D1476517

MODERN SPECTRAL ESTIMATION:

THEORY AND APPLICATION

Steven M. Kay
University of Rhode Island
Kingston, Rhode Island

P T R PRENTICE HALL
Englewood Cliffs, New Jersey 07632

Library of Congress Cataloging-in-Publication Data

KAY, STEVEN M.,
Modern spectral estimation.

Bibliography: p.
Includes index.
1. Spectral theory (Mathematics)—Data processing.
2. Estimation theory—Data processing. I. Title.
QA280.K39 1987 519.5 87-10883
ISBN 0-13-598582-X

Editorial/production supervision
and interior design: Richard Woods
Cover design: 20/20 Services
Manufacturing buyers: S. Gordon Osbourne and Ron Chapman

Published by P T R Prentice Hall
Prentice-Hall, Inc.
A Paramount Communications Company
Englewood Cliffs, New Jersey 07632

If your diskette is defective or damaged in transit, return it directly to Prentice Hall at the address below for a no-charge replacement within 90 days of the date of purchase. Mail the defective diskette together with your name and address.

Prentice Hall
Attention: College Software
Englewood Cliffs, NJ 07632

Printed in the United States of America
10 9 8 7 6

ISBN 0-13-598582-X

Prentice-Hall International (UK) Limited, *London*
Prentice-Hall of Australia Pty. Limited, *Sydney*
Prentice-Hall Canada Inc., *Toronto*
Prentice-Hall Hispanoamericana, S.A., *Mexico*
Prentice-Hall of India Private Limited, *New Delhi*
Prentice-Hall of Japan, Inc., *Tokyo*
Simon & Schuster Southeast Asia Pte. Ltd., *Singapore*
Editora Prentice-Hall do Brasil, Ltda., *Rio de Janeiro*

To Cindy, Lisa, and Ashley

P T R PRENTICE HALL SIGNAL PROCESSING SERIES

Alan V. Oppenheim, Editor

ANDREWS AND HUNT *Digital Image Restoration*
BRIGHAM *The Fast Fourier Transform*
BRIGHAM *The Fast Fourier Transform and its Applications*
BURDIC *Underwater Acoustic System Analysis*
CASTLEMAN *Digital Image Processing*
COWAN AND GRANT *Adaptive Filters*
CROCHIERE AND RABINER *Multirate Digital Signal Processing*
DUDGEON AND MERSEREAU *Multi-Dimensional Digital Signal Processing*
HAMMING *Digital Filters, 2ed.*
HAYKIN, ET AL. *Array Signal Processing*
JAYANT AND NOLL *Digital Coding of Waveforms*
KAY *Modern Spectral Estimation: Theory and Application*
KINO *Acoustic Waves: Devices, Imaging, and Analog Signal Processing*
LEA, ED. *Trends in Speech Recognition*
LIM *Speech Enhancement*
LIM AND OPPENHEIM *Advanced Topics in Signal Processing*
MARPLE *Digital Spectral Analysis with Applications*
MCCLELLAN AND RADER *Number Theory in Digital Processing*
MENDEL *Lessons in Digital Estimation Theory*
OPPENHEIM, ED. *Applications of Digital Signal Processing*
OPPENHEIM AND SCHAFER *Digital Signal Processing*
OPPENHEIM, WILLSKY, WITH YOUNG *Signals and Systems*
RABINER AND GOLD *Theory and Application of Digital Signal Processing*
RABINER AND SCHAFER *Digital Processing of Speech Signals*
ROBINSON AND TREITEL *Geophysical Signal Analysis*
STEARNS AND DAVID *Signal Processing Algorithms*
TRIBOLET *Seismic Applications of Homomorphic Signal Processing*
WIDROW AND STEARNS *Adaptive Signal Processing*

CONTENTS

5 PARAMETRIC MODELING 106

6 AUTOREGRESSIVE SPECTRAL ESTIMATION: GENERAL 153

7 AUTOREGRESSIVE SPECTRAL ESTIMATION: METHODS 217

PREFACE

Power spectral density estimation has traditionally been based on Fourier transform and filtering theory. Within the last decade there has been a flurry of research activity into formulating and comparing alternative means of spectral estimation. The impetus has been the promise of high resolution. Many of these "modern" methods have their roots in such established fields as time series analysis and approximation theory. However, with the rapid increase in computing power available with digital computers, once purely theoretical methods can now be applied in practice. Researchers from diverse fields have contributed to this effort, resulting in a virtual "tower of Babel" of nomenclature, terminology, and symbolism. An important goal of this book, then, is to describe the existing methods in a consistent manner in an effort to unify the various approaches. Another aim is to place the numerous algorithms in a proper perspective.

Since a primary motivation for the recent interest in alternative methods is improved performance, the important but difficult case of short data records is stressed. For longer data records Fourier methods prove to be adequate. It is natural to attempt a definitive comparison of the various spectral estimation methods. However, no judgments have been rendered since the merits of a particular approach tend to be application dependent, the performance critically dependent on the data type. To illustrate the characteristics of each spectral estimate, numerous computer simulations have been included. It is hoped that they will serve as a guide in helping the reader to make intelligent choices for an application. The tools for making those choices have been provided in the form of computer program subroutine implementations for many of the methods described. The subroutines listed within the book as well as the test case data sets (see Appendix 1A) are included on the floppy disk provided with the book and may be read as standard IBM PC-readable files (see also Appendix 3 for a further description). The computer programs have been written in a straightforward manner for ease of understanding. In some cases, computational efficiency has been sacrificed. Standard Fortran 77 is used throughout. To verify the operation of each program, test cases have been provided with the expected output listed within the documentation. A brief description of the available programs is given in Appendix 1. It is felt that "hands-on" experience with the various spectral estimation methods is crucial for understanding the strong points and limitations of each method. To this end an interactive computer program that integrates all the computer subroutines into a menu-driven format is available (although not included with the book). The program may be run on an IBM PC to generate full graphic output or on a mainframe computer, requiring user-supplied graphics. A further description as well as ordering information is contained in Appendix 4. A demonstration program is included on the floppy disk (see Appendix 3).

The philosophical viewpoint adopted toward the problem of spectral estimation is based on classical statistics. Maximum likelihood estimation is stressed as the primary practical means of obtaining good estimators. Also, important statistical results have been included, although for the most part without proof. The reader is referred to the vast body of statistical literature for further details. A shortcoming of the statistical theory is its limitation to large data records (asymptotic results) and real data. Nonetheless, the available theory establishes an important basis for conceptual understanding of spectral estimation methods.

Since it is an impossible task to describe the vast field of modern spectral estimation in a single book, several prominent topics have had to be omitted. Only discrete data are assumed in accordance with the predominant use of digital computers as a means of implementation. Only wide sense stationary discrete random processes (including the degenerate case of sinusoidal processes) are considered. Spectral estimation of transient processes such as damped exponentials in noise is not addressed. Less attention has been given to implementational issues, so that many computationally efficient or "fast" algorithms have been omitted. Similarly, gradient methods such as the LMS algorithm or the gradient lattice are not described since many other books deal specifically with these topics. For a strong focus on implementational concerns as well for descriptions of some of these omitted areas, the interested reader may consult the complementary text, *Digital Spectral Analysis with Applications* by S. L. Marple, Jr. Where proofs of theorems and algorithm developments were essential for critical understanding of the methods, they have been included. Lengthy proofs or developments, however, have been omitted with appropriate references given instead.

This book is the outgrowth of a one-semester graduate course given at the University of Rhode Island. As such, there are numerous homework problems which illustrate the theory, provide developments of properties or proofs of theorems omitted in the text, and derive additional results. A solutions manual is available from the publisher. To enhance the readability and to allow easy access to the spectral estimation algorithms, a summary section has been provided in each chapter. It calls attention to the important results and specific algorithms. The level of maturity assumed is that of a first-year graduate student who has taken basic courses in digital signal processing and probability and random processes. Some exposure to linear algebra and statistics is helpful, although necessary concepts have been summarized. It is believed that the book will be equally useful for the practicing engineer as well as the student.

The book has been divided into two parts. Part I describes the more basic methods of spectral estimation and includes numerous computer programs. Part II discusses more advanced concepts which are still in the early research phase, and therefore it contains no computer programs. The organization is as follows. Chapter 1 is an introduction to the problem of spectral estimation. Chapters 2 and 3 provide necessary background material, and Chapter 4 summarizes classical Fourier methods. Chapter 5 introduces the topic of parametric modeling, which is used in Chapters 6 and 7 for autoregressive spectral estimation, in Chapter 8 for moving average spectral estimation, and in Chapters 9 and 10 for autoregressive moving average spectral estimation. Chapter 11 discusses minimum variance spectral estimation (Capon's method), and Chapter 12 completes Part I by summarizing the popular spectral estimators. Part II is composed of Chapter 13 on sinusoidal parameter estimation, Chapter 14 on multichannel spectral estimation, Chapter 15 on two-dimensional spectral estimation, and Chapter 16, which introduces the reader to some other applications areas that make heavy use of the theory of modern spectral estimation.

The author wishes to acknowlege the contributions of people in the preparation of this book. The close collaboration with S. L. Marple, Jr. of Martin-Marietta on the early version of the manuscript as well as the inclusion of some of his computer programs have greatly enhanced the quality of this book. Also, he has provided expert support on the use of the TEX typesetting program, which was crucial to the preparation of a readable manuscript. Research discussions of the material on which this book is based with L. Jackson, R. Kumaresan, L. Pakula, and D. Tufts of the University of Rhode Island and with L. Scharf of the University of Colorado have been especially helpful. Many people have provided comments on the preliminary manuscript which served to improve the presentation of material. They are W. Gardner of the University of California; M. Kaveh of the University of Minnesota; J. Makhoul of Bolt, Beranek, and Newman; S. L. Marple, Jr.; C. Nikias of Northeastern University; L. Scharf; and D. Sengupta of the University of Rhode Island. Many of the computer simulations were conducted by graduate students at the University of Rhode Island and in particular by D. Sengupta. Finally, thanks are due to the Office of Naval Research, whose continued research support has enabled the author to investigate the field of modern spectral estimation.

A floppy disk has been included with this text. Its contents and hardware/software requirements are described in Appendix 3. Questions regarding the software may be directed to the author at:

Electrical Engineering Dept.
University of Rhode Island
Kingston, R.I. 02881
(401)792-2505

Steven M. Kay
Kingston, Rhode Island

PART I

BASIC METHODS

Chapter 1
Introduction

1.1 THE SPECTRAL ESTIMATION PROBLEM AND ITS APPLICATION

The general problem of spectral estimation is that of determining the spectral content of a random process based on a finite set of observations from that process. Formally, the *power spectral density* (PSD), which will be denoted by $P_{xx}(f)$, of a complex wide sense stationary (WSS) random process $x[n]$ is defined as

$$P_{xx}(f) = \sum_{k=-\infty}^{\infty} r_{xx}[k] \exp(-j2\pi fk) \qquad -\tfrac{1}{2} \leq f \leq \tfrac{1}{2} \tag{1.1}$$

where $r_{xx}[k]$ is the autocorrelation function (ACF) of $x[n]$ defined as

$$r_{xx}[k] = \mathscr{E}(x^*[n]x[n+k]) \tag{1.2}$$

and $\mathscr{E}$ is the expectation operator. The frequency f may either be thought of as the fraction of the sampling frequency used in obtaining the data samples from a continuous random process or as the number of cycles/sample. The PSD function describes the distribution of power with frequency of the random process. Physically, we could determine this distribution by filtering the random process with a bandpass filter that is centered at $f = f_0$ and has a sufficiently narrow filter bandwidth, and then measuring the power at its output. The power is then divided by the filter bandwidth. This procedure would be repeated for all center frequencies $-\frac{1}{2} \leq f_0 \leq \frac{1}{2}$. This method, however, presupposes that the *observed* random process will be of sufficient duration to allow the filter transients to decay. The narrower the filter bandwidth, the longer the observation interval must be.

Fourier methods of spectral estimation implicitly use this approach (see Chapter 4). For many practical applications the observation interval can be quite short. Typically, we have the contiguous observations $\{x[0], x[1], \ldots, x[N-1]\}$ with which to determine the PSD. Since by (1.1) the PSD depends on an infinite number of ACF values, determination of the PSD is in general an impossible task. Any attempt to do so must fail. A more reasonable goal is to obtain a good *estimate* of the PSD. It is the latter problem that is the subject of this book. The spectral estimation problem may now be summarized. *Based on the N contiguous observations* $\{x[0], x[1], \ldots, x[N-1]\}$ *of a single realization of a WSS random process, it is desired to estimate the PSD for* $-\frac{1}{2} \leq f \leq \frac{1}{2}$.

The reason for the restriction of the observation interval to only N data points is a practical one. Many random processes generated by physical mechanisms cannot be characterized as WSS. Hence a PSD cannot be defined since the ACF $\mathcal{E}(x^*[n]x[n+k])$ will depend not only on the lag k between the samples but also on the value of n. In many cases, the nonstationarity is not severe, so that the process may be considered to be *locally WSS*. Locally WSS random processes are ones for which the variation of $\mathcal{E}(x^*[n]x[n+k])$ with n is small over the duration of the observation interval. If the observation interval, which is N samples in length, is sufficiently small, the process may be considered to be locally WSS. An example occurs in speech, where a speech sound or *phoneme* is WSS for about 20 to 80 ms, depending on the particular phoneme [Makhoul et al. 1974]. Other examples are in radar and sonar, in which the received process usually consists of a signal embedded in a background of noise. Because of the doppler effect, the radiated signal of a target will change in center frequency as the target moves [Knight et al. 1981]. Furthermore, the propagation characteristics of the medium may change in time, causing an additional nonstationarity. An example of this behavior is shown in Figure 1.1, in which the spectrogram of a received passive sonar waveform has been plotted [Kay 1980]. The spectrogram was generated by taking 256 point FFTs of successive blocks of data which were overlapped by 50%. A strong sinusoidal component is present but appears to be fading in amplitude, indicative of time-varying multipath propagation through the ocean. Clearly, the data exhibit a mild degree of nonstationarity. However, over short time intervals the change in the estimated PSD is slight, so that the process may be considered to be locally WSS. Hence one must restrict the data analysis window to be short to avoid the smoothing or biasing inherent in estimation of a time-varying PSD. The use of a short data record is not without its problems since any spectral estimate based on a limited data set will exhibit a large variability due to a lack of averaging. This *bias–variance* trade-off is a salient feature of all spectral estimators. In other fields of application a short data record is caused by a genuine lack of data. In economics the data observed are limited by the length of time a particular economic phenomenon has been studied or has been known to exist [Box and Jenkins 1970]. Seismic data tend to be transient since events such as the eruption of a volcano or an earthquake last for only a short period of time [Landers and Lacoss 1977]. Still other applications find the cost of obtaining large amounts of data to be prohibitive. Examples are in optical interferometry, in which each data sample is obtained by mechanical placement of a mirror [Cham-

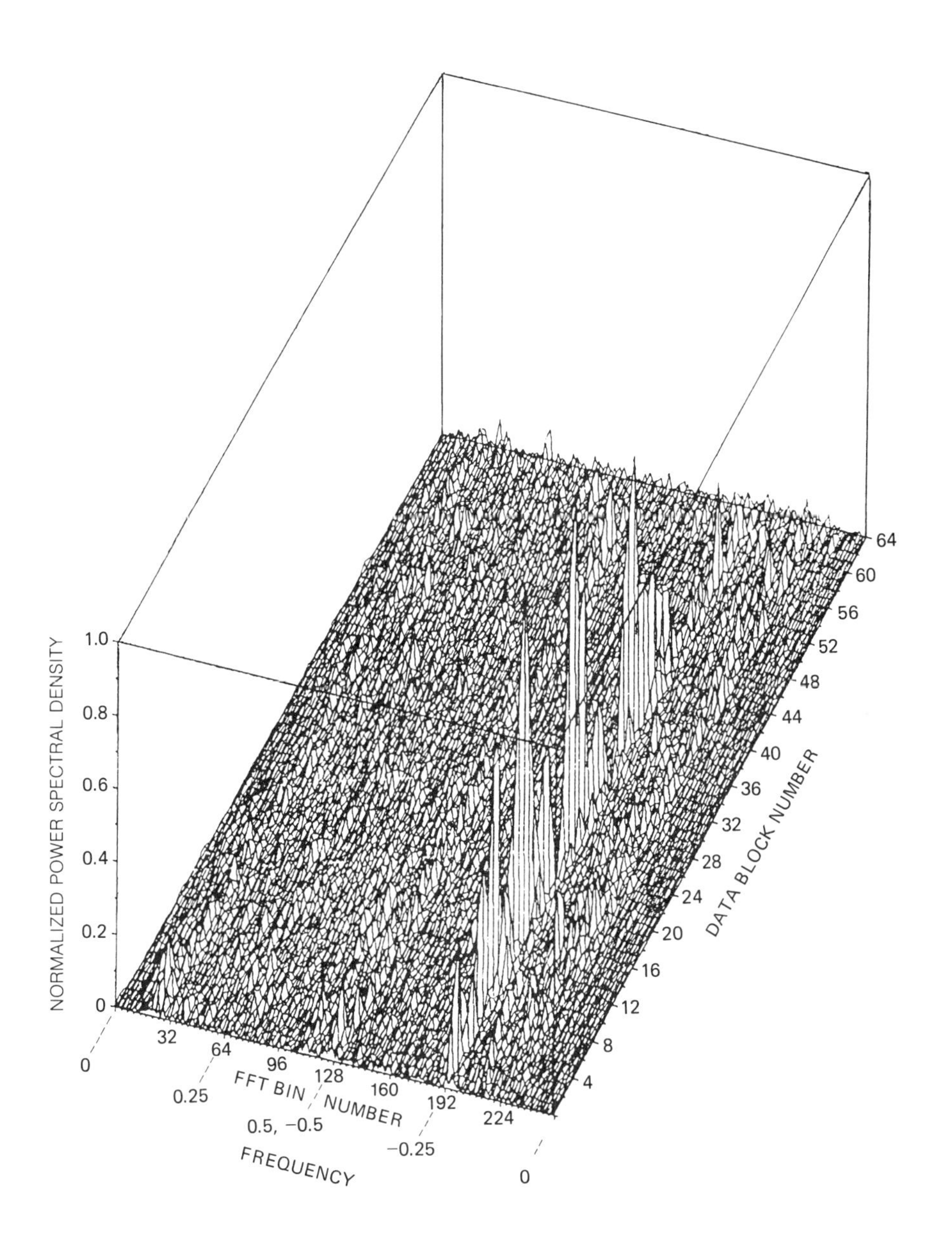

Figure 1.1 Spectrogram of passive sonar data indicating nonstationary sinusoidal signal.

berlain 1979] and in spatial arrays in which each data sample arises from a physical sensing device such as a radar antenna element [Monzingo and Miller 1980]. Clearly, we are driven to base the spectral estimate on as few data points as possible.

Because of the practical constraints spectral estimators are based on finite and usually short data records. It is this case that will be of paramount interest. It should be mentioned that in addition to the fields of application previously described, spectral estimation plays an important role in biomedicine [Gersch 1970], vibration analysis [Bendat and Piersol 1971], image processing [Jain 1981], radio astronomy [Wernecke 1977], oceanography [Holm and Hovem 1979], and ecological systems [Hacker 1978].

1.2 ON THE USES OF SPECTRAL ESTIMATION

Spectral estimation is a *preliminary data analysis tool.* A spectral estimate should not be used to answer *specific* questions about the data, such as whether a resonance is present, but only to suggest possible hypotheses. These hypotheses can then be formulated within a statistical framework. The most probable hypothesis is chosen by employing established methods of hypothesis testing [Kendall and Stuart 1979]. For example, the detection of a positive dc signal in colored Gaussian noise for which the PSD of the noise is unknown might lead one to consider spectral estimation as a preliminary step in designing a practical detector. If a segment of the noise only were available, a spectral estimate might reveal that the PSD is nearly flat at $f = 0$. Based on this information, a good means of detection would be to compare the sample mean to a threshold since this procedure approximates the optimal matched filter [Van Trees 1968]. On the other hand, if the spectral estimate revealed a strong narrowband component near $f = 0$, we would need to use a prewhitener prior to a matched filter.

Implicit in the preceding example is the use of spectral estimation for model building. A principal use, then, is to postulate models that describe the data as broadband or narrowband, stationary or nonstationary, low pass or high pass, and so on. Once the model is chosen and verified, either by physical justification or by further statistical tests of the data, new insights become available. New methods for solving the original problem may suggest themselves.

In many cases the problem of interest is so well defined that the preliminary step of spectral estimation may be eliminated entirely. For instance, if we know or are willing to assume that the data consist of two sinusoids in white Gaussian noise, then the problem of frequency estimation becomes a problem in *parameter estimation* (see Chapter 13). The latter is more adequately handled using the theory of statistical estimation [Kendall and Stuart 1979]. This is not to say that the spectral estimation and parameter estimation problems are mutually exclusive. If the frequencies of the sinusoids are far enough apart so that the sinusoids are well "resolved," then a maximum likelihood estimator corresponds to picking the frequencies as the locations of the two largest peaks in the periodogram (see Chapter 13). If the noise is colored, however, then the periodogram does not provide the best frequency estimates. Maximum likelihood estimates in this case will require knowledge of the noise PSD, which in practice is usually unknown. Hence a periodogram may still be a practical alternative. The salient point is that if the problem of interest is so well defined that the only unknowns are a set of parameters, be they sinusoidal frequencies or the power of white noise as examples, then parameter estimation techniques should be employed. Similarly, if the problem is well defined as a problem in detection, then the principles of hypothesis testing are applicable.

1.3 PRINCIPAL APPROACHES TO SPECTRAL ESTIMATION

Since the PSD is a function of an infinite number of ACF values, the task of estimating the PSD based on a finite data set is an impossible one. At best we could only hope to estimate a subset of the ACF values. The Fourier-based spectral estimators estimate only the most significant values, which are assumed to be for $k = 0$ to $k = M$ (see Chapter 4). The resulting spectral estimator may be severely biased if the process exhibits strong correlations for $k > M$. In an attempt to reduce the spectral estimation problem to one that is more manageable, various models for the PSD can be assumed. The class of time series models or PSDs which are rational functions are typical (see Chapter 5). The models are parameterized by a finite (and hopefully small) set of variables. The incorporation of a model allows one to replace the spectral estimation problem by a parameter estimation problem. *It is critical, however, that the model be an accurate representation of the random process, at least as far as the PSD is concerned.* Inaccurate models produce systematic biases in the spectral estimator which are always present, even for very large data records. Unfortunately, there exists no theory to guide us in the choice of an appropriate model. Rather, it is usually based on the physical constraints of the data generation process. Alternatively, a model may be used that maintains the important spectral characteristics sought. An example is the determination of the location and bandwidth of the spectral peaks or *formants* in speech. An autoregressive model accurately models the peaks in the PSD but not the valleys. Fortunately, the valleys are not important for speech perception [Rabiner and Schafer 1978]. If the model chosen is an accurate representation of the PSD, less biased (higher resolution) and lower-variance spectral estimators will result.

A key benefit of the modeling approach is that the wealth of available theory of statistical parameter estimation may be applied. The theory of maximum likelihood [Kendall and Stuart 1979] provides by far the most important approach to this problem. The maximum likelihood estimator (MLE) is optimal for large data records and in practice usually produces accurate estimates even for short data records. If an MLE of the model parameters is used and then substituted into the theoretical PSD of the model, an MLE of the PSD is obtained. When the MLE or a good approximation to the MLE is not easily computed, which occurs for the moving average (see Chapter 8) and autoregressive moving average models (see Chapters 9 and 10), suboptimal estimators must be tried. The relative loss in performance is assessed by comparing the variance of the estimator to the Cramer–Rao lower bound [Van Trees 1968, Kendall and Stuart 1979].

A major difficulty in obtaining accurate spectral estimators is that no "best" spectral estimator exists. This is partly due to the lack of an accepted definition of optimality. Any measure of optimality must weight the various characteristics of the spectral estimator in some way. For instance, it is not obvious how one should assign weights to the ability of a spectral estimator to reproduce peaks and at the same time reduce sidelobes. Also, any weighting will be application dependent. A second difficulty arises because optimality implies being able to assign a number to the error between a spectral estimate and the true PSD. The true PSD is obviously unknown. A contrasting problem is in digital filter design. One is usually given the exact magnitude of the frequency response and asked to design a filter for which the approximate magnitude response is close to the given one [Oppenheim and Schafer 1975]. But in spectral estimation the desired PSD is *not* given. With a lack of a definition of optimality and the means to describe the error, many spectral estimators have been proposed on the basis of "ad hoc" arguments. As a result, a multitude of spectral estimators can be found in the literature. Only the more promising approaches are discussed in this book.

1.4 COMPARISON OF SPECTRAL ESTIMATORS

Due to a lack of an accepted definition of optimality the comparison of spectral estimators has led to a wild assortment of claims and counterclaims. It is no wonder that the novice is at a loss to sort out the "wheat from the chaff." In comparing the performance of spectral estimators *some objective measure of goodness must be adopted.* The attribute of resolution is a prime example. The intuitive concept of resolution deals with the ability of a spectral estimator to display fine spectral detail. However, the spectral detail that is displayed should actually be there and not be an artifact of the estimator. An example might be the high resolution property of autoregressive spectral estimators. For autoregressive processes the spectral estimator can resolve closely spaced spectral peaks but may produce other peaks, which are nonexistent. Clearly, in comparing spectral estimators it is imperative to state clearly which spectral characteristics are of interest so that "apples and oranges" are not being compared.

Once the important characteristics have been agreed upon, a statistical measure such as bias and variance, mean square error, and so on, must be chosen to quantify the performance. Here a second problem in the comparison of spectral estimators is encountered. Because the spectral estimator is a statistical function of the data, *its performance can only be described statistically*. This requires that we determine the statistics of the estimator. Unfortunately, results are available only for large data records (i.e., asymptotic statistics) [Jenkins and Watts 1968, Box and Jenkins 1970, Anderson 1971, Koopmans 1974]. No exact statistical descriptions are available for finite data records for the Fourier spectral estimators (other than for the special case of white Gaussian noise processes) or any of the other estimators to be described. The use of asymptotic statistics is inadvisable to compare spectral estimators because the length of the data record necessary for the results to apply is never quantified. It is conceivable and quite probable that the minimum data record length required will depend on the PSD of the process as well as the spectral estimator under consideration. At most the asymptotic statistics should be used only as a guide to describe the characteristics of a *single* spectral estimator and not as a means to compare several spectral estimators.

The lack of theory has forced researchers to compare the performance of competing spectral estimators by Monte Carlo computer simulations. In doing so, one must be careful to bless each spectral estimator with the same a priori knowledge so that any one estimator does not enjoy an unfair advantage. Also, comparisons should be based on the statistics obtained from many realizations. In the simulations to be described in Chapters 4, 7, 8, 10, and 11, the mean of the spectral estimator is plotted together with a scatter plot. The scatter plot, which is formed by overlaying many realizations, indicates the variance of the spectral estimator. A few realizations that are visually compared are not sufficient to describe the performance of a spectral estimator. As an example, consider the two spectral estimates shown in Figure 1.2 [Kay and Demeure 1984]. They were obtained by applying two spectral estimators to a single realization of data consisting of two complex sinusoids in complex white noise. We might conclude that the spectral estimate in Figure 1.2b is preferable since it exhibits higher resolution. Actually, the two spectral estimates are related by

$$\hat{P}_{xx}^{B}(f) = \frac{1}{1 - \hat{P}_{xx}^{A}(f)} \tag{1.3}$$

where

$$0 \leq \hat{P}_{xx}^{A}(f) \leq 1.$$

The sharpness of the peaks in Figure 1.2b is due to the nonlinear warping caused by the transformation $1/(1 - x)$. Since the transformation is monotonic, the peak locations in Figure 1.2b are identical to the peak locations in Figure 1.2a, although not clearly visible (see Problem 1.1). If one were to examine the variance of the two spectral estimators, it would be obvious that the apparent increase in resolution of $\hat{P}_{xx}^{B}(f)$ is gained at the expense of a large increase in variance of the spectral peak amplitudes (see Problem 1.2). Due to (1.3), both spectral estimators contain the same information, although clearly displayed differently. This example serves to illustrate many of the potential pitfalls awaiting the practitioner of spectral estimation. A further discussion of these spectral estimators is contained in Section 13.9.2. Several key points must be kept in mind at all times in comparing spectral estimators via computer simulations. The first is that many realizations need to be generated and appropriate statistics computed because simulation results in isolation are misleading. Visual comparison of spectral estimates can be misleading. An objective comparison needs to be made. A comparison of various

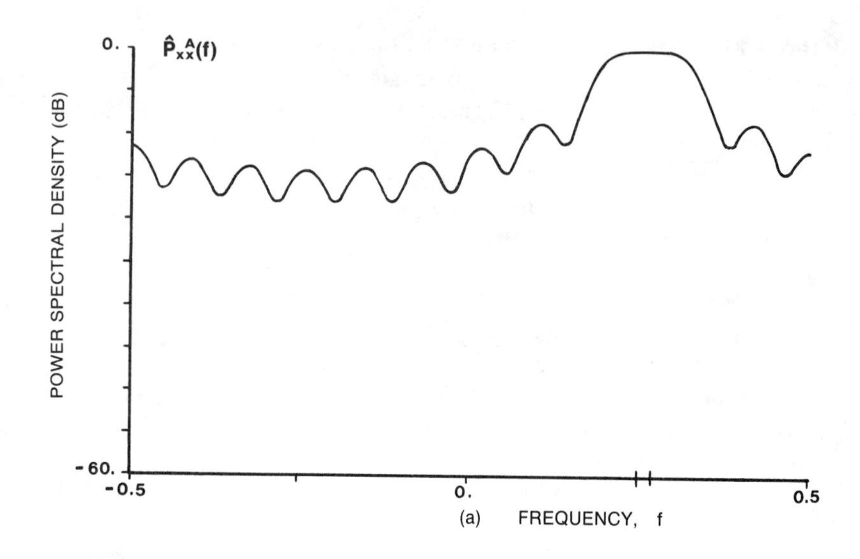

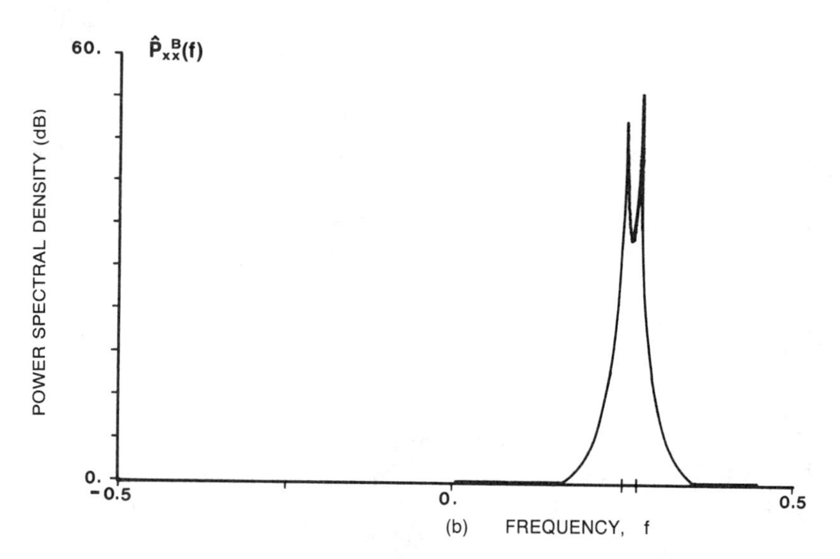

Figure 1.2 Resolution comparison of spectral estimators. (© 1984 IEEE)

spectral estimators can only be made, however, if the a priori assumptions are specified and they are the same for each estimator. If, for example, we knew a priori that either one, two, or no sinusoids were present in the data used to generate the spectral estimates of Figure 1.2, then the assessed merits of the spectral estimators would be radically different. Finally, the promise of "high resolution" which has been responsible for the resurgence in interest in spectral estimation methods can be fulfilled only if resolution is formally defined. It is evident from this example that *the sharpness of the peaks of a spectral estimate is not related to the resolution of a spectral estimator.*

1.5 A COMMON TEST CASE

In an effort to illustrate the typical characteristics of the various spectral estimators to be described and *not as a means of a quantitative comparison*, a test case was developed. The PSD is shown in Figure 1.3. The process consists of narrowband components or sinusoids as well as a broadband component. The heights of the sinusoidal components displayed in Figure 1.3 were arbitrarily chosen since they are theoretically infinite (being Dirac delta functions). Specifically, 32 complex data points have been generated from the process

$$x[n] = 2\cos(2\pi f_1 n) + 2\cos(2\pi f_2 n) + 2\cos(2\pi f_3 n) + z[n] \tag{1.4}$$

for $n = 0, 1, \ldots, 31$. In (1.4), $f_1 = 0.05$, $f_2 = 0.40$, $f_3 = 0.42$, and $z[n]$ is a complex autoregressive process (see Chapter 5) of order 1 or

$$z[n] = -a[1]z[n-1] + u[n]. \tag{1.5}$$

$u[n]$ is a zero-mean complex white Gaussian noise process (see Section 3.5.1) with variance σ^2. It is defined by

$$u[n] = u_R[n] + ju_I[n]$$

where $u_R[n]$ and $u_I[n]$ are each zero-mean, real white Gaussian noise processes with variance $\sigma^2/2$ which are uncorrelated with each other. The values of $a[1]$ and σ^2 are -0.850848 and 0.101043, respectively. Since $a[1]$ is real, the PSD of

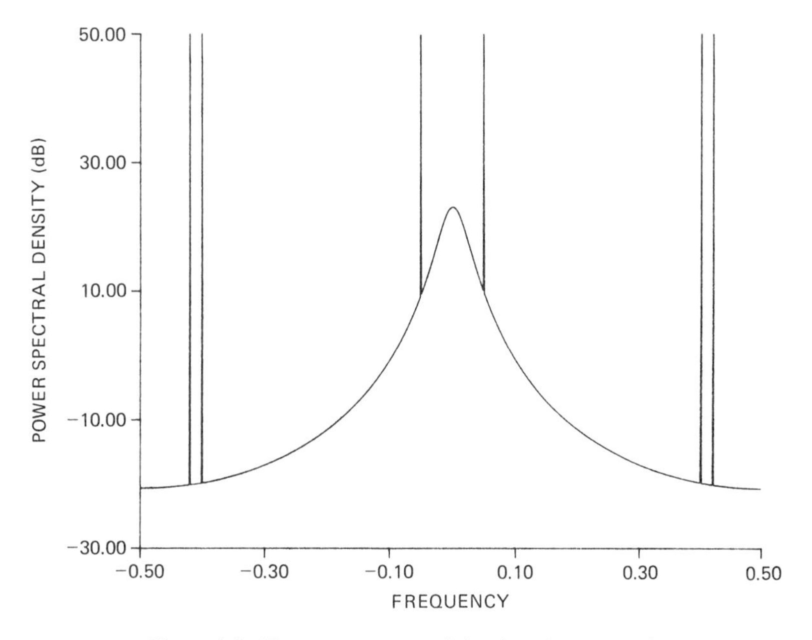

Figure 1.3 True power spectral density of test case data.

$z[n]$ is symmetric and is given by

$$P_{zz}(f) = \frac{\sigma^2}{|1 + a[1]\exp(-j2\pi f)|^2}. \tag{1.6}$$

The choice of a symmetric PSD was deliberate, as will be explained shortly. The 32 complex data points are listed in Appendix 1A.

The sinusoidal components at $f_2 = 0.40$ and $f_3 = 0.42$ will not be resolved by a Fourier spectral estimator since their separation is less than the resolution limit of $1/N = 0.03$. The component at $f_1 = 0.05$ will be well resolved since it is 0.1 cycles/sample apart from its nearest neighbor. The specific values of $a[1]$ and σ^2 were chosen to cause the "local" signal-to-noise ratio (SNR) to be 30 dB in the frequency band near 0.41 and 15 dB in the band near 0.05. The local SNR is defined at f_1 as

$$\text{SNR} = 10\log_{10}\frac{1}{\int_{f_1-1/(2N)}^{f_1+1/(2N)} P_{zz}(f)\,df} \quad \text{dB}. \tag{1.7}$$

This is the power of the sinusoidal component at $f = f_1$ divided by the noise power in a band centered at f_1 and having a width of $1/N$, which is the Fourier resolution. Similarly, for the sinusoids centered about $f = f_0 = 0.41$, the local SNR is

$$\text{SNR} = 10\log_{10}\frac{1}{\int_{f_0-1/(2N)}^{f_0+1/(2N)} P_{zz}(f)\,df} \quad \text{dB}. \tag{1.8}$$

These SNRs may be computed using (1.7) and (1.8) and the result that (see Problem 1.3)

$$\int_{f_L}^{f_H} P_{zz}(f)\,df = \frac{\sigma^2}{\pi(1 - a^2[1])}\arctan\left\{\frac{\dfrac{1 - a[1]}{1 + a[1]}[\tan(\pi f_H) - \tan(\pi f_L)]}{1 + \left(\dfrac{1 - a[1]}{1 + a[1]}\right)^2 \tan(\pi f_L)\tan(\pi f_H)}\right\}. \tag{1.9}$$

Using (1.7) and (1.8), the values of $a[1]$ and σ^2 have been computed to yield the given SNRs. The broadband SNR which is also of interest is defined as

$$\text{SNR} = 10\log_{10}\frac{1}{\dfrac{\sigma^2}{1 - a^2[1]}} \quad \text{dB}$$

and equals 4.4 dB for the test case.

The reason for choosing a complex process for which the PSD is symmetric is that the same PSD may also be realized by a real process. The PSD shown in Figure 1.3 also corresponds to the PSD of the real process (see Problem 1.4)

$$x[n] = 2\cos(2\pi f_1 n) + 2\cos(2\pi f_2 n) + 2\cos(2\pi f_3 n) + \sqrt{2}\,\text{Re}\{z[n]\}. \tag{1.10}$$

Hence for spectral estimators that have only been developed for real data a reasonable qualitative comparison to the spectral estimators based on complex data may be made by using the real data described by (1.10). The same 32 complex points of $z[n]$ used to generate the test case as per (1.4) have also been used to obtain $\sqrt{2}\ \mathrm{Re}\ \{z[n]\}$ for the real data set given by (1.10). The real data set is listed in Appendix 1A. The use of complex data for the spectral estimators which are valid for complex data and the use of real data for the spectral estimators which are restricted to real data is desirable in order for the user to validate the correctness of the computer subroutines listed throughout the book. The results of running the subroutines with the appropriate test data as input are listed within the documentation of each subroutine. Also, the spectral estimates based on the test case for each method are given in the summary section of the chapter describing that method, and a compilation of these results may be found in Chapter 12.

REFERENCES

Anderson, T. W., *The Statistical Analysis of Time Series*, Wiley, New York, 1971.

Bendat, J. S., and A. G. Piersol, *Random Data: Analysis and Measurement Procedures*, Wiley, New York, 1971.

Box, G. E. P., and G. M. Jenkins, *Time Series Analysis: Forecasting and Control*, Holden-Day, San Francisco, 1970.

Chamberlain, J., *The Principles of Interferometric Spectroscopy*, Wiley, New York, 1979.

Gersch, W., "Spectral Analysis of EEG's by Autoregressive Decomposition of Time Series," *Math. Biosci.*, Vol. 7, pp. 205–222, 1970.

Hacker, C. S., "Autoregressive and Transfer Function Models of Mosquito Populations," in *Time Series and Ecological Processes*, H. M. Shugat, ed., SIAM, Philadelphia, 1978, pp. 294–303.

Holm, S., and J. M. Hovem, "Estimation of Scalar Ocean Wave Spectra by the Maximum Entropy Method," *IEEE Trans. Ocean. Eng.*, Vol. OE4, pp. 76–83, July 1979.

Jain, A. K., "Advances in Mathematical Modeling for Image Processing," *Proc. IEEE*, Vol. 69, pp. 502–528, May 1981.

Jenkins, G. M., and D. G. Watts, *Spectral Analysis and Its Applications*, Holden-Day, San Francisco, 1968.

Kay, S., "Autoregressive Spectral Analysis of Narrowband Processes in White Noise with Application to Sonar Signals," Ph.D. dissertation, Georgia Institute of Technology, 1980.

Kay, S., and C. Demeure, "The High Resolution Spectrum Estimator: A Subjective Entity," *Proc. IEEE*, pp. 1815–1816, Dec. 1984.

Kendall, M., and A. S. Stuart, *The Advanced Theory of Statistics*, Vol. 2, Macmillan, New York, 1979.

Knight, W. S., R. G. Pridham, and S. M. Kay, "Digital Signal Processing for Sonar," *Proc. IEEE*, Vol. 69, pp. 1451–1506, Nov. 1981.

Koopmans, L. H., *The Spectral Analysis of Time Series*, Academic Press, New York, 1974.

Landers, T. E., and R. T. Lacoss, "Some Geophysical Applications of Autoregressive Spectral Estimates," *IEEE Trans. Geosci. Electron.*, Vol. GE15, pp. 26–32, Jan. 1977.

Makhoul, J., R. Viswanathan, L. Cosell, and W. Russell, "Natural Communication with Computers: Speech Compression Research at BBN," BBN Rep. 2976 II, Bolt, Beranek, and Newman, Inc., Cambridge, Mass., 1974.

Monzingo, R. A., and T. W. Miller, *Introduction to Adaptive Arrays*, Wiley, New York, 1980.

Oppenheim, A. V., and R. W. Schafer, *Digital Signal Processing*, Prentice-Hall, Englewood Cliffs, N.J., 1975.

Rabiner, L. R., and R. W. Schafer, *Digital Processing of Speech Signals*, Prentice-Hall, Englewood Cliffs, N.J., 1978.

Van Trees, H. L., *Detection, Estimation, and Modulation Theory.* Vol. 1, J. Wiley, New York, 1968.

Wernecke, S., "Two-Dimensional Maximum Entropy Reconstruction of Radio Brightness," *Radio Sci.*, Vol. 12, pp. 831 – 844, Sept.–Oct. 1977.

PROBLEMS

1.1. Prove that

$$g(x) = \frac{1}{1 - x}$$

is a monotonically increasing function over the interval $0 < x < 1$. Show that due to the monotonicity the peaks of the spectral estimate shown in Figure 1.2b must also be present in the spectral estimate of Figure 1.2a.

1.2. Show that if a random variable x is transformed into a random variable y according to the transformation $y = g(x)$, the variance of y is given approximately by

$$\text{var}\,(y) \approx [g'(\mu)]^2\ \text{var}\,(x)$$

where $g'(\mu)$ is the derivative of g evaluated at the mean μ of x. *Hint:* Linearize $g(x)$ about the mean of x. The linearization is valid if $g(x)$ is sufficiently smooth over the range of x values for which the probability density function of x is essentially nonzero. What can you conclude about the relative variances of the spectral estimators related by (1.3)?

1.3. Verify the relationship given by (1.9).

1.4. Show that if $a[1]$ is real, then $z[n]$ as given by (1.5) can be written as $z[n] = z_R[n] + jz_I[n]$, where

$$z_R[n] = -a[1]z_R[n-1] + u_R[n]$$

$$z_I[n] = -a[1]z_I[n-1] + u_I[n].$$

$z_R[n]$, $z_I[n]$ are real processes. Using this result, prove that the PSD of $\sqrt{2}\ \text{Re}\,\{z[n]\}$ is identical to the PSD of $z[n]$ as given by (1.6).

APPENDIX 1A

Listing of Test Data Sets

COMPLEX TEST DATA
(Provided on Floppy Disk)

```
x[0]  = ( 6.3307,     -0.174915)
x[1]  = (-1.33539,    -0.03044)
x[2]  = ( 3.61896,    -0.260459)
x[3]  = ( 1.87513,    -0.323974)
x[4]  = (-1.08561,    -0.136055)
x[5]  = ( 3.99114,    -0.101864)
x[6]  = (-4.10184,     0.130571)
x[7]  = ( 1.55399,     0.0977916)
x[8]  = (-2.1258,     -0.306485)
x[9]  = (-3.27873,    -0.0544436)
x[10] = ( 0.241218,    0.0962379)
x[11] = (-5.74708,     0.0186908)
x[12] = (-0.0165977,   0.237493)
x[13] = (-3.28921,    -0.188478)
x[14] = (-1.31227,    -0.120636)
x[15] = ( 0.745251,   -0.0679575)
x[16] = (-1.77199,    -0.416229)
x[17] = ( 2.56419,    -0.270373)
x[18] = ( 0.21325,    -0.232544)
x[19] = ( 2.23409,     0.236383)
x[20] = ( 2.2949,      0.173061)
x[21] = ( 1.09186,     0.140938)
x[22] = ( 2.29353,     0.442044)
x[23] = ( 0.695823,    0.509325)
x[24] = ( 0.759858,    0.417967)
x[25] = (-0.354267,    0.506891)
x[26] = (-0.594517,    0.39708)
x[27] = (-1.88618,     0.649179)
x[28] = (-1.39041,     0.867086)
x[29] = (-3.06381,     0.422965)
x[30] = (-2.0433,      0.0825514)
x[31] = (-2.1628,     -0.0933218)
```

REAL TEST DATA
(Provided on Floppy Disk)

$x[0] = 6.46768$
$x[1] = -1.28024$
$x[2] = 3.74788$
$x[3] = 1.96092$
$x[4] = -0.768349$
$x[5] = 4.14569$
$x[6] = -4.05277$
$x[7] = 1.65836$
$x[8] = -2.06405$
$x[9] = -3.33397$
$x[10] = 0.085145$
$x[11] = -6.06562$
$x[12] = -0.411658$
$x[13] = -3.61831$
$x[14] = -1.53352$
$x[15] = 0.481522$
$x[16] = -1.93653$
$x[17] = 2.35532$
$x[18] = 0.145624$
$x[19] = 2.21991$
$x[20] = 2.25884$
$x[21] = 1.07373$
$x[22] = 2.26531$
$x[23] = 0.685007$
$x[24] = 0.762859$
$x[25] = -0.501008$
$x[26] = -0.640518$
$x[27] = -1.99263$
$x[28] = -1.60416$
$x[29] = -3.22751$
$x[30] = -2.21946$
$x[31] = -2.42246$

Chapter 2
Review of Linear and Matrix Algebra

2.1 INTRODUCTION

Important results from linear and matrix algebra theory are reviewed in this chapter. In the discussions to follow it is assumed that the reader already has some familiarity with these topics. The specific concepts to be described are used heavily throughout the book. They are for the most part given without proof, although some of the proofs are explored in the problems. For a more comprehensive treatment, the reader is referred to the books *Applied Linear Algebra* [Noble and Daniel 1977] and *Elementary Matrix Algebra* [Hohn 1973]. An excellent summary of matrix theorems may be found in the work by Marcus [1960]. Unless otherwise stated, all matrices and vectors are assumed to be complex with real matrices and vectors as special cases.

2.2 DEFINITIONS

Consider an $m \times n$ matrix $\mathbf{A}$ with elements a_{ij}, $i = 1, 2, \ldots, m; j = 1, 2, \ldots, n$. A shorthand notation for describing $\mathbf{A}$ is

$$[\mathbf{A}]_{ij} = a_{ij}.$$

The *transpose* of $\mathbf{A}$, which is denoted by $\mathbf{A}^T$, is defined as the $n \times m$ matrix with elements a_{ji}, or

$$[\mathbf{A}^T]_{ij} = a_{ji}.$$

Similarly, the *hermitian transpose* (or conjugate transpose) is denoted by $\mathbf{A}^H$ and is defined as the $n \times m$ matrix with elements

$$[\mathbf{A}^H]_{ij} = a_{ji}^*.$$

For a real matrix, $\mathbf{A}^H = \mathbf{A}^T$.

A *square* matrix is one for which $m = n$. A real square matrix is *symmetric* if $\mathbf{A}^T = \mathbf{A}$ and a complex square matrix is *hermitian* if $\mathbf{A}^H = \mathbf{A}$.

The *rank* of a matrix is the number of linearly independent rows or columns, whichever is less. The *inverse* of a square $n \times n$ matrix is the square $n \times n$ matrix $\mathbf{A}^{-1}$, for which

$$\mathbf{A}^{-1}\mathbf{A} = \mathbf{A}\mathbf{A}^{-1} = \mathbf{I}$$

where $\mathbf{I}$ is the $n \times n$ identity matrix. The inverse will exist if and only if the rank of $\mathbf{A}$ is n. If the inverse does not exist, $\mathbf{A}$ is *singular*.

The *determinant* of a square $n \times n$ matrix is denoted by det ($\mathbf{A}$). It is computed as

$$\det(\mathbf{A}) = \sum_{j=1}^{n} a_{ij} C_{ij} \tag{2.1}$$

where

$$C_{ij} = (-1)^{i+j} M_{ij}. \tag{2.2}$$

M_{ij} is the determinant of the submatrix of $\mathbf{A}$ obtained by deleting the ith row and jth column and is termed the *minor* of a_{ij}. C_{ij} is the *cofactor* of a_{ij}. Note that any choice of i for $i = 1, 2, \ldots, n$ will yield the same value for det ($\mathbf{A}$).

A *quadratic form* Q is defined as

$$Q = \sum_{i=1}^{n} \sum_{j=1}^{n} a_{ij} x_i x_j \tag{2.3}$$

for real a_{ij} and x_i. In defining the quadratic form it is assumed that $a_{ji} = a_{ij}$. This entails no loss in generality since any quadratic function may be expressed in this manner. Q may also be expressed as

$$Q = \mathbf{x}^T \mathbf{A} \mathbf{x} \tag{2.4}$$

where $\mathbf{x} = [x_1 x_2 \ldots x_n]^T$ and $\mathbf{A}$ is a square $n \times n$ matrix with $a_{ji} = a_{ij}$ or $\mathbf{A}$ is a symmetric matrix. Similarly, a *hermitian form* Q is defined as

$$Q = \sum_{i=1}^{n} \sum_{j=1}^{n} a_{ij} x_i^* x_j \tag{2.5}$$

for complex a_{ij} and x_i and where it is assumed that $a_{ji} = a_{ij}^*$, so that $\mathbf{A}$ is hermitian. Defining $\mathbf{A}$ as the square $n \times n$ hermitian matrix with elements a_{ij}, (2.5) can alternatively be written as

$$Q = \mathbf{x}^H \mathbf{A} \mathbf{x}. \tag{2.6}$$

Note that since $\mathbf{A}$ is hermitian, Q is real.

A complex square $n \times n$ matrix $\mathbf{A}$ is *positive semidefinite* if it is hermitian and its hermitian form is nonnegative or

$$\mathbf{x}^H \mathbf{A} \mathbf{x} \geq 0 \tag{2.7}$$

for all $\mathbf{x} \neq \mathbf{0}$. If the hermitian form is strictly positive, the matrix is *positive definite*. Similarly, a real square $n \times n$ matrix $\mathbf{A}$ is positive semidefinite if $\mathbf{A}$ is symmetric and

$$\mathbf{x}^T \mathbf{A} \mathbf{x} \geq 0 \tag{2.8}$$

for all real $\mathbf{x} \neq \mathbf{0}$. If the quadratic form is strictly positive, $\mathbf{A}$ is positive definite. When referring to a matrix as positive definite or positive semidefinite, it is always assumed that the matrix is hermitian (symmetric).

A *partitioned* $m \times n$ matrix $\mathbf{A}$ is one which is expressed in terms of its submatrices. An example is the 2×2 partitioning

$$\mathbf{A} = \begin{bmatrix} \mathbf{A}_{11} & \mathbf{A}_{12} \\ \mathbf{A}_{21} & \mathbf{A}_{22} \end{bmatrix}. \tag{2.9}$$

Each "element" $\mathbf{A}_{ij}$ is a submatrix of $\mathbf{A}$. The dimensions of the partitions are given as

$$\begin{bmatrix} k \times l & k \times (n - l) \\ (m - k) \times l & (m - k) \times (n - l) \end{bmatrix}.$$

2.3 SPECIAL MATRICES

A *diagonal* matrix is a square $n \times n$ matrix with $a_{ij} = 0$ for $i \neq j$ or all elements off the principal diagonal are zero. A diagonal matrix appears as

$$\mathbf{A} = \begin{bmatrix} a_{11} & 0 & \cdots & 0 \\ 0 & a_{22} & \cdots & 0 \\ \vdots & \vdots & \ddots & \vdots \\ 0 & 0 & \cdots & a_{nn} \end{bmatrix}. \tag{2.10}$$

A diagonal matrix will sometimes be denoted by diag $(a_{11}, a_{22}, \ldots, a_{nn})$. The inverse of a diagonal matrix is found by simply inverting each element on the principal diagonal.

A generalization of the diagonal matrix is the square $n \times n$ *block diagonal* matrix

$$\mathbf{A} = \begin{bmatrix} \mathbf{A}_{11} & \mathbf{0} & \cdots & \mathbf{0} \\ \mathbf{0} & \mathbf{A}_{22} & \cdots & \mathbf{0} \\ \vdots & \vdots & \ddots & \vdots \\ \mathbf{0} & \mathbf{0} & \cdots & \mathbf{A}_{kk} \end{bmatrix} \tag{2.11}$$

in which all submatrices $\mathbf{A}_{ii}$ are square and the other submatrices are identically zero. The dimensions of the submatrices need not be identical. For instance, if

$k = 2$, $\mathbf{A}_{11}$ might have dimension 2×2 while $\mathbf{A}_{22}$ might be a scalar. If all $\mathbf{A}_{ii}$ are nonsingular, the inverse is easily found as

$$\mathbf{A}^{-1} = \begin{bmatrix} \mathbf{A}_{11}^{-1} & \mathbf{0} & \cdots & \mathbf{0} \\ \mathbf{0} & \mathbf{A}_{22}^{-1} & \cdots & \mathbf{0} \\ \vdots & \vdots & \ddots & \vdots \\ \mathbf{0} & \mathbf{0} & \cdots & \mathbf{A}_{kk}^{-1} \end{bmatrix}. \tag{2.12}$$

Also, the determinant is

$$\det(\mathbf{A}) = \prod_{i=1}^{k} \det(\mathbf{A}_{ii}). \tag{2.13}$$

The *exchange matrix* $\mathbf{J}$ is defined as the square $n \times n$ matrix

$$\mathbf{J} = \begin{bmatrix} 0 & \cdots & 0 & 1 \\ 0 & \cdots & 1 & 0 \\ \vdots & \cdot^{\cdot^{\cdot}} & \vdots & \vdots \\ 1 & \cdots & 0 & 0 \end{bmatrix}. \tag{2.14}$$

The elements on the principal cross-diagonal are unity while all other elements are zero. Note that $\mathbf{J}$ is symmetric and $\mathbf{J}^2 = \mathbf{I}$. The effect of applying the exchange matrix to a vector is to exchange or "flip around" the elements since

$$\mathbf{Jx} = \begin{bmatrix} x_n \\ x_{n-1} \\ \vdots \\ x_1 \end{bmatrix}.$$

A *lower triangular* square $n \times n$ matrix is defined as

$$\mathbf{L} = \begin{bmatrix} a_{11} & 0 & \cdots & 0 \\ a_{21} & a_{22} & \cdots & 0 \\ \vdots & \vdots & \ddots & \vdots \\ a_{n1} & a_{n2} & \cdots & a_{nn} \end{bmatrix}. \tag{2.15}$$

The inverse of a lower triangular matrix is also lower triangular (see Problem 2.1). The determinant of $\mathbf{L}$ is

$$\det(\mathbf{L}) = \prod_{i=1}^{n} a_{ii}. \tag{2.16}$$

Similarly, an *upper triangular* square $n \times n$ matrix is defined as

$$\mathbf{U} = \begin{bmatrix} a_{11} & a_{12} & \cdots & a_{1n} \\ 0 & a_{22} & \cdots & a_{2n} \\ \vdots & \vdots & \ddots & \vdots \\ 0 & 0 & \cdots & a_{nn} \end{bmatrix}. \tag{2.17}$$

The inverse is also upper triangular and the determinant is given by (2.16).

A real square $n \times n$ matrix is *orthogonal* if

$$\mathbf{A}^{-1} = \mathbf{A}^T. \tag{2.18}$$

For a matrix to be orthogonal the columns (and rows) must be orthonormal or if $\mathbf{a}_i$ denotes the ith column (or row), the conditions

$$\mathbf{a}_i^T \mathbf{a}_j = \begin{cases} 0 & \text{for } i \neq j \\ 1 & \text{for } i = j \end{cases} \tag{2.19}$$

must be satisfied. A complex square $n \times n$ matrix is *unitary* if

$$\mathbf{A}^{-1} = \mathbf{A}^H. \tag{2.20}$$

To be unitary the matrix $\mathbf{A}$ must satisfy (see Problem 2.2)

$$\mathbf{a}_i^H \mathbf{a}_j = \begin{cases} 0 & \text{for } i \neq j \\ 1 & \text{for } i = j. \end{cases} \tag{2.21}$$

An important example of a unitary matrix is encountered in the discrete Fourier transform (DFT). Except for a normalization factor the linear transformation from a set of data samples to the DFT coefficients is representable as a unitary matrix (see Problem 2.3). Let the $n \times n$ matrix be

$$\mathbf{W} = \frac{1}{\sqrt{n}} \begin{bmatrix} 1 & 1 & 1 & \cdots & 1 \\ 1 & \exp\left(-j\frac{2\pi}{n}\right) & \exp\left(-j\frac{2\pi}{n}2\right) & \cdots & \exp\left(-j\frac{2\pi}{n}[n-1]\right) \\ \vdots & \vdots & \vdots & \vdots & \vdots \\ 1 & \exp\left(-j\frac{2\pi}{n}[n-1]\right) & \exp\left(-j\frac{2\pi}{n}2[n-1]\right) & \cdots & \exp\left(-j\frac{2\pi}{n}[n-1]^2\right) \end{bmatrix}. \tag{2.22}$$

It is easily shown that the columns (and rows) are orthonormal and hence, $\mathbf{W}$ is unitary. A computer program entitled FFT is provided in Appendix 2A to implement the DFT efficiently via a *fast Fourier transform* (FFT).

A real *dyad* or *outer product* is a real square $n \times n$ matrix defined as

$$\mathbf{A} = \mathbf{x}\mathbf{x}^T \tag{2.23}$$

where $\mathbf{x}$ is a real $n \times 1$ vector. Similarly, a complex dyad is defined as

$$\mathbf{A} = \mathbf{x}\mathbf{x}^H \tag{2.24}$$

where $\mathbf{x}$ is a complex $n \times 1$ vector. Note that the complex dyad is explicitly given by

$$\mathbf{A} = \begin{bmatrix} x_1 x_1^* & x_1 x_2^* & \cdots & x_1 x_n^* \\ x_2 x_1^* & x_2 x_2^* & \cdots & x_2 x_n^* \\ \vdots & \vdots & \ddots & \vdots \\ x_n x_1^* & x_n x_2^* & \cdots & x_n x_n^* \end{bmatrix}.$$

The rank of a dyad is 1.

An *idempotent matrix* is a square $n \times n$ matrix that satisfies

$$\mathbf{A}^2 = \mathbf{A}.$$

This condition implies that $\mathbf{A}^l = \mathbf{A}$ for $l \geq 1$. An example is the *projection matrix* (see Problem 2.4)

$$\mathbf{A} = \mathbf{F}(\mathbf{F}^H\mathbf{F})^{-1}\mathbf{F}^H$$

where $\mathbf{F}$ is an $m \times n$ full rank matrix with $m > n$. This matrix occurs in least squares problems (see Section 3.3.2).

A *circulant matrix* is a square $n \times n$ matrix whose elements are formed from the n values $\{a_0, a_1, \ldots, a_{n-1}\}$ as [Gray 1971]

$$\mathbf{A} = \begin{bmatrix} a_0 & a_1 & \cdots & a_{n-1} \\ a_{n-1} & a_0 & \cdots & a_{n-2} \\ \vdots & \vdots & \ddots & \vdots \\ a_1 & a_2 & \cdots & a_0 \end{bmatrix}. \tag{2.25}$$

Each row is generated by circularly shifting the previous row to the right. The inverse of a circulant matrix is easily found from an eigendecomposition (see Section 2.6) as

$$\mathbf{A}^{-1} = \mathbf{W}\mathbf{\Lambda}^{-1}\mathbf{W}^H. \tag{2.26}$$

$\mathbf{W}$ is given by (2.22) and

$$\mathbf{\Lambda}^{-1} = \operatorname{diag}\left(\frac{1}{\lambda_1}, \frac{1}{\lambda_2}, \ldots, \frac{1}{\lambda_n}\right)$$

where λ_i is given by (2.45). The determinant of a circulant matrix is found from (2.41) and (2.45). An example of a circulant matrix is the square $n \times n$ hermitian matrix (see Problem 2.5)

$$\mathbf{R} = \sigma^2\mathbf{I} + P_1\mathbf{e}_1\mathbf{e}_1^H + P_2\mathbf{e}_2\mathbf{e}_2^H \tag{2.27}$$

where

$$\mathbf{e}_i = [1 \ \exp(j2\pi f_i) \ \exp(j2\pi 2 f_i) \ \cdots \ \exp(j2\pi(n-1)f_i)]^T$$

and $f_1 = k/n$, $f_2 = l/n$ for k, l distinct integers in the range $[-n/2, n/2 - 1]$ for n even and $[-(n-1)/2, (n-1)/2]$ for n odd. This matrix arises in Chapters 6 and 11.

A square $n \times n$ *Toeplitz matrix* is defined as

$$[\mathbf{A}]_{ij} = a_{i-j}$$

or

$$\mathbf{A} = \begin{bmatrix} a_0 & a_{-1} & a_{-2} & \cdots & a_{-(n-1)} \\ a_1 & a_0 & a_{-1} & \cdots & a_{-(n-2)} \\ \vdots & \vdots & \vdots & \ddots & \vdots \\ a_{n-1} & a_{n-2} & a_{n-3} & \cdots & a_0 \end{bmatrix}. \tag{2.28}$$

Each element along a NW-SE diagonal is the same. If, in addition, $a_{-k} = a_k^*$, then $\mathbf{A}$ is *hermitian Toeplitz*. If the matrix is real and $a_{-k} = a_k$, then $\mathbf{A}$ is *symmetric Toeplitz*. The inverse of a hermitian Toeplitz matrix is no longer hermitian Toeplitz but has the *persymmetry* property [Cantoni and Butler 1976]. For complex matrices the inverse is hermitian and also symmetric about its principal cross-diagonal, termed the persymmetry property. The principal cross-diagonal is the principal diagonal running from NE to SW. As an example if $\mathbf{A}$ is a 4×4 hermitian Toeplitz matrix, then

$$\mathbf{A}^{-1} = \begin{bmatrix} b_{11} & b_{12} & b_{13} & b_{14} \\ b_{12}^* & b_{22} & b_{23} & b_{13} \\ b_{13}^* & b_{23}^* & b_{22} & b_{12} \\ b_{14}^* & b_{13}^* & b_{12}^* & b_{11} \end{bmatrix}.$$

The inverse is observed to be hermitian. It is also persymmetric or

$$[\mathbf{A}^{-1}]_{ij} = [\mathbf{A}^{-1}]_{n-j+1,n-i+1}.$$

For real Toeplitz matrices the inverse is symmetric about the principal diagonal and the principal cross-diagonal (see Problem 2.7). An explicit procedure for the computation of the inverse and determinant of a hermitian Toeplitz matrix is given by the Levinson recursion described in Chapter 6.

2.4 MATRIX MANIPULATIONS AND FORMULAS

Some useful formulas for the algebraic manipulation of matrices are summarized in this section. For complex $m \times n$ matrices $\mathbf{A}$ and $\mathbf{B}$ the following relationships are useful. The formulas involving inverses and determinants require the matrices to be square or $m = n$.

$$\begin{aligned}
(\mathbf{AB})^T &= \mathbf{B}^T\mathbf{A}^T \\
(\mathbf{AB})^H &= \mathbf{B}^H\mathbf{A}^H \\
(\mathbf{A}^T)^{-1} &= (\mathbf{A}^{-1})^T \\
(\mathbf{A}^H)^{-1} &= (\mathbf{A}^{-1})^H \\
(\mathbf{AB})^{-1} &= \mathbf{B}^{-1}\mathbf{A}^{-1} \\
\det(\mathbf{A}^T) &= \det(\mathbf{A}) \\
\det(\mathbf{A}^H) &= \det{}^*(\mathbf{A}) \\
\det(c\mathbf{A}) &= c^n \det(\mathbf{A}) \qquad c \text{ a scalar} \\
\det(\mathbf{AB}) &= \det(\mathbf{A})\det(\mathbf{B}) \\
\det(\mathbf{A}^{-1}) &= \frac{1}{\det(\mathbf{A})}
\end{aligned} \tag{2.29}$$

It is frequently necessary to determine the inverse of a matrix analytically. To do so we can make use of the following formulas. The inverse of a square $n \times n$ matrix is

$$\mathbf{A}^{-1} = \frac{\mathbf{C}^T}{\det(\mathbf{A})} \tag{2.30}$$

where $\mathbf{C}$ is the square $n \times n$ matrix of cofactors of $\mathbf{A}$. The cofactor matrix is defined by

$$[\mathbf{C}]_{ij} = (-1)^{i+j} M_{ij}$$

where M_{ij} is the minor of a_{ij} obtained by deleting the ith row and jth column of $\mathbf{A}$.

Another formula that is quite useful is the *matrix inversion lemma*,

$$(\mathbf{A} + \mathbf{BCD})^{-1} = \mathbf{A}^{-1} - \mathbf{A}^{-1}\mathbf{B}(\mathbf{DA}^{-1}\mathbf{B} + \mathbf{C}^{-1})^{-1}\mathbf{DA}^{-1} \tag{2.31}$$

where it is assumed that $\mathbf{A}$ is $n \times n$, $\mathbf{B}$ is $n \times m$, $\mathbf{C}$ is $m \times m$, and $\mathbf{D}$ is $m \times n$ and that the indicated inverses exist. A special case known as *Woodbury's identity* results for $\mathbf{B}$ an $n \times 1$ column vector $\mathbf{u}$, $\mathbf{C}$ a scalar of unity, and $\mathbf{D}$ a $1 \times n$ row vector $\mathbf{u}^H$. Then

$$(\mathbf{A} + \mathbf{u}\mathbf{u}^H)^{-1} = \mathbf{A}^{-1} - \frac{\mathbf{A}^{-1}\mathbf{u}\mathbf{u}^H\mathbf{A}^{-1}}{1 + \mathbf{u}^H\mathbf{A}^{-1}\mathbf{u}}. \tag{2.32}$$

Note that Woodbury's identity may be recursively applied to compute the inverse of a matrix plus a sum of dyads (see Problem 2.8). As an example, if the inverse of $\mathbf{A} + \mathbf{u}_1\mathbf{u}_1^H + \mathbf{u}_2\mathbf{u}_2^H$ is desired, let

$$\mathbf{B} = \mathbf{A} + \mathbf{u}_1\mathbf{u}_1^H.$$

Using the identity yields

$$(\mathbf{A} + \mathbf{u}_1\mathbf{u}_1^H + \mathbf{u}_2\mathbf{u}_2^H)^{-1} = \mathbf{B}^{-1} - \frac{\mathbf{B}^{-1}\mathbf{u}_2\mathbf{u}_2^H\mathbf{B}^{-1}}{1 + \mathbf{u}_2^H\mathbf{B}^{-1}\mathbf{u}_2}$$

where

$$\mathbf{B}^{-1} = \mathbf{A}^{-1} - \frac{\mathbf{A}^{-1}\mathbf{u}_1\mathbf{u}_1^H\mathbf{A}^{-1}}{1 + \mathbf{u}_1^H\mathbf{A}^{-1}\mathbf{u}_1}.$$

Partitioned matrices may be manipulated according to the usual rules of matrix algebra by considering each submatrix as an element. For multiplication of partitioned matrices the submatrices which are multiplied together must be conformable. As an illustration, for 2×2 partitioned matrices

$$\begin{aligned}\mathbf{AB} &= \begin{bmatrix} \mathbf{A}_{11} & \mathbf{A}_{12} \\ \mathbf{A}_{21} & \mathbf{A}_{22} \end{bmatrix} \begin{bmatrix} \mathbf{B}_{11} & \mathbf{B}_{12} \\ \mathbf{B}_{21} & \mathbf{B}_{22} \end{bmatrix} \\ &= \begin{bmatrix} \mathbf{A}_{11}\mathbf{B}_{11} + \mathbf{A}_{12}\mathbf{B}_{21} & \mathbf{A}_{11}\mathbf{B}_{12} + \mathbf{A}_{12}\mathbf{B}_{22} \\ \mathbf{A}_{21}\mathbf{B}_{11} + \mathbf{A}_{22}\mathbf{B}_{21} & \mathbf{A}_{21}\mathbf{B}_{12} + \mathbf{A}_{22}\mathbf{B}_{22} \end{bmatrix}.\end{aligned}$$

The conjugate transposition of a partitioned matrix is formed by transposing the "elements" of the matrix and applying H to each submatrix. For a 2×2 partitioned matrix

$$\begin{bmatrix} \mathbf{A}_{11} & \mathbf{A}_{12} \\ \mathbf{A}_{21} & \mathbf{A}_{22} \end{bmatrix}^H = \begin{bmatrix} \mathbf{A}_{11}^H & \mathbf{A}_{21}^H \\ \mathbf{A}_{12}^H & \mathbf{A}_{22}^H \end{bmatrix}.$$

The extension of these properties to arbitrary partitioning is straightforward. Determination of the inverses and determinants of partitioned matrices is facilitated by employing the following formulas. Let $\mathbf{A}$ be a square $n \times n$ matrix partitioned as

$$\mathbf{A} = \begin{bmatrix} \mathbf{A}_{11} & \mathbf{A}_{12} \\ \mathbf{A}_{21} & \mathbf{A}_{22} \end{bmatrix} = \begin{bmatrix} k \times k & k \times (n-k) \\ (n-k) \times k & (n-k) \times (n-k) \end{bmatrix}.$$

Then

$$\mathbf{A}^{-1} = \begin{bmatrix} (\mathbf{A}_{11} - \mathbf{A}_{12}\mathbf{A}_{22}^{-1}\mathbf{A}_{21})^{-1} & -(\mathbf{A}_{11} - \mathbf{A}_{12}\mathbf{A}_{22}^{-1}\mathbf{A}_{21})^{-1}\mathbf{A}_{12}\mathbf{A}_{22}^{-1} \\ -(\mathbf{A}_{22} - \mathbf{A}_{21}\mathbf{A}_{11}^{-1}\mathbf{A}_{12})^{-1}\mathbf{A}_{21}\mathbf{A}_{11}^{-1} & (\mathbf{A}_{22} - \mathbf{A}_{21}\mathbf{A}_{11}^{-1}\mathbf{A}_{12})^{-1} \end{bmatrix} \tag{2.33}$$

$$\begin{aligned} \det(\mathbf{A}) &= \det(\mathbf{A}_{22}) \det(\mathbf{A}_{11} - \mathbf{A}_{12}\mathbf{A}_{22}^{-1}\mathbf{A}_{21}) \\ &= \det(\mathbf{A}_{11}) \det(\mathbf{A}_{22} - \mathbf{A}_{21}\mathbf{A}_{11}^{-1}\mathbf{A}_{12}) \end{aligned} \tag{2.34}$$

where the inverses of $\mathbf{A}_{11}$ and $\mathbf{A}_{22}$ are assumed to exist.

2.5 IMPORTANT THEOREMS

Some important theorems used throughout the book are summarized in this section.

1. A square $n \times n$ matrix $\mathbf{A}$ is invertible (nonsingular) if and only if its columns (or rows) are linearly independent or equivalently if its determinant is nonzero. In such a case, $\mathbf{A}$ is *full rank*. Otherwise, it is singular.
2. A square $n \times n$ matrix $\mathbf{A}$ is positive definite if and only if
   a. it can be written as

$$\mathbf{A} = \mathbf{C}\mathbf{C}^H \tag{2.35}$$

   where $\mathbf{C}$ is also $n \times n$ and is full rank, or
   b. the principal minors are all positive.
   If $\mathbf{A}$ can be written as in (2.35), but $\mathbf{C}$ is not full rank or the principal minors are only nonnegative, then $\mathbf{A}$ is positive semidefinite.
3. If a square $n \times n$ matrix $\mathbf{A}$ can be written as in (2.35) with $\mathbf{C}$ an $n \times m$ matrix and $m < n$, then $\mathbf{A}$ is positive semidefinite. An example is the weighted sum of $m < n$ dyads, or (see Problem 2.9)

$$\mathbf{A} = \sum_{i=1}^{m} d_i \mathbf{u}_i \mathbf{u}_i^H = [\mathbf{u}_1 \quad \mathbf{u}_2 \quad \cdots \quad \mathbf{u}_m] \begin{bmatrix} d_1 & 0 & \cdots & 0 \\ 0 & d_2 & \cdots & 0 \\ \vdots & \vdots & \ddots & \vdots \\ 0 & 0 & \cdots & d_m \end{bmatrix} \begin{bmatrix} \mathbf{u}_1^H \\ \mathbf{u}_2^H \\ \vdots \\ \mathbf{u}_m^H \end{bmatrix}$$

   where the $\mathbf{u}_i$'s are $n \times 1$ vectors and the d_i's are real and positive.

4. If $\mathbf{A}$ is positive definite, the inverse exists and may be found from (2.35) as $\mathbf{A}^{-1} = (\mathbf{C}^{-1})^H(\mathbf{C}^{-1})$.
5. Let $\mathbf{A}$ be positive definite. If $\mathbf{B}$ is an $m \times n$ matrix of full rank with $m \leq n$, then $\mathbf{BAB}^H$ is also positive definite (see Problem 2.10).
6. If $\mathbf{A}$ is positive definite (positive semidefinite), then
   a. the diagonal elements are positive (nonnegative)
   b. the determinant of $\mathbf{A}$, which is a principal minor, is positive (nonnegative).

2.6 EIGENDECOMPOSITION OF MATRICES

An *eigenvector* of a square $n \times n$ matrix $\mathbf{A}$ is an $n \times 1$ vector $\mathbf{v}$ satisfying

$$\mathbf{Av} = \lambda \mathbf{v} \tag{2.36}$$

for some complex scalar λ. λ is the *eigenvalue* of $\mathbf{A}$ corresponding to the eigenvector $\mathbf{v}$. It is assumed that the eigenvector is normalized to have unit length or $\mathbf{v}^H\mathbf{v} = 1$. If $\mathbf{A}$ is hermitian, one can always find n linearly independent eigenvectors, although they will not in general be unique. An example is the identity matrix for which any vector is an eigenvector with eigenvalue one. As a consequence, the eigenvectors of $\mathbf{A} + \sigma^2\mathbf{I}$ are the same as those of $\mathbf{A}$ and the eigenvalues are $\lambda_i + \sigma^2$, $i = 1, 2, \ldots, n$, where the λ_i's are the eigenvalues of $\mathbf{A}$. If $\mathbf{A}$ is hermitian (symmetric for real $\mathbf{A}$), the eigenvectors corresponding to distinct eigenvalues are orthonormal or $\mathbf{v}_i^H\mathbf{v}_j = \delta_{ij}$ and the eigenvalues are real. If, furthermore, the matrix is positive definite (positive semidefinite), the eigenvalues are positive (nonnegative). For a positive semidefinite matrix the rank is equal to the number of nonzero eigenvalues.

The defining relation of (2.36) can also be written as

$$\mathbf{A}[\mathbf{v}_1 \ \ \mathbf{v}_2 \ \ \cdots \ \ \mathbf{v}_n] = [\lambda_1\mathbf{v}_1 \ \ \lambda_2\mathbf{v}_2 \ \ \cdots \ \ \lambda_n\mathbf{v}_n]$$

or

$$\mathbf{AV} = \mathbf{V\Lambda} \tag{2.37}$$

where

$$\mathbf{V} = [\mathbf{v}_1 \ \ \mathbf{v}_2 \ \ \cdots \ \ \mathbf{v}_n]$$

$$\mathbf{\Lambda} = \text{diag}\,(\lambda_1, \lambda_2, \ldots, \lambda_n).$$

If $\mathbf{A}$ is hermitian so that the eigenvectors corresponding to distinct eigenvalues are orthonormal and the remaining eigenvectors are chosen to yield an orthonormal eigenvector set, then $\mathbf{V}$ is a unitary matrix. As such, its inverse is $\mathbf{V}^H$, so that (2.37) becomes

$$\begin{aligned} \mathbf{A} &= \mathbf{V\Lambda V}^H \\ &= \sum_{i=1}^{n} \lambda_i \mathbf{v}_i \mathbf{v}_i^H. \end{aligned} \tag{2.38}$$

This relationship is referred to as the *spectral decomposition* or the *eigendecomposition* of $\mathbf{A}$. Note that if $\mathbf{A} = \mathbf{I}$, then $\lambda_i = 1$, for $i = 1, 2, \ldots, n$. As a result, (2.38) becomes

$$\mathbf{I} = \sum_{i=1}^{n} \mathbf{v}_i \mathbf{v}_i^H. \tag{2.39}$$

The inverse of a hermitian matrix $\mathbf{A}$ is easily found if all the eigenvalues are nonzero (a necessary and sufficient condition for a nonsingular hermitian matrix). From (2.38) the inverse is

$$\begin{aligned} \mathbf{A}^{-1} &= \mathbf{V}^{H^{-1}} \mathbf{\Lambda}^{-1} \mathbf{V}^{-1} = \mathbf{V} \mathbf{\Lambda}^{-1} \mathbf{V}^H \\ &= \sum_{i=1}^{n} \frac{1}{\lambda_i} \mathbf{v}_i \mathbf{v}_i^H. \end{aligned} \tag{2.40}$$

Also, the determinant follows from (2.38) as

$$\det(\mathbf{A}) = \prod_{i=1}^{n} \lambda_i. \tag{2.41}$$

In general, it is difficult to determine the eigendecomposition of a matrix. A special case of interest occurs when $\mathbf{A}$ is a circulant matrix. Then (2.25) can be written as

$$\mathbf{A} = \sum_{k=0}^{n-1} a_k \mathbf{P}^k \tag{2.42}$$

where $\mathbf{P}$ is the *permutation* matrix

$$\mathbf{P} = \begin{bmatrix} 0 & 1 & 0 & \cdots & 0 \\ 0 & 0 & 1 & \cdots & 0 \\ \vdots & \vdots & \vdots & \ddots & \vdots \\ 0 & 0 & 0 & \cdots & 1 \\ 1 & 0 & 0 & \cdots & 0 \end{bmatrix}. \tag{2.43}$$

$\mathbf{P}^0$ is defined in (2.42) to be the identity matrix. The eigenvectors of $\mathbf{P}$ are

$$\begin{aligned} \mathbf{v}_i = \frac{1}{\sqrt{n}} \Bigg[1 \quad & \exp\left[-j\frac{2\pi}{n}(i-1)\right] \\ & \exp\left[-j\frac{2\pi}{n}2(i-1)\right] \quad \cdots \quad \exp\left[-j\frac{2\pi}{n}(n-1)(i-1)\right] \Bigg]^T \end{aligned} \tag{2.44}$$

with corresponding eigenvalues

$$\lambda_i = \exp\left[-j\frac{2\pi}{n}(\mathrm{i}-1)\right]$$

for $i = 1, 2, \ldots, n$. From the representation of (2.42) it follows that the eigenvectors of a circulant matrix are also given by (2.44) with corresponding eigenvalues (see Problem 2.11)

$$\lambda_i = \sum_{k=0}^{n-1} a_k \exp\left[-j\frac{2\pi}{n}k(i-1)\right] \qquad i = 1, 2, \ldots, n. \tag{2.45}$$

Observe that the eigenvalues are found by taking the DFT of the elements in the first row of the circulant matrix.

2.7 SOLUTIONS OF LINEAR EQUATIONS

The solution of simultaneous linear equations plays an important role in spectral estimation. In matrix notation the equations to be solved take the form

$$\mathbf{Ax} = \mathbf{b} \tag{2.46}$$

where $\mathbf{A}$ is $m \times n$, $\mathbf{x}$ is $n \times 1$, and $\mathbf{b}$ is $m \times 1$. Consider first the case of a square matrix $\mathbf{A}$ so that $m = n$. If $\mathbf{A}$ is full rank, there is a unique solution and it is given by $\mathbf{A}^{-1}\mathbf{b}$. If $\mathbf{A}$ is positive definite, then it is guaranteed to be full rank, so that the unique solution is $\mathbf{A}^{-1}\mathbf{b}$. If $\mathbf{A}$ is not full rank, then no solution may exist or an infinite number of solutions may exist. If the rank of the augmented matrix $[\mathbf{Ab}]$ is equal to the rank of $\mathbf{A}$, an infinite number of solutions exist and otherwise no solution exists. To find the solutions, assuming that they exist, let the rank of $\mathbf{A}$ be $k < n$. Then the general solution is given by

$$\mathbf{x} = \mathbf{x}_p + \sum_{i=1}^{n-k} \xi_i \mathbf{x}_{h_i} \tag{2.47}$$

where $\mathbf{x}_p$ is any solution of (2.46) and $\mathbf{x}_{h_i}$, $i = 1, 2, \ldots, n-k$ are the linearly independent solutions of

$$\mathbf{Ax} = \mathbf{0}. \tag{2.48}$$

The coefficients ξ_i are arbitrary constants. The subspace spanned by the solutions to (2.48) is the *nullspace* of $\mathbf{A}$ and its dimension $n - k$ is the *nullity* of $\mathbf{A}$. The subspace spanned by the columns of $\mathbf{A}$ is the *range space* of $\mathbf{A}$ and its dimension k is the rank of $\mathbf{A}$. Note that the rank + nullity = n.

Assuming that a unique solution exists to (2.46) for $\mathbf{A}$ a square $n \times n$ matrix, there are several ways to find it. We can use

$$\mathbf{x} = \mathbf{A}^{-1}\mathbf{b}$$

where the inverse of $\mathbf{A}$ is computed from (2.30). Alternatively, Cramer's rule finds the solution by using

$$x_i = \frac{\det([\mathbf{a}_1 \;\; \mathbf{a}_2 \;\; \cdots \;\; \mathbf{a}_{i-1} \;\; \mathbf{b} \;\; \mathbf{a}_{i+1} \;\; \cdots \;\; \mathbf{a}_n])}{\det(\mathbf{A})} \qquad i = 1, 2, \ldots, n \tag{2.49}$$

where $\mathbf{A} = [\mathbf{a}_1 \quad \mathbf{a}_2 \quad \cdots \quad \mathbf{a}_n]$. If an eigendecomposition of $\mathbf{A}$ is available, then from (2.40) the solution becomes

$$\mathbf{x} = \sum_{i=1}^{n} \frac{1}{\lambda_i} \mathbf{v}_i \mathbf{v}_i^H \mathbf{b}. \tag{2.50}$$

If $\mathbf{A}$ is a positive definite matrix, then a computationally efficient method known as the *Cholesky decomposition* or *square-root* method may be used to solve (2.46) [Lawson and Hanson 1974]. In the first step of this approach $\mathbf{A}$ is decomposed as (see Problem 2.12)

$$\mathbf{A} = \mathbf{L}\mathbf{D}\mathbf{L}^H \tag{2.51}$$

$\mathbf{L}$ is a lower triangular $n \times n$ matrix with 1's on the principal diagonal and $\mathbf{D}$ is a diagonal $n \times n$ matrix with real and positive elements on the principal diagonal or $\mathbf{D} = \text{diag}\,(d_1, d_2, \ldots, d_n)$ with $d_i > 0$. Equation (2.46) is solved by employing two stages of *back-substitution*. Substituting (2.51) into (2.46) yields

$$\mathbf{L}\mathbf{D}\mathbf{L}^H\mathbf{x} = \mathbf{b}.$$

The first stage proceeds by letting

$$\mathbf{y} = \mathbf{D}\mathbf{L}^H\mathbf{x} \tag{2.52}$$

so that $\mathbf{L}\mathbf{y} = \mathbf{b}$ is to be solved. These equations are explicitly given by

$$\begin{bmatrix} 1 & 0 & \cdots & 0 \\ l_{21} & 1 & \cdots & 0 \\ \vdots & \vdots & \ddots & \vdots \\ l_{n1} & l_{n2} & \cdots & 1 \end{bmatrix} \begin{bmatrix} y_1 \\ y_2 \\ \vdots \\ y_n \end{bmatrix} = \begin{bmatrix} b_1 \\ b_2 \\ \vdots \\ b_n \end{bmatrix}.$$

Due to the lower triangular matrix the solution is easily obtained by the following back-substitution recursion:

$$\begin{aligned} y_1 &= b_1 \\ y_k &= b_k - \sum_{j=1}^{k-1} l_{kj} y_j \qquad k = 2, 3, \ldots, n. \end{aligned} \tag{2.53}$$

Once $\mathbf{y}$ is obtained, the final solution is found from (2.52) by solving

$$\mathbf{L}^H\mathbf{x} = \mathbf{D}^{-1}\mathbf{y}$$

which again employs the back-substitution recursion to yield

$$\begin{aligned} x_n &= \frac{y_n}{d_n} \\ x_k &= \frac{y_k}{d_k} - \sum_{j=k+1}^{n} l_{jk}^* x_j \qquad k = n-1, n-2, \ldots, 1. \end{aligned} \tag{2.54}$$

The elements of **L** and **D** may be found from the recursion (see Problem 2.12)

$$d_1 = a_{11}.$$

For $i = 2, 3, \ldots, n$,

$$l_{ij} = \begin{cases} \dfrac{a_{i1}}{d_1} & \text{for } j = 1 \\ \\ \dfrac{a_{ij}}{d_j} - \displaystyle\sum_{k=1}^{j-1} \dfrac{l_{ik} d_k l_{jk}^*}{d_j} & \text{for } j = 2, 3, \ldots, i-1 \end{cases} \tag{2.55}$$

$$d_i = a_{ii} - \sum_{k-1}^{i-1} d_k \mid l_{ik} \mid^2.$$

A computer program entitled CHOLESKY, which is given in Appendix 2B, solves a set of linear equations using the Cholesky decomposition.

Now consider the solution of (2.46) for $m \neq n$. If there are more equations than unknowns or $m > n$, then in general no solution will exist. The usual approach, then, is to find $\mathbf{x}$ so that the *least squares error*

$$\| \mathbf{Ax} - \mathbf{b} \|^2 = (\mathbf{Ax} - \mathbf{b})^H(\mathbf{Ax} - \mathbf{b}) \tag{2.56}$$

is minimized. The minimization can be effected using results provided in Section 2.8 [see (2.75)] to yield (see Problem 2.13)

$$\mathbf{x} = (\mathbf{A}^H\mathbf{A})^{-1}\mathbf{A}^H\mathbf{b}. \tag{2.57}$$

$\mathbf{A}$ is assumed to be full rank so that the inverse of $\mathbf{A}^H\mathbf{A}$ exists. The $n \times m$ matrix $(\mathbf{A}^H\mathbf{A})^{-1}\mathbf{A}^H$ is referred to as the *pseudoinverse* of $\mathbf{A}$ for the overdetermined problem and is denoted by $\mathbf{A}^\#$. If, on the other hand, there are fewer equations than unknowns or $m < n$, then many solutions exist. A common approach is to choose the solution which has minimum length or for which $\mathbf{x}^H\mathbf{x}$ is minimized. This solution, termed the *minimum norm* solution, is

$$\mathbf{x} = \mathbf{A}^H(\mathbf{AA}^H)^{-1}\mathbf{b}. \tag{2.58}$$

$\mathbf{A}$ is assumed to be full rank so that the inverse of $\mathbf{AA}^H$ exists. The $n \times m$ matrix $\mathbf{A}^H(\mathbf{AA}^H)^{-1}$ is termed the pseudoinverse for the underdetermined problem and is also denoted by $\mathbf{A}^\#$.

For the case previously described in which $\mathbf{A}$ is $n \times n$ but has rank $m < n$, the solution was given by (2.47) for $k = m$. This solution is not unique since the ξ_i's are arbitrary. If $\mathbf{A}$ is hermitian, a minimum norm solution is $\mathbf{x} = \mathbf{A}^\#\mathbf{b}$ where the pseudoinverse is given by (see Problem 2.14)

$$\mathbf{A}^\# = \sum_{i=1}^{m} \frac{1}{\lambda_i} \mathbf{v}_i\mathbf{v}_i^H. \tag{2.59}$$

$\{\lambda_1, \lambda_2, \ldots, \lambda_m\}$ are the nonzero eigenvalues of $\mathbf{A}$ and $\{\mathbf{v}_1, \mathbf{v}_2, \ldots, \mathbf{v}_m\}$ are the corresponding eigenvectors.

2.8 MINIMIZATION OF QUADRATIC AND HERMITIAN FUNCTIONS

The need to solve a set of simultaneous linear equations frequently arises from the desire to minimize a quadratic or hermitian function. Consider first the minimization of a real quadratic function of real variables

$$g(\mathbf{x}) = \mathbf{x}^T\mathbf{A}'\mathbf{x} - 2\mathbf{b}'^T\mathbf{x} + c' \tag{2.60}$$

with respect to the real $n \times 1$ vector $\mathbf{x}$. $\mathbf{A}'$ is a real positive definite $n \times n$ matrix, $\mathbf{b}'$ is a real $n \times 1$ vector, and c' is a real scalar. We can make use of the following formulas [Graybill 1976]:

$$\begin{aligned} \frac{\partial \mathbf{x}^T\mathbf{A}'\mathbf{x}}{\partial \mathbf{x}} &= 2\mathbf{A}'\mathbf{x} \\ \frac{\partial \mathbf{b}'^T\mathbf{x}}{\partial \mathbf{x}} &= \mathbf{b}' \end{aligned} \tag{2.61}$$

where $\partial(\ )/\partial\mathbf{x}$ denotes the gradient of a real function with respect to $\mathbf{x}$ (see Problem 2.15). For (2.61) to hold, $\mathbf{A}'$ need only be symmetric and not positive definite. Setting the gradient of (2.60) equal to zero and making use of (2.61) yields the minimizing value of $\mathbf{x}$ as

$$\mathbf{x}_0 = \mathbf{A}'^{-1}\mathbf{b}' \tag{2.62}$$

and the minimum value of the function as

$$g(\mathbf{x}_0) = c' - \mathbf{b}'^T\mathbf{A}'^{-1}\mathbf{b}'. \tag{2.63}$$

Note that if $\mathbf{A}'$ is positive semidefinite, the minimum will still exist but will not be unique. An important application of this result is the minimization of the weighted least squares error

$$g(\mathbf{x}) = (\mathbf{b} - \mathbf{A}\mathbf{x})^T\mathbf{R}^{-1}(\mathbf{b} - \mathbf{A}\mathbf{x}) \tag{2.64}$$

where $\mathbf{A}$ is an $m \times n$ matrix with $m > n$ and is full rank, $\mathbf{b}$ is an $m \times 1$ vector, $\mathbf{x}$ is an $n \times 1$ vector, and $\mathbf{R}$ is a positive definite $m \times m$ matrix. By expanding $g(\mathbf{x})$ and using (2.62) and (2.63), we obtain

$$\mathbf{x}_0 = (\mathbf{A}^T\mathbf{R}^{-1}\mathbf{A})^{-1}\mathbf{A}^T\mathbf{R}^{-1}\mathbf{b} \tag{2.65}$$

and a minimum value of

$$g(\mathbf{x}_0) = \mathbf{b}^T\mathbf{R}^{-1}\mathbf{b} - \mathbf{b}^T\mathbf{R}^{-1}\mathbf{A}(\mathbf{A}^T\mathbf{R}^{-1}\mathbf{A})^{-1}\mathbf{A}^T\mathbf{R}^{-1}\mathbf{b}. \tag{2.66}$$

Alternatively, we could minimize $g(\mathbf{x})$ by expressing it as

$$g(\mathbf{x}) = (\mathbf{b} - \mathbf{A}\mathbf{x}_0)^T\mathbf{R}^{-1}(\mathbf{b} - \mathbf{A}\mathbf{x}_0) + (\mathbf{x} - \mathbf{x}_0)^T\mathbf{A}^T\mathbf{R}^{-1}\mathbf{A}(\mathbf{x} - \mathbf{x}_0) \tag{2.67}$$

where $\mathbf{x}_0$ is given by (2.65). The second term is a quadratic form with a positive definite matrix. ($\mathbf{R}$ is positive definite and hence $\mathbf{R}^{-1}$ is positive definite. Since $\mathbf{A}$ is full rank, the overall matrix $\mathbf{A}^T\mathbf{R}^{-1}\mathbf{A}$ is positive definite.) Because the second term is nonnegative and the first term does not depend on $\mathbf{x}$, the function is minimized for $\mathbf{x} = \mathbf{x}_0$.

Next consider the minimization of a real hermitian function of complex variables

$$g(\mathbf{x}) = \mathbf{x}^H\mathbf{A}'\mathbf{x} - \mathbf{b}'^H\mathbf{x} - \mathbf{x}^H\mathbf{b}' + c'. \tag{2.68}$$

with respect to the complex $n \times 1$ vector $\mathbf{x}$. $\mathbf{A}'$ is a complex positive definite $n \times n$ matrix, $\mathbf{b}'$ is a complex $n \times 1$ vector, and c' is a real scalar. To find the minimum, we could set the gradient of $g(\mathbf{x})$ with respect to the real and imaginary parts of $\mathbf{x}$ equal to zero, or

$$\frac{\partial g(\mathbf{x})}{\partial \mathbf{x}_R} = \frac{\partial g(\mathbf{x})}{\partial \mathbf{x}_I} = \mathbf{0}.$$

Less complicated equations will result if one defines a *complex gradient* as

$$\frac{\partial g(\mathbf{x})}{\partial \mathbf{x}} = \frac{\partial g(\mathbf{x})}{\partial \mathbf{x}_R} + j\frac{\partial g(\mathbf{x})}{\partial \mathbf{x}_I}. \tag{2.69}$$

If the complex gradient is zero, the real gradients with respect to the real and imaginary parts of $\mathbf{x}$ must also equal zero, and conversely. Furthermore, we can show that (see Problem 2.16)

$$\begin{aligned} \frac{\partial \mathbf{x}^H\mathbf{A}'\mathbf{x}}{\partial \mathbf{x}} &= 2\mathbf{A}'\mathbf{x} \\ \frac{\partial(\mathbf{x}^H\mathbf{b}' + \mathbf{b}'^H\mathbf{x})}{\partial \mathbf{x}} &= 2\mathbf{b}' \end{aligned} \tag{2.70}$$

which is the complex version of (2.61). Note that $\mathbf{x}^H\mathbf{A}'\mathbf{x}$ as well as $\mathbf{x}^H\mathbf{b}' + \mathbf{b}'^H\mathbf{x}$ are real functions of $\mathbf{x}$. Using these relations, it follows that $g(\mathbf{x})$ is minimized for

$$\mathbf{x}_0 = \mathbf{A}'^{-1}\mathbf{b}' \tag{2.71}$$

and has a minimum value of

$$g(\mathbf{x}_0) = c' - \mathbf{b}'^H\mathbf{A}'^{-1}\mathbf{b}'. \tag{2.72}$$

In the weighted least squares problem for complex data, it is desired to minimize

$$g(\mathbf{x}) = (\mathbf{b} - \mathbf{A}\mathbf{x})^H\mathbf{R}^{-1}(\mathbf{b} - \mathbf{A}\mathbf{x}) \tag{2.73}$$

where $\mathbf{A}$ is a complex $m \times n$ matrix with $m > n$ and is full rank, $\mathbf{b}$ is a complex $m \times 1$ vector, $\mathbf{x}$ is a complex $n \times 1$ vector, and $\mathbf{R}$ is a complex positive definite $m \times m$ matrix. The solution may be found by expanding $g(\mathbf{x})$ and applying (2.71) and (2.72). Alternatively, if we take the complex gradient of (2.73) by employing (2.70),

$$\frac{\partial g(\mathbf{x})}{\partial \mathbf{x}} = 2\mathbf{A}^H\mathbf{R}^{-1}(\mathbf{A}\mathbf{x} - \mathbf{b}). \tag{2.74}$$

and set it equal to zero, the minimizing solution becomes

$$\mathbf{x}_0 = (\mathbf{A}^H\mathbf{R}^{-1}\mathbf{A})^{-1}\mathbf{A}^H\mathbf{R}^{-1}\mathbf{b} \tag{2.75}$$

with a minimum value of

$$g(\mathbf{x}_0) = \mathbf{b}^H\mathbf{R}^{-1}\mathbf{b} - \mathbf{b}^H\mathbf{R}^{-1}\mathbf{A}(\mathbf{A}^H\mathbf{R}^{-1}\mathbf{A})^{-1}\mathbf{A}^H\mathbf{R}^{-1}\mathbf{b}. \tag{2.76}$$

We could also minimize $g(\mathbf{x})$ by expressing it as (see Problem 2.17)

$$g(\mathbf{x}) = (\mathbf{b} - \mathbf{A}\mathbf{x}_0)^H\mathbf{R}^{-1}(\mathbf{b} - \mathbf{A}\mathbf{x}_0) + (\mathbf{x} - \mathbf{x}_0)^H\mathbf{A}^H\mathbf{R}^{-1}\mathbf{A}(\mathbf{x} - \mathbf{x}_0) \tag{2.77}$$

where $\mathbf{x}_0$ is given by (2.75). A special case that occurs frequently is when $\mathbf{R} = \mathbf{I}$. Then from (2.73),

$$g(\mathbf{x}) = (\mathbf{b} - \mathbf{A}\mathbf{x})^H(\mathbf{b} - \mathbf{A}\mathbf{x}) = \sum_{i=1}^{m} \left| b_i - \sum_{j=1}^{n} a_{ij}x_j \right|^2. \tag{2.78}$$

To minimize this over the x_j, we can set the complex gradient equal to zero. The complex gradient is, from (2.74),

$$\frac{\partial g(\mathbf{x})}{\partial \mathbf{x}} = -2\mathbf{A}^H(\mathbf{b} - \mathbf{A}\mathbf{x})$$

or

$$\frac{\partial g(\mathbf{x})}{\partial x_k} = -2 \sum_{i=1}^{m} a_{ik}^* \left(b_i - \sum_{j=1}^{n} a_{ij}x_j \right) \qquad k = 1, 2, \ldots, n. \tag{2.79}$$

REFERENCES

Cantoni, A., and P. Butler, "Eigenvalues and Eigenvectors of Symmetric Centrosymmetric Matrices," *Linear Algebra Appl.*, Vol. 13, pp. 275–288, Mar. 1976.

Gray, R. M., "Toeplitz and Circulant Matrices: A Review," Tech. Rep. 6502-1, Information Systems Laboratory, Stanford University, June 1971.

Graybill, F. A., *Theory and Application of the Linear Model*, Duxbury Press, Belmont, Calif., 1976.

Hohn, F. E., *Elementary Matrix Algebra*, Macmillan, New York, 1973.

Lawson, C. L., and R. J. Hanson, *Solving Least-Squares Problems*, Prentice-Hall, Englewood Cliffs, N.J., 1974.

Marcus, M., *Basic Theorems in Matrix Theory*, National Bureau of Standards, Applied Math. Ser. 57, Jan. 22, 1960.

Noble, B., and J. W. Daniel, *Applied Linear Algebra*, Prentice-Hall, Englewood Cliffs, N.J., 1977.

Oppenheim, A. V., and R. W. Schafer, *Digital Signal Processing*, Prentice-Hall, Englewood Cliffs, N.J., 1975.

PROBLEMS

2.1. Prove that the inverse of a lower triangular matrix is also lower triangular. *Hint:* Use (2.30).

2.2. Prove that the rows and columns of a unitary matrix are orthonormal as per (2.21).

2.3. Prove that the normalized DFT matrix given in (2.22) is unitary.

2.4. Let $\mathbf{F}$ be a full rank $m \times n$ matrix with $m > n$, $\mathbf{x}$ be an $m \times 1$ vector, and $\mathbf{y}$ be an $m \times 1$ vector. Show that the effect of the linear transformation

$$\mathbf{y} = \mathbf{F}(\mathbf{F}^H\mathbf{F})^{-1}\mathbf{F}^H\mathbf{x}$$

is to project $\mathbf{x}$ onto the subspace spanned by the columns of $\mathbf{F}$. Specifically, if $\{\mathbf{f}_1, \mathbf{f}_2, \ldots, \mathbf{f}_n\}$ are the columns of $\mathbf{F}$, show that

$$(\mathbf{x} - \mathbf{y})^H\mathbf{f}_i = 0 \qquad i = 1, 2, \ldots, n.$$

Why must $\mathbf{F}(\mathbf{F}^H\mathbf{F})^{-1}\mathbf{F}^H$ be idempotent?

2.5. Prove that the matrix given in (2.27) is circulant.

2.6. Prove that if $\mathbf{A}$ and $\mathbf{B}$ are complex $n \times n$ hermitian matrices and $\mathbf{A} - \mathbf{B}$ is positive semidefinite, then

$$[\mathbf{A}]_{ii} \geq [\mathbf{B}]_{ii} \qquad i = 1, 2, \ldots, n.$$

2.7. Find the inverse of the real symmetric Toeplitz matrix

$$\mathbf{A} = \begin{bmatrix} 1 & -a & a^2 \\ -a & 1 & -a \\ a^2 & -a & 1 \end{bmatrix}$$

and show that it is symmetric and persymmetric.

2.8. Find the inverse of the $n \times n$ hermitian matrix $\mathbf{R}$ given by (2.27) by recursively applying Woodbury's identity. Consider the cases where f_1, f_2 are arbitrary and where $f_1 = k/n$, $f_2 = l/n$ for k, l being distinct integers in the range $[-n/2, n/2 - 1]$ for n even and $[-(n-1)/2, (n-1)/2]$ for n odd.

2.9. Prove that the rank of the complex $n \times n$ matrix

$$\mathbf{A} = \sum_{i=1}^{m} d_i \mathbf{u}_i \mathbf{u}_i^H$$

where the $\mathbf{u}_i$'s are linearly independent complex $n \times 1$ vectors and the d_i's are real and positive, is equal to m if $m \leq n$. What is the rank if $m > n$?

2.10. Prove that if $\mathbf{A}$ is a complex $n \times n$ positive definite matrix and $\mathbf{B}$ is a full rank complex $m \times n$ matrix with $m \leq n$, then $\mathbf{B}\mathbf{A}\mathbf{B}^H$ is positive definite.

2.11. Find the eigenvalues of the circulant matrix given in (2.27).

2.12. Verify the equations given for the Cholesky decomposition, (2.53)–(2.55). Use these equations to find the inverse of the matrix given in Problem 2.7.

2.13. Consider the problem of fitting the data $\{x[0], x[1], \ldots, x[N-1]\}$ by the sum of a dc signal and a sinusoid as

$$\hat{x}[n] = \mu + A_c \exp(j2\pi f_0 n) \qquad n = 0, 1, \ldots, N-1.$$

The complex dc level μ and the complex sinusoidal amplitude A_c are unknown. We may view the determination of μ, A_c as the solution of the overdetermined set of linear equations

$$\begin{bmatrix} 1 & 1 \\ 1 & \exp(j2\pi f_0) \\ \vdots & \vdots \\ 1 & \exp(j2\pi f_0[N-1]) \end{bmatrix} \begin{bmatrix} \mu \\ A_c \end{bmatrix} = \begin{bmatrix} x[0] \\ x[1] \\ \vdots \\ x[N-1] \end{bmatrix}.$$

Find the least squares solution for μ and A_c. Now assume that $f_0 = k/N$, where k is a nonzero integer in the range $[-N/2, N/2 - 1]$ for N even and $[-(N-1)/2, (N-1)/2]$ for N odd and again find the least squares solution.

2.14. Consider the solution to the set of linear equations

$$\mathbf{R}\mathbf{x} = -\mathbf{r}$$

where $\mathbf{R}$ is an $n \times n$ hermitian matrix given by

$$\mathbf{R} = P_1 \mathbf{e}_1 \mathbf{e}_1^H + P_2 \mathbf{e}_2 \mathbf{e}_2^H$$

and $\mathbf{r}$ is a complex $n \times 1$ vector given by

$$\mathbf{r} = P_1 \exp(j2\pi f_1)\mathbf{e}_1 + P_2 \exp(j2\pi f_2)\mathbf{e}_2.$$

The complex vectors $\mathbf{e}_1$, $\mathbf{e}_2$ are defined in (2.27). It is assumed that $f_1 = k/n$, $f_2 = l/n$ for k, l distinct integers in the range $[-n/2, n/2 - 1]$ for n even and $[-(n-1)/2, (n-1)/2]$ for n odd. $\mathbf{x}$ is a complex $n \times 1$ vector. First show that $\mathbf{R}$ is a singular matrix (assuming $n > 2$) and that there are an infinite number of solutions. Find the general solution and also the minimum norm solution. *Hint:* Note that $\mathbf{e}_1/\sqrt{n}$, $\mathbf{e}_2/\sqrt{n}$ are eigenvectors of $\mathbf{R}$ with nonzero eigenvalues and that the remaining eigenvectors have zero eigenvalues. Then assume a solution of the form

$$\mathbf{x} = \xi_1 \mathbf{e}_1 + \xi_2 \mathbf{e}_2 + \sum_{i=3}^{n} \xi_i \mathbf{e}_i$$

where $\mathbf{e}_i^H \mathbf{e}_1 = 0$, $\mathbf{e}_i^H \mathbf{e}_2 = 0$ for $i = 3, 4, \ldots, n$ and solve for ξ_1, ξ_2.

2.15. Verify the formulas given for the gradient of a quadratic and linear form (2.61).

2.16. Verify the formulas given for the complex gradient of a hermitian and a linear form (2.70). To do so, decompose the matrices and vectors into their real and imaginary parts as $\mathbf{x} = \mathbf{x}_R + j\mathbf{x}_I$, $\mathbf{A}' = \mathbf{A}_R + j\mathbf{A}_I$, $\mathbf{b}' = \mathbf{b}_R + j\mathbf{b}_I$. Apply the gradient formulas for real functions as given by (2.61) and note that $\mathbf{A}_R^T = \mathbf{A}_R$, $\mathbf{A}_I^T = -\mathbf{A}_I$ since $\mathbf{A}'$ is hermitian.

2.17. Verify the alternative expression (2.77) for a hermitian function.

2.18. Prove that the inverse of a complex matrix $\mathbf{A} = \mathbf{A}_R + j\mathbf{A}_I$ may be found by first inverting

$$\begin{bmatrix} \mathbf{A}_R & -\mathbf{A}_I \\ \mathbf{A}_I & \mathbf{A}_R \end{bmatrix}$$

to yield

$$\begin{bmatrix} \mathbf{B}_R & -\mathbf{B}_I \\ \mathbf{B}_I & \mathbf{B}_R \end{bmatrix}$$

and then letting $\mathbf{A}^{-1} = \mathbf{B}_R + j\mathbf{B}_I$.

APPENDIX 2A

Computer Program for FFT (Provided on Floppy Disk)

The program contained in this appendix computes the discrete Fourier transform (DFT) $X[k]$ of a data sequence $x[n]$ using the fast Fourier transform (FFT). The DFT is defined as

$$X[k] = \sum_{n=0}^{N-1} x[n] \exp\left(-\frac{j2\pi kn}{N}\right) \qquad k = 0, 1, \ldots, N-1$$

and the inverse DFT is defined as

$$x[n] = \frac{1}{N} \sum_{k=0}^{N-1} X[k] \exp\left(\frac{j2\pi kn}{N}\right) \qquad n = 0, 1, \ldots, N-1.$$

```
      SUBROUTINE FFT(X,M,INVRS)
C
C     This program computes the fast Fourier transform of
C     an array of complex data using a decimation-in-time
C     FFT. See Oppenheim and Schafer [1975] for further details.
C
C     Input Parameters:
C
C       X          -Complex array of dimension 2**Mx1 of data samples
C       M          -Power of two for FFT size
C       INVRS      -Set equal to:  1 for forward Fourier transform
C                                 -1 for inverse Fourier transform
C
C     Output Parameters:
C
C       X          -Complex array of dimension 2**Mx1 of Fourier
C                   transform values
C
C     Notes:
C
C       The calling program must dimension the complex array X
C       greater than or equal to 2**Mx1. The complex array Y must
C       be dimensioned greater than or equal to 2**Mx1.
C
C     Verification Test Case:
C
```

```
C         If M=3, INVRS=1, and
C           X(1)=X(5)=(1., 0.)
C           X(2)=X(6)=(0., 1.)
C           X(3)=X(7)=(-1., 0.)
C           X(4)=X(8)=(0., -1.)
C         then the output should be
C           X(1)=(0.00000, 0.00000)          X(5)=(0.00000,       0.00000)
C           X(2)=(0.00000, 0.00000)          X(6)=(0.00000,       0.00000)
C           X(3)=(8.00000, -1.74845E-07)     X(7)=(0.00000, 1.74845E-07)
C           X(4)=(0.00000, 0.00000)          X(8)=(0.00000,       0.00000)
C         for a DEC VAX 11/780.
C
          COMPLEX X(1), Y(2**M),W,SAVE
          N=2**M
C       Put the data into bit reversed form
          Y(1)=X(1)
          DO 20 I=2,N
          NUM=I-1
          J=0
          L=N
          DO 10 K=1,M
          L=L/2
          IF(NUM.LT.L)GO TO 10
          J=J+2**(K-1)
          NUM=NUM-L
10        CONTINUE
          Y(I)=X(J+1)
20        CONTINUE
C       Begin computation of FFT
          PI=4.*ATAN(1.)
          LDFT=1
          NDFT=N
          DO 50 K=1,M
          LDFT=2*LDFT
          NDFT=NDFT/2
          DO 40 I=1,NDFT
          DO 30 J=1,LDFT/2
C       Determine twiddle factor
          ARG=-(2.*PI*INVRS)*(J-1.)/LDFT
          W=CMPLX(COS(ARG),SIN(ARG))
C       Determine indices for next butterfly to be computed
          NP=J+LDFT*(I-1)
          NQ=NP+LDFT/2
C       Compute butterfly for the Ith DFT in stage K
          SAVE=Y(NP)+W*Y(NQ)
          Y(NQ)=Y(NP)-W*Y(NQ)
```

```
30        Y(NP)=SAVE
40        CONTINUE
50        CONTINUE
C       Divide array values by N if inverse FFT desired
          DO 60 I=1,N
          IF(INVRS.EQ.1)X(I)=Y(I)
60        IF(INVRS.EQ.-1)X(I)=Y(I)/N
          RETURN
          END
```

APPENDIX 2B

Computer Program For Cholesky Solution of Simultaneous Linear Equations (Provided on Floppy Disk)

```
           SUBROUTINE CHOLESKY(A,B,N,EPS,IFLAG)
C       This program solves a hermitian positive definite set of
C       complex linear simultaneous equations using the Cholesky
C       decomposition method. See (2.53)-(2.55).
C
C                 AX=B
C
C       Input Parameters:
C
C         N         -Order of the matrix (number of linear equations)
C         EPS       -Small positive number to test for ill-
C                    conditioning of matrix A
C         A         -Complex array of dimension N*(N+1)/2 of matrix
C                    elements stored columnwise for upper triangular part,
C                     i.e., A(1,1) stored as A(1), A(1,2) stored as A(2),
C                     A(2,2) stored as A(3), etc. (The lower triangular part
C                     is found by hermitian symmetry.)
C         B         -Complex array of dimension Nx1 of right-hand-side
C                    vector elements
C
C       Output Parameters:
C
```

```
C         IFLAG     -Ill-conditioning indicator
C                       =    0 for normal program termination
C                       =  -1 if one of the di's is less than EPS
C                             causing premature termination
C         B         -Complex array of dimension Nx1 containing solution
C                    X vector stored in place of B on output
C
C       Notes:
C
C         The calling program must dimension arrays A,B. The arrays
C         XL,Y,D must be dimensioned greater than or equal to the variable
C         dimensions shown.
C
C       Verification Test Case:
C
C         If A(1)=(2.,0.), A(2)=(-1.,-1.), A(3)=(2.,0.), B(1)=(1.,-1.),
C         B(2)=(0.,0.), N=2, and EPS=1.0E-15, then the output should be
C           B(1)=(0.99999,-0.99999)
C           B(2)=(0.00000,-0.99999)
C           IFLAG=0
C         for a DEC VAX 11/780.
C
          COMPLEX A(1),B(1),XL(N,N),Y(N)
          DIMENSION D(N)
C       Factor into triangular and diagonal form (2.55)
          IFLAG=0
          L=1
          D(1)=REAL(A(L))
          DO 40 I=2,N
          DO 20 J=1,I-1
          L=L+1
          XL(I,J)=CONJG(A(L))/D(J)
          IF(J.EQ.1)GO TO 20
          DO 10 K=1,J-1
10        XL(I,J)=XL(I,J)-XL(I,K)*CONJG(XL(J,K))*D(K)/D(J)
20        CONTINUE
          L=L+1
          D(I)=REAL(A(L))
          DO 30 K=1,I-1
30        D(I)=D(I)-D(K)*CABS(XL(I,K))**2
C       Test for non-positive value of di
          IF(D(I).GT.EPS)GO TO 40
          IFLAG=-1
          RETURN
```

```
40      CONTINUE
C     Solve for intermediate column vector solution (2.53)
        Y(1)=B(1)
        DO 60 K=2,N
        Y(K)=B(K)
        DO 50 J=1,K-1
50      Y(K)=Y(K)-XL(K,J)*Y(J)
60      CONTINUE
C     Solve for final column vector solution (2.54)
        B(N)=Y(N)/D(N)
        DO 80 K=N-1,1,-1
        B(K)=Y(K)/D(K)
        DO 70 J=K+1,N
70      B(K)=B(K)-CONJG(XL(J,K))*B(J)
80      CONTINUE
        RETURN
        END
```

Chapter 3

Review of Probability, Statistics, and Random Processes

3.1 INTRODUCTION

An assumption is made that the reader already has some familiarity with probability theory, elementary statistics, and basic random process theory. This chapter serves as a review of these topics as well as a summary of some more advanced concepts and theorems. For those readers needing a more extensive review or more details of the advanced concepts, the text by Papoulis [1965] on probability and random processes and the treatise on statistics by Kendall and Stuart [1977] are recommended.

3.2 SOME USEFUL PROBABILITY DENSITY FUNCTIONS

3.2.1 Real Random Variables

A probability density function (PDF) that is frequently used to model the statistical behavior of a random variable is the Gaussian distribution. A real random variable x with mean μ_x and variance σ_x^2 is distributed according to a Gaussian or normal distribution if the PDF is given by

$$p(x) = \frac{1}{\sqrt{2\pi}\,\sigma_x} \exp\left[-\frac{1}{2}\left(\frac{x-\mu_x}{\sigma_x}\right)^2\right] \qquad -\infty < x < \infty. \tag{3.1}$$

The shorthand notation $x \sim N(\mu_x, \sigma_x^2)$ is often used, where $\sim$ means "is distributed according to.'' The extension to a set of real random variables or real random vector $\mathbf{x} = [x_0 \quad x_1 \quad \cdots \quad x_{N-1}]^T$ with mean

$$\mathcal{E}(\mathbf{x}) = \boldsymbol{\mu}_x$$

and covariance matrix

$$\mathcal{E}[(\mathbf{x} - \boldsymbol{\mu}_x)(\mathbf{x} - \boldsymbol{\mu}_x)^T] = \mathbf{C}_{xx}$$

is the multivariate Gaussian PDF

$$p(\mathbf{x}) = \frac{1}{(2\pi)^{N/2} \det^{1/2}(\mathbf{C}_{xx})} \exp\left[-\frac{1}{2} (\mathbf{x} - \boldsymbol{\mu}_x)^T \mathbf{C}_{xx}^{-1} (\mathbf{x} - \boldsymbol{\mu}_x) \right]. \quad (3.2)$$

Note that $\mathbf{C}_{xx}$ is $N \times N$ with $[\mathbf{C}_{xx}]_{ij} = \mathcal{E}\{[x_i - \mathcal{E}(x_i)][x_j - \mathcal{E}(x_j)]\}$. If $\mathbf{C}_{xx}$ is a diagonal matrix, the random variables are uncorrelated. In this case (3.2) factors into the product of N univariate Gaussian PDFs of the form of (3.1), and hence the random variables are also independent. If $\mathbf{x}$ is zero mean, the higher-order joint moments are easily computed. In particular, the fourth-order moment is

$$\mathcal{E}(x_i x_j x_k x_l) = \mathcal{E}(x_i x_j)\mathcal{E}(x_k x_l) + \mathcal{E}(x_i x_k)\mathcal{E}(x_j x_l) + \mathcal{E}(x_i x_l)\mathcal{E}(x_j x_k).$$

If $\mathbf{x}$ is linearly transformed as

$$\mathbf{y} = \mathbf{A}\mathbf{x} + \mathbf{b}$$

where $\mathbf{A}$ is $N \times N$ and $\mathbf{b}$ is $N \times 1$, then $\mathbf{y}$ is also distributed according to a multivariate Gaussian distribution with

$$\mathcal{E}(\mathbf{y}) = \boldsymbol{\mu}_y = \mathbf{A}\boldsymbol{\mu}_x + \mathbf{b} \quad (3.3)$$

and

$$\mathcal{E}[(\mathbf{y} - \boldsymbol{\mu}_y)(\mathbf{y} - \boldsymbol{\mu}_y)^T] = \mathbf{C}_{yy} = \mathbf{A}\mathbf{C}_{xx}\mathbf{A}^T. \quad (3.4)$$

Of particular interest is the transformation $\mathbf{A}$, which decorrelates or whitens the elements of $\mathbf{x}$. This whitening occurs if

$$\mathbf{A}\mathbf{C}_{xx}\mathbf{A}^T = \mathbf{D} \quad (3.5)$$

where $\mathbf{D}$ is a diagonal matrix with positive diagonal elements. Many such matrices exist. One possibility follows from the Cholesky decomposition described in Section 2.7. The Cholesky decomposition factors a square positive definite matrix $\mathbf{C}_{xx}$ as

$$\mathbf{C}_{xx} = \mathbf{L}\mathbf{D}\mathbf{L}^T \quad (3.6)$$

where $\mathbf{L}$ is a lower triangular matrix with 1's on the principal diagonal. Comparing (3.6) with (3.5) it is seen that $\mathbf{A} = \mathbf{L}^{-1}$, which is again lower triangular. Hence

$$\mathbf{y} = \mathbf{L}^{-1}\mathbf{x} + \mathbf{b}$$

yields a set of independent Gaussian random variables (see Problem 3.1).

A useful PDF is the χ^2 distribution, which is derived from the Gaussian distribution. If $\mathbf{x}$ is composed of independent and identically distributed random variables with $x_i \sim N(0, 1)$, $i = 0, 1, \ldots, N-1$, then

$$y = \sum_{i=0}^{N-1} x_i^2 \sim \chi_N^2$$

where χ_N^2 denotes a χ^2 random variable with N degrees of freedom. The PDF is given as (see Problem 3.2)

$$p(y) = \begin{cases} \dfrac{1}{2^{N/2}\Gamma(N/2)} y^{N/2-1} \exp\left(-\dfrac{y}{2}\right) & \text{for } y \geq 0 \\ 0 & \text{for } y < 0 \end{cases} \tag{3.7}$$

where $\Gamma(u)$ is the gamma integral. The mean and variance of y are

$$\begin{aligned} \mathscr{E}(y) &= N \\ \operatorname{var}(y) &= 2N. \end{aligned} \tag{3.8}$$

The cumulative distribution function (CDF) is

$$\Phi(y) = \begin{cases} 0 & \text{for } y < 0 \\ \displaystyle\int_0^y p(\xi)\, d\xi & \text{for } y \geq 0 \end{cases} \tag{3.9}$$

and can be shown to be

$$\Phi(y) = \begin{cases} 1 - \exp\left(-\dfrac{y}{2}\right) \displaystyle\sum_{k=0}^{N/2-1} \frac{(y/2)^k}{k!} & \text{for } N \text{ even} \\ 1 - \sqrt{\dfrac{2}{\pi}} \displaystyle\int_{\sqrt{y}}^{\infty} \exp\left(-\frac{t^2}{2}\right) dt & \\ \quad - \dfrac{2}{\sqrt{\pi}} \exp\left(-\dfrac{y}{2}\right) \displaystyle\sum_{k=1}^{(N-1)/2} \frac{k!}{(2k)!} (2y)^{k-1/2} & \text{for } N \text{ odd.} \end{cases}$$

The CDF is useful for certain hypothesis testing procedures. A tabulation of the CDF for different degrees of freedom can be found in the work by Abramowitz and Stegen [1966].

3.2.2 Complex Random Variables

A Gaussian PDF can be defined for a complex random variable $x = u + jv$, where u and v are both real. Let u and v be independent and distributed as

$$u \sim N\left(\mu_u, \frac{\sigma_x^2}{2}\right)$$

$$v \sim N\left(\mu_v, \frac{\sigma_x^2}{2}\right).$$

The joint PDF of u and v is the product of the individual PDFs:

$$p(u, v) = \frac{1}{\pi\sigma_x^2} \exp\left\{-\frac{1}{\sigma_x^2}[(u - \mu_u)^2 + (v - \mu_v)^2]\right\}.$$

Noting that

$$\mathscr{E}(x) = \mathscr{E}(u) + j\mathscr{E}(v) = \mu_u + j\mu_v = \mu_x$$

and

$$\begin{aligned}\text{var}(x) &= \mathscr{E}[|(u + jv) - (\mu_u + j\mu_v)|^2] \\ &= \mathscr{E}[(u - \mu_u)^2] + \mathscr{E}[(v - \mu_v)^2] \\ &= \sigma_x^2\end{aligned}$$

the PDF can be written as

$$p(x) = \frac{1}{\pi\sigma_x^2} \exp\left[-\frac{|x - \mu_x|^2}{\sigma_x^2}\right] \tag{3.10}$$

where μ_x is the mean and σ_x^2 is the variance of x. The notation $x \sim CN(\mu_x, \sigma_x^2)$ is sometimes used. For complex random vectors a multivariate PDF can be defined [Goodman 1963]. If $\mathbf{x} = \mathbf{u} + j\mathbf{v}$ is a complex random vector with mean $\boldsymbol{\mu}_x$ and covariance matrix

$$\mathscr{E}[(\mathbf{x} - \boldsymbol{\mu}_x)(\mathbf{x} - \boldsymbol{\mu}_x)^H] = \mathbf{C}_{xx} \tag{3.11}$$

then $\mathbf{x}$ is distributed according to a complex multivariate Gaussian distribution under the following conditions: $\mathbf{u}$ and $\mathbf{v}$ are jointly distributed according to a real multivariate Gaussian distribution and the real vector $[\mathbf{u}^T \quad \mathbf{v}^T]^T$ has a covariance matrix

$$\begin{bmatrix} \mathscr{E}[(\mathbf{u} - \boldsymbol{\mu}_u)(\mathbf{u} - \boldsymbol{\mu}_u)^T] & \mathscr{E}[(\mathbf{u} - \boldsymbol{\mu}_u)(\mathbf{v} - \boldsymbol{\mu}_v)^T] \\ \mathscr{E}[(\mathbf{v} - \boldsymbol{\mu}_v)(\mathbf{u} - \boldsymbol{\mu}_u)^T] & \mathscr{E}[(\mathbf{v} - \boldsymbol{\mu}_v)(\mathbf{v} - \boldsymbol{\mu}_v)^T] \end{bmatrix} = \frac{1}{2}\begin{bmatrix} \mathbf{E} & -\mathbf{F} \\ \mathbf{F} & \mathbf{E} \end{bmatrix} \tag{3.12}$$

where $\mathbf{F}^T = -\mathbf{F}$. Under these conditions if one lets $\mathbf{C}_{xx} = \mathbf{E} + j\mathbf{F}$, the PDF is given by

$$p(\mathbf{x}) = \frac{1}{\pi^N \det(\mathbf{C}_{xx})} \exp[-(\mathbf{x} - \boldsymbol{\mu}_x)^H \mathbf{C}_{xx}^{-1}(\mathbf{x} - \boldsymbol{\mu}_x)]. \tag{3.13}$$

Note that (3.13) reduces to (3.10) for $N = 1$ (see Problem 3.3). Properties of the complex multivariate Gaussian PDF which are analogous to those for the real case are that linear transformations of complex Gaussian vectors yields a complex Gaussian vector and the fourth-order moments can be factored into sums of products of second-order moments [Reed 1962, Goodman 1963].

3.3 ESTIMATION THEORY

Parameter estimation may be classified as estimation either of unknown but constant parameters, termed the *deterministic parameter* case or of statistically fluctuating parameters, termed the *random parameter* case. Estimation of deterministic parameters is sometimes referred to as *classical* estimation, while the estimation of random parameters is described as *Bayesian* estimation. In either case $\boldsymbol{\theta}$ (a vector in general) will denote the parameter to be estimated and $\hat{\boldsymbol{\theta}}$ will denote its estimator.

3.3.1 Deterministic Parameters

It will be assumed in the following discussion that θ is a *real* parameter to be estimated from the *real* data set $\{x[0], x[1], \ldots, x[N-1]\}$. Extensions to complex data are possible, although in practice the theory for real data is usually adequate. Where the extensions are useful they will be described.

There are many desirable properties that an estimator should possess. As the number of data points N becomes large, a good estimator should converge to the true parameter value. Such an estimator is termed *consistent*. For an estimator of a scalar parameter to be consistent,

$$\lim_{N\to\infty} \Pr(|\hat{\theta} - \theta| > \epsilon) = 0 \tag{3.14}$$

where Pr denotes probability and ϵ is a small positive number. For the estimator of a vector parameter to be consistent, (3.14) must hold with the absolute-value sign replaced by a suitable norm. One way to measure the quality of an estimator of a scalar parameter is by the *mean square error* (MSE)

$$\text{MSE} = \mathcal{E}[(\hat{\theta} - \theta)^2]. \tag{3.15}$$

The MSE can be further decomposed into the sum of the estimator variance and the estimator bias as

$$\text{MSE} = \mathcal{E}\{[\hat{\theta} - \mathcal{E}(\hat{\theta})]^2\} + \{\mathcal{E}(\hat{\theta}) - \theta\}^2.$$

The first term is the variance while the second term is the square of *bias*, which is defined as

$$B(\theta) = \mathcal{E}(\hat{\theta}) - \theta. \tag{3.16}$$

The bias measures the *average* deviation of the estimator from the true value. An estimator is said to be *unbiased* if $B(\theta) = 0$ for all θ. In such a case the MSE reduces to the variance. We can usually trade off bias for variance by a slight modification of the estimator (usually a scaling) but only an increase in N will allow both errors to be reduced simultaneously. For vector parameters all the definitions and results are valid if instead of squaring the various quantities, we use the norm squared.

As an alternative to the MSE for describing the performance of an estimator, we could give an interval (assuming a scalar parameter) which includes the true value of the parameter with high probability. Such an interval is termed a *confidence interval*. Two cases are of primary interest. They are when $\hat{\theta}$ is distributed according to a Gaussian distribution as $\hat{\theta} \sim N(\theta, a\theta^2)$ and when a scaled version of $\hat{\theta}$ is distributed according to a χ^2 distribution as $\nu\hat{\theta}/\theta \sim \chi^2_\nu$. In the first case the $100(1 - \alpha)\%$ confidence interval is

$$\left(\frac{\hat{\theta}}{1 + \sqrt{a}\, z(1 - \alpha/2)}, \frac{\hat{\theta}}{1 - \sqrt{a}\, z(1 - \alpha/2)}\right).$$

$z(\alpha)$ is the α percentage point of a $N(0, 1)$ CDF, or

$$\Pr\{Z \leq z(\alpha)\} = \alpha$$

where $Z \sim N(0, 1)$. For the second case the $100(1 - \alpha)\%$ confidence interval is

$$\left(\frac{\nu\hat{\theta}}{\chi^2_\nu(1 - \alpha/2)}, \frac{\nu\hat{\theta}}{\chi^2_\nu(\alpha/2)}\right)$$

where $\chi^2_\nu(\alpha)$ is the α percentage point of a χ^2_ν CDF, or

$$\Pr\{\chi^2_\nu \leq \chi^2_\nu(\alpha)\} = \alpha.$$

In searching for good estimators it is advantageous to be able to bound the performance of any estimator. As a practical matter we are usually interested in unbiased estimators, so that a figure of merit of the performance of the estimator is the variance (which is identical to the MSE). The variance of any unbiased estimator is bounded from below by the Cramer–Rao (CR) bound, which is computed from the PDF of $\mathbf{x}$. Specifically, the CR bound asserts that for a scalar parameter

$$\operatorname{var}(\hat{\theta}) \geq \frac{1}{\mathcal{E}[(\partial \ln p(\mathbf{x}; \theta)/\partial\theta)^2]} \tag{3.17}$$

where $p(\mathbf{x}; \theta)$ is the PDF of $\mathbf{x}$ with the dependence on θ explicitly noted. The denominator of (3.17) is evaluated at the true value of the parameter θ. Alternatively, the denominator of (3.17) can be computed from the relation

$$\mathcal{E}\left[\left(\frac{\partial \ln p(\mathbf{x}; \theta)}{\partial\theta}\right)^2\right] = -\mathcal{E}\left[\frac{\partial^2 \ln p(\mathbf{x}; \theta)}{\partial\theta^2}\right]. \tag{3.18}$$

An estimator which actually attains the CR bound is said to be *efficient* in that it uses the data efficiently. It is possible that such an estimator does not exist. If a differentiable function of θ, say $g(\theta)$, is to be estimated, then

$$\operatorname{var}(\hat{g}) \geq \frac{(\partial g/\partial\theta)^2}{\mathcal{E}[(\partial \ln p(\mathbf{x}; \theta)/\partial\theta)^2]}.$$

For a $p \times 1$ vector parameter the CR bound asserts that the *covariance matrix* of $\hat{\boldsymbol{\theta}}$ which is denoted by $\mathbf{C}_{\hat{\theta}}$, is always "larger" than some $\mathbf{I}_\theta^{-1}$. The bound is expressed as

$$\mathbf{C}_{\hat{\theta}} \geq \mathbf{I}_\theta^{-1} \tag{3.19}$$

with the $\geq$ sign is interpreted as meaning that the matrix $\mathbf{C}_{\hat{\theta}} - \mathbf{I}_\theta^{-1}$ is positive semidefinite. $\mathbf{I}_\theta$ is evaluated at the true value of $\boldsymbol{\theta}$. In particular, for a positive semidefinite matrix the diagonal elements are all nonnegative, which implies that (see Problem 2.6)

$$\text{var}(\hat{\theta}_i) \geq [\mathbf{I}_\theta^{-1}]_{ii} \qquad i = 1, 2, \ldots, p. \tag{3.20}$$

$\mathbf{I}_\theta$ is termed the *Fisher information matrix* and is defined as

$$\mathbf{I}_\theta = \mathscr{E}\left[\left(\frac{\partial \ln p(\mathbf{x}; \boldsymbol{\theta})}{\partial \boldsymbol{\theta}}\right)\left(\frac{\partial \ln p(\mathbf{x}; \boldsymbol{\theta})}{\partial \boldsymbol{\theta}}\right)^T\right] \tag{3.21}$$

where $\partial \ln p(\mathbf{x}; \boldsymbol{\theta})/\partial \boldsymbol{\theta}$ is the gradient of $\ln p(\mathbf{x}; \boldsymbol{\theta})$ with respect to $\boldsymbol{\theta}$. The $[i, j]$ element of the Fisher information matrix is

$$\begin{aligned} [\mathbf{I}_\theta]_{ij} &= \mathscr{E}\left[\frac{\partial \ln p(\mathbf{x}; \boldsymbol{\theta})}{\partial \theta_i} \frac{\partial \ln p(\mathbf{x}; \boldsymbol{\theta})}{\partial \theta_j}\right] \\ &= -\mathscr{E}\left[\frac{\partial^2 \ln p(\mathbf{x}; \boldsymbol{\theta})}{\partial \theta_i \, \partial \theta_j}\right]. \end{aligned} \tag{3.22}$$

An estimator whose covariance matrix is given by $\mathbf{I}_\theta^{-1}$ is said to be efficient.

If a vector function of $\boldsymbol{\theta}$ is to be estimated, i.e., $\mathbf{g}(\boldsymbol{\theta})$, where $\mathbf{g}$ maps the $p \times 1$ vector $\boldsymbol{\theta}$ into the $r \times 1$ vector $\mathbf{g}(\boldsymbol{\theta})$, then the CR bound is

$$\mathbf{C}_{\hat{g}} \geq \mathscr{J}\mathbf{I}_\theta^{-1}\mathscr{J}^T. \tag{3.23}$$

$\mathscr{J}$ is the Jacobian matrix of dimension $r \times p$ of the transformation and is defined as

$$[\mathscr{J}]_{ij} = \frac{\partial g_i(\boldsymbol{\theta})}{\partial \theta_j}.$$

There are two principal methods for finding estimators of parameters. The first approach, termed the *method of moments*, does not usually lead to efficient estimators but in many instances provides estimators which are readily computed. The second approach, termed the *method of maximum likelihood*, is efficient for large data records but may be difficult to compute. When the maximum likelihood estimator (MLE) can be implemented, it is to be preferred. The method of moments estimates the unknown parameters by equating theoretical moments to sample moments. As an example, to estimate the mean of a Gaussian distribution, we can set the theoretical first moment $\mathscr{E}(x) = \mu_x$ equal to the sample first moment $1/N \sum_{i=0}^{N-1} x_i$. The method of maximum likelihood estimates $\boldsymbol{\theta}$ by choosing that value of $\boldsymbol{\theta}$ which maximizes $p(\mathbf{x}; \boldsymbol{\theta})$, where the *observed value* of $\mathbf{x}$ is substituted into the PDF. The PDF is thus considered as a function of $\boldsymbol{\theta}$ and is referred to as the *likelihood function*. The rationale for this approach is that because $\mathbf{x}$ was observed it must have been very likely. Hence the value of $\boldsymbol{\theta}$ that yields the largest probability for the observed value of $\mathbf{x}$ is probably close to the true value. The MLE has many desirable properties. If an estimator exists that is unbiased and attains the CR bound, then the method of maximum likelihood will produce it. For large data records (N large) the MLE is

1. *Unbiased*, $\mathscr{E}(\hat{\boldsymbol{\theta}}) \rightarrow \boldsymbol{\theta}$
2. *Efficient*, $\mathbf{C}_{\hat{\theta}} \rightarrow \mathbf{I}_{\theta}^{-1}$
3. *Gaussianly distributed*

These properties are referred to as the *asymptotic* properties of the MLE. For moderately sized data records the MLE usually provides a good estimator, although no optimality properties can be ascribed to it. Another important characteristic of the MLE from a practical viewpoint is the *invariance* property. If the MLE of $\mathbf{g}(\boldsymbol{\theta})$ is desired, where as before $\mathbf{g}$ is $r \times 1$, then the MLE is

$$\hat{\mathbf{g}} = \mathbf{g}(\hat{\boldsymbol{\theta}}).$$

The MLE of $\mathbf{g}$ is found by transforming the MLE of $\boldsymbol{\theta}$. Finally, the computation of the likelihood can sometimes be simplified by relying on the concept of a *sufficient statistic*. A statistic $\hat{\boldsymbol{\theta}}$ is said to be sufficient for $\boldsymbol{\theta}$ if the conditional PDF $p(\mathbf{x} \mid \hat{\boldsymbol{\theta}})$ does not depend on $\boldsymbol{\theta}$. This says that knowledge of the sufficient statistic summarizes all the information about the data since the PDF of the data given $\hat{\boldsymbol{\theta}}$ does not depend on $\boldsymbol{\theta}$. It can be shown that a statistic is sufficient if and only if the PDF can be factored as

$$p(\mathbf{x}; \boldsymbol{\theta}) = g(\hat{\boldsymbol{\theta}}; \boldsymbol{\theta})h(\mathbf{x}).$$

g and h are functions that depend only on the variables indicated. This factorization is known as the Neyman–Fisher factorization theorem. Consequently, if a sufficient statistic exists, the likelihood function of the original data can be replaced by the likelihood function of the sufficient statistic $p(\hat{\boldsymbol{\theta}}; \boldsymbol{\theta})$ and the latter maximized over $\boldsymbol{\theta}$ to find the MLE. Problems 3.4–3.8 illustrate many of the important statistical concepts.

3.3.2 An Important Estimation Problem

A model that is frequently employed in practice is the *linear model* for the observed data. Assuming now complex data in general, the linear model is defined as

$$\mathbf{x} = \mathbf{H}\boldsymbol{\theta} + \mathbf{z} \tag{3.24}$$

where $\mathbf{H}$ is an $N \times p$ complex matrix of *known constants*, $\boldsymbol{\theta}$ is a $p \times 1$ complex vector of *unknown deterministic* parameters, and $\mathbf{z}$ is an $N \times 1$ complex vector of zero-mean errors or noise. If $\mathbf{z}$ is a complex Gaussian random vector with zero mean and covariance matrix $\sigma_z^2\mathbf{I}$, then, from (3.13),

$$p(\mathbf{x}; \boldsymbol{\theta}) = \frac{1}{\pi^N \sigma_z^{2N}} \exp\left[-\frac{1}{\sigma_z^2} (\mathbf{x} - \mathbf{H}\boldsymbol{\theta})^H (\mathbf{x} - \mathbf{H}\boldsymbol{\theta}) \right]. \tag{3.25}$$

It is usually assumed that the noise variance σ_z^2 is also unknown. The MLE of $\boldsymbol{\theta}$ is found by maximizing (3.25), or equivalently by minimizing

$$S = (\mathbf{x} - \mathbf{H}\boldsymbol{\theta})^H (\mathbf{x} - \mathbf{H}\boldsymbol{\theta}). \tag{3.26}$$

S is a hermitian function which has one and only one minimum (assuming that $\mathbf{H}$ is full rank). To minimize S, note that it can be written as (see Section 2.8)

$$S = (\mathbf{x} - \mathbf{H}\hat{\boldsymbol{\theta}})^H (\mathbf{x} - \mathbf{H}\hat{\boldsymbol{\theta}}) + (\boldsymbol{\theta} - \hat{\boldsymbol{\theta}})^H \mathbf{H}^H \mathbf{H} (\boldsymbol{\theta} - \hat{\boldsymbol{\theta}}) \tag{3.27}$$

where

$$\hat{\boldsymbol{\theta}} = (\mathbf{H}^H \mathbf{H})^{-1} \mathbf{H}^H \mathbf{x}. \tag{3.28}$$

The matrix $\mathbf{H}^H\mathbf{H}$ is positive definite and hence the second term in (3.27) is nonnegative. Because the first term does not depend on $\boldsymbol{\theta}$, S is minimized by choosing $\boldsymbol{\theta} = \hat{\boldsymbol{\theta}}$. The MLE is given by (3.28). The function S is a sum of squares and therefore $\hat{\boldsymbol{\theta}}$ is called a *least squares estimator*. For the model under consideration the *MLE is identical to the least squares estimator*. The minimum value of S is, from (3.26) and (3.28),

$$\begin{aligned} S_{\text{MIN}} &= [\mathbf{x} - \mathbf{H}(\mathbf{H}^H\mathbf{H})^{-1}\mathbf{H}^H\mathbf{x}]^H [\mathbf{x} - \mathbf{H}(\mathbf{H}^H\mathbf{H})^{-1}\mathbf{H}^H\mathbf{x}] \\ &= \mathbf{x}^H[\mathbf{I} - \mathbf{H}(\mathbf{H}^H\mathbf{H})^{-1}\mathbf{H}^H]\mathbf{x} \end{aligned} \tag{3.29}$$

since the matrix $\mathbf{I} - \mathbf{H}(\mathbf{H}^H\mathbf{H})^{-1}\mathbf{H}^H$ is idempotent. The MLE of σ_z^2 can be found by maximizing (3.25) over σ_z^2 once $\boldsymbol{\theta}$ has been replaced by $\hat{\boldsymbol{\theta}}$ and is readily shown to be

$$\hat{\sigma}_z^2 = \frac{1}{N} S_{\text{MIN}}.$$

It can be shown that $\hat{\boldsymbol{\theta}}$ has mean $\boldsymbol{\theta}$, covariance matrix $\sigma_z^2(\mathbf{H}^H\mathbf{H})^{-1}$, and is distributed according to a complex multivariate Gaussian distribution. For the linear model the MLE is optimal even for finite data records since it can be shown to be efficient. An important point to keep in mind is that the *least squares estimator is not the MLE* if

1. $\mathbf{H}$ is a random matrix, or
2. $\mathbf{z}$ is not Gaussianly distributed with mean zero and covariance matrix $\sigma_z^2\mathbf{I}$.

In such a case *the least squares estimator has no optimality properties associated with it*.

Some extensions to the linear model are possible. If the covariance matrix of $\mathbf{z}$ is $\sigma_z^2\mathbf{V}$, where $\mathbf{V}$ is known, then the MLE is found by minimizing (see Section 2.8)

$$S' = (\mathbf{x} - \mathbf{H}\boldsymbol{\theta})^H \mathbf{V}^{-1} (\mathbf{x} - \mathbf{H}\boldsymbol{\theta})$$

which results in

$$\hat{\boldsymbol{\theta}} = (\mathbf{H}^H\mathbf{V}^{-1}\mathbf{H})^{-1}\mathbf{H}^H\mathbf{V}^{-1}\mathbf{x} \tag{3.30}$$

and

$$\hat{\sigma}_z^2 = \frac{1}{N} S'_{\text{MIN}} \tag{3.31}$$

where

$$S'_{\text{MIN}} = \mathbf{x}^H[\mathbf{V}^{-1} - \mathbf{V}^{-1}\mathbf{H}(\mathbf{H}^H\mathbf{V}^{-1}\mathbf{H})^{-1}\mathbf{H}^H\mathbf{V}^{-1}]\mathbf{x}.$$

Minimizing S' results in the *generalized or weighted least squares estimator*, which is identical to the MLE under the conditions assumed in the linear model. $\hat{\boldsymbol{\theta}}$ has mean $\boldsymbol{\theta}$, covariance matrix $\sigma_z^2(\mathbf{H}^H\mathbf{V}^{-1}\mathbf{H})^{-1}$, and is Gaussianly distributed.

A second extension to the linear model is the nonlinear model of the form

$$\mathbf{x} = \mathbf{H}(\boldsymbol{\alpha})\boldsymbol{\theta} + \mathbf{z}. \tag{3.32}$$

The assumptions are as before except that now $\mathbf{H}$ depends nonlinearly on a vector $\boldsymbol{\alpha}$ of constant but unknown parameters. Considering only the case when the covariance matrix of $\mathbf{z}$ is $\sigma_z^2\mathbf{I}$, the MLE of $\boldsymbol{\alpha}$ and $\boldsymbol{\theta}$ is found in two steps. Step 1 is to minimize S as given by (3.26) over $\boldsymbol{\theta}$. The minimum value S_{MIN} is given by (3.29) and now depends on $\boldsymbol{\alpha}$ through $\mathbf{H}$. Step 2 is to minimize S_{MIN} over $\boldsymbol{\alpha}$. The second step is in general difficult since $\mathbf{H}$ depends nonlinearly on $\boldsymbol{\alpha}$. This type of problem is quite common and is referred to as a *nonlinear least squares* problem. For the particular form of (3.32) the problem decouples into a linear part and a nonlinear part. The linear problem is easily solved so that the final minimization need only be done over $\boldsymbol{\alpha}$.

All the results regarding the linear model are valid for real data if the conjugate transposes are replaced by transposes and the PDF of $\mathbf{z}$ is a real multivariate Gaussian distribution (see Problem 3.9 for an example).

3.3.3 Random Parameters

Many of the properties and results for deterministic parameter estimation can be extended to random parameter estimation. Only the essential background necessary for understanding the linear prediction problem discussed in Chapter 6 is provided here. The random parameter estimation problem is to estimate a realization of the *random parameter* θ, which is assumed to be a complex scalar, based on the observed complex data $\mathbf{x}$. $\mathbf{x}$ and θ are assumed to be zero mean. The estimator $\hat{\theta}$ to be considered is linear in the data

$$\hat{\theta} = -\sum_{i=0}^{N-1} \beta_i^* x_i. \tag{3.33}$$

The choice of a negative sign in (3.33) as well as a conjugation of the β_i's is for consistency with results in Chapter 6. A good estimator chooses the β_i's to minimize the MSE

$$\text{MSE} = \mathscr{E}[|\theta - \hat{\theta}|^2]. \tag{3.34}$$

The MSE as given above differs from that of (3.15) in that the expectation is now taken over $\mathbf{x}$ *and* θ since θ is random. To minimize the MSE, let $\boldsymbol{\beta} = [\beta_0 \quad \beta_1 \quad \cdots \quad \beta_{N-1}]^T$, so that

$$\begin{aligned} \text{MSE} &= \mathscr{E}[|\theta + \boldsymbol{\beta}^H\mathbf{x}|^2] \\ &= \mathscr{E}[(\theta + \boldsymbol{\beta}^H\mathbf{x})(\theta + \boldsymbol{\beta}^H\mathbf{x})^H] \\ &= \sigma_\theta^2 + \mathbf{r}_{\theta x}^H\boldsymbol{\beta} + \boldsymbol{\beta}^H\mathbf{r}_{\theta x} + \boldsymbol{\beta}^H\mathbf{C}_{xx}\boldsymbol{\beta} \end{aligned}$$

where $\sigma_\theta^2 = \mathscr{E}[|\theta|^2]$ and $\mathbf{r}_{\theta x} = \mathscr{E}[\theta^*\mathbf{x}]$. This may be minimized by using the complex gradient formulas of (2.70) or alternatively by rewriting the MSE as

$$\text{MSE} = \sigma_\theta^2 - \mathbf{r}_{\theta x}^H\mathbf{C}_{xx}^{-1}\mathbf{r}_{\theta x} + (\boldsymbol{\beta} + \mathbf{C}_{xx}^{-1}\mathbf{r}_{\theta x})^H\mathbf{C}_{xx}(\boldsymbol{\beta} + \mathbf{C}_{xx}^{-1}\mathbf{r}_{\theta x}). \tag{3.35}$$

In a similar fashion to the development of (3.28), the optimal coefficients are

$$\hat{\boldsymbol{\beta}} = -\mathbf{C}_{xx}^{-1}\mathbf{r}_{\theta x} \tag{3.36}$$

and the minimum MSE is

$$\begin{aligned} \text{MSE}_{\text{MIN}} &= \sigma_\theta^2 - \mathbf{r}_{\theta x}^H\mathbf{C}_{xx}^{-1}\mathbf{r}_{\theta x} \\ &= \sigma_\theta^2 + \mathbf{r}_{\theta x}^H\hat{\boldsymbol{\beta}}. \end{aligned} \tag{3.37}$$

Equations (3.36) are termed the *Wiener–Hopf equations*. If $\mathbf{x}$ and θ are samples from wide sense stationary random processes (to be discussed in Section 3.4), then depending on the relative position of θ with respect to $\mathbf{x}$, (3.36) solves either the prediction, smoothing, or filtering problem (see Problems 3.10 and 3.11).

A useful result follows by rewriting the MSE as

$$\text{MSE} = \mathscr{E}[\theta(\theta + \boldsymbol{\beta}^H\mathbf{x})^H] + \boldsymbol{\beta}^H\mathscr{E}[\mathbf{x}(\theta + \boldsymbol{\beta}^H\mathbf{x})^H]$$

and noting that the minimum MSE is attained when $\boldsymbol{\beta}$ is chosen to make the second term zero. Then

$$\mathscr{E}[\mathbf{x}(\theta + \boldsymbol{\beta}^H\mathbf{x})^H] = \mathscr{E}[\mathbf{x}\epsilon^*] = \mathbf{0}$$

where ϵ is the error $\theta - \hat{\theta}$. Equivalently,

$$\mathscr{E}[\epsilon x_i^*] = 0 \qquad i = 0, 1, \ldots, N - 1. \tag{3.38}$$

We can interpret (3.38) as saying that the optimal estimator chooses the coefficients to make the *error orthogonal to the data*. Two zero-mean random variables are said to be orthogonal if they are uncorrelated. This property, called the *orthogonality principle of linear minimum mean square estimation*, proves to be extremely useful.

3.4 RANDOM PROCESS CHARACTERIZATION

A *discrete random process* $x[n]$ is a sequence of random variables, real or complex, defined for every integer n. If the discrete random process is wide sense stationary (WSS), it has a *mean*

$$\mathscr{E}\{x[n]\} = \mu_x \tag{3.39}$$

which does not depend on n and an *autocorrelation function* (ACF)

$$r_{xx}[k] = \mathcal{E}\{x^*[n]x[n+k]\} \tag{3.40}$$

which depends only on the lag between the two samples, not on their absolute positions. Also, the *autocovariance function* is defined as

$$c_{xx}[k] = \mathcal{E}\{(x^*[n] - \mu_x^*)(x[n+k] - \mu_x)\} = r_{xx}[k] - |\mu_x|^2. \tag{3.41}$$

In a similar manner, two jointly WSS random process $x[n]$ and $y[n]$ have a *cross-correlation function* (CCF)

$$r_{xy}[k] = \mathcal{E}\{x^*[n]y[n+k]\} \tag{3.42}$$

and a *cross-covariance function*

$$c_{xy}[k] = \mathcal{E}\{(x^*[n] - \mu_x^*)(y[n+k] - \mu_y)\} = r_{xy}[k] - \mu_x^*\mu_y. \tag{3.43}$$

Some useful properties of the ACF and CCF are

$$\begin{aligned} r_{xx}[0] &\geq |r_{xx}[k]| \\ r_{xx}[-k] &= r_{xx}^*[k] \\ r_{xy}[-k] &= r_{yx}^*[k]. \end{aligned} \tag{3.44}$$

Note that $r_{xx}[0]$ is real and positive, which follows from (3.40). Also, the ACF satisfies

$$\mathcal{E}\left\{\left|\sum_{n=0}^{M-1} a^*[n]x[n]\right|^2\right\} = \sum_{m=0}^{M-1}\sum_{n=0}^{M-1} a^*[m]a[n]r_{xx}[m-n] \geq 0 \tag{3.45}$$

for any $a[n]$ sequence and any M. Such functions are termed *positive semidefinite*. Equivalently, the autocorrelation matrix

$$\begin{bmatrix} r_{xx}[0] & r_{xx}[-1] & \cdots & r_{xx}[-(M-1)] \\ r_{xx}[1] & r_{xx}[0] & \cdots & r_{xx}[-(M-2)] \\ \vdots & \vdots & \ddots & \vdots \\ r_{xx}[M-1] & r_{xx}[M-2] & \cdots & r_{xx}[0] \end{bmatrix} \tag{3.46}$$

is a positive semidefinite matrix (see Problem 3.14).

The z-transforms of the ACF and CCF, defined as

$$\begin{aligned} \mathcal{P}_{xx}(z) &= \sum_{k=-\infty}^{\infty} r_{xx}[k]z^{-k} \\ \mathcal{P}_{xy}(z) &= \sum_{k=-\infty}^{\infty} r_{xy}[k]z^{-k} \end{aligned} \tag{3.47}$$

lead to the definition of the power spectral density. When evaluated on the unit circle, $\mathcal{P}_{xx}(z)$ and $\mathcal{P}_{xy}(z)$ become the *auto-PSD*, $P_{xx}(f) = \mathcal{P}_{xx}(\exp[j2\pi f])$, and *cross-PSD*, $P_{xy}(f) = \mathcal{P}_{xy}(\exp[j2\pi f])$, or

$$P_{xx}(f) = \sum_{k=-\infty}^{\infty} r_{xx}[k] \exp(-j2\pi fk) \tag{3.48}$$

$$P_{xy}(f) = \sum_{k=-\infty}^{\infty} r_{xy}[k] \exp(-j2\pi fk). \tag{3.49}$$

The relationship that the auto-PSD is the Fourier transform of the ACF as expressed by (3.48) is sometimes referred to as the Wiener–Khinchin theorem. As discussed in Chapter 1, the auto-PSD describes the distribution in frequency of the power of $x[n]$ and as such is real and nonnegative. The cross-PSD, on the other hand, is in general complex. The magnitude of the cross-PSD describes whether frequency components in $x[n]$ are associated with large or small amplitudes at the same frequency in $y[n]$, and the phase of the cross-PSD indicates the phase lag or lead of $x[n]$ with respect to $y[n]$ for a given frequency component. Note that both spectral densities are periodic with period one. The frequency interval $-\frac{1}{2} \le f \le \frac{1}{2}$ will be considered as the fundamental period. When there is no confusion, $P_{xx}(f)$ will be referred to simply as the *power spectral density* (PSD).

A process that will frequently be encountered is discrete white noise. It is defined as having an ACF

$$r_{xx}[k] = \sigma_x^2 \delta[k]$$

where $\delta[k]$ is the discrete impulse function. This says that each sample is uncorrelated with all the others. Using (3.48), the PSD becomes

$$P_{xx}(f) = \sigma_x^2$$

and is observed to be completely flat with frequency. Alternatively, white noise is composed of equipower contributions from all frequencies.

For a linear shift invariant (LSI) system with impulse response $h[n]$ and with a WSS random process input, various relationships between the correlations and spectral density functions of the input process $x[n]$ and output process $y[n]$ hold (see Problem 3.15). The correlation relationships are

$$\begin{aligned}
r_{xy}[k] &= h[k] \star r_{xx}[k] = \sum_{l=-\infty}^{\infty} h[l] r_{xx}[k-l] \\
r_{yx}[k] &= h^*[-k] \star r_{xx}[k] = \sum_{l=-\infty}^{\infty} h^*[-l] r_{xx}[k-l] \\
r_{yy}[k] &= h[k] \star r_{yx}[k] = h[k] \star h^*[-k] \star r_{xx}[k] \\
&= \sum_{m=-\infty}^{\infty} h[k-m] \sum_{l=-\infty}^{\infty} h^*[-l] r_{xx}[m-l]
\end{aligned} \tag{3.50}$$

where $\star$ denotes convolution. Denoting the system function by $\mathcal{H}(z) =$

$\sum_{n=-\infty}^{\infty} h[n]z^{-n}$, the following relations for the PSDs follow from (3.50) (see Problem 3.15):

$$\begin{aligned}\mathcal{P}_{xy}(z) &= \mathcal{H}(z)\mathcal{P}_{xx}(z)\\ \mathcal{P}_{yx}(z) &= \mathcal{H}^*(1/z^*)\,\mathcal{P}_{xx}(z)\\ \mathcal{P}_{yy}(z) &= \mathcal{H}(z)\mathcal{H}^*(1/z^*)\,\mathcal{P}_{xx}(z).\end{aligned} \tag{3.51}$$

If $h[n]$ is real, then $\mathcal{H}^*(1/z^*) = \mathcal{H}(1/z)$. The last relationship in (3.51) is particularly important in that it justifies the interpretation of $P_{xx}(f)$ as a PSD. Specifically, if a random process $x[n]$ is input to a LSI system, the expected power of the output process $y[n]$ is $r_{yy}[0]$. But from (3.48) we note that the ACF is the inverse Fourier transform of the PSD and hence

$$r_{yy}[0] = \int_{-1/2}^{1/2} P_{yy}(f)\,df.$$

Assume now that the LSI system is a narrowband filter with center frequency f_0 and bandwidth B, so that the frequency response is

$$H(f) = \begin{cases} 1 & \text{for } |f - f_0| \le \dfrac{B}{2} \\ 0 & \text{otherwise.}\end{cases}$$

The expected power at the output of the filter is due to frequency components near $f = f_0$ and is measured by $r_{yy}[0]$. Noting from (3.51) that $P_{yy}(f) = |H(f)|^2 P_{xx}(f)$, the expected output power becomes

$$r_{yy}[0] = \int_{-1/2}^{1/2} |H(f)|^2 P_{xx}(f)\,df = \int_{f_0-B/2}^{f_0+B/2} P_{xx}(f)\,df.$$

This says that to obtain the expected power in $x[n]$ over the band $(f_0 - B/2, f_0 + B/2)$ we should integrate the function $P_{xx}(f)$ over these frequencies, leading to the interpretation of a PSD.

For the special case of a white noise input process, the output power spectral density becomes

$$P_{yy}(f) = |H(f)|^2 \sigma_x^2. \tag{3.52}$$

This will form the basis of the rational transfer function spectral estimators discussed in Chapter 5.

3.5 SOME IMPORTANT RANDOM PROCESSES

Two important random process models are the Gaussian random process and the sinusoidal random process. The assumptions behind each process are discussed in this section. Spectral estimators to be described subsequently depend heavily on these processes, so it is critical that the underlying assumptions be clearly understood.

3.5.1 Gaussian Random Processes

A Gaussian random process is one for which any set of samples $\{x[n_0], x[n_1], \ldots, x[n_{N-1}]\}$ is jointed distributed according to a multivariate Gaussian PDF. For a real or complex Gaussian random process the PDF of the samples is given by (3.2) or (3.13), respectively. If the samples are taken at successive times to generate the vector $\mathbf{x} = [x[0] \quad x[1] \quad \cdots \quad x[N-1]]^T$, then assuming a zero-mean WSS random process, the covariance matrix takes on the form

$$\mathbf{C}_{xx} = \mathbf{R}_{xx} = \begin{bmatrix} r_{xx}[0] & r_{xx}[-1] & \cdots & r_{xx}[-(N-1)] \\ r_{xx}[1] & r_{xx}[0] & \cdots & r_{xx}[-(N-2)] \\ \vdots & \vdots & \ddots & \vdots \\ r_{xx}[N-1] & r_{xx}[N-2] & \cdots & r_{xx}[0] \end{bmatrix}. \tag{3.53}$$

The covariance matrix or more appropriately the autocorrelation matrix $\mathbf{R}_{xx}$ has the special hermitian Toeplitz structure, as discussed in Section 2.3. A decorrelation of the samples, which was discussed in Section 3.2, can be effected using the Cholesky decomposition. However, a more efficient method employs the Levinson recursion described in Chapter 6. The Levinson recursion decomposes the complex autocorrelation matrix as (see Section 6.3.6)

$$\mathbf{R}_{xx} = \mathbf{B}^{H^{-1}}\mathbf{P}\mathbf{B}^{-1} \tag{3.54}$$

where $\mathbf{B}^H$ is an $N \times N$ lower triangular matrix and $\mathbf{P}$ is an $N \times N$ diagonal matrix with real and positive diagonal elements. Note that if a complex vector $\boldsymbol{\xi} = [\xi[0] \quad \xi[1] \quad \cdots \quad \xi[N-1]]^T$ of independent $CN(0, 1)$ random variables is transformed as

$$\mathbf{x} = \mathbf{B}^{H^{-1}} \sqrt{\mathbf{P}}\boldsymbol{\xi}$$

where $\sqrt{\mathbf{P}}$ is found from $\mathbf{P}$ by taking the positive square root of all the elements, then $\mathbf{x}$ has the autocorrelation matrix given by (3.54). $\mathbf{x}$ then has the identical PDF as the original Gaussian process. This approach can be used to set the initial conditions of a recursive filter for generation of a time series, as discussed in Appendix 5B.

An important Gaussian random process is the white process. As discussed previously, the ACF for a white process is a discrete delta function. In light of the definition of a Gaussian random process a *real white Gaussian random process* $x[n]$ with mean zero and variance σ_x^2 is one for which

$$x[n] \sim N(0, \sigma_x^2) \qquad -\infty < n < \infty$$

$$r_{xx}[m-n] = \mathscr{E}(x[n]x[m]) = 0 \qquad m \neq n.$$

Because of the Gaussian assumption any two samples are statistically independent. The extension to the complex case defines a *complex white Gaussian random process* as one for which

$$x[n] \sim CN(0, \sigma_x^2) \qquad -\infty < n < \infty$$

$$r_{xx}[m-n] = \mathscr{E}(x^*[n]x[m]) = 0 \qquad m \neq n.$$

The latter equation implies that

$$\mathbf{C}_{xx} = \mathbf{E} + j\mathbf{F} = \sigma_x^2\mathbf{I}$$

or

$$\mathbf{C}_{xx} = \mathbf{E} = \sigma_x^2\mathbf{I}$$

$$\mathbf{F} = \mathbf{0}.$$

Noting that $\mathbf{u}$, $\mathbf{v}$ are zero mean, it follows from (3.12) that

$$\mathscr{E}[\mathbf{u}\mathbf{u}^T] = \mathscr{E}[\mathbf{v}\mathbf{v}^T] = \frac{\mathbf{E}}{2} = \frac{\sigma_x^2}{2}\mathbf{I}$$

$$\mathscr{E}[\mathbf{v}\mathbf{u}^T] = -\mathscr{E}[\mathbf{u}\mathbf{v}^T] = \frac{\mathbf{F}}{2} = \mathbf{0}$$

Consequently, a *complex white Gaussian random process is one for which* $u[n]$ *and* $v[n]$ *are each real white Gaussian random processes which are uncorrelated (independent) of each other.* The variance for each real white Gaussian random process is $\sigma_x^2/2$.

3.5.2 Sinusoidal Random Processes

A sinusoidal process may be modeled in various ways. Consider first a single real sinusoid

$$x[n] = A \cos(2\pi f_0 n + \phi). \tag{3.55}$$

If the amplitude and phase are assumed to be constants, then (3.55) cannot be modeled by a WSS random process since the mean and ACF will both depend on n. If the amplitude is assumed to be constant but the phase is assumed to be a random variable uniformly distributed on $[0, 2\pi)$, the process is WSS with (see Problem 3.16)

$$\mathscr{E}(x[n]) = 0$$

$$r_{xx}[k] = \frac{A^2}{2}\cos(2\pi f_0 k).$$

Furthermore, if the amplitude is assumed to be a random variable with a Rayleigh distribution and independent of the phase, the process will also be Gaussian. For multiple real sinusoids the same conclusions hold if all the amplitudes and phases are independent. Consider next a single complex sinusoid

$$x[n] = A \exp[j(2\pi f_0 n + \phi)]. \tag{3.56}$$

The same conclusions are valid. For $x[n]$ to be WSS the phase ϕ must be uniformly distributed over $[0, 2\pi)$. Then

$$\mathscr{E}(x[n]) = 0$$

$$r_{xx}[k] = A^2 \exp(j2\pi f_0 k).$$

For $x[n]$, also, to be a complex Gaussian random process, A must be a Rayleigh random variable statistically independent of ϕ. It can also be shown that if A is Rayleigh distributed with mean $\sqrt{\pi/2}A_0$ and ϕ is uniformly distributed and independent of A, then

$$A \exp(j\phi) \sim CN(0, A_0^2).$$

The different types of models for sinusoidal processes lead to different estimation procedures for the sinusoidal parameters. As an example, assume that a complex sinusoid $x[n]$ is embedded in complex white Gaussian noise $w[n]$ with zero mean and variance σ_w^2, so that

$$y[n] = x[n] + w[n].$$

The PDF for $\mathbf{y} = [y[0] \quad y[1] \quad \cdots \quad y[N-1]]^T$ will depend on the assumptions made for the sinusoidal amplitude and phase. If the amplitude and phase are deterministic constants, then $x[n]$ represents a shift in the mean of the noise process. The PDF is accordingly given as

$$p(\mathbf{y}) = \frac{1}{\pi^N \sigma_w^{2N}} \exp\left\{-\frac{1}{\sigma_w^2} \sum_{n=0}^{N-1} |y[n] - A \exp[j(2\pi f_0 n + \phi)]|^2\right\}. \tag{3.57}$$

If, on the other hand, the amplitude and phase are both random variables independent of each other and with $w[n]$, and the amplitude and phase are Rayleigh and uniformly distributed random variables, respectively, so as to make $x[n]$ a Gaussian process, then $y[n]$ is also Gaussian, being the sum of two Gaussian processes. $y[n]$ will be zero mean and have an ACF

$$\begin{aligned} r_{yy}[k] &= r_{xx}[k] + r_{ww}[k] \\ &= A^2 \exp(j2\pi f_0 k) + \sigma_w^2 \delta[k]. \end{aligned}$$

The PDF is then

$$p(\mathbf{y}) = \frac{1}{\pi^N \det(\mathbf{R}_{yy})} \exp(-\mathbf{y}^H \mathbf{R}_{yy}^{-1} \mathbf{y}) \tag{3.58}$$

where

$$\mathbf{R}_{yy} = A^2 \mathbf{e}\mathbf{e}^H + \sigma_w^2 \mathbf{I}$$

and

$$\mathbf{e} = [1 \quad \exp(j2\pi f_0) \quad \cdots \quad \exp(j2\pi f_0[N-1])]^T.$$

Hence estimators of sinusoidal parameters will in general be different if different models are used. For example, in the random amplitude/random phase model, the phase cannot be estimated since the PDF is independent of phase. The specific model chosen is usually for mathematical convenience. The same conclusions hold for multiple sinusoids as well as for real data. Some explicit estimation algorithms for sinusoidal parameters are given in Chapter 13.

3.6 ERGODICITY OF THE AUTOCORRELATION FUNCTION

Estimation of the PSD of an arbitrary WSS random process requires one to estimate the ACF. A difficulty arises in that the ACF is defined as the expectation of $x^*[n]x[n + k]$ obtained when averaged over an *ensemble of realizations*. In practice, however, only a *segment of a single realization* is available. Thus it is imperative that a single realization of the random process or the infinite data set $x[n]$ for $-\infty < n < \infty$ be sufficient to determine the ACF. A random process which has this property is said to be *autocorrelation ergodic*. In general, a strictly ergodic process allows one to determine ensemble averages by replacing them with time averages. As an example, consider the estimation of the mean of a WSS random process using the temporal average $[1/(2M + 1)] \sum_{n=-M}^{M} x[n]$. For a consistent estimator based on a single realization, the temporal average should converge to the ensemble mean, or

$$\lim_{M\to\infty} \frac{1}{2M + 1} \sum_{n=-M}^{M} x[n] = \mathscr{E}(x[n]) = \mu_x. \tag{3.59}$$

This limit can be shown to exist if and only if the variance of the time average above approaches zero, that is (see Problem 3.17),

$$\lim_{M\to\infty} \frac{1}{2M + 1} \sum_{k=-2M}^{2M} \left(1 - \frac{|k|}{2M + 1}\right) c_{xx}[k] = 0 \tag{3.60}$$

where $c_{xx}[k]$ is the true ensemble autocovariance of the random process $x[n]$. In this case, $x[n]$ is said to be *ergodic in the mean*. In a similar manner, if we were to observe the product of the observations at two instants separated by k in several places, say $x^*[n_1]x[n_1 + k]$, $x^*[n_2]x[n_2 + k]$ and so on, then on the average, the product should yield the value of $r_{xx}[k]$. In the limit as we average over all product values, it would be expected that the *temporal ACF* would converge to the true ensemble ACF, or

$$\lim_{M\to\infty} \frac{1}{2M + 1} \sum_{n=-M}^{M} x^*[n]x[n + k] = \mathscr{E}(x^*[n]x[n + k]) = r_{xx}[k]. \tag{3.61}$$

This can be shown to be true if and only if the variance of the time average above approaches zero in the limit, that is,

$$\lim_{M\to\infty} \frac{1}{2M + 1} \sum_{k=-2M}^{2M} \left(1 - \frac{|k|}{2M + 1}\right) c_{pp}[k] = 0 \tag{3.62}$$

where $c_{pp}[k]$ is the true ensemble covariance of the random process $p[n] = x^*[n]x[n + k]$, the lagged product of the process $x[n]$. Note that $c_{pp}[k]$ involves a fourth-order statistical moment of the process $x[n]$. If the above is true, $x[n]$ is autocorrelation ergodic (see Problems 3.18 and 3.19). Proving ergodicity of the autocorrelation is a difficult task, especially for non-Gaussian processes. As a matter of practicality, ergodicity must be assumed. *Hereafter, it will be assumed that the measured process is ergodic in the autocorrelation, so that a time average* *can replace an ensemble average*. It should be cautioned, however, that even if the process is autocorrelation ergodic, *it does not follow that the Fourier transform of the temporal ACF will tend to the true or ensemble PSD as the data record length increases*.

3.7 AN ALTERNATIVE DEFINITION OF THE POWER SPECTRAL DENSITY

It is shown in this section that a nearly equivalent definition for the PSD is

$$P_{xx}(f) = \lim_{M\to\infty} \mathcal{E}\left\{\frac{1}{2M+1}\left|\sum_{n=-M}^{M} x[n]\exp(-j2\pi fn)\right|^2\right\}. \tag{3.63}$$

This definition says to take first the magnitude squared of the Fourier transform of the data $x[n]$ and then divide by the data record length. Then, since the Fourier transform will be a random variable (a different value will be obtained for each realization of $x[n]$), the expected value is taken. Finally, since the random process is in general of infinite duration, a limiting operation is required. Equation (3.63) is now proven to be equivalent to the Wiener–Khinchin theorem (3.48).

$$P_{xx}(f) = \lim_{M\to\infty} \mathcal{E}\left\{\frac{1}{2M+1}\sum_{m=-M}^{M}\sum_{n=-M}^{M} x[m]x^*[n]\exp[-j2\pi f(m-n)]\right\}$$

$$= \lim_{M\to\infty} \frac{1}{2M+1}\sum_{m=-M}^{M}\sum_{n=-M}^{M} r_{xx}[m-n]\exp[-j2\pi f(m-n)].$$

But

$$\sum_{m=-M}^{M}\sum_{n=-M}^{M} g[m-n] = \sum_{k=-2M}^{2M}(2M+1-|k|)g[k] \tag{3.64}$$

which may be verified by considering $g[m-n]$ as the $[m, n]$ element of a matrix of dimension $(2M+1)\times(2M+1)$. It follows that

$$P_{xx}(f) = \lim_{M\to\infty}\frac{1}{2M+1}\sum_{k=-2M}^{2M}(2M+1-|k|)r_{xx}[k]\exp(-j2\pi fk)$$

$$= \lim_{M\to\infty}\sum_{k=-2M}^{2M}\left(1-\frac{|k|}{2M+1}\right)r_{xx}[k]\exp(-j2\pi fk)$$

$$= \sum_{k=-\infty}^{\infty} r_{xx}[k]\exp(-j2\pi fk).$$

The last step assumes that the ACF decays sufficiently rapidly. Then (3.63) is equivalent to the Wiener–Khinchin definition of the PSD. For some processes the ACF does not decay quickly enough to make this statement. Examples are random processes with nonzero means or sinusoidal components. It should be observed, however, that the Wiener–Khinchin definition of the PSD is able to accommodate these processes only through the use of Dirac delta functions. Note finally that the definition of the PSD given by (3.63) would have been *incorrect* if the expectation operator had been omitted. In Chapter 4 the nearly equivalent definitions of the PSD given by the (3.48) and (3.63) will be used as a basis for the classical methods of spectral estimation.

REFERENCES

Abramowitz, M., and I. A. Stegen, *Handbook of Mathematical Functions,* National Bureau of Standards, Applied Math. Ser. 55, Aug. 1966.

Goodman, N. R., "Statistical Analysis Based on a Certain Multivariate Complex Distribution," *Ann. Math. Statist.,* Vol. 34, pp. 152–157, Mar. 1963.

Kendall, M., and A. Stuart, *The Advanced Theory of Statistics,* Vols. 1, 2, and 3, Macmillan, New York, 1977.

Papoulis, A., *Probability, Random Variables, and Stochastic Processes,* McGraw-Hill, New York, 1965.

Reed, I. S., "On a Moment Theorem for Complex Gaussian Processes," *IEEE Trans. Inf. Theory,* Vol. IT8, Apr. 1962.

PROBLEMS

3.1. Let $\mathbf{x}$ be a 2×1 real random vector distributed according to a multivariate Gaussian PDF with zero mean and covariance matrix

$$\mathbf{C}_{xx} = \begin{bmatrix} \sigma_1^2 & \sigma_{12} \\ \sigma_{21} & \sigma_2^2 \end{bmatrix}.$$

If

$$y_1 = x_1$$

$$y_2 = ax_1 + x_2$$

so that $\mathbf{y} = \mathbf{L}^{-1}\mathbf{x}$, find a so that y_1 and y_2 are uncorrelated and hence independent. Find the Cholesky decomposition of $\mathbf{C}_{xx}$, which expresses $\mathbf{C}_{xx}$ as $\mathbf{LDL}^T$, where $\mathbf{L}$ is lower triangular with 1's on the principal diagonal and $\mathbf{D}$ is a diagonal matrix with positive diagonal elements.

3.2. Prove that the sum of the squares of N independent and identically distributed $N(0, 1)$ random variables has a χ_N^2distribution. Use the method of characteristic functions.

3.3. Prove that the complex multivariate Gaussian PDF reduces to the complex univariate Gaussian PDF if $N = 1$.

3.4. Given $\{x[0], x[1], \ldots, x[N-1]\}$, which are independent and identically distributed according to a $N(\mu_x, \sigma_x^2)$ distribution, it is desired to estimate μ_x by the *sample mean*

$$\hat{\mu}_x = \frac{1}{N} \sum_{n=0}^{N-1} x[n].$$

Prove that $\hat{\mu}_x$ is an unbiased estimator. Also, find the variance of the estimator.

3.5. For the conditions of Problem 3.4 show that the CR bound is

$$\text{var}(\hat{\mu}_x) \geq \frac{\sigma_x^2}{N}.$$

What can you say about the efficiency of the sample mean estimator?

3.6. Assume now that the variance is to be estimated as well as the mean for the conditions of Problem 3.4. For the vector parameter $\boldsymbol{\theta} = [\mu_x \;\; \sigma_x^2]^T$, prove that the Fisher information matrix is

$$\mathbf{I}_\theta = \begin{bmatrix} \dfrac{N}{\sigma_x^2} & 0 \\ 0 & \dfrac{N}{2\sigma_x^4} \end{bmatrix}.$$

Find the CR bound to determine if the sample mean $\hat{\mu}_x$ is efficient. If the variance is estimated as

$$\hat{\sigma}_x^2 = \frac{1}{N-1} \sum_{n=0}^{N-1} (x[n] - \hat{\mu}_x)^2$$

determine if this estimator is unbiased and efficient. *Hint:* Use the result that

$$\frac{(N-1)\hat{\sigma}_x^2}{\sigma_x^2} \sim \chi_{N-1}^2.$$

3.7. For the conditions of Problem 3.4 find the MLE of μ_x and σ_x^2. Is the MLE of the parameters asymptotically unbiased, efficient, and Gaussianly distributed?

3.8. Prove that the sample mean is a sufficient statistic for the mean under the conditions of Problem 3.4. Assume that σ_x^2 is known. Find the MLE of the mean by maximizing $p(\hat{\mu}_x; \mu_x)$.

3.9. Consider the real linear model

$$x[n] = \alpha + \beta n + z[n] \qquad n = 0, 1, \ldots, N-1.$$

Find the MLE of the slope β and the intercept α by assuming that $z[n]$ is real white Gaussian noise with mean zero and variance σ_x^2. Find the MLE of α if in the linear model we set $\beta = 0$.

3.10. It is desired to predict the complex WSS random process $x[n]$ based on the previous sample $x[n-1]$ by using a *linear predictor*

$$\hat{x}[n] = -\alpha_1 x[n-1].$$

α_1 is to be chosen to minimize the MSE or *prediction error power*

$$\text{MSE} = \mathcal{E}\{|x[n] - \hat{x}[n]|^2\}.$$

Using the orthogonality principle, find the optimal prediction parameter α_1 and the minimum prediction error power.

3.11. Repeat Problem 3.10 for the general case where the predictor is given as

$$\hat{x}[n] = -\sum_{k=1}^{p} \alpha_k x[n-k].$$

Show that the optimal prediction coefficients $\{\alpha_1, \alpha_2, \ldots, \alpha_p\}$ are found by solving (6.4) and the minimum prediction error power is given by (6.5).

3.12. Prove that for a WSS random process

$$r_{xx}[-k] = r_{xx}^*[k].$$

3.13. Prove that the PSD of a real WSS random process is a real even function of frequency.

3.14. The positive semidefinite property of the ACF was defined in (3.45). Using this property, prove that the autocorrelation matrix given by (3.46) is also positive semidefinite.

3.15. Verify the ACF and PSD relationships given in (3.50) and (3.51).

3.16. Show that the random process

$$x[n] = A \cos(2\pi f_0 n + \phi)$$

where ϕ is uniformly distributed on $[0, 2\pi)$, is WSS by finding its mean and ACF. Repeat for a single complex sinusoid

$$x[n] = A \exp[j(2\pi f_0 n + \phi)]$$

using the same assumptions.

3.17. Verify that the variance of the sample mean estimator for the mean of a real WSS random process

$$\frac{1}{2M+1} \sum_{n=-M}^{M} x[n]$$

is given in (3.60). If $x[n]$ is real white noise, what does the variance expression reduce to? *Hint:* Use the relationship given by (3.64).

3.18. For the real sinusoidal random process of Problem 3.16 find the temporal ACF

$$\hat{r}_{xx}[k] = \frac{1}{2M+1} \sum_{n=-M}^{M} x[n]x[n+k]$$

as $M \to \infty$. Is the random process autocorrelation ergodic?

3.19. For the multiple sinusoidal process

$$x[n] = \sum_{i=1}^{p} A_i \cos(2\pi f_i n + \phi_i)$$

where the ϕ_i's are all uniformly distributed random variables on $[0, 2\pi)$ and independent of each other, find the ensemble ACF and the temporal ACF as $M \to \infty$. Is this random process autocorrelation ergodic?

Chapter 4
Classical Spectral Estimation

4.1 INTRODUCTION

In this chapter we discuss two popular spectral estimation methods based on Fourier analysis. They are the periodogram (sometimes called the *sample spectrum*), originally proposed by Schuster in 1898, and the Blackman–Tukey spectral estimator, named after the pioneering work of R. Blackman and J. Tukey in 1958. These methods are also referred to as classical techniques, to distinguish them from the more modern spectral estimation methods discussed in Chapters 6 to 11. The principal conclusions which result from a study of the classical methods are that the bias of the estimator can be reduced if we are willing to accept an increase in variance, and vice versa, but both types of errors cannot be reduced simultaneously. If the bias is too severe or alternatively, the resolution not adequate for a given application, we must resort to other methods. If, however, the resolution is adequate and the variance is at an acceptable level, Fourier methods prove to be satisfactory in practice.

Although many variations of the periodogram and Blackman–Tukey spectral estimators exist, these alternative forms will be mentioned only in passing. This chapter thus serves to summarize the major advantages and disadvantages of the classical methods and to motivate an interest in modern spectral estimation. For further details of classical spectral estimation, the reader is referred to the many excellent texts concerning themselves solely with Fourier based methods: for example, those by Jenkins and Watts [1968], Koopmans [1974], and Bloomfield [1976].

4.2 SUMMARY

The two basic methods of Fourier or classical spectral estimation are described. The first is termed the periodogram and is defined in (4.2). It is shown that the periodogram is an inconsistent estimator in that even though the average value converges to the true value as the data record length becomes large, the variance is a constant, as given by (4.10). To circumvent this problem, the averaged periodogram as defined by (4.11) can be used. For this estimator the data record is segmented into nonoverlapping blocks, which is then followed by an averaging of the periodograms for each block. The variance is then reduced by a factor equal approximately to the number of blocks averaged (4.14), but the bias is increased. A compromise must then be made between bias and variance. The confidence interval for the averaged periodogram is given by (4.17). Next the Blackman–Tukey spectral estimator is defined by (4.23). Again a bias–variance tradeoff is evident, with the mean being given by (4.25) and the variance determined by (4.26). The tradeoff is effected by the choice of a lag window. The confidence interval is given by (4.28). Finally, some computer simulation examples to illustrate typical spectral estimates are given in Section 4.6.

The results of applying the periodogram and Blackman–Tukey spectral estimators to the test case data (see Section 1.5) are shown in Figures 4.1 and 4.2. The lag window used in the Blackman–Tukey estimate was a Bartlett window with $M = 10$. The computer programs PERIODOGRAM (see Appendix 4C) and BLACKTUKEY (see Appendix 4E) were used to obtain the spectral estimates. These results are discussed in Chapter 12.

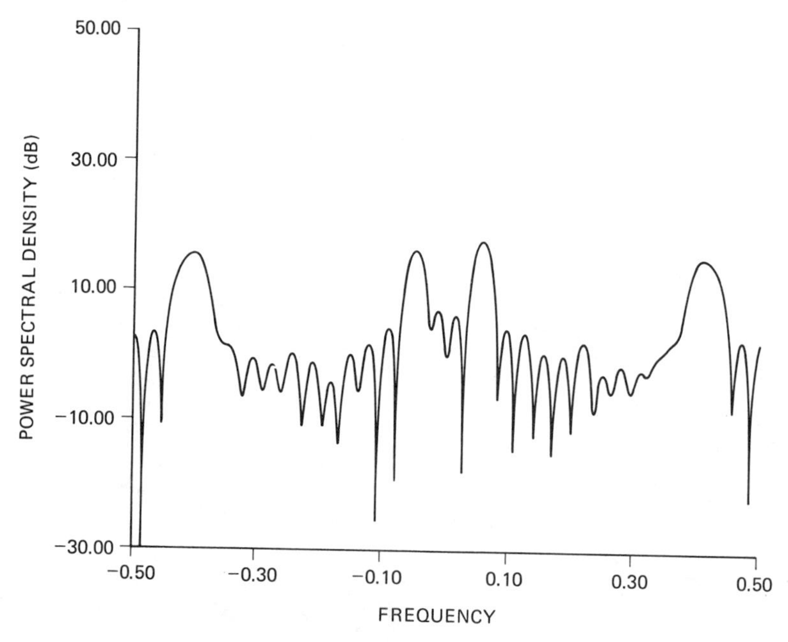

Figure 4.1 Periodogram spectral estimate for complex test case data.

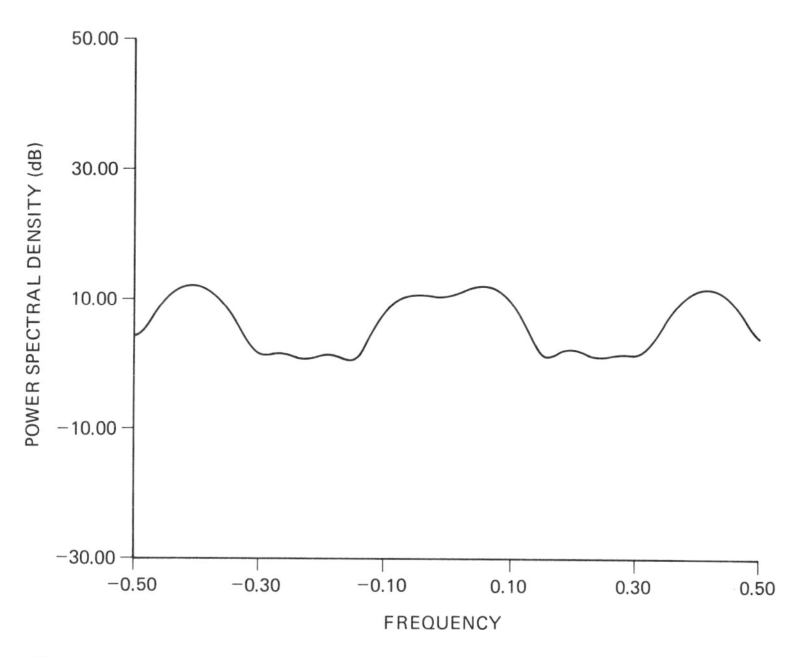

Figure 4.2 Blackman–Tukey spectral estimate using Bartlett window ($M = 10$) for complex test case data.

4.3 PERIODOGRAM

The periodogram spectral estimator relies on the definition of the PSD given by (3.63)

$$P_{xx}(f) = \lim_{M\to\infty} \mathcal{E}\left[\frac{1}{2M+1}\left|\sum_{n=-M}^{M} x[n]\exp(-j2\pi fn)\right|^2\right]. \quad (4.1)$$

By neglecting the expectation operator and using the available data $\{x[0], x[1], \ldots, x[N-1]\}$, the periodogram spectral estimator is defined as

$$\hat{P}_{\text{PER}}(f) = \frac{1}{N}\left|\sum_{n=0}^{N-1} x[n]\exp(-j2\pi fn)\right|^2. \quad (4.2)$$

An interesting interpretation of the periodogram estimator becomes apparent by examining the value at a given frequency f_0. Then the periodogram can be expressed as

$$\hat{P}_{\text{PER}}(f_0) = N\left|\sum_{k=0}^{N-1} h[n-k]x[k]\right|^2_{n=0} \quad (4.3)$$

where

$$h[n] = \begin{cases} \frac{1}{N}\exp(j2\pi f_0 n) & \text{for } n = -(N-1), \ldots, -1, 0 \\ 0 & \text{otherwise.} \end{cases}$$

$h[n]$ is the impulse response of an LSI filter with frequency response

$$\begin{aligned} H(f) &= \sum_{n=-(N-1)}^{0} h[n] \exp(-j2\pi f n) \\ &= \frac{\sin N\pi(f - f_0)}{N \sin \pi(f - f_0)} \exp[j(N-1)\pi(f - f_0)]. \end{aligned} \tag{4.4}$$

This is the frequency response of a bandpass filter with center frequency $f = f_0$. Hence the periodogram estimates the power at f_0 by filtering the data with a bandpass filter, sampling the output, and computing the magnitude squared. The N factor is necessary to account for the bandwidth of the filter in order to yield a spectral *density*. The 3 dB bandwidth can be shown to be approximately $1/N$ (see Problem 4.4). The power when divided by $1/N$ yields the PSD estimate. If a different bandpass filter were used for each f_0 in an effort to optimize the performance, one possible approach is the minimum variance spectral estimator discussed in Chapter 11.

It might be supposed that if enough data are available, say $N \to \infty$, then

$$\hat{P}_{\text{PER}}(f) \to P_{xx}(f)$$

or that the periodogram is a consistent estimator of the PSD. To test this conjecture, consider the periodogram of real zero-mean white Gaussian noise. In Figure 4.3 the periodogram is shown for several data record lengths. Each N point data record was obtained by taking the first N points of a 1024 point data record. It appears that the random fluctuations or variance of the periodogram does not decrease with data record length, prompting us to conclude that the periodogram is *not* a consistent estimator. Although the mean converges to the true PSD as $N \to \infty$, the variance does not tend to zero as $N \to \infty$ (see Problems 4.1 and 4.2). Intuitively, since the filter bandwidth $1/N$ approaches zero as $N \to \infty$, the power out of the filter is due only to spectral components at the frequency of interest for large N. This results in an unbiased spectral estimator. Unfortunately, the variance does not approach zero since only a single output sample of the narrowband filter is used to estimate the power. It is necessary to average the power estimates of many outputs to cause the variance to decrease. Mathematically, for an arbitrary PSD the mean of the periodogram is (see Appendix 4A)

$$\mathscr{E}[\hat{P}_{\text{PER}}(f)] = \int_{-1/2}^{1/2} W_B(f - \xi) P_{xx}(\xi)\, d\xi \tag{4.5}$$

where $W_B(f)$ is the Fourier transform of a triangular or Bartlett window

$$w_B[k] = \begin{cases} 1 - \dfrac{|k|}{N} & \text{for } |k| \leq N - 1 \\ 0 & \text{for } |k| > N \end{cases} \tag{4.6}$$

and is given by

$$W_B(f) = \frac{1}{N} \left(\frac{\sin \pi f N}{\sin \pi f} \right)^2. \tag{4.7}$$

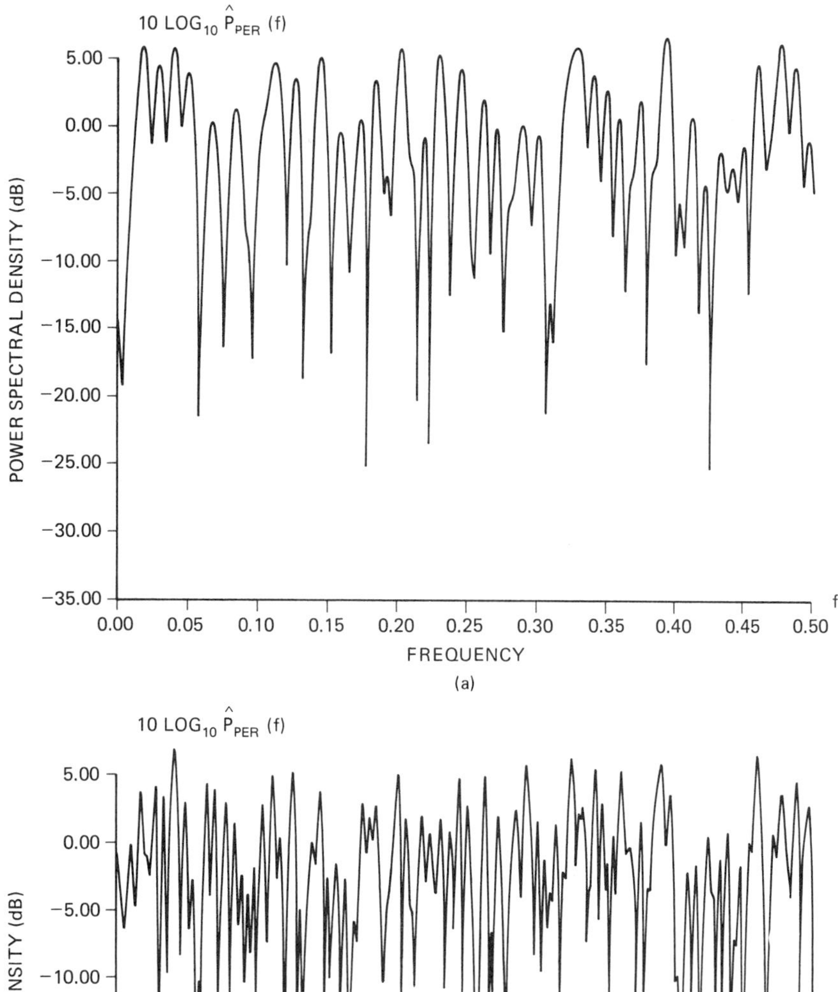

Figure 4.3 Illustration of inconsistency of periodogram for white Gaussian noise ($\sigma_x^2 = 1$). (a) $N = 128$. (b) $N = 256$. (c) $N = 512$. (d) $N = 1024$.

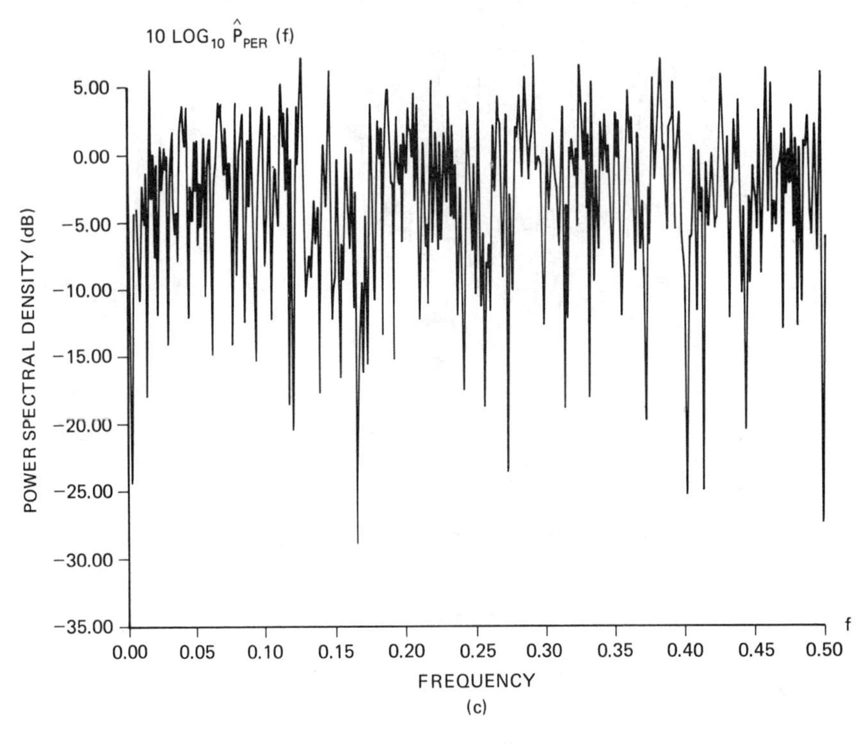

(c)

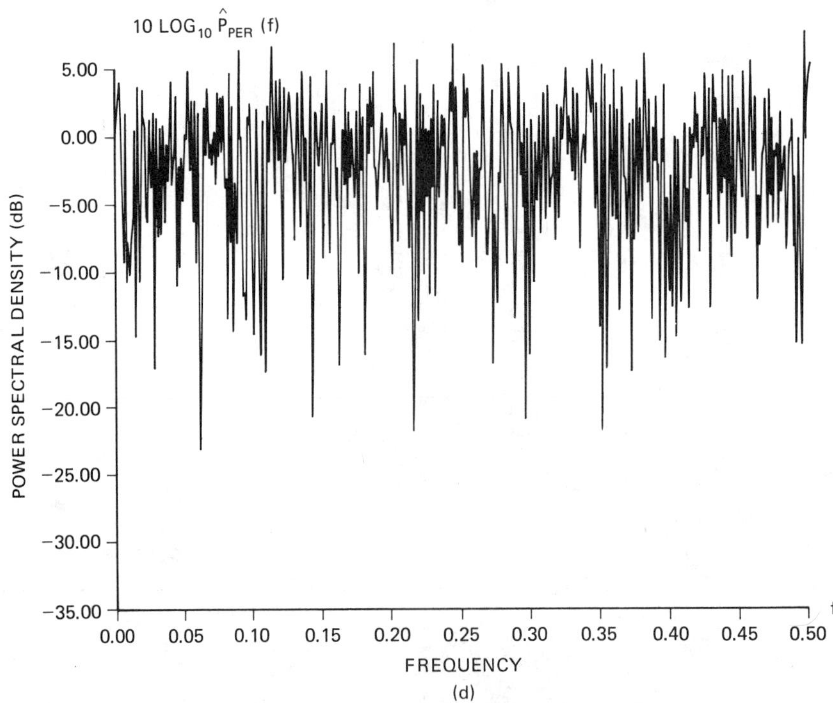

(d)

Figure 4.3 *(Continued)*

This means that *the average periodogram is the convolution of the true PSD with the Fourier transform of a Bartlett window*, yielding on the average a smoothed version of the true PSD. Thus the periodogram is *biased* in general for finite data records but unbiased as $N \to \infty$ since from (4.5)

$$\lim_{N\to\infty} \mathscr{E}[\hat{P}_{\text{PER}}(f)] = P_{xx}(f)$$

which is due to the convergence of $W_B(f)$ to the Dirac delta function, $\delta(f)$ or alternatively to the relationship (4.1). For the special case of white noise it is easily shown that the periodogram is unbiased even for finite data records (see Problem 4.3). From Appendix 4A the covariance is approximately

$$\operatorname{cov}[\hat{P}_{\text{PER}}(f_1), \hat{P}_{\text{PER}}(f_2)] \approx P_{xx}(f_1)P_{xx}(f_2)\left[\left(\frac{\sin N\pi(f_1+f_2)}{N\sin\pi(f_1+f_2)}\right)^2 + \left(\frac{\sin N\pi(f_1-f_2)}{N\sin\pi(f_1-f_2)}\right)^2\right]. \tag{4.8}$$

The variance at any given frequency follows as

$$\operatorname{var}[\hat{P}_{\text{PER}}(f)] = \operatorname{cov}[\hat{P}_{\text{PER}}(f), \hat{P}_{\text{PER}}(f)] \approx P_{xx}^2(f)\left[1 + \left(\frac{\sin 2\pi Nf}{N\sin 2\pi f}\right)^2\right]. \tag{4.9}$$

For frequencies not near 0 or $\pm\frac{1}{2}$, the latter expression approximately reduces to

$$\operatorname{var}[\hat{P}_{\text{PER}}(f)] \approx P_{xx}^2(f). \tag{4.10}$$

It is observed that the *variance is a constant independent of N*, resulting in the important observation that the periodogram estimator is *unreliable* since the standard deviation (square root of the variance) is as large as the mean, which is approximately the quantity to be estimated.

Note that the selection of the harmonic frequencies $f_1 = m/N$ and $f_2 = n/N$, where m and n are distinct integers in the interval $[-N/2, N/2 - 1]$ for N even or $[-(N-1)/2, (N-1)/2]$ for N odd results in a covariance

$$\operatorname{cov}[\hat{P}_{\text{PER}}(f_1), \hat{P}_{\text{PER}}(f_2)] \approx 0.$$

The values of the periodogram separated by integer multiples of $1/N$ are uncorrelated. As the number of data samples N increases, the distance between neighboring frequency samples that are uncorrelated decreases. This feature, coupled with the nondecreasing variance with increasing N, leads to the rapidly fluctuating periodogram as the record length increases. This behavior is illustrated in Figure 4.3. A computer program entitled PERIODOGRAM is provided in Appendix 4C to compute the periodogram.

In many cases of practical interest the data consist of sinusoidal or narrowband signals in white noise. In such a situation it may be advantageous to apply a *data window to* $x[n]$ before computing the periodogram. Without data windowing a lower level signal may be masked by the sidelobes of a higher level signal if the signals are close in frequency. An example is shown in Figure 4.4a, in which the

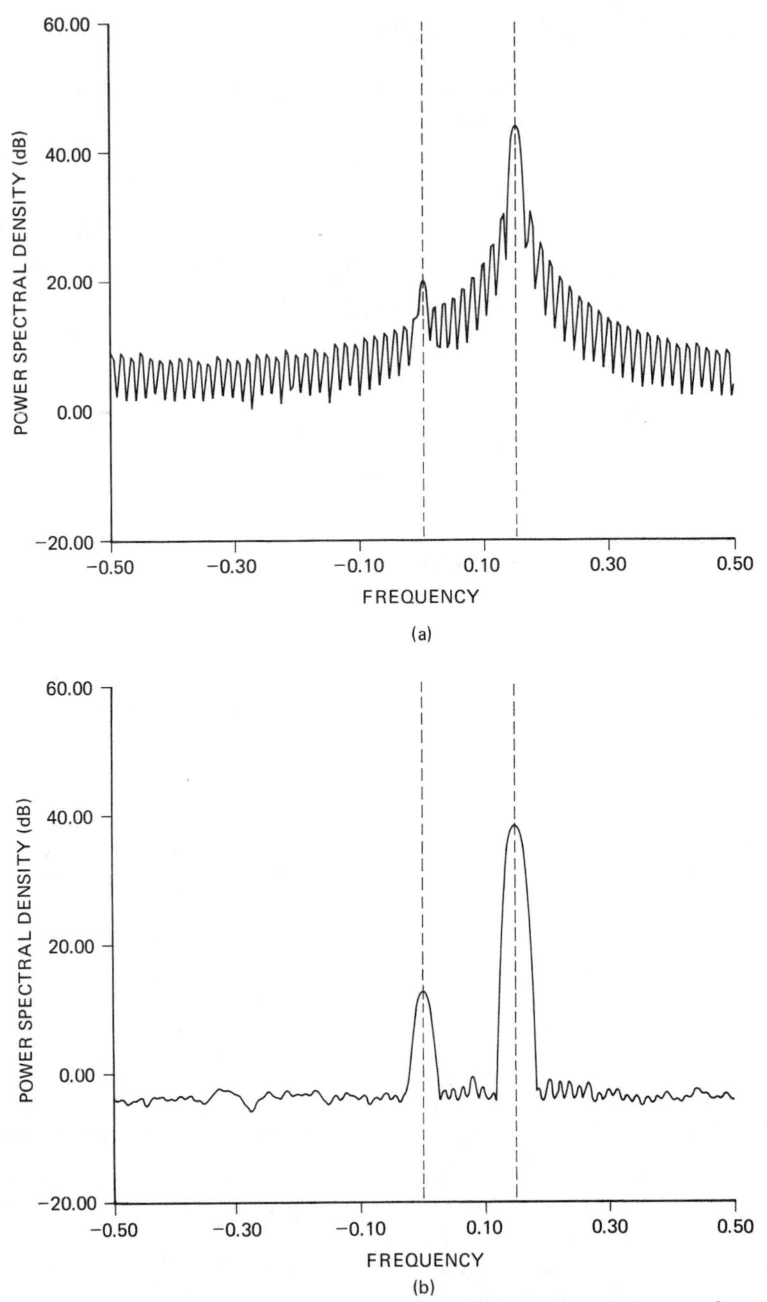

Figure 4.4 Use of data windowing to reduce sidelobes in periodogram of narrowband processes. (a) No data window. (b) Hamming data window.

sidelobes of a sinusoidal signal at $f = 0.12$ nearly mask the presence of a sinusoidal signal at $f = 0$. Data windowing will reduce the magnitude of the periodogram at frequencies not near the signal frequency (sometimes referred to as *the leakage* or the sidelobe level) at the expense of increasing the bandwidth of the mainlobe.

TABLE 4.1 COMMON LAG (DATA[a]) WINDOWS

Name	Definition	Fourier transform
Rectangular	$w[k] = \begin{cases} 1 & \lvert k \rvert \le M \\ 0 & \lvert k \rvert > M \end{cases}$	$W(f) = W_R(f) = \dfrac{\sin \pi f(2M+1)}{\sin \pi f}$
Bartlett	$w[k] = \begin{cases} 1 - \dfrac{\lvert k \rvert}{M} & \lvert k \rvert \le M \\ 0 & \lvert k \rvert > M \end{cases}$	$W(f) = W_B(f) = \dfrac{1}{M}\left(\dfrac{\sin \pi f M}{\sin \pi f}\right)^2$
Hanning	$w[k] = \begin{cases} \frac{1}{2} + \frac{1}{2}\cos \dfrac{\pi k}{M} & \lvert k \rvert \le M \\ 0 & \lvert k \rvert > M \end{cases}$	$W(f) = \frac{1}{4}W_R\left(f - \dfrac{1}{2M}\right) + \frac{1}{2}W_R(f) + \frac{1}{4}W_R\left(f + \dfrac{1}{2M}\right)$
Hamming	$w[k] = \begin{cases} 0.54 + 0.46\cos \dfrac{\pi k}{M} & \lvert k \rvert \le M \\ 0 & \lvert k \rvert > M \end{cases}$	$W(f) = 0.23W_R\left(f - \dfrac{1}{2M}\right) + 0.54W_R(f) + 0.23W_R\left(f + \dfrac{1}{2M}\right)$
Parzen (M even)	$w[k] = \begin{cases} 2\left(1 - \dfrac{\lvert k \rvert}{M}\right)^3 - \left(1 - 2\dfrac{\lvert k \rvert}{M}\right)^3 & \lvert k \rvert \le \dfrac{M}{2} \\ 2\left(1 - \dfrac{\lvert k \rvert}{M}\right)^3 & \dfrac{M}{2} < \lvert k \rvert \le M \\ 0 & \lvert k \rvert > M \end{cases}$	$W(f) = \dfrac{8}{M^3}\left(\dfrac{3}{2}\dfrac{\sin^4 \pi f M/2}{\sin^4 \pi f} - \dfrac{\sin^4 \pi f M/2}{\sin^2 \pi f}\right)$

[a] For the corresponding data window, shift the lag window sequence to the right by M so that $w[k]$ is nonzero over $[0, 2M]$ and replace M by $(N-1)/2$ (N assumed to be odd). The Fourier transform will be that listed after multiplication by $\exp(-j2\pi f M)$.

The application of a Hamming window (see Table 4.1) to the same set of data produces the periodogram shown in Figure 4.4b. The data window has reduced the sidelobe level to more than 40 dB below the peak level. As a result, the lower level sinusoid is now clearly visible. Note, however, that the mainlobe bandwidth has increased due to data windowing. For maximum resolution or the ability to observe two comparable level sinusoidal signals in white noise, no data windowing should be used. A common rule of thumb is that two equiamplitude sinusoids are resolvable if their frequencies are spaced more than $1/N$ cycles/sample apart. It should be emphasized that if the data consist of only one complex sinusoid or several complex sinusoids spaced more than $1/N$ apart in frequency embedded in white Gaussian noise, then the optimal frequency estimator is the periodogram with *no data windowing*, as described more fully in Chapter 13. Some common data windows are given in Table 4.1. Others have been compiled by Harris [1978].

4.4 AVERAGED PERIODOGRAM

The reason why the variance of the periodogram does not decrease with increasing data record length may be attributed to the lack of an expectation operator as required by (4.1). To improve the statistical properties of the periodogram, we can approximate the expectation operator by averaging a set of periodograms together. Assume that K independent data records are available, all for the interval $0 \le n \le L-1$ and all are realizations of the same random process. The data are $\{x_0[n], 0 \le n \le L-1; x_1[n], 0 \le n \le L-1; \ldots ; x_{K-1}[n], 0 \le n \le L-1\}$. Then the *averaged periodogram estimator* is defined as

$$\hat{P}_{\text{AVPER}}(f) = \frac{1}{K}\sum_{m=0}^{K-1} \hat{P}_{\text{PER}}^{(m)}(f) \tag{4.11}$$

where $\hat{P}_{\text{PER}}^{(m)}(f)$ is the periodogram for the mth data set

$$\hat{P}_{\text{PER}}^{(m)}(f) = \frac{1}{L}\left|\sum_{n=0}^{L-1} x_m[n] \exp(-j2\pi f n)\right|^2. \tag{4.12}$$

The mean value of the averaged periodogram will be the same as the mean value of the periodogram based on any of the individual data sets (4.12) since the periodogram for each data set is identically distributed. Hence the mean value is given by (4.5) and (4.7) with N replaced by L or

$$\mathcal{E}[\hat{P}_{\text{AVPER}}(f)] = \int_{-1/2}^{1/2} W_B(f-\xi)P_{xx}(\xi)\,d\xi$$

where

$$W_B(f) = \frac{1}{L}\left(\frac{\sin \pi f L}{\sin \pi f}\right)^2. \tag{4.13}$$

The variance, however, will be decreased by a factor of K. Since the data records are independent, the individual periodograms are also independent. It follows that

$$\text{var}\,[\hat{P}_{\text{AVPER}}(f)] = \frac{1}{K}\,\text{var}\,[\hat{P}^{(m)}_{\text{PER}}(f)] \tag{4.14}$$

for any m. As an example, for the case of white noise with $K = 8$ and $L = 128$ the averaged periodogram estimate is shown in Figure 4.5. A comparison with Figure 4.3a illustrates the reduction in variance. In practice, we seldom have independent data sets but rather, only one data record of length N on which to base the spectral estimator. A common approach, then, is to segment the data into K nonoverlapping blocks of length L, where $N = KL$. In this manner the data sets for use in (4.12) become

$$x_m[n] = x[n + mL] \qquad n = 0,1, \ldots, L-1; \quad m = 0,1, \ldots, K-1.$$

Since the blocks are contiguous, they cannot be uncorrelated let alone independent for any other process but white noise. The variance reduction will in general be less than K. For processes not exhibiting sharp resonances or characterized by ACFs that decay rapidly, the data blocks will be approximately uncorrelated. If the data are Gaussian, the individual periodograms will be nearly independent, so (4.14) will be a good approximation.

It is important to note that in sectioning the data into blocks each block will have a length $L < N$. The bias of the averaged periodogram spectral estimator

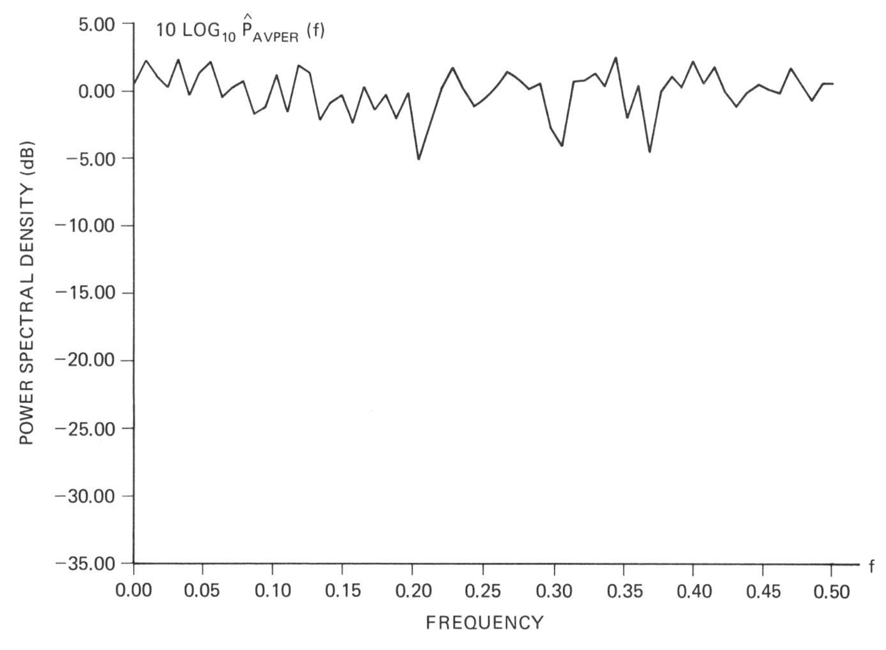

Figure 4.5 Averaged periodogram for white noise with $\sigma_x^2 = 1$, $L = 128$, $K = 8$ illustrating reduction in variance.

will then increase over that of the periodogram since the Bartlett window of (4.6) will be shorter and hence the mainlobe of $W_B(f)$ will be wider. An example of the Fourier transform of the Bartlett window is given in Figure 4.6. The result of the convolution is to produce an average spectral estimate which is smeared. An example is given in Figure 4.7. To avoid the smearing the Bartlett window length L must be chosen so that the bandwidth of the mainlobe of $W_B(f)$ is much less than the bandwidth of the narrowest peak in $P_{xx}(f)$. Since the 3 dB bandwidth of the mainlobe of $W_B(f)$ is about $1/L$ (see Problem 4.4), we cannot resolve details in the PSD finer than $1/L$. The spectral estimator is then said to have a resolution of $1/L$ cycles/sample. Clearly, for maximum resolution L should be chosen as large as possible or $L = N - 1$, which results in the standard periodogram (4.2). However, for some variance reduction we should choose $K = N/L$ large or equivalently L small. Since both goals cannot be met simultaneously, we must trade off bias for variance by adjusting L. In practice, a good strategy is to compute several averaged periodogram estimates, with each successive one having a larger L. If the spectral estimate does not change significantly as L is increased, all the spectral detail has been found. This technique, known as *window closing*, must be used with care since statistical instability increases as L is increased [Jenkins and Watts 1968].

A technique which is useful for reducing the bias is that of prewhitening the data prior to spectral estimation. The bias of the averaged periodogram (or the periodogram) is zero for white noise since if $P_{xx}(f) = \sigma_x^2$,

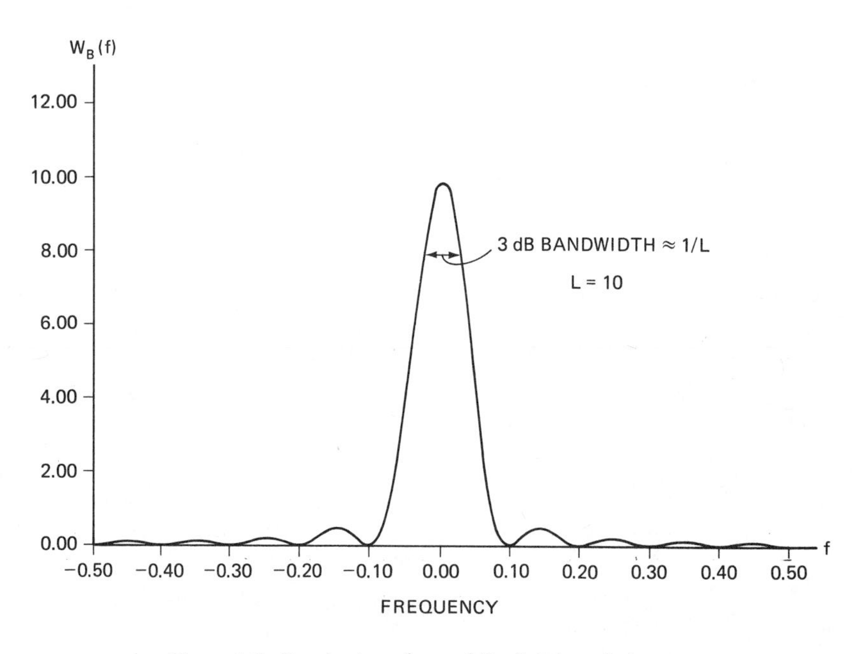

Figure 4.6 Fourier transform of Bartlett lag window.

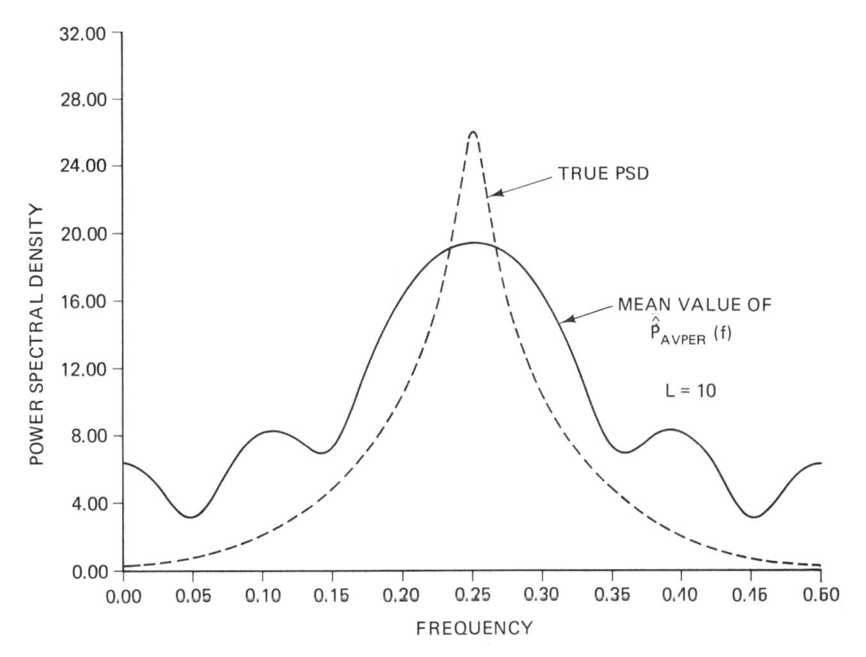

Figure 4.7 Example of mean value of averaged periodogram.

$$
\begin{aligned}
\mathcal{E}[\hat{P}_{\mathrm{AVPER}}(f)] &= \int_{-1/2}^{1/2} W_B(f - \xi)P_{xx}(\xi)\, d\xi \\
&= \sigma_x^2 \int_{-1/2}^{1/2} W_B(f - \xi)\, d\xi \\
&= \sigma_x^2 w_B[0] = \sigma_x^2 = P_{xx}(f).
\end{aligned} \tag{4.15}
$$

Regardless of the value of L, the estimator is unbiased. This is due, of course, to the lack of peaks and valleys in the PSD and hence the lack of smoothing introduced by the convolution operation. If this is the case, L can be made small to reduce the variance, and yet not increase the bias. In practice, if we have some idea of the general shape of the PSD but possibly not the details, a good technique is to filter the data with an LSI filter to yield a PSD at the output of the filter which is flatter than at the input. The FIR filter, which is termed a prediction error or autoregressive filter and is described in Chapters 6 and 16, has been proposed for this purpose [Blackman and Tukey 1958, Thomson 1977]. Such an approach is termed *prewhitening the data* [Blackman and Tukey 1958, Jenkins and Watts 1968].

It is extremely important in interpreting spectral estimates to be able to ascertain whether spectral detail is due to statistical fluctuation or is actually present. In other words, we need some measure of confidence in the spectral estimate. Assuming that the spectral estimator is approximately unbiased, we can derive a confidence interval for the averaged periodogram estimator for real data

[Jenkins and Watts 1968]. The unbiased assumption requires the bandwidth of the narrowest peak or valley of the PSD to be much larger than the mainlobe of the spectral window $W_B(f)$. Under these conditions it can be shown that a $(1 - \alpha) \times 100\%$ confidence interval is (see also Section 3.3.1)

$$\left(\frac{2K\hat{P}_{\text{AVPER}}(f)}{\chi^2_{2K}(1 - \alpha/2)}, \frac{2K\hat{P}_{\text{AVPER}}(f)}{\chi^2_{2K}(\alpha/2)} \right) \tag{4.16}$$

which is valid for $f \neq 0, \pm\frac{1}{2}$. $\chi^2_{2K}(\alpha)$ is the α percentage point of a χ^2_{2K} CDF, or

$$\Phi(\chi^2_{2K}(\alpha)) = \Pr\{\chi^2_{2K} \leq \chi^2_{2K}(\alpha)\} = \alpha.$$

If one plots the PSD in decibels, the confidence interval becomes *a constant length interval for all frequencies*, or

$$10 \log_{10} \hat{P}_{\text{AVPER}}(f) \begin{cases} +10 \quad \log_{10} \dfrac{2K}{\chi^2_{2K}(\alpha/2)} \\ \\ -10 \quad \log_{10} \dfrac{\chi^2_{2K}(1 - \alpha/2)}{2K}. \end{cases} \tag{4.17}$$

As an example, for a 95% confidence interval with $K = 10$, $\alpha = 0.05$, $\chi^2_{20}(0.025) = 10.85$, $\chi^2_{20}(0.975) = 31.41$, the confidence interval becomes

$$10 \log_{10} \hat{P}_{\text{AVPER}}(f) \begin{cases} +2.65 \text{ dB} \\ -1.96 \text{ dB}. \end{cases}$$

Hence spectral peaks and valleys more than a few decibels should be considered as actually being present and not due to statistical fluctuation.

A variation of the averaged periodogram has been proposed by Welch [1967]. Its contrasting features are the use of a *data window* for each data block and the use of *overlapped* data blocks. The data window reduces leakage by reducing the sidelobe levels of the spectral estimates for narrowband components. By overlapping the blocks of data, usually by 50 or 75%, some extra variance reduction is achieved.

Before concluding the discussion of periodogram spectral estimators, it is important to note the use of the FFT in computing them. Since $\hat{P}_{\text{PER}}(f)$ cannot be computed for a continuum of frequencies, we are forced to sample it. Although the fundamental frequency interval of the PSD has been assumed to be $[-\frac{1}{2}, \frac{1}{2}]$, it is more convenient for the present discussion to use the frequency interval [0,1]. Typically, then, equally spaced frequency samples are taken so that $f_k = k/N$ for $k = 0, 1, \ldots, N - 1$, to yield

$$\begin{aligned} \hat{P}_{\text{PER}}(f_k) &= \frac{1}{N} \left| \sum_{n=0}^{N-1} x[n] \exp(-j2\pi f_k n) \right|^2 \\ &= \frac{1}{N} \left| \sum_{n=0}^{N-1} x[n] \exp\left(\frac{-j2\pi kn}{N} \right) \right|^2 \qquad k = 0, 1, \ldots, N - 1. \end{aligned} \tag{4.18}$$

The latter expression is in the form of a DFT and hence may be efficiently computed using an FFT. The samples of the periodogram may be recovered for $-\frac{1}{2} \leq f < 0$ by making use of the periodicity of the PSD to obtain $\hat{P}_{\text{PER}}(f_k - 1) = \hat{P}_{\text{PER}}(f_k)$ for $f_k = k/N$ and $k = N/2, N/2 + 1, \ldots, N - 1$, assuming that N is even. Equivalently, the two halves of the FFT output array must be transposed. To approximate $\hat{P}_{\text{PER}}(f)$ more closely, we may need to have a finer frequency spacing. This is accomplished by zero padding the data with $N' - N$ zeros and then taking an N' point FFT. The effective data set becomes

$$x'[n] = \begin{cases} x[n] & \text{for } n = 0, 1, \ldots, N - 1 \\ 0 & \text{for } n = N, N + 1, \ldots, N' - 1. \end{cases}$$

The frequency spacing will then be $1/N' < 1/N$. *No extra resolution is afforded by zero padding but only a better evaluation of the periodogram.* See Figure 4.8 for an example.

4.5 BLACKMAN–TUKEY SPECTRAL ESTIMATION

In Section 4.4 it was shown that the periodogram is a poor estimator of the PSD. This poor performance can, alternatively, be explained by the following equivalent form for the periodogram (see Problem 4.5):

$$\hat{P}_{\text{PER}}(f) = \sum_{k=-(N-1)}^{N-1} \hat{r}_{xx}[k] \exp(-j2\pi f k) \tag{4.19}$$

where

$$\hat{r}_{xx}[k] = \begin{cases} \dfrac{1}{N} \displaystyle\sum_{n=0}^{N-1-k} x^*[n]x[n+k] & \text{for } k = 0, 1, \ldots, N - 1 \\ \hat{r}^*_{xx}[-k] & \text{for } k = -(N-1), -(N-2), \ldots, -1. \end{cases} \tag{4.20}$$

The periodogram as given by (4.19) is seen to be an estimator based on the Wiener–Khinchin theorem discussed in Chapter 3 with the ACF replaced by the estimate of (4.20). The poor performance of the periodogram may be attributed to the poor performance of the ACF estimator. To illustrate this point, consider the estimate of $r_{xx}[N-1]$. This sample is estimated by $x^*[0]x[N-1]/N$, which will be highly variable due to the lack of averaging, as well as being biased no matter how large N becomes. In general, the mean of $\hat{r}_{xx}[k]$ is, from (4.20),

$$\mathscr{E}[\hat{r}_{xx}[k]] = \frac{N - |k|}{N} r_{xx}[k] \qquad |k| \leq N - 1 \tag{4.21}$$

or the mean value is equal to the true value weighted by a Bartlett window. We could use an unbiased ACF estimator in (4.19) by replacing the $1/N$ factor in (4.20) by $1/(N - |k|)$. (A computer program called CORRELATION [Marple 1987],

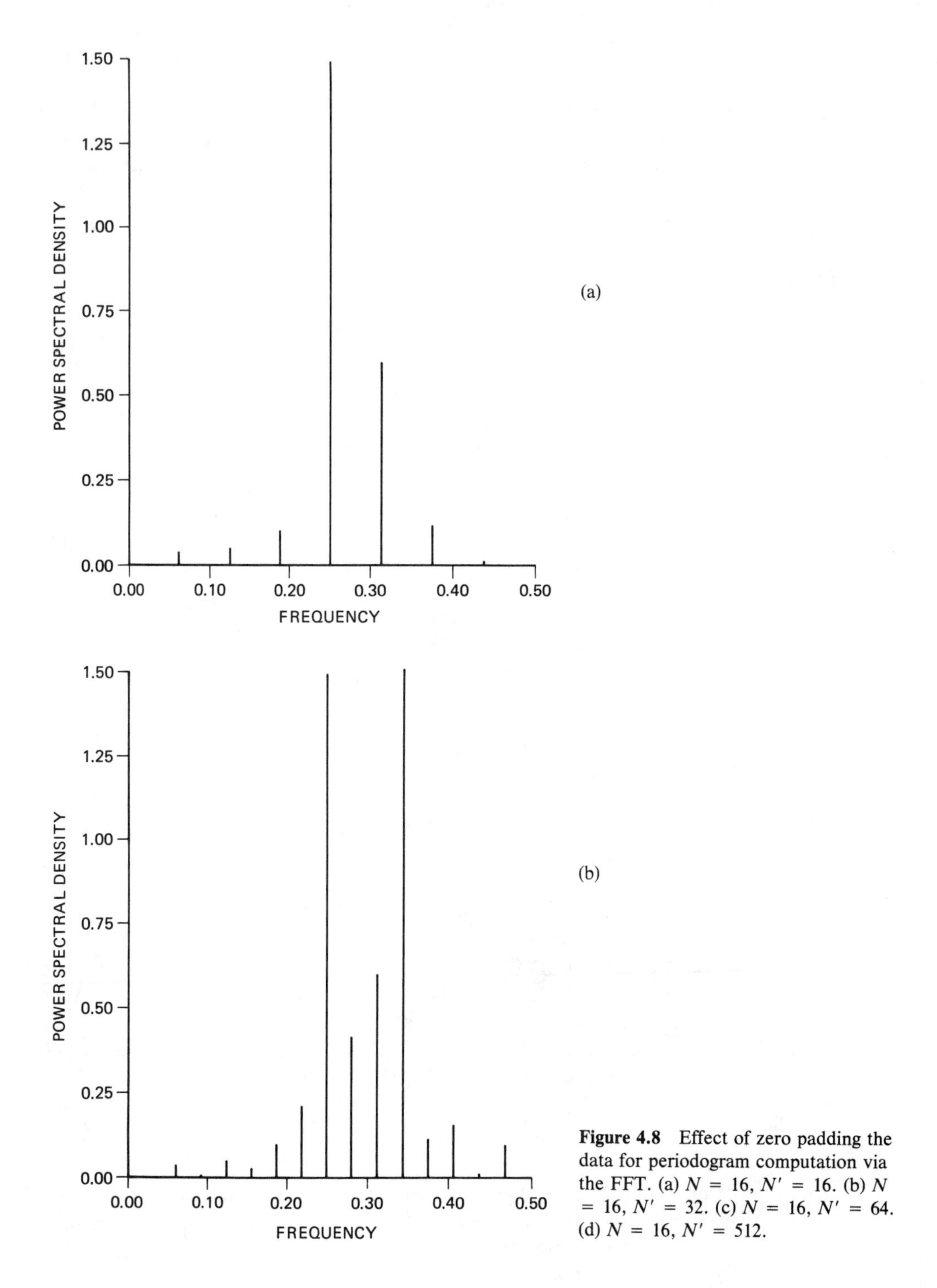

Figure 4.8 Effect of zero padding the data for periodogram computation via the FFT. (a) $N = 16$, $N' = 16$. (b) $N = 16$, $N' = 32$. (c) $N = 16$, $N' = 64$. (d) $N = 16$, $N' = 512$.

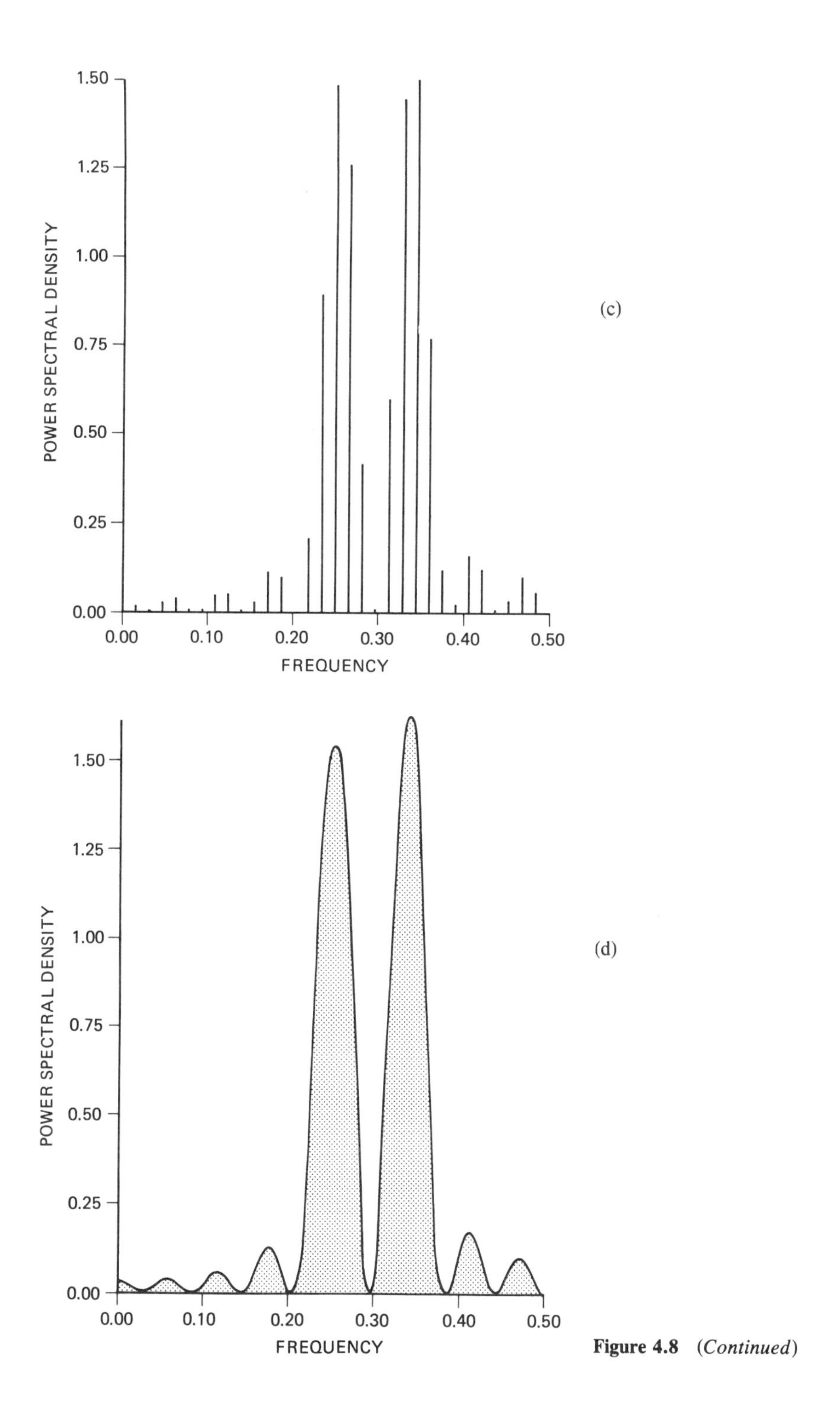

Figure 4.8 *(Continued)*

given in Appendix 4D, can be used to compute either the biased or unbiased ACF estimator.) However, this choice may lead to a negative spectral estimate since the unbiased ACF estimator does not guarantee a positive semidefinite sequence (see Problems 4.6 and 4.7). This point will be discussed in more detail shortly. The poorer estimates of the ACF at higher lags is a result of the fewer number of lag products averaged (see also Problems 4.8 and 4.9). One way to avoid this problem is to weight the ACF estimates at higher lags less or to use the spectral estimator

$$\hat{P}_{\text{BT}}(f) = \sum_{k=-(N-1)}^{N-1} w[k]\hat{r}_{xx}[k] \exp(-j2\pi fk) \tag{4.22}$$

where $w[k]$ is a real sequence termed a *lag window* with the following properties:

1. $0 \le w[k] \le w[0] = 1$
2. $w[-k] = w[k]$
3. $w[k] = 0 \quad \text{for } |k| > M$

where $M \le N - 1$. Due to the last property of the lag window, (4.22) may be written as

$$\hat{P}_{\text{BT}}(f) = \sum_{k=-M}^{M} w[k]\hat{r}_{xx}[k] \exp(-j2\pi fk). \tag{4.23}$$

(4.23) is called the *Blackman–Tukey (BT) spectral estimator* after the researchers who pioneered its use [Blackman and Tukey 1958]. As described previously, it is equivalent to the periodogram if $w[k] = 1$ for $|k| \le M = N - 1$. The BT spectral estimator is sometimes called a *weighted covariance* estimator. The weighting of the ACF estimator will reduce the variance of the spectral estimator at the expense of increasing the bias (unless the process is white noise, for which the bias is zero for any lag window). Many windows are available. Table 4.1 lists a few of the popular ones. Many others are available [Harris 1978]. We must be careful, however, to ensure that the window chosen will always lead to a non-negative spectral estimate. To see how a negative spectral estimate may occur as a result of windowing, note from (4.23) that

$$\begin{aligned} \hat{P}_{\text{BT}}(f) &= \mathcal{F}\{w[k]\hat{r}_{xx}[k]\} \\ &= \int_{-1/2}^{1/2} W(f - \xi)\hat{P}_{\text{PER}}(\xi)\, d\xi \end{aligned} \tag{4.24}$$

since $\mathcal{F}\{\hat{r}_{xx}[k]\} = \hat{P}_{\text{PER}}(f)$ ($\mathcal{F}$ denotes the Fourier transform operator). Although $\hat{P}_{\text{PER}}(f) \ge 0$, $W(f)$ may be negative enough to cause $\hat{P}_{\text{BT}}(f)$ to become negative (see Problem 4.10 for an example). To ensure that this will not happen, the lag window should have a Fourier transform that is nonnegative or equivalently $w[k]$ must be a positive semidefinite sequence. A window that satisfies these requirements is the Bartlett window, whose frequency response is given in Figure 4.6. Of the windows listed in Table 4.1, only the Bartlett and Parzen windows have nonnegative Fourier transforms (see Problem 4.11). Also, it is apparent from (4.24) that the use of the unbiased ACF estimator may result in a negative spectral estimate even if a positive semidefinite lag window is employed. This is because the autocorrelation sequence may not be positive semidefinite, which implies that its Fourier transform may be negative.

The bias and variance of the BT spectral estimator are derived in Appendix 4B. The results are now summarized. If the periodogram is approximately an unbiased estimator or the true PSD is smooth over any frequency interval of length $1/N$, then the mean is

$$\mathcal{E}[\hat{P}_{\mathrm{BT}}(f)] \approx \int_{-1/2}^{1/2} W(f - \xi)P_{xx}(\xi)\, d\xi. \tag{4.25}$$

The mean is a smeared version of the true PSD. It is said that $W(f)$ acts as a *spectral window*. For most lag windows the BT spectral estimator will be able to resolve spectral detail to about $1/M$ cycles/sample or to the bandwidth of the mainlobe of the spectral window. Details finer than this will, after the convolution operation of (4.25), appear smeared and so will not be visible. In general, heavy biases are incurred for those spectra having a large dynamic range or nonflat spectra. As in the case of the periodogram, prewhitening can be used to reduce this bias. The variance is derived under the assumption that the PSD is smooth over any frequency interval which is equal to the bandwidth of the mainlobe of the spectral window ($\approx 1/M$). For frequencies not near 0, $\pm\frac{1}{2}$, the variance is

$$\mathrm{var}\,[\hat{P}_{\mathrm{BT}}(f)] \approx \frac{P_{xx}^2(f)}{N} \sum_{k=-M}^{M} w^2[k]. \tag{4.26}$$

Again a bias–variance trade-off is evident by examining (4.25) and (4.26). For a small bias M should be chosen to be large since that choice will cause the spectral window to behave as a Dirac delta function. On the other hand, for a small variance M should be chosen to be small, as per (4.26). A maximum value of $M = N/5$ is usually recommended. As an example, for the Bartlett window

$$\mathrm{var}\,[\hat{P}_{\mathrm{BT}}(f)] \approx \frac{2M}{3N} P_{xx}^2(f) \tag{4.27}$$

so that $M = N/5$ would result in a variance reduction of 7.5 relative to the periodogram. Much of the art in classical spectral estimation is in choosing an appropriate window, both the type and the length.

A confidence interval for the BT spectral estimator may be derived in a similar fashion to that of the averaged periodogram [Koopmans 1974]. The results are

$$10 \log_{10} \hat{P}_{\mathrm{BT}}(f) \begin{cases} +10 \log_{10} \dfrac{\nu}{\chi_\nu^2(\alpha/2)} \\[2ex] -10 \log_{10} \dfrac{\chi_\nu^2(1 - \alpha/2)}{\nu} \end{cases} \tag{4.28}$$

where

$$\nu = \frac{2N}{\sum_{k=-M}^{M} w^2[k]} \tag{4.29}$$

which represents the degrees of freedom of a χ^2 distribution. The assumption made in deriving (4.28) is that the spectral estimator is approximately unbiased. As an example, for the Bartlett window given by (4.6) with N replaced by M, ν is approximately $3N/M$. A computer program entitled BLACKTUKEY is provided in Appendix 4E to compute the Blackman–Tukey spectral estimate.

A variant of the BT estimator which is also related to the periodogram is the *smoothed periodogram* estimator. From (4.24) the BT spectral estimator may be interpreted as a convolution of the periodogram with a spectral window in frequency. This suggests computing the periodogram and smoothing or averaging it in frequency. Assuming that an N' point FFT is used to compute the periodogram, a discrete version of (4.24) for uniform spectral weighting is

$$\hat{P}_{\text{DAN}}(f_i) = \frac{1}{2J+1} \sum_{j=-J}^{J} \hat{P}_{\text{PER}}\left(f_i + \frac{j}{N'}\right) \tag{4.30}$$

where $f_i = i/N'$, $i = -N'/2, -N'/2 + 1, \ldots, N'/2 - 1$. This is called Daniell's spectral estimator [Koopmans 1974]. Many other spectral estimators may be generated by choosing various spectral weightings.

4.6 COMPUTER SIMULATION EXAMPLES

Some computer simulation examples of the periodogram and BT spectral estimators are described in this section to illustrate the properties of these estimators. In all the examples 50 realizations of the spectral estimator were generated. Each figure displays the 50 realizations in an overlaid format to indicate the variability of the estimator as well as the average of the realizations to indicate the bias.

The first example consists of sinusoidal signals in white noise:

$$x[n] = \sqrt{10} \exp\,[j2\pi(0.15)n] + \sqrt{20} \exp\,[j2\pi(0.20)n] + z[n] \tag{4.31}$$

where $z[n]$ is complex white Gaussian noise with variance equal to unity (see Section 3.5.1). For a data record length of $N = 20$, the periodogram spectral estimate with no data windowing is shown in Figure 4.9a. The true positions of the sinusoids are indicated by dashed lines. The two sinusoids are just resolved since the frequency spacing is at the resolution limit of $1/N$. The corresponding BT spectral estimator which employs a Bartlett window with length $M = 4$ is plotted in Figure 4.9b. The use of $M = N/5 = 4$ is the maximum value recommended. The BT estimate is unable to resolve the sinusoids and hence exhibits a large bias but much less variance than the periodogram. If the data record length is increased to $N = 100$ with $M = 20$ used for the Bartlett window, the spectral estimates appear as shown in Figure 4.10. The periodogram yields a better representation of the PSD since the peaks become narrower. However, the variance does not decrease, as may be observed by examining the estimate at frequencies not near the sinusoidal locations. Also, it is interesting to note that near the sinusoidal locations the variance is small. This is because the response of the narrowband filters of the periodogram [see (4.4)] is due mainly to the sinusoids in this frequency region. The nonrandom nature of the sinusoids results in a spectral estimate with low variability. The spectral estimate at frequencies not near the sinusoids is due solely to the white noise component of the data. As expected from the discussion of Section 4.3, the variability is large and does not decrease with an increase in the data record length. The BT spectral estimate is nearly able to resolve the sinusoids since the resolution of the BT spectral estimator as indicated by (4.25) is about $1/M$ cycles/sample. The variance is about the same as for the BT estimate with $N = 20$ and $M = 4$, which is in accordance with (4.27). Next, the averaged periodogram spectral estimator (4.11) was applied to the data given by (4.31) for $N = 256$ points. In Figure 4.11 the results of averaging the periodograms for $K = 4$ nonoverlapping blocks of 64 data points (part a) and $K = 16$ nonoverlapping blocks of 16 data points (part b) are shown. As expected, the variance decreases as more periodograms are averaged, but so also does the resolution.

The periodogram and BT spectral estimators were next applied to *real* processes which may be characterized as having either narrowband or broadband spectral features. The exact character of the random processes is discussed in Section 10.7, where they are described as the ARMA1 and ARMA3 processes (see Table 10.1). The spectral estimates were all based on an $N = 256$ point real data record. The results obtained for the broadband process are shown in Figure 4.12. Note that the vertical axis scaling is not the same for both estimates. The reason for this will be explained shortly. The periodogram estimate did not window the data and the BT estimate used a Bartlett window with $M = 51$. In each case the dashed curve represents the true PSD. The average spectral estimate for either the periodogram or BT estimate appears to yield the true PSD so that very little bias is observed. Because of the broadband nature of the PSD the resolution of the periodogram ($1/N = 0.004$) and the BT ($1/M = 0.02$) spectral estimators are adequate to resolve the spectral detail. The variance of the periodogram, however, is substantially larger than that of the BT spectral estimator. The periodogram spectral values vary over a range of about 20 dB, while those of the BT estimate vary only over about a 7 dB range. The large variability of the periodogram is the reason for the difference in vertical scaling used in Figure 4.12. In fact, from (4.10) and (4.27) the ratio of the variances would be predicted to be about $3N/2M = 7.5$. Hence for broadband PSDs the variance of the BT spectral estimator may be made small by choosing M small without incurring a large bias. This is because for this type of PSD the ACF damps out rapidly so that truncation of the ACF will not cause a significant smearing of the average spectral estimate. On the other hand, for narrowband PSDs this will not be the case. An example is given in Figure 4.13, in which the spectral estimates for a narrowband PSD based on a $N = 256$ real data point record are shown. As before, the BT spectral estimate employed a Bartlett window with $M = 51$. The true PSD is shown as a dashed

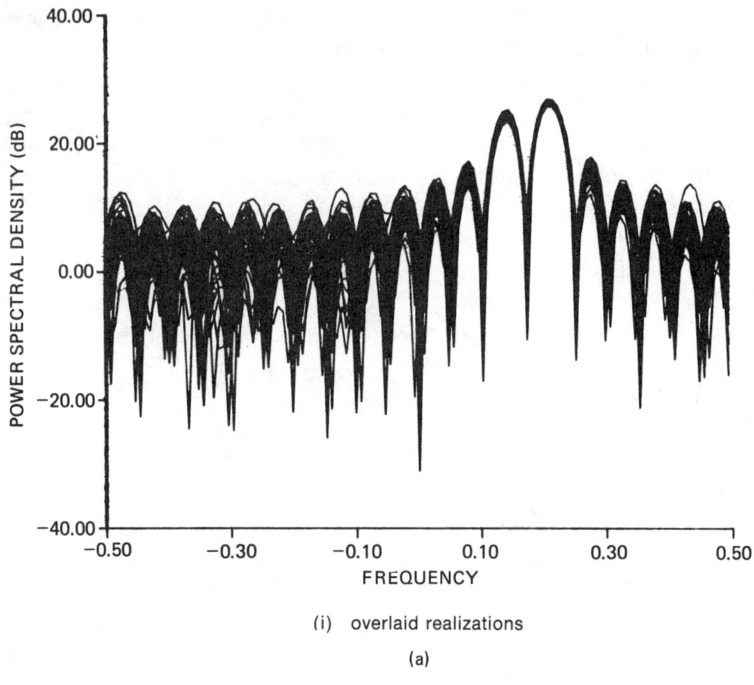

(i) overlaid realizations

(a)

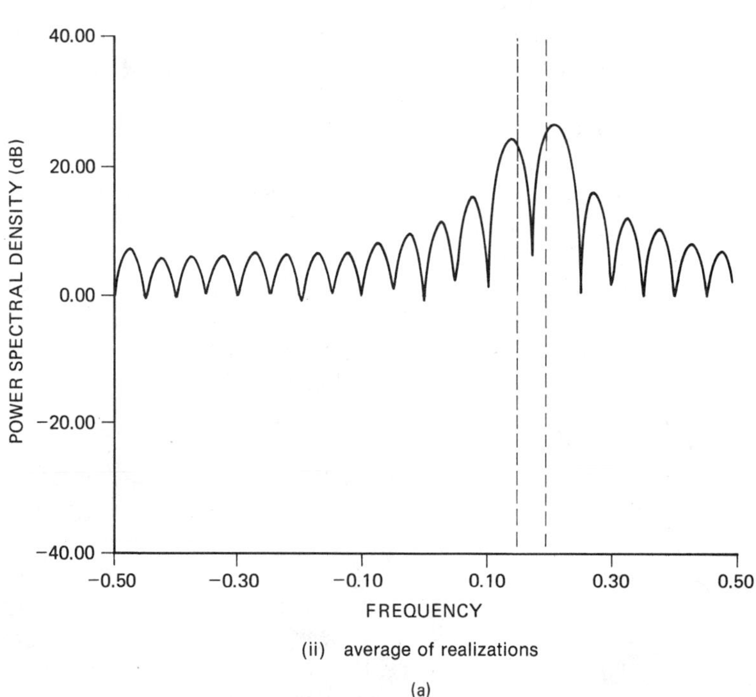

(ii) average of realizations

(a)

Figure 4.9 (a) Periodogram spectral estimate for two complex sinusoids in white noise ($N = 20$). (b) Blackman–Tukey spectral estimate for two complex sinusoids in white noise ($N = 20$).

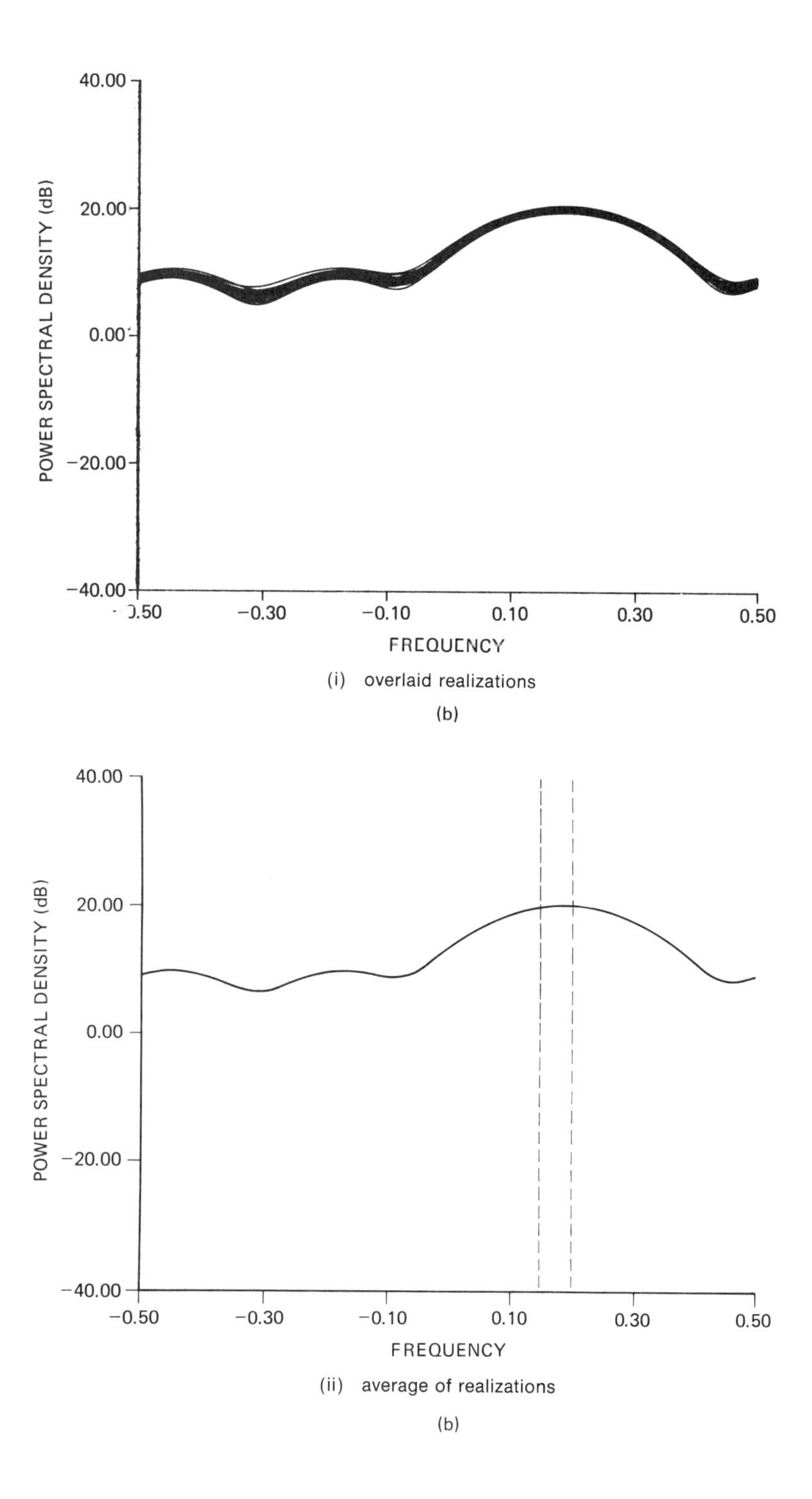

Figure 4.9 *(Continued)*

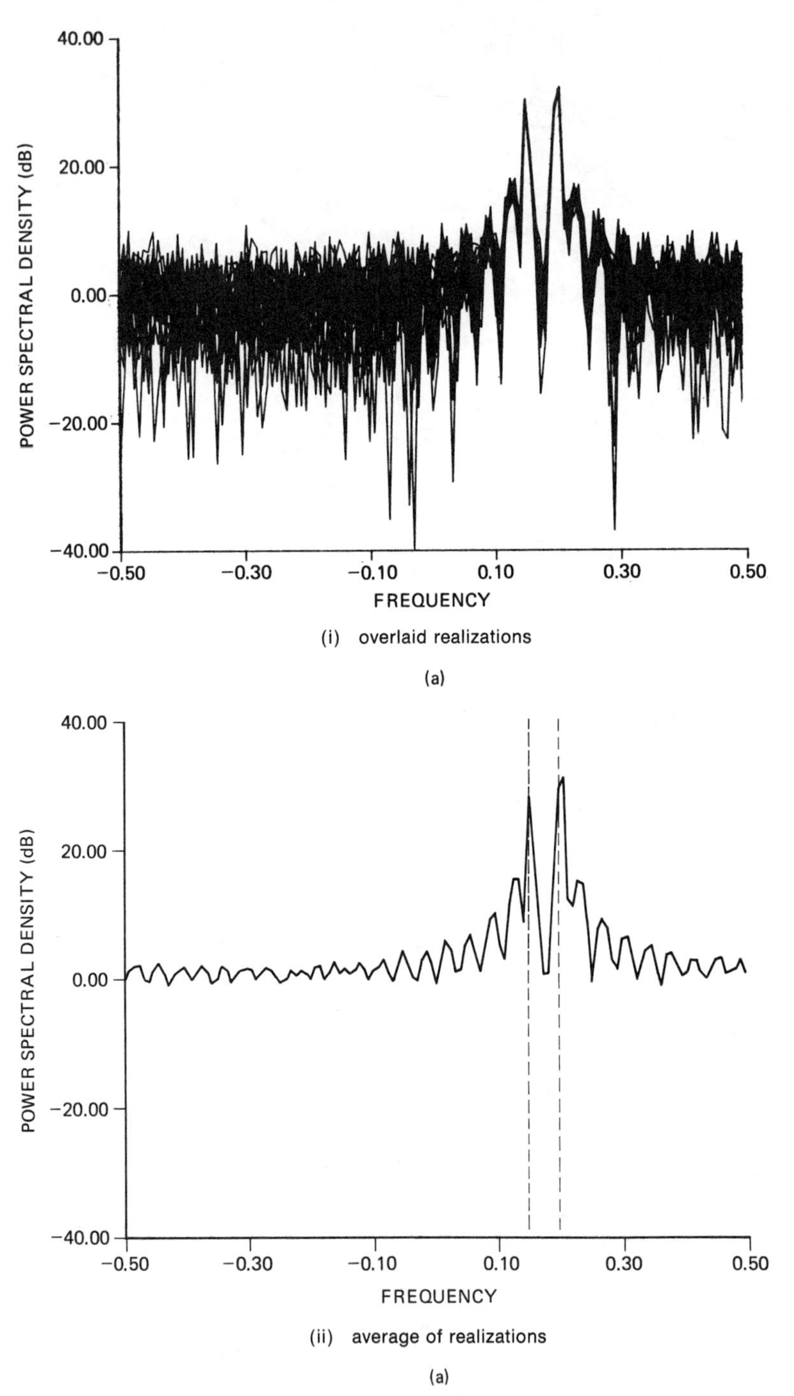

Figure 4.10 (a) Periodogram spectral estimate for two complex sinusoids in white noise (N = 100). (b) Blackman–Tukey spectral estimate using a Bartlett window (M = 20) for two complex sinusoids in white noise (N = 100).

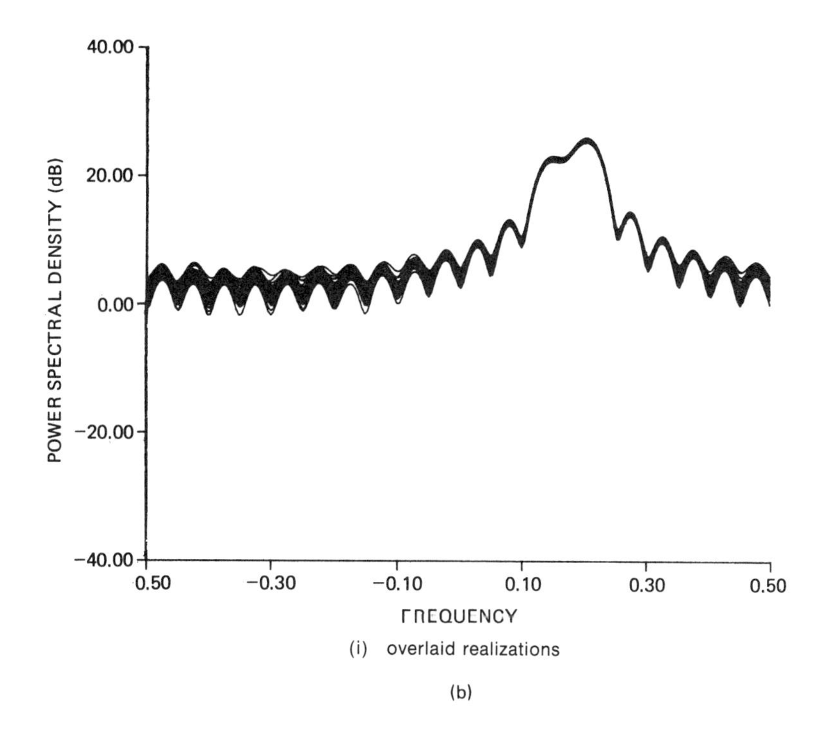

(i) overlaid realizations

(b)

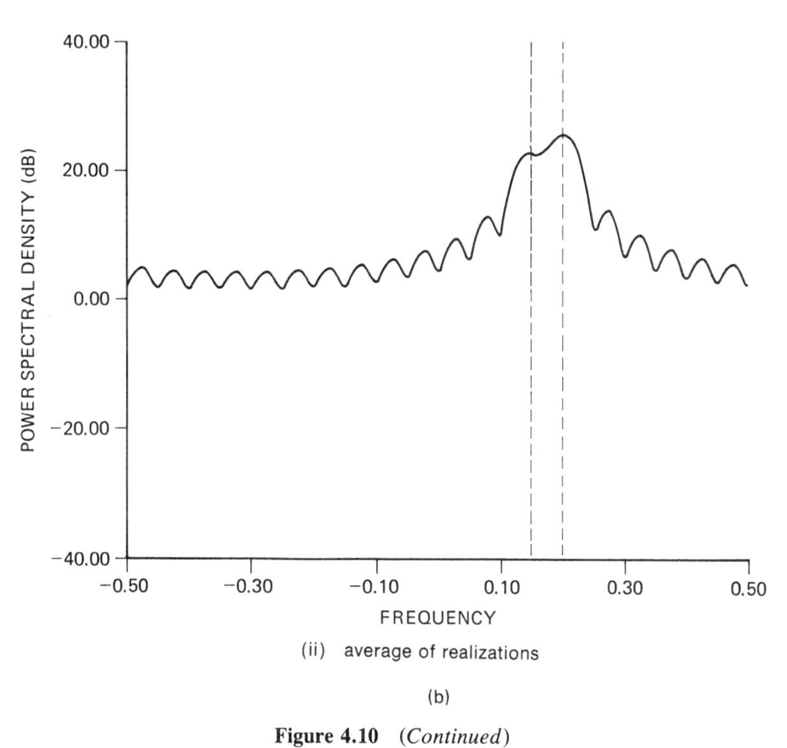

(ii) average of realizations

(b)

Figure 4.10 (*Continued*)

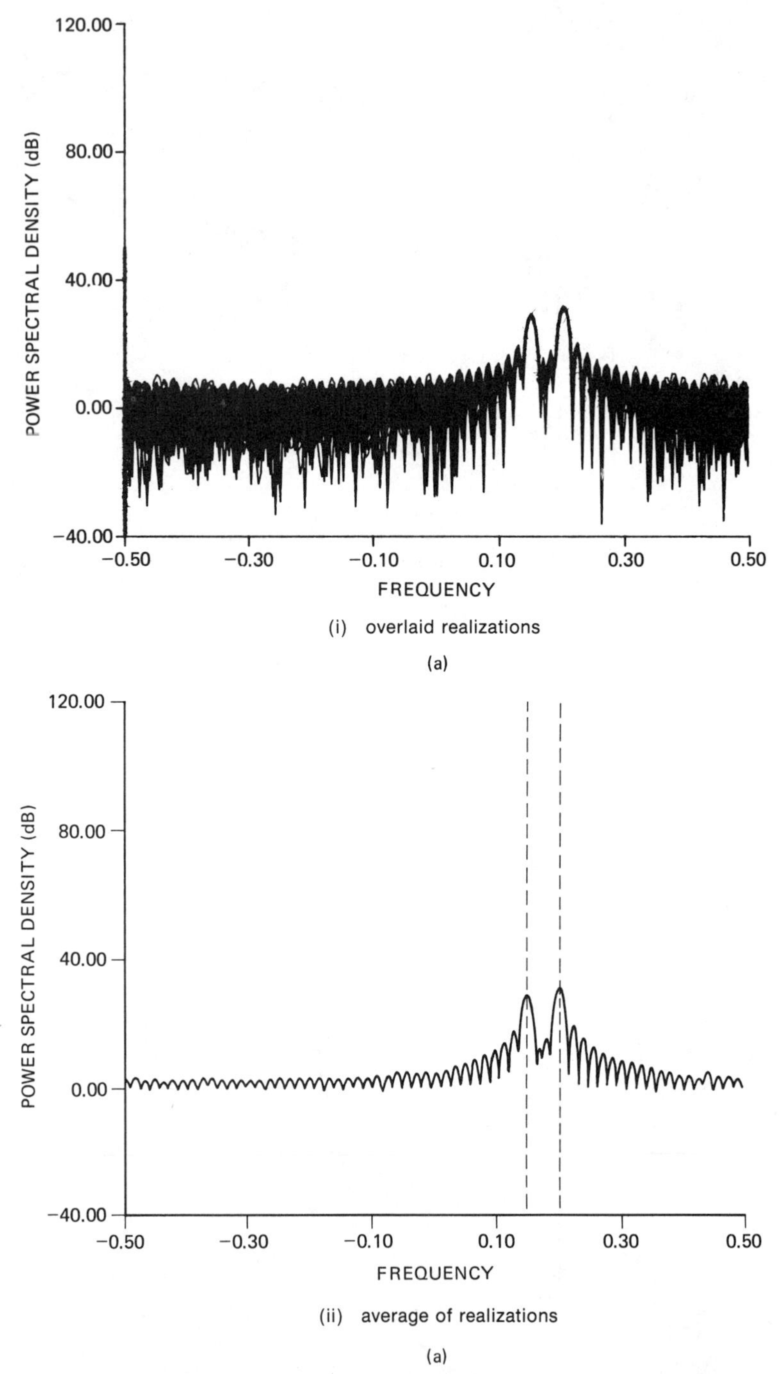

Figure 4.11 Averaged periodogram spectral estimate for two complex sinusoids in white noise. (a) $K = 4$ periodograms averaged with each one based on $L = 64$ point data block. (b) $K = 16$ periodograms averaged with each one based on $L = 16$ point data block.

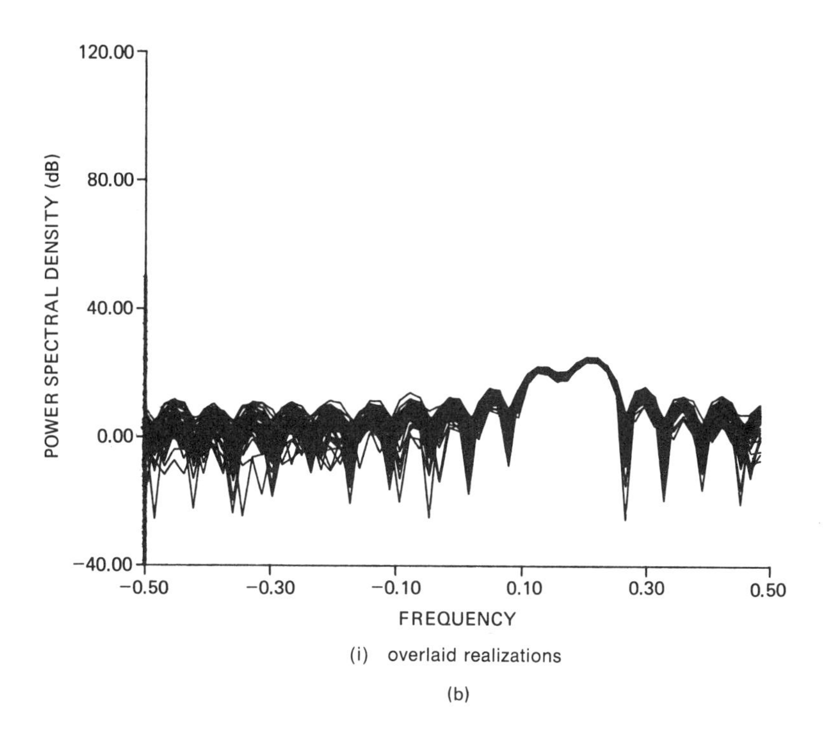

(i) overlaid realizations

(b)

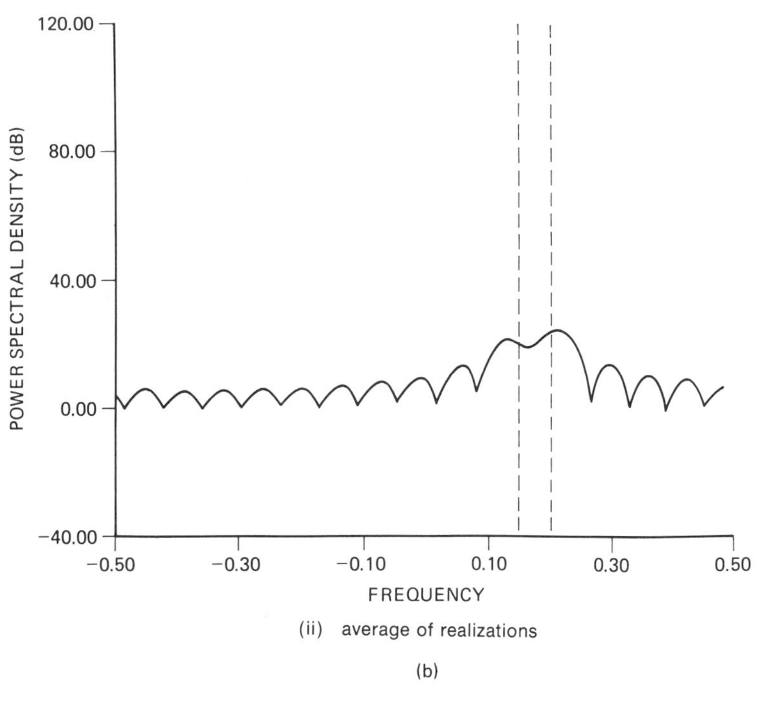

(ii) average of realizations

(b)

Figure 4.11 (*Continued*)

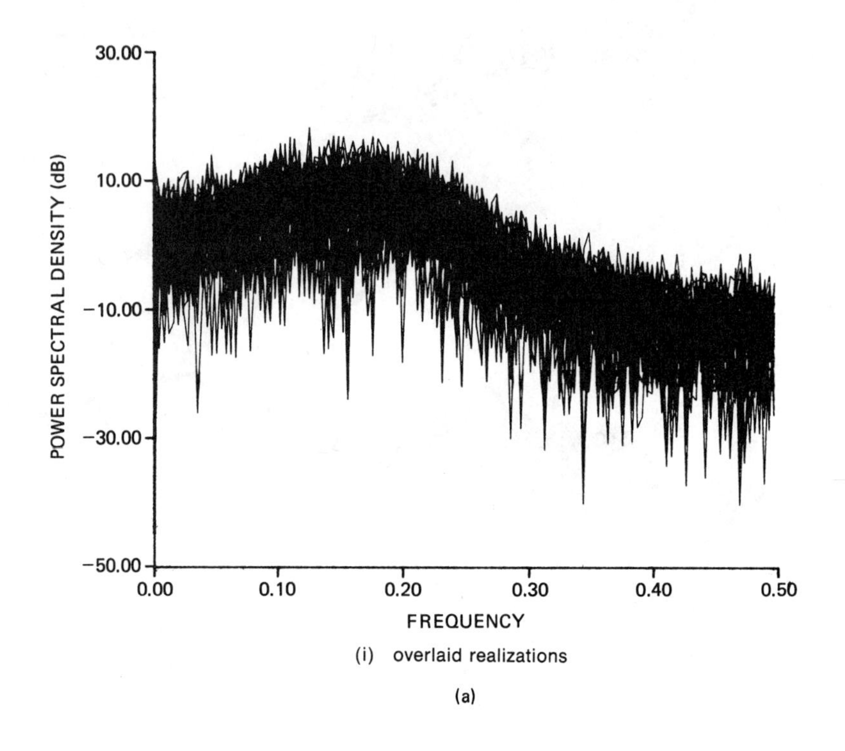

(i) overlaid realizations

(a)

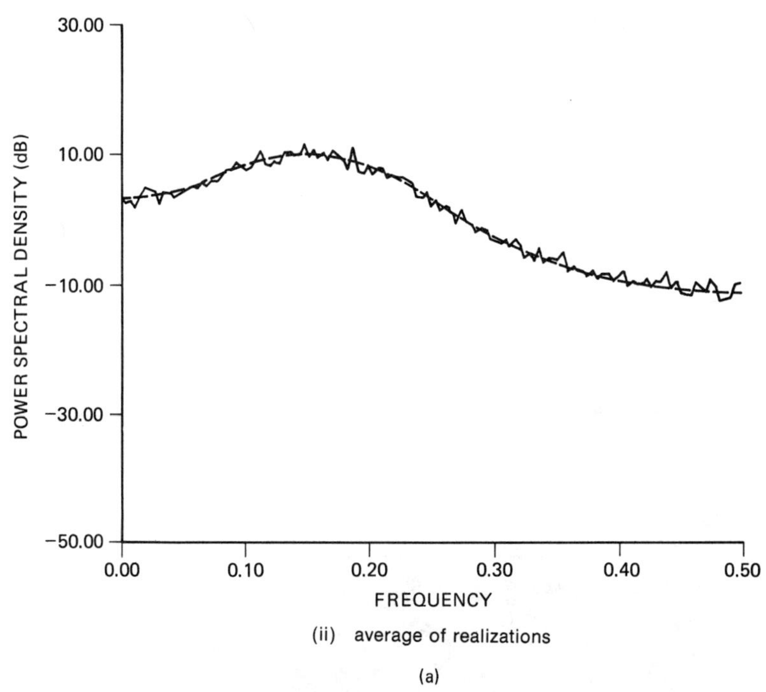

(ii) average of realizations

(a)

Figure 4.12 (a) Periodogram spectral estimate for broadband PSD. (b) Blackman–Tukey spectral estimate using a Bartlett window ($M = 51$) for broadband PSD.

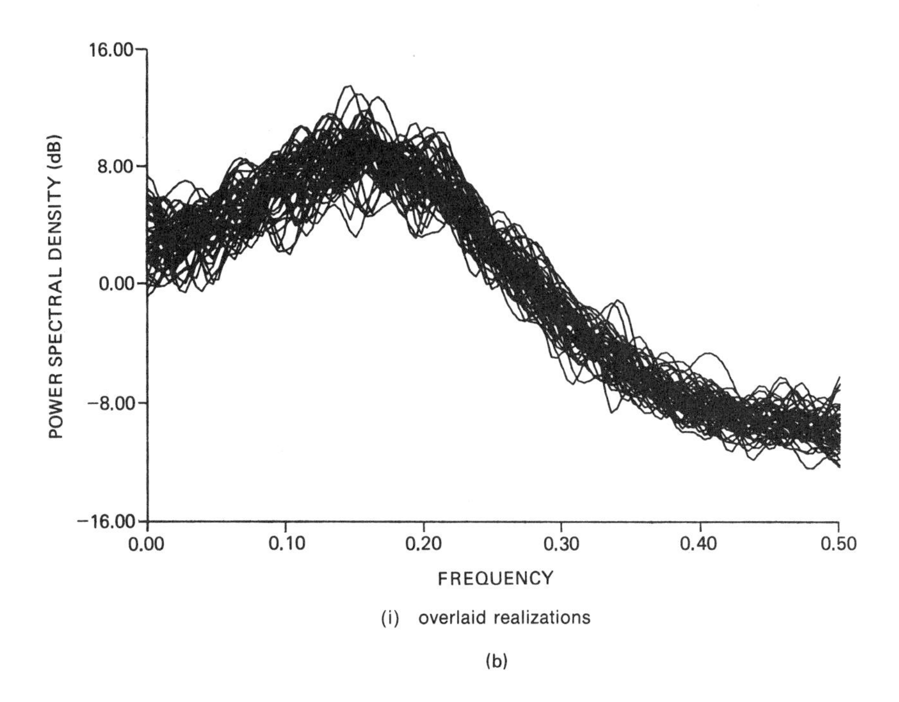

(i) overlaid realizations

(b)

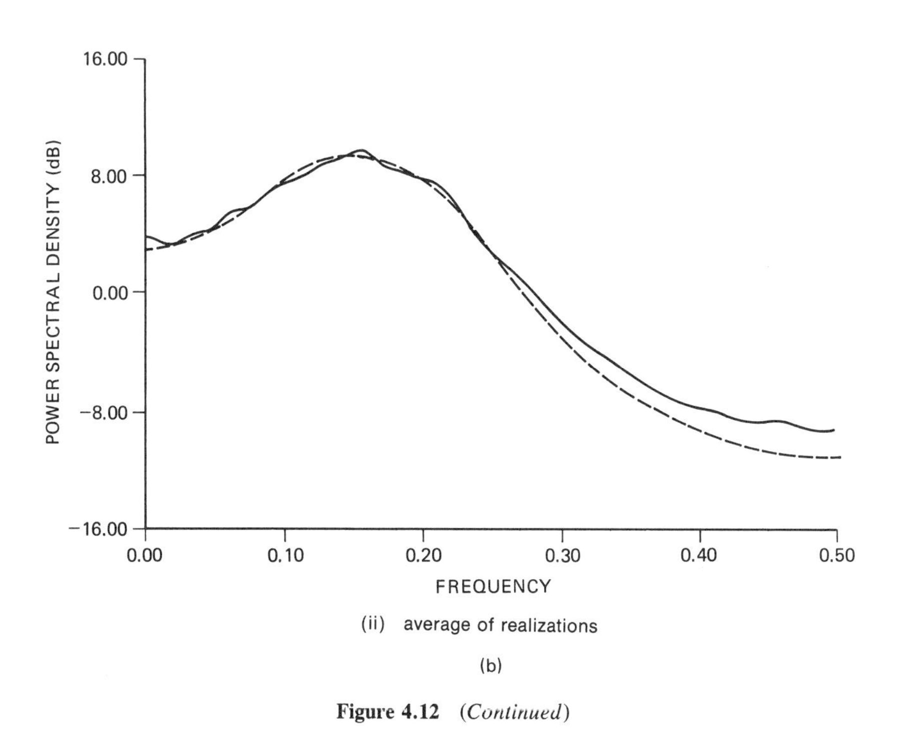

(ii) average of realizations

(b)

Figure 4.12 *(Continued)*

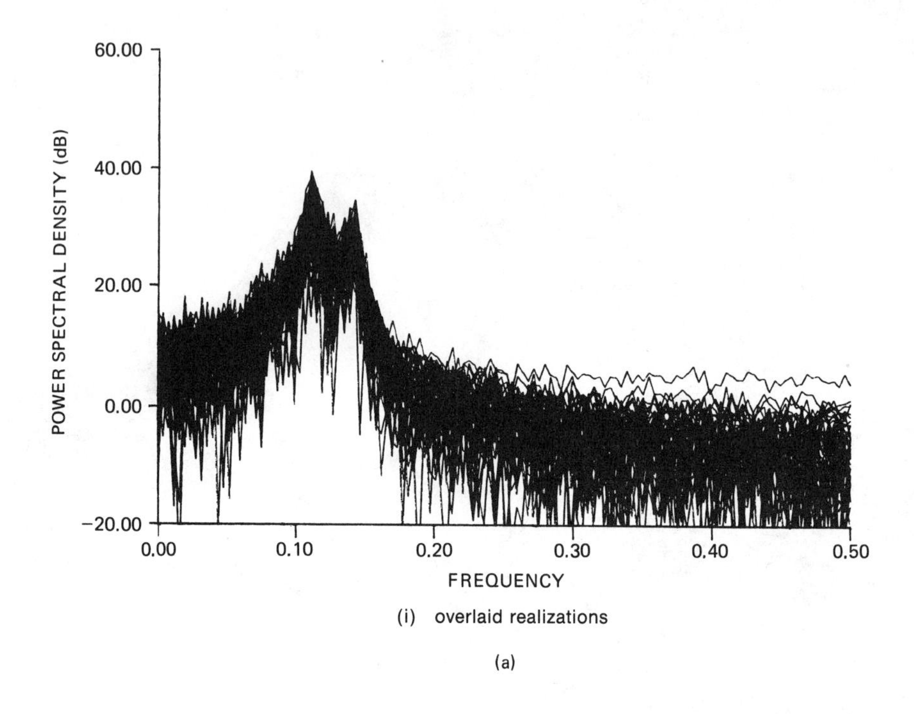

(i) overlaid realizations

(a)

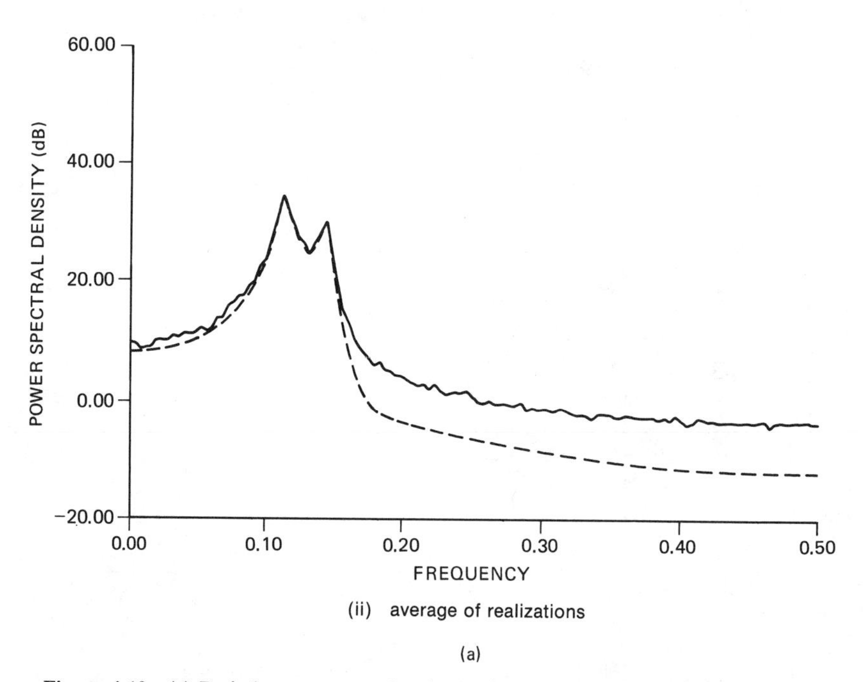

(ii) average of realizations

(a)

Figure 4.13 (a) Periodogram spectral estimate for narrowband PSD. (b) Blackman–Tukey spectral estimate using a Bartlett window ($M = 51$) for narrowband PSD.

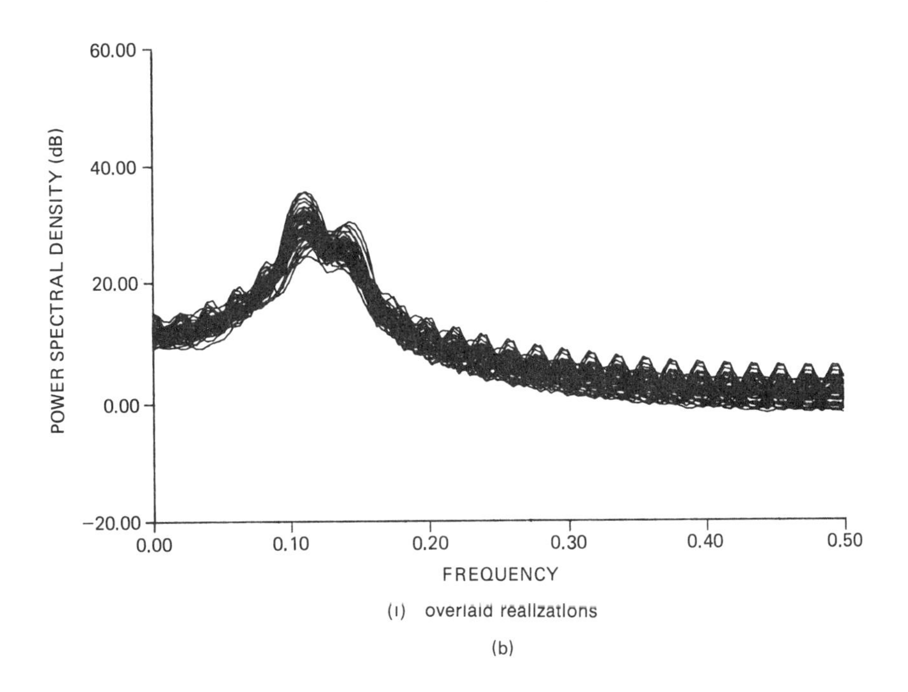

(i) overlaid realizations

(b)

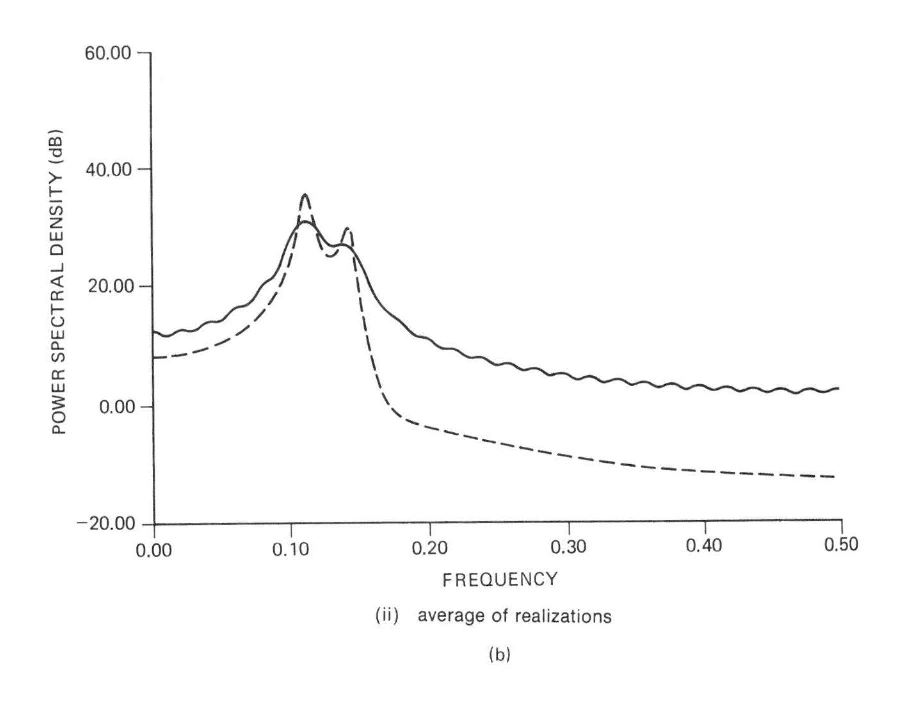

(ii) average of realizations

(b)

Figure 4.13 *(Continued)*

curve. It is seen that the BT estimate exhibits significant bias due to the windowing of the ACF, while the periodogram exhibits less bias. The tradeoff as usual is the reduced variance of the BT spectral estimator.

REFERENCES

Blackman, R. B., and J. W. Tukey, *The Measurement of Power Spectra from the Point of View of Communications Engineering*, Dover, New York, 1958.

Bloomfield, P., *Fourier Analysis of Time Series: An Introduction*, Wiley, New York, 1976.

Harris, F. J., "On the Use of Windows for Harmonic Analysis with the Discrete Fourier Transform," *Proc. IEEE*, Vol. 66, pp. 51–83, Jan. 1978.

Jenkins, G. M., and D. G. Watts, *Spectral Analysis and Its Applications*, Holden-Day, San Francisco, 1968.

Koopmans, L. H., *The Spectral Analysis of Time Series*, Academic Press, New York, 1974.

Marple, S. L., Jr., *Digital Spectral Analysis with Applications,* Prentice-Hall, Englewood Cliffs, N.J., 1987.

Schuster, A., "On the Investigation of Hidden Periodicities with Application to a Supposed 26 Day Period of Meteorological Phenomena," *Terr. Magn.*, Vol. 3, pp. 13–41, Mar. 1898.

Thomson, D. J., "Spectrum Estimation Techniques for Characterization and Development of WT4 Waveguide-I," *Bell Syst. Tech. J.*, Vol. 56, pp. 1769–1815, Nov. 1977.

Welch, P. D., "The Use of Fast Fourier Transform for Estimation of Power Spectra: A Method Based on Time Averaging over Short, Modified Periodograms," *IEEE Trans. Audio Electroacoust.*, Vol. AU15, pp. 70–73, June 1967.

PROBLEMS

4.1. In this problem we show that the periodogram is an inconsistent estimator by examining the estimator at $f = 0$, or

$$\hat{P}_{\text{PER}}(0) = \frac{1}{N}\left(\sum_{n=0}^{N-1} x[n]\right)^2.$$

If $x[n]$ is a real white Gaussian noise process with PSD

$$P_{xx}(f) = \sigma_x^2$$

find the mean and variance of $\hat{P}_{\text{PER}}(0)$. Does the variance converge to zero as $N \to \infty$? *Hint:* Note that

$$\hat{P}_{\text{PER}}(0) = \sigma_x^2 \left(\sum_{n=0}^{N-1} \frac{x[n]}{\sigma_x \sqrt{N}}\right)^2$$

where the quantity inside the parentheses is $\sim N(0,1)$.

4.2. Consider the estimator

$$\hat{P}_{\text{AVPER}}(0) = \frac{1}{N} \sum_{m=0}^{N-1} \hat{P}_{\text{PER}}^{(m)}(0)$$

where

$$\hat{P}_{\text{PER}}^{(m)}(0) = x^2[m]$$

for the process of Problem 4.1. This estimator may be viewed as an *averaged periodogram*. In essence the data record is sectioned into blocks (in this case, of length 1) and the periodograms for each block are averaged. Find the mean and variance of $\hat{P}_{\text{AVPER}}(0)$. Compare this result to that obtained in Problem 4.1.

4.3. Prove that the periodogram is an unbiased estimator for the case of a white noise process.

4.4. Show that the 3 dB bandwidth of the filter frequency response given by (4.4) is approximately $0.88/N$. What is the 3 dB bandwidth of the Fourier transform of a Bartlett window as given by (4.7)? *Hint:* Assume that N is large.

4.5. Prove that the BT spectral estimator is identical to the periodogram if the biased ACF estimator as given by (4.20) is used in conjunction with a rectangular lag window, or

$$w[k] = \begin{cases} 1 & \text{for } |k| \leq N - 1 \\ 0 & \text{otherwise.} \end{cases}$$

4.6. Prove that the sequence generated by using the biased ACF estimator (4.20) is positive semidefinite. Note that $\hat{r}_{xx}[k]$ may be written as

$$\frac{1}{N} \sum_{n=-\infty}^{\infty} x^*[n]x[n + k]$$

where $x[n] = 0$ for $n < 0$ and $n > N - 1$.

4.7. For the real data set $\{x[0], x[1], x[2]\}$ the unbiased ACF estimator yields

$$\hat{r}'_{xx}[0] = \tfrac{1}{3}(x^2[0] + x^2[1] + x^2[2])$$

$$\hat{r}'_{xx}[1] = \tfrac{1}{2}(x[0]x[1] + x[1]x[2]).$$

Show that the autocorrelation sequence is not positive semidefinite by giving a numerical example where the matrix

$$\begin{bmatrix} \hat{r}'_{xx}[0] & \hat{r}'_{xx}[1] \\ \hat{r}'_{xx}[1] & \hat{r}'_{xx}[0] \end{bmatrix}$$

is not positive semidefinite.

4.8. Find the variance of the unbiased ACF estimator

$$\hat{r}'_{xx}[k] = \frac{1}{N - k} \sum_{n=0}^{N-1-k} x[n]x[n + k] \qquad 0 \leq k \leq N - 1$$

for real data which is a zero-mean white Gaussian process with variance σ_x^2. What happens as the lag k increases? Repeat the problem for the biased ACF estimator and explain your results. *Hint:* First prove that for any real zero-mean Gaussian process the variance of the unbiased ACF estimator is

$$\text{var}\,[\hat{r}'_{xx}[k]] = \frac{1}{N-k} \sum_{j=-(N-1-k)}^{N-1-k} \left(1 - \frac{|j|}{N-k}\right) (r_{xx}^2[j] + r_{xx}[j+k]r_{xx}[j-k]).$$

4.9. Find the variance of the unbiased and biased ACF estimators for real data if N is large and $N \gg k$ using the results of Problem 4.8.

4.10. For the sinusoidal data $x[n] = \exp(j2\pi f_0 n)$ for $0 \le n \le N-1$ find the BT spectral estimate using the biased ACF estimator if the lag window is rectangular and the length is $M = 1$. Also, find the estimate for a Bartlett lag window with $M = 1$ and compare the results. Do the lag windows result in nonnegative spectra?

4.11. Find a lag window which has a nonnegative Fourier transform by convolving a rectangular window with itself. Compare your result to the Bartlett window. How could other lag windows with nonnegative Fourier transforms be generated?

APPENDIX 4A

Bias and Variance of Periodogram

A statistical analysis of the periodogram for the general case is mathematically intractable. The analysis that follows derives the mean for any real process but only the variance for the special case of a real zero-mean white Gaussian process with PSD $P_{xx}(f) = \sigma_x^2$. The variance result may, however, be used as a guideline to the behavior of the variance of the periodogram spectral estimate in more general situations [Jenkins and Watts 1968].

A periodogram of the finite-length sequence $x[n]$ for $n = 0$ to $n = N-1$ is

$$\hat{P}_{\text{PER}}(f) = \frac{1}{N} \left| \sum_{n=0}^{N-1} x[n] \exp(-j2\pi f n) \right|^2.$$

The periodogram can be shown to be identical to the Fourier transform of an estimated autocorrelation sequence, or (see Problem 4.5)

$$\hat{P}_{\text{PER}}(f) = \sum_{k=-(N-1)}^{N-1} \hat{r}_{xx}[k] \exp(-j2\pi f k)$$

where $\hat{r}_{xx}[k]$ is the biased autocorrelation estimate, which is defined as

$$\hat{r}_{xx}[k] = \frac{1}{N} \sum_{n=0}^{N-1-|k|} x[n]x[n+|k|].$$

The expected value of the periodogram is simply

$$\begin{aligned}
\mathscr{E}\{\hat{P}_{\text{PER}}(f)\} &= \sum_{k=-(N-1)}^{N-1} \mathscr{E}\{\hat{r}_{xx}[k]\} \exp(-j2\pi fk) \\
&= \sum_{k=-(N-1)}^{N-1} \left[\frac{N - |k|}{N}\right] r_{xx}[k] \exp(-j2\pi fk) \\
&= \mathscr{F}\{w_B[k]r_{xx}[k]\} \\
&= \int_{-1/2}^{1/2} W_B(f - \xi)P_{xx}(\xi)\, d\xi
\end{aligned}$$

where $W_B(f)$ is the Fourier transform of the Bartlett window

$$w[k] = \begin{cases} 1 - \dfrac{|k|}{N} & \text{for } |k| \le N - 1 \\ 0 & \text{for } |k| > N - 1 \end{cases}$$

and is given by (4.7). For the specific case of a white Gaussian process, $r_{xx}[k] = \sigma_x^2\delta[k]$, which results in (see Problem 4.3)

$$\mathscr{E}[\hat{P}_{\text{PER}}(f)] = \sigma_x^2 = P_{xx}(f)$$

or the periodogram is unbiased.

The second moment of the periodogram is given by

$$\begin{aligned}
&\mathscr{E}\{\hat{P}_{\text{PER}}(f_1)\hat{P}_{\text{PER}}(f_2)\} \\
&\quad = \left(\frac{1}{N}\right)^2 \sum_{k=0}^{N-1} \sum_{l=0}^{N-1} \sum_{m=0}^{N-1} \sum_{n=0}^{N-1} \mathscr{E}\{x[k]x[l]x[m]x[n]\} \\
&\qquad\qquad \exp(j2\pi[f_1(k - l) + f_2(m - n)]).
\end{aligned}$$

This expression involves fourth-order moments, which are difficult to evaluate in the general case. However, since $x[n]$ is white and Gaussian, the fourth-order statistical moment reduces to a sum of second-order moments (see Section 3.2.1):

$$\begin{aligned}
\mathscr{E}\{x[k]x[l]x[m]x[n]\} &= \sigma_x^4(\delta[k - l]\delta[m - n] \\
&\qquad + \delta[k - m]\delta[l - n] + \delta[k - n]\delta[l - m]) \\
&= \begin{cases} \sigma_x^4 & \text{if } k = l, m = n, \text{ or } k = m, l = n, \text{ or } k = n, l = m \\ 0 & \text{otherwise.} \end{cases}
\end{aligned}$$

This means that

$$\mathscr{E}\{\hat{P}_{\text{PER}}(f_1)\hat{P}_{\text{PER}}(f_2)\} = \sigma_x^4 \left[1 + \left(\frac{\sin N\pi(f_1 + f_2)}{N \sin \pi(f_1 + f_2)}\right)^2 + \left(\frac{\sin N\pi(f_1 - f_2)}{N \sin \pi(f_1 - f_2)}\right)^2\right].$$

The covariance is then

$$\text{cov}[\hat{P}_{\text{PER}}(f_1), \hat{P}_{\text{PER}}(f_2)] = \mathscr{E}\{\hat{P}_{\text{PER}}(f_1)\hat{P}_{\text{PER}}(f_2)\} - \mathscr{E}\{\hat{P}_{\text{PER}}(f_1)\}\mathscr{E}\{\hat{P}_{\text{PER}}(f_2)\}.$$

Noting that $\mathscr{E}\{\hat{P}_{\text{PER}}(f)\} = \sigma_x^2 = P_{xx}(f)$ for a white Gaussian process, we have

$$\operatorname{cov}[\hat{P}_{\text{PER}}(f_1), \hat{P}_{\text{PER}}(f_2)]$$
$$= P_{xx}(f_1)P_{xx}(f_2)\left[\left(\frac{\sin N\pi(f_1 + f_2)}{N \sin \pi(f_1 + f_2)}\right)^2 + \left(\frac{\sin N\pi(f_1 - f_2)}{N \sin \pi(f_1 - f_2)}\right)^2\right]. \quad \text{(4A.1)}$$

In particular, the variance at any given frequency is

$$\begin{aligned} \operatorname{var}\{\hat{P}_{\text{PER}}(f)\} &= \operatorname{cov}[\hat{P}_{\text{PER}}(f), \hat{P}_{\text{PER}}(f)] \\ &= P_{xx}^2(f)\left[1 + \left(\frac{\sin 2\pi N f}{N \sin 2\pi f}\right)^2\right]. \end{aligned} \quad \text{(4A.2)}$$

In general, it can be shown that (4A.1) and (4A.2) are approximately true for any nonwhite process for a large enough data record.

APPENDIX 4B

Bias and Variance of Blackman–Tukey Spectral Estimator

The bias and variance of the Blackman–Tukey spectral estimator for a real data process are derived in this appendix [Jenkins and Watts 1968]. The definition of the Blackman–Tukey spectral estimator as given by (4.22) is

$$\begin{aligned} \hat{P}_{\text{BT}}(f) &= \sum_{k=-(N-1)}^{N-1} w[k]\hat{r}_{xx}[k] \exp(-j2\pi f k) \\ &= \mathscr{F}\{w[k]\hat{r}_{xx}[k]\}. \end{aligned}$$

Since $\mathscr{F}\{\hat{r}_{xx}[k]\}$ is just the periodogram,

$$\hat{P}_{\text{BT}}(f) = \int_{-1/2}^{1/2} W(f - \xi)\hat{P}_{\text{PER}}(\xi)\, d\xi. \quad \text{(4B.1)}$$

The mean value follows as

$$\mathscr{E}[\hat{P}_{\text{BT}}(f)] = \int_{-1/2}^{1/2} W(f - \xi)\mathscr{E}[\hat{P}_{\text{PER}}(\xi)]\, d\xi$$

which when evaluated for large data records leads to

$$\mathscr{E}[\hat{P}_{\text{BT}}(f)] \approx \int_{-1/2}^{1/2} W(f - \xi)P_{xx}(\xi)\, d\xi$$

since the periodogram is unbiased as $N \rightarrow \infty$ [see (4.1)].

The derivation of the variance relies heavily on the results for the periodogram. Specifically, in Appendix 4A it was shown that

$$\operatorname{cov}[\hat{P}_{\mathrm{PER}}(f_1), \hat{P}_{\mathrm{PER}}(f_2)$$

$$\approx P_{xx}(f_1)P_{xx}(f_2)\left[\left(\frac{\sin N\pi(f_1+f_2)}{N\sin\pi(f_1+f_2)}\right)^2 + \left(\frac{\sin N\pi(f_1-f_2)}{N\sin\pi(f_1-f_2)}\right)^2\right]. \tag{4B.2}$$

Using (4B.1) gives us

$$\begin{aligned}\operatorname{var}[\hat{P}_{\mathrm{BT}}(f)] &= \mathscr{E}\left[\left\{\int_{-1/2}^{1/2} W(f-\xi)\,[\hat{P}_{\mathrm{PER}}(\xi) - \mathscr{E}[\hat{P}_{\mathrm{PER}}(\xi)]]\right\}^2\right] \\ &= \int_{-1/2}^{1/2}\int_{-1/2}^{1/2} W(f-\xi)W(f-\nu)\operatorname{cov}[\hat{P}_{\mathrm{PER}}(\xi), \hat{P}_{\mathrm{PER}}(\nu)]\,d\xi\,d\nu.\end{aligned}$$

For large data records the covariance as given by (4B.2) can be approximated as

$$\operatorname{cov}[\hat{P}_{\mathrm{PER}}(f_1), \hat{P}_{\mathrm{PER}}(f_2)] \approx \frac{P_{xx}(f_1)P_{xx}(f_2)}{N}[\delta(f_1+f_2) + \delta(f_1-f_2)]$$

where $\delta(f)$ is the Dirac delta function. This is because

$$\frac{1}{N}\left(\frac{\sin N\pi f}{\sin \pi f}\right)^2$$

is the Fourier transform of a Bartlett window. As $N \to \infty$, the Bartlett window approaches the sequence $w[n] = 1$ for all n, which has the Fourier transform $\delta(f)$. Using this approximation, the variance becomes

$$\operatorname{var}[\hat{P}_{\mathrm{BT}}(f)] \approx \frac{1}{N}\left[\int_{-1/2}^{1/2} W^2(f-\xi)P_{xx}^2(\xi)\,d\xi \right.$$

$$\left. + \int_{-1/2}^{1/2} W(f-\xi)W(f+\xi)P_{xx}^2(\xi)\,d\xi\right].$$

The second term in brackets will be small with respect to the first if f is not near $0, \pm\frac{1}{2}$ since the shifted mainlobes of the spectral windows $W(f-\xi)$, $W(f+\xi)$ will not overlap and hence their product will be small. As a result, this term will be neglected. Furthermore, if the assumption is made that the PSD is smooth over the bandwidth of the mainlobe of the spectral window, then $P_{xx}^2(\xi)$ may be considered to be a constant equal to $P_{xx}^2(f)$ in the integral expression. The final result follows as

$$\begin{aligned}\operatorname{var}[\hat{P}_{\mathrm{BT}}(f)] &\approx \frac{P_{xx}^2(f)}{N}\int_{-1/2}^{1/2} W^2(\xi)\,d\xi \\ &= \frac{P_{xx}^2(f)}{N}\sum_{k=-M}^{M} w^2[k].\end{aligned}$$

APPENDIX 4C

Computer Program for the Periodogram (Provided on Floppy Disk)

```
      SUBROUTINE PERIODOGRAM(X,N,W,NEXP,PPER)
C
C     This program computes a data windowed periodogram. For
C     the case where no data windowing is used the periodogram
C     is given by (4.2). The periodogram is evaluated at the
C     frequencies F=-1/2+(I-1)/L where I=1,2,...,L. The number
C     of frequencies is given by L=2**NEXP.
C
C     Input Parameters:
C
C       X        -Complex array of dimension 2**NEXPx1 with first
C                 N locations containing the data points
C       N        -Number of data points
C       W        -Real array of dimension Nx1 of data window weights;
C                 the weights W(1),W(2),...,W(N) should be
C                 symmetric about the point (N+1)/2. If no windowing
C                 is desired, then set W(I)=1 for I=1,2,...,N.
C       NEXP     -Power of two which determines number of frequency
C                 samples desired; L=2**NEXP must be chosen so that
C                 L is greater than or equal to N
C
C     Output Parameters:
C
C       PPER     -Real array of dimension L=2**NEXPx1 of samples
C                 of the periodogram, where PPER(I) corresponds to
C                 the spectral estimate at the frequency F=-1/2+(I-1)/L
C
C     External Subroutines:
C
C       FFT      -see Appendix 2A
C
C     Notes:
C
C       The calling program must dimension the arrays X,W,PPER.
C       The array Y in FFT must be dimensioned greater than or
```

```
C        equal to 2**NEXP.
C
C      Verification Test Case:
C
C        If the complex data test case listed in Appendix 1A is
C        input as X, with N=32, W(I)=1 for I=1,2,. . .,32 and NEXP=9,
C        then the output should be
C          PPER(100) = 0.89216
C          PPER(200) = 0.15918
C          PPER(300) = 0.76762
C          PPER(400) = 0.76319
C          PPER(500) = 1.35960
C        for a DEC VAX 11/780.
C
         COMPLEX X(1)
         DIMENSION W(1),PPER(1)
C      Window the data
         DO 10 I=1,N
10       X(I)=X(I)*W(I)
C      Compute FFT of windowed data
         L=2**NEXP
         IF(L.EQ.N)GO TO 25
         DO 20 I=N+1,L
20       X(I)=(0.,0.)
25       INVRS=1
         CALL FFT(X,NEXP,INVRS)
C      Compute samples of periodogram and transpose halves of
C      FFT output so that first PSD sample is at a frequency of -1/2
         DO 30 I=1,L/2
         PPER(I+L/2)=(CABS(X(I)))**2/N
30       PPER(I)=(CABS(X(I+L/2)))**2/N
         RETURN
         END
```

APPENDIX 4D

Computer Program for the Correlation Estimate (Provided on Floppy Disk)

```
      SUBROUTINE CORRELATION(N,LAG,MODE,X,Y,R)
C     This program computes either the unbiased or biased complex
C     correlation estimates between arrays X and Y of complex data
C     samples. If X=Y, then the estimate of the autocorrelation,
C     rxx[k] for k=0,1,...,LAG-1, is computed (see (4.20)); otherwise
C     the estimate of the crosscorrelation, rxy[k] for k=0,1,...,LAG-1,
C     is computed. If the crosscorrelation estimate is desired for
C     negative lags, then the relation rxy[-k]=ryx[k]* may be used
C     (see (3.44)). Hence, the X and Y arrays need only be interchanged
C     before calling this program and the output crosscorrelation
C     samples complex conjugated.
C
C     Input Parameters:
C
C       N       -Number of data samples in complex arrays X and Y
C       LAG     -Number of correlation samples desired
C       MODE    -Set equal to 0 for unbiased estimates; otherwise
C                biased estimates are computed
C       X       -Complex array of dimension Nx1 of data samples
C       Y       -Complex array of dimension Nx1 of data samples
C
C     Output Parameters:
C
C       R       -Complex array of dimension NLAGx1 of correlation
C                estimates where R(1),R(2),...,R(LAG) correspond to
C                r[0],r[1],...,r[LAG-1]
C
C     Notes:
C
C       The calling program must dimension arrays X,Y,R.
C
C     Verification Test Case:
C
C       If N=5, LAG=3, MODE=1, and
C         X(1)=(  1.,  0.) Y(1)=(  1.,0.)
```

```
C          X(2)=(  0.,   1.) Y(2)=(-1.,0.)
C          X(3)=(-1.,   0.) Y(3)=(  1.,0.)
C          X(4)=(  0.,-1.) Y(4)=(-1.,0.)
C          X(5)=(  1.,   0.) Y(5)=(  1.,0.)
C        then the output should be
C          R(1)=(0.20000,0.00000)
C          R(2)=(0.00000,0.00000)
C          R(3)=(0.00000,0.20000)
C        for a DEC VAX 11/780.
C
         COMPLEX X(1),Y(1),R(1),SUM
         DO 20 K=0,LAG-1
         NK=N-K
         SUM=(0.,0.)
         DO 10 J=1,NK
10       SUM=SUM+CONJG(X(J))*Y(J+K)
         IF(MODE.EQ.0)R(K+1)=SUM/(N-K)
         IF(MODE.NE.0)R(K+1)=SUM/N
20       CONTINUE
         RETURN
         END
```

APPENDIX 4E

Computer Program for the Blackman–Tukey Spectral Estimator (Provided on Floppy Disk)

```
          SUBROUTINE BLACKTUKEY(X,N,MODE,W,M,NEXP,PBT)
C       This program computes the Blackman-Tukey spectral estimator
C       as given by (4.23). Either the biased or unbiased
C       autocorrelation estimator may be used as well as a lag window.
C       The spectral estimate is evaluated at the frequencies
C       F=-1/2+(I-1)/L for I=1,2,...,L. The number of frequencies is
C       given by L=2**NEXP.
C
C       Input Parameters:
C
```

```
C     X              -Complex array of dimension N×1 of data points
C     N              -Number of data points
C     MODE           -Set equal to zero for unbiased autocorrelation
C                     estimator; otherwise, biased estimator used
C     W              -Real array of dimension 2M+1 of lag window weights;
C                     W(1),...,W(M+1),...,W(2M+1) correspond
C                     to w[-M],...,w[0],...,w[M].
C     M              -Largest lag desired
C     NEXP           -Power of two which determines number of frequency
C                     samples desired, L=2**NEXP must be chosen so that
C                     L is greater than or equal to 2*M+2
C
C   Output Parameters:
C
C     PBT            -Real array of dimension L=2**NEXP×1 of samples of
C                     the Blackman-Tukey spectral estimate, where PBT(I)
C                     corresponds to the spectral estimate at the
C                     frequency F=-1/2+(I-1)/L
C
C   External Subroutines:
C
C     FFT            -see Appendix 2A
C     CORRELATION    -see Appendix 4D
C
C   Notes:
C
C     The calling program must dimension the arrays X,W,PBT.
C     The arrays R,RCORR,P must be dimensioned greater than
C     or equal to the variable dimension shown. The array Y
C     in FFT must be dimensioned greater than or equal to 2**NEXP.
C
C   Verification Test Case:
C
C     If the complex data test case listed in Appendix 1A is
C     input as X, with N=32, MODE=1, W(I)=1. -ABS(I-11.)/10.
C     for I=1,2,...,21, M=10, and NEXP=9, then the output should be
C       PBT(100) =  1.54635
C       PBT(200) =  4.55061
C       PBT(300) = 12.39030
C       PBT(400) =  1.47874
C       PBT(500) =  4.95506
C     for a DEC VAX 11/780.
C
```

```
      COMPLEX X(1),R(2**NEXP),RCORR(M+1)
      DIMENSION W(1),P(2**NEXP),PBT(1)
      PI=4.*ATAN(1.)
C     Compute the autocorrelation estimates from the data
        M1=M+1
        CALL CORRELATION(N,M1,MODE,X,X,RCORR)
C     Window the M+1 autocorrelation estimates and insert them
C     into last M+1 locations of RCORR. Then, fill first
C     M points of RCORR array with the complex conjugate of
C     the last M points (shift autocorrelation sequence to
C     the right by M samples so that FFT may be used)
        R(M1)=W(M1)*RCORR(1)
        DO 10 I=1,M
        R(M1+I)=W(M1+I)*RCORR(I+1)
10      R(M1-I)=W(M1-I)*CONJG(RCORR(I+1))
C     Zero pad array of windowed autocorrelation samples to
C     obtain an array of dimension equal to L
        L=2**NEXP
        DO 20 I=2*M+2,L
20      R(I)=(0.,0.)
C     Compute FFT of autocorrelation sequence
        INVRS=1
        CALL FFT(R,NEXP,INVRS)
C     Compensate for shifting the autocorrelation sequence to the
C     right by M samples
        DO 30 I=1,L
        F=(I-1.)/L
        ARG=2.*PI*F*M
30      P(I)=REAL(R(I)*CEXP(CMPLX(0.,ARG)))
C     Transpose halves of FFT output so that first PSD sample is
C     at a frequency of -1/2
        DO 40 I=1,L/2
        PBT(I+L/2)=P(I)
40      PBT(I)=P(I+L/2)
        RETURN
        END
```

Chapter 5

Parametric Modeling

5.1 INTRODUCTION

In Chapter 3 it was shown that the second-order statistics of a random process could be represented alternatively by either the autocorrelation function (ACF) or the power spectral density (PSD), both of which are nonparametric descriptions. In this chapter an alternative approach is introduced to describe the random process by means of a parametric model. The parameters of the model are obtainable from the ACF with the explicit relationships provided in this chapter. In subsequent chapters we describe how one actually estimates these parameters from a limited data set.

The rationale for using parametric modeling of random processes will now be discussed. The classical methods of spectral estimation as presented in Chapter 4 used Fourier transform operations on either windowed data or windowed ACF estimates. Windowing of data or ACF values makes the implicit assumption that the unobserved data or ACF values outside the window are zero, which is normally an unrealistic assumption. A smeared spectral estimate is a consequence of the windowing. Often, we have more knowledge about the process from which the data samples are taken, or at least we are able to make a more reasonable assumption other than to assume the data or ACF values are zero outside the window. Use of a priori information (or assumptions) may permit the selection of an exact model for the process that generated the data samples, or at least a model that is a good approximation to the actual underlying process. It is then usually possible to obtain a better spectral estimate by basing it on the model and estimating the parameters of the model from the observations. Thus, spectral estimation, in the context of modeling, becomes a three-step procedure. The first step is to select a model. The second step is to estimate the parameters of the assumed model using the available data samples. The third step is to obtain the spectral estimate by substituting the estimated model parameters into the theoretical PSD implied by the model. One major motivation for the current interest in the modeling approach to spectral estimation is the apparent higher resolution achievable with these modern techniques over that achievable with the classical techniques discussed in Chapter 4. The degree of improvement in resolution and spectral fidelity, if any, will be determined by the ability to fit an assumed model with a small number of parameters. Any inaccuracy in the model will result in a systematic or bias error in the spectral estimate. As an example, if the PSD of a real low-pass process is modeled by a Gaussian function

$$P_{xx}(f) = \frac{r_{xx}[0]}{\sqrt{2\pi}\sigma_f} \exp\left[-\frac{1}{2}\left(\frac{f}{\sigma_f}\right)^2\right] \qquad |f| \leq \tfrac{1}{2} \tag{5.1}$$

we need only estimate $r_{xx}[0]$ and σ_f to estimate the PSD. (It is assumed that σ_f is small so that $P_{xx}(f)$ is essentially zero at $f = \pm\frac{1}{2}$.) However, if the true PSD does not fit the Gaussian model, the spectral estimator will be highly biased, an example of which is shown in Figure 5.1 (see also Problem 5.1). Even for large data records for which the variability of the spectral estimator is small, a persistent bias will be present. Clearly, *the accuracy of the model is of utmost importance*.

The selection of a model and hence a spectral estimator is intimately related to the identification techniques employed in linear systems theory [Hsia 1977]. One key feature of the modeling approach to spectral estimation that differentiates it from the general identification problem is that *only the output process of the model is available for analysis*; the input driving process is not assumed available as it is for general system identification. This restriction precludes the direct application of the myriad of system identification techniques to spectral estimation. However, based on the ability to *estimate* the input process, some system identification algorithms have been adapted for spectral estimation, as discussed in Chapter 10.

One of the promising aspects of the modeling approach to spectral estimation

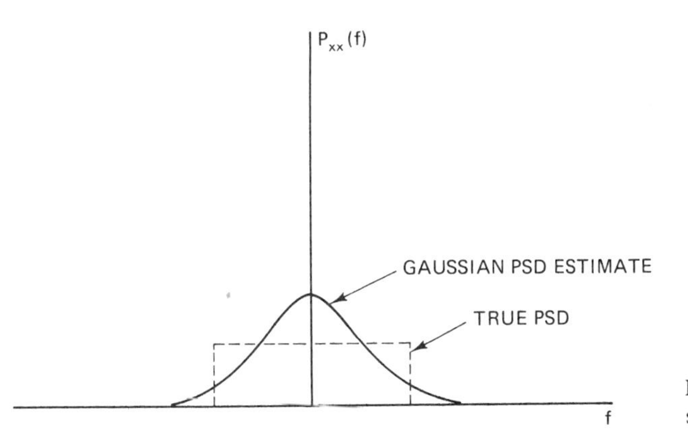

Figure 5.1 Example of inaccurate spectral modeling.

is that one can make more realistic assumptions concerning the nature of the measured process outside the measurement interval, other than to assume it is zero or cyclic. Thus the need for window functions can be eliminated, together with their distorting impact. As a result, the improvement over the conventional FFT spectral estimator can be quite dramatic, especially for short data records [Ulrych 1972].

The classification of spectral estimation techniques into parametric or nonparametric methods is not straightforward. The philosophy will be adopted that a parametric technique is one that assumes more knowledge about the PSD other than the validity of the Wiener–Khinchin theorem. With this definition the Fourier based methods of Chapter 4 are nonparametric spectral estimators, as is the minimum variance spectral estimator of Chapter 11, while the autoregressive, moving average, and autoregressive moving average approaches of Chapters 6 to 10 are parametric. Techniques which assume the data to consist of sinusoids in white noise as in the methods of Chapter 13 are parametric. The distinction is important since comparison of a nonparametric spectral estimator to a parametric one is by definition unfair in that the latter is blessed with additional knowledge about the form of the PSD. Care must be exercised in assessing the relative merits of parametric versus nonparametric spectral estimators or even different parametric spectral estimators based on varying assumptions.

5.2 SUMMARY

Parametric modeling for spectral estimation consists of choosing an appropriate model, estimating the parameters of the model, and then substituting these estimated values into the theoretical PSD expressions. The models to be discussed are the time series or rational transfer function models. They are the autoregressive (AR) model (5.8), the moving average (MA) model (5.6), and the autoregressive-moving average (ARMA) model (5.2). The theoretical PSDs associated with these models are given by (5.9), (5.7), and (5.5), respectively. Estimation of the parameters, which is described in Chapters 6 to 10, of each model usually begins with the estimation of the ACF. Thus it is important to relate the ACF to the model parameters. For an AR process the relationship is given by (5.17), while for an MA process it is given by (5.20), and for the ARMA process (5.16) is the relationship.

Some examples of AR, MA, and ARMA process ACFs and PSDs are given in Section 5.5. To model a time series appropriately for spectral estimation, we desire a model with as few parameters as necessary. Through the examples presented it is shown that the AR model is appropriate for spectra with sharp peaks but not deep valleys. The MA model can be used for spectra with deep valleys but not sharp peaks. The more general ARMA model will be able to represent both these extremes. Another consideration is the roll-off of the PSD. For too rapid a spectral roll-off, neither of the models may yield an accurate representation of the PSD. Next, the use of an AR model for sinusoidal data is shown to be valid in the context of spectral estimation. A sinusoid can be viewed as a very narrowband random process so that the ACF of a sinusoid is obtained from that of an AR process by a limiting argument. For M real sinusoids (M complex sinusoids) an AR model of order $2M$ (M) is appropriate.

The form of the Blackman–Tukey spectral estimator and the MA PSD are shown to be identical in Section 5.7. Consequently, characteristics of MA modeled spectra are also shared by Blackman–Tukey spectral estimators. Finally, in Section 5.8 the question of determining the model parameters given the PSD or ACF is addressed. In general, the uniqueness of the model requires filter causality and invertibility.

5.3 RATIONAL TRANSFER FUNCTION MODELS

5.3.1 Definitions

Many discrete-time random processes encountered in practice are well approximated by a *time series or rational transfer function* model. In this model, an input driving sequence $u[n]$ and the output sequence $x[n]$ that is to model the data are related by the linear difference equation,

$$x[n] = -\sum_{k=1}^{p} a[k]x[n-k] + \sum_{k=0}^{q} b[k]u[n-k]. \tag{5.2}$$

This most general linear model is termed an ARMA model and is shown in Figure 5.2. The interest in these models stems from their relationship to linear filters with rational transfer functions.

It is important to distinquish between the driving noise of the model $u[n]$ and any observation noise. The ARMA model noise is not an additive or observation noise which is typically encountered in signal processing applications. $u[n]$ is an innate part of the model and gives rise to the random nature of the observed process $x[n]$. Any observation noise then needs to be modeled within the ARMA process by modification of its parameters. More will be said about this issue in Section 5.6.

The system function $\mathcal{H}(z)$ between the input $u[n]$ and the output $x[n]$ for the ARMA process of (5.2) is the rational function

$$\mathcal{H}(z) = \frac{\mathcal{B}(z)}{\mathcal{A}(z)} \tag{5.3}$$

$$\text{where } \mathcal{A}(z) = z\text{-transform of AR branch} = \sum_{k=0}^{p} a[k]z^{-k}$$

$$\mathcal{B}(z) = z\text{-transform of MA branch} = \sum_{k=0}^{q} b[k]z^{-k}.$$

It is assumed that $\mathcal{A}(z)$ has all its zeros within the unit circle of the z-plane. This guarantees that $\mathcal{H}(z)$ is a stable and causal filter. Without this assumption it can

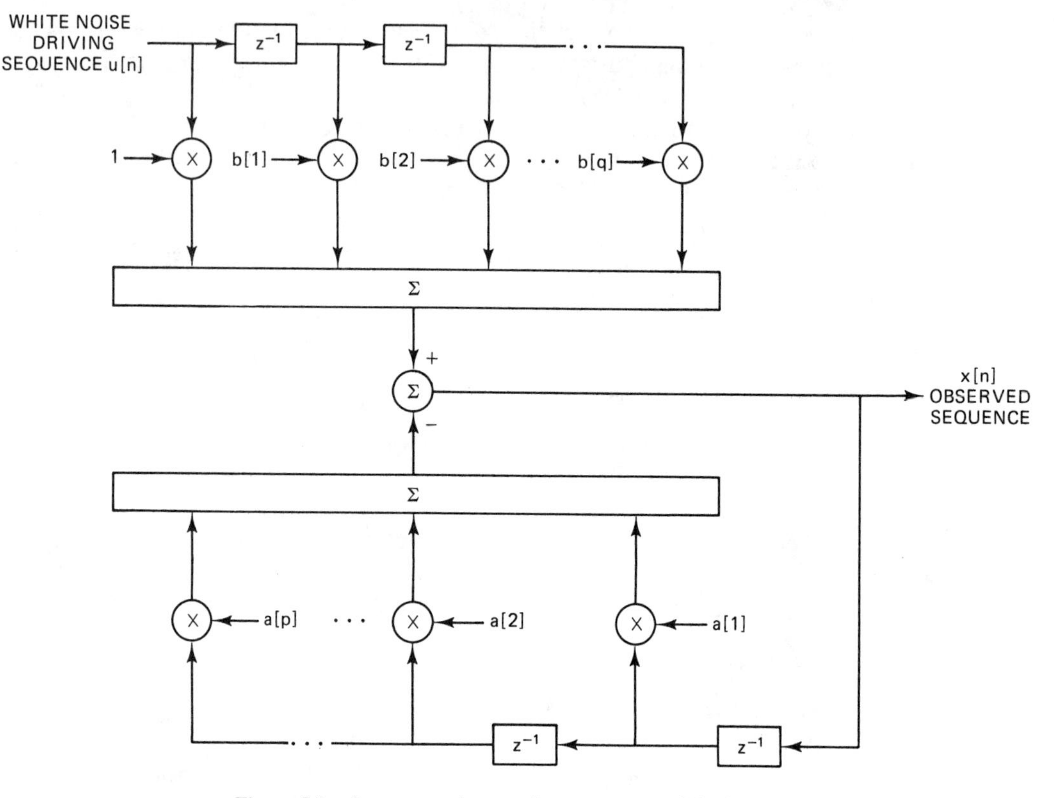

Figure 5.2 Autoregressive moving average model of random process.

be shown (see Problem 5.2) that $x[n]$ as given by (5.2) would not be a valid description of a WSS random process.

It is well known that the z-transform of the ACF at the output of a linear filter, $\mathcal{P}_{xx}(z)$, is related to that at the input, $\mathcal{P}_{uu}(z)$, as follows [see (3.51)]:

$$\mathcal{P}_{xx}(z) = \mathcal{H}(z)\mathcal{H}^*(1/z^*)\,\mathcal{P}_{uu}(z) = \frac{\mathcal{B}(z)\mathcal{B}^*(1/z^*)}{\mathcal{A}(z)\mathcal{A}^*(1/z^*)}\,\mathcal{P}_{uu}(z). \tag{5.4}$$

When (5.4) is evaluated along the unit circle, $z = \exp(j2\pi f)$ for $-\frac{1}{2} \le f \le \frac{1}{2}$, it becomes the PSD $P_{xx}(f)$. Often the driving process is assumed to be a white noise sequence of zero mean and variance σ^2. The PSD of the noise is then σ^2. The PSD of the ARMA output process becomes

$$P_{\text{ARMA}}(f) = P_{xx}(f) = \sigma^2 \left|\frac{B(f)}{A(f)}\right|^2 \tag{5.5}$$

where $A(f)$ denotes $\mathcal{A}(\exp[j2\pi f])$ and $B(f)$ denotes $\mathcal{B}(\exp[j2\pi f])$. Specification of the parameters $a[k]$ (termed the autoregressive coefficients), $b[k]$ (termed the moving average coefficients), and σ^2 is equivalent to specifying the PSD of the process $x[n]$. Without loss of generality, we can assume that $a[0] = 1$ and $b[0]$

$= 1$ since any filter gain can be incorporated into σ^2. This model is sometimes referred to as a *pole–zero model* and is denoted as an ARMA(p, q) process.

If all the $a[k]$ coefficients except $a[0] = 1$ vanish for the ARMA parameters, then

$$x[n] = \sum_{k=0}^{q} b[k]u[n - k] \tag{5.6}$$

and the process is strictly an MA process of order q, and

$$P_{\mathrm{MA}}(f) = \sigma^2 \mid B(f) \mid^2. \tag{5.7}$$

This model is sometimes termed an *all-zero* model and is shown in Figure 5.3. It is denoted as an MA(q) process. If all the $b[k]$ coefficients except $b[0] = 1$ are zero in the ARMA model, then

$$x[n] = -\sum_{k=1}^{p} a[k]x[n - k] + u[n] \tag{5.8}$$

and the process is strictly an AR process of order p. The process is termed an autoregression in that the sequence $x[n]$ is a linear regression on itself with $u[n]$ representing the error. With this model, the present value of the process is expressed as a weighted sum of past values plus a noise term. The PSD is

$$P_{\mathrm{AR}}(f) = \frac{\sigma^2}{\mid A(f) \mid^2}. \tag{5.9}$$

This model is sometimes termed an *all-pole* model and is shown in Figure 5.4. It is denoted as an AR(p) process.

A computer program entitled WGN is provided in Appendix 5A to generate real white Gaussian noise. AR, MA, or ARMA time series data can be generated with the computer program GENDATA given in Appendix 5B. To compute the

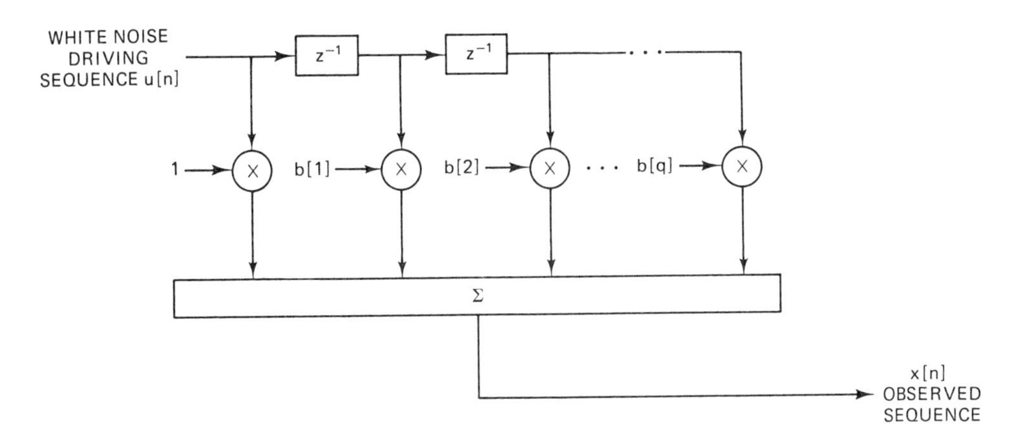

Figure 5.3 Moving average model of random process.

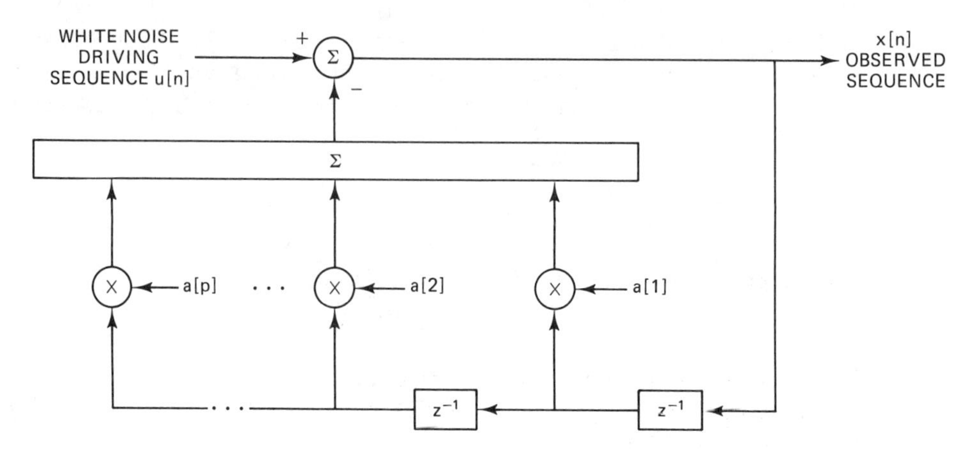

Figure 5.4 Autoregressive model of random process.

PSD of an AR, MA, or ARMA process given the coefficients, we can use the computer program called ARMAPSD [Marple 1985] given in Appendix 5C.

5.3.2 Relating the Time Series Models

The Wold decomposition theorem [Wold 1954] relates the AR, MA, and ARMA models. The basic theorem states that any WSS random process can be decomposed into a component that is completely random and one that is deterministic. A deterministic random process is one that is perfectly predictable based on the infinite past. This is to say that $x[n]$ may be expressed as $-\sum_{k=1}^{\infty} \alpha_k x[n-k]$ without error for some constants α_k (see Section 6.3). An example of this would be the decomposition of a process consisting of a pure sinusoid (randomly phased to ensure WSS) and white noise into a purely random component (the white noise) and a deterministic component (the sinusoid). Alternatively, we may view this decomposition in the spectral domain as one that separates the PSD into a continuous component representing the white noise and a discrete component (impulsive behavior) representing the sinusoid. A corollary of Wold's theorem states that if the PSD is purely continuous, any ARMA or AR process can be represented by a unique MA model of infinite order. A theorem due to Kolmogorov [1941] states that, similarly, any ARMA or MA process can be represented by an AR process of infinite order. These theorems are important because if we choose the wrong model among the three, a reasonable approximation may still be obtained by using a high enough model order.

To illustrate these theorems, it is now shown how an ARMA(1, 1) process can be modeled by an AR(∞) or by an MA(∞) process. From (5.3) the transfer function for an ARMA(1, 1) process is

$$\mathcal{H}(z) = \frac{1 + b[1]z^{-1}}{1 + a[1]z^{-1}}.$$

To represent the ARMA process as an AR process of infinite order, let

$$\mathcal{H}(z) = \frac{1}{1 + c[1]z^{-1} + c[2]z^{-2} + \cdots}$$

where

$$\begin{aligned} \mathcal{C}(z) &= 1 + \sum_{k=1}^{\infty} c[k]z^{-k} \\ &= \frac{1 + a[1]z^{-1}}{1 + b[1]z^{-1}} \end{aligned}$$

and $c[k]$ is the kth AR coefficient of the AR(∞) model. The inverse z-transform of $\mathcal{C}(z)$ is readily shown to be

$$c[k] = \begin{cases} 1 & \text{if } k = 0 \\ (a[1] - b[1])(-b[1])^{k-1} & \text{if } k \geq 1. \end{cases}$$

If we wish to use a finite order AR model, say AR(L), then L should be chosen so that $c[L + 1] \approx 0$ or, equivalently, $b[1]^L \approx 0$. Thus, as the zero of the ARMA process gets closer to the unit circle, a larger order AR model will be required. In general, L should be chosen so that the impulse response of $1/\mathcal{B}(z)$ has decayed to zero for an index greater than L.

In a similar fashion, to represent an ARMA(1, 1) process as an MA process of infinite order, let

$$\mathcal{H}(z) = \sum_{k=0}^{\infty} d[k]z^{-k}$$

so that

$$d[k] = \mathcal{Z}^{-1}\left\{\frac{1 + b[1]z^{-1}}{1 + a[1]z^{-1}}\right\}.$$

$d[k]$ is easily shown to be

$$d[k] = \begin{cases} 1 & \text{if } k = 0 \\ (b[1] - a[1])(-a[1])^{k-1} & \text{if } k \geq 1. \end{cases}$$

For a finite order MA model, say MA(L), the order L should be chosen so that $d[L + 1] \approx 0$ or equivalent $a[1]^L \approx 0$. Hence, as the pole of the ARMA process gets closer to the unit circle, a larger order MA model will be required. In general, L should be chosen so that the impulse response of $1/\mathcal{A}(z)$ has decayed to zero for an index greater than L.

For a general ARMA(p, q) process the AR(∞) model parameters are easily obtained as the inverse z-transform of $\mathcal{A}(z)/\mathcal{B}(z)$ or by the recursive difference equation

$$c[n] = -\sum_{k=1}^{q} b[k]c[n - k] + \sum_{k=0}^{p} a[k]\delta[n - k] \tag{5.10}$$

for $n \geq 0$. The initial conditions $\{c[-q], c[-q+1], \ldots, c[-1]\}$ are set equal to zero. It is also straightforward to obtain the ARMA parameters given the AR(∞) model parameters. Since

$$\mathscr{C}(z)\mathscr{B}(z) = \mathscr{A}(z)$$

on taking the inverse z-transform,

$$\sum_{k=0}^{q} b[k]c[n-k] = a[n]$$

for $n \geq 0$. But $a[n] = 0$ for $n > p$. Hence

$$\sum_{k=0}^{q} b[k]c[n-k] = 0 \qquad \text{for} \quad n \geq p+1. \tag{5.11}$$

This expression for $n = p+1, p+2, \ldots, p+q$ may be written in matrix form as

$$\begin{bmatrix} c[p] & c[p-1] & \cdots & c[p+1-q] \\ c[p+1] & c[p] & \cdots & c[p+2-q] \\ \vdots & \vdots & \ddots & \vdots \\ c[p+q-1] & c[p+q-2] & \cdots & c[p] \end{bmatrix} \begin{bmatrix} b[1] \\ b[2] \\ \vdots \\ b[q] \end{bmatrix} = - \begin{bmatrix} c[p+1] \\ c[p+2] \\ \vdots \\ c[p+q] \end{bmatrix}. \tag{5.12}$$

Once $\{b[1], b[2], \ldots, b[q]\}$ are found, then $\{a[1], a[2], \ldots, a[p]\}$ may be found from

$$\sum_{k=0}^{q} b[k]c[n-k] = a[n] \qquad \text{for} \quad n = 1, 2, \ldots, p \tag{5.13}$$

with $\{c[1-q], c[2-q], \ldots, c[-1]\}$ set equal to zero. In matrix form this is

$$\begin{bmatrix} a[1] \\ a[2] \\ \vdots \\ a[p] \end{bmatrix} = \begin{bmatrix} c[1] & 1 & 0 & \cdots & 0 \\ c[2] & c[1] & 1 & \cdots & 0 \\ \vdots & \vdots & \vdots & \vdots & \vdots \\ c[p] & c[p-1] & c[p-2] & \cdots & c[p-q] \end{bmatrix} \begin{bmatrix} 1 \\ b[1] \\ \vdots \\ b[q] \end{bmatrix}. \tag{5.14}$$

For the MA(∞) approximation (5.10)–(5.14) can still be used if we replace $a[k]$ by $b[k]$, and vice versa. In this case $\mathscr{D}(z) = \mathscr{B}(z)/\mathscr{A}(z)$.

5.4 MODEL PARAMETER RELATIONSHIPS TO AUTOCORRELATION

It is frequently desirable to have an explicit relationship between the parameters of the model and the ACF. These relationships are obtained by taking the inverse z-transform of $\mathscr{P}_{xx}(z)$. Also, they can be derived directly from the time series models. Since these relationships play a fundamental role in spectral estimation they are derived by both methods.

From (5.4) it follows that

$$\begin{aligned}\mathcal{P}_{xx}(z)\mathcal{A}(z) &= \frac{\mathcal{B}^*(1/z^*)}{\mathcal{A}^*(1/z^*)}\mathcal{B}(z)\sigma^2 \\ &= \mathcal{H}^*(1/z^*)\mathcal{B}(z)\sigma^2\end{aligned} \tag{5.15}$$

where it is assumed that the driving noise process is white. Taking the inverse z-transform of (5.15) yields

$$\mathcal{Z}^{-1}[\mathcal{P}_{xx}(z)\mathcal{A}(z)] = r_{xx}[k] \star a[k] = \sum_{l=0}^{p} a[l]r_{xx}[k-l]$$

$$\mathcal{Z}^{-1}[\mathcal{H}^*(1/z^*)\mathcal{B}(z)\sigma^2] = \sigma^2 \sum_{l=0}^{q} b[l]q[k-l]$$

where $q[l] = \mathcal{Z}^{-1}[\mathcal{H}^*(1/z^*)]$. But $q[l] = h^*[-l]$, so that

$$\sum_{l=0}^{p} a[l]r_{xx}[k-l] = \sigma^2 \sum_{l=0}^{q} b[l]h^*[l-k].$$

Using the causality of $\mathcal{H}(z)$, $h[k] = 0$ for $k < 0$, and the final result for an ARMA process becomes

$$r_{xx}[k] = \begin{cases} -\sum_{l=1}^{p} a[l]r_{xx}[k-l] + \sigma^2 \sum_{l=0}^{q-k} h^*[l]b[l+k] & \text{for } k = 0, 1, \ldots, q \\ -\sum_{l=1}^{p} a[l]r_{xx}[k-l] & \text{for } k \geq q+1. \end{cases} \tag{5.16}$$

It should be noted that the relationship between the parameters of an ARMA process and the ACF is a nonlinear one. Given the ACF, we must solve a set of nonlinear equations to find the model parameters. This is due to the $\sum_{l=0}^{q-k} h^*[l]b[l+k]$ term. Specializing the result to an AR process by letting $b[l] = \delta[l]$ results in

$$r_{xx}[k] = -\sum_{l=1}^{p} a[l]r_{xx}[k-l] + \sigma^2 h^*[-k].$$

Since $h^*[-k] = 0$ for $k > 0$ and

$$h^*[0] = \left[\lim_{z\to\infty} \mathcal{H}(z)\right]^* = 1$$

it follows that

$$r_{xx}[k] = \begin{cases} -\sum_{l=1}^{p} a[l] r_{xx}[k-l] & \text{for } k \geq 1 \\ \\ -\sum_{l=1}^{p} a[l] r_{xx}[-l] + \sigma^2 & \text{for } k = 0. \end{cases} \tag{5.17}$$

Equations (5.17) have been termed the Yule–Walker equations. They define a nonlinear relationship between the parameters of an AR process and the ACF. However, *given* the ACF one may determine the AR parameters by solving a set of *linear* equations. To see this, first express the Yule–Walker equations in matrix form as

$$\underbrace{\begin{bmatrix} r_{xx}[0] & r_{xx}[-1] & \cdots & r_{xx}[-(p-1)] \\ r_{xx}[1] & r_{xx}[0] & \cdots & r_{xx}[-(p-2)] \\ \vdots & \vdots & \ddots & \vdots \\ r_{xx}[p-1] & r_{xx}[p-2] & \cdots & r_{xx}[0] \end{bmatrix}}_{\mathbf{R}_{xx}} \begin{bmatrix} a[1] \\ a[2] \\ \vdots \\ a[p] \end{bmatrix} = - \begin{bmatrix} r_{xx}[1] \\ r_{xx}[2] \\ \vdots \\ r_{xx}[p] \end{bmatrix}. \tag{5.18}$$

Note that the autocorrelation matrix $\mathbf{R}_{xx}$ is hermitian ($\mathbf{R}_{xx}^H = \mathbf{R}_{xx}$) and it is Toeplitz since the elements along any diagonal are equal (see Section 2.3). Also, the matrix is positive semidefinite and will be positive *definite* if $x[n]$ does not consist purely of $p - 1$ or fewer sinusoids. This follows from the positive semidefinite property of the ACF (see Section 3.4 and Problem 5.4). Equation (5.18) may be efficiently solved in $O(p^2)$ operations using the Levinson algorithm, which is discussed in Chapter 6. A computer program entitled LEVINSON for implementing the Levinson algorithm is given in Appendix 6B.

Equation (5.18) can also be augmented by incorporating the σ^2 equation to yield

$$\begin{bmatrix} r_{xx}[0] & r_{xx}[-1] & \cdots & r_{xx}[-p] \\ r_{xx}[1] & r_{xx}[0] & \cdots & r_{xx}[-(p-1)] \\ \vdots & \vdots & \ddots & \vdots \\ r_{xx}[p] & r_{xx}[p-1] & \cdots & r_{xx}[0] \end{bmatrix} \begin{bmatrix} 1 \\ a[1] \\ \vdots \\ a[p] \end{bmatrix} = \begin{bmatrix} \sigma^2 \\ 0 \\ \vdots \\ 0 \end{bmatrix} \tag{5.19}$$

which follows from (5.17). This form will be useful later.

Finally, the relationships for an MA process are obtained from (5.16) by letting $a[l] = \delta[l]$ and $h[l] = b[l]$. Then

$$r_{xx}[k] = \begin{cases} \sigma^2 \sum_{l=0}^{q-k} b^*[l] b[l+k] & \text{for } k = 0, 1, \ldots, q \\ \\ 0 & \text{for } k \geq q + 1. \end{cases} \tag{5.20}$$

It is seen that the relationship between the parameters of an MA process and the ACF is a nonlinear one.

The Yule–Walker equations for an ARMA process can also be derived directly from (5.2). Subsequent specializations to the AR and MA processes proceed exactly as in the preceding derivation. In this derivation, an additional relationship (5.21) is found which will be useful later. Multiply (5.2) by $x^*[n - k]$ and take the expectation to yield

$$r_{xx}[k] = -\sum_{l=1}^{p} a[l] r_{xx}[k - l] + \sum_{l=0}^{q} b[l] r_{xu}[k - l]$$

where

$$r_{xu}[k] = \mathscr{E}(u[n]x^*[n - k]) = \mathscr{E}(x^*[n]u[n + k]).$$

But $r_{xu}[k] = 0$ for $k > 0$. This is because $x[n]$, which is the output of a causal filter, is $\sum_{l=-\infty}^{n} h[n - l]u[l]$ and clearly depends on $\{u[n], u[n - 1], \ldots\}$, which is uncorrelated with $u[n + k]$ for $k > 0$. Therefore,

$$r_{xx}[k] = \begin{cases} -\sum_{l=1}^{p} a[l] r_{xx}[k - l] + \sum_{l=k}^{q} b[l] r_{xu}[k - l] & \text{for } k = 0, 1, \ldots, q \\ -\sum_{l=1}^{p} a[l] r_{xx}[k - l] & \text{for } k \geq q + 1. \end{cases} \tag{5.21}$$

But

$$\begin{aligned} r_{xu}[k] &= \mathscr{E}(x^*[n]u[n + k]) \\ &= \mathscr{E}\left(\sum_{l=-\infty}^{n} h^*[n - l]u^*[l]u[n + k]\right) \\ &= \sum_{l=-\infty}^{n} h^*[n - l]\sigma^2\delta[n + k - l] = \sigma^2 h^*[-k] \end{aligned}$$

so that (5.21) becomes (5.16). As alluded to previously, (5.21) is useful for ARMA processes if one knows $r_{xu}[k]$. In this case (5.21) allows one to find the ARMA parameters from the ACF $r_{xx}[k]$ and the CCF $r_{xu}[k]$ by solving a set of linear equations. In Chapter 10, one ARMA estimation method that makes use of (5.21) and an additional relationship estimates the input noise sequence and then uses this to estimate the CCF (see Section 10.6). Another ARMA estimation method discussed in Section 10.4 estimates the AR parameters directly by solving the set of linear equations that are given in (5.16) for $k = q + 1, q + 2, \ldots, p + q$. To find the AR parameters we need to solve

$$\underbrace{\begin{bmatrix} r_{xx}[q] & r_{xx}[q-1] & \cdots & r_{xx}[q-p+1] \\ r_{xx}[q+1] & r_{xx}[q] & \cdots & r_{xx}[q-p+2] \\ \vdots & \vdots & \ddots & \vdots \\ r_{xx}[q+p-1] & r_{xx}[q+p-2] & \cdots & r_{xx}[q] \end{bmatrix}}_{\mathbf{R}'_{xx}} \begin{bmatrix} a[1] \\ a[2] \\ \vdots \\ a[p] \end{bmatrix} = - \begin{bmatrix} r_{xx}[q+1] \\ r_{xx}[q+2] \\ \vdots \\ r_{xx}[q+p] \end{bmatrix}. \tag{5.22}$$

These equations have been called the *extended or modified* Yule–Walker equations. The matrix is Toeplitz, although not hermitian, and is therefore not guaranteed to be positive semidefinite. In fact, $\mathbf{R}'_{xx}$ can be singular for nonsinusoidal processes (see Problem 5.5). An algorithm requiring $O(p^2)$ operations has been developed for solving (5.22). The computer program MYWE given in Appendix 10D implements this algorithm.

5.5 EXAMPLES OF ARMA, AR, AND MA PROCESSES

In this section the results obtained in Sections 5.3 and 5.4 are applied to generate the ACF and PSD for low order (i.e., p and q small) ARMA, AR, and MA processes. Similar results are obtainable for higher order processes but are extremely tedious to derive. The following discussion is restricted to real processes.

5.5.1 AR Processes

For an AR(1) process it follows from (5.17) that

$$r_{xx}[k] = -a[1]r_{xx}[k-1] \qquad k \geq 1 \tag{5.23}$$

which leads to

$$r_{xx}[k] = r_{xx}[0](-a[1])^{|k|}. \tag{5.24}$$

$r_{xx}[0]$ is found to be $\sigma^2/(1 - a^2[1])$ since, from (5.17),

$$\begin{aligned} \sigma^2 &= r_{xx}[0] + a[1]r_{xx}[-1] \\ &= r_{xx}[0] + a[1]r_{xx}[1] \\ &= r_{xx}[0] + a[1](-a[1])r_{xx}[0] \end{aligned}$$

which implies that

$$r_{xx}[0] = \frac{\sigma^2}{1 - a^2[1]}$$

and hence

$$r_{xx}[k] = \frac{\sigma^2}{1 - a^2[1]}(-a[1])^{|k|}. \tag{5.25}$$

$r_{xx}[k]$ is plotted in Figure 5.5 for $a[1] < 0$ and $a[1] > 0$ (see also Problems 5.6 and 5.7). The corresponding PSDs are given by

$$P_{AR}(f) = \frac{\sigma^2}{|1 + a[1] \exp(-j2\pi f)|^2} \tag{5.26}$$

and are plotted in decibels. Note that $a[1] < 0$ corresponds to a low-pass process while $a[1] > 0$ corresponds to a high-pass process. Thus a real AR(1) process cannot model a bandpass process. To do so, we must use an AR(2) process whose ACF can be shown to be (see Problem 5.8)

$$r_{xx}[k] = \sigma^2 \frac{\frac{1 + r^2}{1 - r^2}\sqrt{1 + \left(\frac{1 - r^2}{1 + r^2}\right)^2 \cot^2 2\pi f_0}}{1 - 2r^2 \cos(4\pi f_0) + r^4} r^{|k|} \cos(2\pi f_0 |k| - \psi) \tag{5.27}$$

where

$$\psi = \arctan\left[\frac{1 - r^2}{1 + r^2} \cot 2\pi f_0\right]$$

and $a[1] = -2r \cos 2\pi f_0$, $a[2] = r^2$. Here $r \exp(\pm j2\pi f_0)$ are the poles of $1/\mathcal{A}(z)$, which have been assumed to be complex. The PSD is

$$\begin{aligned} P_{AR}(f) &= \frac{\sigma^2}{|1 + a[1] \exp(-j2\pi f) + a[2] \exp(-j4\pi f)|^2} \\ &= \frac{\sigma^2}{|1 - r \exp(-j2\pi(f - f_0))|^2 \, |1 - r \exp(-j2\pi(f + f_0))|^2} . \end{aligned} \tag{5.28}$$

Examples of the ACF and PSD are given in Figure 5.6. As $a[2] = r^2 \to 1$, the PSD becomes more peaked about $f = f_0$ and the ACF becomes more nearly sinusoidal. It is also possible that the poles of $1/\mathcal{A}(z)$ may be real. In this case the PSD will peak at $f = 0$ or $f = \frac{1}{2}$ or both, depending on the pole locations.

5.5.2 ARMA Processes

Next consider a real ARMA(1, 1) process. The ACF is

$$r_{xx}[k] = \begin{cases} \sigma^2 \left[\dfrac{1 + b^2[1] - 2a[1]b[1]}{1 - a^2[1]}\right] & \text{for } k = 0 \\[2ex] \sigma^2 \left[\dfrac{(1 - a[1]b[1])(b[1] - a[1])}{1 - a^2[1]}\right] (-a[1])^{|k|-1} & \text{for } k \geq 1. \end{cases} \tag{5.29}$$

To derive these relationships, use (5.16).

$$r_{xx}[0] = -a[1]r_{xx}[-1] + \sigma^2(1 + b[1]h[1])$$

$$r_{xx}[1] = -a[1]r_{xx}[0] + \sigma^2 b[1]$$

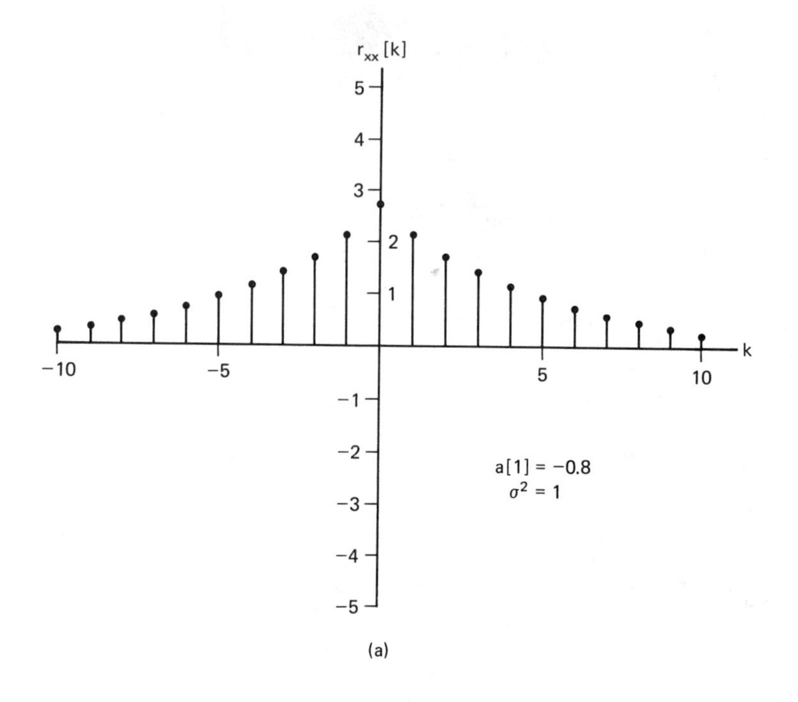

(a)

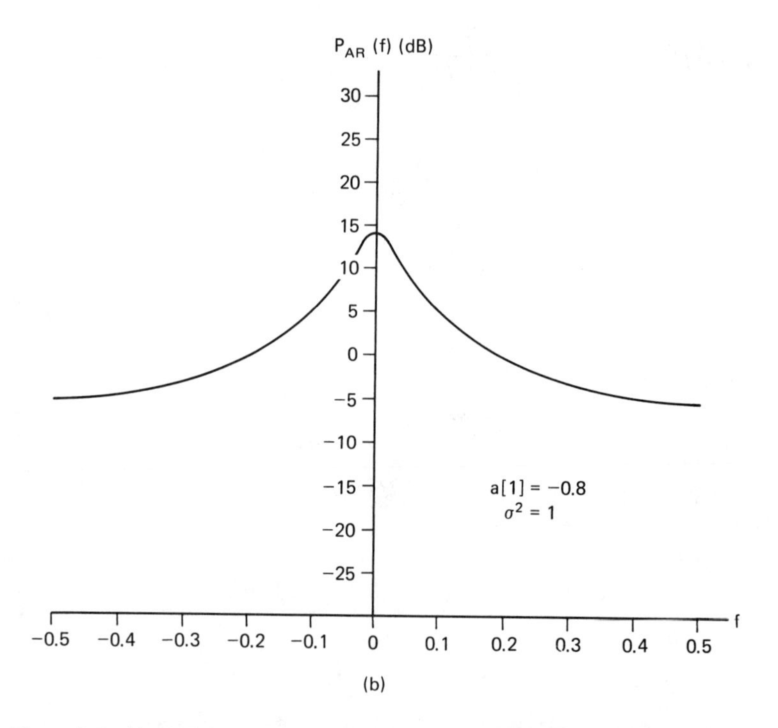

(b)

Figure 5.5 (a),(c) Autocorrelation of real AR(1) process. (b),(d) Power spectral density of real AR(1) process.

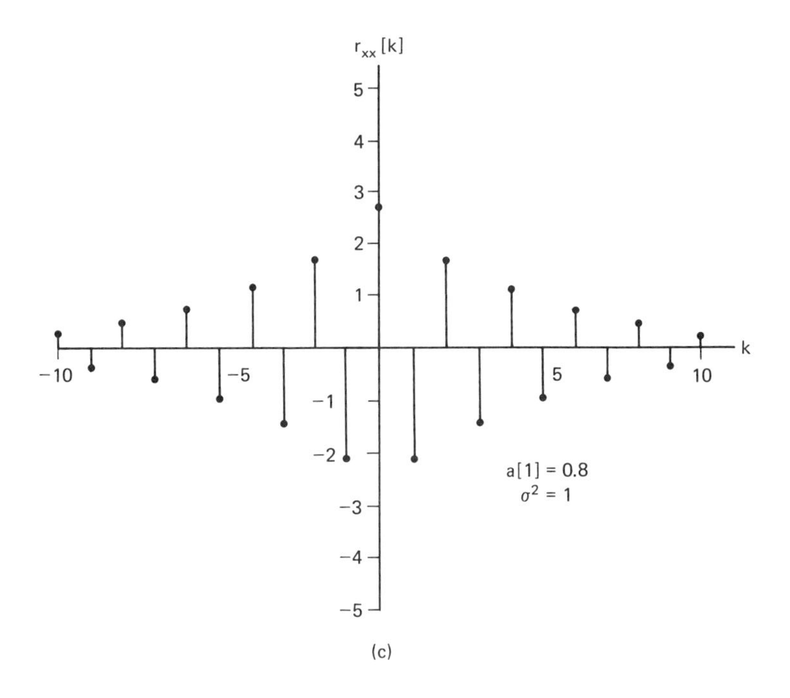

(c)

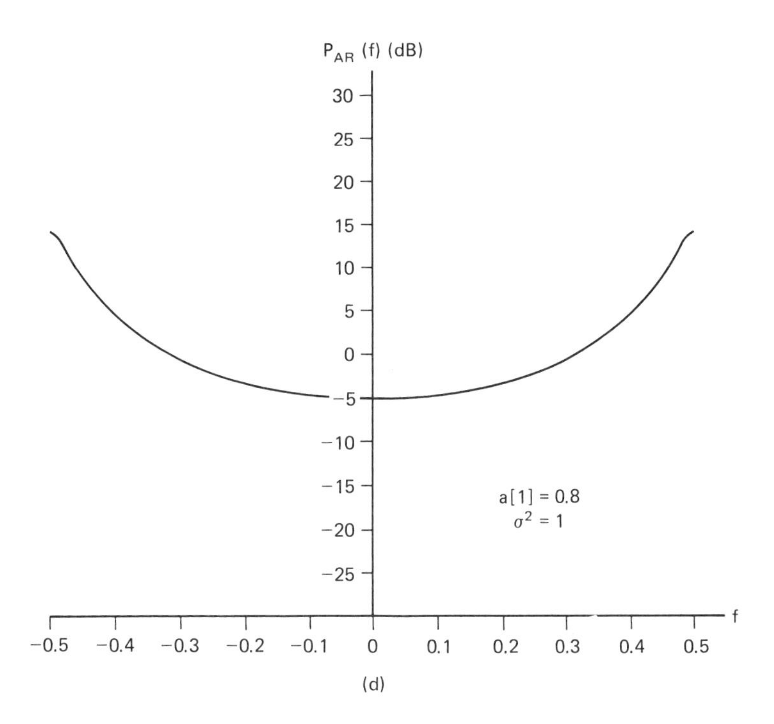

(d)

Figure 5.5 *(Continued)*

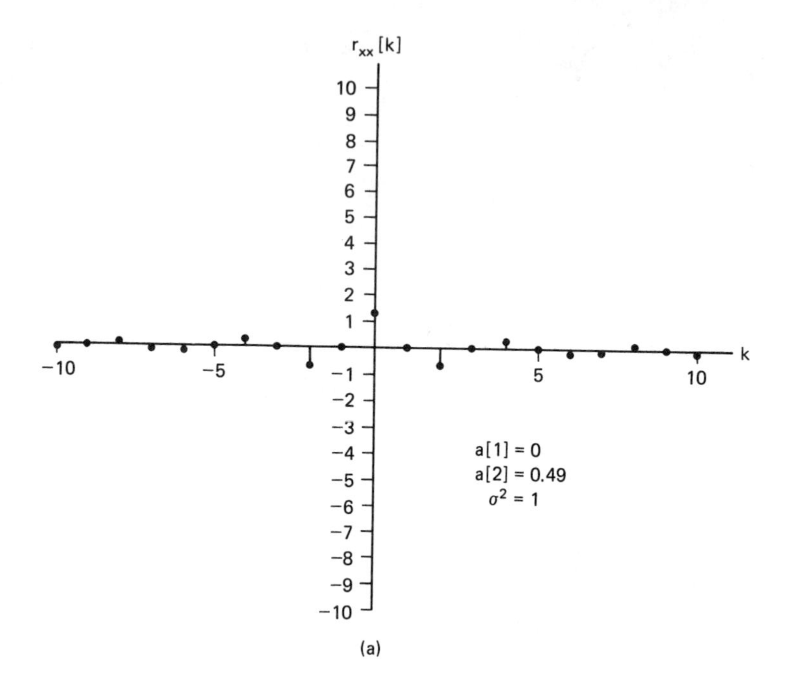

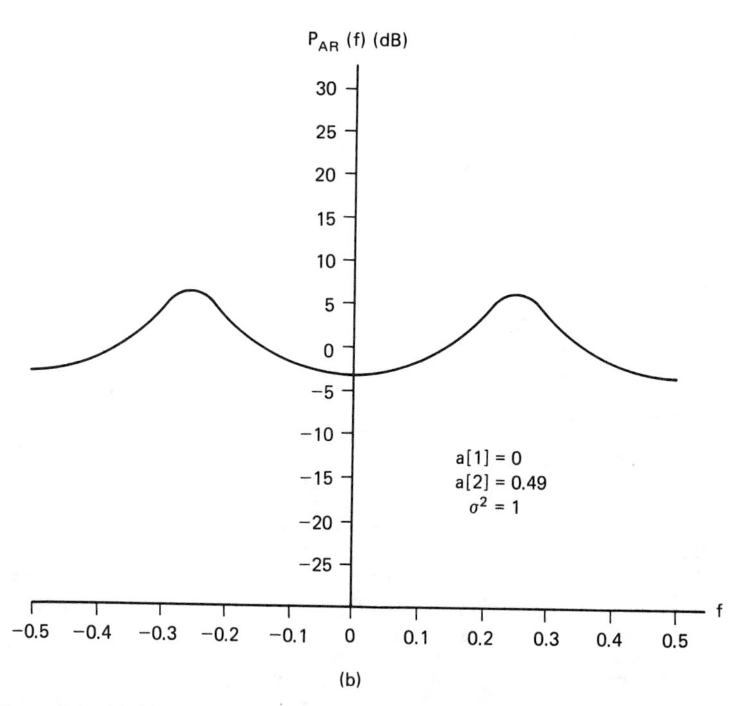

Figure 5.6 (a),(c) Autocorrelation of real AR(2) process with complex conjugate poles. (b),(d) Power spectral density of real AR(2) process with complex conjugate poles.

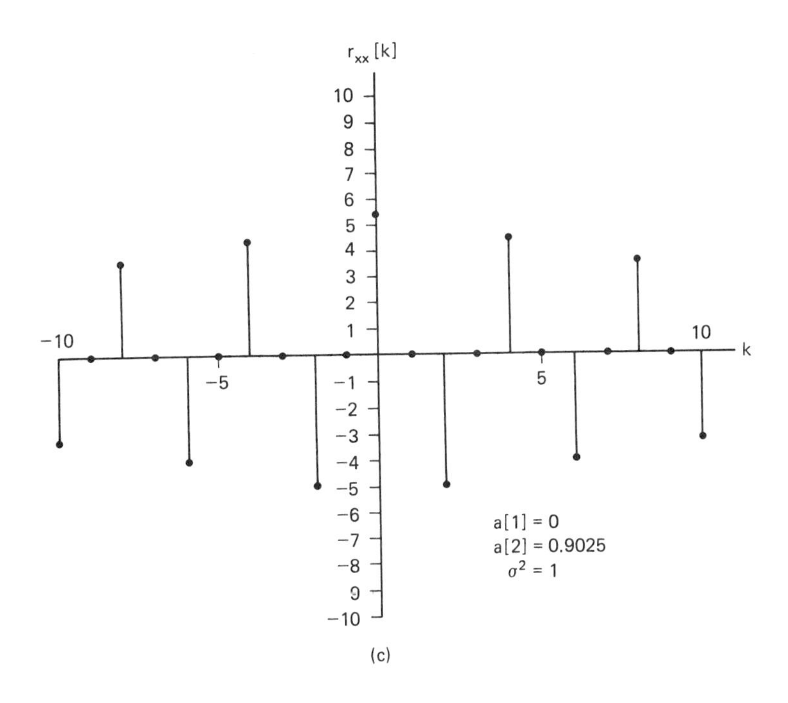

(c)

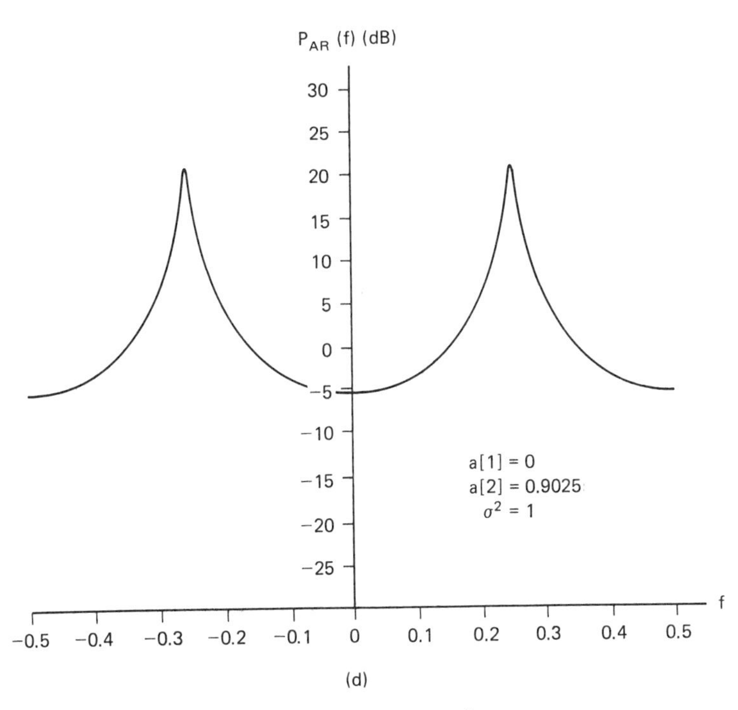

(d)

Figure 5.6 *(Continued)*

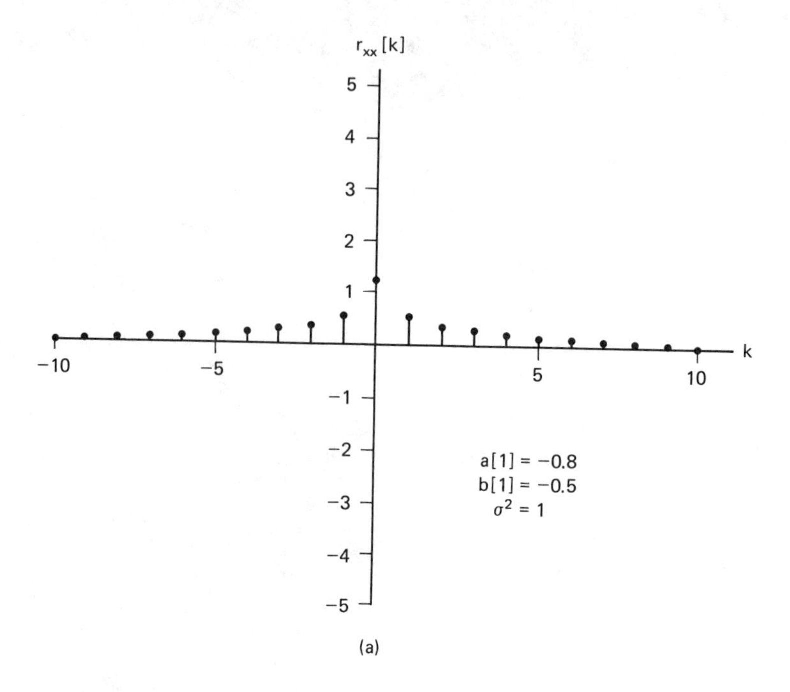

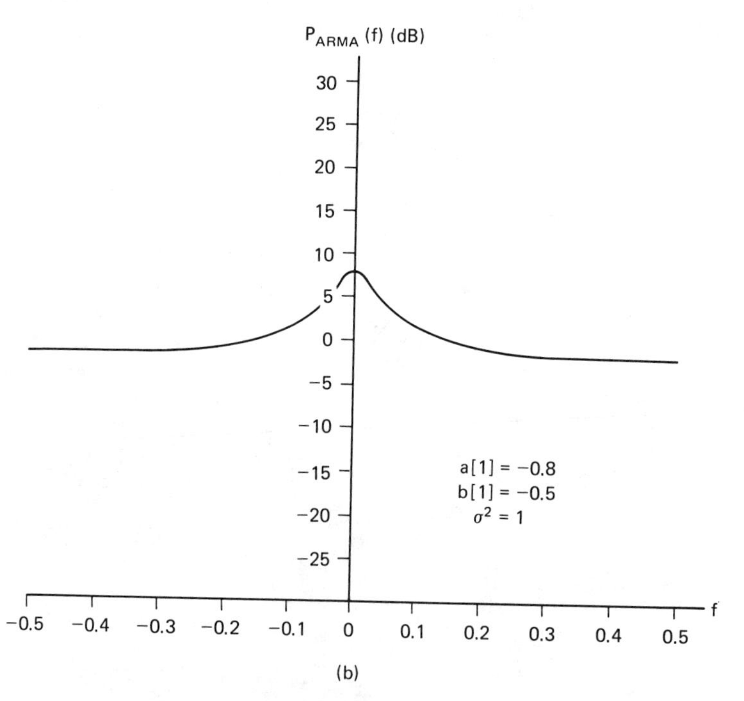

Figure 5.7 (a),(c) Autocorrelation of real ARMA(1, 1) process. (b),(d) Power spectral density of real ARMA(1, 1) process.

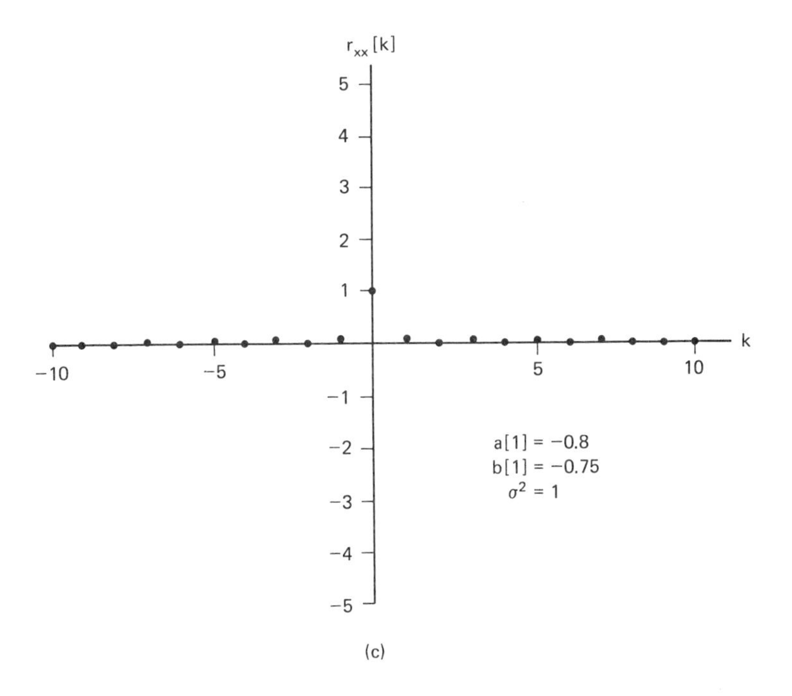

(c)

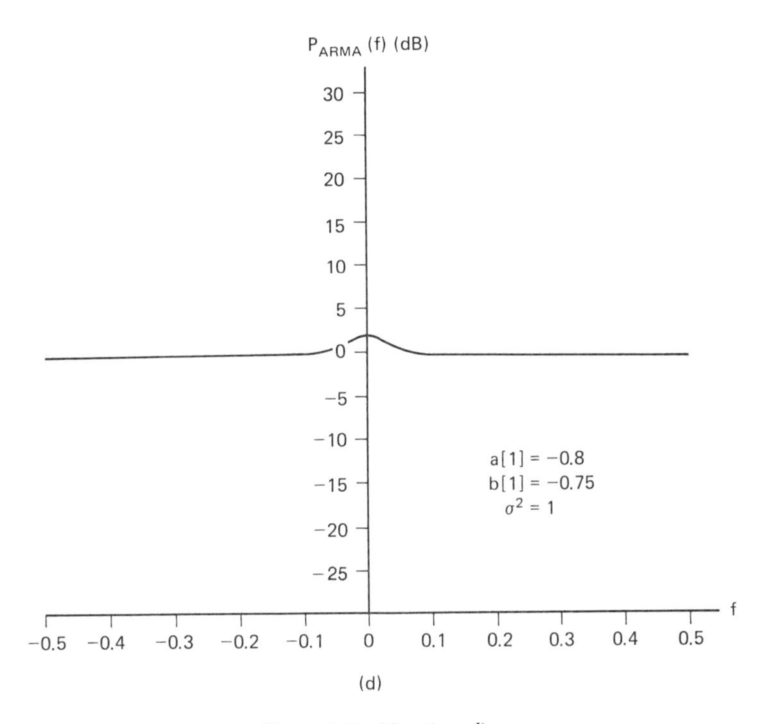

(d)

Figure 5.7 *(Continued)*

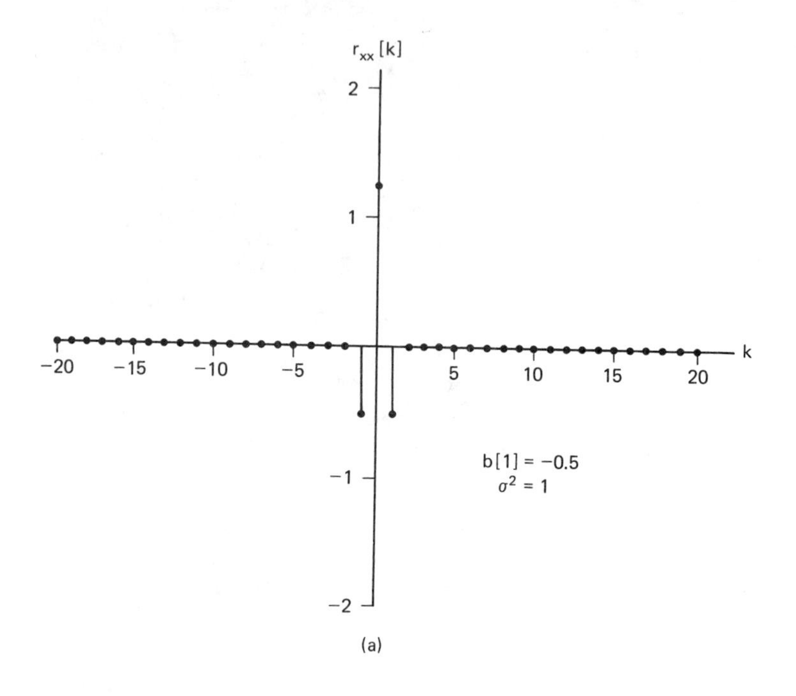

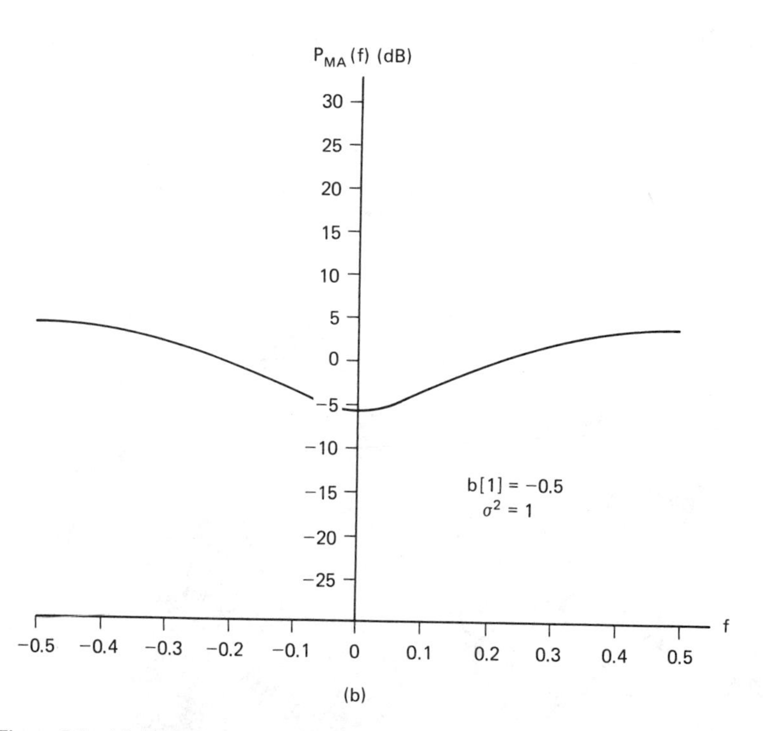

Figure 5.8 (a),(c) Autocorrelation of real MA(1) process. (b),(d) Power spectral density of real MA(1) process.

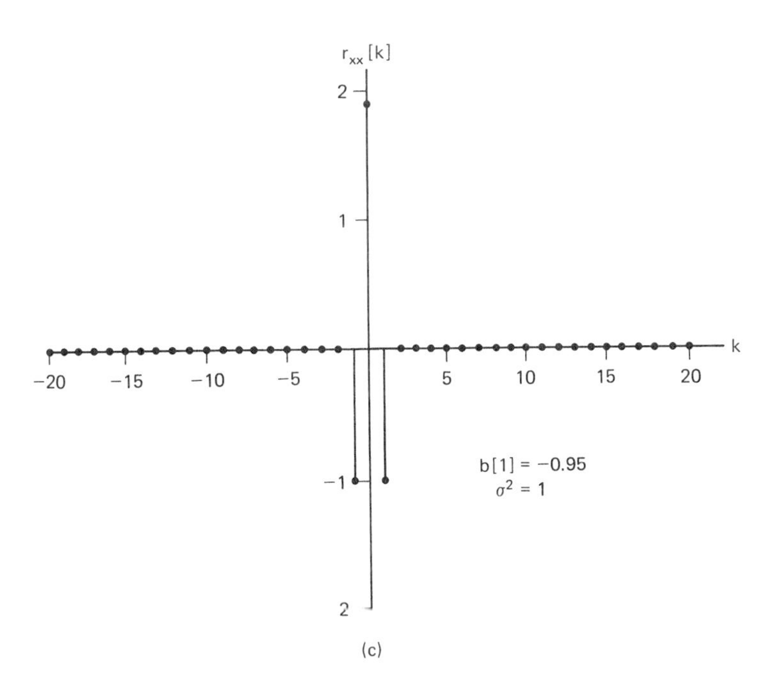

(c)

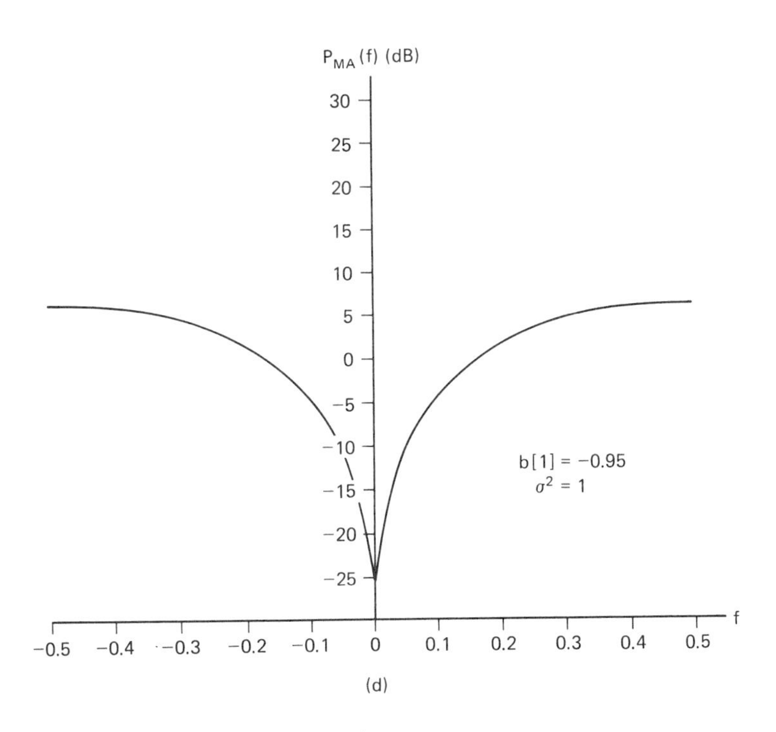

(d)

Figure 5.8 *(Continued)*

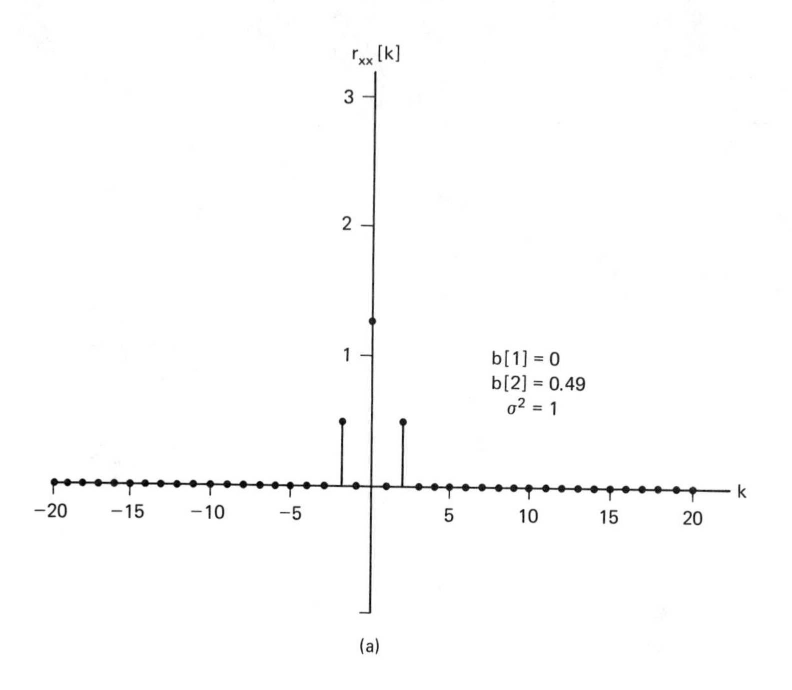

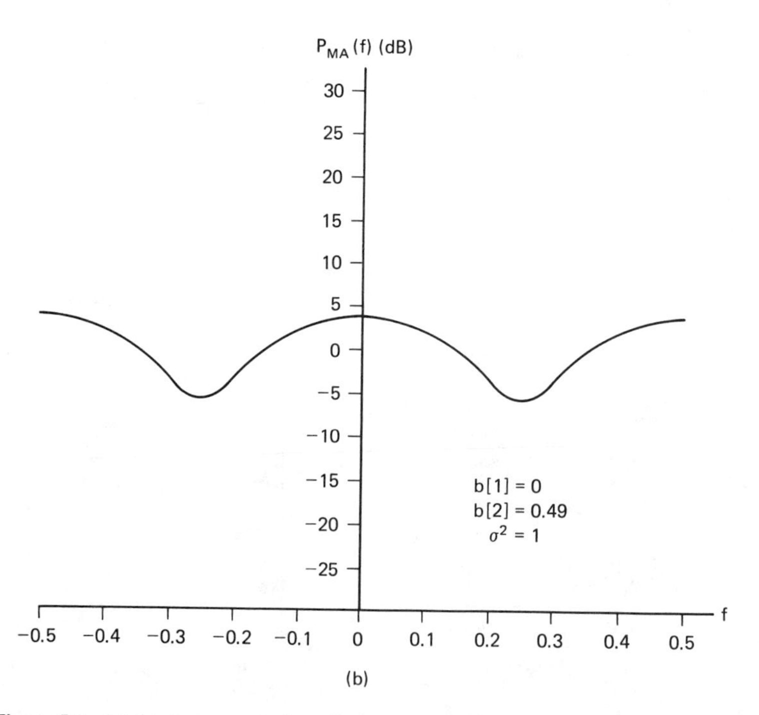

Figure 5.9 (a),(c) Autocorrelation of real MA(2) process. (b),(d) Power spectral density of real MA(2) process.

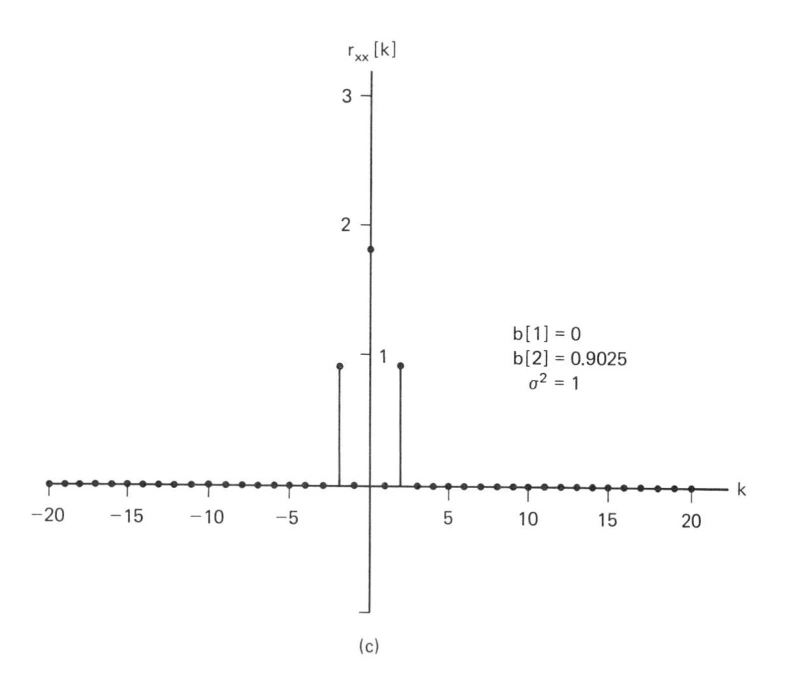

(c)

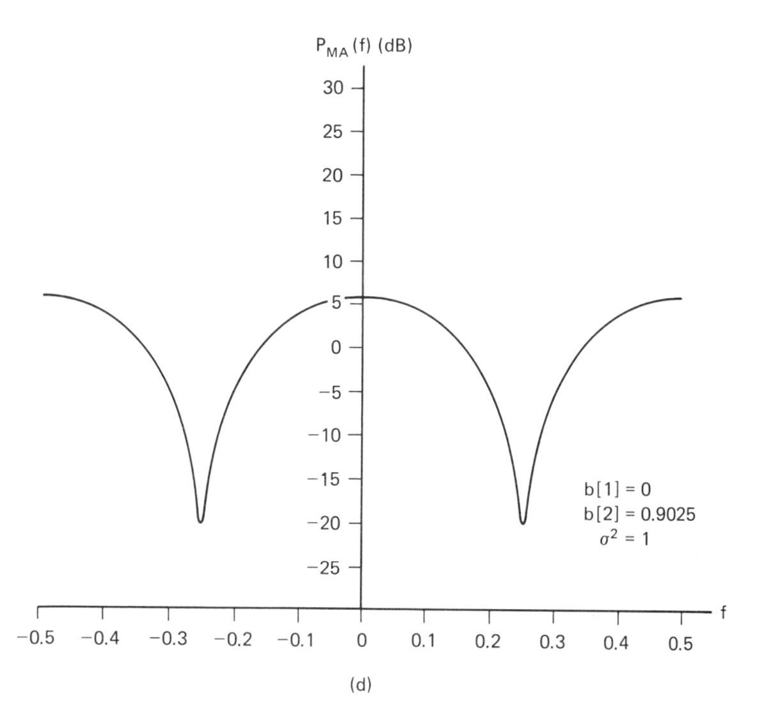

(d)

Figure 5.9 *(Continued)*

where the identities $b[0] = 1$, $h[0] = 1$ have been used. It is easily shown that $h[1] = -a[1] + b[1]$ by noting that

$$h[n] = -a[1]h[n-1] + \delta[n] + b[1]\delta[n-1]$$

so that

$$\begin{bmatrix} 1 & a[1] \\ a[1] & 1 \end{bmatrix} \begin{bmatrix} r_{xx}[0] \\ r_{xx}[1] \end{bmatrix} = \begin{bmatrix} \sigma^2(1 + b^2[1] - a[1]b[1]) \\ \sigma^2 b[1] \end{bmatrix}$$

which is easily solved to yield $r_{xx}[0]$, $r_{xx}[1]$. $r_{xx}[k]$ for $k \geq q + 1 = 2$ is also found from (5.16) using $r_{xx}[k] = -a[1]r_{xx}[k-1]$.

The PSD is given by

$$P_{\text{ARMA}}(f) = \frac{\sigma^2 \mid 1 + b[1] \exp(-j2\pi f) \mid^2}{\mid 1 + a[1] \exp(-j2\pi f) \mid^2}. \tag{5.30}$$

Some examples of the ACF and PSD are shown in Figure 5.7. The character of the plots are almost the same as for the AR(1) process. Although when $b[1] \approx a[1]$, the pole and zero "cancel," yielding a flat spectrum. When $b[1] \to \pm 1$, a null appears in the spectrum.

5.5.3 MA Processes

Consider real MA(1) and MA(2) processes. From (5.20) the ACF for an MA(1) process is given by (see Problem 5.9)

$$r_{xx}[k] = \begin{cases} \sigma^2(1 + b^2[1]) & \text{for } k = 0 \\ \sigma^2 b[1] & \text{for } k = 1 \\ 0 & \text{for } k \geq 2 \end{cases} \tag{5.31}$$

with a corresponding PSD from (5.7) of

$$P_{\text{MA}}(f) = \sigma^2 \mid 1 + b[1] \exp(-j2\pi f) \mid^2. \tag{5.32}$$

For the MA(2) process the relationships are

$$r_{xx}[k] = \begin{cases} \sigma^2(1 + b^2[1] + b^2[2]) & \text{for } k = 0 \\ \sigma^2(b[1] + b[1]b[2]) & \text{for } k = 1 \\ \sigma^2 b[2] & \text{for } k = 2 \\ 0 & \text{for } k \geq 3 \end{cases} \tag{5.33}$$

and

$$P_{\text{MA}}(f) = \sigma^2 \mid 1 + b[1] \exp(-j2\pi f) + b[2] \exp(-j4\pi f) \mid^2. \tag{5.34}$$

Some examples of the ACF and PSD are given in Figures 5.8 and 5.9. It is observed that the PSD for MA processes tends to be broad but may exhibit nulls if the zeros are close to the unit circle.

5.6 MODEL FITTING

In practical spectral estimation, we do not usually know a priori which model to choose. Furthermore, once a model, either AR, MA or ARMA, has been chosen, we also need to specify the model orders p and/or q. In choosing a model, ideally the model selected should have as few parameters as necessary. This is because to estimate the PSD, we will ultimately need to estimate the model parameters. Since the data set is usually limited, the estimation accuracy will be poor if too many model parameters are estimated. The inclusion of as few parameters as possible is sometimes referred to as the principle of *parsimony* [Box and Jenkins 1970]. Among competing models, however, it may not be computationally efficient to estimate the parameters of the model with the fewest number of parameters. The principle of parsimony then needs to be tempered with practical reality.

Some of the considerations involved in choosing a model are the ability of the model to represent spectral peaks, valleys, and roll-offs. For spectra with sharp peaks it is necessary to employ a model which has poles, either an AR or an ARMA model. As an example, consider the data as having been generated by an AR process so that the PSD is given by (5.9). If one attempts to use an MA model to estimate the PSD, very poor results may be expected. To illustrate this, the PSD of a real AR(2) process with complex-conjugate poles and that of an approximating MA process are compared. The PSD obtained from the MA model is

$$P_{\mathrm{MA}}(f) = \sum_{k=-q}^{q} w[k] r_{xx}[k] \exp(-j2\pi f k)$$

where $r_{xx}[k]$ is given by (5.27) and $w[k] = 1 - |k|/q$, which is chosen to ensure a nonnegative PSD (see Section 4.5). The results are shown in Figure 5.10. It is seen that for a true PSD with a sharp peak very large model orders must be chosen if an MA model is employed. This is due to the lack of poles in the MA model. Similarly, if the true PSD is characterized by poles *and zeros*, then the use of an AR model, which has only poles, would again be expected to yield poor results. As an example, consider the data as having been generated by the sum of a real AR(2) process with complex-conjugate poles and a white noise process, uncorrelated with the AR(2) process. Then

$$\begin{aligned} P_{xx}(f) &= P_{\mathrm{AR}}(f) + \sigma_w^2 \\ &= \frac{\sigma^2}{|A(f)|^2} + \sigma_w^2 \\ &= \frac{(\sigma^2 + \sigma_w^2 |A(f)|^2)}{|A(f)|^2} \\ &= P_{\mathrm{ARMA}}(f) \end{aligned} \tag{5.35}$$

where $A(f) = 1 + a[1] \exp(-j2\pi f) + a[2] \exp(-j4\pi f)$. It is observed that the PSD corresponds to that of an ARMA(2, 2) model. An AR(2) model would be

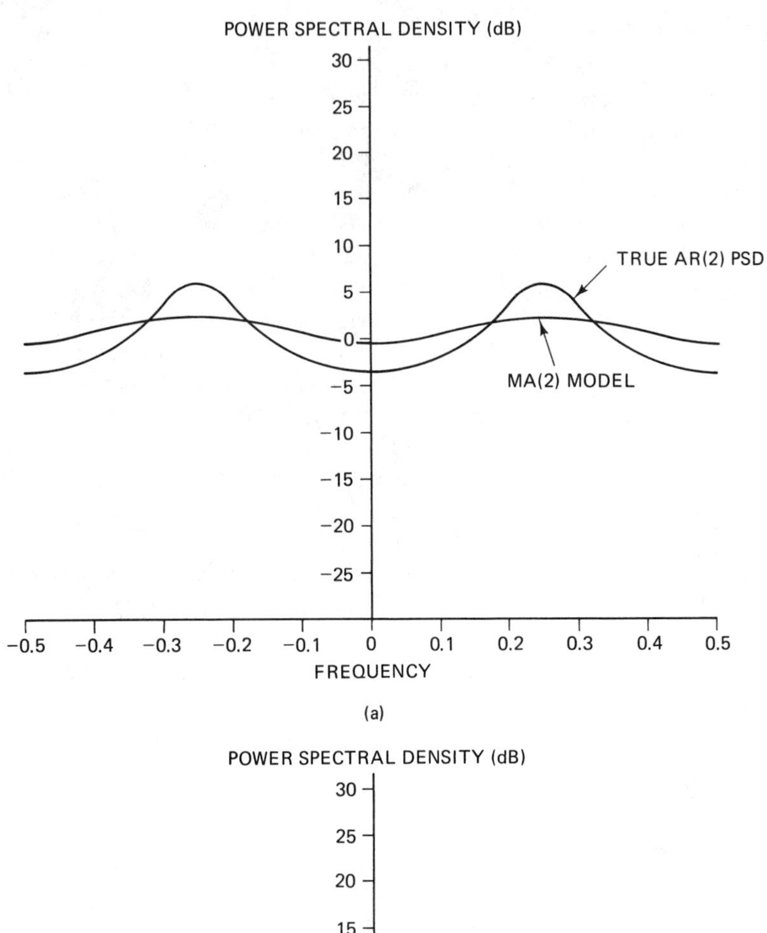

(a)

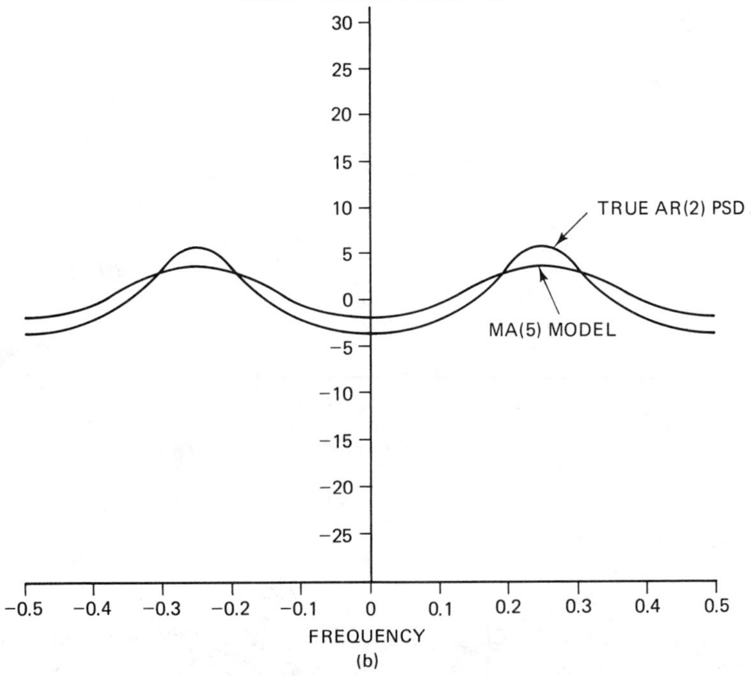

(b)

Figure 5.10 Modeling of AR(2) PSD. (a),(b),(c) MA(2), MA(5), and MA(10) modeling of broadband AR(2) PSD; (d),(e),(f) MA(2), MA(5), and MA(10) modeling of narrowband AR(2) PSD.

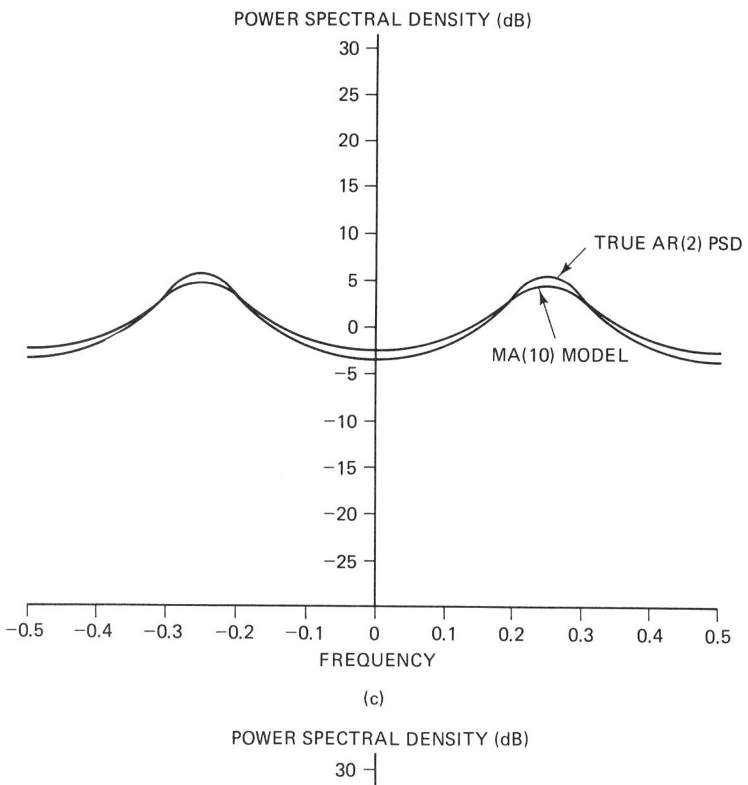

(c)

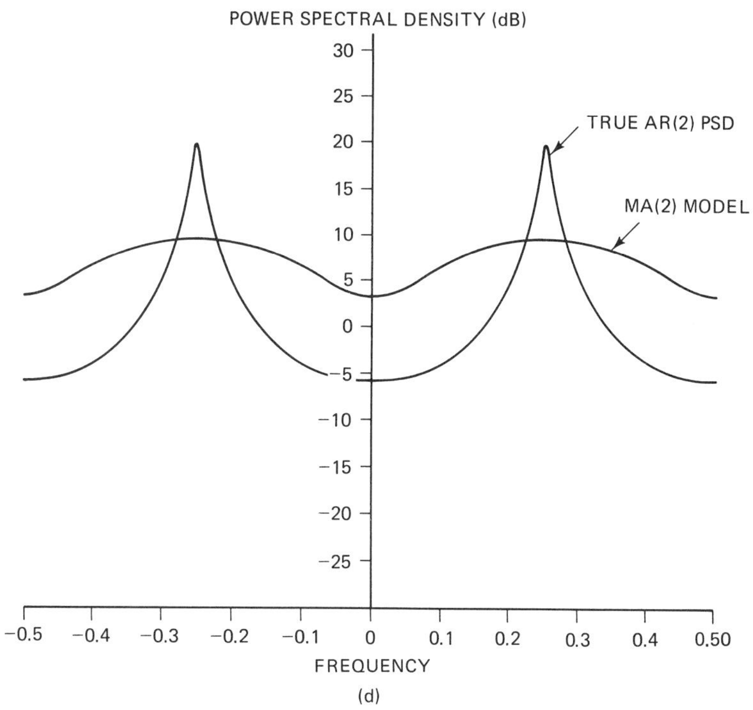

(d)

Figure 5.10 *(Continued)*

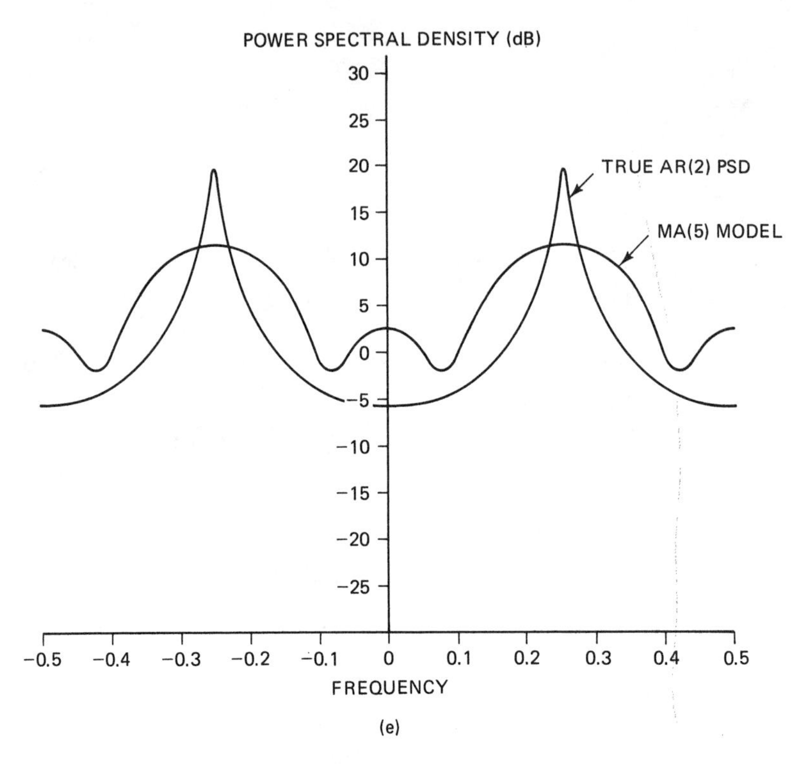

(e)

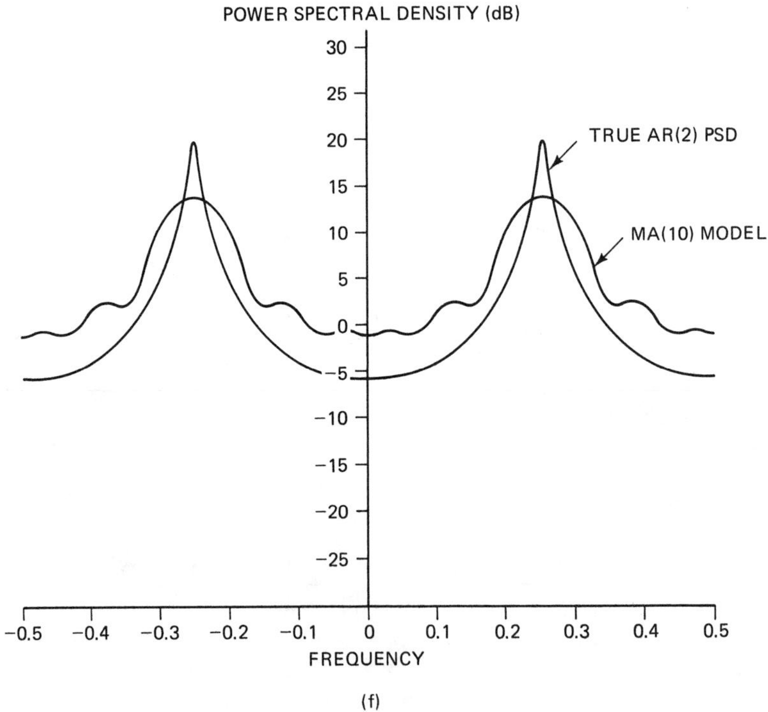

(f)

Figure 5.10 *(Continued)*

appropriate only if $\sigma_w^2 \rightarrow 0$ (see Problems 5.10 and 5.11). Otherwise, the AR(2) model would not be able to represent the PSD and a biased spectral estimator will result. To illustrate this point further, assume that the parameters of an AR model are found by solving the Yule–Walker equations (5.17) using the ACF

$$\begin{aligned} r_{yy}[k] &= \mathscr{F}^{-1}[P_{\text{ARMA}}(f)] \\ &= r_{xx}[k] + \sigma_w^2\delta[k] \end{aligned} \tag{5.36}$$

where $r_{xx}[k]$ is given by (5.27). The results are shown in Figure 5.11 for various AR model orders and SNRs. For σ_w^2 small or a high SNR, the AR(2) model is appropriate and good results are obtained (Figure 5.11c). For σ_w^2 larger or a low SNR, the resultant ARMA(2, 2) process can be shown to have zeros near the unit circle [Kay 1979] and hence the all-pole AR(2) model is inappropriate (Figure 5.11a). Note, however, that as the AR model order p increases, the AR model approximation becomes better (Figure 5.11a and b). This is not purely coincidental but is a manifestation of the Kolmogorov theorem. A careful examination of the manner in which the AR PSD estimate was obtained reveals that an autocorrelation matching process has occurred. By using (5.36), the true ACF, in the Yule–Walker equations with $p = 2$ as an illustration, the first three samples ($k = 0, 1, 2$) of the ACF of the AR(2) PSD model are equal to the true ones. The remaining samples of the ACF of the AR(2) PSD model are given by (5.17) as

$$\hat{r}_{xx}[k] = -a[1]\hat{r}_{xx}[k-1] - a[2]\hat{r}_{xx}[k-2] \qquad k \geq 3.$$

The initial conditions are $\hat{r}_{xx}[1] = r_{xx}[1]$ and $\hat{r}_{xx}[2] = r_{xx}[2]$. This is the so-called "correlation matching" property of the AR model (see also Section 6.4.1). It is therefore apparent that as the AR(p) model order $p \rightarrow \infty$, more of the ACF samples will be matched and consequently, the PSD of the AR(p) model must approach the true PSD. The general conclusion to be drawn is that an AR(∞) model is appropriate for *any* PSD, although clearly impractical.

Another characteristic of the true PSD important in the choice of a model is the spectral roll-off. Considering an AR(p) model it is well known from analog filter design that a filter having p poles will roll off at $-6p$ dB/octave. Hence for PSDs with large roll-offs, a large order AR model must be chosen. As an example, consider a PSD with a nonrational form. In particular, a Gaussian PSD model for a real low-pass process is

$$P_{xx}(f) = \frac{r_{xx}[0]}{\sqrt{2\pi}\sigma_f} \exp\left[-\frac{1}{2}\left(\frac{f}{\sigma_f}\right)^2\right] \qquad -1/2 \leq f \leq 1/2 \tag{5.1}$$

where it is assumed that the value of the Gaussian PSD is nearly zero at $f = \pm\frac{1}{2}$. The roll-off over the octave band of frequencies from f_0 to $2f_0$ is

$$\text{roll-off} = 10\log_{10} P_{xx}(2f_0) - 10\log_{10} P_{xx}(f_0) \qquad \text{dB/octave} \tag{5.37}$$

which from (5.1) becomes

$$\text{roll-off} = \frac{-30\log_{10} e}{2\sigma_f^2} f_0^2. \tag{5.38}$$

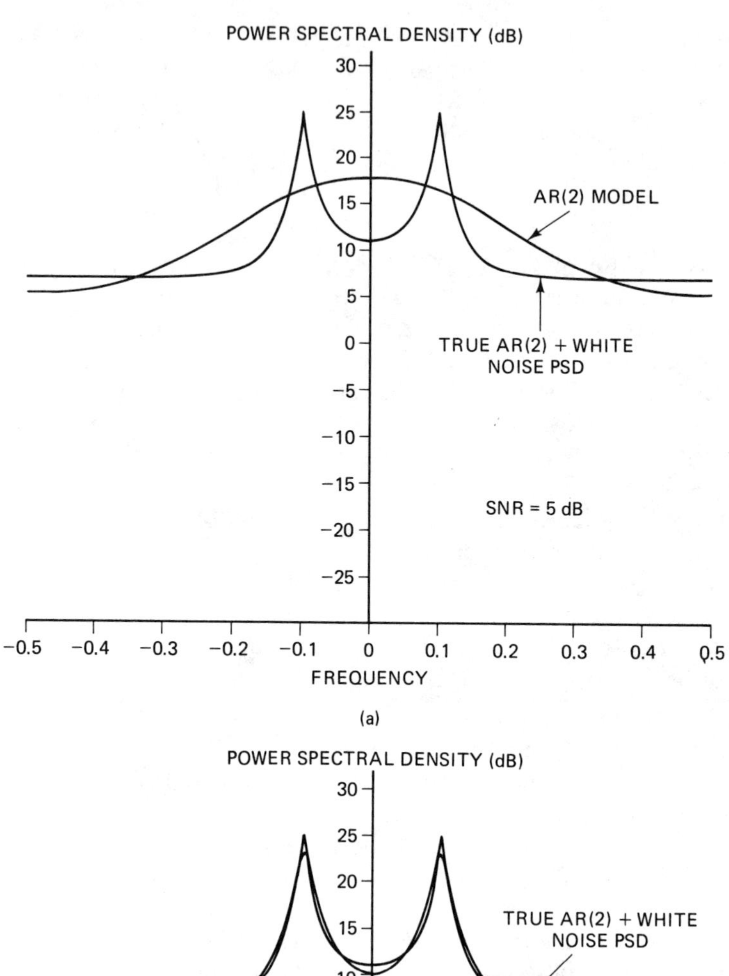

(a)

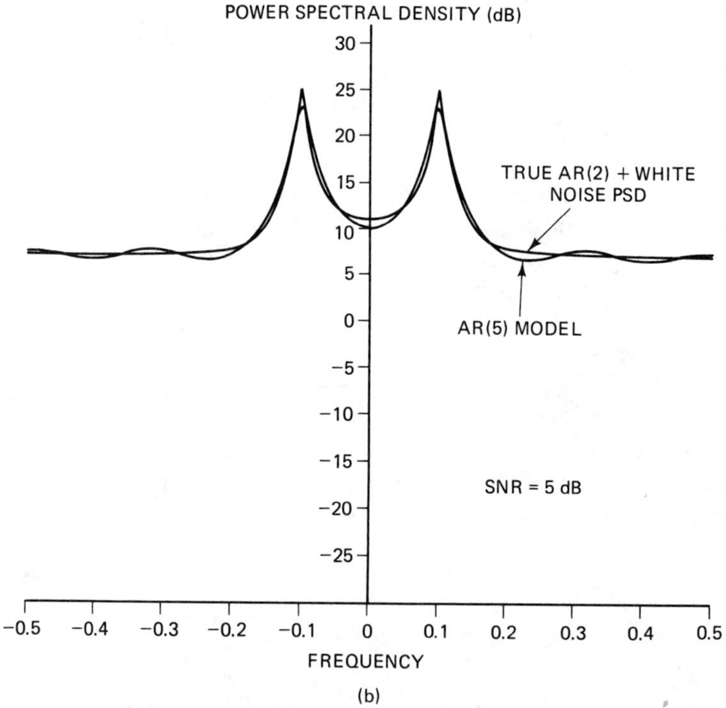

(b)

Figure 5.11 Modeling of noise corrupted AR(2) PSD. (a),(b) AR(2) and AR(5) modeling for low SNR PSD; (c),(d) AR(2) and AR(5) modeling for high SNR PSD.

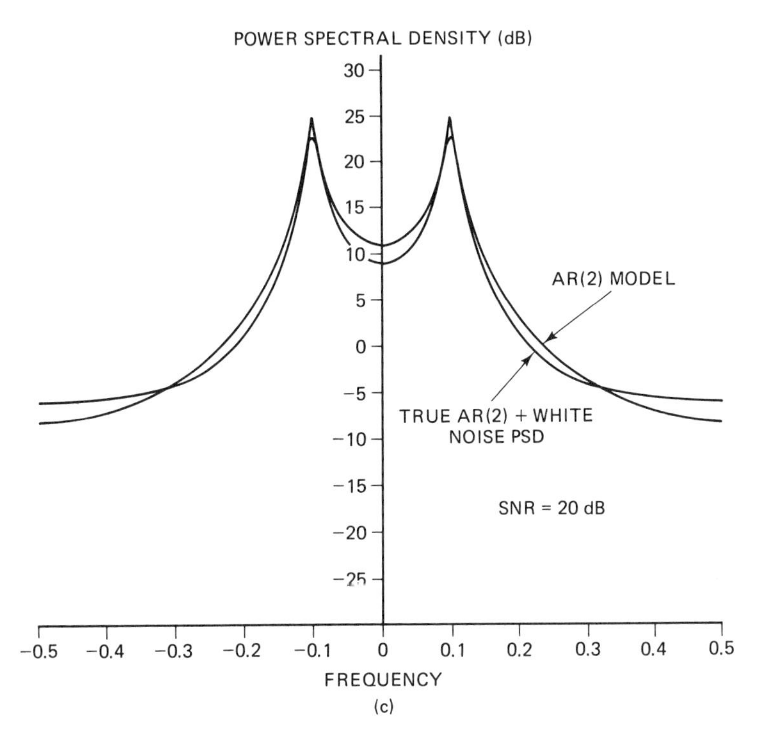

(c)

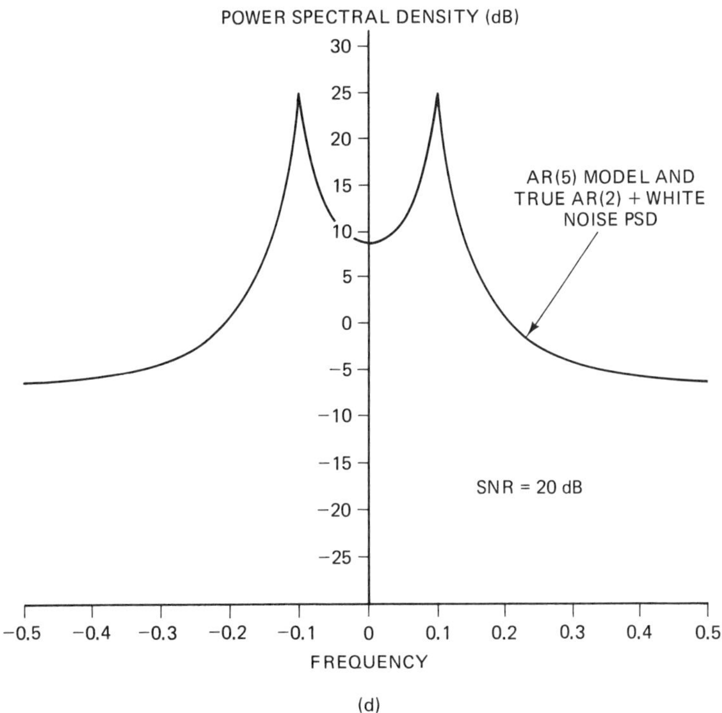

(d)

Figure 5.11 *(Continued)*

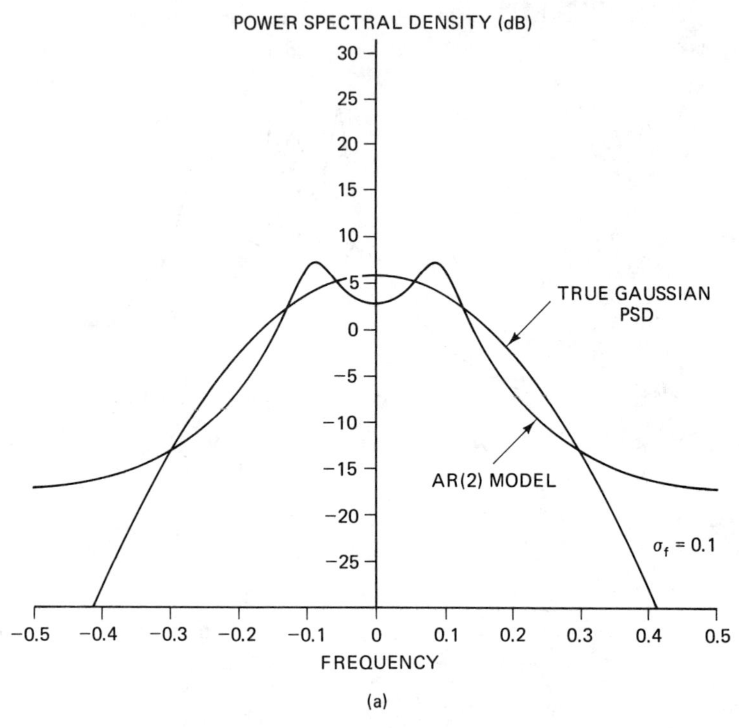

(a)

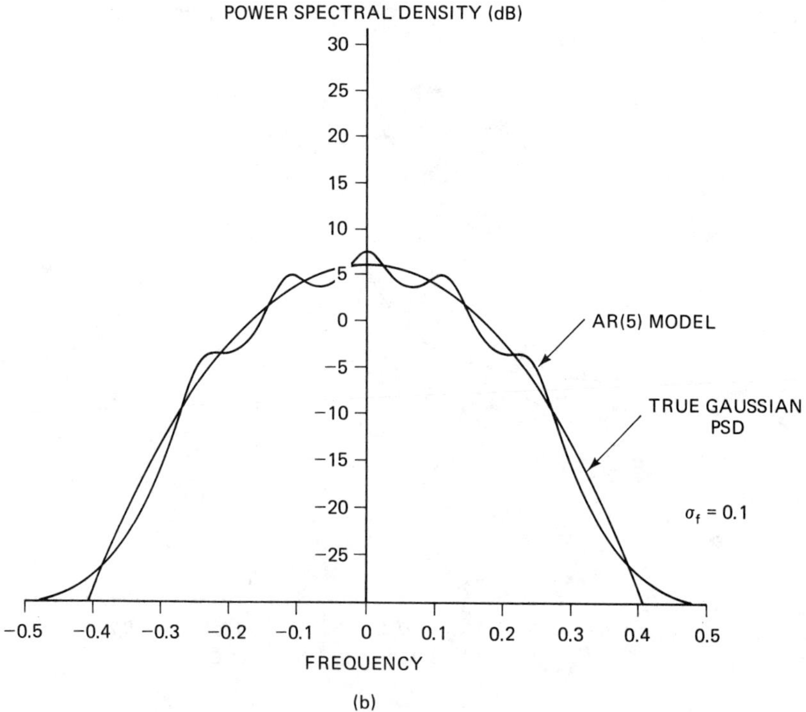

(b)

Figure 5.12 Modeling of Gaussian PSD. (a) AR(2) model. (b) AR(5) model. (c) AR(10) model.

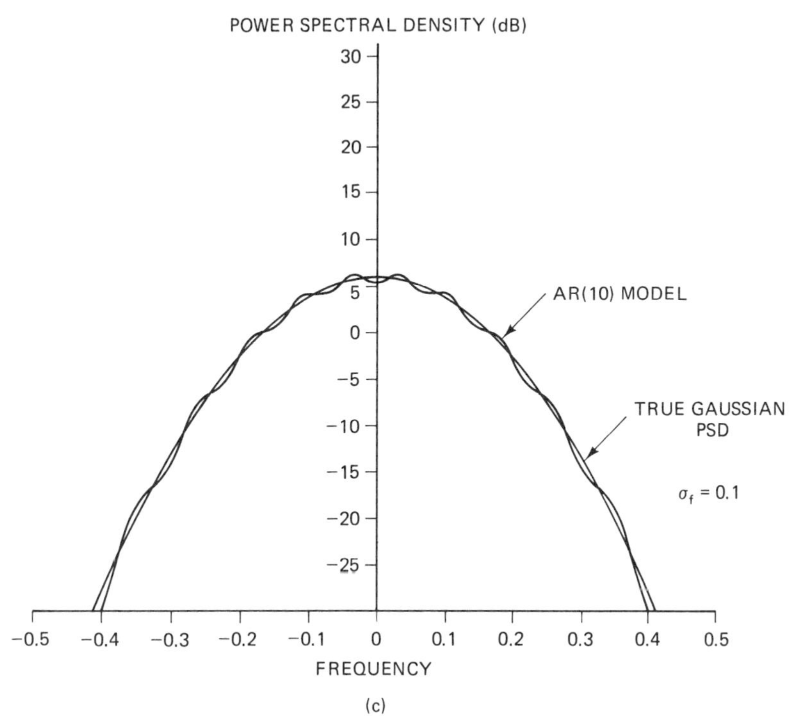

(c)

Figure 5.12 *(Continued)*

For σ_f small the roll-off will be large and hence a large-order AR model will be necessary. Figure 5.12 confirms this observation. $P_{xx}(f)$ as given by (5.1) has been plotted together with some AR(p) model PSDs. The AR(p) PSDs have been obtained from (5.9) and (5.17) with an ACF

$$r_{xx}[k] = \mathcal{F}^{-1}[P_{xx}(f)] = r_{xx}[0] \exp(-2\pi^2\sigma_f^2k^2). \tag{5.39}$$

Further examples of the poor spectral estimates obtained with inappropriate time-series models have been provided by Gutowski et al. [1978].

Finally, some comments regarding the use of time series models for sinusoidal processes should be made. Consider the use of the real AR(2) model for the real sinusoid

$$x[n] = A\cos(2\pi f_0 n + \phi). \tag{5.40}$$

Strictly speaking, the use of any time series model for a sinusoid (or in general a sum of sinusoids) is inappropriate. The AR(2) model

$$x[n] = -a[1]x[n-1] - a[2]x[n-2] + u[n] \tag{5.41}$$

is not capable of generating a sinusoidal time series. A recursive difference equation that will generate (5.40) for $n \geq 0$ is

$$x[n] = 2\cos 2\pi f_0\, x[n-1] - x[n-2] + A\cos\phi\,\delta[n] - A\cos(2\pi f_0 - \phi)\delta[n-1] \tag{5.42}$$

where $x[-2] = x[-1] = 0$ are the initial conditions. In addition to the different forms of the excitation, white noise versus discrete impulses, the poles z_p of the all-pole filter are from (5.42) the solutions to

$$\mathcal{A}(z) = 1 - 2 \cos 2\pi f_0 z^{-1} + z^{-2} = 0$$

or

$$z_p = \exp(\pm j 2\pi f_0) \tag{5.43}$$

so that (5.42) describes the output of an unstable filter. In the AR model all the poles must lie *within* the unit circle for $x[n]$ to be a WSS random process.

In spite of these difficulties it is still possible to model the ACF of (5.40) as the limiting form of the ACF of a real AR(2) process. First, to ensure that $x[n]$ is WSS, we assume that ϕ is a random variable uniformly distributed over $[0, 2\pi)$. The ACF as discussed in Section 3.5.2 is

$$r_{xx}[k] = \frac{A^2}{2} \cos(2\pi f_0 k). \tag{5.44}$$

This ACF can be approximated by the ACF of an AR(2) process with poles at $r \exp(\pm j2\pi f_0)$ if $r \to 1$ and $\sigma^2 \to 0$ in such a way that $r_{xx}[0] = A^2/2$. From (5.27) this approximation yields

$$r_{xx}[k] \to \frac{\sigma^2 \left(\dfrac{2}{1 - r^2}\right)}{2 - 2\cos(4\pi f_0)} \cos(2\pi f_0 k).$$

To maintain $r_{xx}[0] = A^2/2$, let

$$\lim_{\substack{r \to 1 \\ \sigma^2 \to 0}} \frac{\sigma^2}{(1 - r^2)[1 - \cos(4\pi f_0)]} = \frac{A^2}{2}.$$

In this way a sinusoid may be modeled by an AR(2) process, at least as far as the ACF is concerned. Since only the PSD is of interest, such a modeling viewpoint is reasonable and has proven quite useful. A convenient way to view the sinusoidal process is as the limiting form of a narrowband random process, such that the bandwidth of the PSD approaches zero as the peak of the PSD approaches infinity such that the area under the PSD remains constant. The foregoing argument can be generalized to the multiple sinusoid case. For M real sinusoids a real AR($2M$) model is required and for M complex sinusoids a complex AR(M) model is needed.

5.7 MA MODELING AND THE BLACKMAN–TUKEY SPECTRAL ESTIMATOR

The Blackman–Tukey (BT) spectral estimator was defined in Chapter 4 to be

$$\hat{P}_{\mathrm{BT}}(f) = \sum_{k=-M}^{M} w[k] \hat{r}_{xx}[k] \exp(-j2\pi f k) \tag{4.23}$$

where $\hat{r}_{xx}[k]$ is the biased ACF estimator and $w[k]$ is a lag window. The form of the BT spectral estimator with a rectangular lag window ($w[k] = 1$) is identical to that of the MA PSD since if we use $q = M$ in (5.7),

$$P_{\mathrm{MA}}(f) = \sum_{k=-M}^{M} r_{xx}[k] \exp(-j2\pi f k) \tag{5.45}$$

where

$$r_{xx}[k] = \sigma^2 \sum_{l=0}^{M-k} b^*[l]b[l+k] \qquad k = 0, 1, \ldots, M \tag{5.46}$$

In practice, if an MA(M) model is assumed, the MA parameters would be estimated and then used in (5.46) to yield the ACF estimates. The *form* of the PSD will be identical to that of the BT spectral estimator. The only difference will be in the way the ACF is estimated from the data. The BT spectral estimator will thus be subject to the same problems as an inappropriate MA model. A further comparison of the BT and MA spectral estimators may be found in Chapter 12.

5.8 MODEL PARAMETER DETERMINATION BASED ON PSD OR ACF

The definitions of the AR and ARMA processes have assumed that the poles of the process lie within the unit circle of the z-plane. This condition is necessary to ensure that $\mathcal{H}(z)$ is a stable and causal filter. Clearly, if $\mathcal{H}(z)$ is anticausal, then for stability the poles would need to be outside the unit circle. In either case stability is critical since otherwise $x[n]$ would have infinite variance (see Problem 5.2). To illustrate these two possibilities, consider an AR(1) process. A causal filter $\mathcal{H}(z) = 1/\mathcal{A}(z) = 1/(1 + a[1]z^{-1})$ will lead to the time series model

$$x[n] = -a[1]x[n-1] + u[n]. \tag{5.47}$$

However, an anticausal filter $\mathcal{H}(z) = 1/\mathcal{A}^*(1/z^*) = 1/(1 + a^*[1]z)$ results in the time series model

$$x[n] = -a^*[1]x[n+1] + u[n] \tag{5.48}$$

which generates $x[n]$ based on the "future" sample $x[n+1]$. In the first case the pole of $\mathcal{H}(z)$ is at $z = -a[1]$, while for the second case the pole is at $z = -1/a^*[1]$. For $|a[1]| < 1$ the poles are, respectively, inside and outside the unit circle. The z-transform of the ACF of the time series is

$$\mathcal{P}_{xx}(z) = \frac{\sigma^2}{\mathcal{A}(z)\mathcal{A}^*(1/z^*)}. \tag{5.49}$$

Hence, for a given PSD or equivalently for a given ACF, two distinct AR time series models are possible. The causal model given by (5.47) is termed the *forward model,* while the anticausal one of (5.48) is called the *backward model.*

If $\mathcal{P}_{xx}(z)$ is given by (5.49) but $\mathcal{A}(z)$ does not have all its zeros within the unit circle (i.e., it is not *minimum phase*), then to determine the parameters of a stable AR model from $\mathcal{P}_{xx}(z)$ one needs to perform a *spectral factorization*. The polynomial $\mathcal{A}(z)\mathcal{A}^*(1/z^*) = \sum_{k=-p}^{p} r_{aa}[k]z^{-k}$ is rooted to find the $2p$ zeros. The zeros within the unit circle are used to determine the AR parameters of the forward model, while the zeros outside the unit circle are used for the backward model. Other techniques for spectral factorization can be found in the works by Riddle and Anderson [1966], Wilson [1969], Rino [1970], Oppenheim and Schafer [1975], and Friedlander [1983]. Note that in general the forward and backward AR models are given by

$$x[n] = -\sum_{k=1}^{p} a[k]x[n-k] + u[n] \tag{5.50}$$

and

$$x[n] = -\sum_{k=1}^{p} a^*[k]x[n+k] + u[n] \tag{5.51}$$

(see Problem 5.12). The forward and backward models will be used later in discussing various AR estimation methods (see Chapter 7).

For an ARMA or MA process a similar problem occurs with respect to the determination of the MA parameters from the PSD. Considering the MA process, the z-transform of the ACF is

$$\mathcal{P}_{xx}(z) = \sigma^2\mathcal{B}(z)\mathcal{B}^*(1/z^*).$$

There are 2^q sets of MA parameters which yield the same PSD. The different sets correspond to the 2^q possible locations of the zeros. Any zero of $\mathcal{B}(z)$, call it z_i, can be replaced by $1/z_i^*$ or moved to its conjugate reciprocal location to cause a simple scaling of the PSD. The original PSD may be recovered by adjusting σ^2 (see Problem 5.13). For the MA model stability is not in question since all $\mathcal{H}(z)$ filters are stable. In many cases it is useful to choose $\mathcal{B}(z)$ as the minimum-phase filter so as to guarantee a stable and causal *inverse* filter, which is $1/\mathcal{B}(z)$. Such a choice results in an *invertible* process. Filtering by $1/\mathcal{B}(z)$ can be useful if we wish to convert an ARMA process into an AR process by filtering out the zeros as some common ARMA estimation schemes require (see Chapter 11).

In summary, if we are given a PSD or ACF, the ARMA, AR, or MA parameters are not unique but can only be made so by restricting the process to be causal and/or invertible. For causality and invertibility both $\mathcal{A}(z)$ and $\mathcal{B}(z)$ need to be minimum phase. If this is not the case, spectral factorization may be employed to convert the filters to minimum-phase form. For spectral estimation the question of whether we should constrain the parameter estimator to yield only minimum-phase filter estimates has not been satisfactorily answered. Only some of the estimators to be discussed in Chapters 7 to 11 satisfy the minimum-phase constraint.

REFERENCES

Box, G. E. P., and G. M. Jenkins, *Time Series Analysis: Forecasting and Control,* Holden-Day, San Francisco, 1970.

Friedlander, B., "A Lattice Algorithm for Factoring the Spectrum of a Moving Average Process," *IEEE Trans. Autom. Control,* Vol. AC28, pp. 1051–1055, Nov. 1983.

Gutowski, P. R., E. A. Robinson, and S. Treitel, "Spectral Estimation: Fact or Fiction," *IEEE Trans. Geosci. Electron.,* Vol. GE16, pp. 80–84, Apr. 1978.

Hsia, T. C., *System Identification,* Lexington Books, D.C. Heath, Lexington, Mass., 1977.

Kay, S., "The Effects of Noise on the Autoregressive Spectral Estimator," *IEEE Trans. Acoust. Speech Signal Process.,* Vol. ASSP27, pp. 478–485, Oct. 1979.

Kay, S., "Autoregressive Spectral Analysis of Narrowband Signals in White Noise with Application to Sonar Signals," Ph.D. dissertation, Georgia Institute of Technology, 1980.

Kay, S., "Efficient Generation of Colored Noise," *Proc. IEEE,* Vol. 69, pp. 480–481, Apr. 1981.

Kolmogorov, A. N., "Interpolation und Extrapolation von Stationären Zufälligen Folgen," *Bull. Acad. Sci. USSR Ser. Math,* Vol. 5, pp. 3–14, 1941.

Marple, S. L., Jr., original version provided in private communication, 1985.

Oppenheim, A. V., and R. W. Schafer, *Digital Signal Processing,* Prentice-Hall, Englewood Cliffs, N.J., 1975.

Riddle, A. C., and B. D. O. Anderson, "Spectral Factorization, Computational Aspects," *IEEE Trans. Autom. Control,* Vol. AC11, pp. 764–765, Oct. 1966.

Rino, C. L., "Factorization of Spectra by Discrete Fourier Transforms," *IEEE Trans. Inf. Theory,* Vol. IT16, pp. 484–485, July 1970.

Ulrych, T. J., "Maximum Entropy Power Spectrum of Truncated Sinusoids," *J. Geophys. Res.,* Vol. 77, pp. 1396–1400, Mar. 10, 1972.

Wilson, G. T., "Factorization of the Covariance Generating Function of a Pure Moving-Average Process," *SIAM J. Numer. Anal.,* Vol. 6, pp. 1–7, Mar. 1969.

Wold, H., *A Study in the Analysis of Stationary Time Series,* Almqvist & Wiksell, Stockholm, 1954.

PROBLEMS

5.1. If we wish to model the true PSD

$$P_{xx}(f) = \begin{cases} 2 & \text{for } |f| \le 0.25 \\ 0 & \text{for } 0.25 < |f| \le 0.5 \end{cases}$$

by the Gaussian PSD (5.1), then $r_{xx}[0]$ and σ_f must be estimated. Assume that enough data are available so that the ACF estimates $\hat{r}_{xx}[0]$, $\hat{r}_{xx}[1]$ are equal to the true ACF samples. Find the estimates of the unknown parameters by letting

$$\begin{aligned} r_{xx}[0] &= \hat{r}_{xx}[0] \\ r_{xx}[1] &= \hat{r}_{xx}[1] \end{aligned}$$

where $r_{xx}[0]$, $r_{xx}[1]$ are the ACF samples corresponding to the Gaussian PSD model. Plot the estimated PSD.

5.2. Consider a zero mean complex AR(1) process with parameters $a[1]$ and $\sigma^2 > 0$. Find the variance of the process for $|a[1]| < 1$ and $|a[1]| \geq 1$. Where does the pole of the AR filter have to lie for the process to be WSS?

5.3. Consider the problem of generating data from a zero mean complex AR(1) process $x[n] = -a[1]x[n-1] + u[n]$. On a computer we usually let $x[-1] = 0$ and use the difference equation for $n \geq 0$. Using this method, find the variance of $x[n]$ for $n \geq 0$. Is the process WSS? Now let $x[-1] = c$, where c is a zero mean complex random variable uncorrelated with $u[n]$ for $n \geq 0$. If the variance of c is σ_c^2, how should σ_c^2 be chosen so that $x[n]$ is WSS for $n \geq 0$? Is this plausible? (See also Appendix 5B.)

5.4. Prove that the autocorrelation matrix given in (5.18) is positive semidefinite. *Hint:* Use the result that $\text{var}(\mathbf{w}^H\mathbf{x}) \geq 0$ for any $p \times 1$ vector $\mathbf{w}$ and the zero mean vector $\mathbf{x} = [x[0]\ x[1]\ \cdots\ x[p-1]]^T$. Under what conditions will the matrix be singular [or $\text{var}(\mathbf{w}^H\mathbf{x}) = 0$]?

5.5. Give an example where $\mathbf{R}'_{xx}$ as given in (5.22) is singular. *Hint:* Let $p = q = 1$.

5.6. Find the ACF of a real AR(1) process by evaluating the inverse z-transform of

$$\mathcal{P}_{xx}(z) = \frac{\sigma^2}{(1 + a[1]z^{-1})(1 + a[1]z)}.$$

Use contour integration along the unit circle and assume that $|a[1]| < 1$. Do the results agree with (5.25)?

5.7. Find the ACF and PSD for a complex AR(1) process. How does it differ from the results for a real AR(1) process as given by (5.25) and (5.26)?

5.8. Derive the ACF for a real AR(2) process with complex poles as given by (5.27) [Kay 1980].

5.9. By the method of moments (see Section 3.3.1) the filter parameter for a real MA(1) process may be estimated as the solution of

$$\frac{\hat{r}_{xx}[1]}{\hat{r}_{xx}[0]} = \frac{b[1]}{1 + b^2[1]}$$

where $\hat{r}_{xx}[0]$, $\hat{r}_{xx}[1]$ are estimates of the ACF samples. Find the estimator of $b[1]$ under the assumption that the zero of the MA process lies within the unit circle.

5.10. The filter parameter of a real AR(1) process embedded in white noise of variance σ_w^2 is estimated using

$$\hat{a}[1] = -\frac{r_{yy}[1]}{r_{yy}[0]}$$

where $y[n]$ denotes the noise corrupted process. Show that

$$\hat{a}[1] = a[1]\frac{\eta}{\eta + 1}$$

where $\eta = r_{xx}[0]/\sigma_w^2$ is the SNR. Generalize the result to a real AR(p) process in white noise to show that

$$\hat{\mathbf{a}} = (\mathbf{R}_{xx} + \sigma_w^2\mathbf{I})^{-1}\mathbf{R}_{xx}\mathbf{a}$$

where $\mathbf{a}$ and $\hat{\mathbf{a}}$ are the true and estimated AR parameter vectors and $\mathbf{R}_{xx}$ is given in (5.18) (see also Kay [1979]).

5.11. Show that the sum of a real AR(1) process in white noise can be modeled as an ARMA(1, 1) process where the MA filter parameter is given as the solution of

$$\frac{1 + b^2[1]}{b[1]} = \frac{\sigma^2 + \sigma_w^2(1 + a^2[1])}{a[1]\sigma_w^2}.$$

What is $b[1]$ for large and small σ^2/σ_w^2?

5.12. Prove that the forward and backward AR processes given by (5.50) and (5.51) have the same PSD.

5.13. Consider the complex MA(2) process

$$x[n] = u[n] - (\tfrac{3}{2} + j)u[n-1] + \tfrac{1}{2}(1 + j)u[n-2].$$

Find the parameters of three other MA processes that have the identical PSD.

APPENDIX 5A

Computer Program to Generate Real White Gaussian Noise (Provided on Floppy Disk)

```
      SUBROUTINE WGN(N,VAR,ISEED,W)
C
C     This program generates N samples of real white Gaussian
C     noise with zero mean and a specified variance. M samples of
C     complex white Gaussian noise may be generated by calling
C     this program to generate 2*M real samples and then concatenating
C     the output samples as (W(2*I-1),W(2*I)) for I=1,2, . . . ,M.
C
C     The random number generator used in this program is an
C     intrinsic function on the DEC VAX 11/780 and requires a seed
C     number to start the generation. The seed should be a large
C     positive and odd integer such as 69069. For non-VAX
C     computers this program may be used if the random number
C     generator is replaced (See comments in code).
C
C     Input Parameters:
C
C       N       -Number of real white noise samples desired
C       VAR     -Variance of white noise samples desired
C       ISEED   -Seed number for random number generator
C
```

```
C       Output Parameters:
C
C         W         -Real array of dimension N×1 of white noise
C                    samples
C
C       External Subroutines:
C
C         RAN       -Intrinsic function on DEC VAX 11/780. For other
C                    machines replace with random number generator
C                    which generates uniformly distributed random
C                    numbers on the interval [0, 1].
C
C       Notes:
C
C         The calling program must dimension the array W.
C
C       Verification Test Case:
C
C         If N=5, VAR=2., and ISEED=69069, the output should be
C           W(1)=-1.70822, W(2)=-2.42586, W(3)=-1.59297,
C           W(4)=0.14068, W(5)=-2.16533
C         for a DEC VAX 11/780. For non-VAX machines one should
C         replace line 10 of the subroutine RANDOM to input the
C         random numbers
C           W(1)=0.11072, W(2)=0.65235, W(3)=0.52763,
C           W(4)=0.48598, W(5)=0.16694, W(6)=0.40004
C         to test the WGN subroutine.
C
          DIMENSION W(1)
          PI=4.*ATAN(1.)
C       Add one to desired number of samples if N is odd
          M=N
          IF(MOD(N,2).NE.0)M=N+1
C       Generate M independent and uniformly distributed random
C       variables on [0,1]
          CALL RANDOM(M,VAR, ISEED, W)
          L=M/2
C       Convert uniformly distributed random variables to Gaussian
C       ones using Box-Mueller transform
          DO 10 I=1,L
          U1=W(2*I-1)
          U2=W(2*I)
          TEMP=SQRT(-2.*ALOG(U1))
          W(2*I-1)=TEMP*COS(2.*PI*U2)*SQRT(VAR)
10        W(2*I)=TEMP*SIN(2.*PI*U2)*SQRT(VAR)
          RETURN
          END
          SUBROUTINE RANDOM(M,VAR,ISEED,W)
          DIMENSION W(1)
          DO 10 I=1,M
```

```
C       For machines other than DEC VAX 11/780 replace RAN(ISEED)
C       with random number generator.
10        W(I)=RAN(ISEED)
          RETURN
          END
```

APPENDIX 5B

Computer Program to Generate Time Series (Provided on Floppy Disk)

The subroutine GENDATA generates a complex AR, MA, or ARMA time series. In order to eliminate the starting transient caused by the lack of an input sequence before $k = 0$, the initial conditions of the filter are specified so as to cause the filter output to be in statistical steady state at $k = 0$. This is accomplished for an AR process by setting

$$x[k] = \begin{cases} \sqrt{\rho_0}v[0] & \text{for } k = 0 \\ -\sum_{l=1}^{k} a_k[l]x[k-l] + \sqrt{\rho_k}v[k] & \text{for } k = 1, 2, \ldots, p-1 \end{cases}$$

where $v[k]$ is a zero mean complex white noise sequence with a variance of 1. Specifically, $v[k] = v_R[k] + jv_I[k]$, where $v_R[k]$, $v_I[k]$ are each zero mean real random variables with variance of $\frac{1}{2}$ and uncorrelated with each other. (For a Gaussian time series $v_R[k]$, $v_I[k]$ for all k should be Gaussian random variables as described in Section 3.5.1.) $a_j[i]$ and ρ_j are the prediction coefficients and prediction error power for a jth-order linear predictor obtained from the given AR parameters using the step-down procedure as described in Section 6.3.4. For an ARMA process the same approach is employed followed by an FIR filtering with $\mathcal{B}(z)$. If a real time series is desired, we need only set all imaginary parts equal to zero and let $v_R[k]$ have a variance of 1. See the work by Kay [1981] for further details.

```
          SUBROUTINE GENDATA(IP,IQ,A,B,SIG2,N,V,X)
C
C         This program generates a complex AR, MA, or ARMA time series
C         given the filter parameters, excitation noise variance, and
C         a complex array of zero mean, unit variance uncorrelated
C         random variables (complex white noise). For an AR or ARMA
C         process the starting transient is eliminated because the
C         initial conditions of the filter are specified to place the
```

```
C      filter output in statistical steady state.
C
C      For a real AR, MA, or ARMA time series set the imaginary parts
C      of the filter parameters (complex arrays A,B) equal to zero
C      as well as the imaginary part of the white noise input (complex
C      array V). The real part of the samples in the V array should now
C      have a variance of one. On output take the real part of the
C      complex array X.
C
C      Input Parameters:
C
C        IP      -AR model order (for MA process IP=0)
C        IQ      -MA model order (for AR process IQ=0)
C        A       -Complex array of dimension IP×1 of AR filter parameters
C                 arranged as A(1) to A(IP)
C        B       -Complex array of dimension IQ×1 of MA filter parameters
C                 arranged as B(1) to B(IQ)
C        SIG2    -Variance of excitation noise
C        N       -Number of data points desired
C        V       -Complex array of dimension (N+IQ)×1 of white noise
C                 samples; each complex noise sample should have real (Vr[k]) and
C                 imaginary (Vi[k]) parts with zero means and variances of 1/2.
C                 The random variables {Vr[1],Vi[1], . . . ,Vr[N+IQ],Vi[N+IQ]}
C                 should all be uncorrelated with each other. For a
C                 Gaussian time series each Vr[k] and Vi[k] should be
C                 Gaussian. Program WGN may be used to generate Gaussian
C                 Vr[n],Vi[n] samples. For a real time series let Vi[n]=0
C                 and set the variance of each Vr[n] equal to one.
C
C      Output Parameters:
C
C        X       -Complex array of dimension N×1 of time series samples
C
C      External Subroutines:
C
C        STEPDOWN    -see Appendix 6C
C
C      Notes:
C
C        The calling program must dimension arrays V,X and also
C        A,B even if IP=0 or IQ=0. The arrays X1,AA,RHO must be
C        dimensioned greater than or equal to the variable dimensions
C        shown. If IP=0, set the dimensions of AA and RHO as AA(1,1)
C        and RHO(1). The array AA in STEPDOWN must be dimensioned
C        greater than or equal to (IP,IP) if IP>0 or to (1,1) if IP=0.
C        The user must insure that the AR filter parameters chosen
C        result in a stable all-pole filter.
C
C      Verification Test Case:
C
```

```
C       If IP=2, IQ=2, A(1)=(-0.0707,-1.2), A(2)=(-0.72,0.),
C       B(1)=(0.,-0.707), B(2)=(-0.25,0.), SIG2=2., N=5, and
C       V(I)=(I,I) for I=1,2, . . . ,7, then the output should be
C         X(1)=(2.43056,7.50579), X(2)=(1.79000,8.79596),
C         X(3)=(1.33120,10.18507), X(4)=(1.23115,10.7224),
C         X(5)=(2.30939,11.70134)
C       for a DEC VAX 11/780.
C
        COMPLEX A(1),B(1),V(1),X(1),X1(N+IQ),AA(IP,IP)
        DIMENSION RHO(IP)
        IF(IP.NE.0)GO TO 20
        DO 10 K=1,N+IQ
10      X1(K)=SQRT(SIG2)*V(K)
        GO TO 80
C     Set initial conditions for AR filtering
20      IF(IP.GT.1)CALL STEPDOWN(IP,A,SIG2,AA, RHO0, RHO)
        IF(IP.EQ.1)RHO0=SIG2/(1.-CABS(A(1))**2)
        X1(1)=SQRT(RHO0)*V(1)
        IF(IP.EQ.1)GO TO 50
        DO 40 K=2,IP
        X1(K)=SQRT(RHO(K-1))*V(K)
        DO 30 L=1,K-1
30      X1(K)=X1(K)-AA(L,K-1)*X1(K-L)
40      CONTINUE
C     Generate AR part of time series
50      DO 70 K=IP+1,N+IQ
        X1(K)=SQRT(SIG2)*V(K)
        DO 60 L=1,IP
60      X1(K)=X1(K)-A(L)*X1(K-L)
70      CONTINUE
80      IF(IQ.NE.0)GO TO 100
        DO 90 K=1,N
90      X(K)=X1(K)
        RETURN
C     Generate MA part of time series
100     DO 120 K=1,N
        X(K)=X1(K+IQ)
        DO 110 L=1,IQ
110     X(K)=X(K)+B(L)*X1(K+IQ-L)
120     CONTINUE
        RETURN
        END
```

APPENDIX 5C

Computer Program to Compute PSD Values (Provided on Floppy Disk)

```
      SUBROUTINE ARMAPSD(IP,IQ,A,B,SIG2,NEXP,PSD)
C
C     This subroutine computes a set of PSD values across the
C     frequency band [-1/2,1/2), given the parameters of a complex or real
C     ARMA model. The FFT is used to evaluate the numerator
C     and denominator polynomials of the ARMA PSD function. PSD(I) corre-
C     sponds to the PSD at frequency F=-1/2+(I-1)/L, where L=2**NEXP is
C     the number of frequency samples desired.
C
C     The PSD for a pure AR model may be obtained as a special case
C     by setting IQ to zero. Similarly, the PSD for a pure
C     MA model may be obtained as a special case by setting IP to zero.
C
C     For a real model set the imaginary parts of the ARMA filter
C     parameters equal to zero.
C
C     Input parameters:
C
C       IP       -AR model order (for MA process IP=0)
C       IQ       -MA model order (for AR process IQ=0)
C       A        -Complex array of dimension IPx1 of AR parameters
C                 arranged as A(1) to A(IP)
C       B        -Complex array of dimension IQx1 of MA parameters
C                 arranged as B(1) to B(IQ)
C       SIG2     -Variance of excitation noise
C       NEXP     -Power of two which determines number of frequency
C                 samples desired, L=2**NEXP (L must be greater than
C                 or equal to the maximum of IP+2 and IQ+2.)
C
C     Output parameters:
C
C       PSD      -Real array of dimension 2**NEXPx1 of PSD values
C
C     External subroutines:
C
C       FFT      -see Appendix 2A
C
C
C     Notes:
C
C       The calling program must dimension arrays PSD and A,B even
C       if IP=0 or IQ=0. The arrays DEN,XNUM,P must be dimensioned
```

```
C          to be greater than or equal to the variable dimensions shown.
C          The array Y in FFT must be dimensioned greater than or equal
C          to 2**NEXP.
C
C        Verification Test Case:
C
C          If IP=2, IQ=2, A(1)=(-0.0707,-1.2), A(2)=(-0.72,0.),
C          B(1)=(0.,-0.707), B(2)=(-0.25,0.), SIG2=2., NEXP=9,
C          then the output should be
C            PSD(100) = 0.90657
C            PSD(200) = 0.97848
C            PSD(300) = 6.04913
C            PSD(400) = 2.27597
C            PSD(500) = 1.64754
C          for a DEC VAX 11/780.
C
           COMPLEX A(1),B(1)
           COMPLEX DEN(2**NEXP),XNUM(2**NEXP)
           DIMENSION P(2**NEXP),PSD(1)
           L=2**NEXP
C        PSD computation begins
5          IF(IP.EQ.0)GO TO 30
C        Compute denominator frequency function
           DEN(1)=(1.,0.)
           DO 10 I=1,IP
10         DEN(I+1)=A(I)
           DO 20 I=IP+2,L
20         DEN(I)=(0.,0.)
           INVRS=1
           CALL FFT(DEN,NEXP,INVRS)
30         IF(IQ.EQ.0)GO TO 60
C        Compute numerator frequency function
           XNUM(1)=(1.,0.)
           DO 40 I=1,IQ
40         XNUM(I+1)=B(I)
           DO 50 I=IQ+2,L
50         XNUM(I)=(0.,0.)
           INVRS=1
           CALL FFT(XNUM,NEXP,INVRS)
C        Compute PSD values
60         IF(IP.EQ.0)GO TO 80
           IF(IQ.EQ.0)GO TO 100
           DO 70 I=1,L
70         P(I)=SIG2*(REAL(XNUM(I))**2+AIMAG(XNUM(I))**2)/
         *  (REAL(DEN(I))**2+AIMAG(DEN(I))**2)
           GO TO 120
80         DO 90 I=1,L
90         P(I)=SIG2*(REAL(XNUM(I))**2+AIMAG(XNUM(I))**2)
           GO TO 120
100        DO 110 I=1,L
```

```
110     P(I)=SIG2/(REAL(DEN(I))**2+AIMAG(DEN(I))**2)
C     Transpose halves of FFT outputs so that first PSD
C     value is at a frequency of -1/2
120     DO 130 I=1,L/2
        PSD(I+L/2)=P(I)
130     PSD(I)=P(I+L/2)
        RETURN
        END
```

Chapter 6

Autoregressive Spectral Estimation: General

6.1 INTRODUCTION

The most popular of the time series modeling approaches to spectral estimation is the AR spectral estimator. This is because accurate estimates of the AR parameters can be found by solving a set of linear equations. For accurate estimation of the parameters of ARMA or MA processes, we need to solve a set of highly nonlinear equations. When the AR modeling assumption is valid, spectral estimators are obtained which are less biased and have a lower variability than conventional Fourier based spectral estimators. Other names by which the AR spectral estimator is known are the *maximum entropy spectral estimator* and the *linear prediction spectral estimator*. Although the theoretical foundations for the latter two spectral estimators differ from those of the AR spectral estimator, in practice all the approaches are identical. The difficulty with adopting either the maximum entropy or linear prediction philosophies is that the all-pole filter assumption implicit in both approaches is not highlighted. As a consequence, application to non-AR time series usually results in poor quality spectral estimates with no clues provided as to the reasons why. The AR modeling approach is the vehicle used to describe this class of high resolution spectral estimators, although the maximum entropy and linear prediction philosophies also are discussed.

For a thorough understanding of the AR spectral estimator, a knowledge of the properties of AR processes and PSDs as well as the basic approaches to estimation of the AR parameters is needed. This chapter provides that background material. Based on the properties of AR processes and PSDs, explicit algorithms used to estimate the AR PSD are described in Chapter 7.

6.2 SUMMARY

Section 6.3 describes important properties of AR processes. The connection between linear prediction theory and AR modeling is first discussed. If $x[n]$ is an AR(p) process, the optimal one-step linear prediction coefficients are given by the solution of the Yule–Walker equations. The optimal predictor, one that minimizes the mean square error, is given by (6.1) with the coefficients given as the solution of (6.4). Hence the solution of the linear prediction equations (or Wiener–Hopf equations), as given by (6.4), will produce the AR parameters if the order of the AR process and the linear predictor are identical. By this equivalence many properties of the Yule–Walker equations are proven. Specifically, the solution produces a filter $\mathcal{A}(z)$ which is guaranteed to have its zeros inside the unit circle if the ACF samples are positive definite. Next, the important Levinson algorithm is derived using a vector space approach. The algorithm that is summarized in Figure 6.4 computes the solution of the Yule–Walker equations recursively. That is, the Yule–Walker equations of order 1 are solved and based on that solution the equations of order 2 are solved and in like fashion until the final solution is achieved. In this way AR model parameters can be found recursively for AR(1), AR(2), . . . , AR(p) models. Alternatively, the process may be viewed as one of determining the optimal first order predictor, followed by the optimal second-order predictor, and so on. The recursion is essentially a Gram–Schmidt orthogonalization of the random variables $\{x[n-1], x[n-2], \ldots, x[n-p]\}$ used to predict $x[n]$. An important practical consequence of the Levinson algorithm is that it allows one to solve the Yule–Walker equations in $O(p^2)$ operations as opposed to $O(p^3)$ operations for a general Gaussian elimination routine. A computer program entitled LEVINSON, which implements the Levinson algorithm, is provided in Appendix 6B.

The Levinson algorithm gives rise to a set of parameters known as the reflection coefficients $\{k_1, k_2, \ldots, k_p\}$. The kth reflection coefficient is shown to be the negative of the partial correlation coefficient between $x[n]$ and $x[n-k]$, with the dependence due to the intervening samples removed by linear estimation [see (6.43)]. As such the reflection coefficients are bounded by 1 in magnitude. An equivalent representation for an AR process is shown to be based on $r_{xx}[0]$ and the set of reflection coefficients as described in Figure 6.5. Conversions between these representations are based on the Levinson algorithm. For example, to find the reflection coefficients from the AR parameters, one can make use of the step-down procedure given in (6.51). A computer program to do so is described in Appendix 6C. An explicit relationship between the reflection coefficients and the ACF is given in (6.53).

Because the AR filter parameters can be expressed as a function of the reflection coefficients, the filter $\mathcal{A}(z)$ can be implemented with the reflection coefficients as its coefficients. The resulting lattice filter, which is shown in Figure 6.6, consists of p stages, each of which can be considered to be one step in the Gram–Schmidt orthogonalization. Forward prediction errors and backward prediction errors for predictors of length 1 through p are generated as a by-product of the filter structure [see (6.58)]. Additionally, the coefficients which are bounded by 1 in magnitude are easily quantized for storage or transmission.

Next, the inverse of the autocorrelation matrix for an arbitrary process is derived. The inverse, which is based on the application of the Levinson algorithm to the autocorrelation samples, is given by (6.61). Some examples for low order AR processes as well as a general formula for an AR(p) process are described in Section 6.3.6.

The AR spectral estimator is examined in Section 6.4. It is first shown that the high resolution property of the AR spectral estimator is due to an implied extrapolation of the ACF samples by the assumed AR model [see (6.68)]. This is illustrated in Figures 6.7 and 6.8. In contrast to the Fourier methods of spectral estimation, the AR spectral estimator does not assume the unknown ACF samples to be zero. The maximum entropy spectral estimator (MESE) is also based on the desire to eliminate the ACF windowing. It assumes that the ACF for $k = 0, 1, \ldots, p$ is known exactly and attempts to extrapolate the ACF for $k > p$. Since the extrapolation is not unique, the criterion chosen is to maximize the entropy of the time series characterized by the extrapolated ACF. The MESE which results from this approach is shown to be identical to the pth order AR spectral estimator. The AR parameters are found by the usual procedure of solving the Yule–Walker equations using the known ACF samples. In practice, then, there is no difference between the MESE and the AR spectral estimator.

The spectral flatness measure (6.73) is introduced next. Minimizing the power out of a prediction error filter is shown to be equivalent to maximizing the spectral flatness of the PSD at the output of the filter. A link is thus established between the use of linear prediction and spectral estimation for arbitrary processes. Unfortunately, examination of the measure reveals that the AR spectral estimator is appropriate only when the output of the prediction error filter is a white process. Otherwise, poor spectral estimates may result. Finally, bounds on the dynamic range of the AR PSD (6.77) conclude the discussion of the properties of the AR spectral estimator.

The problem of estimation of AR parameters or reflection coefficients of an AR(p) process based on a finite set of data is investigated in Section 6.5. Using some simplifying assumptions which are valid for large data records and poles not too close to the unit circle, the approximate MLEs are derived. The estimators, (6.89) and (6.91), are equivalent to solving the Yule–Walker equations with the theoretical ACF samples replaced by estimates. This estimator is sometimes referred to as the *covariance method of linear prediction*. The resulting matrix is not Toeplitz but can be made so by a slight modification. This leads to the *autocorrelation method of linear prediction* (6.95), which is efficiently implemented by the Levinson algorithm. The approximate MLEs for the reflection coefficients are found by using the values generated by the Levinson algorithm. The statistics of the AR parameter and reflection coefficient estimators are not known for finite-length data records. Based on the theory of the MLE, for large data records the estimators for real data are all distributed according to a real multivariate Gaussian PDF with means equal to the true values. The covariance matrix for the real AR parameter estimator is given by (6.97), while that for the real reflection coefficient estimator is given by (6.100).

Estimation of the AR PSD is described in Section 6.6. The statistics of the AR PSD estimator are valid only for large data records. Under the assumption that the AR model order is very large but small relative to the data record length, the estimator is Gaussian with mean equal to the true value and variance given by (6.104). For statistically reliable spectral estimates it is required that $N \gg p$. Based on maximum likelihood theory, other statistics are given which do not assume a large order AR model (6.106) and (6.107).

In Section 6.7 we discuss the effect of noise on the AR spectral estimator. The high resolution properties of the AR spectral estimator are highly dependent on the presence, if any, of observation noise. As the SNR ratio decreases, so does the resolution [see (6.108)]. Although this effect can be partially compensated for by increasing the AR model order, too large an order will cause the spectral estimate to be unreliable. Principally, spurious peaks will begin to appear (see Figure 6.11) if the model order is not very much smaller than the data record length. The effect of observation noise is to reduce the dynamic range of the estimated AR PSD. Quantitative confirmation of this behavior is obtained by proving that the spectral flatness measure of the AR spectral estimator increases in the presence of white observation noise [see (6.117)]. Further insight is gained by examining the analytical form of the AR spectral estimator for sinusoids in white noise. For one complex sinusoid in complex white noise, the AR spectral estimate is given by (6.122). It is seen that the peak of the PSD occurs at the sinusoidal frequency and has a 3 dB bandwidth which decreases with increasing model order and SNR (6.125). The peak value is proportional, however, to the square of the sinusoidal power (6.124). For multiple sinusoids the analytical form of the AR spectral estimator is extremely complicated but may be found by solving a reduced set of equations given by (6.134) and (6.135). The number of equations to be solved is equal to the number of sinusoids. Examination of the AR spectral estimate for multiple sinusoids in white noise indicates that in general the peaks are not at the sinusoidal frequency locations. Also, because of the property that the peaks are proportional to the square of the power, lower level sinusoids tend to be masked by higher level ones.

Concluding the chapter is a brief discussion of model order selection in Section 6.8. As in all spectral estimation methods a trade-off must be effected in practice between the conflicting goals of minimizing the bias and also the variance. In AR spectral estimation the trade-off is based on the selection of a model order.

6.3 PROPERTIES OF AR PROCESSES

6.3.1 Linear Prediction of AR Processes

Since the AR PSD depends on the AR parameters, insights gained by examining the properties of AR processes are valuable in interpreting the AR spectral estimator. In this section some fundamental properties of AR processes that rely on the linear prediction viewpoint are examined. In particular, an alternative set of parameters, termed the reflection coefficients, used to describe an AR process and the lattice filter, are introduced. The important Levinson algorithm, which is an efficient means for solving the Yule–Walker equations, is shown to evolve naturally from the linear prediction approach.

It is henceforth assumed that $x[n]$ is an AR(p) process. The problem of linear prediction is to predict the unobserved sample $x[n]$ based on the observed data set $\{x[n-1], x[n-2], \ldots, x[n-p]\}$ (i.e., the previous p samples). Assuming a predictor that is a linear combination of the past samples,

$$\hat{x}[n] = -\sum_{k=1}^{p} \alpha_k x[n-k] \tag{6.1}$$

the prediction coefficients $\{\alpha_1, \alpha_2, \ldots, \alpha_p\}$ are chosen to minimize the power of the prediction error $e[n]$:

$$\rho = \mathscr{E}[|e[n]|^2] = \mathscr{E}[|x[n] - \hat{x}[n]|^2]. \tag{6.2}$$

Although $x[n]$ has specifically been chosen to be predicted, the optimal prediction coefficients are independent of the value of n. This is because $x[n]$ is assumed to be WSS, so that the prediction coefficients, which will be a function of the ACF, are independent of n. Proceeding to minimize ρ by employing the orthogonality principle (see Section 3.3.3 and Problem 3.11), we have

$$\mathscr{E}[x^*[n-k](x[n] - \hat{x}[n])] = 0 \qquad k = 1, 2, \ldots, p \tag{6.3}$$

or

$$r_{xx}[k] = -\sum_{l=1}^{p} \alpha_l r_{xx}[k-l] \qquad k = 1, 2, \ldots, p. \tag{6.4}$$

The minimum prediction error power is found by making use of (6.3) to yield

$$\begin{aligned} \rho_{\mathrm{MIN}} &= \mathscr{E}[x^*[n](x[n] - \hat{x}[n])] \\ &= r_{xx}[0] + \sum_{k=1}^{p} \alpha_k r_{xx}[-k]. \end{aligned} \tag{6.5}$$

Because these equations are identical to the Yule–Walker equations for AR processes [see (5.17)], it must be true that $\alpha_k = a[k]$ for $k = 1, 2, \ldots, p$ and $\rho_{\mathrm{MIN}} = \sigma^2$. The optimal linear prediction coefficients are just the AR parameters, and the resulting minimum prediction error power is just the excitation noise variance. This will only be true, however, if the *order of the AR process and the order of the linear predictor are identical.* The equations given by (6.4) are called the Wiener–Hopf equations in the theory of linear prediction [Papoulis 1965]. Furthermore, the prediction error $e[n]$ is just $u[n]$ since

$$e[n] = x[n] - \hat{x}[n] = x[n] - \left[- \sum_{k=1}^{p} \alpha_k x[n-k] \right]$$

$$= x[n] + \sum_{k=1}^{p} a[k]x[n-k]$$

$$= u[n]. \tag{6.6}$$

If we now attempt to predict $x[n]$ on the basis of the previous p samples for successive values of n, a filtering interpretation is possible, as shown in Figure 6.1. The optimal prediction error filter (PEF), which produces the prediction error at the output for $x[n]$ at the input, is the inverse filter $\mathcal{A}(z) = 1 + \sum_{k=1}^{p} a[k]z^{-k}$. Clearly, the prediction error time series is $u[n]$, so that the PEF may be viewed as a whitening filter. In fact, with these observations an AR(p) process can alternatively be defined as

$$x[n] = \hat{x}[n] + u[n] \tag{6.7}$$

where $\hat{x}[n] = -\sum_{k=1}^{p} a[k]x[n-k]$ is the optimal one-step linear prediction based on the previous p samples (see also Problems 6.1 and 6.2).

6.3.2 Minimum-Phase Property of Prediction Error Filter

In defining the AR process it has been assumed that all the poles of $1/\mathcal{A}(z)$ are inside the unit circle. This condition is necessary to ensure that $x[n]$ is a WSS process. Indeed, if any pole is on or outside the unit circle, the variance of $x[n]$

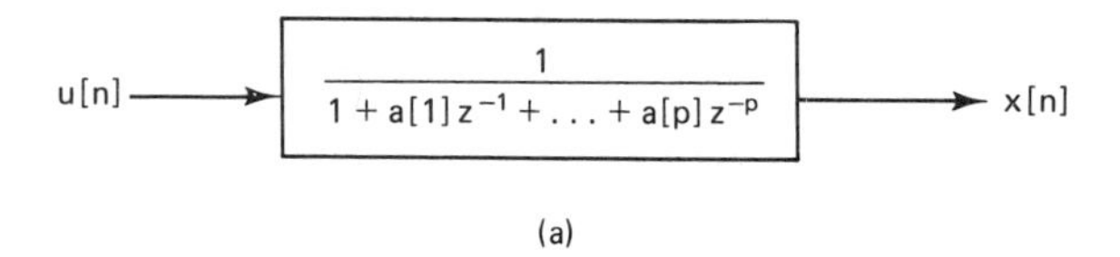

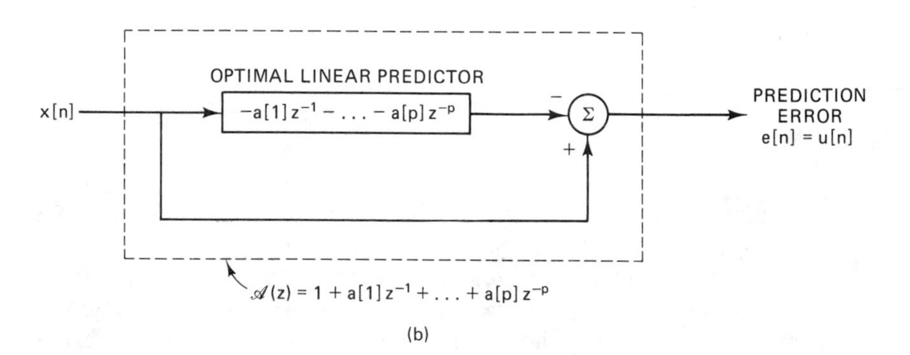

Figure 6.1 Filtering interpretation of linear prediction. (a) AR model of order p. (b) Prediction error filter.

will be infinite. On the other hand, if the AR parameters are obtained by solving the Yule–Walker equations, it is not obvious that the poles will be inside the unit circle. That the poles are guaranteed to be inside the unit circle follows from the observation that the optimal pth order linear prediction coefficients are identical to the AR parameters. With the latter result it is now shown that the solution of the Yule–Walker equations yields a stable all-pole filter $1/\mathcal{A}(z)$ or a minimum-phase $\mathcal{A}(z)$ if the autocorrelation sequence $\{r_{xx}[0], r_{xx}[1], \ldots, r_{xx}[p]\}$ is a valid one. By valid it is meant that the $(p+1) \times (p+1)$ autocorrelation matrix

$$\mathbf{R}_{xx}^{(p+1)} = \begin{bmatrix} r_{xx}[0] & r_{xx}[-1] & \cdots & r_{xx}[-p] \\ r_{xx}[1] & r_{xx}[0] & \cdots & r_{xx}[-(p-1)] \\ \vdots & \vdots & \ddots & \vdots \\ r_{xx}[p] & r_{xx}[p-1] & \cdots & r_{xx}[0] \end{bmatrix} \tag{6.8}$$

is a positive definite matrix. Although many proofs of the minimum-phase property exist [Grenander and Szego 1958, Burg 1975, Markel and Gray 1976, Lang and McClellan 1979], the one that follows is perhaps the simplest [Pakula and Kay 1983].

Because the solution of the Yule–Walker equations yields the optimal one step linear predictor for an AR(p) process, the solution minimizes

$$\begin{aligned} \rho &= \mathcal{E}[\,|\, x[n] - \hat{x}[n] \,|^2] \\ &= \mathcal{E}\left[\left| \sum_{k=0}^{p} \alpha_k x[n-k] \right|^2\right] \end{aligned} \tag{6.9}$$

where $\alpha_0 = 1$. The minimum prediction error power ρ_{MIN} can be written as

$$\rho_{\text{MIN}} = \int_{-1/2}^{1/2} |\, \mathcal{A}(\exp[j2\pi f]) \,|^2 \, P_{xx}(f)\, df \tag{6.10}$$

where

$$\mathcal{A}(\exp[j2\pi f]) = 1 + \sum_{k=1}^{p} a[k] \exp(-j2\pi f k)$$

and the $a[k]$'s are the optimal linear prediction coefficients found by solving the Yule–Walker equations. This follows by observing that (6.9) is the power at the output of the PEF with $x[n]$ at the input. It is now shown that if any zero is outside the unit circle or $|\, z_i \,| > 1$, then ρ_{MIN} can be decreased by replacing z_i by $1/z_i^*$. Hence ρ_{MIN} could not have been the minimum prediction error power and $\mathcal{A}(z)$ could not have been the optimal PEF. By contradiction, then, $\mathcal{A}(z)$ must have all its zeros on or within the unit circle. Assume that $|\, z_i \,| > 1$ for some i. Then

$$\begin{aligned} \mathcal{A}(z) &= \prod_{j=1}^{p} (1 - z_j z^{-1}) = (1 - z_i z^{-1}) \prod_{\substack{j=1 \\ j \neq i}}^{p} (1 - z_j z^{-1}) \\ &= (1 - z_i z^{-1}) \mathcal{A}'(z). \end{aligned}$$

Using this in (6.10) yields

$$\rho_{\text{MIN}} = \int_{-1/2}^{1/2} | 1 - z_i \exp(-j2\pi f) |^2 \, | \mathcal{A}'(\exp(j2\pi f)) |^2 \, P_{xx}(f) \, df.$$

The integrand may be reduced for all f by replacing the zero of $\mathcal{A}(z)$ outside the unit circle by one at the conjugate reciprocal location (i.e., at $1/z_i^*$). This follows from

$$\begin{aligned} | 1 - z_i \exp(-j2\pi f) |^2 &= | z_i |^2 \left| \frac{1}{z_i} - \exp(-j2\pi f) \right|^2 \\ &= | z_i |^2 \left| \exp(j2\pi f) - \frac{1}{z_i^*} \right|^2 \\ &= | z_i |^2 \left| 1 - \frac{1}{z_i^*} \exp(-j2\pi f) \right|^2 \\ &> \left| 1 - \frac{1}{z_i^*} \exp(-j2\pi f) \right|^2 \end{aligned}$$

since $| z_i | > 1$ by assumption—hence the assumption that $| z_i | > 1$ leads to a contradiction so that all zeros of the PEF must be on or within the unit circle. It can further be shown [Burg 1975] that if $\mathbf{R}_{xx}^{(p+1)}$ as given by (6.8) is positive definite, then all the zeros must be *within* the unit circle. If, however, $\mathbf{R}_{xx}^{(p)}$ is positive definite but $\mathbf{R}_{xx}^{(p+1)}$ is singular (and hence $\mathbf{R}_{xx}^{(p+1)}$ is only positive semidefinite), the PEF will have all its zeros *on* the unit circle. The latter will occur if $x[n]$ consists of p sinusoids (see Problem 6.3). In this case, $x[n]$ is perfectly predictable and hence $\rho_{\text{MIN}} = 0$. More generally, if the data consist of $k \leq p$ sinusoids, the minimum prediction error power will be zero. To see this, note that from (6.10) if

$$P_{xx}(f) = \sum_{i=1}^{k} P_i \delta(f - f_i) \tag{6.11}$$

then

$$\rho_{\text{MIN}} = \sum_{i=1}^{k} P_i \, | \mathcal{A}(\exp(j2\pi f_i)) |^2. \tag{6.12}$$

ρ_{MIN} can be made to be zero for $k \leq p$ if $\mathcal{A}(z)$ has zeros on the unit circle at $f = f_1, f_2, \ldots, f_k$. Because $\rho_{\text{MIN}} \geq 0$, $\mathcal{A}(z)$ must then be the optimal PEF. For fewer than p sinusoids ($k < p$), the remaining zeros $\{z_{k+1}, z_{k+2}, \ldots, z_p\}$ can be located arbitrarily while still maintaining $\rho_{\text{MIN}} = 0$. As a result, the prediction coefficients are not unique or equivalently the Wiener–Hopf or Yule–Walker equations have an infinite number of solutions.

The practical consequences of the minimum-phase property is that if the biased estimator of the ACF is used in the Yule–Walker equations, the estimated AR parameters will be minimum-phase. This method of estimation is termed the *autocorrelation method of linear prediction*. It is discussed more fully in Section 6.5.1 and in Chapter 7. Whether AR parameter estimators should be constrained to produce minimum-phase estimates is still in question. It may be argued that by allowing the AR parameter estimator to produce non-minimum-phase estimates, we increase the variance of the estimator since the known constraint is not incorporated. In applications other than spectral estimation, the minimum-phase constraint is important to ensure the stability of the estimated all-pole filter (see Chapter 16).

6.3.3 The Levinson Algorithm

The solution of the Yule–Walker equations for an AR(p) process was shown to produce the optimal one-step linear prediction coefficients. One can use any standard method to solve the set of linear equations. For instance, Gaussian elimination could be used but would require $O(p^3)$ operations. The Yule–Walker equations, however, are a special set of equations which can be solved in $O(p^2)$ operations by the Levinson algorithm. Although appearing at first to be just an efficient computational algorithm, it in fact reveals fundamental properties of AR processes. The concepts of reflection coefficient representations and lattice filters all have their origins in the Levinson algorithm. To make these connections apparent, it becomes necessary to employ a vector space approach [Narasimha et al. 1974] to optimal prediction. Although somewhat lengthy, the vector space development that follows provides valuable insights into the prediction problem and hence AR modeling. The important results of this section have been summarized in Section 6.2.

The Yule–Walker or Wiener–Hopf equations are now rederived using a vector space viewpoint. Let the linear vector space be composed of random variables with zero mean. The inner product is defined as

$$\langle x, y \rangle = \mathscr{E}(x^* y) \tag{6.13}$$

so that the squared norm of a vector is

$$\| x \|^2 = \langle x, x \rangle = \mathscr{E}(| x |^2) = \text{var}(x). \tag{6.14}$$

The linear prediction problem is to find the optimal set of coefficients $\{a[1], a[2], \ldots, a[p]\}$ such that

$$\hat{x}[n] = -\sum_{k=1}^{p} a[k]x[n-k] \tag{6.15}$$

is the "best" predictor of $x[n]$ given $\{x[n-1], x[n-2], \ldots, x[n-p]\}$. In anticipation of the result that the linear prediction coefficients are equal to the AR(p) parameters, $a[k]$ has been used to denote the prediction coefficients. "Best" means that the mean square error

$$\rho = \mathscr{E}(| x[n] - \hat{x}[n] |^2) = \| x[n] - \hat{x}[n] \|^2 \tag{6.16}$$

is minimized. By the orthogonality principle (see Section 3.3.3) the optimal predictor is found by requiring the error vector $x[n] - \hat{x}[n]$ to be orthogonal to the subspace spanned by $\{x[n-1], x[n-2], \ldots, x[n-p]\}$ or

$$\langle x[n-k], x[n] - \hat{x}[n] \rangle = 0 \qquad k = 1, 2, \ldots, p. \tag{6.17}$$

By using (6.15) in (6.17) and standard properties of inner products, we obtain

$$\langle x[n-k], x[n] + \sum_{l=1}^{p} a[l]x[n-l] \rangle = 0$$

$$\sum_{l=1}^{p} a[l] \langle x[n-k], x[n-l] \rangle = -\langle x[n-k], x[n] \rangle.$$

Evaluating the inner products as

$$\sum_{l=1}^{p} a[l]\mathscr{E}(x^*[n-k]x[n-l]) = -\mathscr{E}(x^*[n-k]x[n])$$

results in

$$\sum_{l=1}^{p} a[l]r_{xx}[k-l] = -r_{xx}[k] \qquad k = 1, 2, \ldots, p. \tag{6.18}$$

To find the minimum prediction error power, we begin with

$$\begin{aligned} \rho_{\text{MIN}} &= \langle x[n] - \hat{x}[n], x[n] - \hat{x}[n] \rangle \\ &= \langle x[n], x[n] - \hat{x}[n] \rangle - \langle \hat{x}[n], x[n] - \hat{x}[n] \rangle. \end{aligned}$$

But $\langle \hat{x}[n], x[n] - \hat{x}[n] \rangle = 0$ from (6.17), so that

$$\begin{aligned} \rho_{\text{MIN}} &= \langle x[n], x[n] \rangle - \langle x[n], \hat{x}[n] \rangle \\ &= \langle x[n], x[n] \rangle + \sum_{k=1}^{p} a[k] \langle x[n], x[n-k] \rangle \\ &= r_{xx}[0] + \sum_{k=1}^{p} a[k]r_{xx}[-k]. \end{aligned} \tag{6.19}$$

Equations (6.18) and (6.19) are the Yule–Walker equations. The solution of (6.18) provides the optimal set of coefficients to predict $x[n]$ as a linear combination of $\{x[n-1], x[n-2], \ldots, x[n-p]\}$ (i.e., the optimal pth order linear predictor). If we wish to determine not only the pth order linear predictor but also the linear predictors of orders $p-1, p-2, \ldots, 1$, one possibility is to solve (6.18) for the various assumed model orders. The result will be sets of prediction coefficients $\{a_1[1]\}, \{a_2[1], a_2[2]\}, \ldots, \{a_p[1], a_p[2], \ldots, a_p[p]\}$, where $a_j[i]$ is the ith coefficient of the jth order linear predictor. Clearly, $a_p[i] = a[i]$ for $i = 1, 2, \ldots, p$. This procedure, although straightforward, proves to be computationally burdensome and is altogether unnecessary. An alternative approach is to recursively update the predictor of order $k-1$ to order k. This requires that we perform a Gram–Schmidt orthogonalization of the data $\{x[n-1], x[n-2], \ldots, x[n -$

$p]\}$ into orthogonal or uncorrelated random variables [Luenberger 1969]. To see how this is done, let $\hat{x}_{k-1}[n]$ be the optimal $(k-1)$st order linear predictor of $x[n]$ based on the previous $k-1$ samples or

$$\hat{x}_{k-1}[n] = -\sum_{i=1}^{k-1} a_{k-1}[i]x[n-i]. \tag{6.20}$$

The subscript on $\hat{x}[n]$ indicates the number of previous samples used in the prediction. Consider a first order linear predictor so that $k-1=1$. Then

$$\hat{x}_1[n] = -a_1[1]x[n-1] \tag{6.21}$$

and $a_1[1]$ is found by minimizing

$$\rho_1 = \| x[n] - \hat{x}_1[n] \|^2. \tag{6.22}$$

The solution, depicted geometrically in Figure 6.2a, can be obtained using the orthogonality principle as

$$\langle x[n-1], x[n] - \hat{x}_1[n]\rangle = 0$$

which yields

$$a_1[1] = -\frac{\langle x[n-1], x[n]\rangle}{\langle x[n-1], x[n-1]\rangle} \tag{6.23}$$

so that

$$\hat{x}_1[n] = \frac{\langle x[n-1], x[n]\rangle}{\langle x[n-1], x[n-1]\rangle} x[n-1]. \tag{6.24}$$

Now let

$$\bar{e}_0^b[n-1] = \frac{x[n-1]}{\| x[n-1] \|}. \tag{6.25}$$

$\bar{e}_0^b[n-1]$ is a zero-mean random variable with the overbar denoting that it is also unit variance. An overbar will henceforth denote a random variable that is "normalized" or has unit variance. The "0" subscript and "b" superscript will be explained shortly. The optimal first order predictor then becomes, from (6.24),

$$\begin{aligned}\hat{x}_1[n] &= \frac{\langle x[n-1], x[n]\rangle}{\| x[n-1] \|} \frac{x[n-1]}{\| x[n-1] \|} \\ &= \langle \bar{e}_0^b[n-1], x[n]\rangle\, \bar{e}_0^b[n-1].\end{aligned} \tag{6.26}$$

It is seen that the optimal first order linear predictor is found by projecting $x[n]$ along the $x[n-1]$ "direction," where the "unit vector" along the $x[n-1]$ direction is $\bar{e}_0^b[n-1]$.

Now consider a second order or updated linear predictor with $k-1=2$:

$$\hat{x}_2[n] = -a_2[1]x[n-1] - a_2[2]x[n-2].$$

Referring to Figure 6.2b, we observe that $x[n-2]$ is in general not orthogonal

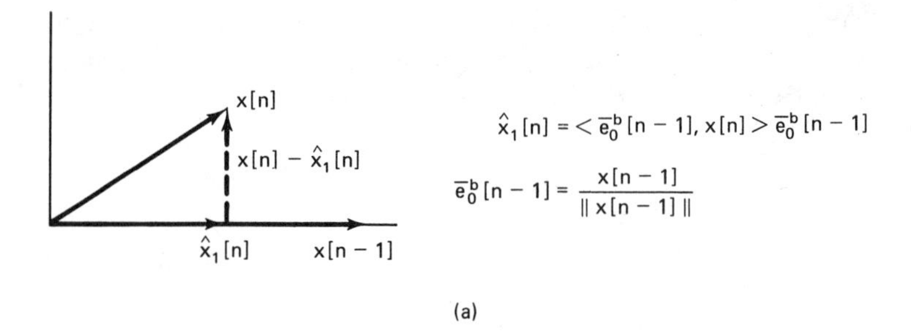

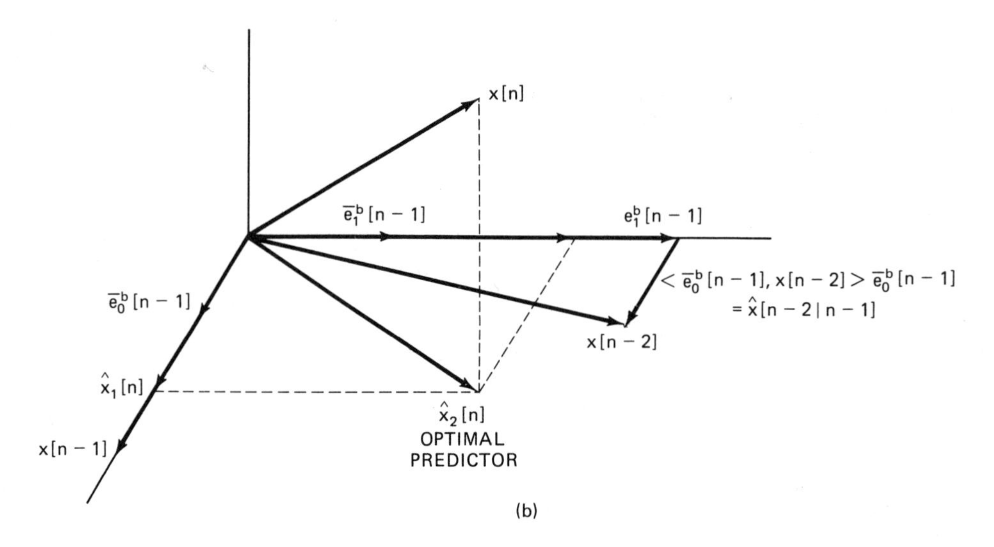

Figure 6.2 Vector space interpretation of linear prediction. (a) First order predictor. (b) Second order predictor.

to $x[n-1]$. This means that $x[n-2]$ is correlated with $x[n-1]$, so that not all of the information provided by $x[n-2]$ about $x[n]$ is new information. The optimal predictor $\hat{x}_2[n]$ can be decomposed into the sum of two vectors in orthogonal directions. One of the directions has already been specified by $x[n-1]$. The second direction will be that which is orthogonal to $x[n-1]$. The optimal second order predictor combines the first order predictor with the best prediction of $x[n]$ based on the *new information* provided by $x[n-2]$, or

$$\hat{x}_2[n] = \hat{x}_1[n] + \text{best prediction of } x[n] \text{ based on part of } x[n-2] \text{ in new orthogonal direction.} \tag{6.27}$$

To find the new information or part of $x[n-2]$ orthogonal to $x[n-1]$, recall that if we "predict" $x[n-2]$ based on $x[n-1]$, the error will be orthogonal to $x[n-1]$. Let $\hat{x}[n-2 \mid n-1]$ be the prediction of $x[n-2]$ based on $x[n-1]$

and let $e_1^b[n-1]$ be the error. Then, referring to Figure 6.2b, we have

$$\begin{aligned} e_1^b[n-1] &= x[n-2] - \hat{x}[n-2 \mid n-1] \\ &= x[n-2] - \langle \bar{e}_0^b[n-1], x[n-2] \rangle \, \bar{e}_0^b[n-1]. \end{aligned} \tag{6.28}$$

$\hat{x}[n-2 \mid n-1]$ is called the *backward* prediction since it is an estimate of $x[n-2]$ based on the future sample $x[n-1]$. Also, $e_1^b[n-1]$ is called the backward prediction error of order 1. The subscript denotes the order of the prediction error or number of future samples used in the prediction. The b subscript has been added to distinguish it from the usual *forward* prediction error. The $(k-1)$st order forward prediction error, $x[n] - \hat{x}_{k-1}[n]$, will be denoted by $e_{k-1}^f[n]$. From Figure 6.2b, $e_1^b[n-1]$ is orthogonal to $x[n-1]$ and so represents the new information in $x[n-2]$ about $x[n]$ not already provided by $x[n-1]$. For this reason $e_1^b[n-1]$ is sometimes referred to as the *innovation*.

If $e_1^b[n-1]$ is normalized,

$$\bar{e}_1^b[n-1] = \frac{e_1^b[n-1]}{\| e_1^b[n-1] \|} \tag{6.29}$$

then, from (6.27),

$$\hat{x}_2[n] = \hat{x}_1[n] + \langle \bar{e}_1^b[n-1], x[n] \rangle \, \bar{e}_1^b[n-1]. \tag{6.30}$$

The evaluation of the inner products will produce the equivalent form

$$\hat{x}_2[n] = -a_2[1]x[n-1] - a_2[2]x[n-2].$$

In general, if $\hat{x}_{k-1}[n]$ has been found, then $\hat{x}_k[n]$ is given by

$$\hat{x}_k[n] = \hat{x}_{k-1}[n] + \langle \bar{e}_{k-1}^b[n-1], x[n] \rangle \, \bar{e}_{k-1}^b[n-1] \tag{6.31}$$

where $e_{k-1}^b[n-1]$ is the backward prediction error if $x[n-k]$ is predicted on the basis of $\{x[n-(k-1)], x[n-(k-2)], \ldots, x[n-1]\}$. See Figure 6.3 for

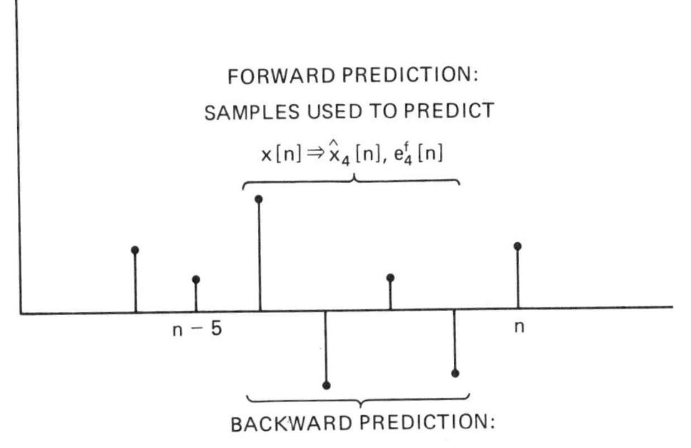

Figure 6.3 Illustration of forward and backward prediction.

an illustration. Note that $e^b_{k-1}[n-1]$ has been defined so that the time index $n-1$ refers to the latest sample used in the prediction, *not* to the sample to be "predicted." Using (6.20), the kth order predictor becomes (see also Problem 6.4)

$$\hat{x}_k[n] = -\sum_{i=1}^{k-1} a_{k-1}[i]x[n-i] + \frac{\langle e^b_{k-1}[n-1], x[n]\rangle}{\|e^b_{k-1}[n-1]\|^2} e^b_{k-1}[n-1]. \quad (6.32)$$

Now let

$$k_k = -\frac{\langle e^b_{k-1}[n-1], x[n]\rangle}{\| e^b_{k-1}[n-1]\|^2} \quad (6.33)$$

where k_k is termed the kth *reflection coefficient*. The backward prediction error may be written explicitly as

$$e^b_{k-1}[n-1] = x[n-k] - \left[-\sum_{i=0}^{k-2} b_{k-1}[i]x[n-1-i]\right] \quad (6.34)$$

where $b_{k-1}[i]$ are the optimal backward prediction coefficients. If we define $b_{k-1}[k-1] = 1$, the backward prediction error becomes

$$e^b_{k-1}[n-1] = \sum_{i=0}^{k-1} b_{k-1}[i]x[n-1-i]. \quad (6.35)$$

Substituting (6.33) and (6.35) into (6.32) results in

$$\hat{x}_k[n] = -\sum_{i=1}^{k-1} a_{k-1}[i]x[n-i] - k_k \sum_{i=0}^{k-1} b_{k-1}[i]x[n-1-i] \quad (6.36)$$

which must be identical to

$$\hat{x}_k[n] = -\sum_{i=1}^{k} a_k[i]x[n-i].$$

The optimal prediction coefficients for the kth order predictor will be the sum of the coefficients for the $(k-1)$st order predictor and a correction term due to $k_k b_{k-1}[i]$.

Before proceeding further it is of interest to examine the relationship between the backward prediction, which is based on the $k-1$ future samples, and the forward prediction, which is based on the $k-1$ past samples, an example of which is shown in Figure 6.3. Considering $n = 0$ for simplicity, in forward prediction we predict $x[0]$ based on $\{x[-1], x[-2], \ldots, x[-(k-1)]\}$, while in backward prediction we predict $x[-k]$ based on $\{x[-(k-1)], x[-(k-2)], \ldots, x[-1]\}$. The two problems are nearly equivalent except for the reversal of time, so that it is not surprising that the optimal backward prediction coefficients are the same as the optimal forward prediction coefficients except *reversed in time and complex conjugated*. This relationship is also apparent if we consider the coefficients of the forward and backward AR models discussed in Section 5.8.

Specifically, it is now shown that

$$b_{k-1}[i] = a^*_{k-1}[k - 1 - i] \qquad i = 0, 1, \ldots, k - 1. \tag{6.37}$$

The optimal backward predictor coefficients are given by the orthogonality principle as the solution of the linear equations

$$\langle x[n - j], \sum_{i=0}^{k-1} b_{k-1}[i]x[n - 1 - i]\rangle = 0 \qquad j = 1, 2, \ldots, k - 1$$

or

$$\sum_{i=0}^{k-1} b_{k-1}[i]r_{xx}[j - i - 1] = 0.$$

Letting $l = k - 1 - i$ in the summation produces

$$\sum_{l=0}^{k-1} b_{k-1}[k - 1 - l]r_{xx}[j - k + l] = 0$$

or taking the complex-conjugate yields

$$\sum_{l=0}^{k-1} b^*_{k-1}[k - 1 - l]r_{xx}[k - j - l] = 0 \qquad j = 1, 2, \ldots, k - 1.$$

Finally, these equations may be rewritten as the Yule–Walker equations,

$$\sum_{l=0}^{k-1} b^*_{k-1}[k - 1 - l]r_{xx}[m - l] = 0 \qquad m = 1, 2, \ldots, k - 1$$

so that

$$a_{k-1}[l] = b^*_{k-1}[k - 1 - l] \qquad l = 0, 1, \ldots, k - 1$$

from which (6.37) follows.

Next, an explicit form for the kth order prediction coefficients is found. Using (6.36) and (6.37) gives us

$$\begin{aligned}\hat{x}_k[n] &= -\sum_{i=1}^{k-1} a_{k-1}[i]x[n - i] - k_k\left(x[n - k]\right. \\ &\qquad \left. + \sum_{i=0}^{k-2} a^*_{k-1}[k - 1 - i]x[n - 1 - i]\right) \\ &= -\sum_{i=1}^{k-1} (a_{k-1}[i] + k_k a^*_{k-1}[k - i])x[n - i] - k_k x[n - k].\end{aligned}$$

Because this is the optimal kth order predictor, it may also be expressed as

$$\hat{x}_k[n] = -\sum_{i=1}^{k} a_k[i]x[n - i]$$

and consequently, equating the two expressions yields

$$a_k[i] = \begin{cases} a_{k-1}[i] + k_k a^*_{k-1}[k-i] & i = 1, 2, \ldots, k-1 \\ k_k & i = k. \end{cases} \tag{6.38}$$

Equations (6.38) are the model order update relations for the prediction coefficients for the kth order predictor. Note that the new coefficients are computed recursively based on the coefficients of the previous lower order predictor and the new reflection coefficient. Also, the $a_k[k]$ coefficient is just the reflection coefficient.

To complete the recursion of (6.38), we need to compute the reflection coefficient sequence. From (6.33),

$$k_k = -\frac{\langle e^b_{k-1}[n-1], x[n]\rangle}{\| e^b_{k-1}[n-1] \|^2}.$$

The backward prediction error is, from (6.35) and (6.37),

$$e^b_{k-1}[n-1] = x[n-k] + \sum_{i=1}^{k-1} a^*_{k-1}[i]x[n-k+i] \tag{6.39}$$

so that

$$\langle e^b_{k-1}[n-1], x[n]\rangle = r_{xx}[k] + \sum_{i=1}^{k-1} a_{k-1}[i]r_{xx}[k-i].$$

Also, $\| e^b_{k-1}[n-1] \|^2 = \langle e^b_{k-1}[n-1], e^b_{k-1}[n-1]\rangle = \langle e^b_{k-1}[n-1], x[n-k]\rangle$ since the backward prediction error is orthogonal to the data. Evaluation of the inner products produces

$$\| e^b_{k-1}[n-1] \|^2 = r_{xx}[0] + \sum_{i=1}^{k-1} a_{k-1}[i]r_{xx}[-i].$$

The reflection coefficient is found to be

$$k_k = -\frac{r_{xx}[k] + \sum_{i=1}^{k-1} a_{k-1}[i]r_{xx}[k-i]}{r_{xx}[0] + \sum_{i=1}^{k-1} a_{k-1}[i]r_{xx}[-i]}. \tag{6.40}$$

Note that the reflection coefficients depend on the ACF as well as the lower order PEFs.

The interpretation of k_k is as a correlation coefficient. The first step in making this correspondence is to realize that $\| e^b_{k-1}[n-1] \|^2$, which is the prediction error power for the $(k-1)$st order backward predictor, is the same as that for the $(k-1)$st order forward predictor. That is (see Problem 6.5),

$$\| e^b_{k-1}[n-1] \|^2 = \| e^f_{k-1}[n-1] \|^2 = \| e^f_{k-1}[n] \|^2 = \rho_{k-1}. \tag{6.41}$$

This is a consequence of the hermitian symmetry property of the ACF of a WSS process. With this result we can show that the reflection coefficients are actually the negated correlation coefficients between the forward and backward prediction errors or between $e^f_{k-1}[n]$ and $e^b_{k-1}[n-1]$. Since $e^b_{k-1}[n-1]$ is the prediction error based on $x[n-(k-1)], x[n-(k-2)], \ldots, x[n-1]$,

$$\langle e^b_{k-1}[n-1], x[n-i]\rangle = 0 \qquad i = 1, 2, \ldots, k-1$$

which allows us to write

$$\begin{aligned}\langle e^b_{k-1}[n-1], x[n]\rangle &= \langle e^b_{k-1}[n-1], x[n] + \sum_{i=1}^{k-1} a_{k-1}[i]x[n-i]\rangle \\ &= \langle e^b_{k-1}[n-1], e^f_{k-1}[n]\rangle.\end{aligned}$$

Hence, from (6.33) and (6.41),

$$\begin{aligned}k_k &= -\frac{\langle e^b_{k-1}[n-1], e^f_{k-1}[n]\rangle}{\| e^f_{k-1}[n] \| \, \| e^b_{k-1}[n-1] \|} \\ &= -\frac{\text{cov}\,(e^b_{k-1}[n-1], e^f_{k-1}[n])}{\sqrt{\text{var}\,(e^f_{k-1}[n])}\sqrt{\text{var}\,(e^b_{k-1}[n-1])}}\end{aligned} \tag{6.42}$$

where cov denotes the covariance. k_k is bounded by 1 in magnitude by the Cauchy–Schwartz inequality. The reflection coefficient is readily seen to be the negative of the correlation coefficient between the forward and backward prediction errors. Also, it may be interpreted as the negative of the *partial* correlation (PARCOR) coefficient [Kendall and Stuart 1979] between $x[n-k]$ and $x[n]$, or

$$k_k = -\gamma_{x[n-k]x[n]\cdot x[n-(k-1)],\ldots,x[n-1]}. \tag{6.43}$$

The partial correlation coefficient between two random variables x_1 and x_3 both of which are correlated with a third random variable x_2 is denoted as $\gamma_{x_1x_3\cdot x_2}$. It is defined as the usual correlation coefficient between $\tilde{x}_1$ and $\tilde{x}_3$, where $\tilde{x}_1 = x_1 - ax_2$ is the error if x_1 is linearly estimated from x_2 and $\tilde{x}_3 = x_3 - bx_2$ is the error if x_3 is linearly estimated from x_2. $\tilde{x}_1, \tilde{x}_3$ are the components of x_1, x_3 which are uncorrelated with x_2. We may thus interpret the partial correlation coefficient as the correlation between x_1 and x_3 after the correlation due to x_2 has been removed. Applying this definition to the negative of the kth reflection coefficient, the latter is seen to be the correlation coefficient between two samples of a time series separated by k samples after the correlation due to the intervening $k-1$ samples has been removed.

Finally, to complete the development of the Levinson algorithm, the simple recursive expression for ρ_k, the prediction error power for the kth order linear predictor,

$$\rho_k = (1 - |k_k|^2)\rho_{k-1} \tag{6.44}$$

is derived. From (6.31) and (6.33) the kth order linear predictor may be written as

$$\hat{x}_k[n] = \hat{x}_{k-1}[n] - k_k e^b_{k-1}[n-1].$$

Adding $-x[n]$ to both sides of this expression produces

$$e_k^f[n] = e_{k-1}^f[n] + k_k e_{k-1}^b[n-1]. \qquad (6.45)$$

Use standard properties of inner products and (6.41), we have

$$\begin{aligned}\rho_k &= \| e_k^f[n] \|^2 \\ &= \langle e_{k-1}^f[n] + k_k e_{k-1}^b[n-1], e_{k-1}^f[n] + k_k e_{k-1}^b[n-1]\rangle \\ &= \rho_{k-1} + k_k \langle e_{k-1}^f[n], e_{k-1}^b[n-1]\rangle \\ &\quad + k_k^* \langle e_{k-1}^b[n-1], e_{k-1}^f[n]\rangle + | k_k |^2 \rho_{k-1}.\end{aligned}$$

But from (6.41) and (6.42) $\langle e_{k-1}^b[n-1], e_{k-1}^f[n]\rangle = -k_k \rho_{k-1}$, which upon substitution in the equation above results in (6.44). As expected, the prediction error power decreases as the order of the predictor increases (assuming that $k_k \neq 0$) and is nonnegative since $| k_k | < 1$.

In summary, the Levinson algorithm recursively computes the parameter sets $\{a_1[1], \rho_1\}$, $\{a_2[1], a_2[2], \rho_2\}$, . . . , $\{a_p[1], a_p[2], \ldots, a_p[p], \rho_p\}$. The final set at order p is the desired solution of the Yule–Walker equations. If $x[n]$ is an AR(p) process, then $a_p[i] = a[i]$ for $i = 1, 2, \ldots, p$ and $\rho_p = \sigma^2$ as described in Section 6.3.1. The recursive algorithm is initialized by

$$a_1[1] = -\frac{r_{xx}[1]}{r_{xx}[0]}$$

$$\rho_1 = (1 - | a_1[1] |^2) r_{xx}[0]$$

with the recursion for $k = 2, 3, \ldots, p$ given by

$$a_k[k] = -\frac{r_{xx}[k] + \sum_{l=1}^{k-1} a_{k-1}[l] r_{xx}[k-l]}{\rho_{k-1}} \qquad (6.46)$$

$$a_k[i] = a_{k-1}[i] + a_k[k] a_{k-1}^*[k-i] \qquad i = 1, 2, \ldots, k-1 \qquad (6.47)$$

$$\rho_k = (1 - | a_k[k] |^2) \rho_{k-1}. \qquad (6.48)$$

The reflection coefficients are given by $k_k = a_k[k]$. The Levinson algorithm is summarized in Figure 6.4. The form of the algorithm given in (6.46)–(6.48) is due to Levinson [1947], who formulated an efficient means of solving a hermitian Toeplitz set of equations, and to Durbin [1960], who refined the algorithm to take advantage of the special form of the right-hand-side vector. A computer program entitled LEVINSON can be found in Appendix 6B to implement the Levinson algorithm. It is important to note that $\{a_j[1], a_j[2], \ldots, a_j[j], \rho_j\}$, as obtained from the Levinson algorithm, is the same as would be obtained by solving (6.4) and (6.5) with $p = j$. The algorithm provides the AR parameters for all lower order AR model fits to the ACF as well as the desired model. This is a useful

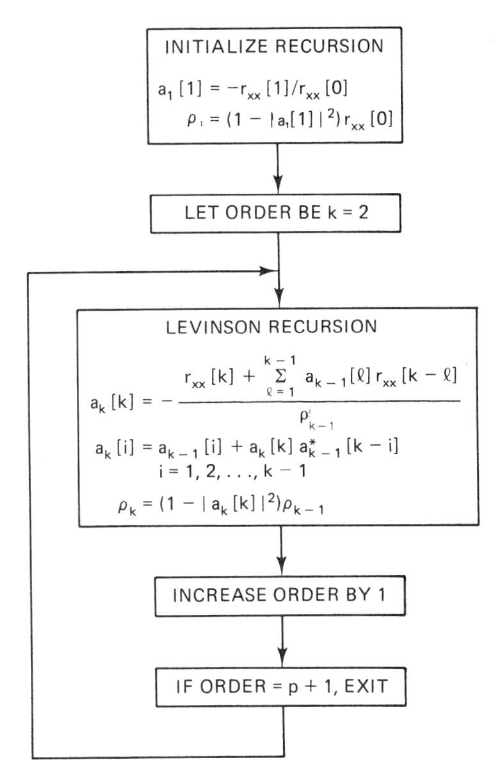

Figure 6.4 Summary of Levinson recursion.

property when we do not know a priori the correct model order. Using the Levinson recursion, successively higher order models can be generated until the modeling error ρ_k is reduced to a desired value. If the process is actually an AR(p) process, then $a_{p+1}[k] = a_p[k]$ for $k = 1, 2, \ldots, p$ and $a_{p+1}[p+1] = k_{p+1} = 0$. In general, for an AR($p$) process, $a_k[k] = k_k = 0$ for $k > p$ and hence $\rho_k = \rho_p$ for $k > p$. This says that the variance of the excitation noise in the model is a constant for a model order equal to or greater than the true order. Hence the point at which ρ_k does not change would appear to be a good indicator of the correct model order. The property that $|a_k[k]| = |k_k| < 1$ leads to $\rho_{k+1} \leq \rho_k$, which furthermore implies that ρ_k first attains its minimum value at the correct model order. A further discussion of this point is given in Section 6.8 on model order determination. If it occurs that $|k_k| = 1$ for some k, the recursion must terminate since $\rho_k = 0$. This case will only occur, however, if the process consists solely of k sinusoids.

6.3.4 Alternative Representations for AR Processes

An AR process is completely characterized by its AR parameters $\{a[1], a[2], \ldots, a[p], \sigma^2\}$. Alternatively, from the Levinson algorithm the AR filter parameters $\{a[1], a[2], \ldots, a[p]\}$ can be found from the reflection coefficients via

(6.47). The variance of the white noise σ^2 can be determined from $r_{xx}[0]$ and the reflection coefficients as

$$\sigma^2 = r_{xx}[0] \prod_{i=1}^{p} (1 - |k_i|^2) \tag{6.49}$$

which follows from the recursive application of (6.48) and $\sigma^2 = \rho_p$. As a result, an AR(p) process has the following equivalent characterizations: $\{a[1], a[2], \ldots, a[p], \sigma^2\}$, $\{r_{xx}[0], k_1, k_2, \ldots, k_p\}$, and $\{r_{xx}[0], r_{xx}[1], \ldots, r_{xx}[p]\}$. We can convert between these representations as shown in Figure 6.5. Conversion A to C is accomplished using the Levinson algorithm, which transforms the ACF samples up to lag p by the AR parameters, where $a[k] = a_p[k]$, $\sigma^2 = \rho_p$. As noted earlier, the reflection coefficients are obtained as $k_i = a_i[i]$, so that conversion A to B is also obtained. Conversion B to C is also effected via the Levinson algorithm. We first use (6.49) to obtain σ^2 and then (6.47) to find $a[k] = a_p[k]$, $k = 1, 2, \ldots, p$. Conversion B to A is accomplished by using the Levinson recursion to generate all the prediction error filter coefficients via (6.47) and the prediction error powers via (6.48). Once this step is completed, $\{r_{xx}[0], r_{xx}[1], \ldots, r_{xx}[p]\}$ can be generated from (6.46) as

$$r_{xx}[k] = \begin{cases} -a_1[1]r_{xx}[0] & \text{for } k = 1 \\ -\sum_{l=1}^{k-1} a_{k-1}[l]r_{xx}[k-l] - a_k[k]\rho_{k-1} & \text{for } k = 2, 3, \ldots, p. \end{cases} \tag{6.50}$$

Conversions C to A and C to B require an additional algorithm referred to as the step-down (SD) procedure. It converts the AR parameters to the reflection coefficients. Consider (6.47):

$$a_k[i] = a_{k-1}[i] + a_k[k]a^*_{k-1}[k-i] \qquad i = 1, 2, \ldots, k-1$$

and let i be replaced by $k - i$ and the entire expression complex-conjugated to yield

$$a^*_k[k-i] = a^*_{k-1}[k-i] + a^*_k[k]a_{k-1}[i].$$

Solving for $a^*_{k-1}[k-i]$ and substituting in (6.47) gives us

$$a_k[i] = a_{k-1}[i] + a_k[k](a^*_k[k-i] - a^*_k[k]a_{k-1}[i])$$

which results in

$$a_{k-1}[i] = \frac{a_k[i] - a_k[k]a^*_k[k-i]}{1 - |a_k[k]|^2} \tag{6.51}$$

$$i = 1, 2, \ldots, k-1; \quad k = p, p-1, \ldots, 2.$$

Equation (6.51) is the first part of the SD procedure which allows us to find the coefficients of the $(k-1)$st order PEF from those of the kth order PEF. In the process the reflection coefficients as well as all the lower order PEFs are found (see Problem 6.6). The prediction error powers for each model order are easily found by reversing the recursion of (6.48) to yield

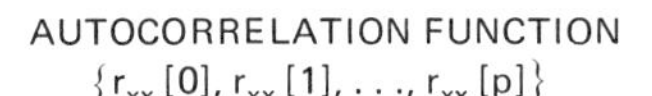
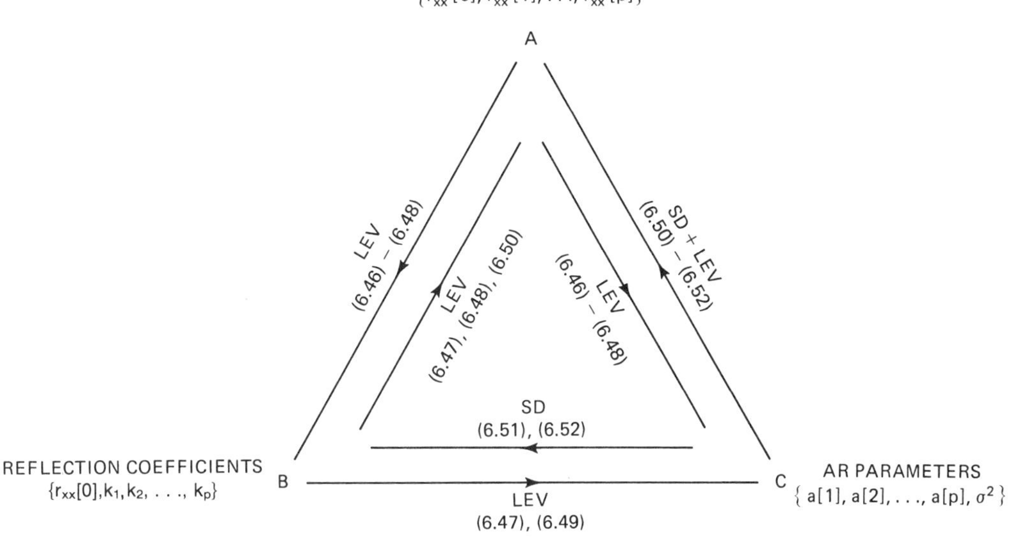

Figure 6.5 Alternative AR representations.

$$\rho_{k-1} = \frac{\rho_k}{1 - |a_k[k]|^2} \qquad k = p, p-1, \ldots, 1. \tag{6.52}$$

$r_{xx}[0]$ is equal to ρ_0. The SD procedure is initialized with $a_p[k] = a[k]$ for $k = 1, 2, \ldots, p$ and $\rho_p = \sigma^2$. It accomplishes the conversion from C to B. Finally, if $\{r_{xx}[0], r_{xx}[1], \ldots, r_{xx}[p]\}$ is desired (the conversion from C to A), (6.50) is used following the SD procedure. A computer program entitled STEPDOWN is provided in Appendix 6C to implement the SD algorithm.

Explicit relationships between the various AR characterizations can also be obtained. For example, since the reflection coefficients satisfy $k_k = a_k[k]$ and $a_k[k]$ is the solution of the Yule–Walker equations for an kth order model, by Cramer's rule,

$$k_k = -\frac{\det\left(\begin{bmatrix} r_{xx}[0] & r_{xx}[-1] & \cdots & r_{xx}[-(k-2)] & r_{xx}[1] \\ r_{xx}[1] & r_{xx}[0] & \cdots & r_{xx}[-(k-3)] & r_{xx}[2] \\ \vdots & \vdots & \vdots & \vdots & \vdots \\ r_{xx}[k-1] & r_{xx}[k-2] & \cdots & r_{xx}[1] & r_{xx}[k] \end{bmatrix}\right)}{\det\left(\begin{bmatrix} r_{xx}[0] & r_{xx}[-1] & \cdots & r_{xx}[-(k-2)] & r_{xx}[-(k-1)] \\ r_{xx}[1] & r_{xx}[0] & \cdots & r_{xx}[-(k-3)] & r_{xx}[-(k-2)] \\ \vdots & \vdots & \vdots & \vdots & \vdots \\ r_{xx}[k-1] & r_{xx}[k-2] & \cdots & r_{xx}[1] & r_{xx}[0] \end{bmatrix}\right)}. \tag{6.53}$$

Some examples of the ACF to reflection coefficient transformation (conversion A to B) are for an AR(1) process,

$$k_1 = -\frac{r_{xx}[1]}{r_{xx}[0]} \tag{6.54}$$

and for an AR(2) process,

$$\begin{aligned} k_1 &= -\frac{r_{xx}[1]}{r_{xx}[0]} \\ k_2 &= \frac{r_{xx}^2[1] - r_{xx}[2]r_{xx}[0]}{r_{xx}^2[0] - |r_{xx}[1]|^2} \end{aligned} \tag{6.55}$$

The characterization of an arbitrary process requires us to specify $\{r_{xx}[0], k_1, k_2, \ldots, k_p, k_{p+1}, \ldots\}$ since the reflection coefficient sequence will not be identically zero after k_p. However, $k_n \to 0$ as $n \to \infty$ and hence, as expected from the Kolmogorov theorem, any process can be modeled as an AR process of sufficiently large order. In practice, we can determine how well an AR process models a given process by computing the reflection coefficient sequence and choosing the AR model order p so that $k_{p+1} \approx 0$ (see Section 7.8).

Finally, it can be shown that the various representations of an AR process must satisfy the following equivalent properties (see Problem 6.7):

1. $\mathcal{A}(z) = 1 + \sum_{k=1}^{p} a[k]z^{-k}$ is minimum-phase and $\sigma^2 > 0$.
2. $r_{xx}[0] > 0$ and $|k_i| < 1$ for $i = 1, 2, \ldots, p$.
3. $\{r_{xx}[0], r_{xx}[1], \ldots, r_{xx}[p]\}$ is a positive definite sequence (see Burg [1975]).

6.3.5 Lattice Filters

Using the results obtained in the derivation of the Levinson algorithm, we can derive an alternative but equivalent form for the PEF. The so-called *lattice* filter has as its coefficients the reflection coefficients. The filter is based on the recursion given by (6.45) for the forward prediction error,

$$e_k^f[n] = e_{k-1}^f[n] + k_k e_{k-1}^b[n-1]$$

and a similar relationship for the backward prediction error,

$$e_k^b[n] = e_{k-1}^b[n-1] + k_k^* e_{k-1}^f[n]. \tag{6.56}$$

The latter relationship is found as follows. The backward prediction error is, from (6.39),

$$e_k^b[n] = x[n-k] + \sum_{i=1}^{k} a_k^*[i]x[n-k+i] \tag{6.57}$$

where $a_k[0] = 1$. But using (6.47) from the Levinson algorithm, we obtain

$$
\begin{aligned}
e_k^b[n] &= x[n-k] + \sum_{i=1}^{k-1} (a_{k-1}^*[i] + k_k^* a_{k-1}[k-i]) x[n-k+i] + k_k^* x[n] \\
&= x[n-k] + \sum_{i=1}^{k-1} a_{k-1}^*[i] x[n-k+i] + k_k^* \sum_{i=1}^{k} a_{k-1}[k-i] x[n-k+i] \\
&= \sum_{i=0}^{k-1} a_{k-1}^*[i] x[n-k+i] + k_k^* \sum_{i=0}^{k-1} a_{k-1}[i] x[n-i] \\
&= e_{k-1}^b[n-1] + k_k^* e_{k-1}^f[n].
\end{aligned}
$$

Thus the lattice filter relationships, which can be used to compute the forward and backward prediction errors for the kth order predictor based on those for the $(k-1)$st order predictor and k_k, are

$$
\begin{aligned}
e_k^f[n] &= e_{k-1}^f[n] + k_k e_{k-1}^b[n-1] \\
e_k^b[n] &= e_{k-1}^b[n-1] + k_k^* e_{k-1}^f[n]
\end{aligned}
\tag{6.58}
$$

where $e_0^f[n] = e_0^b[n] = x[n]$. The lattice filter is shown in Figure 6.6 and is equivalent to the standard PEF $\mathcal{A}(z) = 1 + \sum_{k=1}^{p} a_p[k] z^{-k}$. The lattice filter has the interesting and useful property that its coefficients are all less than 1 in magnitude since $|k_i| < 1$, $i = 1, 2, \ldots, p$. This property becomes important when the filter coefficients need to be quantized for storage or transmission (see Chapter 16). It should be noted that the lattice filter relationships are valid for any process, not just an AR(p) process. If, however, the process is AR(p), then $k_i = 0$ for $i \geq p+1$. Also, since the backward prediction errors were determined as a result of a Gram–Schmidt orthogonalization, they are all uncorrelated at the same time instant:

$$
\mathcal{E}(e_i^b[n]^* e_j^b[n]) = 0 \qquad \text{for} \quad i \neq j \tag{6.59}
$$

and from (6.41),

$$
\text{var}(e_k^b[n]) = \text{var}(e_k^f[n]) = \rho_k. \tag{6.60}
$$

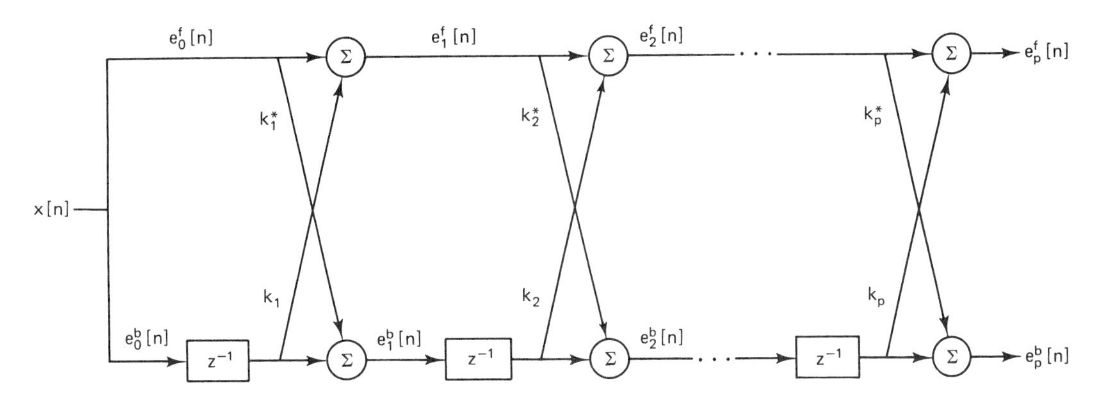

Figure 6.6 Lattice filter implementation of prediction error filter.

6.3.6 Inverting the Autocorrelation Matrix

At times it is useful to be able to compute the inverse of the autocorrelation matrix, not just explicitly solve for a set of prediction coefficients. With the results of the preceding sections, we can easily determine the inverse after first applying the Levinson algorithm to the autocorrelation sequence $\{r_{xx}[0], r_{xx}[1], \ldots, r_{xx}[N]\}$ to yield $\{a_j[i], 1 \le i \le j, j = 1, 2, \ldots, N\}$ and $\{\rho_i, i = 1, 2, \ldots, N\}$. The resultant decomposition of the $(N+1) \times (N+1)$ inverse autocorrelation matrix $\mathbf{R}_{xx}^{-1}$, sometimes referred to as a Cholesky decomposition (see Section 2.7), is [Burg 1975]

$$\mathbf{R}_{xx}^{-1} = \mathbf{B}\mathbf{P}^{-1}\mathbf{B}^H \tag{6.61}$$

where $\mathbf{B}$ is the $(N+1) \times (N+1)$ matrix

$$\mathbf{B} = \begin{bmatrix} 1 & a_1^*[1] & a_2^*[2] & \cdots & a_N^*[N] \\ 0 & 1 & a_2^*[1] & \cdots & a_N^*[N-1] \\ \vdots & \vdots & \vdots & \vdots & \vdots \\ 0 & 0 & 0 & \cdots & 1 \end{bmatrix}$$

and $\mathbf{P}$ is the $(N+1) \times (N+1)$ diagonal matrix

$$\mathbf{P} = \operatorname{diag}(r_{xx}[0], \rho_1, \rho_2, \ldots, \rho_N).$$

The result is valid for any process, not just an AR process. However, if the process is AR(p), then for $j \ge p$,

$$a_j[i] = \begin{cases} a_p[i] & \text{for } i = 1, 2, \ldots, p \\ 0 & \text{for } i > p \end{cases}$$

$$\rho_j = \rho_p.$$

To show that (6.61) is indeed valid, consider $\mathbf{B}^H\mathbf{R}_{xx}\mathbf{B}$ and let $\mathbf{x} = [x[n-1] \;\; x[n-2] \;\; \cdots \;\; x[n-(N+1)]]^T$. Since $\mathbf{R}_{xx}^* = \mathcal{E}[\mathbf{x}\mathbf{x}^H]$,

$$\mathbf{B}^H\mathbf{R}_{xx}\mathbf{B} = \mathcal{E}[(\mathbf{B}^T\mathbf{x})(\mathbf{B}^T\mathbf{x})^H]^*.$$

Expanding $\mathbf{B}^T\mathbf{x}$ and using (6.39) yields

$$\mathbf{B}^T\mathbf{x} = \begin{bmatrix} x[n-1] \\ x[n-2] + a_1^*[1]x[n-1] \\ x[n-3] + a_2^*[1]x[n-2] + a_2^*[2]x[n-1] \\ \vdots \\ x[n-N-1] + \sum_{i=1}^{N} a_N^*[i]x[n-N-1+i] \end{bmatrix} = \begin{bmatrix} e_0^b[n-1] \\ e_1^b[n-1] \\ e_2^b[n-1] \\ \vdots \\ e_N^b[n-1] \end{bmatrix}.$$

Using the property that the backward prediction errors are uncorrelated, that is, (6.59) and (6.60), or

$$\mathcal{E}(e_i^b[n-1]e_j^b[n-1]^*) = \rho_i\delta[i-j]$$

we can write

$$\mathbf{B}^H\mathbf{R}_{xx}\mathbf{B} = \mathcal{E}[(\mathbf{B}^T\mathbf{x})(\mathbf{B}^T\mathbf{x})^H]^* = \mathbf{P}^* = \mathbf{P}.$$

Finally, (6.61) follows from

$$\mathbf{R}_{xx} = \mathbf{B}^{H^{-1}}\mathbf{P}\mathbf{B}^{-1}$$

which implies that

$$\mathbf{R}_{xx}^{-1} = \mathbf{B}\mathbf{P}^{-1}\mathbf{B}^{H}.$$

Note that $\mathbf{B}^{-1}$ exists since $\det(\mathbf{B}) = 1$.

As a corollary to (6.61), one can express the determinant of the autocorrelation matrix as a function of the prediction errors powers. Since $\mathbf{R}_{xx} = \mathbf{B}^{H^{-1}}\mathbf{P}\mathbf{B}^{-1}$, it follows that

$$\det(\mathbf{R}_{xx}) = \frac{\det(\mathbf{P})}{\det(\mathbf{B}^H)\det(\mathbf{B})}$$

However, since $\mathbf{B}$ is an upper triangular matrix [see (2.16)],

$$\det(\mathbf{B}) = \det(\mathbf{B}^H) = 1$$

so that

$$\det(\mathbf{R}_{xx}) = \det(\mathbf{P}) = \prod_{i=0}^{N} \rho_i \tag{6.62}$$

where $\rho_0 = r_{xx}[0]$. A recursive expression for the determinant of the autocorrelation matrix follows as

$$\det(\mathbf{R}_{xx}^{(N+1)}) = \rho_N \det(\mathbf{R}_{xx}^{(N)}) \tag{6.63}$$

where $\mathbf{R}_{xx}^{(N)}$ is the autocorrelation matrix of dimension $N \times N$.

Some examples of the inverse of the $N \times N$ autocorrelation matrix for AR processes will now be given. It is assumed that $N \geq 2p + 1$. First, consider a complex AR(1) process. Then

$$\mathbf{B} = \begin{bmatrix} 1 & a^*[1] & 0 & 0 & \cdots & 0 \\ 0 & 1 & a^*[1] & 0 & \cdots & 0 \\ 0 & 0 & 1 & a^*[1] & \cdots & 0 \\ \vdots & \vdots & \vdots & \vdots & \vdots & \vdots \\ 0 & 0 & 0 & 0 & \cdots & 1 \end{bmatrix}$$

$$\mathbf{P} = \operatorname{diag}(r_{xx}[0], \sigma^2, \sigma^2, \ldots, \sigma^2) = \sigma^2 \operatorname{diag}\left(\frac{1}{1 - |a[1]|^2}, 1, 1, \ldots, 1\right).$$

Performing the matrix multiplication results in

$$\mathbf{R}_{xx}^{-1} = \frac{1}{\sigma^2}\begin{bmatrix} 1 & a^*[1] & 0 & 0 & \cdots & 0 \\ a[1] & 1 + |a[1]|^2 & a^*[1] & 0 & \cdots & 0 \\ 0 & a[1] & 1 + |a[1]|^2 & a^*[1] & \cdots & 0 \\ \vdots & \vdots & \ddots & \ddots & \ddots & \vdots \\ 0 & 0 & \cdots & a[1] & 1 + |a[1]|^2 & a^*[1] \\ 0 & 0 & \cdots & 0 & a[1] & 1 \end{bmatrix}. \tag{6.64}$$

It is interesting to note that while $\mathbf{R}_{xx}$ is a hermitian Toeplitz matrix, $\mathbf{R}_{xx}^{-1}$ is not. $\mathbf{R}_{xx}^{-1}$ does, however, have the persymmetry property (see Section 2.3), which asserts that the matrix is symmetric about the principal cross-diagonal (running from NE to SW).

Similarly, for an AR(2) process, $\mathbf{R}_{xx}^{-1}$ can be shown to be

$$\mathbf{R}_{xx}^{-1} = \frac{1}{\sigma^2} \begin{bmatrix} 1 & a^*[1] & a^*[2] & 0 & \cdots & 0 \\ a[1] & 1+|a[1]|^2 & a^*[1]+a[1]a^*[2] & a^*[2] & \cdots & 0 \\ a[2] & a[1]+a^*[1]a[2] & 1+|a[1]|^2+|a[2]|^2 & a^*[1]+a[1]a^*[2] & \cdots & 0 \\ \vdots & \ddots & \ddots & \ddots & \ddots & \vdots \\ 0 & \cdots & a[1]+a^*[1]a[2] & 1+|a[1]|^2+|a[2]|^2 & a^*[1]+a[1]a^*[2] & a^*[2] \\ 0 & \cdots & a[2] & a[1]+a^*[1]a[2] & 1+|a[1]|^2 & a^*[1] \\ 0 & \cdots & 0 & a[2] & a[1] & 1 \end{bmatrix} \tag{6.65}$$

In general, for an AR(p) process the inverse of the $N \times N$ autocorrelation matrix is, for $i \geq j$ [Siddiqui 1958, Kailath et al. 1978],

$$[\mathbf{R}_{xx}^{-1}]_{ij} = \frac{1}{\sigma^2} \sum_{k=1}^{N} (a[i-k]a^*[j-k] - a^*[N-i+k]a[N-j+k])$$

$$i = 1, 2, \ldots, N; \quad j = 1, 2, \ldots, N \tag{6.66}$$

where $a[k] = 0$ for $k < 0$ and $k > p$.

6.4 PROPERTIES OF THE AR SPECTRAL ESTIMATOR

6.4.1 The Implied Autocorrelation Function Extension

The high resolution property of the AR spectral estimator is due to an implicit extension of the measured autocorrelation sequence. Assume that an AR(p) spectral estimate is desired and that the known ACF samples are $\{r_{xx}[0], r_{xx}[1], \ldots, r_{xx}[p]\}$. (In practice, the ACF samples would be estimated from the data.) The AR parameter estimates are found by substituting the known ACF samples into the Yule–Walker equations and solving. Then the AR spectral estimator as defined on the z-plane is

$$\hat{\mathcal{P}}_{AR}(z) = \frac{\hat{\sigma}^2}{\hat{\mathcal{A}}(z)\hat{\mathcal{A}}^*(1/z^*)} = \sum_{k=-\infty}^{\infty} \hat{r}_{xx}[k]z^{-k} \tag{6.67}$$

where the "hat" or caret denotes the estimated quantities. It is now shown that the implied ACF $\hat{r}_{xx}[k]$, which is given by the inverse z-transform of $\hat{\mathcal{P}}_{AR}(z)$, is

$$\hat{r}_{xx}[k] = \begin{cases} r_{xx}[k] & \text{for } 0 \le k \le p \\ -\sum_{l=1}^{p} \hat{a}[l]\hat{r}_{xx}[k-l] & \text{for } k > p. \end{cases} \tag{6.68}$$

Hence the estimate of the ACF *matches* the known ACF up to lag p and the remaining samples are extrapolated by a recursive difference equation (see also Section 5.6). This is the so-called *correlation matching* property of the AR spectral estimator. In contrast to the BT spectral estimator, *no windowing* of the ACF occurs. Thus AR spectra do not exhibit sidelobes due to windowing. Also, it is the implied extension of (6.68) which is responsible for the high resolution property of the AR spectral estimator. An example of the ACF extension property is given in Figures 6.7 and 6.8. The BT spectral estimator windows the known autocorrelation sequence and then extrapolates the sequence by appending zeros, thus giving rise to the usual smearing of the spectral estimator. The AR spectral estimator, on the other hand, extrapolates the autocorrelation sequence according to (6.68). As seen in Figures 6.7 and 6.8, the resultant spectral estimate is a less biased version of the true one.

To verify (6.68), first multiply both sides of (6.67) by $\hat{\mathcal{A}}(z)$ and then take the inverse z-transform to yield

$$\mathcal{Z}^{-1}\left[\frac{\hat{\sigma}^2}{\hat{\mathcal{A}}^*(1/z^*)}\right] = \mathcal{Z}^{-1}\left[\hat{\mathcal{A}}(z) \sum_{k=-\infty}^{\infty} \hat{r}_{xx}[k]z^{-k}\right]$$

However,

$$\mathcal{Z}^{-1}\left[\frac{\hat{\sigma}^2}{\hat{\mathcal{A}}^*(1/z^*)}\right] = \hat{\sigma}^2\hat{h}^*[-k] = \hat{\sigma}^2\delta[k] \qquad \text{for } k \ge 0$$

since the filter is assumed to be causal and as shown in Section 5.4, $h[0] = 1$. Then

$$\begin{aligned} \hat{\sigma}^2\delta[k] &= \hat{a}[k] \bigstar \hat{r}_{xx}[k] && \text{for } k \ge 0 \\ &= \sum_{l=0}^{p} \hat{a}[l]\hat{r}_{xx}[k-l] && \text{for } k \ge 0. \end{aligned} \tag{6.69}$$

Therefore, (6.68) is satisfied for $k > p$. For $k = 0, 1, \ldots, p$ (6.69) corresponds to the Yule–Walker equations used to find $\hat{a}[k]$ and $\hat{\sigma}^2$ except that $r_{xx}[k]$ has been

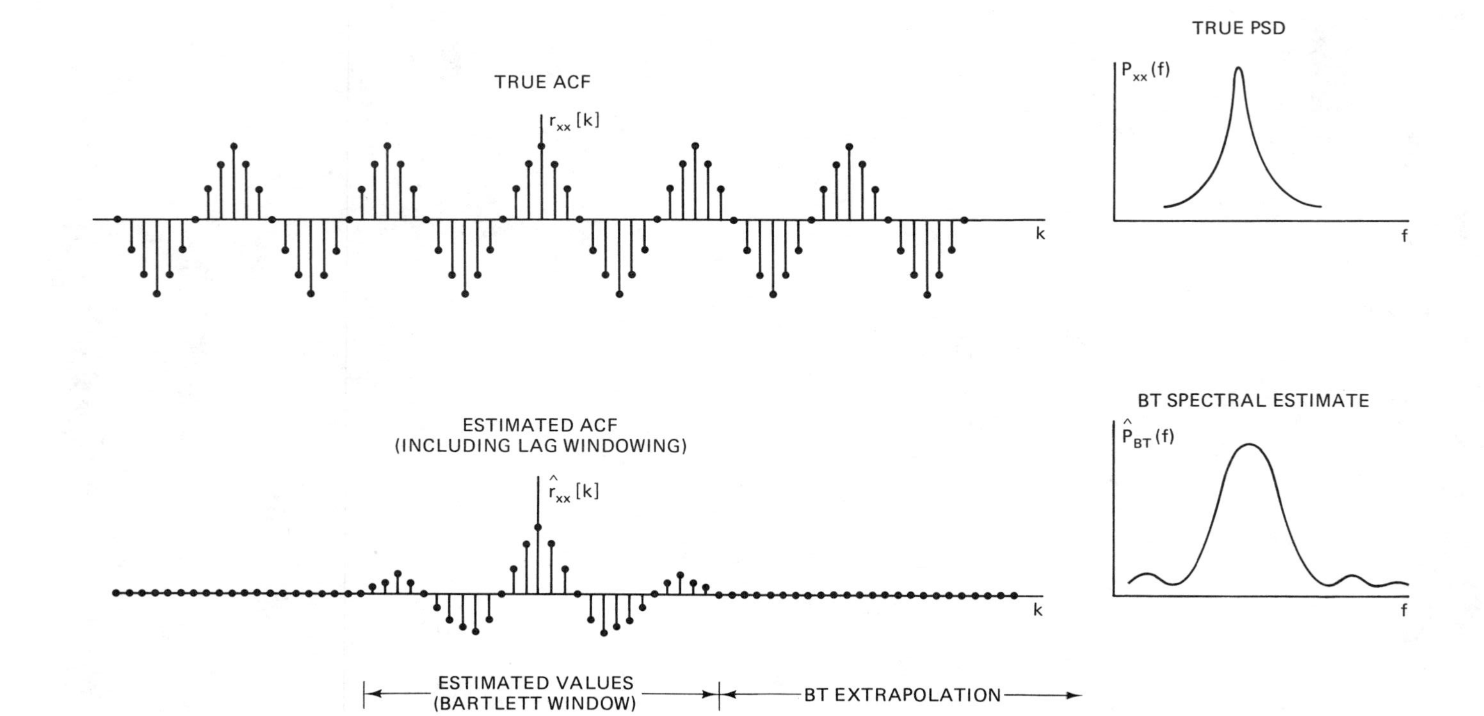

Figure 6.7 Implied autocorrelation extrapolation of Blackman–Tukey spectral estimator.

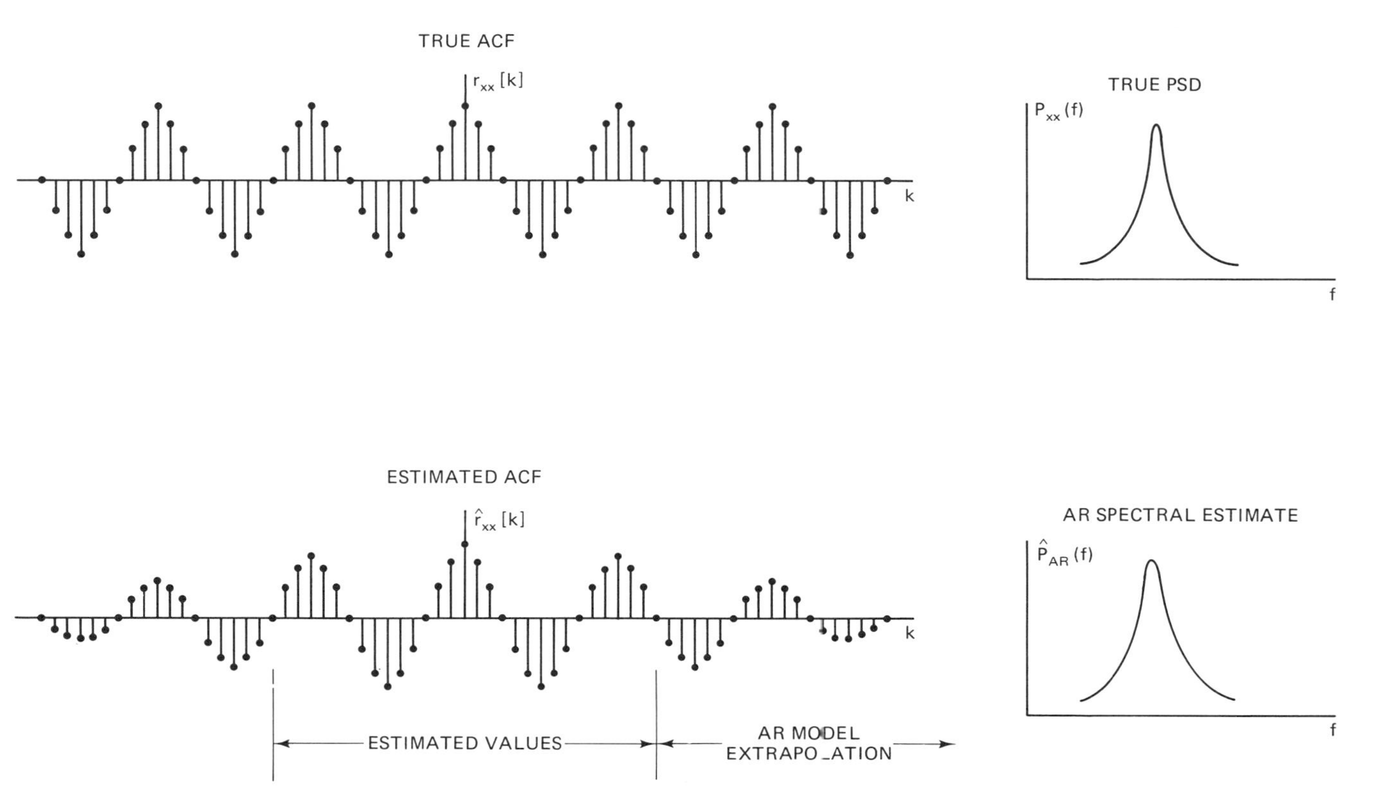

Figure 6.8 Implied autocorrelation extrapolation of autcregressive spectral estimator.

replaced by $\hat{r}_{xx}[k]$. Because the autocorrelation samples $r_{xx}[k]$ and $\hat{r}_{xx}[k]$ for $k = 0, 1, \ldots, p$ both yield the same set of AR parameters, they must be equal due to the one-to-one mapping depicted in Figure 6.5 as conversion A to C.

6.4.2 Maximum Entropy Spectral Estimation

Maximum Entropy spectral estimation (MESE) is based on an explicit extrapolation of a segment of a known ACF for the samples which are not known [Burg 1975]. In this way the characteristic smearing of the estimated PSD due to truncation of the ACF is alleviated. If $\{r_{xx}[0], r_{xx}[1], \ldots, r_{xx}[p]\}$ is known, the question arises as to how $\{r_{xx}[p+1], r_{xx}[p+2], \ldots\}$ should be specified to guarantee that the entire ACF is positive definite. In general, there are an infinite number of possible extrapolations, all of which yield valid ACFs. Burg argued that the extrapolation should be chosen so that the time series characterized by the extrapolated ACF has maximum entropy. The time series will then be the most random one which has the known autocorrelation samples for its first $p + 1$ lag values. Alternatively, the PSD will be the one with the flattest (whitest) spectrum of all spectra for which the first $p + 1$ autocorrelation samples are equal to the known ones. The resultant spectral estimator is termed the MESE. The rationale for choosing the maximum entropy criterion is that it imposes the fewest constraints on the unknown time series by maximizing its randomness, thereby producing a minimum bias solution.

In particular, if we assume a Gaussian random process, then the entropy per sample is proportional to

$$\int_{-1/2}^{1/2} \ln P_{xx}(f)\, df. \tag{6.70}$$

The MESE is found by maximizing (6.70) subject to the constraint that the ACF corresponding to $P_{xx}(f)$ has as its first $p + 1$ lags the known samples, or

$$\int_{-1/2}^{1/2} P_{xx}(f) \exp(j2\pi fk)\, df = r_{xx}[k] \qquad \text{for } k = 0, 1, \ldots, p. \tag{6.71}$$

By applying the technique of Lagrangian multipliers, the solution becomes

$$P_{xx}(f) = \frac{1}{\sum_{k=-p}^{p} \lambda_k \exp(-j2\pi fk)}$$

where the λ_k's are the Lagrangian multipliers to be found by imposing the constraints of (6.71). Evaluating the Lagrangian multipliers results in the MESE

$$P_{xx}(f) = \frac{\sigma^2}{\left| 1 + \sum_{k=1}^{p} a[k] \exp(-j2\pi fk) \right|^2} \tag{6.72}$$

where $\{a[1], a[2], \ldots, a[p], \sigma^2\}$ are found by solving the Yule–Walker equations using the known samples of the ACF. Hence, with knowledge of $\{r_{xx}[0], r_{xx}[1],$

$\ldots, r_{xx}[p]\}$, the MESE is equivalent to the AR spectral estimator. This equivalence, however, is maintained only for Gaussian random processes and *known* ACF samples.

6.4.3 The Spectral Flatness Measure

It was seen in Section 6.3 that the AR(p) parameters could be obtained as the coefficients that minimized the prediction error power of a pth order linear predictor. The PEF acted as a whitener which decorrelated the AR process to produce white noise at its output. In this section the concept of whitening is extended to show that the AR parameters also result by maximizing the *spectral flatness* of the process at the output of an FIR filter or PEF [Gray and Markel 1974]. With the spectral flatness measure the AR spectral estimator can be viewed as the result of an optimal whitening operation.

The spectral flatness measure for a PSD is defined as

$$\xi_x = \frac{\exp\left[\int_{-1/2}^{1/2} \ln P_{xx}(f)\, df\right]}{\int_{-1/2}^{1/2} P_{xx}(f)\, df}. \tag{6.73}$$

ξ_x is the geometric mean of $P_{xx}(f)$ divided by the arithmetic mean of $P_{xx}(f)$ and therefore can be shown to satisfy

$$0 \leq \xi_x \leq 1. \tag{6.74}$$

If $P_{xx}(f)$ is very peaky (i.e., if it has a large dynamic range), then $\xi_x \approx 0$ (see Problem 6.9). On the other hand, if $P_{xx}(f)$ is a constant for all f (i.e., if it has a zero dynamic range), then $\xi_x = 1$, as can be verified directly from (6.73). Thus the flatter the PSD, the larger is the spectral flatness measure. It is interesting to observe that the MESE corresponds to a constrained maximization of ξ_x. This is because in the MESE formulation the numerator of (6.73) is maximized. Since $r_{xx}[0]$ and hence the denominator of (6.73) is constrained, this is equivalent to maximizing the spectral flatness. The MESE therefore maximizes the spectral flatness of the estimated PSD subject to the constraints that $\{r_{xx}[0], r_{xx}[1], \ldots, r_{xx}[p]\}$ are fixed.

Now consider the problem of *maximizing the spectral flatness* of the PSD of the time series at the output of a minimum-phase PEF $\mathcal{A}(z) = 1 + \sum_{k=1}^{p} \alpha_k z^{-k}$, by suitably choosing the prediction coefficients. This procedure will be shown to be equivalent to *minimizing the prediction error power*. The input time series is not restricted to be an AR process but may be arbitrary. The following result is needed. If $A(f)$ is minimum phase, then [Markel and Gray 1976] (see Problem 6.8)

$$\int_{-1/2}^{1/2} \ln |A(f)|^2\, df = 0. \tag{6.75}$$

Let $e[n]$ denote the prediction error time series. Then

$$\int_{-1/2}^{1/2} \ln P_{ee}(f)\, df = \int_{-1/2}^{1/2} \ln [| A(f) |^2 P_{xx}(f)]\, df$$

$$= \int_{-1/2}^{1/2} \ln P_{xx}(f)\, df$$

where (6.75) has been employed. Now exponentiate and divide both sides by $\int_{-1/2}^{1/2} P_{ee}(f)\, df$ to yield

$$\xi_e = \frac{\exp\left[\int_{-1/2}^{1/2} \ln P_{xx}(f)\, df\right]}{\int_{-1/2}^{1/2} P_{ee}(f)\, df}$$

$$= \xi_x \frac{\int_{-1/2}^{1/2} P_{xx}(f)\, df}{\int_{-1/2}^{1/2} P_{ee}(f)\, df}$$

$$= \xi_x \frac{r_{xx}[0]}{r_{ee}[0]} \, .$$

To maximize ξ_e we must minimize $r_{ee}[0]$ since $P_{xx}(f)$ and hence $\xi_x r_{xx}[0]$ is fixed. It is concluded that maximizing the spectral flatness of the prediction error PSD is equivalent to minimizing the prediction error power of a pth order linear predictor. (It is now obvious that the assumption of a minimum-phase PEF is not restrictive since the prediction error power is always minimized by a minimum-phase filter, as shown in Section 6.3.)

In the case of an AR(p) process, maximizing the spectral flatness at the output of a pth order PEF will produce a PEF with coefficients equal to the AR parameters and an output time series $e[n]$ equal to white noise $u[n]$ with variance σ^2. Hence

$$P_{ee}(f) = \sigma^2 = | A(f) |^2 P_{xx}(f)$$

which indicates that the PSD of the AR process must have been

$$P_{xx}(f) = \frac{\sigma^2}{| A(f) |^2} \, .$$

If, however, the process is *not* AR(k), where $k \leq p$, then the prediction error *will not be white* and

$$P_{xx}(f) = \frac{P_{ee}(f)}{| A(f) |^2} \qquad (6.76)$$

where $A(f)$ is the optimal pth order PEF. Equation (6.76) has an important implication for AR spectral estimation of non-AR processes. By using an AR spectral estimator with the AR parameters estimated as the solution of the Yule–Walker equations, we are assuming $P_{ee}(f)$ to be a constant and so any important spectral detail contributed by $P_{ee}(f)$ is lost. Some examples of the loss in spectral resolution were given in Section 5.6. As a simple example of the nonwhiteness of the prediction error time series, consider the optimal first order linear predictor for a real MA(1) process. From (6.4),

$$\alpha_1 = -\frac{r_{xx}[1]}{r_{xx}[0]} = -\frac{b[1]}{1 + b^2[1]}$$

and from (6.76),

$$P_{ee}(f) = \sigma^2 \mid 1 + b[1] \exp(-j2\pi f) \mid^2 \left| 1 - \frac{b[1]}{1 + b^2[1]} \exp(-j2\pi f) \right|^2 .$$

The prediction error time series will be white only if $b[1] = 0$ or if $x[n]$ is a white noise process to begin with.

6.4.4 Bounds on AR Spectral Estimators

It is sometimes useful to be able to determine the dynamic range of the AR spectral estimator without having to explicitly compute the PSD. The PSD can be bounded above and below if we know $r_{xx}[0]$ and the magnitudes of the reflection coefficients. It has been shown that [Burg 1975]

$$r_{xx}[0] \prod_{i=1}^{p} \frac{1 - \mid k_i \mid}{1 + \mid k_i \mid} \leq P_{\text{AR}}(f) \leq r_{xx}[0] \prod_{i=1}^{p} \frac{1 + \mid k_i \mid}{1 - \mid k_i \mid} . \tag{6.77}$$

The lower bound will be smaller and the upper bound larger when the magnitude of any of the reflection coefficients is close to 1. This is just a restatement of the property that AR processes with large reflection coefficient magnitudes or equivalently with poles near the unit circle are characterized by peaky spectra.

6.5 ESTIMATION OF AR PARAMETERS AND REFLECTION COEFFICIENTS

6.5.1. Maximum Likelihood Estimators

To estimate the PSD using an AR model we need to estimate the parameters of the model. The theoretical PSD is given from (5.9) as

$$P_{\text{AR}}(f) = \frac{\sigma^2}{\mid 1 + a[1] \exp(-j2\pi f) + \cdots + a[p] \exp(-j2\pi fp) \mid^2} . \tag{6.78}$$

The estimate of the PSD is formed by replacing the theoretical AR parameters by their estimates to yield

$$\hat{P}_{\text{AR}}(f) = \frac{\hat{\sigma}^2}{\mid 1 + \hat{a}[1] \exp(-j2\pi f) + \cdots + \hat{a}[p] \exp(-j2\pi fp) \mid^2} \tag{6.79}$$

where the "hats" denote estimated quantities. For good estimates of the AR parameters a maximum likelihood estimator (MLE) is usually employed. The desirable properties of the MLE are that for large data records the estimator is unbiased and efficient, in that it attains the Cramer–Rao lower bound on variance (see Section 3.3.1 for a summary of important MLE properties). Assuming that $x[n]$ is a *real* Gaussian random process, it is now shown that the approximate MLE of the AR parameters is found by solving the Yule–Walker equations using a suitable estimate of the ACF [Jenkins and Watts 1968]. The complex Gaussian case proceeds analogously to the derivation to be presented but requires the concept of a complex multivariate Gaussian distribution [Goodman 1963] (see also Section 3.5.1). It is assumed that $\{x[0], x[1], \ldots, x[N-1]\}$ is observed, so that the joint PDF of a real AR(p) process is denoted by

$$p(x[0], x[1], \ldots, x[N-1]; a[1], a[2], \ldots, a[p], \sigma^2). \tag{6.80}$$

The dependence of the PDF on the AR parameters is noted. Letting $\mathbf{a} = [a[1] \;\; a[2] \;\; \cdots \;\; a[p]]^T$, the PDF can also be written

$$\begin{aligned} p(x[p], x[p+1], \ldots, x[N-1] \mid x[0], x[1], \ldots, x[p-1]; \mathbf{a}, \sigma^2) \\ p(x[0], x[1], \ldots, x[p-1]; \mathbf{a}, \sigma^2). \end{aligned} \tag{6.81}$$

The exact maximization of (6.81) with respect to the AR parameters produces a set of highly nonlinear equations [Box and Jenkins 1970]. For large data records the maximization of the PDF or likelihood function can be effected approximately by maximizing only the conditional PDF in (6.81). [See also Problems 6.10 and 6.11 for the exact MLE of the filter parameter of an AR(1) process.] The effect of the PDF of the initial conditions $p(x[0], x[1], \ldots, x[p]; \mathbf{a}, \sigma^2)$ on the MLE will be small as long as the poles are not too close to the unit circle [Box and Jenkins 1970]. With this approximation the conditional PDF can be maximized by assuming that $x[n]$ is a Gaussian random process. Since $u[n]$ is the output of the linear filter $\mathcal{A}(z)$ excited by $x[n]$, $u[n]$ will also be a Gaussian random process. Using the definition of an AR(p) process,

$$u[n] = x[n] + \sum_{k=1}^{p} a[k]x[n-k] \tag{6.82}$$

it follows that the transformation from the observed data samples to the input noise samples is

$$\begin{aligned} u[p] &= x[p] + a[1]x[p-1] + \cdots + a[p]x[0] \\ u[p+1] &= x[p+1] + a[1]x[p] + \cdots + a[p]x[1] \\ &\;\vdots \\ u[N-1] &= x[N-1] + a[1]x[N-2] + \cdots + a[p]x[N-p-1]. \end{aligned} \tag{6.83}$$

The PDF of $\mathbf{u} = [u[p] \;\; u[p+1] \;\; \cdots \;\; u[N-1]]^T$ is

$$p(\mathbf{u}) = \prod_{n=p}^{N-1} \frac{1}{\sqrt{2\pi\sigma^2}} \exp\left(-\frac{u^2[n]}{2\sigma^2}\right)$$

$$= \frac{1}{(2\pi\sigma^2)^{(N-p)/2}} \exp\left(-\frac{1}{2\sigma^2}\sum_{n=p}^{N-1} u^2[n]\right). \tag{6.84}$$

Next, use is made of the transformation of (6.83) from $\{x[p], x[p+1], \ldots, x[N-1]\}$ to $\{u[p], u[p+1], \ldots, u[N-1]\}$, which is rewritten in matrix form as

$$\mathbf{u} = \underbrace{\begin{bmatrix} 1 & 0 & 0 & \cdots & 0 \\ a[1] & 1 & 0 & \cdots & 0 \\ a[2] & a[1] & 1 & \cdots & 0 \\ \vdots & \vdots & \vdots & \ddots & \vdots \\ 0 & \cdots & a[p] & \cdots & 1 \end{bmatrix}}_{\mathcal{J}} \mathbf{x} + \begin{bmatrix} \sum_{n=1}^{p} a[n]x[p-n] \\ \sum_{n=2}^{p} a[n]x[p+1-n] \\ \vdots \\ a[p]x[p-1] \\ 0 \\ \vdots \\ 0 \end{bmatrix} \tag{6.85}$$

where $\mathbf{x} = [x[p] \quad x[p+1] \quad \cdots \quad x[N-1]]^T$. The Jacobian of the transformation $\partial\mathbf{u}/\partial\mathbf{x}$ is just the matrix $\mathcal{J}$ and hence $\det(\mathcal{J}) = 1$. Using the standard formula for the PDF of a set of transformed random variables [Papoulis 1965] yields

$$p(\mathbf{x}|x[0], x[1], \ldots, x[p-1]; \mathbf{a}, \sigma^2)$$

$$= p(\mathbf{u}(\mathbf{x}))\left|\det\left(\frac{\partial\mathbf{u}}{\partial\mathbf{x}}\right)\right| = p(\mathbf{u}(\mathbf{x}))$$

$$= \frac{1}{(2\pi\sigma^2)^{(N-p)/2}} \exp\left[-\frac{1}{2\sigma^2}\sum_{n=p}^{N-1}\left(x[n] + \sum_{j=1}^{p} a[j]x[n-j]\right)^2\right] \tag{6.86}$$

This is maximized over $\mathbf{a}$ by minimizing

$$S_1(\mathbf{a}) = \sum_{n=p}^{N-1}\left(x[n] + \sum_{j=1}^{p} a[j]x[n-j]\right)^2. \tag{6.87}$$

Since (6.87) is a quadratic form in $\mathbf{a}$, differentiating it will produce a global minimum, although as will be discussed shortly, it may not be unique. Performing the differentiation

$$\sum_{j=1}^{p} \hat{a}[j] \sum_{n=p}^{N-1} x[n-j]x[n-k] = -\sum_{n=p}^{N-1} x[n]x[n-k] \qquad k = 1, 2, \ldots, p \tag{6.88}$$

or in matrix form the MLE of $\mathbf{a}$ is found from

$$\underbrace{\begin{bmatrix} c_{xx}[1,1] & c_{xx}[1,2] & \cdots & c_{xx}[1,p] \\ c_{xx}[2,1] & c_{xx}[2,2] & \cdots & c_{xx}[2,p] \\ \vdots & \vdots & \ddots & \vdots \\ c_{xx}[p,1] & c_{xx}[p,2] & \cdots & c_{xx}[p,p] \end{bmatrix}}_{\mathbf{C}} \underbrace{\begin{bmatrix} \hat{a}[1] \\ \hat{a}[2] \\ \vdots \\ \hat{a}[p] \end{bmatrix}}_{\hat{\mathbf{a}}} = -\underbrace{\begin{bmatrix} c_{xx}[1,0] \\ c_{xx}[2,0] \\ \vdots \\ c_{xx}[p,0] \end{bmatrix}}_{\mathbf{c}} \tag{6.89}$$

where

$$c_{xx}[j,k] = \frac{1}{N-p} \sum_{n=p}^{N-1} x[n-j]x[n-k].$$

The factor $1/(N-p)$ has been obtained by dividing both sides of (6.88) by $N - p$. This allows $c_{xx}[j, k]$ to be interpreted as an autocorrelation estimator at lag $j - k$.

To find $\hat{\sigma}^2$, (6.86) is maximized by equivalently taking the logarithm, differentiating, and setting the result equal to zero, or

$$\frac{\partial \ln p}{\partial \sigma^2} = \frac{\partial}{\partial \sigma^2}\left[-\frac{N-p}{2}\ln 2\pi - \frac{N-p}{2}\ln \sigma^2 - \frac{1}{2\sigma^2}S_1(\mathbf{a})\right] = 0.$$

We thus obtain

$$\begin{aligned} \hat{\sigma}^2 &= \frac{1}{N-p} S_1(\hat{\mathbf{a}}) \\ &= \frac{1}{N-p} \sum_{n=p}^{N-1} \left(x[n] + \sum_{j=1}^{p} \hat{a}[j]x[n-j]\right)^2. \end{aligned} \tag{6.90}$$

An alternative expression for $\hat{\sigma}^2$ is found by observing from (6.88) that

$$\sum_{n=p}^{N-1} \left(x[n] + \sum_{j=1}^{p} \hat{a}[j]x[n-j]\right) \sum_{k=1}^{p} \hat{a}[k]x[n-k] = 0$$

so that (6.90) may be rewritten as

$$\hat{\sigma}^2 = \frac{1}{N-p} \sum_{n=p}^{N-1} x^2[n] + \sum_{j=1}^{p} \hat{a}[j] \frac{1}{N-p} \sum_{n=p}^{N-1} x[n]x[n-j]$$

and finally

$$\hat{\sigma}^2 = c_{xx}[0,0] + \sum_{j=1}^{p} \hat{a}[j]c_{xx}[0,j]. \tag{6.91}$$

It should be noted that $c_{xx}[j, k] = c_{xx}[k, j]$, and therefore $\mathbf{C}$ is a symmetric matrix.

Furthermore, $\mathbf{C}$ is positive semidefinite since

$$\begin{aligned}\boldsymbol{\beta}^T\mathbf{C}\boldsymbol{\beta} &= \sum_{j=1}^{p}\sum_{k=1}^{p} \beta_j\beta_k c_{xx}[j, k] \\ &= \frac{1}{N-p}\sum_{n=p}^{N-1}\sum_{j=1}^{p}\sum_{k=1}^{p} \beta_j\beta_k x[n-j]x[n-k] \\ &= \frac{1}{N-p}\sum_{n=p}^{N-1}\left(\sum_{j=1}^{p} \beta_j x[n-j]\right)^2 \geq 0. \end{aligned} \tag{6.92}$$

The quadratic form in (6.92) will be strictly positive and hence $\mathbf{C}$ will be positive definite, and therefore invertible, if $x[n]$ does not consist solely of sinusoids and/or damped exponentials. Otherwise, we could have

$$\sum_{j=1}^{p} \beta_j x[n-j] = 0 \qquad n = p, p+1, \ldots, N-1 \tag{6.93}$$

and thus $\mathbf{C}$ would not be invertible. Equivalently, the solution of (6.89) would not be unique. (Specifically, if $x[n]$ were composed of L real sinusoids, M real damped sinusoids, and J real damped exponentials, then $\mathbf{C}$ would not be invertible if $2L + 2M + J < p$.)

The MLE of the AR parameters for complex $x[n]$ is identical to that given in (6.89) except that $c_{xx}[j, k]$ is defined as

$$c_{xx}[j, k] = \frac{1}{N-p}\sum_{n=p}^{N-1} x^*[n-j]x[n-k]. \tag{6.94}$$

The approximate MLE as given by (6.89) and (6.91) is called the *covariance method* of linear prediction [Makhoul 1975]. It is also discussed in Section 7.4 and a computer program entitled COVMCOV, which is given in Appendix 7C, implements the method. Since the covariance method is only an approximate MLE, we can introduce many variants by redefining the estimator of the ACF, $c_{xx}[j, k]$. These AR parameter estimators are detailed in Chapter 7, which by the foregoing discussion are all MLEs for large data records. As an example, if we modify the range of summation in (6.94) to $n = 0$ to $n = N - 1 + p$, change the $N - p$ divisor to N, and assume that $x[n] = 0$ for n not in the range 0 to $N - 1$, then

$$c_{xx}[j, k] = \frac{1}{N}\sum_{n=0}^{N-1+p} x^*[n-j]x[n-k].$$

Letting $m = n - j$ yields

$$\begin{aligned} c_{xx}[j, k] &= \frac{1}{N}\sum_{m=-j}^{N-1+p-j} x^*[m]x[m+j-k] \\ &= \frac{1}{N}\sum_{m=0}^{N-1+p-j} x^*[m]x[m+j-k] \end{aligned}$$

since $x[n] = 0$ for $n < 0$. Assuming that $j \geq k$ and noting that $x[n] = 0$ for $n > N - 1$, we have

$$\begin{aligned} c_{xx}[j, k] &= \frac{1}{N} \sum_{m=0}^{N-1-(j-k)} x^*[m]x[m + j - k] \\ &= \hat{r}_{xx}[j - k] \end{aligned} \tag{6.95}$$

which is the biased estimator of the ACF at lag $j - k$. With this modification (6.89) becomes the Yule–Walker equations with the biased ACF estimator. This estimator is called the *autocorrelation method* of linear prediction [Makhoul 1975]. It is also discussed in Section 7.3 and a computer program entitled AUTOCORR, which is given in Appendix 7B, implements this method. Note that for large data records the inclusion of the extra $2p$ terms in $c_{xx}[j, k]$ will be negligible.

Maximum likelihood estimation of the reflection coefficients follows directly from the estimators for the AR parameters. Although the reflection coefficients can be estimated by various methods, the following discussion assumes that they are estimated by the autocorrelation method, which solves the Yule–Walker equations using the Levinson algorithm. The Levinson algorithm will produce AR parameter estimates as well as reflection coefficient estimates. Since the reflection coefficients are functions of the AR parameters, by the invariance property of MLEs (see Section 3.3.1), the reflection coefficient estimates obtained via the Levinson algorithm will also be MLEs for large data records. As an example, for a real AR(2) process it is easily shown that

$$\begin{aligned} k_1 &= \frac{a[1]}{1 + a[2]} \\ k_2 &= a[2] \end{aligned}$$

which follows directly from (6.51). The other methods for reflection coefficient estimation described in Chapter 7 all produce the same numerical values for large data records and hence are also approximate MLEs [Kay and Makhoul 1983].

6.5.2 Statistics of AR Parameter and Reflection Coefficient Estimators

Determination of the statistics of the AR parameter and reflection coefficient estimators is a formidable, if not impossible task. No exact results are available for even the simplest case of $p = 1$. We must therefore resort to approximations based on large sample theory, and in particular the large sample properties of the MLE as described in Section 3.3.1. The case of a *real* AR process is now considered. Although the statistics for the MLEs based on complex data can probably be determined in a similar manner, no results appear to be available.

It is well known that for large data records (N large) MLEs are unbiased estimators and have variances that attain the Cramer–Rao (CR) lower bound. Furthermore, for large data records the MLE is also distributed as a Gaussian random variable. For these asymptotic results to apply for finite data records, it is necessary and sufficient that:

1. The maximization of the conditional likelihood function as given by (6.86) be a good approximation to that of the unconditional likelihood function as given by (6.80).
2. The maximum of (6.86) not be close to a stability boundary in the AR parameter space.
3. The ACF estimation errors are small.

These requirements will be met for moderately sized data records if the poles of $1/\mathcal{A}(z)$ are not near the unit circle of the z-plane. Assuming this to be the case, the statistics of the AR parameter and reflection coefficient estimators are now summarized.

From the asymptotic properties of the MLE, $[\mathbf{a}^T \quad \sigma^2]^T$ will be distributed according to a real multivariate Gaussian PDF with mean [Box and Jenkins 1970]

$$\boldsymbol{\mu} = \mathcal{E}\begin{bmatrix} \hat{\mathbf{a}} \\ \hat{\sigma}^2 \end{bmatrix} - \begin{bmatrix} \mathbf{a} \\ \sigma^2 \end{bmatrix} \tag{6.96}$$

and covariance matrix (see Appendix 6A)

$$\begin{aligned} \mathbf{C}_{a,\sigma^2} &= \mathcal{E}\left\{\begin{bmatrix} \hat{\mathbf{a}} - \mathcal{E}(\hat{\mathbf{a}}) \\ \hat{\sigma}^2 - \mathcal{E}(\hat{\sigma}^2) \end{bmatrix} \begin{bmatrix} \hat{\mathbf{a}} - \mathcal{E}(\hat{\mathbf{a}}) \\ \hat{\sigma}^2 - \mathcal{E}(\hat{\sigma}^2) \end{bmatrix}^T\right\} \\ &= \begin{bmatrix} \dfrac{\sigma^2}{N}\mathbf{R}_{xx}^{-1} & \mathbf{0} \\ \mathbf{0}^T & \dfrac{2\sigma^4}{N} \end{bmatrix} \end{aligned} \tag{6.97}$$

where $\mathbf{R}_{xx}$ is the $p \times p$ autocorrelation matrix and $\mathbf{0}$ is a vector of dimension $p \times 1$ of all zeros. The covariance matrix is the inverse of the Fisher information matrix and thus is the CR lower bound. Hence, even if (6.97) is not a good approximation to the covariance of the AR parameter estimator, it still provides a lower bound in that

$$\begin{aligned} \text{var}\,(\hat{a}[i]) &\geq \frac{\sigma^2}{N}[\mathbf{R}_{xx}^{-1}]_{ii} \qquad i = 1, 2, \ldots, p \\ \text{var}\,(\hat{\sigma}^2) &\geq \frac{2\sigma^4}{N} \end{aligned} \tag{6.98}$$

assuming that $\hat{\mathbf{a}}$, $\hat{\sigma}^2$ are unbiased. It is interesting to note that $\hat{\mathbf{a}}$ and $\hat{\sigma}^2$ are asymptotically uncorrelated and consequently, independent (since they are jointly Gaussian), even though $\hat{\sigma}^2$ clearly depends on $\mathbf{a}$. These results were rigorously proven by Mann and Wald [1943].

The corresponding results for the reflection coefficients are found by applying the properties of MLEs as described in Section 3.3.1. Denoting

$[k_1 \quad k_2 \quad \cdots \quad k_p]^T$ by $\mathbf{k}$, for large data records $\mathbf{k}$ will be distributed according to a multivariate Gaussian PDF with mean

$$\boldsymbol{\mu} = \mathcal{E}[\hat{\mathbf{k}}] = \mathbf{k} \tag{6.99}$$

and covariance matrix given by (3.23) as

$$\begin{aligned} \mathbf{C}_k &= \mathcal{E}\{[\hat{\mathbf{k}} - \mathcal{E}(\hat{\mathbf{k}})][\hat{\mathbf{k}} - \mathcal{E}(\hat{\mathbf{k}})]^T\} \\ &= \mathbf{A}\mathbf{C}_a\mathbf{A}^T \end{aligned} \tag{6.100}$$

where $\mathbf{C}_a = \sigma^2\mathbf{R}_{xx}^{-1}/N$ is the covariance matrix of the AR parameter estimator and $\mathbf{A}$ is the $p \times p$ Jacobian matrix with $[\mathbf{A}]_{ij} = \partial k_i/\partial a[j]$ [Barndorf-Nielsen and Schou 1973]. $\mathbf{C}_k$ is the inverse of the Fisher information matrix for $\mathbf{k}$. As an example, for $p = 1$,

$$\mathbf{C}_k = \text{var}(\hat{k}_1) = \frac{1}{N}(1 - k_1^2)$$

and for $p = 2$,

$$\mathbf{C}_k = \frac{1}{N}\begin{bmatrix} \dfrac{(1 - k_1^2)(1 - k_2)}{1 + k_2} & 0 \\ 0 & 1 - k_2^2 \end{bmatrix}.$$

For arbitrary p the calculation of $\mathbf{A}$ is tedious so that the following recursive algorithm may be used for the computation of the covariance matrix [Kay and Makhoul 1983].

1. For $n = 1$,

$$\begin{aligned} \mathbf{a}_1 &= a_1[1] = k_1 \\ \mathbf{A}_1 &= \mathbf{B}_1 = \mathbf{E}_1 = \rho_0 = 1 \\ \rho_1 &= (1 - k_1^2)\rho_0 \\ \mathbf{C}_k(1) &= \frac{1}{N}(1 - k_1^2) \end{aligned}$$

2. Increment n by one and compute

$$\begin{aligned} \rho_n &= (1 - k_n^2)\rho_{n-1} \\ \mathbf{D}_{n-1} &= \frac{\mathbf{A}_{n-1}(\mathbf{I}_{n-1} - k_n\mathbf{J}_{n-1})\mathbf{B}_{n-1}}{1 - k_n^2} \\ \mathbf{C}_k(n) &= \frac{1}{N}\begin{pmatrix} \rho_n\mathbf{D}_{n-1}\mathbf{E}_{n-1}^{-1}\mathbf{D}_{n-1}^T & \mathbf{0} \\ \mathbf{0}^T & 1 - k_n^2 \end{pmatrix} \end{aligned}$$

3. If $n = p$, then $\mathbf{C}_k = \mathbf{C}_k(n)$ and exit. Otherwise,

4.

$$\mathbf{E}_n^{-1} = \begin{bmatrix} \mathbf{E}_{n-1}^{-1} & \mathbf{0} \\ \mathbf{0}^T & \dfrac{1}{\rho_{n-1}} \end{bmatrix}$$

$$\mathbf{a}_{n-1} = \begin{bmatrix} (\mathbf{I}_{n-2} + k_{n-1}\mathbf{J}_{n-2})\mathbf{a}_{n-2} \\ k_{n-1} \end{bmatrix}$$

$$\mathbf{B}_n = \begin{bmatrix} \mathbf{B}_{n-1} & \mathbf{J}_{n-1}\mathbf{a}_{n-1} \\ \mathbf{0}^T & 1 \end{bmatrix}$$

$$\mathbf{A}_n = \begin{bmatrix} \dfrac{\mathbf{A}_{n-1}(\mathbf{I}_{n-1} - k_n\mathbf{J}_{n-1})}{1 - k_n^2} & -\dfrac{\mathbf{A}_{n-1}(\mathbf{I}_{n-1} - k_n\mathbf{J}_{n-1})\mathbf{J}_{n-1}\mathbf{a}_{n-1}}{1 - k_n^2} \\ \mathbf{0}^T & 1 \end{bmatrix}$$

5. Go to step 2.

Note that $\mathbf{a}_{n-1}$ and $\mathbf{0}$ have the dimensions $(n - 1) \times 1$ and $\mathbf{I}_n$, $\mathbf{J}_n$ are the $n \times n$ identity and exchange matrices, respectively. The dimensions of the $\mathbf{A}_n$, $\mathbf{B}_n$, $\mathbf{D}_n$, and $\mathbf{E}_n$ matrices are $n \times n$. The algorithm computes the covariance matrix of the reflection coefficients for an AR process with parameters $\{k_1, k_2, \ldots, k_n\}$ in a recursive manner until the desired one $(n = p)$ is attained. It is seen from the form of the covariance matrix that for an AR(p) process, $\hat{k}_p$ is independent of the lower order reflection coefficients and

$$\operatorname{var}(\hat{k}_p) = \frac{1}{N}(1 - k_p^2). \tag{6.101}$$

Also, by repeated application of the algorithm, it can be shown that since $k_n = 0$ for an AR(p) process, if $n > p$,

$$\operatorname{var}(\hat{k}_n) = \frac{1}{N} \qquad n > p. \tag{6.102}$$

The latter result can be used to test the order of an AR process as described in Section 7.8.

6.6 ESTIMATION OF THE AR POWER SPECTRAL DENSITY

6.6.1 Maximum Likelihood Estimation

Once the MLE for the AR parameters has been obtained it is a simple matter to determine the MLE for the AR PSD [Parzen 1968]. We need only substitute the estimated AR parameters obtained from (6.89) and (6.91) into the theoretical PSD to obtain

$$\hat{P}_{AR}(f) = \frac{\hat{\sigma}^2}{|1 + \hat{a}[1] \exp(-j2\pi f) + \cdots + \hat{a}[p] \exp(-j2\pi fp)|^2}. \tag{6.103}$$

That this is the MLE for the AR PSD follows from the invariance property of the MLE discussed in Section 3.3.1. Any of the AR parameter estimation methods discussed so far and those to be described in Chapter 7 will result in an approximate MLE for the AR PSD.

6.6.2 Statistics of AR Power Spectral Density Estimator

As might be expected, exact results for the statistics of the AR spectral estimator are not available. Approximations based on large sample theory will be given. All the results in this section are valid only for *real AR processes*. It has been shown that for large data records $\hat{P}_{AR}(f)$ is distributed according to a Gaussian PDF with [Kromer 1970, Berk 1974]

$$\begin{aligned}
\mathscr{E}[\hat{P}_{AR}(f)] &= P_{AR}(f) \\
\operatorname{var}[\hat{P}_{AR}(f)] &= \begin{cases} \dfrac{4p}{N} P_{AR}^2(f) & \text{for } f = 0, \pm\frac{1}{2} \\[2ex] \dfrac{2p}{N} P_{AR}^2(f) & \text{otherwise} \end{cases} \\
\operatorname{cov}[\hat{P}_{AR}(f_1), \hat{P}_{AR}(f_2)] &= 0 \qquad f_1 \neq f_2.
\end{aligned} \tag{6.104}$$

These results were derived assuming not only that $N \to \infty$ but also that $p \to \infty$, so it is not known how well the approximation performs in practice. Nonetheless, it is of interest to observe that the variance of the AR spectral estimator decreases as N/p increases. Also, the mean to standard deviation ratio, which should be large for a reliable spectral estimator, is, from (6.104),

$$\frac{\mathscr{E}[\hat{P}_{AR}(f)]}{\sqrt{\operatorname{var}(\hat{P}_{AR}(f))}} = \begin{cases} \sqrt{\dfrac{N}{4p}} & \text{for } f = 0, \pm\frac{1}{2} \\[3ex] \sqrt{\dfrac{N}{2p}} & \text{otherwise.} \end{cases} \tag{6.105}$$

Finally, because of the form of the variance, if we plot the logarithm of $\hat{P}_{AR}(f)$ versus f, the confidence interval will be the same length for any f. It may be shown that the $100(1 - \alpha)\%$ confidence interval for $f \neq 0, \pm\frac{1}{2}$ is

$$10 \log_{10} \hat{P}_{AR}(f) \begin{cases} -10 \log_{10}\left[1 + z\left(1 - \dfrac{\alpha}{2}\right)\sqrt{\dfrac{2p}{N}}\right] \\[3ex] -10 \log_{10}\left[1 - z\left(1 - \dfrac{\alpha}{2}\right)\sqrt{\dfrac{2p}{N}}\right] \end{cases}$$

where $z(\alpha)$ is the α percentage point of a $N(0, 1)$ CDF (see Section 3.3.1). Consequently, it behooves us to plot the PSD estimate on a logarithmic (dB) rather than a linear scale.

Statistics may also be obtained for the AR spectral estimator by relying on the large data record properties of the MLE. In this case no assumption about a large AR model order need be made. The general results are stated in Chapter 9 for an ARMA process. For the special case of an AR process, if

$$\mathbf{p} = [\hat{P}_{\text{AR}}(f_1) \quad \hat{P}_{\text{AR}}(f_2) \quad \cdots \quad \hat{P}_{\text{AR}}(f_M)]^T$$

then for large data records $\mathbf{p}$ is distributed according to a multivariate Gaussian PDF with mean equal to the true value and covariance matrix (see Section 9.4)

$$\mathbf{C}_p = \mathbf{D}\mathbf{C}_{a,\sigma^2}\mathbf{D}^T \tag{6.106}$$

where $\mathbf{D}$ is the $M \times (p + 1)$ Jacobian matrix of the transformation from $\mathbf{a}$, σ^2 to $\mathbf{p}$.

$$[\mathbf{D}]_{mn} = \begin{cases} -2P_{\text{AR}}(f_m)\,\text{Re}\left[\dfrac{\exp(-j2\pi f_m n)}{A(f_m)}\right] & \text{for } m = 1, 2, \ldots, M;\ n = 1, 2, \ldots, p \\ \dfrac{P_{\text{AR}}(f_m)}{\sigma^2} & \text{for } m = 1, 2, \ldots, M;\ n = p + 1 \end{cases}$$

and $\mathbf{C}_{a,\sigma^2}$ is given by (6.97). In particular, at any frequency the variance is given by (6.106) for $M = 1$ or

$$\text{var}\,[\hat{P}_{\text{AR}}(f)] = \frac{4\sigma^2}{N} P_{\text{AR}}^2(f)\mathbf{e}^T\mathbf{R}_{xx}^{-1}\mathbf{e} + \frac{2}{N} P_{\text{AR}}^2(f) \tag{6.107)}$$

where $[\mathbf{e}]_n = -\text{Re}\,[\exp(-j2\pi f n)/A(f)]$ for $n = 1, 2, \ldots, p$ and $\mathbf{R}_{xx}$ is the $p \times p$ autocorrelation matrix (see Problem 6.12). Similar results can be found in Akaike [1969]. Note that a confidence interval based on these results will have a length that is frequency dependent.

6.7 EFFECT OF NOISE ON THE AR SPECTRAL ESTIMATOR

A very important problem with the AR spectral estimator, which limits its utility, is its sensitivity to the addition of observation noise to the AR process. An example is given in Figure 6.9 for data consisting of two real sinusoids with powers separated by 6 dB in white noise. Using an AR model order of $p = 4$, it is seen that the spectral peaks are broadened, displaced from their true positions (indicated by dashed lines), and one peak is completely lost at low SNR. It would appear that the resolution of the AR spectral estimator decreases as the SNR decreases. Numerous reported simulations attest to this property [Lacoss 1971, Chen and Stegen 1974, Marple 1977]. Methods for estimating AR parameters in the presence of observation noise are given in Section 7.9.

One means of assessing the resolution of the AR spectral estimator is by measuring the minimum frequency separation of two equiamplitude sinusoids embedded in white noise for which the two sinusoids can be discerned. If the data record is large enough to ignore the variance of the spectral estimator and thus to consider only the bias as being significant, then it is appropriate to consider the ACF samples as known. The results to follow make use of this assumption.

The resolution in cycles/sample of the AR spectral estimator, assuming known ACF samples, has been found to be approximately [Marple 1976]

$$\delta f_{AR} = \frac{1.03}{p[\eta(p+1)]^{0.31}} \tag{6.108}$$

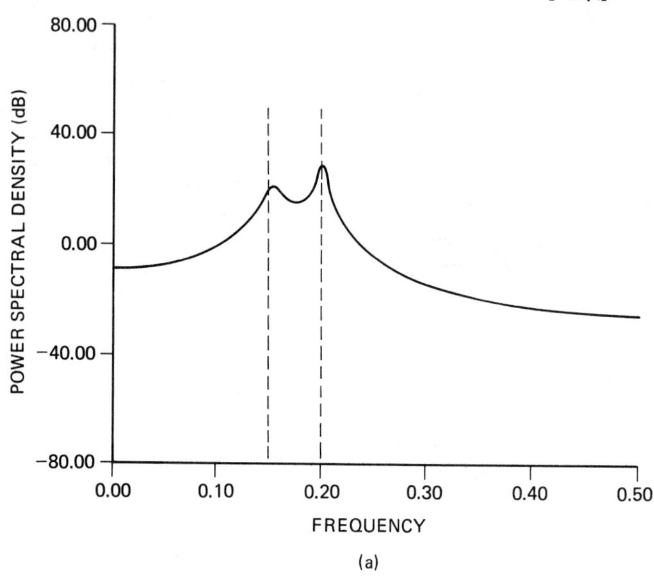

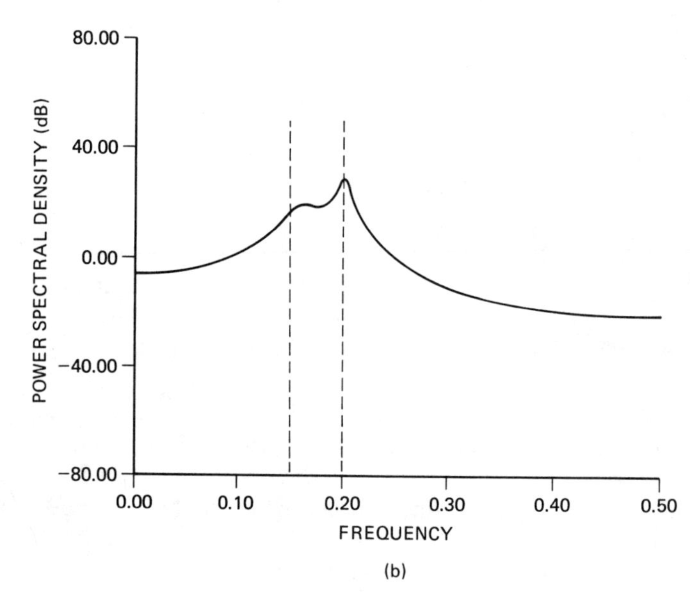

Figure 6.9 Degradation of resolution of AR spectral estimator with SNR. (a) SNR = 27, 33 dB. (b) SNR = 23, 29 dB. (c) SNR = 20, 26 dB. (d) SNR = 17, 23 dB.

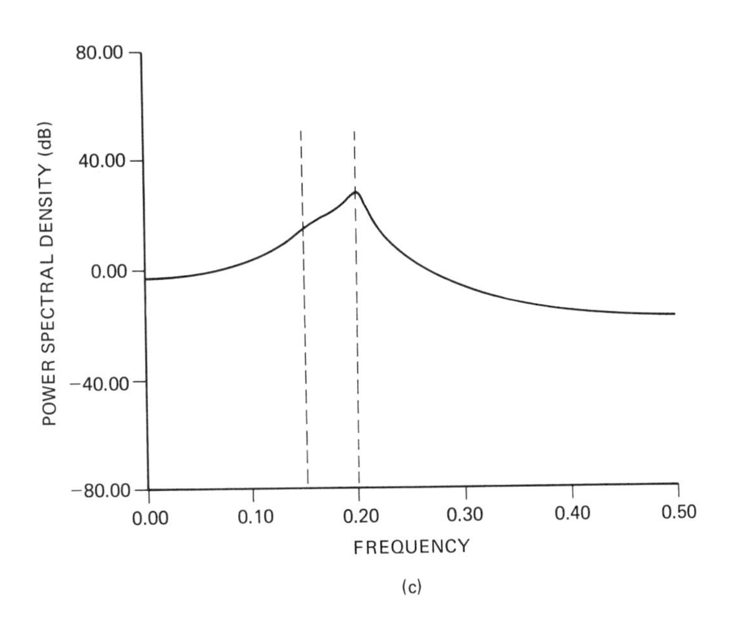

(c)

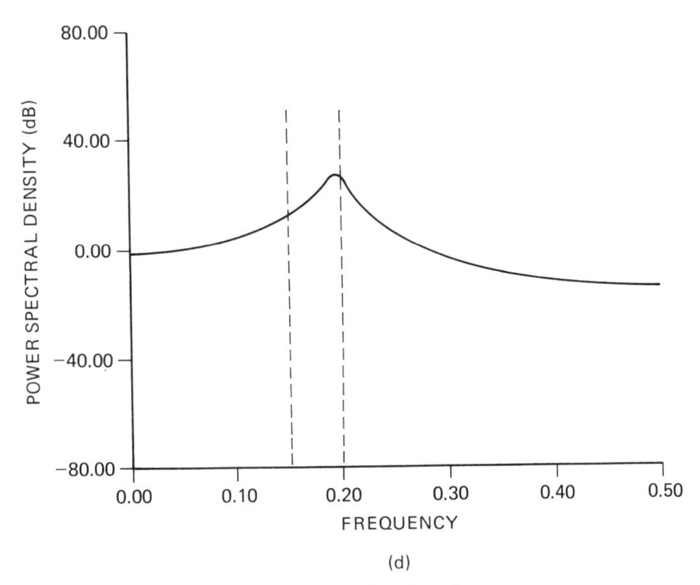

(d)

Figure 6.9 *(Continued)*

for $p\eta > 10$, where η is the SNR of one sinusoid. As expected, the resolution decreases with decreasing SNR but also increases with increasing model order. The resolution of the periodogram can similarly be shown to be [Marple 1976]

$$\delta f_p = \frac{0.86}{N} \tag{6.109}$$

where N is the number of data samples. As has frequently been observed in practice, there is no dependence on SNR. Now assume a model order of $p = 4$, which would be the appropriate order for two real sinusoids and compare the resolutions as

$$\gamma = \frac{1/\delta f_{AR}}{1/\delta f_p}. \tag{6.110}$$

A value of γ greater than 1 indicates a superior resolution for the AR spectral estimator. From (6.108) and (6.109), γ becomes

$$\gamma = \frac{5.5\eta^{0.31}}{N}. \tag{6.111}$$

For the two spectral estimators to have the same resolution or $\gamma = 1$, the SNR needs to be

$$10 \log_{10} \eta = -23.9 + 32.2 \log_{10} N \quad \text{dB} \tag{6.112}$$

for $10 \log_{10} \eta > 4$ dB. This is plotted in Figure 6.10. It is seen that very high SNRs may be necessary for the AR spectral estimator to have higher resolution than the periodogram. For a given SNR the resolution of the AR spectral estimator may be improved by increasing the model order. However, as p increases relative to the data record length, the variance increases according to (6.104). If p is chosen too large, spurious peaks in the spectral estimate will appear. This is due to the extra or "noise" poles, which have situated themselves too close to the unit circle. An example of this type of behavior is given in Figure 6.11 for data consisting of two real sinusoids in white noise. Clearly, the issue of model order selection is a critical one. For low SNR the resolution of the AR spectral estimator is no better than that of the periodogram. The reason for the degradation is that the all-pole model assumed in AR spectral estimation is no longer valid when obser-

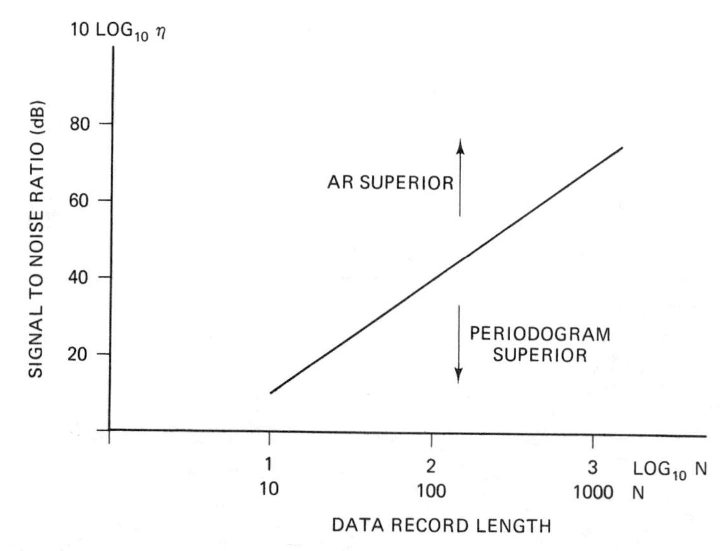

Figure 6.10 Comparison of resolution of autoregressive and periodogram spectral estimators.

vation noise is present. If $y[n]$ denotes the noise corrupted AR process $x[n]$, and $w[n]$ denotes observation noise, the observed process is

$$y[n] = x[n] + w[n]. \tag{6.113}$$

Assuming that the noise is white with variance σ_w^2 and is uncorrelated with $x[n]$,

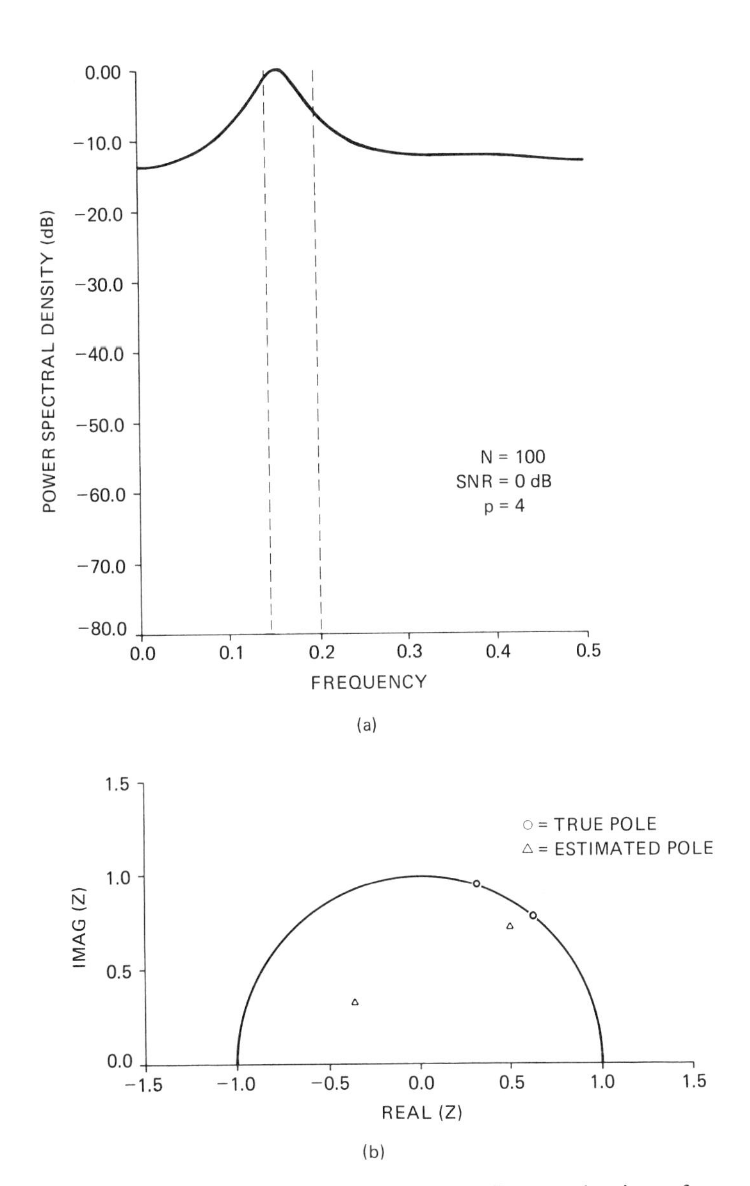

Figure 6.11 Effect of increasing model order on AR spectral estimate for two real sinusoids in white noise. (a),(c) Spectral estimates. (b),(d) Pole plots.

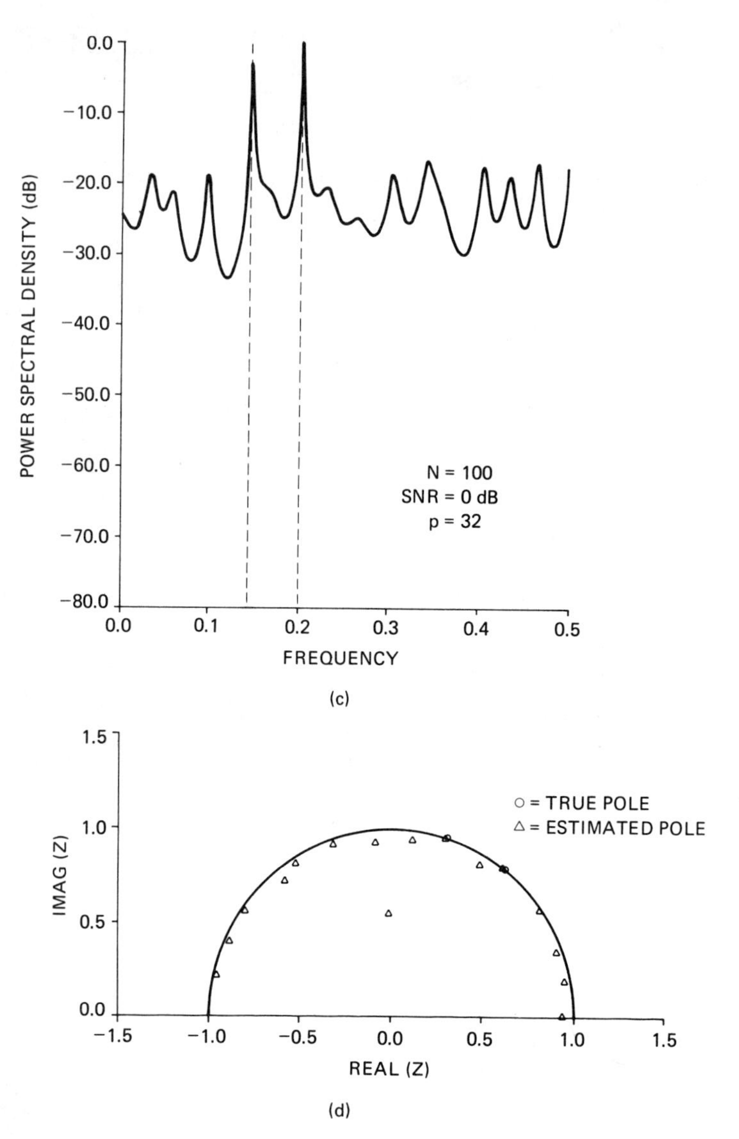

Figure 6.11 (*Continued*)

the PSD is

$$\begin{aligned}\mathcal{P}_{yy}(z) &= \frac{\sigma^2}{\mathcal{A}(z)\mathcal{A}^*(1/z^*)} + \sigma_w^2 \\ &= \frac{[\sigma^2 + \sigma_w^2 \mathcal{A}(z)\mathcal{A}^*(1/z^*)]}{\mathcal{A}(z)\mathcal{A}^*(1/z^*)}.\end{aligned} \tag{6.114}$$

Thus the PSD of $y[n]$ is characterized by *zeros* as well as poles. That is, the appropriate model for the observed process $y[n]$ is an ARMA(p, p) process, as has already been discussed in Section 5.6. The inconsistency of the AR model for a noise corrupted AR process leads to the degradation observed in Figure 6.9.

The phenomenon is explained as follows. The effect of noise is to reduce the dynamic range of the PSD of the observed process. Since the PEF $\mathcal{A}(z)$ attempts to whiten the PSD, it is not surprising that for low SNR, the zeros of $\mathcal{A}(z)$ are located near the origin of the z-plane, so that $\mathcal{A}(z) \approx 1$. This is because the PSD of $y[n]$ is already relatively flat due to noise, and hence any subsequent filtering operations by a PEF will not significantly whiten the PSD further. The AR spectral estimate in the presence of white observation noise will thus be a smoothed version of the AR spectral estimate which would have been obtained if no noise were present. More quantitatively, the spectral flatness for the AR spectral estimate based on $y[n]$ exceeds that based on $x[n]$ or in terms of the spectral flatness measure $\xi_y > \xi_x$ [Kay 1979]. To prove this property, let ρ_x, ρ_y denote the minimum prediction error power at the output of the optimal pth order PEFs $A_x(f)$, $A_y(f)$ for $x[n]$ and $y[n]$, respectively. Then

$$\hat{P}_{xx}(f) = \frac{\rho_x}{|A_x(f)|^2} \tag{6.115}$$

$$\hat{P}_{yy}(f) = \frac{\rho_y}{|A_y(f)|^2} \tag{6.116}$$

where $\hat{P}_{xx}(f)$ and $\hat{P}_{yy}(f)$ are the pth order AR spectral estimators based on $x[n]$ and $y[n]$, respectively. It is now shown that

$$\xi_y = \frac{\exp\left[\int_{-1/2}^{1/2} \ln \hat{P}_{yy}(f)\, df\right]}{\int_{-1/2}^{1/2} \hat{P}_{yy}(f)\, df} > \frac{\exp\left[\int_{-1/2}^{1/2} \ln \hat{P}_{xx}(f)\, df\right]}{\int_{-1/2}^{1/2} \hat{P}_{xx}(f)\, df} = \xi_x \tag{6.117}$$

regardless of whether or not $x[n]$ is truly an AR(p) process. It is concluded that the *effect of white observation noise is to produce a flattened AR spectral estimate, regardless of the nature of the process under consideration.* Using (6.75) and noting the correlation matching property of the AR spectral estimator (see Section 6.4.1) yields

$$\xi_x = \frac{\rho_x}{r_{xx}[0]} \tag{6.118}$$

and similarly for ξ_y. Next it is proven that

$$\rho_y > \rho_x + \sigma_w^2. \tag{6.119}$$

Consider the minimization of

$$J = \int_{-1/2}^{1/2} |A(f)|^2 P_{yy}(f)\, df$$

over all PEFs. Since

$$J = \int_{-1/2}^{1/2} |A(f)|^2 P_{xx}(f)\, df + \int_{-1/2}^{1/2} |A(f)|^2 \sigma_w^2\, df$$

it is clear that any *one* $A(f)$ cannot produce a minimum of J which is equal to or less than that which would be obtained if each integral were minimized separately.

Performing the minimization of each integral separately produces

$$\rho_y > \rho_x + \sigma_w^2.$$

The second integral, which is equivalent to $\sigma_w^2 \sum_{k=0}^{p} | a[k] |^2$, is minimized for $a[k] = 0$, $k = 1, 2, \ldots, p$. This leads to the desired result,

$$\begin{aligned}
\xi_y &= \frac{\rho_y}{r_{yy}[0]} = \frac{\rho_y}{r_{xx}[0] + \sigma_w^2} > \frac{\rho_x + \sigma_w^2}{r_{xx}[0] + \sigma_w^2} \\
&= \frac{\xi_x r_{xx}[0] + \sigma_w^2}{r_{xx}[0] + \sigma_w^2} \\
&> \frac{\xi_x r_{xx}[0] + \xi_x \sigma_w^2}{r_{xx}[0] + \sigma_w^2} = \xi_x.
\end{aligned}$$

The last inequality assumes that $x[n]$ is not white noise, so that $\xi_x < 1$, completing the proof.

Some further insight may be gained by considering the special cases of one and two complex sinusoids in complex white noise [Lacoss 1971]. First, consider one complex sinusoid, so that the ACF is

$$r_{yy}[k] = P \exp(j2\pi f_0 k) + \sigma_w^2 \delta[k]. \tag{6.120}$$

Then using the Yule–Walker equations, we find that [see (6.134) and (6.135)]

$$\begin{aligned}
a[i] &= -\frac{P}{pP + \sigma_w^2} \exp(j2\pi f_0 i) \qquad i = 1, 2, \ldots, p \\
\sigma^2 &= \sigma_w^2 \left(1 + \frac{P}{pP + \sigma_w^2}\right)
\end{aligned} \tag{6.121}$$

and therefore

$$P_{\text{AR}}(f) = \frac{\sigma_w^2 \left(1 + \dfrac{P}{pP + \sigma_w^2}\right)}{\left| 1 - \dfrac{P}{pP + \sigma_w^2} \sum_{i=1}^{p} \exp[-j2\pi(f - f_0)i] \right|^2}.$$

Letting

$$B_p(f) = \frac{1}{p} \sum_{n=0}^{p-1} \exp(j2\pi f n) = \frac{\sin p\pi f}{p \sin \pi f} \exp[j\pi f(p - 1)]$$

the final form for the AR spectral estimator is

$$P_{\text{AR}}(f) = \frac{\sigma_w^2 \left(1 + \dfrac{P}{pP + \sigma_w^2}\right)}{\left| 1 - \dfrac{pP}{pP + \sigma_w^2} \exp[j2\pi(f - f_0)] B_p(f - f_0) \right|^2}. \tag{6.122}$$

Note that the peak of the PSD occurs at $f = f_0$ (see Problem 6.13) and can easily be shown to be

$$P_{AR}(f)_{MAX} = P_{AR}(f_0) = \sigma_w^2[1 + \eta p][1 + \eta(p + 1)] \tag{6.123}$$

where $\eta = P/\sigma_w^2 = \text{SNR}$. For $\eta p \gg 1$ and $p \gg 1$ the peak value becomes

$$P_{AR}(f)_{MAX} = \sigma_w^2(\eta p)^2. \tag{6.124}$$

The peak of the PSD is not proportional to the power of the sinusoid as is the case for a Fourier based spectral estimator but to the square of the power. We must therefore be careful in interpreting the spectral estimate since relative peak heights do not directly relate to sinusoidal powers. The 3 dB bandwidth of the spectral peak can be shown to be [Lacoss 1971]

$$\text{BW} = \frac{2}{\pi(p + 1)^2\eta}. \tag{6.125}$$

For the case of two complex sinusoids in white noise the ACF is

$$r_{yy}[k] = P_1 \exp(j2\pi f_1 k) + P_2 \exp(j2\pi f_2 k) + \sigma_w^2\delta[k]. \tag{6.126}$$

Using this in the Yule–Walker equations produces an AR spectral estimate that is extremely complicated. In general, it is found that the peaks of the PSD will not occur at the sinusoidal frequency locations and the 3 dB bandwidths will increase with decreasing model order and decreasing SNR (see also Problem 6.14). For the special case when the frequencies satisfy $f_1 = k/p$, $f_2 = l/p$, where k, l are distinct integers in the interval $[-p/2, p/2]$ for p even and $[-(p - 1)/2, (p - 1)/2]$ for p odd, we can show that

$$a[i] = -\left[\frac{P_1}{pP_1 + \sigma_w^2}\exp(j2\pi f_1 i) + \frac{P_2}{pP_2 + \sigma_w^2}\exp(j2\pi f_2 i)\right] \qquad i = 1, 2, \ldots, p \tag{6.127}$$

$$\sigma^2 = \sigma_w^2\left(1 + \frac{P_1}{pP_1 + \sigma_w^2} + \frac{P_2}{pP_2 + \sigma_w^2}\right)$$

so that

$$P_{AR}(f) = \sigma_w^2\left(1 + \frac{P_1}{pP_1 + \sigma_w^2} + \frac{P_2}{pP_2 + \sigma_w^2}\right) \left[\left|1 - \frac{pP_1}{pP_1 + \sigma_w^2}\exp[j2\pi(f - f_1)]B_p(f - f_1) - \frac{pP_2}{pP_2 + \sigma_w^2}\exp[j2\pi(f - f_2)]B_p(f - f_2)\right|^2\right]^{-1} \tag{6.128}$$

Equation (6.128) may be verified by noting that when the sinusoidal frequencies are distinct multiples of $1/p$, the autocorrelation matrix becomes a circulant matrix for which the inverse may be found in closed form (see Problem 2.5) or by applying the results given in (6.134) and (6.135). The analytical results can be used to explain why AR spectral estimates for sinusoids in white noise tend to accentuate the higher level sinusoid. From (6.128),

$$P_{\mathrm{AR}}(f_1) = \frac{\text{numerator}}{\left[\dfrac{\sigma_w^2}{pP_1 + \sigma_w^2}\right]^2}$$

since $B_p(f_1 - f_2) = 0$. Letting $\eta_1 = P_1/\sigma_w^2$ and assuming that $\eta_1 p \gg 1$ yields

$$P_{\mathrm{AR}}(f_1) \approx \text{numerator} \times (\eta_1 p)^2.$$

Similarly,

$$P_{\mathrm{AR}}(f_2) \approx \text{numerator} \times (\eta_2 p)^2$$

or

$$P_{\mathrm{AR}}(f_2) \approx \left(\frac{\eta_2}{\eta_1}\right)^2 P_{\mathrm{AR}}(f_1). \tag{6.129}$$

If, for example, the sinusoid at f_1 is 10 dB below that at f_2 ($\eta_1 = 0.1\eta_2$), then from (6.129), the AR spectral estimate will produce a peak at f_1 which is 20 dB below that at f_2.For two complex sinusoids of arbitrary frequencies in white noise the exact expression for the AR spectral estimate may be found in Lacoss [1971].

To reduce the computation required to solve the Yule–Walker equations for M complex sinusoids in white noise as well as to obtain some useful analytical results, we can make use of the following approach. If the ACF

$$r_{yy}[k] = \sum_{i=1}^{M} P_i \exp(j2\pi f_i k) + \sigma_w^2\delta[k] \tag{6.130}$$

is substituted into the Yule–Walker equations, it can be shown that the solution must satisfy

$$a[k] = \sum_{i=1}^{M} \gamma_i \exp[j2\pi f_i(k - 1)] \tag{6.131}$$

for some complex constants γ_i. Then, for $M < p$, (6.131) can be substituted into the Yule–Walker equations and solved for the unknown constants, thereby resulting in the solution of a smaller set of M equations. To do so, we solve the Yule–Walker equations, $\mathbf{R}_{yy}\mathbf{a} = -\mathbf{r}$, by noting that

$$\mathbf{R}_{yy} = \sigma_w^2\mathbf{I} + \sum_{i=1}^{M} P_i\mathbf{e}_i\mathbf{e}_i^H = \sigma_w^2\mathbf{I} + \mathbf{EPE}^H$$

$$\mathbf{a} = \sum_{i=1}^{M} \gamma_i\mathbf{e}_i = \mathbf{E}\boldsymbol{\gamma} \tag{6.132}$$

$$\mathbf{r} = \sum_{i=1}^{M} P_i \exp(j2\pi f_i)\mathbf{e}_i = \mathbf{EPh}.$$

The various matrices and vectors are defined as

$$\mathbf{e}_i = [1 \; \exp(j2\pi f_i) \cdots \exp[j2\pi f_i(p-1)]]^T$$

$$\mathbf{E} = [\mathbf{e}_1 \quad \mathbf{e}_2 \quad \cdots \quad \mathbf{e}_M] \qquad (p \times M)$$

$$\mathbf{P} = \operatorname{diag}(P_1, P_2, \ldots, P_M)$$

$$\boldsymbol{\gamma} = [\gamma_1 \quad \gamma_2 \quad \cdots \quad \gamma_M]^T$$

$$\mathbf{h} = [\exp(j2\pi f_1) \quad \exp(j2\pi f_2) \quad \cdots \quad \exp(j2\pi f_M)]^T.$$

Thus the Yule–Walker equations become

$$(\sigma_w^2 \mathbf{I} + \mathbf{E}\mathbf{P}\mathbf{E}^H)\mathbf{E}\boldsymbol{\gamma} = -\mathbf{E}\mathbf{P}\mathbf{h}$$

or

$$\mathbf{E}(\sigma_w^2 \boldsymbol{\gamma} + \mathbf{P}\mathbf{E}^H\mathbf{E}\boldsymbol{\gamma} + \mathbf{P}\mathbf{h}) = \mathbf{0}.$$

Since $\mathbf{E}$ is full rank, the vector within parentheses must be zero or

$$\sigma_w^2 \boldsymbol{\gamma} + \mathbf{P}\mathbf{E}^H\mathbf{E}\boldsymbol{\gamma} = -\mathbf{P}\mathbf{h} \tag{6.133}$$

which represents a set of M simultaneous equations. Because $[\mathbf{E}^H\mathbf{E}]_{mn} = \mathbf{e}_m^H\mathbf{e}_n$ and $[\mathbf{Ph}]_m = P_m \exp(j2\pi f_m)$, the equations that need to be solved can be explicitly written as

$$\sigma_w^2 \gamma_m + P_m \sum_{n=1}^{M} \mathbf{e}_m^H \mathbf{e}_n \gamma_n = -P_m \exp(j2\pi f_m) \qquad m = 1, 2, \ldots, M$$

or, after some manipulations,

$$\gamma_m + \sum_{\substack{n=1 \\ n \neq m}}^{M} c_{mn}\gamma_n = -\frac{P_m \exp(j2\pi f_m)}{pP_m + \sigma_w^2} \qquad m = 1, 2, \ldots, M \tag{6.134}$$

where

$$c_{mn} = \frac{P_m}{pP_m + \sigma_w^2} \sum_{i=0}^{p-1} \exp[j2\pi(f_n - f_m)i] = \frac{pP_m}{pP_m + \sigma_w^2} B_p(f_n - f_m).$$

The white noise variance can be found once (6.134) has been solved. Since

$$\sigma^2 = r_{yy}[0] + \mathbf{r}^H\mathbf{a}$$

and from (6.132),

$$\mathbf{r}^H\mathbf{a} = \mathbf{h}^H\mathbf{P}\mathbf{E}^H\mathbf{E}\boldsymbol{\gamma}$$

it follows from (6.133) that

$$\begin{aligned} \sigma^2 &= r_{yy}[0] + \mathbf{h}^H(-\sigma_w^2\boldsymbol{\gamma} - \mathbf{P}\mathbf{h}) \\ &= r_{yy}[0] - \sum_{i=1}^{M} P_i - \sigma_w^2 \sum_{i=1}^{M} \gamma_i \exp(-j2\pi f_i). \end{aligned}$$

The final result is

$$\sigma^2 = \sigma_w^2 \left[1 - \sum_{i=1}^{M} \gamma_i \exp(-j2\pi f_i) \right]. \tag{6.135}$$

The use of (6.134) and (6.135) for one sinusoid will verify (6.121). For multiple sinusoids with distinct frequencies at multiples of $1/p$, it is easily shown that $c_{mn} = 0$ for $m \neq n$, so the solution of (6.134) is particularly simple:

$$\gamma_m = -\frac{P_m \exp(j2\pi f_m)}{pP_m + \sigma_w^2} \qquad m = 1, 2, \ldots, M$$

which can be used to verify (6.127). An analogous result for real sinusoids is given by Satorius and Zeidler [1978] (see Problem 6.15).

6.8 CONSIDERATIONS IN MODEL ORDER SELECTION

Because the best choice of the AR model order is usually not known a priori, it is necessary in practice to postulate several model orders. Based on these we then compute some error criterion that indicates which model order to choose. These criteria, as well as specific model order estimation methods, are discussed in Section 7.8. If the process is truly AR(p), then a model order less than p will result in a smoothed spectral estimate. As an example, consider the AR(4) PSD shown in Figure 6.12 as a dashed curve and the AR spectral estimate based on

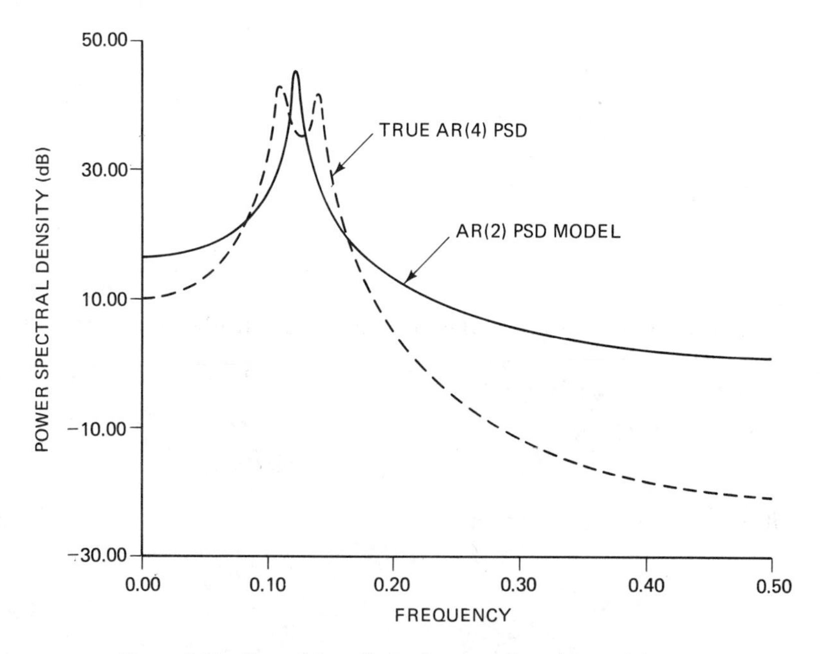

Figure 6.12 Smoothing effect of too small an AR model order.

a model order of 2, both for real data. We should thus choose a model order equal to or greater than p. If an order $k > p$ is chosen and if the ACF samples are estimated without error, the estimated AR parameters for the AR(k) model would be

$$\hat{a}_k[i] = \begin{cases} a_p[i] & \text{for } i = 1, 2, \ldots, p \\ 0 & \text{for } i = p + 1, p + 2, \ldots, k \end{cases}$$

and the correct PSD would result. In practice, however, spurious peaks may occur due to estimation errors, as was observed in Figure 6.11. Furthermore, if an AR spectral estimator is applied to a process that is not AR (AR in white noise as an example), then the true AR model would be one of infinite order. Any finite order AR model will introduce bias errors due to modeling inaccuracies. A trade-off must be effected between the desires to choose p large to reduce the bias and to choose p small to reduce the estimation errors. As will be discussed in Section 7.8, various measures for model order determination have been proposed which are based on the estimated prediction error power. They generally attempt to choose a model order that minimizes the prediction error power but at the same time uses as small a model order as possible. It is reasonable that the prediction error power be central to model order selection. This is because from the Levinson recursion ρ_k decreases monotonically with k and first reaches a minimum for $k = p$ for an AR(p) process.

REFERENCES

Akaike, H., "Power Spectrum Estimation through Autoregressive Model Fitting," *Ann. Inst. Statist. Math.*, Vol. 21, pp. 407–419, 1969.

Anderson, T. W., *The Statistical Analysis of Time Series*, Wiley, New York, 1971.

Barndorf-Nielsen, O., and G. Schou, "On the Parametrization of Autoregressive Models by Partial Autocorrelations," *J. Multivar. Anal.*, Vol. 3, pp. 408–419, 1973.

Berk, K. N., "Consistent Autoregressive Spectral Estimates," *Ann. Statist.*, Vol. 2, pp. 489–502, 1974.

Box, G. E. P., and G. M. Jenkins, *Time Series Analysis: Forecasting and Control*, Holden-Day, San Francisco, 1970.

Burg, J., "Maximum Entropy Spectral Analysis," Ph.D. dissertation, Stanford University, 1975.

Chen, W. Y., and G. R. Stegen, "Experiments with Maximum Entropy Power Spectra of Sinusoids," *J. Geophys. Res.*, Vol. 79, pp. 3019–3022, July 10, 1974.

Durbin, J., "The Fitting of Time-Series Models," *Rev. Inst. Int. Statist.*, Vol. 28, pp. 233–243, 1960.

Goodman, N. R., "Statistical Analysis Based on a Certain Multivariate Complex Gaussian Distribution," *Ann. Math. Statist.*, Vol. 34, pp. 152–157, 1963.

Gray, A. H., Jr., and J. D. Markel, "A Spectral Flatness Measure for Studying the Autocorrelation Method of Linear Prediction of Speech," *IEEE Trans. Acoust. Speech Signal Process.*, Vol. ASSP22, pp. 207–217, 1974.

Grenander, U., and G. Szego, *Toeplitz Forms and Their Applications,* University of California Press, Berkeley, Calif., 1958.

Jenkins, G. M., and D. G. Watts, *Spectral Analysis and Its Applications,* Holden-Day, San Francisco, 1968.

Kailath, T., A. Vieira, and M. Morf, "Inverses of Toeplitz Operators, Innovations, and Orthogonal Polynomials," *SIAM Rev.,* Vol. 20, pp. 106–119, 1978.

Kay, S., "The Effects of Noise on the Autoregressive Spectral Estimator," *IEEE Trans. Acoust. Speech Signal Process.,* ASSP27, pp. 478–485, Oct. 1979.

Kay, S., "Recursive Maximum Likelihood Estimation of Autoregressive Processes," *IEEE Trans. Acoust. Speech Signal Process.,* Vol. ASSP31, pp. 56–65, Feb. 1983.

Kay, S., and J. Makhoul, "On the Statistics of the Estimated Reflection Coefficients of an Autoregressive Process," *IEEE Trans. Acoust. Speech Signal Process.,* Vol. ASSP31, pp. 1447–1455, Dec. 1983.

Kendall, M. G., and A. Stuart, *The Advanced Theory of Statistics,* Vol. 2, Macmillan, New York, 1979.

Kromer, R. E., "Asymptotic Properties of the Autoregressive Spectral Estimator," Ph.D. dissertation, Stanford University, 1970.

Lacoss, R. T., "Data Adaptive Spectral Analysis Methods," *Geophysics,* Vol. 36, pp. 661–675, Aug. 1971.

Lang, S. W., and J. H. McClellan, "A Simple Proof of Stability for All-Pole Linear Prediction Models," *Proc. IEEE,* Vol. 67, pp. 860–861, May 1979.

Levinson, N., "The Wiener RMS (Root Mean Square) Error Criterion in Filter Design and Prediction," *J. Math. Phys.,* Vol. 25, pp. 261–278, 1947.

Luenberger, D. G., *Optimization by Vector Space Methods,* Wiley, New York, 1969.

Makhoul, J., "Linear Prediction: A Tutorial Review," *Proc. IEEE,* Vol. 63, pp. 561–580, Apr. 1975.

Mann, H. B., and A. Wald, "On the Statistical Treatment of Linear Stochastic Difference Equations," *Econometrica,* Vol. 11, pp. 173–220, 1943.

Markel, J. D., and A. H. Gray, Jr., *Linear Prediction of Speech,* Springer-Verlag, New York, 1976.

Marple, S. L., Jr., "Conventional Fourier, Autoregressive, and Special ARMA Methods of Spectrum Analysis," Engineer's dissertation, Stanford University, 1976.

Marple, S. L., Jr., "Resolution of Conventional Fourier, Autoregressive, and Special ARMA Methods of Spectral Analysis," in *Rec. 1977 IEEE Int. Conf. Acoust. Speech Signal Process.,* pp. 74–77. May 9–11, 1977.

Narasimha, M. J., K. Shenoi, and A. M. Peterson, "A Hilbert Space Approach to Linear Predictive Analysis of Speech Signals," Tech. Rep. 3606-10, Radioscience Lab., Stanford Electronics Lab., Stanford University, Stanford, Calif., 1974.

Pakula, L., and S. Kay, "Simple Proofs of the Minimum Phase Property of the Prediction Error Filter," *IEEE Trans. Acoust. Speech Signal Process.,* Vol. ASSP31, pp. 501–502, Apr. 1983.

Papoulis, A., *Probability, Random Variables, and Stochastic Processes,* McGraw-Hill, New York, 1965.

Parzen, E., "Statistical Spectral Analysis (Single Channel Case) in 1968," Tech. Rep. 11, Department of Statistics, Stanford University, Stanford, Calif., June 10, 1968.

Satorius, E. H., and J. Zeidler, "Maximum Entropy Spectral Analysis of Multiple Sinusoids in Noise," *Geophysics,* Vol. 43, pp. 1111–1118, Oct. 1978.

Siddiqui, M. M., "On the Inversion of the Sample Covariance Matrix in a Stationary Autoregressive Process," *Ann. Math. Statist.,* Vol. 58, pp. 585–588, 1958.

Tufts, D. W., F. Giannella, I. Kirsteins, and L. L. Scharf, "Cramer–Rao Bounds on the Accuracy of Autoregressive Parameter Estimators," *Proc. 2nd ASSP Spectrum Estimation Workshop,* pp. 12–16, 1983.

PROBLEMS

6.1. Show that the optimal one-step linear predictor for an AR(p) process based on the infinite past is the same as one based on the previous p samples. *Hint:* Show that the solution of the Wiener–Hopf (Yule–Walker) equations yields $a[k] = 0$ for $k > p$ for an AR(p) process.

6.2. Show that the optimal one-step linear predictor for an ARMA(1, 1) process based on the infinite past can be written as

$$\hat{x}[n] = -a[1]x[n-1] + b[1]u[n-1].$$

If $\hat{x}[n]$ is written as $-\sum_{k=1}^{\infty} \alpha_k x[n-k]$, find the α_k's. Generalize your results to an ARMA(p, q) process and comment on the use of linear prediction to estimate the AR parameters of an ARMA process. *Hint:* Replace the ARMA process by an AR(∞) model.

6.3. Consider the linear prediction of two complex sinusoids by a second order predictor. For an ACF given by $r_{xx}[k] = P_1 \exp(j2\pi f_1 k) + P_2 \exp(j2\pi f_2 k)$ find the prediction coefficients α_1, α_2 and the minimum prediction error power ρ_{MIN}. Where are the zeros of the PEF located?

6.4. Show that the optimal pth order linear predictor may be written as

$$\hat{x}[n] = -\sum_{i=1}^{p} k_i e_{i-1}^b[n-1]$$

by making use of (6.31) and (6.33).

6.5. Prove that the minimum forward prediction error power is equal to the minimum backward prediction error power, or

$$\mathcal{E}[|e_k^f[n]|^2] = \mathcal{E}[|e_k^b[n]|^2]$$

by making use of the property that the backward prediction coefficients are related to the forward prediction coefficients via (6.37).

6.6. The Levinson recursion effects a mapping of the set of p reflection coefficients k_i into the set of p filter coefficients $a[i]$. Derive the mapping for $p = 3$. Also, derive the inverse mapping for $p = 3$.

6.7. Prove that a necessary and sufficient condition for $\mathcal{A}(z) = 1 + a[1]z^{-1} + a[2]z^{-2}$ to be minimum-phase is that $|k_1| < 1$, $|k_2| < 1$. Restrict the proof to the case of real $a[1]$, $a[2]$ coefficients.

6.8. Prove that if $\mathcal{A}(z)$ is minimum-phase, then [Markel and Gray 1976]

$$\int_{-1/2}^{1/2} \ln | A(f) |^2 \, df = 0.$$

Hint: First show that the integral may be rewritten as

$$2 \operatorname{Re} \left\{ \int_{-1/2}^{1/2} \ln \mathcal{A}(\exp(j2\pi f)) \, df \right\} = 2 \operatorname{Re} \left\{ \frac{1}{2\pi j} \oint \ln \mathcal{A}(z) \frac{dz}{z} \right\}$$

where the contour of integration is the unit circle in the z-plane.

6.9. Find the spectral flatness measure for an AR(p) process in terms of the reflection coefficients. Based on your result, find a process for which the measure is zero. *Hint:* use the results of Problem 6.8.

6.10. Show that the exact likelihood function for a real Gaussian AR(p) process is [Box and Jenkins 1970]

$$p(\mathbf{x}; \mathbf{a}, \sigma^2) = \frac{1}{(2\pi\sigma^2)^{N/2} \det^{1/2}(\overline{\mathbf{R}}_{xx})} \exp\left[-\frac{1}{2\sigma^2} S(\mathbf{a}) \right]$$

where

$$S(\mathbf{a}) = \sum_{n=p}^{N-1} \left(\sum_{k=0}^{p} a[k]x[n-k] \right)^2 + \mathbf{x}_0^T \overline{\mathbf{R}}_{xx}^{-1} \mathbf{x}_0$$

and $\mathbf{x}_0 = [x[0] \;\; x[1] \;\; \cdots \;\; x[p-1]]^T$. The autocorrelation matrix $\overline{\mathbf{R}}_{xx}$ is the usual autocorrelation matrix $\mathbf{R}_{xx}$ [see (5.18)] divided by σ^2. Note that $\overline{\mathbf{R}}_{xx}$ depends only on the AR filter parameters.

6.11. Using the exact likelihood function given in Problem 6.10, show that the exact MLE of $a[1]$ for a real AR(1) process is found as a solution of [Anderson 1971, Kay 1983]

$$a^3[1] + \frac{N-2}{N-1} \frac{S_{01}}{S_{11}} a^2[1] - \frac{S_{00} + NS_{11}}{(N-1)S_{11}} a[1] - \frac{NS_{01}}{(N-1)S_{11}} = 0$$

where

$$S_{ij} = \sum_{n=0}^{N-1-i-j} x[n+i]x[n+j].$$

6.12. For a real Gaussian AR(1) process find the large data record variance of the MLEs for $a[1]$, σ^2, and $P_{\text{AR}}(f)$ by using (6.97) and (6.107). Compute $\mathcal{E}[\hat{P}_{\text{AR}}(f)]/\sqrt{\operatorname{var}[\hat{P}_{\text{AR}}(f)]}$ to determine which processes are most easily estimated.

6.13. Show that the peak of the AR spectral estimator for one complex sinusoid in complex white noise as given by (6.123) occurs at the sinusoidal frequency location.

6.14. For M complex sinusoids in white noise the AR spectral estimator can be found by applying (6.134) and (6.135). Show that as $pP_m/\sigma_w^2 \to \infty$, the spectral estimate exhibits peaks of infinite amplitudes at the sinusoidal frequency locations.

6.15. Perform an analysis similar to the one used to derive (6.134) to show that for the case of M real sinusoids in real white noise the solution of the Yule–Walker equations is given by [Satorius and Zeidler 1978]

$$\beta_m + \sum_{\substack{n=1 \\ n \neq m}}^{2M} c_{mn}\beta_n = -\frac{P_m \exp(j2\pi f_m)}{pP_m + 2\sigma_w^2} \qquad m = 1, 2, \ldots, 2M$$

where

$$P_{m+M} = P_m \qquad m = 1, 2, \ldots, M$$

$$c_{mn} = \frac{P_m}{pP_m + 2\sigma_w^2} \sum_{i=0}^{p-1} \exp[j2\pi(f_n - f_m)i].$$

The ACF is assumed to be given by

$$r_{yy}[k] = \sum_{m=1}^{M} P_m \cos(2\pi f_m k) + \sigma_w^2 \delta[k]$$

and the solution is by assumption

$$a[k] = \sum_{i=1}^{2M} \beta_i \exp[j2\pi f_i(k-1)] \qquad k = 1, 2, \ldots, p.$$

The β_i's are in general complex.

APPENDIX 6A

Derivation of Cramer–Rao Lower Bounds for AR Parameter Estimators

The CR bound that will be derived is approximate since it is based on the conditional PDF as given by (6.86). [See the paper by Tufts et al. [1983] for the exact CR bounds for a real AR(1) process.] The Fisher information matrix is defined as (see Section 3.3.1)

$$[\mathbf{I}_\theta]_{ij} = -\mathscr{E}\left[\frac{\partial^2 \ln p(\mathbf{x}; \boldsymbol{\theta})}{\partial\theta_i \, \partial\theta_j}\right] \qquad i, j = 1, 2, \ldots, p+1$$

where $\boldsymbol{\theta} = [\mathbf{a}^T \;\; \sigma^2]^T$. $p(\mathbf{x}; \mathbf{a}, \sigma^2)$, as given by the approximation of (6.86), will be used to evaluate the matrix. The first-order partial derivatives are

$$\frac{\partial \ln p(\mathbf{x}; \mathbf{a}, \sigma^2)}{\partial a[k]} = -\frac{1}{\sigma^2} \sum_{n=p}^{N-1} \left(x[n] + \sum_{j=1}^{p} a[j]x[n-j]\right) x[n-k]$$

$$\frac{\partial \ln p(\mathbf{x}; \mathbf{a}, \sigma^2)}{\partial \sigma^2} = -\frac{N-p}{2\sigma^2} + \frac{1}{2\sigma^4} \sum_{n=p}^{N-1} \left(x[n] + \sum_{j=1}^{p} a[j]x[n-j]\right)^2.$$

The second order partial derivatives are

$$\frac{\partial^2 \ln p(\mathbf{x}; \mathbf{a}, \sigma^2)}{\partial a[k]\, \partial a[l]} = -\frac{1}{\sigma^2} \sum_{n=p}^{N-1} x[n-l]x[n-k]$$

$$\frac{\partial^2 \ln p(\mathbf{x}; \mathbf{a}, \sigma^2)}{\partial a[k]\, \partial \sigma^2} = \frac{1}{\sigma^4} \sum_{n=p}^{N-1} \left(x[n] + \sum_{j=1}^{p} a[j]x[n-j] \right) x[n-k]$$

$$\frac{\partial^2 \ln p(\mathbf{x}; \mathbf{a}, \sigma^2)}{\partial \sigma^2\, \partial \sigma^2} = \frac{N-p}{2\sigma^4} - \frac{1}{\sigma^6} \sum_{n=p}^{N-1} \left(x[n] + \sum_{j=1}^{p} a[j]x[n-j] \right)^2.$$

Taking expectations yields

$$\mathcal{E}\left[\frac{\partial^2 \ln p(\mathbf{x}; \mathbf{a}, \sigma^2)}{\partial a[k]\, \partial a[l]} \right] = -\frac{N-p}{\sigma^2} r_{xx}[k-l]$$

$$\mathcal{E}\left[\frac{\partial^2 \ln p(\mathbf{x}; \mathbf{a}, \sigma^2)}{\partial a[k]\, \partial \sigma^2} \right] = \frac{N-p}{\sigma^4} \left[r_{xx}[k] + \sum_{j=1}^{p} a[j] r_{xx}[k-j] \right] = 0$$

since the quantity inside the brackets are the Yule–Walker equations for $k = 1, 2, \ldots, p$. Also,

$$\mathcal{E}\left[\frac{\partial^2 \ln p(\mathbf{x}; \mathbf{a}, \sigma^2)}{\partial \sigma^2\, \partial \sigma^2} \right] = \frac{N-p}{2\sigma^4} - \frac{1}{\sigma^6} \mathcal{E}\left(\sum_{n=p}^{N-1} u^2[n] \right)$$

making use of the definition of an AR process. It follows that

$$\mathcal{E}\left[\frac{\partial^2 \ln p(\mathbf{x}; \mathbf{a}, \sigma^2)}{\partial \sigma^2\, \partial \sigma^2} \right] = \frac{N-p}{2\sigma^4} - \frac{N-p}{\sigma^4} = -\frac{N-p}{2\sigma^4}.$$

Hence the Fisher information matrix is given as

$$\mathbf{I}_{\mathbf{a},\sigma^2} = \begin{bmatrix} \frac{N-p}{\sigma^2} \mathbf{R}_{xx} & \mathbf{0} \\ \mathbf{0}^T & \frac{N-p}{2\sigma^4} \end{bmatrix}$$

where $\mathbf{R}_{xx}$ is the $p \times p$ autocorrelation matrix. Because it was assumed that the data record was large in order to ignore the effect of the likelihood function on the initial conditions, the condition $N \gg p$ must be satisfied. Using this approximation, the CR lower bound on the variance becomes

$$\mathbf{I}^{-1}_{\mathbf{a},\sigma^2} = \begin{bmatrix} \frac{\sigma^2}{N} \mathbf{R}^{-1}_{xx} & \mathbf{0} \\ \mathbf{0}^T & \frac{2\sigma^4}{N} \end{bmatrix}.$$

Note that the same result may be derived as a special case of the CR lower bound for the parameters of an ARMA process. The latter is derived in Appendix 9A using a frequency domain approach.

APPENDIX 6B

Computer Program for the Levinson Recursion (Provided on Floppy Disk)

```
      SUBROUTINE LEVINSON(R,IP,A,AA,RHO)
C
C     This program implements the Levinson recursion.
C     See (6.46)-(6.48) and Figure 6.1.
C
C     Input Parameters:
C
C       R      -Complex array of dimension (IP+1)x1 of autocorrelation
C               samples where R(1),R(2), . . . ,R(IP+1) correspond to
C               rxx[0],rxx[1], . . . ,rxx[p]
C       IP     -Order or dimension of Yule-Walker equations to be
C               solved
C
C     Output Parameters:
C
C       A       -Complex array of dimension IPx1 containing solution,
C                arranged as A(1) to A(IP)
C       AA      -Complex array of dimension IPxIP containing all lower
C                order solutions as well as desired one; AA(I,J)
C                corresponds to the Ith coefficient for the Jth order
C                solution (AA(I,IP)=A(I))
C       RHO     -Real array of dimension IPx1 containing prediction
C                error powers where RHO(J) is the prediction error
C                power for the Jth order predictor (RHO(IP)=white noise
C                variance)
C
C     Notes:
C
C       The calling program must dimension the arrays R,A,AA,RHO.
C       The array AA must be dimensioned in the calling program
C       to be greater than or equal to the variable dimensions
C       shown and in this program to agree with calling program
C       dimensions.
```

```
C
C     Verification Test Case:
C
C       If R(1)=(2.,0.), R(2)=(-1.,1.), R(3)=(0.,0.), and IP=2, then
C       the output should be
C         A(1)=AA(1,2)=(0.99999,-0.99999), A(2)=AA(2,2)=(0.00000,-0.99999),
C         AA(1,1)=(0.50000,-0.50000), RHO(1)=1.00000, RHO(2)=2.38418E-07
C       for a DEC VAX 11/780.
C
        COMPLEX R(1),AA(IP,IP),A(1),B
        DIMENSION RHO(1)
C     Initialization of Levinson recursion
        RHO0=R(1)
        AA(1,1)=-R(2)/R(1)
        RHO(1)=(1.-CABS(AA(1,1))**2)*R(1)
        A(1)=AA(1,1)
        IF(IP.EQ.1)GO TO 50
C     Begin recursion by computing reflection coefficient (6.46)
        DO 30 I=2,IP
        B=-R(I+1)
        DO 10 K=1,I-1
10      B=B-AA(K,I-1)*R(I+1-K)
        AA(I,I)=B/RHO(I-1)
C     Update in order the prediction error filter and prediction
C     error power, (6.47) and (6.48)
        DO 20 K=1,I-1
20      AA(K,I)=AA(K,I-1)+AA(I,I)*CONJG(AA(I-K,I-1))
        RHO(I)=(1.-CABS(AA(I,I))**2)*RHO(I-1)
30      CONTINUE
C     Store final solution of Yule-Walker equations
        DO 40 I=1,IP
40      A(I)=AA(I,IP)
50      RETURN
        END
```

APPENDIX 6C

Computer Program for the Step-Down Procedure (Provided on Floppy Disk)

```
      SUBROUTINE STEPDOWN(IP,A,SIG2,AA,RHO0,RHO)
C
C     This program implements the step-down procedure to find the
C     coefficients and prediction error powers for all the lower
```

```
C       order predictors given the coefficients and prediction
C       error power for the IPth order linear predictor or
C       equivalently given the filter parameters and white noise
C       variance of an IPth order AR model. See (6.51) and (6.52).
C
C
C       Input Parameters:
C
C         IP          -AR model order (must be 2 or greater)
C         A           -Complex array of dimension IP×1 of AR parameters
C                      arranged as A(1) to A(IP)
C         SIG2        -Variance of excitation noise of AR model of order
C                      IP or equivalently the prediction error power of
C                      the optimal linear predictor of order IP
C
C       Output Parameters:
C
C         AA          -Complex array of dimension IP×IP of prediction
C                      coefficients where AA(I,J) is the Ith coeff-
C                      icient of the Jth order predictor
C         RHO0        -Zeroth lag of ACF
C         RHO         -Real vector of dimension IP×1 of prediction error
C                      powers where RHO(J) is the prediction error power
C                      for the Jth order predictor
C       Notes:
C
C         The calling program must dimension the arrays A,AA,RHO.
C         The array AA must be dimensioned in the calling program
C         to be greater than or equal to the variable dimensions
C         shown and in this program to agree with calling program
C         dimensions.
C
C       Verification Test Case:
C
C         If IP=3, A(1)=(0.625,0.375), A(2)=(0.5,-0.375), A(3)=(0.5,0.),
C         and SIG2=0.328125, then the output should be
C           AA(1,1)=(0.50000,0.50000), AA(1,2)=(0.50000,0.25000),
C           AA(2,2)=(0.25000,-0.25000), RHO0=0.99999, RHO(1)=0.50000,
C           RHO(2)=0.43750, RHO(3)=0.328125
C         for a DEC VAX 11/780.
C
          COMPLEX A(1),AA(IP,IP)
          DIMENSION RHO(1)
C       Initialize step-down by equating AR parameters to IPth
C       order prediction coefficients and prediction error power
          DO 10 I=1,IP
10        AA(I,IP)=A(I)
          RHO(IP)=SIG2
C       Begin step-down
          IP2=IP-1
```

```
      DO 30 J=1,IP2
      K=IP+1-J
      DEN=1.-CABS(AA(K,K))**2
C     Compute lower order prediction error power (6.52)
      RHO(K-1)=RHO(K)/DEN
C     Compute lower order prediction coefficients (6.51)
      K1=K-1
      DO 20 I=1,K1
      AA(I,K-1)=(AA(I,K)-AA(K,K)*CONJG(AA(K-I,K)))/DEN
20    CONTINUE
30    CONTINUE
C     Complete step-down by computing zeroth lag of ACF
      RHO0=RHO(1)/(1.-CABS(AA(1,1))**2)
      RETURN
      END
```

Chapter 7

Autoregressive Spectral Estimation: Methods

7.1 INTRODUCTION

General properties of AR processes and the estimation of the parameters and PSD have been examined in Chapter 6. A variety of methods currently used to estimate the AR PSD are described in this chapter. They include the autocorrelation, covariance, modified covariance, Burg, and recursive MLE methods. All the approaches are approximate MLEs. As such, their performance for large data records is comparable. However, for short data records some marked differences exist between the various approaches. Generally, the quality of the spectral estimates is quite good, even for short data records. Also, the computational burden of the AR spectral estimators is rather modest. For many of the methods the computational requirements may be reduced even further by taking advantage of the structure of the equations to be solved in order to implement "fast" algorithms. Because of the good performance of the AR spectral estimation methods as well as the computational efficiency, many of the algorithms to be described are widely used in practice.

The AR spectral estimators are based on estimation of either the AR parameters or the reflection coefficients. Excepting the recursive MLE, the techniques estimate the parameters by minimizing an estimate of the prediction error power. As described in Section 6.5.1, minimization of the prediction error power arose from maximum likelihood considerations, although subject to certain approximations. The recursive MLE, on the other hand, attempts to obtain a better approximation to the MLE by maximizing the exact likelihood function in a recursive manner.

In describing the characteristics of the various AR spectral estimators, it will frequently be assumed that the data consist of sinusoids in white noise. This is due to the overwhelming application of AR methods to this specialized data set in an attempt to capitalize on the "high resolution" property. It should be emphasized, however, that if the data are *known to consist of sinusoids in white noise* and it is desired to estimate the sinusoidal parameters such as amplitude, frequency, and phase, the techniques of Chapter 13 should be investigated first. This is because those methods use the a priori knowledge that the data are composed of sinusoids in white noise and so are "tailored" to that problem. Poorer results in terms of frequency estimation accuracy may be expected for the more general AR spectral estimator. Computer simulation results will be described to illustrate the behavior of the AR spectral estimators for nonsinusoidal data.

Finally, many other approximate maximum likelihood methods for estimation of the AR parameters exist. Some of these may be found in Box and Jenkins [1970], Pagano [1972], and Dickinson [1978]. Also, algorithms for sequential estimation of AR parameters, in which the estimates can be updated as new data points become available, are frequently of interest. They are described and computer subroutine implementations given in the book by Marple [1987].

7.2 SUMMARY

AR parameter estimation methods are described in Sections 7.3 to 7.7. The autocorrelation or Yule–Walker method requires that we solve the Yule–Walker equations given by (7.3) using a biased autocorrelation estimator. The estimate of the white noise variance is given by (7.5) or alternatively by (7.6) if the Levinson recursion is used to solve the equations. The covariance method estimates the AR filter parameters by solving (7.8) with the estimate of the white noise variance given by (7.9). It is shown to be identical to the extended Prony method used to estimate the pole positions of exponential signals in noise. The modified covariance method is identical to the covariance method except that the definition of the autocorrelation estimator is modified to be (7.22). The Burg method estimates the reflection coefficients for use in the Levinson recursion. The entire algorithm is summarized by (7.39)–(7.41). Some other reflection coefficient estimators are also briefly described. Finally, the recursive MLE estimates the AR parameters by a recursive maximization of the exact likelihood function. The algorithm is described by (7.46)–(7.52). All these methods have been implemented as computer subroutines, which can be found in Appendices 7B to 7E. The problem of model order selection is discussed in Section 7.8. The final prediction error is defined by (7.53) while the Akaike information criterion is given by (7.54). The order is selected as the value of k that minimizes these functions. For large data records they yield identical results due to (7.56). Another method is the criterion autoregressive transfer function defined in (7.57). A few other model order selection approaches are described to complete the section. Spectral estimation of noise

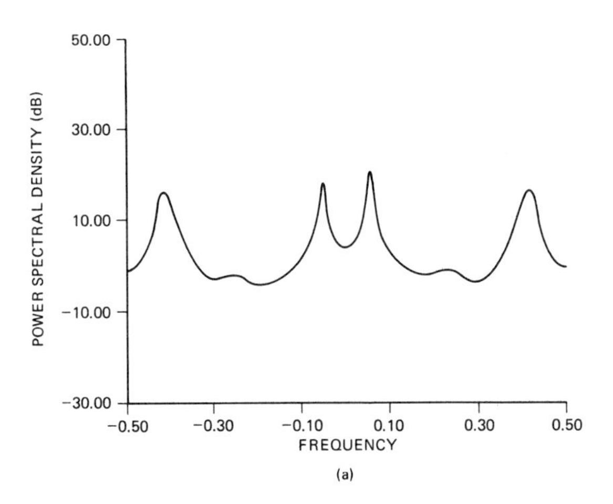

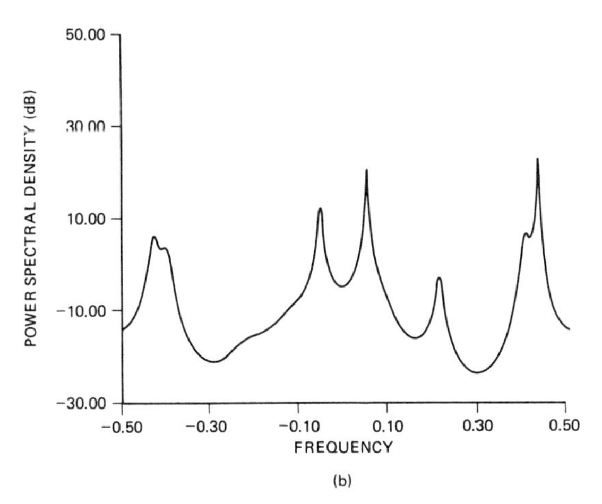

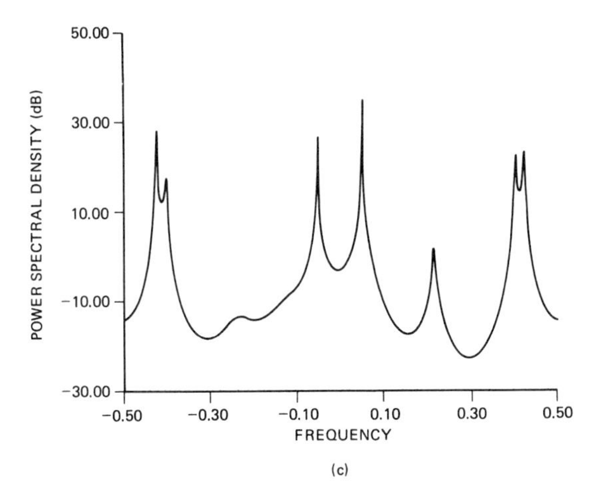

Figure 7.1 Autoregressive spectral estimates for test case data ($p = 10$). (a) Autocorrelation method. (b) Covariance method. (c) Modified covariance method. (d) Burg method. (e) Recursive MLE.

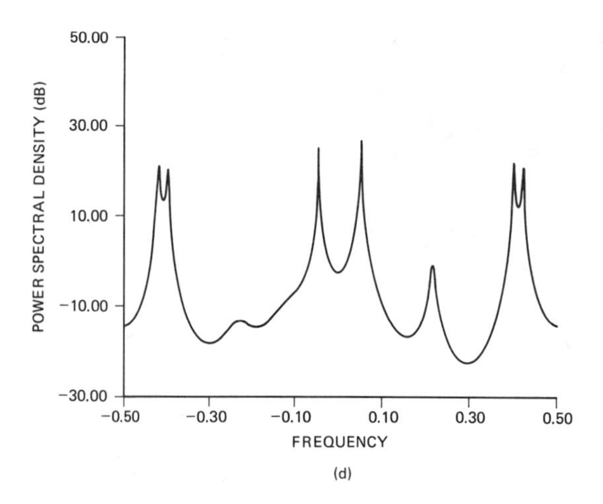

(d)

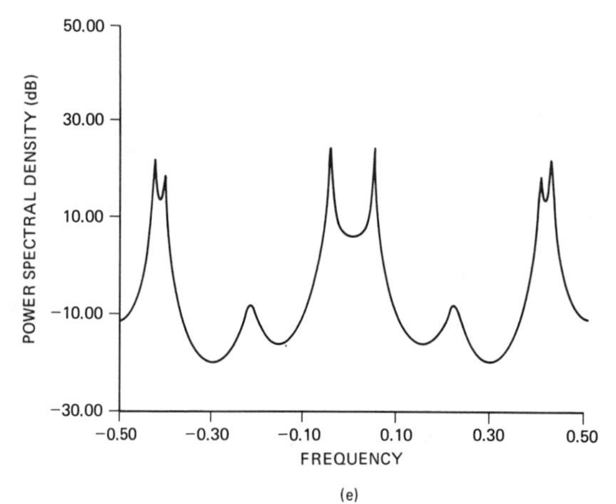

(e)

Figure 7.1 *(Continued)*

corrupted AR processes is discussed in Section 7.9. Whereas MLE methods are analytically intractable, several suboptimum approaches are described. None of the methods, though, has met with great success. Finally, some computer simulations are described in Section 7.10 which illustrate the characteristics of the AR spectral estimators.

The application of the AR spectral estimation methods to the test case data described in Section 1.5 produces the spectral estimates shown in Figure 7.1. The computer programs AUTOCORR (Appendix 7B), COVMCOV (Appendix 7C), BURG (Appendix 7D), and RMLE (Appendix 7E) were used to obtain the results. For all the methods a model order of $p = 10$ was chosen. The recursive MLE, which is valid only for real data, used the real data set, while the remaining methods were applied to the complex test data. The results are discussed in Chapter 12.

7.3 AUTOCORRELATION METHOD

As usual, it is assumed that the data $\{x[0], x[1], \ldots, x[N-1]\}$ are observed. The *autocorrelation method* [Makhoul 1975], or the *Yule–Walker* approach as it is sometimes referred to, has been described in Section 6.5.1. The AR parameters are estimated by minimizing an estimate of the prediction error power

$$\hat{\rho} = \frac{1}{N} \sum_{n=-\infty}^{\infty} \left| x[n] + \sum_{k=1}^{p} a[k]x[n-k] \right|^2. \tag{7.1}$$

The samples of the $x[n]$ process which are not observed (i.e., those not in the range $0 \le n \le N-1$) are set equal to zero in (7.1). The estimated prediction error power is minimized by differentiating (7.1) with respect to the real and imaginary parts of the $a[k]$'s. This may be done by using the complex gradient [see (2.79)] to yield

$$\frac{1}{N} \sum_{n=-\infty}^{\infty} \left(x[n] + \sum_{k=1}^{p} a[k]x[n-k] \right) x^*[n-l] = 0 \qquad l = 1,2,\ldots,p. \tag{7.2}$$

In matrix form this set of equations becomes

$$\begin{bmatrix} \hat{r}_{xx}[0] & \hat{r}_{xx}[-1] & \cdots & \hat{r}_{xx}[-(p-1)] \\ \hat{r}_{xx}[1] & \hat{r}_{xx}[0] & \cdots & \hat{r}_{xx}[-(p-2)] \\ \vdots & \vdots & \ddots & \vdots \\ \hat{r}_{xx}[p-1] & \hat{r}_{xx}[p-2] & \cdots & \hat{r}_{xx}[0] \end{bmatrix} \begin{bmatrix} \hat{a}[1] \\ \hat{a}[2] \\ \vdots \\ \hat{a}[p] \end{bmatrix} = - \begin{bmatrix} \hat{r}_{xx}[1] \\ \hat{r}_{xx}[2] \\ \vdots \\ \hat{r}_{xx}[p] \end{bmatrix} \tag{7.3}$$

where

$$\hat{r}_{xx}[k] = \begin{cases} \dfrac{1}{N} \displaystyle\sum_{n=0}^{N-1-k} x^*[n]x[n+k] & \text{for } k = 0, 1, \ldots, p \\ \hat{r}^*_{xx}[-k] & \text{for } k = -(p-1), -(p-2), \ldots, -1 \end{cases} \tag{7.4}$$

which is recognized as the biased ACF estimator. The matrix in (7.3) is hermitian ($\hat{r}_{xx}[-k] = \hat{r}^*_{xx}[k]$) and Toeplitz, and furthermore can be shown (see Problem 7.1) to be positive definite. The alternative name *Yule–Walker method* is due to the equivalence of the autocorrelation method to the use of the Yule–Walker equations with a biased ACF estimator. As such, the Levinson recursion may be used to solve the equations and the resulting estimated poles are guaranteed to be within the unit circle by the minimum-phase theorem of Section 6.3.2 (see Problem 7.2).

The estimate of the white noise variance σ^2 is found as $\hat{\rho}_{\text{MIN}}$, which is given by

$$\hat{\sigma}^2 = \hat{\rho}_{\text{MIN}} = \frac{1}{N} \sum_{n=-\infty}^{\infty} \left| x[n] + \sum_{k=1}^{p} \hat{a}[k]x[n-k] \right|^2$$

$$= \frac{1}{N} \sum_{n=-\infty}^{\infty} \left[\left(x[n] + \sum_{k=1}^{p} \hat{a}[k]x[n-k] \right) x^*[n] \right.$$

$$\left. + \left(x[n] + \sum_{k=1}^{p} \hat{a}[k]x[n-k] \right) \sum_{l=1}^{p} \hat{a}^*[l]x^*[n-l] \right].$$

From (7.2) the second term in the summation over n is zero, leading to the final result that

$$\hat{\sigma}^2 = \hat{r}_{xx}[0] + \sum_{k=1}^{p} \hat{a}[k]\hat{r}_{xx}[-k]. \tag{7.5}$$

$\hat{\sigma}^2$ may also be found in the last step of the Levinson recursion as the pth order prediction error power or in the alternative form as

$$\hat{\sigma}^2 = \hat{r}_{xx}[0] \prod_{i=1}^{p} (1 - |\hat{k}_i|^2) \tag{7.6}$$

where $\hat{k}_i$ is the estimate of the ith order reflection coefficient generated within the Levinson recursion. The autocorrelation method given by (7.3) and (7.5), which uses the Levinson algorithm, has been implemented in the computer program AUTOCORR, which is given in Appendix 7B.

The autocorrelation method has been found to produce poorer resolution spectral estimates than the other estimators to be described [Nuttall 1976]. Computer simulation results in Section 7.10 illustrate this behavior. For this reason it is not usually recommended for short data records. A variant of this approach is to use the unbiased autocorrelation estimator in the Yule–Walker equations. With this modification it may be shown that the autocorrelation matrix in (7.3) is no longer guaranteed to be positive definite (see Problem 7.3). As a consequence, the matrix is frequently observed to be singular or nearly singular, causing the spectral estimator to exhibit a large variance [Nuttall 1976, Kay 1980]. The use of the unbiased ACF estimator is therefore not recommended.

7.4 COVARIANCE METHOD

The *covariance method* [Makhoul 1975] was derived in Section 6.5.1 for real data as an approximate MLE. For complex data the analogous estimator may be found by minimizing the estimate of the prediction error power

$$\hat{\rho} = \frac{1}{N-p} \sum_{n=p}^{N-1} \left| x[n] + \sum_{k=1}^{p} a[k]x[n-k] \right|^2. \tag{7.7}$$

Note that the only difference between the covariance method and the autocorrelation method is the range of summation in the prediction error power estimate.

In the covariance method all the data points needed to compute $\hat{\rho}$ have been observed. No zeroing of the data is necessary.

The minimization of (7.7) may be effected by applying the complex gradient [see (2.79)] to yield the AR parameter estimates as the solution of the equations [see also the development of modified covariance equations (7.23)]

$$\begin{bmatrix} c_{xx}[1,1] & c_{xx}[1,2] & \cdots & c_{xx}[1,p] \\ c_{xx}[2,1] & c_{xx}[2,2] & \cdots & c_{xx}[2,p] \\ \vdots & \vdots & \ddots & \vdots \\ c_{xx}[p,1] & c_{xx}[p,2] & \cdots & c_{xx}[p,p] \end{bmatrix} \begin{bmatrix} \hat{a}[1] \\ \hat{a}[2] \\ \vdots \\ \hat{a}[p] \end{bmatrix} = - \begin{bmatrix} c_{xx}[1,0] \\ c_{xx}[2,0] \\ \vdots \\ c_{xx}[p,0] \end{bmatrix} \tag{7.8}$$

where

$$c_{xx}[j,k] = \frac{1}{N-p} \sum_{n=p}^{N-1} x^*[n-j]x[n-k].$$

The white noise variance is estimated as

$$\hat{\sigma}^2 = \hat{\rho}_{\text{MIN}} = c_{xx}[0,0] + \sum_{k=1}^{p} \hat{a}[k]c_{xx}[0,k]. \tag{7.9}$$

The matrix in (7.8) is hermitian ($c_{xx}[k,j] = c_{xx}^*[j,k]$) and positive semidefinite. It may be shown to be singular if the data consist of $p-1$ or fewer complex sinusoids (see Problem 7.4). Any observation noise, however, will cause the matrix to be nonsingular. The equations may be solved using the Cholesky decomposition (see Section 2.7). A computer program entitled COVMCOV and given in Appendix 7C implements the covariance method. A more computationally efficient solution has been developed by Morf et al. [1977]. The estimated poles using the covariance method are not guaranteed to lie within the unit circle. As an example, if $p = 1$ and $N = 2$ [Nuttall 1976],

$$\hat{a}[1] = -\frac{x[1]}{x[0]}$$

which may be greater than or equal to 1 in magnitude. In practice, however, the poles are seldom observed to fall on or outside the unit circle.

As implied from the definition, $c_{xx}[j,k]$ is readily seen to be an estimate of $r_{xx}[j-k]$, although a different estimate than that encountered in the autocorrelation method. $c_{xx}[j,k]$ uses the sum of only $N-p$ lag products to estimate the ACF for each lag even though more are available. As an example, in the estimation of $r_{xx}[0]$ the biased autocorrelation estimator of the autocorrelation method uses all N data points, while the covariance method uses only $N-p$ data points in the summation. For large data records in which $N \gg p$, these "end effects" are negligible and consequently, the autocorrelation and covariance methods will yield similar spectral estimates. A second contrasting feature is that for data consisting of pure sinusoids, the covariance method may be used to perfectly extract the frequencies. This is due to the relationship of the covariance method to Prony's method, which is described next. This property is not shared by the autocorrelation method. Methods to estimate sinusoidal frequencies as described more fully in Chapter 13 are usually based on the covariance method or a variant thereof, the modified covariance method, described in the next section.

The covariance method is identical to the modern version of Prony's method for pole estimation. As originally proposed, Prony's method [Prony 1795] is a technique for identifying the frequencies f_i, damping factors α_i, amplitudes A_i, and phases ϕ_i of real exponential signals, or

$$x[n] = \sum_{i=1}^{p} A_{c_i} \exp\left[(\alpha_i + j2\pi f_i)n\right] \qquad n \geq 0 \tag{7.10}$$

where $A_{c_i} = A_i \exp(j\phi_i)$ is a complex amplitude. For real signals the *modes* $z_i = \exp(\alpha_i + j2\pi f_i)$ and the complex amplitudes A_{c_i} must occur in complex-conjugate pairs. Since the data may be written as

$$x[n] = \sum_{i=1}^{p} A_{c_i} z_i^n \qquad n \geq 0$$

it is well known [Oppenheim and Schafer 1975] that $x[n]$ may be generated by the recursive difference equation (see Problem 7.5)

$$x[n] = -\sum_{k=1}^{p} a[k]x[n-k] \qquad n \geq p \tag{7.11}$$

with appropriate initial conditions $\{x[0], x[1], \ldots, x[p-1]\}$. The signal may also be viewed as the natural response of an all-pole filter with the given initial conditions. The coefficients $a[k]$ are related to the modes or *poles* z_i by

$$\mathcal{A}(z) = 1 + \sum_{k=1}^{p} a[k]z^{-k} = \prod_{i=1}^{p} (1 - z_i z^{-1}). \tag{7.12}$$

If we are given the data $\{x[0], x[1], \ldots, x[2p-1]\}$, the poles may be determined *exactly* by solving the set of linear equations given by (7.11) for $n = p, p+1, \ldots, 2p-1$ to find the $a[k]$'s and then rooting the polynomial of (7.12).

When noise is present the original Prony method performs poorly [Van Blaricum and Mittra 1978] since (7.11) no longer holds. Attempts have been made to extend Prony's method to the case of exponential signals in noise. If $y[n]$ denotes the noise corrupted process and $w[n]$ denotes white observation noise so that

$$y[n] = x[n] + w[n]$$

then (7.11) becomes

$$y[n] - w[n] = -\sum_{k=1}^{p} a[k](y[n-k] - w[n-k])$$

or

$$y[n] = -\sum_{k=1}^{p} a[k]y[n-k] + w[n] + \sum_{k=1}^{p} a[k]w[n-k]. \tag{7.13}$$

Note that $y[n]$ has the form of an ARMA(p, p) process and in fact for a pure sinusoid, a special case of (7.10), in white noise, (7.13) may be considered as the limiting form of an ARMA process (see Section 9.7). However, for exponential signals that are not WSS, no such interpretation is possible. If we now let

$$\epsilon[n] = w[n] + \sum_{k=1}^{p} a[k]w[n-k] \tag{7.14}$$

which may be thought of as an error due to noise, a least squares estimator of the $a[k]$'s is found by minimizing

$$\frac{1}{N-p}\sum_{n=p}^{N-1} |\epsilon[n]|^2 = \frac{1}{N-p}\sum_{n=p}^{N-1} \left| y[n] + \sum_{k=1}^{p} a[k]y[n-k] \right|^2. \tag{7.15}$$

This method has been termed the *extended Prony method*. Note that the minimization of (7.15) is *identical* to that encountered in the covariance method and thus the *extended Prony method and covariance method are identical*. The underlying assumptions, however, are radically different. From the results in Chapter 6 which illustrated the poor performance of AR spectral estimators for sinusoids in noise (see Figure 6.11 for large errors in the pole position estimates), it would be expected that even the extended Prony method would perform poorly in the presence of noise. This has been demonstrated by Kumaresan [1982], although the extended Prony method is clearly superior to the original method. It should also be mentioned that since (7.14) is not a white Gaussian process, the minimization of (7.15) *does not produce* an approximate MLE of the $a[k]$'s. An approximate MLE that is based on the representation of (7.11) is described in Section 13.5 for sinusoids in white Gaussian noise. For exponential signals in white noise, improved Prony-type methods may be found in Kumaresan [1982]. Finally, if the data are noiseless, consisting of p complex sinusoids, the extended Prony method or covariance method will yield the correct pole positions. This is because (7.15) will be zero and hence minimized for the true $a[k]$'s due to (7.11) (see Problem 13.9). Further discussions of Prony's method as well as some computer subroutine implementations may be found in the book by Marple [1987].

Some computer simulation results that illustrate the behavior of the covariance method are included in Section 7.10. They generally indicate that the resolution of the covariance method exceeds that of the autocorrelation method.

7.5 MODIFIED COVARIANCE METHOD

In Chapter 6 it was shown that for an AR(p) process the optimal forward predictor is

$$\hat{x}[n] = -\sum_{k=1}^{p} a[k]x[n-k] \tag{7.16}$$

while the optimal backward predictor is [see (6.57)]

$$\hat{x}[n] = - \sum_{k=1}^{p} a^*[k]x[n+k] \tag{7.17}$$

where the $a[k]$'s are the AR filter parameters. In either case the minimum prediction error power is just the white noise variance σ^2. The modified covariance method estimates the AR parameters by minimizing the *average* of the estimated forward and backward prediction error powers, or

$$\hat{\rho} = \frac{1}{2}(\hat{\rho}^f + \hat{\rho}^b) \tag{7.18}$$

where

$$\begin{aligned} \hat{\rho}^f &= \frac{1}{N-p} \sum_{n-p}^{N-1} \left| x[n] + \sum_{k=1}^{p} a[k]x[n-k] \right|^2 \\ \hat{\rho}^b &= \frac{1}{N-p} \sum_{n=0}^{N-1-p} \left| x[n] + \sum_{k=1}^{p} a^*[k]x[n+k] \right|^2 . \end{aligned} \tag{7.19}$$

As in the covariance method the summations are only over the prediction errors that involve observed data samples. Note that an alternative way of viewing this estimator is to recognize that $\hat{\rho}^b$ is the prediction error power estimate obtained by "flipping the data record around" and complex conjugating it (i.e., letting $x'[0] = x^*[N-1]$, $x'[1] = x^*[N-2]$, etc.) and applying a forward predictor to this new data set. In this manner we obtain some "extra" data points and hence more prediction errors over which to average. Note that for any set of $a[k]$'s the forward and backward prediction error estimates will be slightly different due to the range of the summations. This procedure is also equivalent to combining the prediction errors of the forward and backward AR models described in Section 5.8.

To minimize (7.18) we can differentiate $\hat{\rho}$ with respect to the real and imaginary parts of $a[k]$ for $k = 1, 2, \ldots, p$. Alternatively, as shown in Section 2.8, we can take advantage of the complex gradient relationship given by (2.79) to yield

$$\begin{aligned} \frac{\partial \hat{\rho}}{\partial a[l]} = \frac{1}{N-p} \Bigg[& \sum_{n=p}^{N-1} \left(x[n] + \sum_{k=1}^{p} a[k]x[n-k] \right) x^*[n-l] \\ & + \sum_{n=0}^{N-1-p} \left(x^*[n] + \sum_{k=1}^{p} a[k]x^*[n+k] \right) x[n+l] \Bigg] \\ = 0 \quad & l = 1, 2, \ldots, p. \end{aligned} \tag{7.20}$$

After some simplification this becomes

$$\begin{aligned} \sum_{k=1}^{p} \hat{a}[k] & \left(\sum_{n=p}^{N-1} x[n-k]x^*[n-l] + \sum_{n=0}^{N-1-p} x^*[n+k]x[n+l] \right) \\ & = - \left(\sum_{n=p}^{N-1} x[n]x^*[n-l] + \sum_{n=0}^{N-1-p} x^*[n]x[n+l] \right) \end{aligned} \tag{7.21}$$

for $l = 1, 2, \ldots, p$. Letting

$$c_{xx}[j, k] = \frac{1}{2(N-p)}\left(\sum_{n=p}^{N-1} x^*[n-j]x[n-k] + \sum_{n=0}^{N-1-p} x[n+j]x^*[n+k]\right) \tag{7.22}$$

(7.21) can be written in the identical matrix form as (7.8), or

$$\begin{bmatrix} c_{xx}[1, 1] & c_{xx}[1, 2] & \cdots & c_{xx}[1, p] \\ c_{xx}[2, 1] & c_{xx}[2, 2] & \cdots & c_{xx}[2, p] \\ \vdots & \vdots & \ddots & \vdots \\ c_{xx}[p, 1] & c_{xx}[p, 2] & \cdots & c_{xx}[p, p] \end{bmatrix} \begin{bmatrix} \hat{a}[1] \\ \hat{a}[2] \\ \vdots \\ \hat{a}[p] \end{bmatrix} = -\begin{bmatrix} c_{xx}[1, 0] \\ c_{xx}[2, 0] \\ \vdots \\ c_{xx}[p, 0] \end{bmatrix}. \tag{7.23}$$

The estimate of the white noise variance is

$$\hat{\sigma}^2 = \hat{\rho}_{\text{MIN}} = \frac{1}{2(N-p)}\left[\sum_{n=p}^{N-1}\left(x[n] + \sum_{k=1}^{p} \hat{a}[k]x[n-k]\right)x^*[n] + \sum_{n=0}^{N-1-p}\left(x^*[n] + \sum_{k=1}^{p} \hat{a}[k]x^*[n+k]\right)x[n]\right]$$

where (7.20) has been used, or finally,

$$\hat{\sigma}^2 = c_{xx}[0, 0] + \sum_{k=1}^{p} \hat{a}[k]c_{xx}[0, k] \tag{7.24}$$

where $c_{xx}[j, k]$ is defined by (7.22). It is observed that the modified covariance method is identical to the covariance method except for the definition of $c_{xx}[j, k]$, the autocorrelation estimator. The matrix in (7.23) is hermitian ($c_{xx}[k, j] = c_{xx}^*[j, k]$) and positive definite (excluding the pure sinusoid case), so that the Cholesky decomposition may be used to solve the linear equations (see Section 2.7). A computer program entitled COVMCOV, which is given in Appendix 7C, implements the modified covariance method. A more computationally efficient means of solving the equations derived by Marple [1980] takes advantage of the special structure of the equations. As in the covariance method a singular matrix will occur for data consisting of $p - 1$ or fewer sinusoids (see Problem 7.4). For p sinusoids the modified covariance method may be used to obtain perfect estimates of the frequencies. In Chapter 13 we present more details on the frequency estimation problem. The modified covariance method does not guarantee a stable all-pole filter, although in practice the estimated poles usually fall inside the unit circle. This method of AR parameter estimation was originally proposed by Nuttall [1976], who termed it the *forward-backward* method and also, independently by Ulrych and Clayton [1976], who called it the *least squares* approach.

The modified covariance method appears to yield statistically stable spectral estimates with high resolution [Nuttall 1976, Kay 1983, Shon and Mehrota 1984]. For data consisting of sinusoids in white noise a number of desirable properties of the estimator have been observed from computer simulations. The usual shifting of the peaks of an AR spectral estimate from the true frequency locations due to additive observation noise appears to be less pronounced than for many of the other AR spectral estimators [Swingler 1979A]. Also, the peak location dependence on initial sinusoidal phase [Chen and Stegen 1974] is considerably reduced [Ulrych and Clayton 1976]. Spectral line splitting in which a single sinusoidal component gives rise to two distinct spectral peaks [Fougere et al. 1976, Herring 1980] has never been observed with the modified covariance method [Kay and Marple 1979, Marple 1980]. Illustrations of spectral estimation performance for nonsinusoidal processes are provided in Section 7.10.

7.6 BURG METHOD

In contrast to the autocorrelation, covariance, and modified covariance methods, which estimate the AR parameters directly, the Burg method estimates the reflection coefficients and then uses the Levinson recursion to obtain the AR parameter estimates. The reflection coefficient estimates are obtained by minimizing estimates of the prediction error power for different order predictors in a recursive manner. Specifically, based on the Levinson algorithm (see Figure 6.4), if estimates of the reflection coefficients $\{k_1, k_2, \ldots, k_p\}$ are available, the AR parameters may be estimated as follows:

$$\hat{r}_{xx}[0] = \frac{1}{N} \sum_{n=0}^{N-1} |x[n]|^2$$

$$\hat{a}[1] = \hat{k}_1$$

$$\hat{\rho}_1 = (1 - |\hat{a}_1[1]|^2)\hat{r}_{xx}[0].$$

For $k = 2, 3, \ldots, p$,

$$\hat{a}_k[i] = \begin{cases} \hat{a}_{k-1}[i] + \hat{k}_k \hat{a}^*_{k-1}[k-i] & \text{for } i = 1, 2, \ldots, k-1 \\ \hat{k}_k & \text{for } i = k \end{cases} \tag{7.25}$$

$$\hat{\rho}_k = (1 - |\hat{a}_k[k]|^2)\hat{\rho}_{k-1}. \tag{7.26}$$

The estimates of the AR filter parameters are $\{\hat{a}_p[1], \hat{a}_p[2], \ldots, \hat{a}_p[p]\}$ and the white noise variance estimates is $\hat{\rho}_p$. It remains only to obtain estimates of the reflection coefficients. In deriving the kth reflection coefficient estimate, Burg assumed that the $(k-1)$st order prediction error filter coefficients had already been estimated as $\{\hat{a}_{k-1}[1], \hat{a}_{k-1}[2], \ldots, \hat{a}_{k-1}[k-1]\}$, having been obtained by minimizing the $(k-1)$st order prediction error power. Using the Levinson recursion the coefficients of the kth order prediction error filter depend only on k_k according to (7.25), and hence the kth order prediction error power estimate also depends only on k_k. Burg proposed to estimate k_k by minimizing the *average* of the estimates of the forward and backward prediction error powers. This approach, which preceded the modified covariance method, is a *constrained minimization* of $\hat{\rho}$ as given by (7.18) in contrast to the modified covariance method, which minimizes $\hat{\rho}$ in an unconstrained manner. The constrained or recursive minimization will not in general produce a global minimum. Hence, to obtain the estimate of k_k, we minimize

$$\hat{\rho}_k = \tfrac{1}{2}(\hat{\rho}_k^f + \hat{\rho}_k^b) \tag{7.27}$$

where

$$\hat{\rho}_k^f = \frac{1}{N-k} \sum_{n=k}^{N-1} \left| x[n] + \sum_{i=1}^{k} a_k[i]x[n-i] \right|^2 \tag{7.28}$$

$$\hat{\rho}_k^b = \frac{1}{N-k} \sum_{n=0}^{N-1-k} \left| x[n] + \sum_{i=1}^{k} a_k^*[i]x[n+i] \right|^2 \tag{7.29}$$

and

$$a_k[i] = \begin{cases} \hat{a}_{k-1}[i] + k_k\hat{a}_{k-1}^*[k-i] & \text{for } i = 1, 2, \ldots, k-1 \\ k_k & \text{for } i = k. \end{cases} \tag{7.30}$$

$\hat{\rho}_k^f$ and $\hat{\rho}_k^b$ are functions *only* of k_k since the $(k-1)$st order prediction coefficients are assumed to have already been estimated by minimizing $\hat{\rho}_{k-1}$. Defining estimated forward and backward prediction errors in a similar fashion to those presented in Chapter 6 [see (6.15) and (6.57)],

$$\hat{e}_k^f[n] = x[n] + \sum_{i=1}^{k} a_k[i]x[n-i] \tag{7.31}$$

$$\hat{e}_k^b[n] = x[n-k] + \sum_{i=1}^{k} a_k^*[i]x[n-k+i] \tag{7.32}$$

the forward prediction error power estimate becomes

$$\hat{\rho}_k^f = \frac{1}{N-k} \sum_{n=k}^{N-1} |\hat{e}_k^f[n]|^2 \tag{7.33}$$

while the backward prediction error power estimate becomes

$$\hat{\rho}_k^b = \frac{1}{N-k} \sum_{n=k}^{N-1} |\hat{e}_k^b[n]|^2. \tag{7.34}$$

The lattice filter relations which describe the model order update of the forward and backward prediction error time series as given by (6.58) are applicable here. [Alternatively, substitute (7.30) in (7.31) and (7.32).] These are

$$\hat{e}_k^f[n] = \hat{e}_{k-1}^f[n] + k_k\hat{e}_{k-1}^b[n-1] \tag{7.35}$$

$$\hat{e}_k^b[n] = \hat{e}_{k-1}^b[n-1] + k_k^*\hat{e}_{k-1}^f[n] \tag{7.36}$$

where

$$\hat{e}_0^f[n] = \hat{e}_0^b[n] = x[n].$$

When these relations are substituted into (7.33) and (7.34) and then into (7.27), the average estimated prediction error power becomes

$$\hat{\rho}_k = \frac{1}{2(N-k)} \sum_{n=k}^{N-1} [\,|\,\hat{e}^f_{k-1}[n] + k_k \hat{e}^b_{k-1}[n-1]\,|^2 + |\,\hat{e}^b_{k-1}[n-1] + k_k^* \hat{e}^f_{k-1}[n]\,|^2\,]. \tag{7.37}$$

Differentiating $\hat{\rho}_k$ with respect to the real and imaginary parts of k_k by using the complex gradient [see (2.79)] and setting the result equal to zero yields

$$\frac{\partial \hat{\rho}_k}{\partial k_k} = \frac{1}{N-k} \sum_{n=k}^{N-1} \{(\hat{e}^f_{k-1}[n] + k_k \hat{e}^b_{k-1}[n-1])\hat{e}^b_{k-1}[n-1]^* + (\hat{e}^b_{k-1}[n-1]^* + k_k \hat{e}^f_{k-1}[n]^*)\hat{e}^f_{k-1}[n]\} = 0.$$

Solving for k_k, we have

$$\hat{k}_k = \frac{-2 \sum_{n=k}^{N-1} \hat{e}^f_{k-1}[n]\hat{e}^b_{k-1}[n-1]^*}{\sum_{n=k}^{N-1} (\,|\,\hat{e}^f_{k-1}[n]\,|^2 + |\,\hat{e}^b_{k-1}[n-1]\,|^2)} \tag{7.38}$$

which is the Burg method for estimation of the kth reflection coefficient. It can be shown that $|\,\hat{k}_k\,| \le 1$ (see Problem 7.6), which agrees with the theoretical constraint that a partial correlation coefficient should be less than 1 in magnitude [see (6.42)]. In summary, the Burg method for estimation of the AR parameters is

$$\begin{aligned} \hat{r}_{xx}[0] &= \frac{1}{N} \sum_{n=0}^{N-1} |\,x[n]\,|^2 \\ \hat{\rho}_0 &= \hat{r}_{xx}[0] \\ \hat{e}^f_0[n] &= x[n] \qquad n = 1, 2, \ldots, N-1 \\ \hat{e}^b_0[n] &= x[n] \qquad n = 0, 1, \ldots, N-2. \end{aligned} \tag{7.39}$$

For $k = 1, 2, \ldots, p$,

$$\begin{aligned} \hat{k}_k &= \frac{-2 \sum_{n=k}^{N-1} \hat{e}^f_{k-1}[n]\hat{e}^b_{k-1}[n-1]^*}{\sum_{n=k}^{N-1} (\,|\,\hat{e}^f_{k-1}[n]\,|^2 + |\,\hat{e}^b_{k-1}[n-1]\,|^2)} \\ \hat{\rho}_k &= (1 - |\,\hat{k}_k\,|^2)\hat{\rho}_{k-1} \\ \hat{a}_k[i] &= \begin{cases} \hat{a}_{k-1}[i] + \hat{k}_k \hat{a}^*_{k-1}[k-i] & \text{for } i = 1, 2, \ldots, k-1 \\ \hat{k}_k & \text{for } i = k. \end{cases} \end{aligned} \tag{7.40}$$

(If $k = 1$, $\hat{a}_1[1] = \hat{k}_1$.)

$$\hat{e}_k^f[n] = \hat{e}_{k-1}^f[n] + \hat{k}_k \hat{e}_{k-1}^b[n-1] \qquad n = k+1, k+2, \ldots, N-1 \tag{7.41}$$
$$\hat{e}_k^b[n] = \hat{e}_{k-1}^b[n-1] + \hat{k}_k^* \hat{e}_{k-1}^f[n] \qquad n = k, k+1, \ldots, N-2$$

The estimates are given as $\{\hat{a}_p[1], \hat{a}_p[2], \ldots, \hat{a}_p[p], \hat{\rho}_p\}$. A computer program entitled BURG, which is given in Appendix 7D, implements this algorithm.

The Burg method yields estimated poles that are on or inside the unit circle. This is due to the property $|\hat{k}_k| \le 1$. To reduce the necessary computation of the Burg algorithm, we can recursively compute the denominator of (7.38) by using the lattice recursions of (7.41). It may be shown (see Problem 7.7) that if DEN(k) denotes the denominator of (7.38), then [Anderson 1978]

$$\text{DEN}(k) = (1 - |\hat{k}_{k-1}|^2)\,\text{DEN}(k-1) - |\hat{e}_{k-1}^f[k-1]|^2 - |\hat{e}_{k-1}^b[N-1]|^2. \tag{7.42}$$

In general, the Burg algorithm produces accurate AR spectral estimates for data which are truly AR [Nuttall 1976]. For sinusoidal data, however, some difficulties have been observed. The Burg algorithm is subject to line splitting [Fougere et al. 1976, Herring 1980] and peak locations are strongly dependent on phase [Chen and Stegen 1974]. To reduce the phase dependence, we can use a modified reflection coefficient estimate,

$$\hat{k}_k^W = \frac{-2 \sum_{n=k}^{N-1} w_k[n] \hat{e}_{k-1}^f[n] \hat{e}_{k-1}^b[n-1]^*}{\sum_{n=k}^{N-1} w_k[n] \left(|\hat{e}_{k-1}^f[n]|^2 + |\hat{e}_{k-1}^b[n-1]|^2 \right)} \tag{7.43}$$

where $w_k[n]$ is a suitably chosen window with nonnegative weights. This modified reflection coefficient estimate, which was originally proposed by Burg [1975], has been found to reduce phase dependence effects [Swingler 1979B, Kaveh and Lippert 1983].

The Burg estimate of the reflection coefficient is only one of a large class of estimates that maintain the minimum-phase property. One that was proposed by Itakura and Saito [1971] replaces the theoretical ensemble average [see (6.42)] by their time averages to yield

$$\hat{k}_k^I = -\frac{\sum_{n=k}^{N-1} \hat{e}_{k-1}^f[n] \hat{e}_{k-1}^b[n-1]^*}{\sqrt{\sum_{n=k}^{N-1} |\hat{e}_{k-1}^f[n]|^2 \sum_{n=k}^{N-1} |\hat{e}_{k-1}^b[n-1]|^2}}. \tag{7.44}$$

It can be shown (see Problem 7.8) that

$$|\hat{k}_k^B| \le |\hat{k}_k^I| \tag{7.45}$$

where $\hat{k}_k^B$ denotes the Burg estimate given by (7.38). Many other possible estimators exist [Makhoul 1977], although in practice the differences in spectral estimation performance appear to be minor. Some illustrations of the Burg spectral estimator are given in Section 7.10.

7.7 RECURSIVE MLE

The previous AR parameter estimators were based on minimizing the prediction error power. For large data records this procedure is equivalent to maximizing the likelihood function as shown in Section 6.5.1. However, for short data records which are mainly of interest, this is not so. Since an exact maximization of the likelihood function is not possible analytically [Box and Jenkins 1970], a recursive maximization, *which is also approximate*, has been proposed [Kay 1983]. This approach allows us to retain terms in the likelihood function which would appear to constrain the estimated poles from becoming too close to the unit circle and hence, avoiding spurious peaks. In addition, since a closer approximation to the true MLE is found, more accurate parameter estimates are to be expected. Some limited results support these conjectures [Kay 1983, Shon and Mehrota 1984].

The *recursive MLE* (RMLE) first proceeds by maximizing the exact likelihood function of an AR(1) process (see Problem 6.11). Using these estimates in the Levinson recursion, the AR parameters for an AR(2) process may be generated, which now depend only on k_2. Since the likelihood function depends only on k_2, it can be maximized with respect to this parameter. In this manner a recursive maximization of the likelihood function continues until it is maximized with respect to all the reflection coefficients. Since each maximization assumes that the lower order AR filter has already been estimated, it is only an approximate maximization and hence the estimate obtained is not the true MLE [unless the process is AR(1)]. The recursive MLE is similar in spirit to Burg's method except that the likelihood function is maximized at each step as opposed to a minimization of the prediction error power. The algorithm has been developed only for *real* data. It is summarized below.

$$\epsilon_0 = S_{00}$$

$$c_1 = S_{01}$$

$$d_1 = S_{11}$$

To kind $\hat{k}_1$, solve

$$k_1^3 + \frac{(N-2)c_1}{(N-1)d_1}k_1^2 - \frac{\epsilon_0 + Nd_1}{(N-1)d_1}k_1 - \frac{Nc_1}{(N-1)d_1} = 0 \tag{7.46}$$

and choose the single root within $[-1, 1]$.

$$\hat{a}_1[1] = \hat{k}_1$$

$$\epsilon_1 = S_{00} + 2\hat{k}_1 S_{01} + \hat{k}_1^2 S_{11} \tag{7.47}$$

$$\hat{\rho}_1 = \frac{1}{N}\epsilon_1.$$

For $k = 2, 3, \ldots, p$,

$$\begin{aligned} c_k &= \mathbf{a}_{k-1}'^T \mathscr{C}_k \mathbf{b}_{k-1}' \\ d_k &= \mathbf{b}_{k-1}'^T \mathscr{D}_k \mathbf{b}_{k-1}'. \end{aligned} \tag{7.48}$$

To find $\hat{k}_k$, solve

$$k_k^3 + \frac{(N-2k)c_k}{(N-k)d_k} k_k^2 - \frac{k\epsilon_{k-1} + Nd_k}{(N-k)d_k} k_k - \frac{Nc_k}{(N-k)d_k} = 0 \qquad (7.49)$$

and choose the root within $[-1, 1]$. If more than one root lies within this interval, choose the one that maximizes

$$\frac{(1-k_k^2)^{k/2}}{\left[\frac{1}{N}(\epsilon_{k-1} + 2c_k k_k + d_k k_k^2)\right]^{N/2}}. \qquad (7.50)$$

$$\hat{a}_k[i] = \begin{cases} \hat{a}_{k-1}[i] + \hat{k}_k \hat{a}_{k-1}[k-i] & \text{for } i = 1, 2, \ldots, k-1 \\ \hat{k}_k & \text{for } i = k \end{cases}$$

$$\epsilon_k = \epsilon_{k-1} + 2c_k\hat{k}_k + d_k\hat{k}_k^2 \qquad (7.51)$$

$$\hat{\rho}_k = \frac{1}{N}\epsilon_k.$$

The various vectors and matrices are defined as follows:

$$\mathbf{a}'_{k-1} = [1 \quad \hat{a}_{k-1}[1] \quad \hat{a}_{k-1}[2] \quad \cdots \quad \hat{a}_{k-1}[k-1]]^T$$

$$\mathbf{b}'_{k-1} = [\hat{a}_{k-1}[k-1] \quad \hat{a}_{k-1}[k-2] \quad \cdots \quad \hat{a}_{k-1}[1] \; 1]^T.$$

$\mathscr{C}_k$, $\mathscr{D}_k$ are defined as the $k \times k$ partitions of $\mathbf{S}$:

$$[\mathscr{C}_k]_{ij} = S_{i-1,j} \qquad i, j = 1, 2, \ldots, k$$

$$[\mathscr{D}_k]_{ij} = S_{ij} \qquad i, j = 1, 2, \ldots, k$$

where the matrix $\mathbf{S}$ is $(p+1) \times (p+1)$ with elements

$$[\mathbf{S}]_{ij} = S_{ij} = \sum_{n=0}^{N-1-i-j} x[n+i]x[n+j] \qquad i, j = 0, 1, \ldots, p. \qquad (7.52)$$

The estimates of the AR parameters are given as $\{\hat{a}_p[1], \hat{a}_p[2], \ldots, \hat{a}_p[p], \hat{\rho}_p\}$.

It may be shown that for nonsinusoidal data at least one solution of the cubic equation (7.49) will produce $|\hat{k}_k| < 1$ [Kay 1983], and hence the estimated poles will lie inside the unit circle. Also, for large data records the solutions will be

$$k_k = -1, +1, -\frac{c_k}{d_k}$$

(see Problem 7.9).

A computer program entitled RMLE which implements this algorithm is provided in Appendix 7E. Some computer simulation results that illustrate its performance are given in Section 7.10.

7.8 MODEL ORDER SELECTION

The selection of the model order in AR spectral estimation is a critical one. As described in Section 6.8, too low an order results in a smoothed estimate (see Figure 6.12), while too large an order causes spurious peaks and general statistical instability (see Figure 6.11). Many techniques have been proposed to estimate model order. These approaches have been derived by statistical analysis of *real* data. It is probable that these model order estimators may be applied directly to complex data; however, the extensions to complex data are not available. Caution must therefore be exercised in doing so.

For data observed from a pure AR process the model order estimators produce acceptable spectral estimates if the data record length is not extremely short [Akaike 1974]. It has been observed that for noise corrupted data the AR model order chosen is usually not sufficient to resolve spectral details [Landers and Lacoss 1977]. Of course, the true AR model for noisy data is of infinite order so that this result is not unexpected. It should also be emphasized that different estimates of the model order will be obtained if different AR parameter estimators are used in conjunction with the same model order estimator. No detailed studies are available which assess the *spectral estimation* performance of the various AR spectral estimators when the model order must be estimated in addition to the AR parameters. In comparing the strengths and weaknesses of the model order estimators, we should keep in mind that it is the quality of the spectral estimate which is of importance. For example, an estimator that underestimates the true AR model order for broadband PSDs, which are smooth in appearance, may well be preferable to one that indicates the true order of the broadband AR process but which, when combined with an AR parameter estimator, gives rise to spurious peaks. This situation is possible if, for example, the data record is short.

As was alluded to in Section 6.8, nearly all model order estimators are based on the estimated prediction error power. The estimated prediction error power is guaranteed to decrease or stay the same as the model order increases for all the AR parameter estimation methods described. Hence we cannot simply monitor the decrease in power as a means of determining model order but must also account for the increase in variance of a spectral estimate based on an increasing number of parameters. Two methods proposed by Akaike adhere to this philosophy. The first one, termed the *final prediction error* (FPE), estimates the model order as the value that minimizes [Akaike 1970]

$$\text{FPE}(k) = \frac{N + k}{N - k}\hat{\rho}_k \tag{7.53}$$

where $\hat{\rho}_k$ is the estimate of the white noise variance (prediction error power) for the kth order AR model. It is seen that whereas $\hat{\rho}_k$ decreases with k, the term $(N + k)/(N - k)$ increases with k. The FPE is an estimate of the prediction error power when the prediction coefficients must be estimated from the data. The term $(N + k)/(N - k)$ accounts for the increase in the variance of the prediction error power estimator due to the inaccuracies in the prediction coefficient estimates.

A second criterion, which appears to be in more general usage, is the *Akaike information criterion* (AIC). It is defined as [Akaike 1974]

$$\mathrm{AIC}(k) = N \ln \hat{\rho}_k + 2k. \tag{7.54}$$

As before, the order selected is the one that minimizes the AIC. A development of the AIC is provided in Appendix 7A. It is shown there that the AIC is an estimate of the Kullback–Leibler distance between an assumed PDF and the true PDF of the data. The method is not limited to AR model order determination but may be used more generally (see Appendix 7A) for choosing a model among competing models. Consequently, the AIC is useful for MA model order determination (see Chapter 8) and ARMA model order determination (see Chapter 9). An example is given in Figure 7.2. A real AR(2) process with parameters $a[1] = -1.34$, $a[2] = 0.9025$, $\sigma^2 = 1$ was simulated and $N = 100$ data points used for estimation of the AR parameters by the Burg algorithm. The minimum occurred at an order of 4, although the true order is 2. A larger data record would probably improve the estimate, although Kashyap [1980] has pointed out that the AIC is *not* a consistent estimator. This tendency to overestimate the true model order has led researchers to propose methods that replace the $2k$ factor in (7.54) by one that increases faster with N, typically $k \ln N$ [Rissanen 1978, 1983].

The performance of the AIC and FPE is similar. For short data records the use of the AIC is recommended [Ulrych and Ooe 1979]. For larger data records ($N \to \infty$) the two estimators will yield identical model order estimates since they are functionally related to each other. To prove this [Ulrych and Ooe 1979], consider

$$\ln \mathrm{FPE}(k) = \ln \hat{\rho}_k + \ln \frac{1 + k/N}{1 - k/N}. \tag{7.55}$$

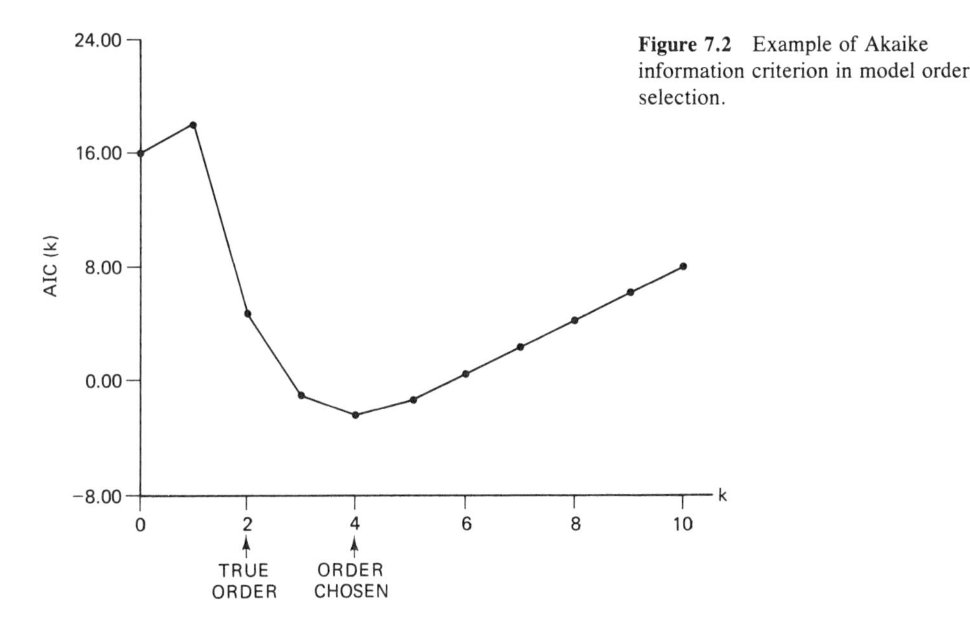

Figure 7.2 Example of Akaike information criterion in model order selection.

For large N, $k/N \ll 1$, so that

$$\ln \frac{1 + x}{1 - x} \approx 2x$$

and hence

$$\ln \text{FPE}(k) = \ln \hat{\rho}_k + \frac{2k}{N}$$

or finally,

$$N \ln \text{FPE}(k) = \text{AIC}(k). \tag{7.56}$$

Minimizing the AIC is therefore equivalent to minimizing the FPE.

A third method has been proposed by Parzen [1976] and is termed the *criterion autoregressive transfer function* (CAT). It is defined as

$$\text{CAT}(k) = \frac{1}{N} \sum_{i=1}^{k} \frac{1}{\tilde{\rho}_i} - \frac{1}{\tilde{\rho}_k} \tag{7.57}$$

where

$$\tilde{\rho}_i = \frac{N}{N - i} \hat{\rho}_i.$$

The value of k that minimizes (7.57) is chosen as the model order estimate. The CAT chooses the AR model order as the one that produces an estimated prediction error filter which is closest to the optimal infinite-length filter. The optimal filter has infinite length because the CAT is derived for AR spectral estimation based on an arbitrary data set, and not just a pure AR process. The criterion takes into account estimation errors in the prediction coefficients. The performance of the CAT has been found to be similar to that of the AIC and FPE [Parzen 1974, Landers and Lacoss 1977].

Some other methods have been proposed which are now briefly described. Jenkins and Watts [1968] suggest choosing the model order as the value of k which causes the estimated prediction error power

$$\hat{\rho}_k = \frac{1}{N - 2k} \sum_{n=k}^{N-1} \left(x[n] + \sum_{i=1}^{k} a_k[i] x[n - i] \right)^2 \tag{7.58}$$

to flatten out or be minimized. Note that the normalizing factor $(N - k) - k$ accounts for the k degrees of freedom lost in estimating the AR filter parameters. Another approach is to examine the sequence of estimated reflection coefficients. As described in Section 6.5.2 for an AR(p) process, the MLEs of the higher order reflection coefficients are distributed as [Box and Jenkins 1970]

$$\hat{k}_k \sim N\left(0, \frac{1}{N}\right) \qquad k \geq p + 1. \tag{7.59}$$

This means that, as an example, with 95% probability $|\hat{k}_k| \leq 1.96/\sqrt{N}$ for $k \geq p + 1$. The model order chosen is the value of k for which all higher order reflection coefficients are less than $1.96/\sqrt{N}$ in magnitude. A final method is based on the whiteness of the estimated prediction error

$$\hat{e}_k^f[n] = x[n] + \sum_{i=1}^{k} \hat{a}_k[i]x[n-i] \qquad n = k, k+1, \ldots, N-1.$$

Denote the autocorrelation estimator of $\hat{e}_k^f[n]$ by $\hat{r}_{ee}^{(k)}[m]$. If the process is truly AR(k), then [Box and Pierce 1970]

$$Q = N \sum_{m=1}^{M} \left[\frac{\hat{r}_{ee}^{(k)}[m]}{\hat{r}_{ee}^{(k)}[0]} \right]^2 \sim \chi^2_{M-k}. \tag{7.60}$$

Hence, if Q is computed versus k, the appropriateness of each model can be tested by comparing Q to a threshold. Assume that it is desired to test the appropriateness of various model orders at a 95% significance level. Thresholds for the orders are computed from (see Section 2.1)

$$\Pr\{\chi^2_{M-k} > \alpha_k\} = 0.05.$$

If Q for a given k falls below the computed threshold α_k, that value of k can be considered as a candidate for the correct model order. If several values of k produce Q that fall below the threshold, it is not clear which model order should be chosen. It is also possible that all values of k may produce Q which exceed their respective thresholds. This type of test may not produce a good indication of model order. M should be chosen so that the impulse response of $1/\mathcal{A}(z)$ is effectively zero for samples greater than M. Because $\mathcal{A}(z)$ is unknown, the choice of M is not obvious.

In summary, many model order estimators are available, but few guidelines as to their use in practical situations. For the purposes of spectral estimation, a rule of thumb is that the AR model order should be chosen to be $N/3 < p < N/2$ [Ulrych and Bishop 1975] for good spectral resolution with few spurious peaks. The reader may consult the papers by Ulrych and Bishop [1975] and Landers and Lacoss [1977] for more performance comparisons of the estimators just described and also Quenoille [1947], Anderson [1971], and Rissanen [1978] for descriptions of other model order estimators.

7.9 SPECTRAL ESTIMATION OF NOISY AR PROCESSES

As discussed in detail in Section 6.7, the effect of observation noise is to produce a smoothed AR spectral estimate. This sensitivity, which manifests itself as a severe bias, limits the practical utility of AR spectral estimation. The various methods for AR parameter estimation which were derived based on the maximum likelihood principle are no longer MLEs when observation noise is present. To find the MLE for real data, assume that the observed process is of the form

$$y[n] = x[n] + w[n]$$

where $x[n]$ is an AR(p) process and $w[n]$ is white Gaussian noise with variance

σ_w^2. Then for large data records we must minimize the following function [Hosoya 1979]:

$$\int_{-1/2}^{1/2} \left[\ln P_{yy}(f) + \frac{I(f)}{P_{yy}(f)} \right] df \tag{7.61}$$

where

$$P_{yy}(f) = \frac{\sigma^2}{|A(f)|^2} + \sigma_w^2 \tag{7.62}$$

over $\{a[1], a[2], \ldots, a[p], \sigma^2, \sigma_w^2\}$. In (7.61), $I(f)$ is the periodogram as defined by (4.2). The different notation for the periodogram in (7.61) is meant to indicate that it is to be considered as a given function of the data, and not necessarily a spectral estimate. This is a difficult nonlinear minimization problem (see Problem 7.10). Hosoya has proposed a Newton–Raphson approach but provides no results for finite data records. An alternative strategy which searches for the minimum over a slightly different set of parameters has been proposed by Pagano [1974]. It, too, employs a Newton–Raphson iteration. Either method will suffer from the usual problems of convergence to only a local minimum or lack of convergence (see the discussion of the Akaike MLE for ARMA processes in Section 10.3 for a further description of the difficulties associated with nonlinear optimization).

Due to the difficulties of an MLE approach, several suboptimum estimators have been proposed. They involve:

1. Applying general ARMA estimators
2. Filtering the data to reduce the observation noise
3. Compensating the AR parameters or reflection coefficient estimates for the biasing effect of noise
4. Using a large order AR model.

The first approach recognizes that the true model for an AR(p) process embedded in white noise is an ARMA(p, p) model as discussed in Section 6.7. The AR filter parameters of the ARMA(p, p) model are identical to the AR filter parameters of the AR(p) process. Hence we can use any of the ARMA estimators described in Chapter 10 to estimate the AR filter parameters. A common method is the use of the modified Yule–Walker equations (see Section 10.4) [Walker 1960, Gersch 1970, Done 1978]. This procedure, however, does not make use of the linkage between the AR and MA parameters of the ARMA(p, p) model. From (6.114) this relationship is

$$\sigma_\eta^2 \mathcal{B}(z)\mathcal{B}^*(1/z^*) = \sigma^2 + \sigma_w^2 \mathcal{A}(z)\mathcal{A}^*(1/z^*) \tag{7.63}$$

where $\mathcal{B}(z) = 1 + \sum_{k=1}^{p} b[k]z^{-k}$. $\{b[1], b[2], \ldots, b[p]\}$ are the MA filter parameters and σ_η^2 is the driving white noise variance of the equivalent ARMA(p, p) model. Once the AR filter parameters have been obtained, σ^2 and σ_w^2 may be estimated using the least squares estimator of Parzen [1961]. This estimator is based on (7.62), or

$$|A(f)|^2 P_{yy}(f) = \sigma^2 + \sigma_w^2 |A(f)|^2. \tag{7.64}$$

The PSD of $y[n]$ is replaced by a periodogram estimate and the AR filter parameters are replaced by the estimates already obtained. Letting $E(f)$ represent the error due to replacements of the theoretical quantities by their estimates, (7.64) becomes

$$|\hat{A}(f)|^2 I(f) = \sigma^2 + \sigma_w^2 |\hat{A}(f)|^2 + E(f). \tag{7.65}$$

Solving (7.65) in a least-squares sense (see Problem 7.11) leads to

$$\begin{bmatrix} 1 & \sum_{k=0}^{p} |\hat{a}[k]|^2 \\ \sum_{k=0}^{p} |\hat{a}[k]|^2 & \sum_{k=0}^{2p} |\hat{a}[k] \star \hat{a}[k]|^2 \end{bmatrix} \begin{bmatrix} \hat{\sigma}^2 \\ \hat{\sigma}_w^2 \end{bmatrix} = \begin{bmatrix} \frac{1}{N-p} \sum_{n=p}^{N-1} |x[n] \star \hat{a}[n]|^2 \\ \frac{1}{N-2p} \sum_{n=2p}^{N-1} |x[n] \star \hat{a}[n] \star \hat{a}[n]|^2 \end{bmatrix}. \tag{7.66}$$

A second approach is to filter the data with a Wiener filter to enhance the signal [the AR(p) process] from the noise. Then, any of the standard AR parameter estimators may be used. Because the optimal Wiener filter frequency response depends on the PSD of the signal (see Problem 7.12), which is to be estimated, the filter must be designed in a bootstrap or adaptive manner. It can be shown that the Wiener filter is an implicit part of the MLE, so that this approach is reasonable. An example is the estimator proposed by Lim [1978], which has met with some success for speech data. Others can be found in the survey paper by Lim and Oppenheim [1979] as well as in Sambur [1979].

A third method attempts to remove the bias of the AR parameters due to observation noise and is termed *noise compensation*. A simple example occurs in the application of the autocorrelation method to noisy data. Because

$$r_{yy}[k] = \begin{cases} r_{xx}[k] + \sigma_w^2 & \text{for } k = 0 \\ r_{xx}[k] & \text{for } k \neq 0 \end{cases}$$

we could subtract an estimate of the observation noise variance from the estimate of the ACF at lag $k = 0$. Thus, in the Yule–Walker equations the autocorrelation estimator $\{\hat{r}_{yy}[0] - \hat{\sigma}_w^2, \hat{r}_{yy}[1], \ldots, \hat{r}_{yy}[p]\}$ is used. A serious deficiency of this approach is that subtraction of too large a noise variance will result in a spectral estimate that is too "peaky" [Marple 1978] (see Problem 7.13). Additionally, the matrix may become ill-conditioned, leading to a spectral estimator with a large variance. A similar approach may be applied to the estimated reflection coefficients [Kay 1980]. More variations may be found in Lim and Oppenheim [1979].

A final approach to AR spectral estimation in noise is to use a large order AR model in an attempt to reduce the bias due to the model mismatch. That this is feasible is guaranteed by the Kolmogorov theorem, which states that an AR(∞) is adequate to model any WSS process. The principal shortcoming is the possibility of spurious peaks as the model order is increased. A maximum order of $p = N/2$ has been recommended [Ulrych and Bishop 1975] to guard against this possibility. If the data consist of sinusoids in white noise, the occurrence of spurious peaks may be substantially reduced by use of the principal component solution as described in Chapter 13.

Model order estimation for AR processes in white noise poses additional difficulties in an otherwise nontrivial problem. The reader is referred to the papers by Tong [1975, 1977] in which the AIC is used in conjunction with an MLE of the AR parameters. Very little is known about the statistics of AR spectral estimates obtained from noise corrupted data. Some recent work that attempts to address this problem for the case where the AR parameters are estimated by using the modified Yule–Walker equations (see Section 10.4) is that of Gingras [1984, 1985].

7.10 COMPUTER SIMULATION EXAMPLES

Some examples of the performance of the various AR spectral estimation methods are now described. Two PSDs have been chosen to represent the contrasting possibilities of spectra with narrowband and broadband features. Real AR(4) processes with true PSDs shown in Figure 7.3 were used. The AR(4) process that produced the PSD of Figure 7.3a will be referred to as the broadband AR process, while the one that produced the PSD shown in Figure 7.3b will be termed the narrowband AR process. The AR parameters for each case are listed in Table 7.1. Also, the corresponding pole positions are illustrated in Figure 7.3. The data record length for all the simulations was $N = 256$ real data points and the AR model order used was the true model order $p = 4$. For each method 50 realizations of the spectral estimator were obtained by computer simulation using the computer programs listed in Appendices 7B to 7E. The average of the 50 estimates is shown for each method together with the true PSD, which is illustrated by a dashed curve. Also, the 50 estimates are plotted in an overlaid fashion to indicate the variability of each spectral estimator. The results are shown in Figure 7.4 for the broadband AR process and in Figure 7.5 for the narrowband AR process. Considering first the broadband process, there appears to be very little bias in the spectral estimator for any of the methods. The variance of each method is nearly the same, as is evident by examining the overlaid plots. Thus, for broadband processes all the spectral estimators yield nearly identical results. This is to be expected since all the methods are approximate MLEs, differing only in the "end effects" which become negligible for large data records. For this example the data record length of $N = 256$ was therefore adequate to ensure the applicability of the asymptotic results. For the narrowband process all the estimators except the autocorrelation method are unbiased and have a comparable variance. The autocorrelation method, on the other hand, displays a large bias, being unable to resolve the spectral peaks, and also a slightly larger variance. From these simulations we may conclude that the covariance, modified covariance, Burg, and recursive MLE methods are comparable in performance, while the autocorrelation method is inferior. It is also of interest to compare these results with those for the periodogram and Blackman–Tukey spectral estimators. As an example, con-

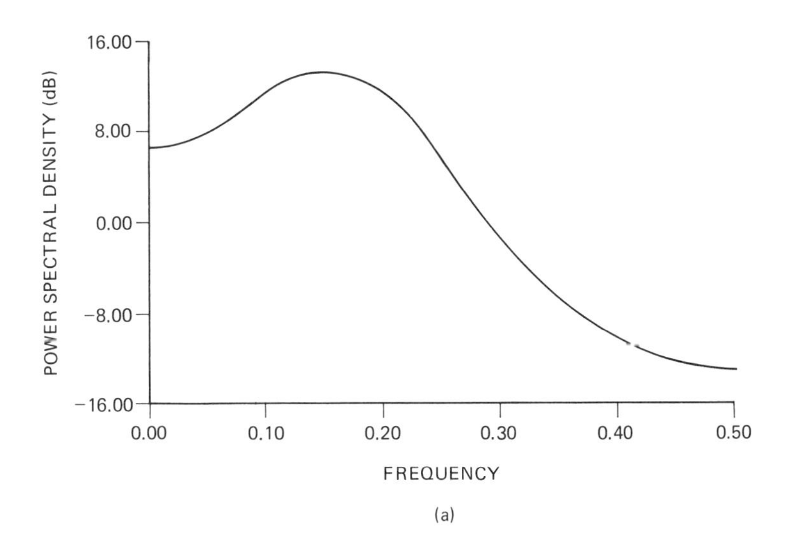

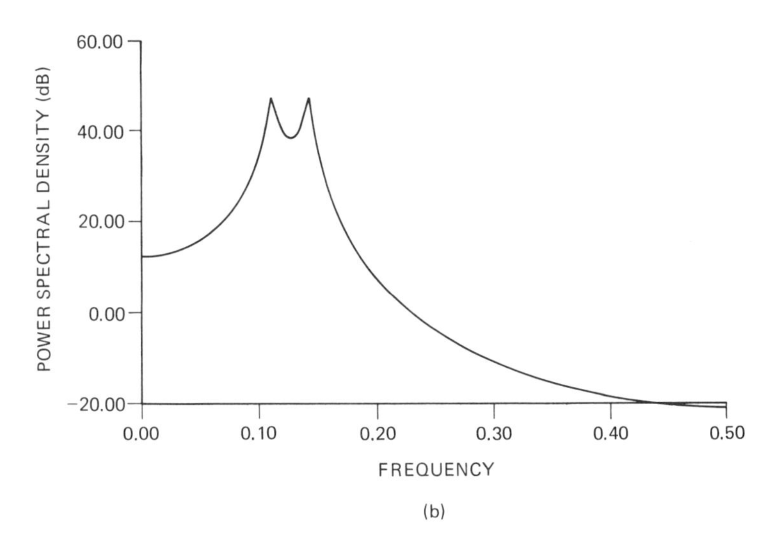

Figure 7.3 True power spectral densities and pole plots a) broadband power spectral density b) narrowband power spectral density c) pole plot for broadband process d) pole plot for narrowband process.

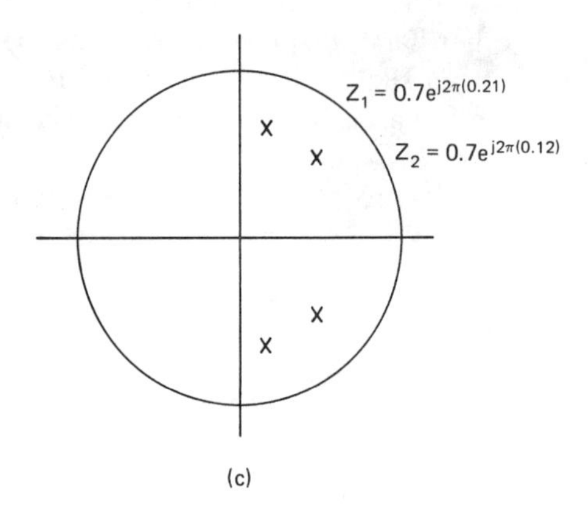

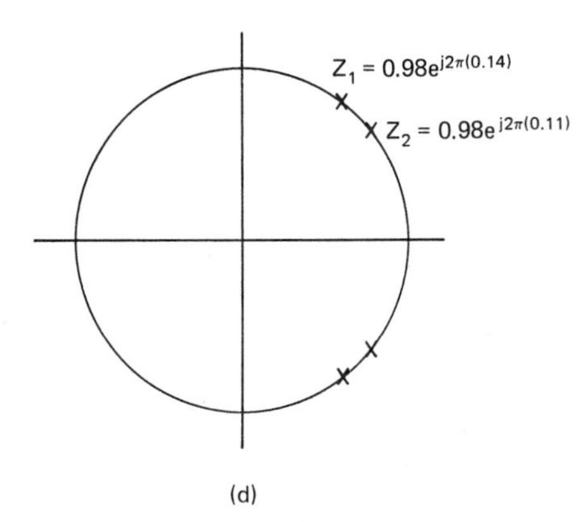

Figure 7.3 (*Continued*)

TABLE 7.1 PARAMETER VALUES OF AR PROCESSES USED FOR COMPUTER SIMULATIONS

	Parameter				
	$a[1]$	$a[2]$	$a[3]$	$a[4]$	σ^2
Broadband process	−1.352	1.338	−0.662	0.240	1
Narrowband process	−2.760	3.809	−2.654	0.924	1

sider the spectral estimates for the broadband process given in Figures 4.12 and 7.4. (Be careful to note that the scaling in Figure 4.12a is not the same as for the other plots.) It is immediately clear from this comparison that if the data are obtained from an AR process and the AR spectral estimator is based on the true model order, the AR spectral estimator is superior in performance to any Fourier based spectral estimator. Further comparisons via computer simulations are available from the works of Nuttall [1976], Kaveh and Cooper [1976], and Shon and Mehrota [1984].

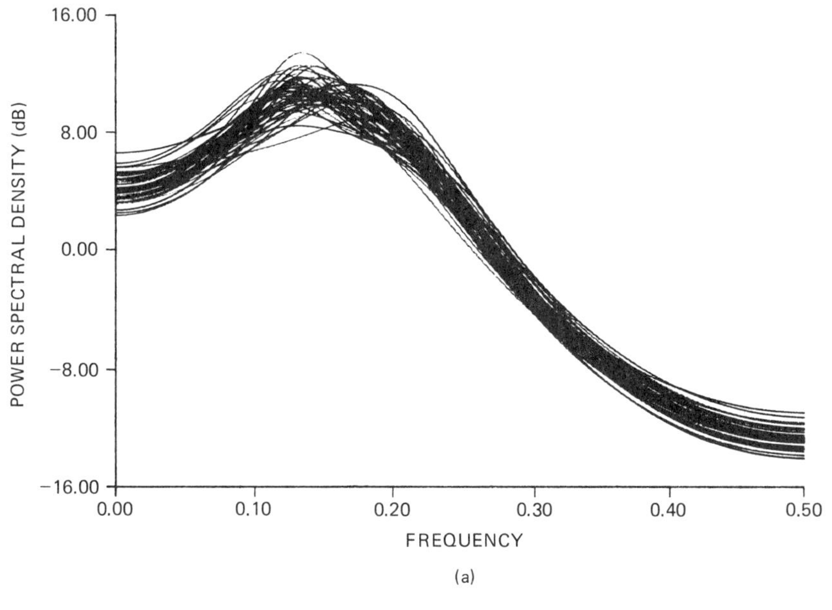

i) overlaid realizations

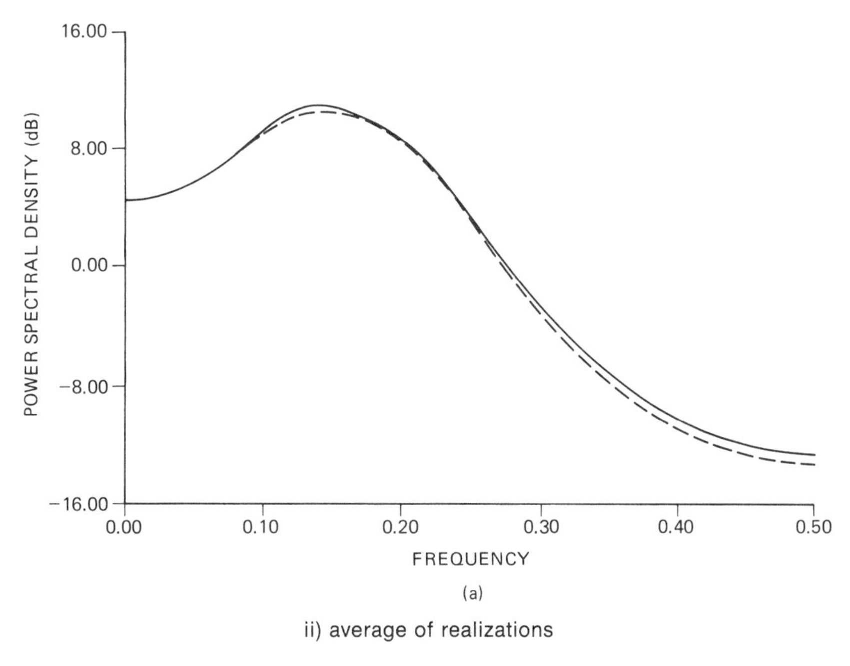

ii) average of realizations

Figure 7.4 Spectral estimation performance of AR methods for broadband AR process. (a) Autocorrelation method. (b) Covariance method. (c) Modified covariance method. (d) Burg method. (e) Recursive MLE.

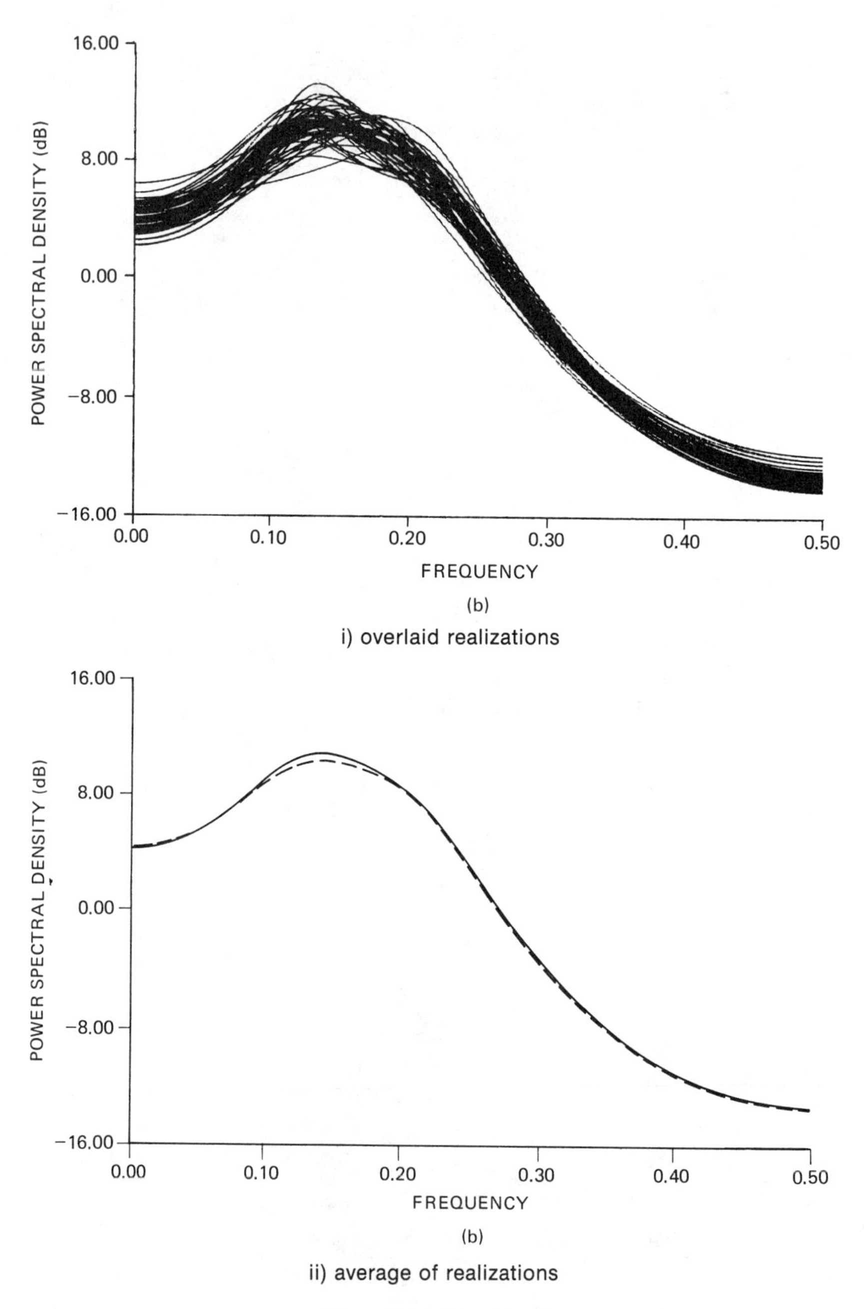

Figure 7.4 (*Continued*)

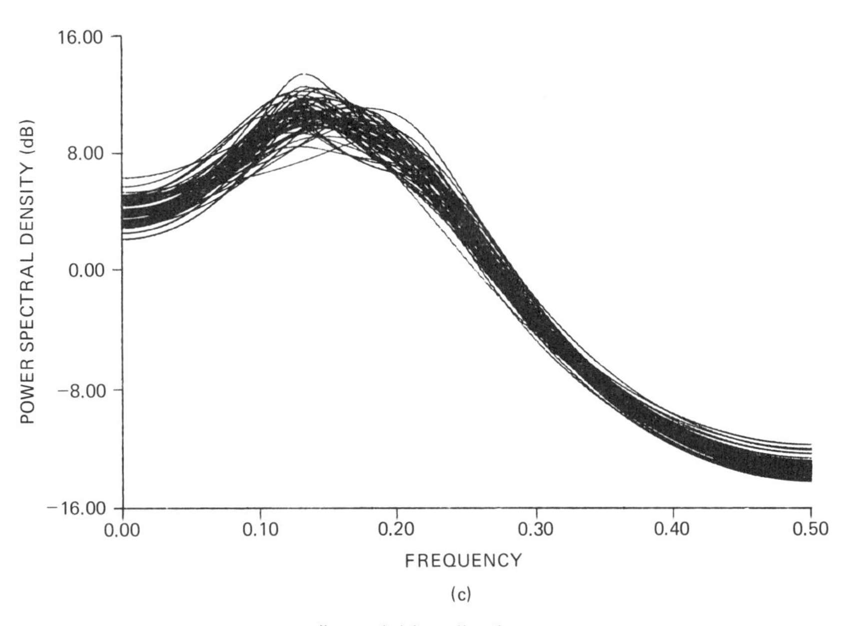

i) overlaid realizations

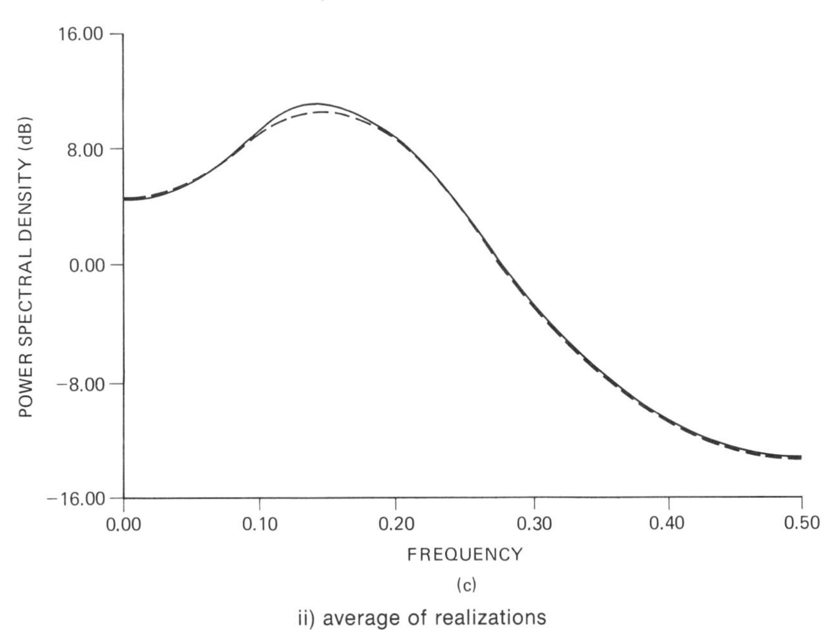

ii) average of realizations

Figure 7.4 *(Continued)*

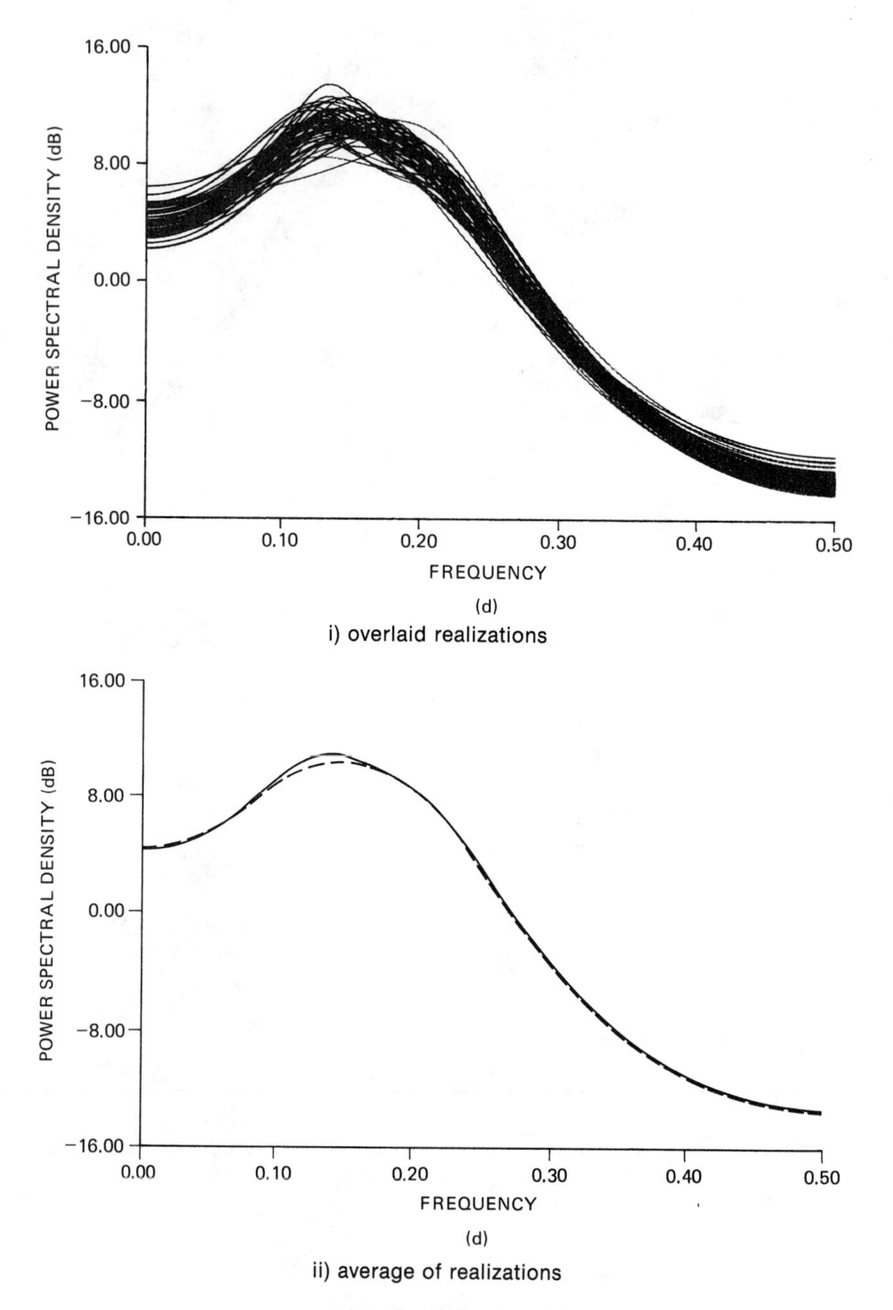

Figure 7.4 (*Continued*)

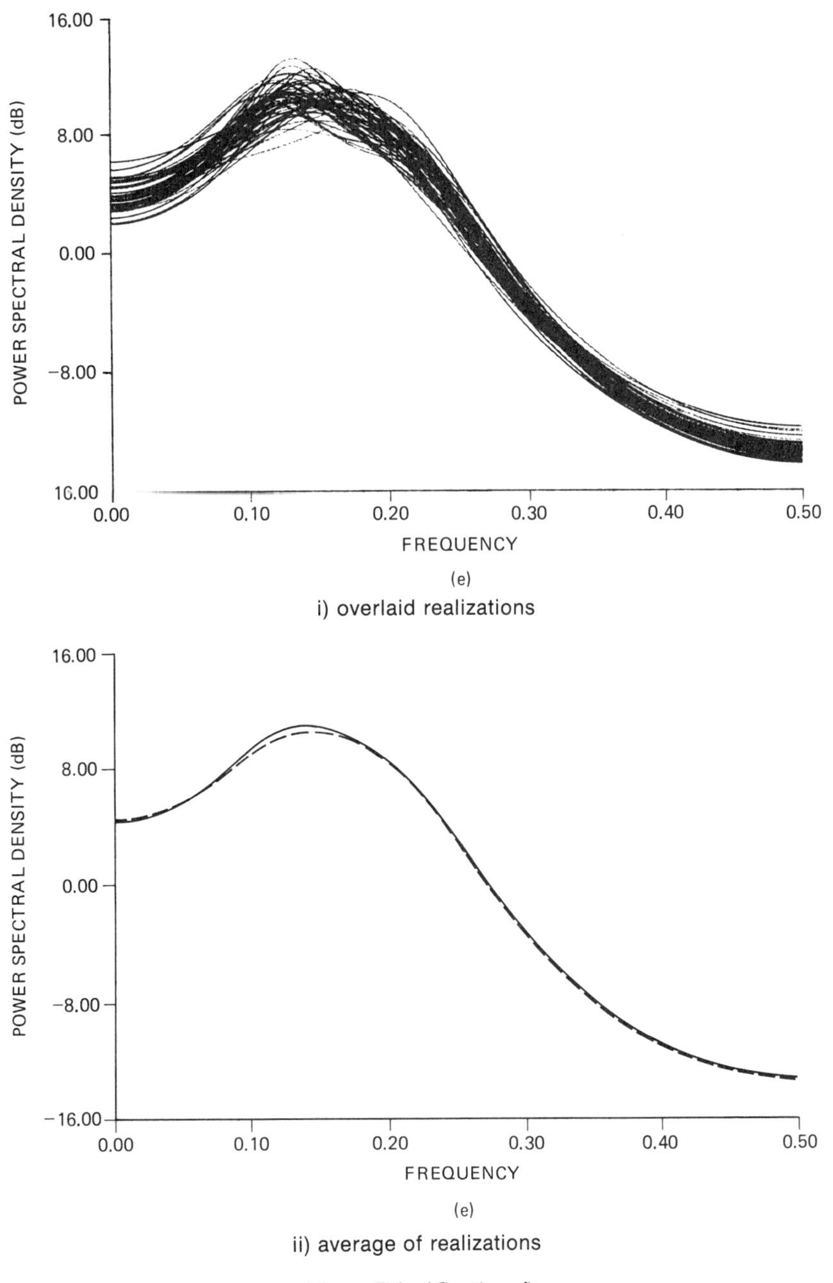

Figure 7.4 (*Continued*)

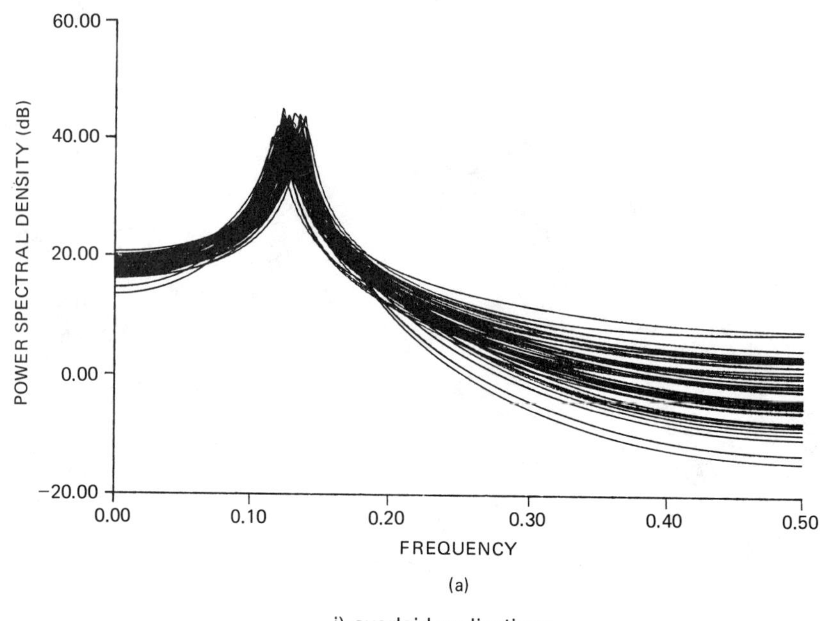

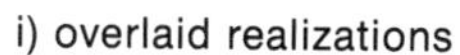

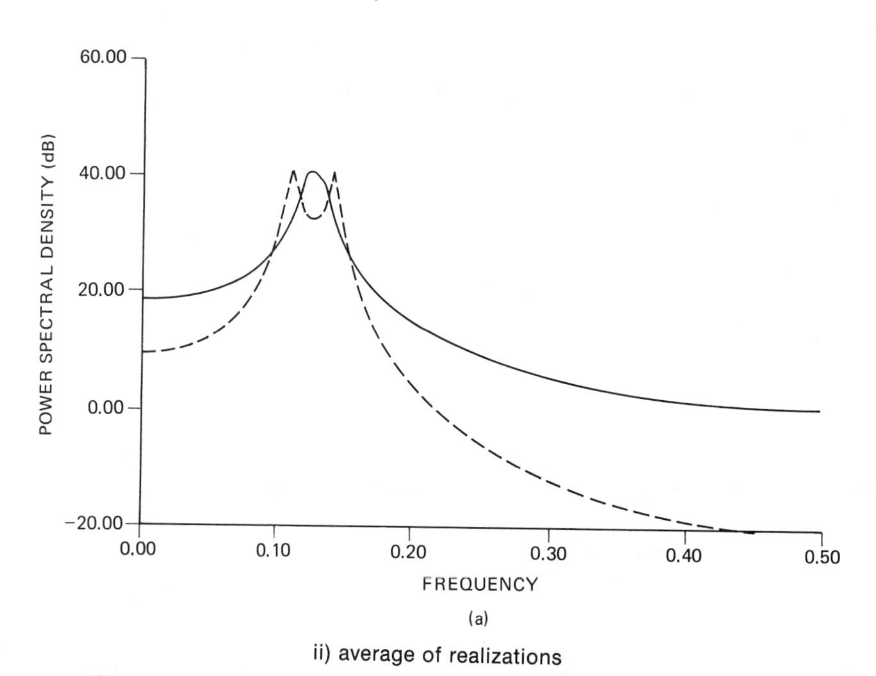

Figure 7.5 Spectral estimation performance of AR methods for narrowband AR process. (a) Autocorrelation method. (b) Covariance method. (c) Modified covariance method. (d) Burg method. (e) Recursive MLE.

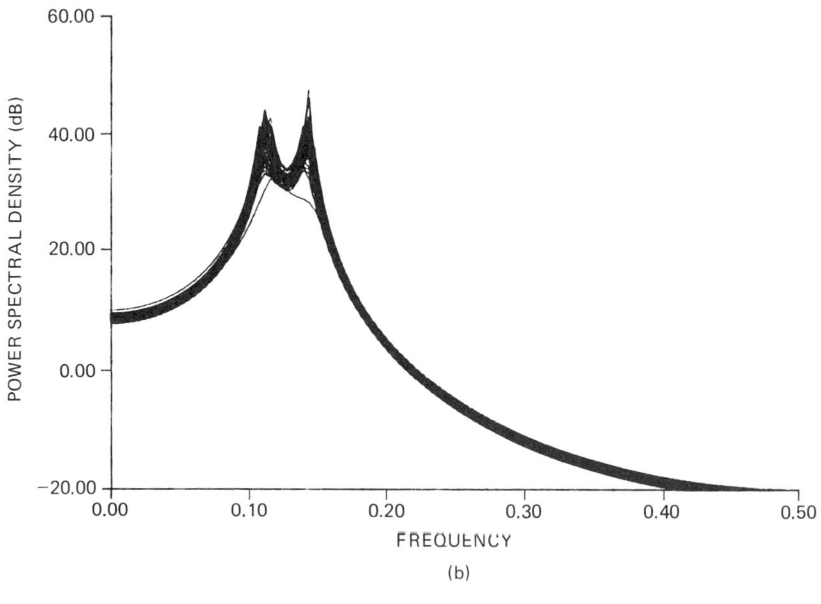

i) overlaid realizations

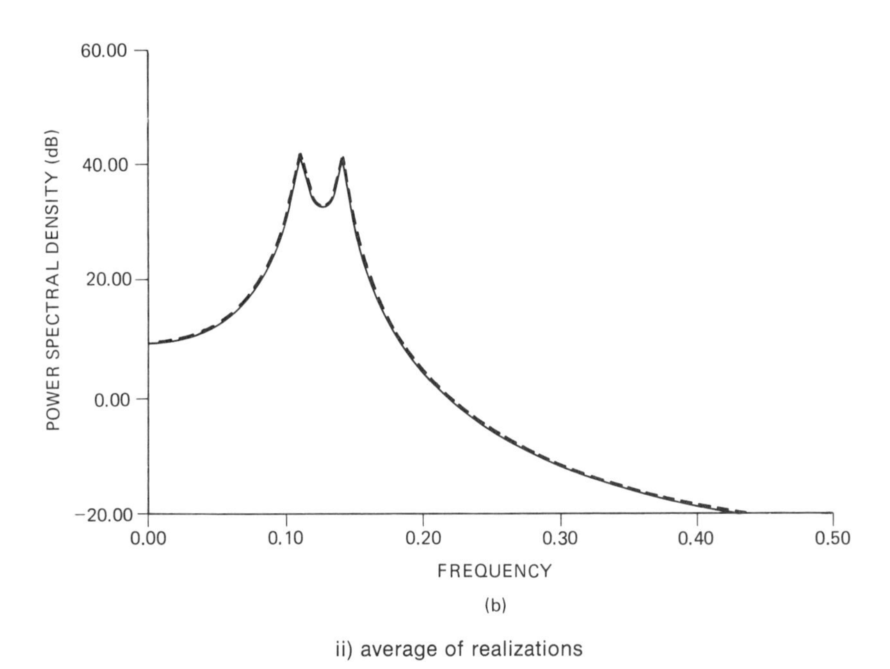

ii) average of realizations

Figure 7.5 *(Continued)*

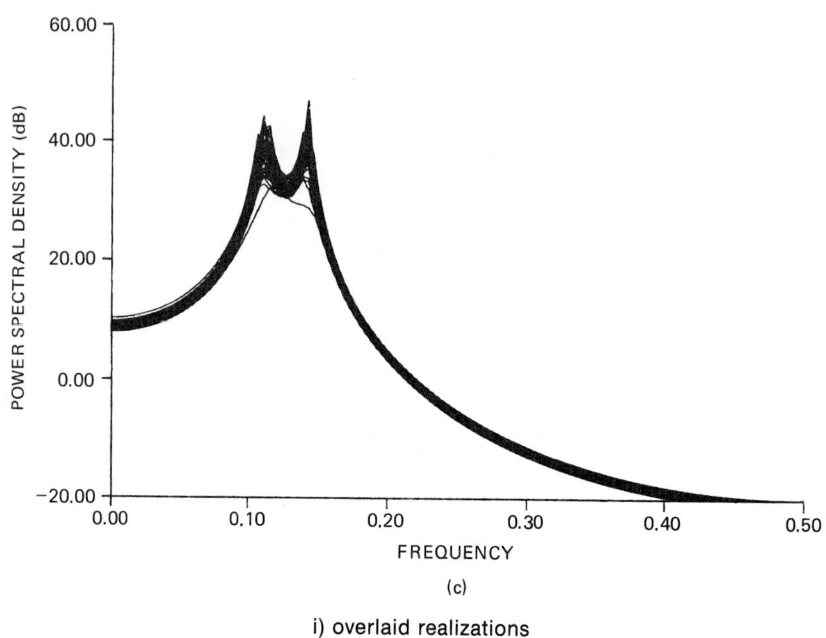

(c)

i) overlaid realizations

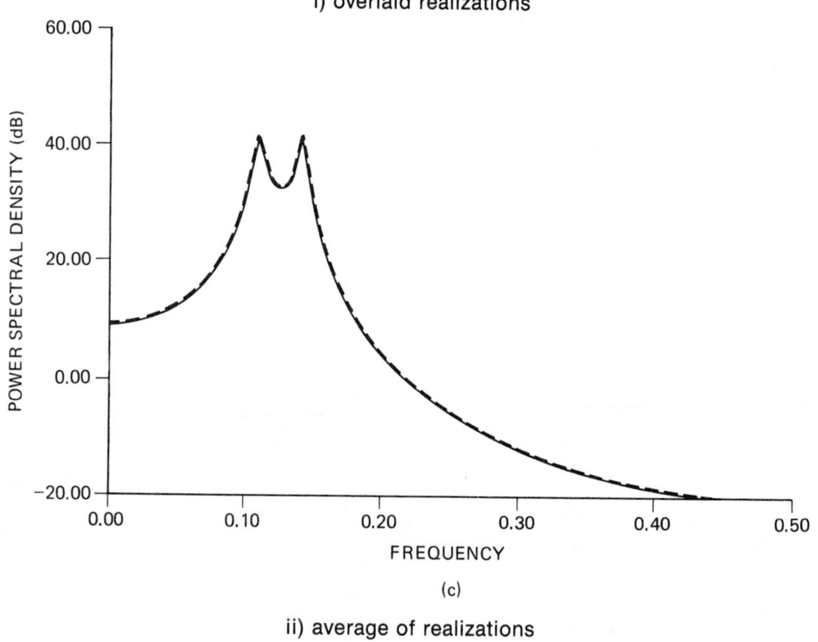

(c)

ii) average of realizations

Figure 7.5 (*Continued*)

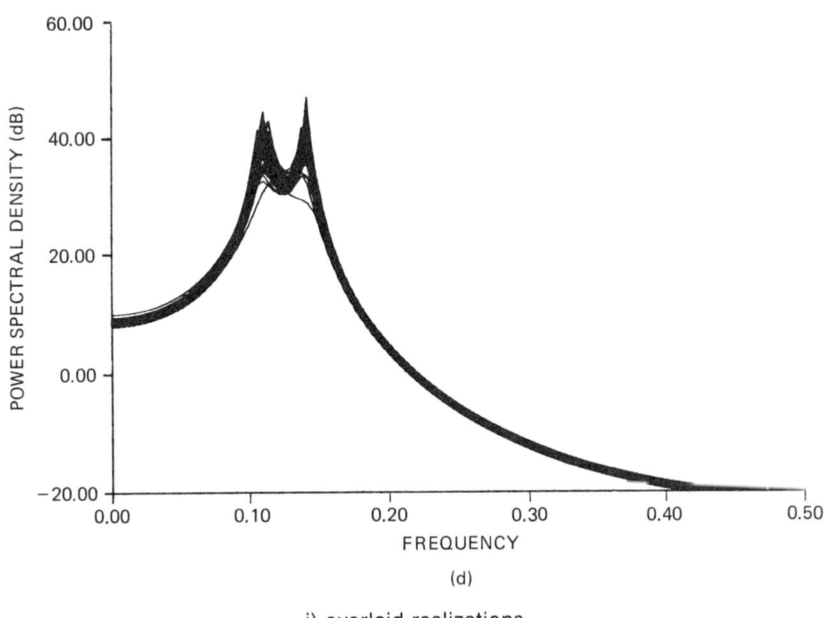

i) overlaid realizations

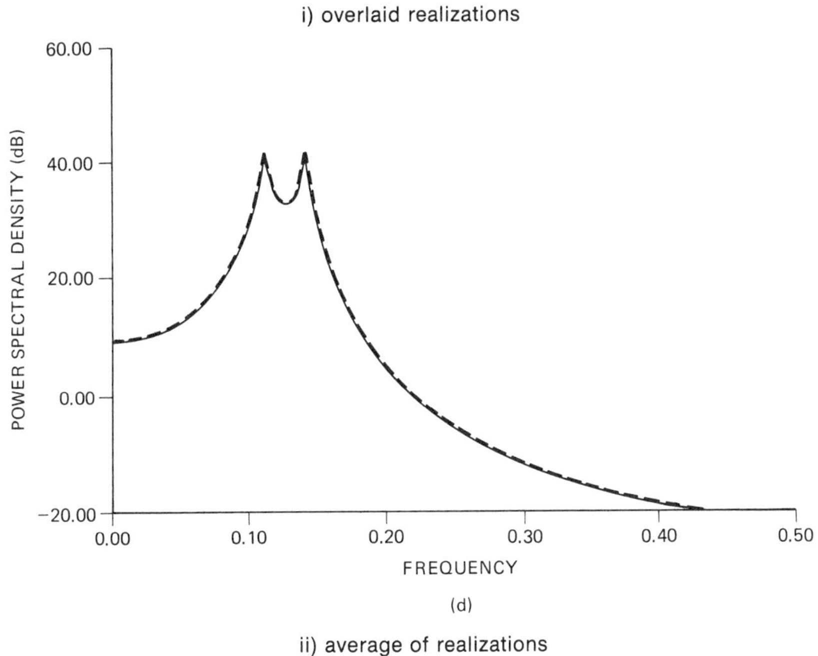

ii) average of realizations

Figure 7.5 (*Continued*)

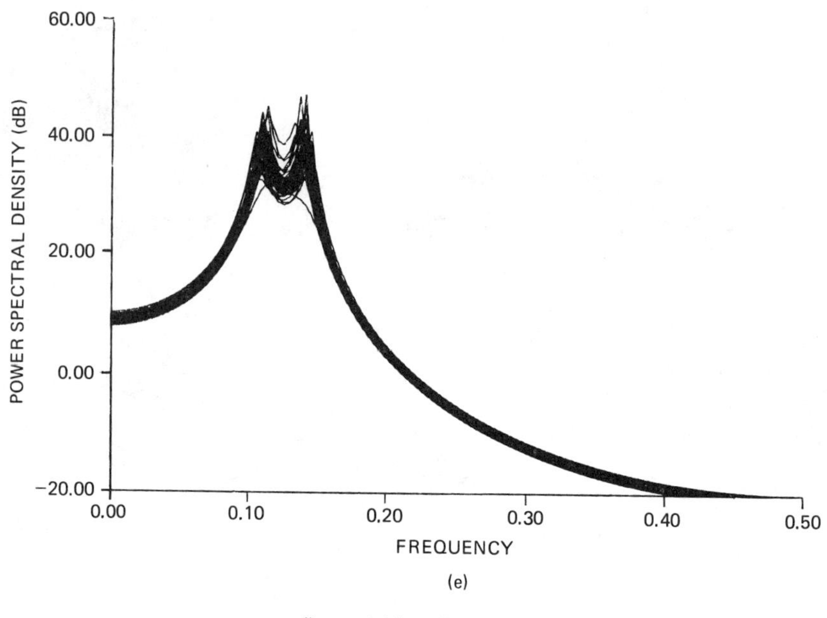

(e)

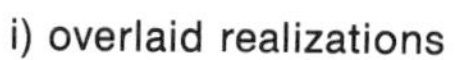

i) overlaid realizations

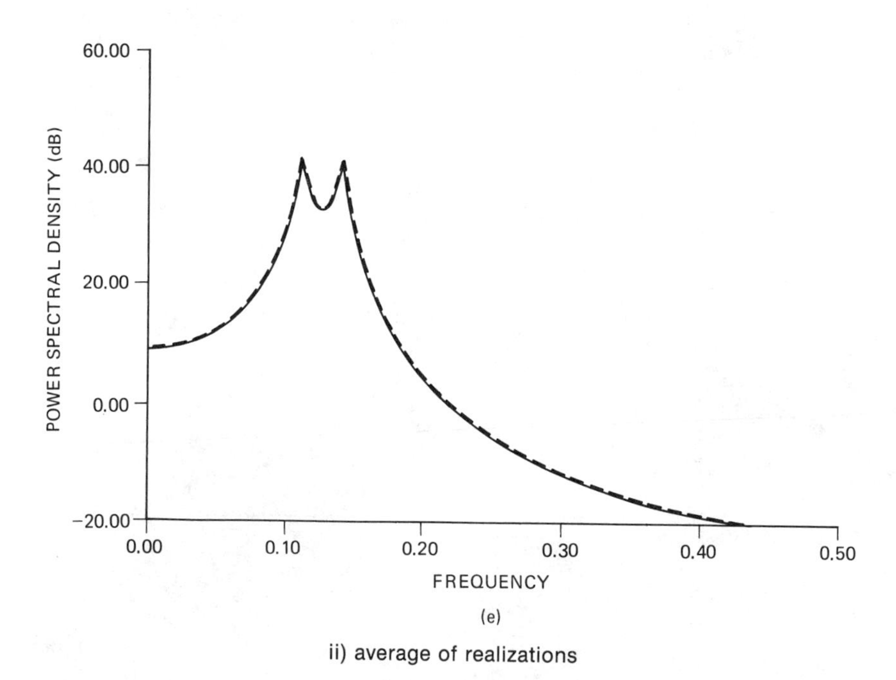

(e)

ii) average of realizations

Figure 7.5 *(Continued)*

REFERENCES

Akaike, H., "Statistical Predictor Identification," *Ann. Inst. Statist. Math.*, Vol. 22, pp. 203–217, 1970.

Akaike, H., "A New Look at the Statistical Model Identification," *IEEE Trans. Autom. Control*, Vol. AC19, pp. 716–723, Dec. 1974.

Anderson, T. W., *The Statistical Analysis of Time Series*, Wiley, New York, 1971.

Anderson, N. O., "Comments on the Performance of the Maximum Entropy Algorithm," *Proc. IEEE*, Vol. 66, pp. 1581–1582, Nov. 1978.

Box, G. E. P., and G. M. Jenkins, *Time Series Analysis: Forecasting and Control*, Holden-Day, San Francisco, 1970.

Box, G. E. P., and D. A. Pierce, "Distribution of Residual Autocorrelations in Autoregressive-Integrated Moving Average Time Series," *J. Amer. Stat. Assoc.*, Vol. 64, 1970.

Burg, J. P., "Maximum Entropy Spectral Analysis," Ph.D. dissertation, Stanford University, May 1975.

Chen, W. Y., and G. R. Stegen, "Experiments with Maximum Entropy Power Spectra of Sinusoids," *J. Geophys. Res.*, Vol. 79, pp. 3019–3022, July 1974.

Cox, D. R., and D. V. Hinkley, *Theoretical Statistics*, Chapman & Hall, New York, 1974.

Dickinson, B., "Two Recursive Estimates of Autoregressive Models Based on Maximum Likelihood," *J. Statist. Comput. Simulation*, Vol. 7, pp. 85–92, 1978.

Done, W. J., "Estimation of the Parameters of an Autoregressive Process in the Presence of Additive White Noise," Rep. UTEC-CSC-79-021, Computer Science Department, University of Utah, Dec. 1978.

Fougere, P. F., E. J. Zawalick, and H. R. Radoski, "Spontaneous Line Splitting in Maximum Entropy Power Spectrum Analysis," *Phys. Earth Planet. Inter.*, Vol. 12, pp. 201–207, Aug. 1976.

Gersch, W., "Estimation of the Autoregressive Parameters of a Mixed Autoregressive Moving-Average Time Series," *IEEE Trans. Autom. Control*, Vol. AC15, pp. 583–588, Oct. 1970.

Gingras, D., "On the Asymptotic Normality of Autoregressive Spectral Density Estimates for the Noise Corrupted Case," *Rec. 1984 IEEE Int. Conf. Acoust. Speech Signal Process.*, pp. 14.9.1–14.9.4.

Gingras, D., "Asymptotic Properties of High-Order Yule–Walker Estimates of the AR Parameters of an ARMA Time Series," *IEEE Trans. Acoust. Speech Signal Process.*, Vol. ASSP33, pp. 1095–1101, Oct. 1985.

Herring, R. W., "The Cause of Line Splitting in Burg Maximum-Entropy Spectral Analysis," *IEEE Trans. Acoust. Speech Signal Process.*, Vol. ASSP28, pp. 692–701, Dec. 1980.

Hosoya, Y., "Efficient Estimation of a Model with an Autoregressive Signal with White Noise," Tech. Rep. 37, Department of Statistics, Stanford University, Mar. 1979.

Itakura, F., and S. Saito, "Digital Filtering Techniques for Speech Analysis and Synthesis," *7th Int. Cong. Acoust.*, Budapest, Paper 25-C-1, 1971.

Jenkins, G. M., and D. G. Watts, *Spectral Analysis and Its Applications*, Holden-Day, San Francisco, 1968.

Kashyap, R. L., "Inconsistency of the AIC Rule for Estimating the Order of Autoregressive Models," *IEEE Trans. Autom. Control*, Vol. AC25, pp. 996–998, Oct. 1980.

Kaveh, M., and G. R. Cooper, "An Empirical Investigation of the Properties of the Autoregressive Spectral Estimator," *IEEE Trans. Inf. Theory*, Vol. IT22, pp. 313–323, May 1976.

Kaveh, M., and G. A. Lippert, "An Optimum Tapered Burg Algorithm for Linear Prediction and Spectral Analysis," *IEEE Trans. Acoust. Speech Siganl Process.*, Vol. ASSP31, pp. 438–444, Apr. 1983.

Kay, S. M., "Noise Compensation for Autoregressive Spectral Estimates," *IEEE Trans. Acoust. Speech Signal Process.*, Vol. ASSP28, pp. 292–303, June 1980.

Kay, S. M., "Recursive Maximum Likelihood Estimation of Autoregressive Processes," *IEEE Trans. Acoust. Speech Signal Process.*, Vol. ASSP31, pp. 56–65, Feb. 1983.

Kay, S. M., and S. L. Marple, Jr., "Sources of and Remedies for Spectral Line Splitting in Autoregressive Spectrum Analysis," *Rec. 1979 IEEE Int. Conf. Acoust. Speech Signal Process.*, pp. 151–154.

Kullback, S., *Information Theory and Statistics*, Wiley, New York, 1959.

Kumaresan, R., "Estimating the Parameters of Exponentially Damped or Undamped Sinusoidal Signals in Noise," Ph.D. dissertation, University of Rhode Island, 1982.

Landers, T. E., and R. T. Lacoss, "Some Geophysical Applications of Autoregressive Spectral Estimates," *IEEE Trans. Geosci. Electron.*, Vol. GE15, pp. 26–32, Jan. 1977.

Lim, J. S., "All Pole Modeling of Degraded Speech," *IEEE Trans. Acoust. Speech Signal Process.*, Vol. ASSP26, pp. 197–209, June 1978.

Lim, J. S., and A. V. Oppenheim, "Enhancement and Bandwidth Compression of Noisy Speech," *Proc. IEEE*, Vol. 67, pp. 1586–1604, Dec. 1979.

Makhoul, J., "Linear Prediction: A Tutorial Review," *Proc. IEEE*, Vol. 63, pp. 561–580, Apr. 1975.

Makhoul, J., "Stable and Efficient Lattice Methods for Linear Prediction," *IEEE Trans. Acoust. Speech Signal Process.*, Vol. ASSP25, pp. 423–428, Oct. 1977.

Makhoul, J., "Corrections to 'Stable and Efficient Lattice Methods for Linear Prediction'," *IEEE Trans. Acoust. Speech Signal Process.*, Vol. ASSP26, p. 111, Feb. 1978.

Marple, S. L., Jr., "High Resolution Autoregressive Spectrum Analysis Using Noise Power Cancelation," *Rec. 1978 IEEE Int. Conf. Acoust. Speech Signal Process.*, pp. 345–348.

Marple, S. L., Jr., "A New Autoregressive Spectrum Analysis Algorithm," *IEEE Trans. Acoust. Speech Signal Process.*, Vol. ASSP28, pp. 441–454, Aug. 1980.

Marple, S. L., Jr., *Digital Spectral Analysis with Applications,* Prentice-Hall, Englewood Cliffs, N.J., 1987.

Matsuoka, T., and T. J. Ulrych, "Information Theory Measures with Application to Model Identification," *IEEE Trans. Acoust. Speech Signal Process.*, Vol. ASSP34, pp. 511–517, June 1986.

Morf, M., B. Dickinson, T. Kailath, and A. Vieira, "Efficient Solution of Covariance Equations for Linear Prediction," *IEEE Trans. Acoust. Speech Signal Process.*, Vol. ASSP25, pp. 429–433, Oct. 1977.

Nuttall, A. H., "Spectral Analysis of a Univariate Process with Bad Data Points, via Maximum Entropy and Linear Predictive Techniques," Tech. Rep. TR-5303, Naval Underwater Systems Center, New London, Conn., Mar. 26, 1976.

Oppenheim, A. V., and R. W. Schafer, *Digital Signal Processing*, Prentice-Hall, Englewood Cliffs, N.J., 1975.

Pagano, M., "An Algorithm for Fitting Autoregressive Schemes," *J. Roy. Statist. Soc.*, Vol. 21, pp. 274–281, 1972.

Pagano, M., "Estimation of Models of Autoregressive Signal plus White Noise," *Ann. Statist.*, Vol. 2, pp. 99–108, 1974.

Parzen, M., "An Approach to Time Series Analysis," *Ann. Math. Statist.*, Vol. 32, pp. 951–989, 1961.

Parzen, E., "Some Recent Advances in Time Series Modeling," *IEEE Trans. Autom. Control*, Vol. AC19, pp. 723–730, Dec. 1974.

Parzen, E., "An Approach to Time Series Modelling and Forecasting Illustrated by Hourly Electricity Demands," Tech. Rep. 37, Statistical Science Division, State University of New York, Jan. 1976.

Prony, G. R. B., "Essai expérimental et analytique, etc.," *J. Ec. Polytech.*, Vol. 1, Cashier 2, pp. 24–76, 1795.

Quenouille, M. H., "A Large-Sample Test for the Goodness of Fit of Autoregressive Schemes," *J. Roy. Statist. Soc.*, Vol. 110, p. 123, 1947.

Rissanen, J., "Modeling by Shortest Data Description," *Automatica*, Vol. 14, pp. 465–471, 1978.

Rissanen, J., "A Universal Prior for the Integers and Estimation by Minimum Description Length," *Ann. Statist.*, Vol. 11, pp. 417–431, 1983.

Sambur, M. R., "A Preprocessing Filter for Enhancing LPC Analysis/Synthesis of Noisy Speech," *Rec. 1979 IEEE Int. Conf. Acoust. Speech Signal Process.*, pp. 971–974.

Shon, S., and K. Mehrota, "Performance Comparisons of Autoregressive Estimation Methods," *Rec. 1984 IEEE Int. Conf. Acoust. Speech Signal Process.*, pp. 14.3.1–14.3.4.

Swingler, D. N., "A Comparison between Burg's Maximum Entropy Method and a Non-recursive Technique for the Spectral Analysis of Deterministic Signals," *J. Geophys. Res.*, Vol. 84, pp. 679–685, Feb. 1979A.

Swingler, D., "A Modified Burg Algorithm for Maximum Entropy Spectral Analysis," *Proc. IEEE*, Vol. 67, pp. 1368–1369, Sept. 1979B.

Tong, H., "Autoregressive Model Fitting with Noisy Data by Akaike's Information Criterion," *IEEE Trans. Inf. Theory*, Vol. IT21, pp. 476–480, July 1975.

Tong, H., "More on Autoregressive Model Fitting with Noisy Data by Akaike's Information Criterion," *IEEE Trans. Inf. Theory*, Vol. IT23, pp. 409–410, May 1977.

Ulrych, T. J., and T. N. Bishop, "Maximum Entropy Spectral Analysis and Autoregressive Decomposition," *Rev. Geophys. Space Phys.*, Vol. 13, pp. 183–200, Feb. 1975.

Ulrych, T. J., and R. W. Clayton, "Time Series Modeling and Maximum Entropy," *Phys. Earth Planet. Inter.*, Vol. 12, pp. 188–200, Aug. 1976.

Ulrych, T. J., and M. Ooe, "Autoregressive and Mixed Autoregressive–Moving Average Models and Spectra," in *Nonlinear Methods of Spectral Analysis*, Springer-Verlag, New York, 1979.

Van Blaricum, M. L., and R. Mittra, "Problems and Solutions Associated with Prony's Method for Processing Transient Data," *IEEE Trans. Antennas Propag.*, Vol. AP26, pp. 174–182, Jan. 1978.

Walker, A. M., "Some Consequences of Superimposed Error in Time Series Analysis," *Biometrika*, Vol. 47, pp. 33–43, 1960.

PROBLEMS

7.1. Prove that the matrix given in the autocorrelation method (7.3) is positive definite.

7.2. Using a proof similar to the one given in Section 6.3.2 for the minimum-phase property of the optimal prediction error filter, prove that the estimated poles obtained using the autocorrelation method must lie inside the unit circle. *Hint:* First show that $\hat{\rho}$ as given by (7.1) can be rewritten as

$$\hat{\rho} = \int_{-1/2}^{1/2} I(f) \mid A(f) \mid^2 df$$

where

$$I(f) = \frac{1}{N} \left| \sum_{n=0}^{N-1} x[n] \exp(-j2\pi fn) \right|^2$$

is the periodogram and

$$A(f) = 1 + \sum_{k=1}^{p} a[k] \exp(-j2\pi fk).$$

7.3. Give a counterexample to show that the use of the unbiased ACF estimator in the autocorrelation method will not guarantee a positive definite matrix. *Hint:* Consider the simple case of $p = 2$, $N = 2$.

7.4. Prove that the matrices encountered in the covariance and modified covariance methods will be singular if the data consist of $p - 1$ or fewer complex sinusoids when the dimension of the matrix is $p \times p$.

7.5. Show that the exponential signal

$$x[n] = \sum_{i=1}^{p} A_{c_i} z_i^n$$

may be generated by the recursive difference equation

$$x[n] = -\sum_{k=1}^{p} a[k]x[n - k]$$

for $n \geq p$ under the following conditions:
(1) If the $a[k]$'s are chosen so that the zeros of the polynomial

$$\mathcal{A}(z) = 1 + \sum_{k=1}^{p} a[k]z^{-k}$$

are $\{z_1, z_2, \ldots, z_p\}$, and
(2) If the initial conditions of the difference equations are chosen as

$$\begin{bmatrix} x[0] \\ x[1] \\ \vdots \\ x[p-1] \end{bmatrix} = \begin{bmatrix} 1 & 1 & \cdots & 1 \\ z_1 & z_2 & \cdots & z_p \\ \vdots & \vdots & \vdots & \vdots \\ z_1^{p-1} & z_2^{p-1} & \cdots & z_p^{p-1} \end{bmatrix} \begin{bmatrix} A_{c1} \\ A_{c2} \\ \vdots \\ A_{c_p} \end{bmatrix}.$$

7.6. Prove that the Burg estimate of the reflection coefficient is always less than or equal to 1 in magnitude.

7.7. Derive Anderson's relation as given by (7.42) for the recursive update of the denominator of the Burg reflection coefficient estimate.

7.8. Prove that the magnitude of the Burg reflection coefficient estimate is always less than or equal to the magnitude of the Itakura estimate.

7.9. Prove that the solutions of the cubic equation (7.49) required in the RMLE method are approximately -1, $+1$, $-c_k/d_k$ if N is large. Interpret $-c_k/d_k$ as a reflection coefficient estimator by rewriting it as an explicit function of the estimated forward and backward prediction errors, assuming N is large.

7.10. The approximate MLE for the parameters of an AR process in white noise requires us to minimize (7.61). If $\sigma_w^2 = 0$, show that minimization of (7.61) reduces to the autocorrelation method. Explain why the minimization becomes much more difficult when $\sigma_w^2 \neq 0$.

7.11. Verify (7.66) for the least squares estimator of σ^2 and σ_w^2 by minimizing

$$\int_{-1/2}^{1/2} (\,|\,\hat{A}(f)\,|^2 I(f) - \sigma^2 - \sigma_w^2\,|\,\hat{A}(f)\,|^2)^2\,df.$$

In doing so, make any necessary approximations to cause the filtering operations in the time domain to involve only observed data samples.

7.12. In this problem the Wiener filter for noise smoothing is derived. Assume that the real process $y[n] = x[n] + w[n]$ is observed for $-\infty < n < \infty$. $x[n]$ is a zero-mean real WSS random process with PSD $P_{xx}(f)$ and $w[n]$ is a zero-mean real white noise process with variance σ_w^2 which is uncorrelated with $x[n]$. The Wiener estimate of $x[n]$ for $-\infty < n < \infty$ is found as

$$\hat{x}[n] = \sum_{k=-\infty}^{\infty} h[k]y[n-k]$$

where the impulse response $h[n]$ is chosen to minimize the mean square error

$$\text{MSE} = \mathcal{E}\{(x[n] - \hat{x}[n])^2\}.$$

Show that the impulse response must satisfy

$$r_{yx}[m] = \sum_{k=-\infty}^{\infty} h[k]r_{yy}[m-k] \qquad -\infty < m < \infty.$$

Use this result to show that the frequency response of the Wiener filter is given by

$$H(f) = \frac{P_{xx}(f)}{P_{xx}(f) + \sigma_w^2}.$$

Comment on the use of a Wiener filter for noise reduction in AR spectral estimation.

7.13. Consider the positive definite autocorrelation matrix encountered in the autocorrelation method (7.3). Subtract $\hat{\sigma}_w^2 I$ from the matrix to effect noise compensation and find the largest value of $\hat{\sigma}_w^2$ that will still result in a nonsingular matrix. *Hint:* Use the properties of eigenvalues of positive definite matrices.

APPENDIX 7A

Development of Akaike Information Criterion

The development of the Akaike information criterion is based on *arguments* presented in Akaike [1974]. Since no rigorous derivation appears to be available, we should view the material presented as only a justification for the AIC. The reader is also referred to the paper by Matsuoka and Ulrych [1986] for a similar argument.

For simplicity, assume that the real observations, $\{x[0], x[1], \ldots, x[N-1]\}$, are independent and identically distributed. The development for the non-independent case proceeds in a similar fashion but is much more complicated. Let $p(\mathbf{x}; \boldsymbol{\theta}_0)$ denote the *true* PDF of $\mathbf{x} = [x[0] \;\; x[1] \;\; \cdots \;\; x[N-1]]^T$, which is dependent on $\boldsymbol{\theta}_0$, a real $p \times 1$ vector of unknown parameters, and let $p(\mathbf{x}; \boldsymbol{\theta})$ denote a family of model PDFs parameterized by a $k \times 1$ vector $\boldsymbol{\theta}$, where $k \neq p$ in general. Then a measure sensitive to the difference between $p(\mathbf{x}; \boldsymbol{\theta}_0)$ and $p(\mathbf{x}; \boldsymbol{\theta})$ is the Kullback–Leibler information measure [Kullback 1959]

$$I(\boldsymbol{\theta}_0, \boldsymbol{\theta}) = \int p(\mathbf{x}; \boldsymbol{\theta}_0) \ln \frac{p(\mathbf{x}; \boldsymbol{\theta}_0)}{p(\mathbf{x}; \boldsymbol{\theta})} \, d\mathbf{x}. \tag{7A.1}$$

It can be shown that $I(\boldsymbol{\theta}_0, \boldsymbol{\theta}) \geq 0$ and the measure equals zero if and only if $p(\mathbf{x}; \boldsymbol{\theta}) = p(\mathbf{x}; \boldsymbol{\theta}_0)$. For competing models we should choose the model that minimizes $I(\boldsymbol{\theta}_0, \boldsymbol{\theta})$. Since $p(\mathbf{x}; \boldsymbol{\theta}_0)$ is fixed, being the true PDF, we should choose the model that minimizes

$$\begin{aligned} I'(\boldsymbol{\theta}_0, \boldsymbol{\theta}) &= -\int p(\mathbf{x}; \boldsymbol{\theta}_0) \ln p(\mathbf{x}; \boldsymbol{\theta}) \, d\mathbf{x} \\ &= -\mathscr{E}_{\boldsymbol{\theta}_0}[\ln p(\mathbf{x}; \boldsymbol{\theta})]. \end{aligned} \tag{7A.2}$$

The subscript on the expectation emphasizes that the underlying PDF is the true one. Assume now that we have an MLE of $\boldsymbol{\theta}$ for each model in question, where the "true value" of $\boldsymbol{\theta}$ is the one obtained from the MLE procedure as $N \to \infty$. (Note that $\boldsymbol{\theta}_0$ is obtained as the "true value" only if the model under consideration is the correct one. This is due to the consistency of the MLE.) Also assume that the "true value" of $\boldsymbol{\theta}$ is close to $\boldsymbol{\theta}_0$. Then, if the MLE $\hat{\boldsymbol{\theta}}$ is close to $\boldsymbol{\theta}$ (which will be the case for large data records), $\ln p(\mathbf{x}; \boldsymbol{\theta})$ may be approximated by

$$\ln p(\mathbf{x}; \boldsymbol{\theta}) \approx \ln p(\mathbf{x}; \hat{\boldsymbol{\theta}}) - \tfrac{1}{2}(\boldsymbol{\theta} - \hat{\boldsymbol{\theta}})^T \mathbf{I}_{\boldsymbol{\theta}} (\boldsymbol{\theta} - \hat{\boldsymbol{\theta}}) \tag{7A.3}$$

where $\mathbf{I}_{\boldsymbol{\theta}}$ is the Fisher information matrix of dimension $k \times k$. This relationship will be justified shortly. Using (7A.3) in (7A.2) results in

$$I'(\boldsymbol{\theta}_0, \boldsymbol{\theta}) = -\mathscr{E}_{\boldsymbol{\theta}_0}[\ln p(\mathbf{x}; \hat{\boldsymbol{\theta}})] + \tfrac{1}{2}\mathscr{E}_{\boldsymbol{\theta}_0}[(\boldsymbol{\theta} - \hat{\boldsymbol{\theta}})^T \mathbf{I}_{\boldsymbol{\theta}} (\boldsymbol{\theta} - \hat{\boldsymbol{\theta}})]. \tag{7A.4}$$

However, if the model under consideration is correct, then for large data records the MLE $\hat{\boldsymbol{\theta}}$ is Gaussian with mean $\boldsymbol{\theta}$ and covariance matrix $\mathbf{I}_{\boldsymbol{\theta}}^{-1}$, which implies that

$$(\hat{\boldsymbol{\theta}} - \boldsymbol{\theta})^T \mathbf{I}_{\boldsymbol{\theta}} (\hat{\boldsymbol{\theta}} - \boldsymbol{\theta}) \sim \chi_k^2 \tag{7A.5}$$

so that the expected value with respect to $p(\mathbf{x}; \boldsymbol{\theta})$ is k (see Section 3.2.1). Since the desired expectation is with respect to the true PDF $p(\mathbf{x}; \boldsymbol{\theta}_0)$, the use of k results in a downward bias for the expectation. This is compensated for in an ad hoc fashion by adding an additional k to yield a value of $2k$. Also,

$$\begin{aligned}\frac{1}{N} \ln p(\mathbf{x}; \hat{\boldsymbol{\theta}}) &= \frac{1}{N} \sum_{n=0}^{N-1} \ln p(x[n]; \hat{\boldsymbol{\theta}}) \\ &\rightarrow \mathscr{E}_{\boldsymbol{\theta}_0}[\ln p(x[n]; \hat{\boldsymbol{\theta}})] \\ &= \frac{1}{N} \mathscr{E}_{\boldsymbol{\theta}_0}[\ln p(\mathbf{x}; \hat{\boldsymbol{\theta}})].\end{aligned} \tag{7A.6}$$

$p(x[n]; \boldsymbol{\theta})$ is the PDF of a single observation. These results follow from the independent and identically distributed assumption as well as the law of large numbers. Finally, using (7A.6) in (7A.4) yields

$$I'(\boldsymbol{\theta}_0, \boldsymbol{\theta}) = -\ln p(\mathbf{x}; \hat{\boldsymbol{\theta}}) + k$$

or multiplying by 2, the AIC is defined as

$$\text{AIC}(k) = -2 \ln p(\mathbf{x}; \hat{\boldsymbol{\theta}}) + 2k. \tag{7A.7}$$

The approximation of the logarithm of the likelihood function as given by (7A.3) is justified by the following arguments [Cox and Hinkley 1974]. If the MLE $\hat{\boldsymbol{\theta}}$ is close to the true value $\boldsymbol{\theta}$, then a truncated Taylor expansion about the point $\boldsymbol{\theta} = \hat{\boldsymbol{\theta}}$ will be a good approximation or

$$\begin{aligned}\ln p(\mathbf{x}; \boldsymbol{\theta}) = \ln p(\mathbf{x}; \hat{\boldsymbol{\theta}}) &+ \left.\frac{\partial \ln p(\mathbf{x}; \boldsymbol{\theta})}{\partial \boldsymbol{\theta}}\right|_{\boldsymbol{\theta}=\hat{\boldsymbol{\theta}}} (\boldsymbol{\theta} - \hat{\boldsymbol{\theta}}) \\ &+ \frac{1}{2}(\boldsymbol{\theta} - \hat{\boldsymbol{\theta}})^T \left.\frac{\partial^2 \ln p(\mathbf{x}; \boldsymbol{\theta})}{\partial \boldsymbol{\theta}^2}\right|_{\boldsymbol{\theta}=\hat{\boldsymbol{\theta}}} (\boldsymbol{\theta} - \hat{\boldsymbol{\theta}})\end{aligned} \tag{7A.8}$$

where $\partial \ln p(\mathbf{x}; \boldsymbol{\theta})/\partial \boldsymbol{\theta}$ is the gradient and $\partial^2 \ln p(\mathbf{x}; \boldsymbol{\theta})/\partial \boldsymbol{\theta}^2$ is the Hessian of $\ln p(\mathbf{x}; \boldsymbol{\theta})$. Now by the definition of the MLE the term of (7A.8) involving the gradient is zero since

$$\left.\frac{\partial \ln p(\mathbf{x}; \boldsymbol{\theta})}{\partial \boldsymbol{\theta}}\right|_{\boldsymbol{\theta}=\hat{\boldsymbol{\theta}}} = \mathbf{0}. \tag{7A.9}$$

Due to the independence of the observed samples,

$$\begin{aligned}\frac{1}{N} \frac{\partial^2 \ln p(\mathbf{x}; \boldsymbol{\theta})}{\partial \boldsymbol{\theta}^2} &= \frac{1}{N} \sum_{n=0}^{N-1} \frac{\partial^2 \ln p(x[n]; \boldsymbol{\theta})}{\partial \boldsymbol{\theta}^2} \\ &\rightarrow \mathscr{E}\left[\frac{\partial^2 \ln p(x[n]; \boldsymbol{\theta})}{\partial \boldsymbol{\theta}^2}\right].\end{aligned} \tag{7A.10}$$

The convergence to the expected value is again due to the law of large numbers.

Now

$$\mathcal{E}\left[\frac{\partial^2 \ln p(x[n]; \boldsymbol{\theta})}{\partial \boldsymbol{\theta}^2}\right] = -\frac{1}{N}\mathbf{I}_\theta \tag{7A.11}$$

by the definition of the Fisher information matrix and the property that the Fisher information matrix for N independent samples is N times that for one sample. Finally, for large data records the effect of evaluating the quadratic term in (7A.8) at $\boldsymbol{\theta}$ instead of $\hat{\boldsymbol{\theta}}$ will be negligible, so that using (7A.9)–(7A.11) in the truncated Taylor expansion yields (7A.3).

The AIC given by (7A.7) is now applied to AR model order estimation. It was shown in Section 6.5.1 that for large data records [see (6.86) and (6.87) with $N \gg p$]

$$\ln p(\mathbf{x}; \boldsymbol{\theta}) = -\frac{N}{2}\ln 2\pi - \frac{N}{2}\ln \sigma^2 - \frac{1}{2\sigma^2}S_1(\mathbf{a}) \tag{7A.12}$$

where $\boldsymbol{\theta} = [\mathbf{a}^T \, \sigma^2]^T$. Letting $\hat{\mathbf{a}}$ denote the MLE of $\mathbf{a}$ (any of the methods in this chapter are approximate MLEs) and recalling that the MLE of the white noise variance is

$$\hat{\sigma}^2 = \frac{1}{N}S_1(\hat{\mathbf{a}})$$

(7A.12) becomes

$$\ln p(\mathbf{x}; \hat{\boldsymbol{\theta}}) = -\frac{N}{2}\ln 2\pi - \frac{N}{2}\ln \hat{\sigma}^2 - \frac{N}{2}.$$

Ignoring the constant terms the AIC becomes

$$\mathrm{AIC}(k) = N \ln \hat{\rho}_k + 2k \tag{7.54}$$

where $\hat{\rho}_k$ is the MLE of the white noise variance for the AR(k) model.

APPENDIX 7B

Computer Program for Autocorrelation Method (Provided on Floppy Disk)

```
      SUBROUTINE AUTOCORR(X,N,IP,A,SIG2)
C     This program implements the autocorrelation method
C     of linear prediction for estimation of the AR parameters.
C     The equations are solved using the Levinson recursion.
```

```
C     See (7.3)-(7.5).
C
C     Input Parameters:
C
C       X         -Complex array of dimension N×1 of data points
C       N         -Number of data points
C       IP        -AR model order desired
C
C     Output Parameters:
C
C       A         -Complex array of dimension IP×1 of AR filter
C                  parameter estimates arranged as A(1) to A(IP)
C       SIG2      -Excitation white noise variance estimate
C
C     External Subroutines:
C
C       CORRELATION      -see Appendix 4D
C       LEVINSON         -see Appendix 6B
C
C     Notes:
C
C       The calling program must dimension the arrays X,A.
C       The arrays R,AA,RHO must be dimensioned greater than or equal
C       to the variable dimensions shown. The array AA in LEVINSON
C       must be dimensioned to agree with dimensions of AA in this
C       program.
C
C     Verification Test Case:
C
C       If the complex test data listed in Appendix 1A is input
C       as X, with N=32, and IP=10, then the output should be
C         A(1)  = ( 6.03004E-02,-4.09581E-02)
C         A(2)  = (-0.66814     ,-6.09021E-02)
C         A(3)  = (-0.36291     , 2.92395E-02)
C         A(4)  = ( 0.43211     , 8.43889E-02)
C         A(5)  = (-0.11293     , 7.85218E-03)
C         A(6)  = ( 0.27237     ,-7.58821E-02)
C         A(7)  = ( 0.13614     ,-3.52657E-02)
C         A(8)  = ( 1.56604E-02, 2.54710E-02)
C         A(9)  = (-5.53038E-02, 2.44507E-02)
C         A(10) = ( 0.15153     ,-1.92422E-02)
C         SIG2  = 1.86565
C       for a DEC VAX 11/780.
C
        COMPLEX X(1),A(1),R(IP+1),AA(IP,IP)
        DIMENSION RHO(IP)
C     Compute samples of biased autocorrelation estimator (7.4)
        MODE=1
        LAG=IP+1
        CALL CORRELATION(N,LAG,MODE,X,X,R)
```

```
C     Solve Yule-Walker equations using Levinson recursion (7.3), (7.5)
        CALL LEVINSON(R,IP,A,AA,RHO)
        SIG2=RHO(IP)
        RETURN
        END
```

APPENDIX 7C

Computer Program for Covariance and Modified Covariance Methods (Provided on Floppy Disk)

```
        SUBROUTINE COVMCOV(X,N,IP,MODE,EPS,IFLAG,A,SIG2)
C     This program implements either the covariance method
C     (7.8) and (7.9) or the modified covariance method
C     (7.22)-(7.24) for estimation of the AR parameters.
C
C     Input Parameters:
C
C       X          -Complex array of dimension Nx1 of data points
C       N          -Number of data points
C       IP         -AR model order desired
C       MODE       -Set =1 for modified covariance method, otherwise
C                   covariance method implemented
C       EPS        -Small positive number to test for ill-conditioning
C                   of matrix, required in CHOLESKY used to solve linear
C                   equations
C
C     Output Parameters:
C
C       IFLAG      -Ill-conditioning indicator
C                     =   0 for normal program termination
C                     = -1 if one of the di's in Cholesky decomposition
C                          is less than EPS causing premature termination
C       A          -Complex array of dimension IPx1 of AR filter
C                   parameter estimates arranged as A(1) to A(IP)
C       SIG2       -Excitation white noise variance estimate
C
C     External Subroutines:
C
C       CHOLESKY     -see Appendix 2B
C
C     Notes:
```

```
C
C        The calling program must dimension arrays X,A.
C        The CC array must be dimensioned greater than or
C        equal to the variable dimensions shown. The arrays
C        XL,Y,D in CHOLESKY must be dimensioned greater than
C        or equal to IP×IP, IP, and IP, respectively.
C
C      Verification Test Case:
C
C        If the complex test data listed in Appendix 1A is input
C        as X, with N=32, IP=10, MODE=0, and EPS=1.0E-15, then the
C        output should be
C          IFLAG=0
C          A(1)  = (-0.22160,-1.87537E-02)
C          A(2)  = (-1.08574,-4.21279E-02)
C          A(3)  = ( 0.14958, 4.17785E-03)
C          A(4)  = ( 0.83520, 0.14371     )
C          A(5)  = (-0.69723,-0.22038     )
C          A(6)  = ( 0.27110,-0.21128     )
C          A(7)  = ( 0.59138, 0.21102     )
C          A(8)  = (-0.21453, 0.22058     )
C          A(9)  = (-0.38712,-0.14536     )
C          A(10) = ( 0.33533,-8.34980E-02)
C          SIG2  = 0.11476
C        for a DEC VAX 11/780.
C        If the complex test data listed in Appendix 1A is input
C        as X, with N=32, IP=10, MODE=1, and EPS=1.0E-15, then the
C        output should be
C          IFLAG=0
C          A(1)  = (-0.10112, 0.12298     )
C          A(2)  = (-1.07629,-6.32359E-02)
C          A(3)  = (-0.14683,-8.97811E-02)
C          A(4)  = ( 0.86527, 9.84762E-02)
C          A(5)  = (-0.61774,-7.41049E-02)
C          A(6)  = ( 0.38049,-0.32719     )
C          A(7)  = ( 0.63081, 0.22249     )
C          A(8)  = (-0.21936, 0.34745     )
C          A(9)  = (-0.52164,-7.32829E-02)
C          A(10) = ( 0.38135,-0.19113     )
C          SIG2 = 0.16222
C        for a DEC VAX 11/780.
C
         COMPLEX X(1),A(1),CC(IP*(IP+1)/2),C,SUM
C      Compute autocorrelation estimates and insert them in
C      matrix (stored in symmetric mode in CC)
         L=1
         DO 50 K=1,IP
         DO 40 J=1,K
         CC(L)=(0.,0.)
C      For covariance method use only single
```

```
C       autocorrelation term (7.8)
          DO 10 I=IP+1,N
10        CC(L)=CC(L)+CONJG(X(I-J))*X(I-K)
C       For modified covariance method add additional
C       autocorrelation term (7.22)
          IF(MODE.NE.1)GO TO 30
          DO 20 I=1,N-IP
20        CC(L)=CC(L)+X(I+J)*CONJG(X(I+K))
30        L=L+1
40        CONTINUE
50        CONTINUE
          DO 80 J=1,IP
C       Compute right-hand-vector of covariance equations
          A(J)=(0.,0.)
C       For covariance method use only single autocorrelation
C       term (7.8)
          DO 60 I=IP+1,N
60        A(J)=A(J)-CONJG(X(I-J))*X(I)
C       For modified covariance method add additional
C       autocorrelation term (7.23)
          IF(MODE.NE.1)GO TO 80
          DO 70 I=1,N-IP
70        A(J)=A(J)-X(I+J)*CONJG(X(I))
80        CONTINUE
C       Compute AR filter parameter estimate
          CALL CHOLESKY(CC,A,IP,EPS,IFLAG)
          IF(IFLAG.EQ.-1)RETURN
C       Compute estimate of white noise variance
          SUM=(0.,0.)
          DO 120 KK=1,IP+1
          C=(0.,0.)
          K=KK-1
C       For covariance method use only single autocorrelation
C       term (7.9)
          DO 90 I=IP+1,N
90        C=C+CONJG(X(I))*X(I-K)
C       For modified covariance method add additional
C       autocorrelation term (7.24)
          IF(MODE.NE.1)GO TO 110
          DO 100 I=1,N-IP
100       C=C+X(I)*CONJG(X(I+K))
110       IF(K.EQ.0)SUM=SUM+C
          IF(K.NE.0)SUM=SUM+C*A(K)
120       CONTINUE
C       Complete estimate of white noise variance
          IF(MODE.EQ.1)SIG2=REAL(SUM)/(2.*(N-IP))
          IF(MODE.NE.1)SIG2=REAL(SUM)/(N-IP)
          RETURN
          END
```

APPENDIX 7D

Computer Program for Burg Method (Provided on Floppy Disk)

```
      SUBROUTINE BURG(X,N,IP,A,SIG2)
C     This program implements the Burg method for
C     estimation of the AR parameters (7.39)-(7.41).
C
C     Input Parameters:
C
C       X       -Complex array of dimension N×1 of data points
C       N       -Number of data points
C       IP      -AR model order desired
C
C     Output Parameters:
C
C       A       -Complex array of dimension IP×1 of AR filter
C                parameter estimates arranged as A(1) to A(IP)
C       SIG2    -Excitation white noise variance estimate
C
C     Notes:
C
C       The calling program must dimension the X,A arrays.
C       The arrays EFK,EFK1,EBK,EBK1,AA,RHO must be
C       dimensioned greater than or equal to the variable
C       dimensions shown.
C
C     Verification Test Case:
C
C       If the complex test data listed in Appendix 1A is input
C       as X, with N=32, and IP=10, then the output should be
C         A(1)  = (-0.10964,  9.80833E-02)
C         A(2)  = (-1.07989, -8.23723E-02)
C         A(3)  = (-0.16670, -7.09936E-02)
C         A(4)  = ( 0.86892,  0.14021    )
C         A(5)  = (-0.60076, -3.46992E-02)
C         A(6)  = ( 0.34997, -0.34456    )
C         A(7)  = ( 0.59898,  0.15030    )
C         A(8)  = (-0.20271,  0.34466    )
C         A(9)  = (-0.50447, -3.14139E-02)
C         A(10) = ( 0.37516, -0.18268    )
C         SIG2  =  0.16930
C       for a DEC VAX 11/780.
C
      COMPLEX X(1),A(1),EFK(N),EBK(N),EFK1(N),EBK1(N),
```

```
     * AA(IP,IP),SUMN,SUMD
       DIMENSION RHO(IP)
C     Compute the estimate of the autocorrelation at lag zero (7.39)
       RHO0=0.
       DO 10 I=1,N
10     RHO0=RHO0+CABS(X(I))**2/N
C     Initialize the forward and backward prediction errors (7.39)
       DO 20 I=2,N
       EFK1(I)=X(I)
20     EBK1(I-1)=X(I-1)
C     Begin recursion
       DO 80 K=1,IP
C     Compute the reflection coefficient estimate (7.40)
       SUMN=(0.,0.)
       SUMD=(0.,0.)
       DO 30 I=K+1,N
       SUMN=SUMN+EFK1(I)*CONJG(EBK1(I-1))
30     SUMD=SUMD+CABS(EFK1(I))**2+CABS(EBK1(I-1))**2
       AA(K,K)=-2.*SUMN/SUMD
C     Update the prediction error power (7.40)
       IF(K.EQ.1)RHO(K)=(1.-CABS(AA(K,K))**2)*RHO0
       IF(K.GT.1)RHO(K)=(1.-CABS(AA(K,K))**2)*RHO(K-1)
       IF(IP.EQ.1)GO TO 90
       IF(K.EQ.1)GO TO 50
C     Update the prediction error filter coefficients (7.40)
       DO 40 J=1,K-1
40     AA(J,K)=AA(J,K-1)+AA(K,K)*CONJG(AA(K-J,K-1))
C     Update the forward and backward prediction errors (7.41)
50     DO 60 I=K+2,N
       EFK(I)=EFK1(I)+AA(K,K)*EBK1(I-1)
60     EBK(I-1)=EBK1(I-2)+CONJG(AA(K,K))*EFK1(I-1)
       DO 70 I=K+2,N
       EFK1(I)=EFK(I)
70     EBK1(I-1)=EBK(I-1)
80     CONTINUE
C     Find final values of the prediction error power,
C     which is the white noise variance estimate, and the
C     prediction coefficients, which are the AR filter
C     parameter estimates
90     SIG2=RHO(IP)
       DO 100 I=1,IP
100    A(I)=AA(I,IP)
       RETURN
       END
```

APPENDIX 7E

Computer Program for Recursive MLE method (Provided on Floppy Disk)

```
      SUBROUTINE RMLE(X,N,IP,A,SIG2)
C     This program implements the recursive maximum likelihood
C     estimator for the AR parameters (7.46)-(7.52). The data
C     must be real.
C
C     Input Parameters:
C
C       X        -Real array of dimension N×1 of data points
C       N        -Number of data points
C       IP       -AR model order desired
C
C     Output Parameters:
C
C       A        -Real array of dimension IP×1 of AR filter
C                 parameter estimates arranged as A(1) to A(IP)
C       SIG2     -Excitation white noise variance estimate
C
C     Notes:
C
C       The calling program must dimension the X,A arrays.
C       The arrays S,AA,RHO in the main subroutine RMLE
C       and the arrays AP,B in the called subroutine COEFF
C       must be dimensioned greater than or equal to the variable
C       dimensions shown. The arrays S,AA in the called
C       subroutines COEFF, ARUPT, and SMAT must be dimensioned
C       to agree with dimensions of S,AA in the main subroutine
C       RMLE.
C
C     Verification Test Case:
C
C       If the real test data listed in Appendix 1A is input
C       as X, with N=32, and IP=10, then the output should be
C         A(1)  =  -7.64353E-02
C         A(2)  =  -1.26934
C         A(3)  =  -0.30781
C         A(4)  =   0.85033
C         A(5)  =  -0.49144
C         A(6)  =   0.36946
C         A(7)  =   0.75698
C         A(8)  =  -0.18951
C         A(9)  =  -0.62877
C         A(10) =   0.21280
```

```
C          SIG2 =     0.21543
C        for a DEC VAX 11/780.
C
         DIMENSION X(1),S(IP+1,IP+1),AA(IP,IP),A(1),RHO(IP),ROOT(3)
C       Compute elements of S matrix (7.52)
         CALL SMAT(X,N,IP,S)
C       Begin recursion
         DO 10 I=1,IP
C       Obtain coefficients of cubic equation to be solved
C       for next reflection coefficient (7.48),(7.51)
         CALL COEFF(I,IP,S,N,AA,P,Q,R,C,D,E)
C       Solve cubic equation (7.49)
         CALL CUBIC(P,Q,R,NREAL,ROOT)
C       If there is only one real root, then set reflection
C       coefficient equal to it.
         AA(I,I)=ROOT(1)
C       If there are three real roots, find the one which is
C       within the [-1,1] interval and which maximizes the
C       likelihood function (7.50)
         IF(NREAL.EQ.3)CALL TEST(ROOT,C,D,E,I,N,AA(I,I))
C       Update the likelihood function (7.51)
         CALL ARUPT(AA,I,S,N,C,D,E,RHO(I))
10       CONTINUE
C       Obtain final AR parameter estimates
         DO 20 I=1,IP
20       A(I)=AA(I,IP)
         SIG2=RHO(IP)
         RETURN
         END
         SUBROUTINE COEFF(I,IP,S,N,AA,P,Q,R,C,D,E)
C       This program computes the coefficients of the cubic
C       equation to be solved for the reflection coefficient.
         DIMENSION S(IP+1,IP+1),B(IP),AA(IP,IP),AP(IP)
         IF(I.EQ.1)GO TO 60
         I1=I-1
         I2=I+1
         AP(1)=1.
         DO 20 J=1,I1
20       AP(J+1)=AA(J,I1)
         DO 35 J=1,I
35       B(J)=AP(I+1-J)
         C=0.
         D=0.
         DO 40 L=1,I
         DO 30 K=1,I
         C=C+AP(K)*B(L)*S(K,L+1)
         D=D+B(K)*B(L)*S(K+1,L+1)
30       CONTINUE
40       CONTINUE
         GO TO 70
```

```
60        C=S(1,2)
          D=S(2,2)
          E=S(1,1)
70        P=((N-2.*I)/(N-I))*(C/D)
          Q=(I*E+N*D)/(D*(I-N))
          R=N*C/(D*(I-N))
          RETURN
          END
          SUBROUTINE ARUPT(AA,I,S,N,C,D,E,POW)
C       This program updates in order the prediction coefficients
C       and the prediction error power.
          DIMENSION AA(IP,IP),S(IP+1,IP+1)
          IF(I.EQ.1)GO TO 15
          I1=I-1
          DO 10 J=1,I1
10        AA(J,I)=AA(J,I1)+AA(I,I)*AA(I-J,I1)
15        E=E+2.*C*AA(I,I)+D*AA(I,I)**2
          POW=E/FLOAT(N)
          RETURN
          END
          SUBROUTINE SMAT(X,N,IP,S)
C       This program computes the S data matrix required in the
C       likelihood function.
          DIMENSION S(IP+1,IP+1),X(1)
          IP1=IP+1
          DO 30 J=1,IP1
          DO 20 I=1,IP1
          S(I,J)=0.
          DO 10 K=1,N-(I-1)-(J-1)
10        S(I,J)=S(I,J)+X(K+I-1)*X(K+J-1)
20        CONTINUE
30        CONTINUE
          RETURN
          END
          SUBROUTINE CUBIC(P,Q,R,NREAL,ROOT)
C       This program determines the real roots of a cubic equation.
          DIMENSION ROOT(3)
          PI=4.*ATAN(1.)
          A=(3.*Q-P*P)/3.
          B=(2.*P**3-9.*P*Q+27.*R)/27.
          D=B*B/4.+A**3/27.
          IF(D.GT.0.)NREAL=1
          IF(D.LE.0.)NREAL=3
          IF(NREAL.EQ.3)GO TO 10
          ARG1=-B/2.+SQRT(D)
          SARG1=ARG1/ABS(ARG1)
          ARG2=-B/2.-SQRT(D)
          SARG2=ARG2/ABS(ARG2)
          AA=SARG1*((ABS(ARG1))**(1./3.))
          BB=SARG2*((ABS(ARG2))**(1./3.))
```

```
      ROOT(1)=AA+BB-P/3.
      GO TO 20
10    C=2.*SQRT(-A/3.)
      Z=(3.*B)/(A*C)
      ALPHA=ATAN((SQRT(1.-Z*Z))/ABS(Z))
      IF(Z.LT.0)ALPHA=PI-ALPHA
      Y=(1./3.)*ALPHA
      ROOT(1)=C*COS(Y)-P/3.
      ROOT(2)=C*COS(Y+2.*PI/3.)-P/3.
      ROOT(3)=C*COS(Y+4.*PI/3.)-P/3.
20    RETURN
      END
      SUBROUTINE TEST(ROOT,C,D,E,I,N,XKEST)
C     This program checks the three real roots of a cubic
C     equation to determine which one is the maximum likelihood
C     estimate of the reflection coefficient.
      DIMENSION ROOT(3)
      FMAX=-10000000.
      DO 30 J=1,3
      IF(ROOT(J).LT.-1..OR.ROOT(J).GT.1.)GO TO 30
      Q=E+2.*ROOT(J)*C+ROOT(J)*ROOT(J)*D
      F=(I/2.)*ALOG10(1.-ROOT(J)*ROOT(J))-(N/2.)*ALOG10(Q/N)
      IF(F.GT.FMAX)XKEST=ROOT(J)
      IF(F.GT.FMAX)FMAX=F
30    CONTINUE
      RETURN
      END
```

Chapter 8

Moving Average Spectral Estimation

8.1 INTRODUCTION

Moving average (MA) spectral estimation is valuable when the PSD is characterized by broad peaks and/or sharp nulls. Since the MA PSD is based on an all-zero model of the data, it is not possible to use it to estimate PSDs with sharp peaks. The latter case is more easily accommodated using an AR or ARMA model. Because the MA spectral estimator is not a high resolution spectral estimator for processes with narrowband spectral features, investigations of its properties have been somewhat limited. It should be noted, however, that when the underlying process is actually an MA process, the MA spectral estimator will be more accurate than Fourier based spectral estimators.

8.2 SUMMARY

The MA spectral density is defined by (8.1) and may alternatively be expressed as in (8.2). Although identical in form to the BT spectral estimator, the MA spectral estimator is more accurate due to the imposed restriction that the ACF be identically zero beyond a given lag. An algorithm due to Durbin for estimating the MA parameters is derived in Section 8.4 and is summarized at the end of that section. Statistics of the estimator which are valid for real processes and large data records are given in Section 8.5. The MA parameter estimator is distributed according to a multivariate Gaussian PDF with mean equal to the true value and covariance matrix given by (8.11). The spectral estimator is also distributed ac-

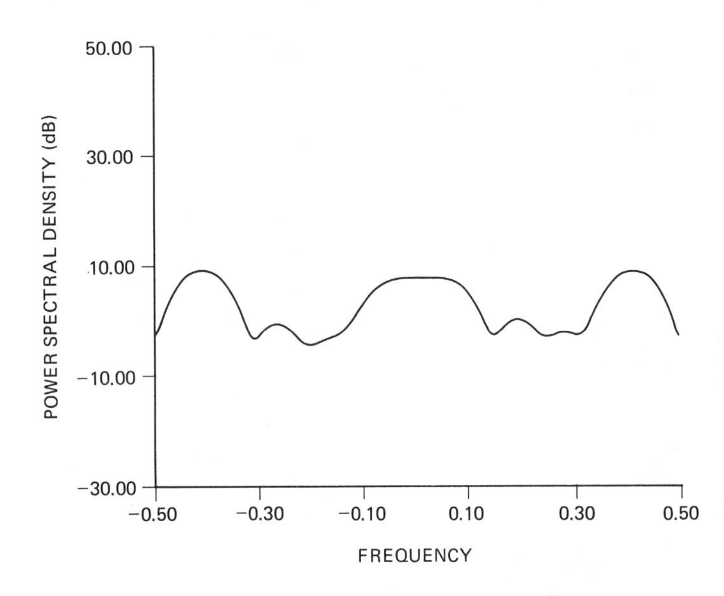

Figure 8.1 MA spectral estimate of complex test case data ($L = 15$, $q = 10$).

cording to a multivariate Gaussian PDF with mean equal to the true value and covariance matrix given by (8.12). The variance at any frequency is given by (8.13). Various methods to determine the model order are given in Section 8.6. Notably, Durbin's method can be used in conjunction with the AIC (8.14). Some other MA estimators are described briefly in Section 8.7. Finally, typical spectral estimates using Durbin's algorithm are given in Section 8.8.

The application of Durbin's algorithm to the complex test data listed in Appendix 1A results in the spectral estimate shown in Figure 8.1. The computer program DURBIN, which is provided in Appendix 8A, was used to obtain the parameter estimates. As might be expected, the spectral estimate is a broadband approximation to the true PSD. A further discussion of this result is provided in Chapter 12.

8.3 THE MA SPECTRAL ESTIMATOR

In Chapter 5 the PSD for an MA(q) process was shown to be

$$P_{\mathrm{MA}}(f) = \sigma^2 \mid B(f) \mid^2 \tag{8.1}$$

where

$$B(f) = 1 + \sum_{k=1}^{q} b[k] \exp(-j2\pi f k).$$

Alternatively, the PSD may be written in terms of the Fourier transform of the MA ACF as

$$P_{\mathrm{MA}}(f) = \sum_{k=-q}^{q} r_{xx}[k] \exp(-j2\pi f k) \tag{8.2}$$

where

$$r_{xx}[k] = \begin{cases} \sigma^2 \sum_{l=0}^{q-k} b^*[l]b[l+k] & \text{for } k = 0, 1, \ldots, q \\ r_{xx}^*[-k] & \text{for } k = -q, -(q-1), \ldots, -1. \end{cases}$$

A natural estimator of the PSD for an MA process would be, from (8.2),

$$\hat{P}_{\text{MA}}(f) = \sum_{k=-q}^{q} \hat{r}_{xx}[k] \exp(-j2\pi fk) \tag{8.3}$$

where $\hat{r}_{xx}[k]$ is a suitable estimator of the ACF (see also Section 8.7). Note that (8.3) is exactly of the form of the BT spectral estimator. A subtle difference exists, however, in the assumptions behind each estimator. Equation (8.3) assumes an MA process of order q so that the theoretical ACF is zero for $k > q$. In fact, if the unbiased ACF estimator is used, then

$$\mathscr{E}(\hat{P}_{\text{MA}}(f)) - P_{\text{MA}}(f)$$

so that there is no bias in the estimator. The BT spectral estimator does not assume an MA process and therefore can be applied to data with an arbitrary PSD. The price paid is a bias due to a zeroing of the unmeasured ACF samples.

Henceforth, it is assumed that $x[n]$ is an MA(q) process. The problem is to estimate $\{b[1], b[2], \ldots, b[q], \sigma^2\}$ for use in (8.1). For reliable estimates of the MA parameters the MLE will be employed. Equivalently, we could obtain the MLEs for $\{r_{xx}[0], r_{xx}[1], \ldots, r_{xx}[q]\}$ and use (8.3) to estimate the PSD. We must be careful, however, to ensure that the estimated ACF is positive semidefinite if the estimated PSD is always to be nonnegative. It should be emphasized that the usual biased or unbiased ACF estimators are *not* the MLEs of the ACF.

In the next section approximate MLEs for the MA parameters and the MA PSD are derived. The algorithm, due to Durbin, first converts the MA(q) process into an AR process and then uses the Yule–Walker equations to estimate the MA parameters. This procedure leads to an asymptotically efficient estimator (see Section 3.3.1).

8.4 MAXIMUM LIKELIHOOD ESTIMATION: DURBIN'S METHOD

Durbin's method is an approximate MLE [Durbin 1959]. It is derived for real data, but the extension to complex data is straightforward and is given. The first step is to replace the MA(q) process by an approximate AR(L) process. As discussed in Section 5.3.2, an MA process

$$x[n] = \sum_{k=0}^{q} b[k]u[n-k]$$

is equivalent to the AR(∞) process

$$x[n] = -\sum_{k=1}^{\infty} a[k]x[n-k] + u[n]$$

if $a[k]$ is the impulse response of $1/\mathcal{B}(z)$. This is immediately observed if we let

$$\mathcal{H}(z) = \mathcal{B}(z) = \frac{1}{\mathcal{A}(z)}$$

so that

$$\mathcal{A}(z) = \frac{1}{\mathcal{B}(z)}.$$

If the impulse response of $1/\mathcal{B}(z)$ has decayed to zero for an index greater than L, then an AR(L) process will be a good approximation to the MA(q) process. Now instead of considering the likelihood function for the data directly, we can use the likelihood function for the AR parameter estimates. This is because the usual AR parameter estimator is a sufficient statistic for the AR parameters for large data records (see Section 3.3.1). Let $\hat{\mathbf{a}}$, $\hat{\sigma}^2$ be the AR parameter estimates obtained by any of the methods of Chapter 7 (i.e., any of the approximate MLE techniques) using an AR(L) model. Then, as discussed in Section 6.5.2, for large data records $\hat{\boldsymbol{\theta}} = [\hat{\mathbf{a}}^T \ \ \hat{\sigma}^2]^T$ is distributed according to a multivariate Gaussian PDF with mean

$$\mathcal{E}(\hat{\boldsymbol{\theta}}) = \boldsymbol{\theta} = \begin{bmatrix} \mathbf{a} \\ \sigma^2 \end{bmatrix}$$

and covariance matrix

$$\mathbf{C}_{a,\sigma^2} = \begin{bmatrix} \frac{\sigma^2}{N}\mathbf{R}_{xx}^{-1} & \mathbf{0} \\ & \\ \mathbf{0}^T & \frac{2\sigma^4}{N} \end{bmatrix}$$

where $\mathbf{R}_{xx}$ is the $L \times L$ autocorrelation matrix of the MA(q) or equivalent AR(L) process. Hence the PDF for $\hat{\boldsymbol{\theta}}$ is

$$p(\hat{\boldsymbol{\theta}}; \mathbf{b}, \sigma^2) = \frac{1}{(2\pi)^{(L+1)/2}\det^{1/2}(\mathbf{C}_{a,\sigma^2})}\exp\left[-\tfrac{1}{2}(\hat{\boldsymbol{\theta}} - \boldsymbol{\theta})^T\mathbf{C}_{a,\sigma^2}^{-1}(\hat{\boldsymbol{\theta}} - \boldsymbol{\theta})\right]. \quad (8.4)$$

Note that the likelihood function depends on $\mathbf{b}$ through its relationship with $\mathbf{a}$. The determinant of the covariance matrix is, from (2.13) and (2.29),

$$\det(\mathbf{C}_{a,\sigma^2}) = \frac{2\sigma^4}{N}\left(\frac{\sigma^2}{N}\right)^L \det^{-1}(\mathbf{R}_{xx}).$$

It may be shown (see Problem 8.1) that for large L,

$$\det(\mathbf{R}_{xx}) \approx \sigma^{2L}$$

so that

$$\det(\mathbf{C}_{a,\sigma^2}) = \frac{2\sigma^4}{N^{L+1}}. \tag{8.5}$$

Also,

$$(\hat{\boldsymbol{\theta}} - \boldsymbol{\theta})^T \mathbf{C}_{a,\sigma^2}^{-1}(\hat{\boldsymbol{\theta}} - \boldsymbol{\theta}) = (\hat{\mathbf{a}} - \mathbf{a})^T \left(\frac{N\mathbf{R}_{xx}}{\sigma^2}\right)(\hat{\mathbf{a}} - \mathbf{a}) + \frac{N}{2\sigma^4}(\hat{\sigma}^2 - \sigma^2)^2 \tag{8.6}$$

due to the block diagonal nature of $\mathbf{C}_{a,\sigma^2}$. Now let $\overline{\mathbf{R}}_{xx} = \mathbf{R}_{xx}/\sigma^2$ so that $\overline{\mathbf{R}}_{xx}$ depends only on the MA *filter* parameters. Using (8.5) and (8.6) in (8.4) yields

$$p(\hat{\boldsymbol{\theta}}; \mathbf{b}, \sigma^2) = \underbrace{\frac{c}{\sigma^2}\exp\left\{-\left[\frac{N}{4\sigma^4}(\hat{\sigma}^2 - \sigma^2)^2\right]\right\}}_{p_1}\underbrace{\exp\left\{-\left[\frac{N}{2}(\hat{\mathbf{a}} - \mathbf{a})^T\overline{\mathbf{R}}_{xx}(\hat{\mathbf{a}} - \mathbf{a})\right]\right\}}_{p_2}$$

where c is a constant not depending on $\mathbf{b}$ or σ^2. Since p_1 depends only on σ^2 and p_2 depends only on $\mathbf{b}$ through $\mathbf{a}$ and $\overline{\mathbf{R}}_{xx}$, each function can be maximized separately. To find the MLE of σ^2, note that

$$\ln p_1 = \ln c - \ln \sigma^2 - \frac{N}{4\sigma^4}(\hat{\sigma}^2 - \sigma^2)^2$$

which is dominated by the last term for $N \gg 1$. Thus the MLE is just $\hat{\sigma}^2$, which is the estimate obtained using the AR(L) model. Assuming that the autocorrelation method has been used for real data, then from (7.5),

$$\hat{\sigma}^2 = \hat{r}_{xx}[0] + \sum_{k=1}^{L} \hat{a}[k]\hat{r}_{xx}[k]. \tag{8.7}$$

To find the MLE of $\mathbf{b}$, we need to maximize p_2 or equivalently to minimize

$$\begin{aligned} Q &= (\hat{\mathbf{a}} - \mathbf{a})^T\overline{\mathbf{R}}_{xx}(\hat{\mathbf{a}} - \mathbf{a}) \\ &= \hat{\mathbf{a}}^T\overline{\mathbf{R}}_{xx}\hat{\mathbf{a}} - 2\hat{\mathbf{a}}^T\overline{\mathbf{R}}_{xx}\mathbf{a} + \mathbf{a}^T\overline{\mathbf{R}}_{xx}\mathbf{a}. \end{aligned}$$

But $\overline{\mathbf{R}}_{xx} = \mathbf{R}_{xx}/\sigma^2$ and for an AR(L) process the Yule–Walker equations yield [see (5.18)]

$$\overline{\mathbf{R}}_{xx}\mathbf{a} = \frac{\mathbf{R}_{xx}\mathbf{a}}{\sigma^2} = -\bar{\mathbf{r}}_{xx}$$

where

$$\bar{\mathbf{r}}_{xx} = [\bar{r}_{xx}[1] \quad \bar{r}_{xx}[2] \quad \cdots \quad \bar{r}_{xx}[L]]^T$$

and $\bar{r}_{xx}[k] = r_{xx}[k]/\sigma^2$ is the normalized ACF. It follows that Q may be written as

$$Q = \hat{\mathbf{a}}^T\overline{\mathbf{R}}_{xx}\hat{\mathbf{a}} + 2\hat{\mathbf{a}}^T\bar{\mathbf{r}}_{xx} - \mathbf{a}^T\bar{\mathbf{r}}_{xx}.$$

Also, for an AR(L) process [see (5.17)],

$$\sigma^2 = r_{xx}[0] + \sigma^2 \mathbf{a}^T \bar{\mathbf{r}}_{xx}$$

which then implies that

$$\begin{aligned} Q &= \hat{\mathbf{a}}^T \bar{\mathbf{R}}_{xx} \hat{\mathbf{a}} + 2\hat{\mathbf{a}}^T \bar{\mathbf{r}}_{xx} + \frac{r_{xx}[0]}{\sigma^2} - 1 \\ &= \begin{bmatrix} 1 \\ \hat{\mathbf{a}} \end{bmatrix}^T \begin{bmatrix} \bar{r}_{xx}[0] & \bar{\mathbf{r}}_{xx}^T \\ \bar{\mathbf{r}}_{xx} & \bar{\mathbf{R}}_{xx} \end{bmatrix} \begin{bmatrix} 1 \\ \hat{\mathbf{a}} \end{bmatrix} - 1. \end{aligned}$$

Expanding the quadratic form, we obtain

$$Q = \sum_{i=0}^{L} \sum_{j=0}^{L} \hat{a}[i]\hat{a}[j]\bar{r}_{xx}[i-j] - 1 \tag{8.8}$$

where $\hat{a}[0] = 1$. Now the normalized ACF for an MA(q) process may be written as

$$\begin{aligned} \bar{r}_{xx}[i-j] &= \sum_{n=-\infty}^{\infty} b[n]b[n+i-j] \\ &= \sum_{n=-\infty}^{\infty} b[n-i]b[n-j] \end{aligned}$$

where $b[0] = 1$ and $b[n] = 0$ for $n < 0$ and $n > q$. Substituting into (8.8) yields

$$\begin{aligned} Q &= \sum_{i=0}^{L} \sum_{j=0}^{L} \hat{a}[i]\hat{a}[j] \sum_{n=-\infty}^{\infty} b[n-i]b[n-j] - 1 \\ &= \sum_{n=-\infty}^{\infty} \left(\sum_{k=0}^{L} \hat{a}[k]b[n-k] \right)^2 - 1 \\ &= \sum_{n=-\infty}^{\infty} \left(\sum_{k=-\infty}^{\infty} \hat{a}[k]b[n-k] \right)^2 - 1 \\ &= \sum_{n=-\infty}^{\infty} \left(\sum_{k=-\infty}^{\infty} b[k]\hat{a}[n-k] \right)^2 - 1 \end{aligned}$$

where $\hat{a}[k] = 0$ for $k < 0$ and $k > L$. Finally,

$$Q = \sum_{n=-\infty}^{\infty} \left(\hat{a}[n] + \sum_{k=1}^{q} b[k]\hat{a}[n-k] \right)^2 - 1.$$

To minimize Q over $\{b[1], b[2], \ldots, b[q]\}$, note that $(Q+1)/(L+1)$ is the prediction error power estimate for the "data" $\{1, \hat{a}[1], \ldots, \hat{a}[L]\}$. Minimizing Q is therefore accomplished via the autocorrelation method of AR parameter estimation [see (7.3) and (7.4)]. The approximate MLE for the MA filter parameters is therefore

$$\hat{\mathbf{b}} = -\hat{\mathbf{R}}_{aa}^{-1}\hat{\mathbf{r}}_{aa} \tag{8.9}$$

where

$$[\hat{\mathbf{R}}_{aa}]_{ij} = \frac{1}{L+1}\sum_{n=0}^{L-|i-j|} \hat{a}[n]\hat{a}[n+|i-j|] \qquad i, j = 1, 2, \ldots, q$$

$$[\hat{\mathbf{r}}_{aa}]_{i} = \frac{1}{L+1}\sum_{n=0}^{L-i} \hat{a}[n]\hat{a}[n+i] \qquad i = 1, 2, \ldots, q.$$

Equation (8.9) is Durbin's method for MA parameter estimation. The Levinson algorithm may be used to solve the equations for $\mathbf{b}$. Because of the minimum-phase property of the autocorrelation method the estimated zeros of $\mathcal{B}(z)$ will be inside the unit circle. Many variants of Durbin's method may be generated by replacing the autocorrelation method of AR parameter estimation by any of the techniques described in Chapter 7.

In summary, Durbin's algorithm for the estimation of the MA parameters of an MA(q) process proceeds as follows (see Problem 8.2 for an example):

1. Using the data $\{x[0], x[1], \ldots, x[N-1]\}$, fit a large order AR model using the autocorrelation method. For an AR model order of L, where $q \ll L \ll N$, the white noise variance estimator $\hat{\sigma}^2$ is given by (8.7).
2. Using the AR parameter estimates obtained from step 1 as the data (i.e., $\{1, \hat{a}[1], \hat{a}[2], \ldots, \hat{a}[L]\}$), use the autocorrelation method with an order of q to find $\{\hat{b}[1], \hat{b}[2], \ldots, \hat{b}[q]\}$ as given by (8.9).

For complex data the same steps apply if we use the complex AR parameter estimators. A computer program implementing Durbin's algorithm for complex data, which is entitled DURBIN, is given in Appendix 8A.

8.5 STATISTICS OF THE MA PARAMETER AND SPECTRAL ESTIMATORS

The statistics which are now described assume that the data are real and have been obtained from an MA(q) process. Since Durbin's algorithm is an approximate MLE, the statistics are obtainable from the asymptotic properties of MLEs as discussed in Section 3.3.1. For large data records the MLE $[\hat{\mathbf{b}}^T \ \hat{\sigma}^2]^T$ will be distributed according to a multivariate Gaussian PDF with mean

$$\boldsymbol{\mu} = \mathcal{E}\begin{bmatrix}\hat{\mathbf{b}} \\ \hat{\sigma}^2\end{bmatrix} = \begin{bmatrix}\mathbf{b} \\ \sigma^2\end{bmatrix} \tag{8.10}$$

and covariance matrix [see Appendix 9A for a derivation of the more general result for an ARMA(p, q) process]

$$\mathbf{C}_{b,\sigma^2} = \mathscr{E}\left\{\begin{bmatrix}\hat{\mathbf{b}} - \mathscr{E}(\hat{\mathbf{b}}) \\ \hat{\sigma}^2 - \mathscr{E}(\hat{\sigma}^2)\end{bmatrix}\begin{bmatrix}\hat{\mathbf{b}} - \mathscr{E}(\hat{\mathbf{b}}) \\ \hat{\sigma}^2 - \mathscr{E}(\hat{\sigma}^2)\end{bmatrix}^T\right\}$$
$$= \begin{bmatrix}\frac{\sigma^2}{N}\mathbf{R}_{zz}^{-1} & \mathbf{0} \\ \mathbf{0}^T & \frac{2\sigma^4}{N}\end{bmatrix} \tag{8.11}$$

where $\mathbf{R}_{zz}$ is the $q \times q$ autocorrelation matrix of the process $z[n]$. $z[n]$ is defined as an AR(q) process with parameters $\{b[1], b[2], \ldots, b[q], \sigma^2\}$. Hence the ijth element of $\mathbf{R}_{zz}$ is $r_{zz}[i-j]$, where $r_{zz}[k]$ is the ACF corresponding to the PSD

$$P_{zz}(f) = \frac{\sigma^2}{\left|1 + \sum_{k=1}^{q} b[k]\exp(-j2\pi fk)\right|^2}.$$

It is interesting to note that $\hat{\mathbf{b}},\hat{\sigma}^2$ are asymptotically uncorrelated and hence independent (since they are jointly Gaussian). The covariance matrix as given by (8.11) is the inverse of the appropriate partition of the Fisher information matrix as derived in Appendix 9A. Even if (8.11) is not a good approximation to the covariance matrix of the MA parameter estimators as may occur for shorter-length data records, it still provides a lower bound in that

$$\mathrm{var}(\hat{b}[i]) \geq \frac{\sigma^2}{N}[\mathbf{R}_{zz}^{-1}]_{ii} \qquad i = 1, 2, \ldots, q$$

$$\mathrm{var}(\hat{\sigma}^2) \geq \frac{2\sigma^4}{N}$$

assuming that $\hat{\mathbf{b}}$, $\hat{\sigma}^2$ are unbiased estimators.

The form of the CR bound of (8.11) is identical to that of an AR(p) process [see (6.97)]. In essence, it says that the MA parameters can be estimated as accurately as the AR parameters of an AR process (see Problem 8.3). It also suggests that if we could convert an MA(q) process into an AR(q) process with parameters $\{b[1], b[2], \ldots, b[q], \sigma^2\}$, we could attain the CR bound for large data records by using a linear prediction estimator for the parameters (which is an approximate MLE). This approach is at the heart of Durbin's algorithm.

The statistics of the MA spectral estimator can be found by applying the standard properties of the MLE as discussed in Section 3.3.1. Let

$$\mathbf{p} = [\hat{P}_{\mathrm{MA}}(f_1) \quad \hat{P}_{\mathrm{MA}}(f_2) \quad \cdots \quad \hat{P}_{\mathrm{MA}}(f_M)]^T$$

Then, for large data records $\mathbf{p}$ is distributed according to a multivariate Gaussian PDF with mean equal to the true PSD and covariance matrix (see Section 9.4)

$$\mathbf{C}_p = \mathbf{D}\mathbf{C}_{b,\sigma^2}\mathbf{D}^T \tag{8.12}$$

where $\mathbf{D}$ is the $M \times (q+1)$ Jacobian matrix of the transformation from $\mathbf{b}$, σ^2 to $\mathbf{p}$, or

$$[\mathbf{D}]_{mn} = \begin{cases} 2P_{\mathrm{MA}}(f_m)\ \mathrm{Re}\left[\dfrac{\exp(-j2\pi f_m n)}{B(f_m)}\right] & \text{for } m = 1, 2, \ldots, M; \\ & n = 1, 2, \ldots, q \\ \dfrac{P_{\mathrm{MA}}(f_m)}{\sigma^2} & \text{for } m = 1, 2, \ldots, M; \\ & n = q + 1 \end{cases}$$

and $\mathbf{C}_{b,\sigma^2}$ is given by (8.11). In particular, at any frequency the variance is

$$\mathrm{var}\,[\hat{P}_{\mathrm{MA}}(f)] = \frac{4\sigma^2 P_{\mathrm{MA}}^2(f)}{N}\,\mathbf{e}^T\mathbf{R}_{zz}^{-1}\mathbf{e} + \frac{2}{N}P_{\mathrm{MA}}^2(f) \tag{8.13}$$

where $[\mathbf{e}]_n = \mathrm{Re}\,[\exp(-j2\pi f n)/B(f)]$, $n = 1, 2, \ldots, q$. Note that (8.13) may be used to establish a confidence interval for the spectral estimate.

8.6 MODEL ORDER SELECTION

Before describing several model order estimators, it should be mentioned that the prediction error power which formed the basis for the AR model order estimators cannot be applied to MA processes. This is because it decreases monotonically with the order of the linear predictor. No theoretical minimum of the prediction error power of a linear predictor occurs for an order equal to the MA model order. Equivalently, the reflection coefficient sequence is not zero after a certain index but is generally composed of a sum of damped exponentials (see Problems 8.4 and 8.5).

Several techniques for MA model order determination are now described. None of the techniques have been thoroughly tested so that a comparison of their relative merits is not available. The first one uses the AIC as previously described in Appendix 7A. For an MA process it is defined as

$$\mathrm{AIC}(i) = N \ln \hat{\sigma}_i^2 + 2i \tag{8.14}$$

where i is the assumed MA model order and $\hat{\sigma}_i^2$ is the MLE of the white noise variance based on an ith order model. For possible model orders the AIC is computed and the model order yielding the minimum is chosen. If Durbin's algorithm is used to estimate the MA parameters, then all the lower order MA models are available. $\hat{\sigma}_i^2$ can be found by filtering the data with an estimate of the ith order inverse MA filter $1/\mathcal{B}_i(z)$, which is guaranteed to be stable, and estimating the power at the output.

A second approach which relies on the statistical properties of Durbin's method is to examine Q_{MIN} versus i. It can be shown that if the MA(i) model is correct, then [Durbin 1959]

$$Q_{\mathrm{MIN}} \sim \chi_{l-i}^2,$$

Hence, if Q_{MIN} is computed versus i, the appropriateness of each model can be tested by comparing Q_{MIN} to a threshold. A large value of Q_{MIN} indicates that the model order is probably incorrect. Assume that it is desired to test the appropriateness of various model orders at a 95% significance level. Thresholds for the orders are computed from (see Section 2.1)

$$\Pr\{\chi^2_{L-i} > \alpha_i\} = 0.05.$$

If Q_{MIN} for a given i falls below the computed threshold α_i, that value of i can be considered as a candidate for the correct model order. If several values of i produce Q_{MIN} which fall below the threshold, it is not clear which model order should be chosen. It is also possible that all values of i may produce Q_{MIN} that exceed their respective thresholds. This type of test may not produce a good indication of model order. It should be noted that the Q_{MIN} are readily available from the Levinson solution of (8.9). Specifically, for an ith order model

$$Q_{\text{MIN}} = (L+1)\left(\hat{r}_{aa}[0] + \sum_{k=1}^{i} b_i[k]\hat{r}_{aa}[k]\right) - 1 \tag{8.15}$$

where the $b_i[k]$'s are the MA filter parameter estimates obtained from the Levinson recursion and

$$\hat{r}_{aa}[k] = \frac{1}{L+1}\sum_{n=0}^{L-k} \hat{a}[n]\hat{a}[n+k].$$

A third model order selection method tests the adequacy of an MA(i) model by testing whether the ACF samples for $k > i$ are zero [Chow 1972]. Let

$$\hat{\mathbf{r}}'_{xx} = [\hat{r}_{xx}[i+1] \quad \hat{r}_{xx}[i+2] \quad \cdots \quad \hat{r}_{xx}[i+M]]^T.$$

Then the MA(i) model is accepted if

$$\hat{\mathbf{r}}'^T_{xx}\mathbf{C}_r^{-1}\hat{\mathbf{r}}'_{xx} < \gamma. \tag{8.16}$$

Here $\mathbf{C}_r$ is the covariance matrix for $\hat{\mathbf{r}}'_{xx}$ based on the assumption that $x[n]$ is an MA(i) process. The threshold γ is chosen to ensure with high probability that if the MA(i) is correct, the test will indicate this. It is not known how M should be chosen nor how the correct model among competing MA models may be chosen. The exact form for the covariance matrix is given in the work by Chow [1971]. This approach is similar in spirit to the test proposed by Box and Pierce [1970], which is described for AR model order determination in Section 7.8 [see (7.60)].

8.7 OTHER MA ESTIMATORS

A common technique in time series analysis for estimating the MA parameters of an MA(q) process is to use the method of moments [Kendall and Stuart 1979] (see also Section 3.3.1). This approach estimates the unknown parameters by solving a set of equations relating the theoretical moments to the estimated moments. We first estimate $r_{xx}[0]$ through $r_{xx}[q]$. Then the MA spectral estimator is formed as

$$\hat{P}_{\mathrm{MA}}(f) = \sum_{k=-q}^{q} \hat{r}_{xx}[k] \exp(-j2\pi f k). \tag{8.17}$$

Since theoretically (8.17) is equal to $\hat{\sigma}^2 \mid \hat{B}(f) \mid^2$, the MA parameters can be found via spectral factorization (see Section 5.8). To ensure a nonnegative PSD and hence a factorable spectrum, the estimated ACF using a biased estimator needs to be multiplied by a positive semidefinite window such as the Bartlett window (see Section 4.5). The method does not in general yield good estimators. Furthermore, the spectral estimator obtained in this fashion is identical to the BT spectral estimator, so that there seems little point in adopting this approach. The observations of Section 8.3 are also applicable here.

An MLE of the MA parameters based on the likelihood function of the *data* as opposed to the likelihood function of the AR parameter estimates as in Durbin's method requires a nonlinear optimization. Typically, a Newton–Raphson iteration is performed on the gradient of an approximate log-likelihood function. For a real ARMA(p, q) process the Newton–Raphson approach is described in Chapter 10. The special case of a real MA process leads us to the iteration (see Section 10.3)

$$\mathbf{b}_{k+1} = \mathbf{b}_k - \mathbf{H}^{-1}(\mathbf{b}_k) \left. \frac{\partial Q(\mathbf{b})}{\partial \mathbf{b}} \right|_{b=b_k} \tag{8.18}$$

where $\mathbf{b}_k$ is the vector of MA parameters $\{b[1], b[2], \ldots, b[q]\}$ for the kth iterate,

$$\frac{\partial Q(\mathbf{b})}{\partial b[l]} = -2 \sum_{i=0}^{q} b[i]\hat{r}_{zz}[l-i] \qquad l = 1, 2, \ldots, q$$

$$[\mathbf{H}(\mathbf{b})]_{kl} = 2\hat{r}_{zz}[k-l] \qquad k, l = 1, 2, \ldots, q$$

and

$$\hat{r}_{zz}[k] = \frac{1}{N} \sum_{n=0}^{N-1-|k|} z[n]z[n + |k|]$$

$$z[n] = \mathcal{Z}^{-1}\left\{\frac{\mathcal{X}(z)}{\mathcal{B}^2(z)}\right\}.$$

$z[n]$ is the sequence obtained by filtering $x[n]$ with $1/\mathcal{B}^2(z)$. Unfortunately, the iteration is not guaranteed to converge or may converge to only a local minimum of the likelihood function. Other problems are detailed in Chapter 10.

A technique which like Durbin's method attempts to replace the MA(q) process by an AR(q) process is due to Cleveland [1972]. To accomplish this transformation, he proposes that an ACF estimator be computed as

$$\hat{r}_i[k] = \frac{1}{M} \sum_{m=0}^{M-1} \frac{\exp(j2\pi km/M)}{\hat{P}_{\text{AVPER}}(m/M)} \qquad k = 0, 1, \ldots, q \tag{8.19}$$

where $\hat{P}_{\text{AVPER}}(f)$ is an averaged periodogram as defined in Section 4.4 with $\hat{P}_{\text{AVPER}}(f) = \hat{P}_{\text{AVPER}}(f - 1)$ for $\frac{1}{2} \le f < 1$ and M is chosen to be suitably large. $\hat{r}_i[k]$ is readily seen to be an estimate of

$$\int_0^1 \frac{\exp(j2\pi fk)}{P_{\text{MA}}(f)}\, df = \int_0^1 \frac{\exp(j2\pi fk)}{\sigma^2 \,|\, B(f) \,|^2}\, df = \int_{-1/2}^{1/2} \frac{\exp(j2\pi fk)}{\sigma^2 \,|\, B(f) \,|^2}\, df$$

which is termed the *inverse ACF* and represents the ACF of an AR(q) process. The conversion to an AR process is by inverting the periodogram. Using (8.19), the Yule–Walker equations are solved for the MA filter parameters.

Finally, a method due to Clevenson [1970] produces an approximate MLE. The algorithm estimates the ACF samples and then uses a spectral factorization to obtain the MA parameter estimates. Once again a "spectral inversion" is implicit in the method.

8.8 COMPUTER SIMULATION EXAMPLES

The spectral estimates obtained using Durbin's method for two real MA(4) processes are now presented. The MA processes have true PSDs shown in Figure 8.2. The MA parameters for each process are listed in Table 8.1. The first process has its zeros sufficiently displaced from the unit circle so as to give rise to a PSD with no sharp nulls. The second process has its zeros close to the unit circle, resulting in a PSD with sharp nulls. The specific zero locations are given in Figure 8.2. For all the simulations $N = 256$ real data points were used with a large AR model order of $L = 51 \approx N/5$ and a true MA model order of $q = 4$. 30 realizations of the spectral estimator for each process were generated via a computer simulation. The average of the 30 realizations is displayed together with the true PSD

TABLE 8.1 PARAMETER VALUES OF MA PROCESSES USED FOR COMPUTER SIMULATIONS

	Parameter				
	$b[1]$	$b[2]$	$b[3]$	$b[4]$	σ^2
Process 1: zeros not near unit circle	−1.352	1.338	−0.662	0.240	1
Process 2: zeros near unit circle	−2.760	3.809	−2.654	0.924	1

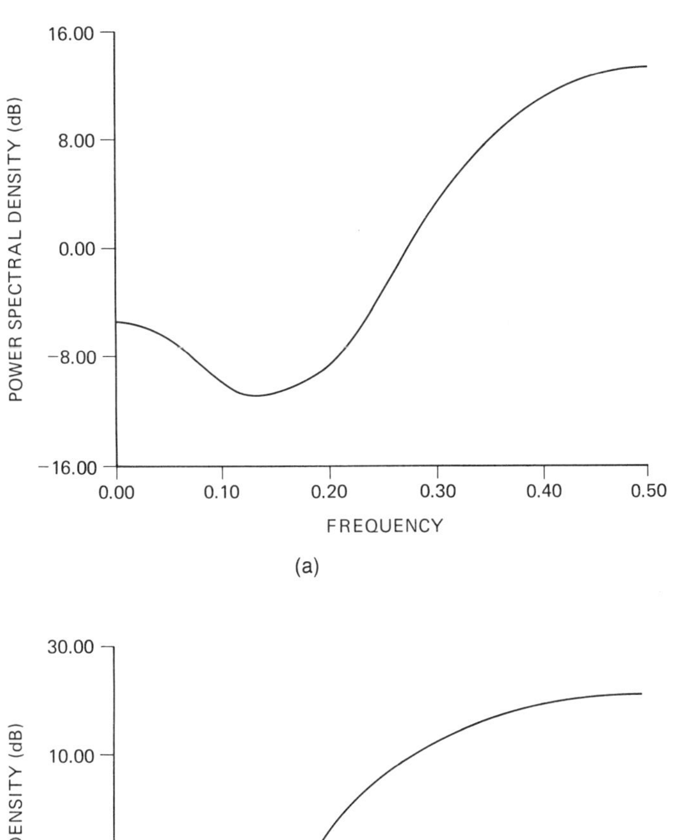

Figure 8.2 True power spectral densities and pole plots. (a), (c) power spectral density and pole plot for process 1 (b), (d) power spectral density and pole plot for process 2.

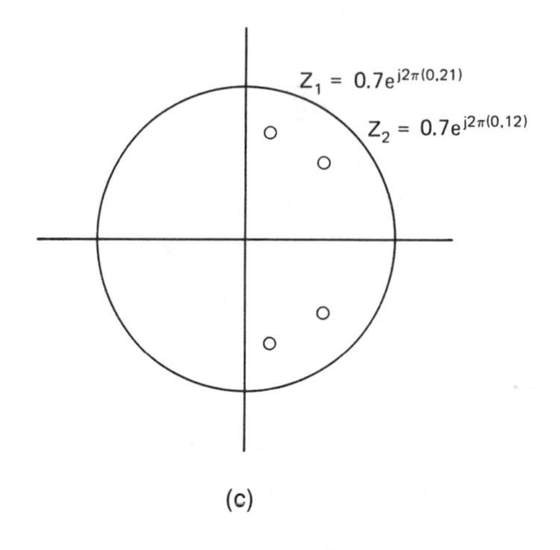

(c)

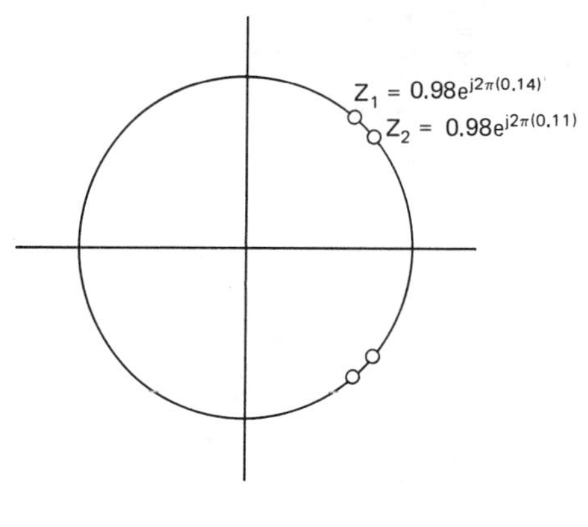

(d)

Figure 8.2 *(Continued)*

(shown as a dashed curve) to indicate the bias of the spectral estimator. Also, the 30 realizations are plotted in an overlaid fashion to measure the variability of the spectral estimator. The results are shown in Figures 8.3 and 8.4. It is seen that for the MA process with a PSD not exhibiting spectral nulls the performance is generally good. Only a slight bias is observed and the variance is not excessive. For the MA process with sharp nulls in the PSD a significant bias is observed for the frequency region in the vicinity of the nulls. In this case, the radii of the complex-conjugate zeros are all equal to 0.98. Hence this result is to be expected since the impulse response of $1/\mathcal{B}(z)$ has not sufficiently decayed for an index of $L = 51$ $(0.98^{51} = 0.36)$. A larger order AR model would be necessary to adequately represent the MA PSD. Doing so, however, would run the risk of a statistically unstable spectral estimator. In summary, Durbin's method works well if the large order AR model provides an accurate representation of the MA PSD.

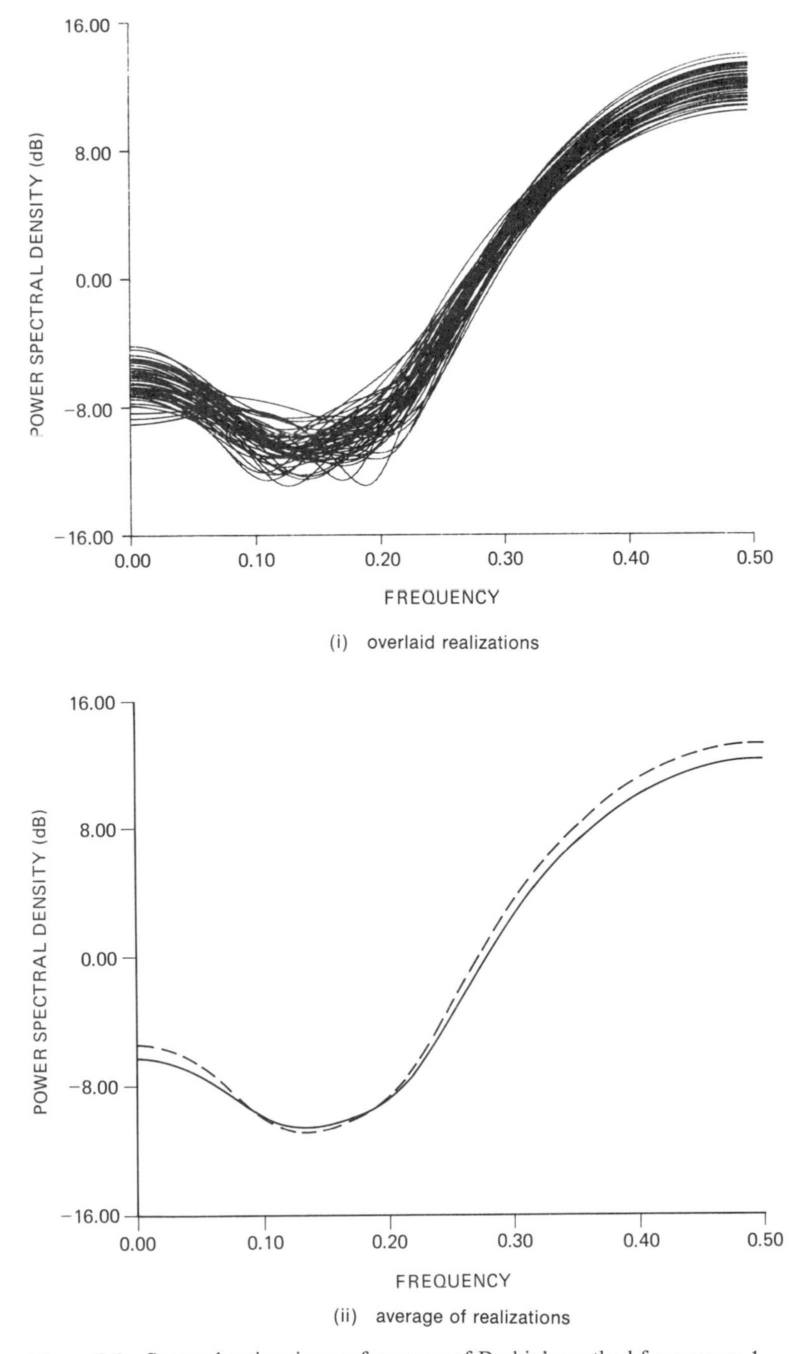

Figure 8.3 Spectral estimation performance of Durbin's method for process 1.

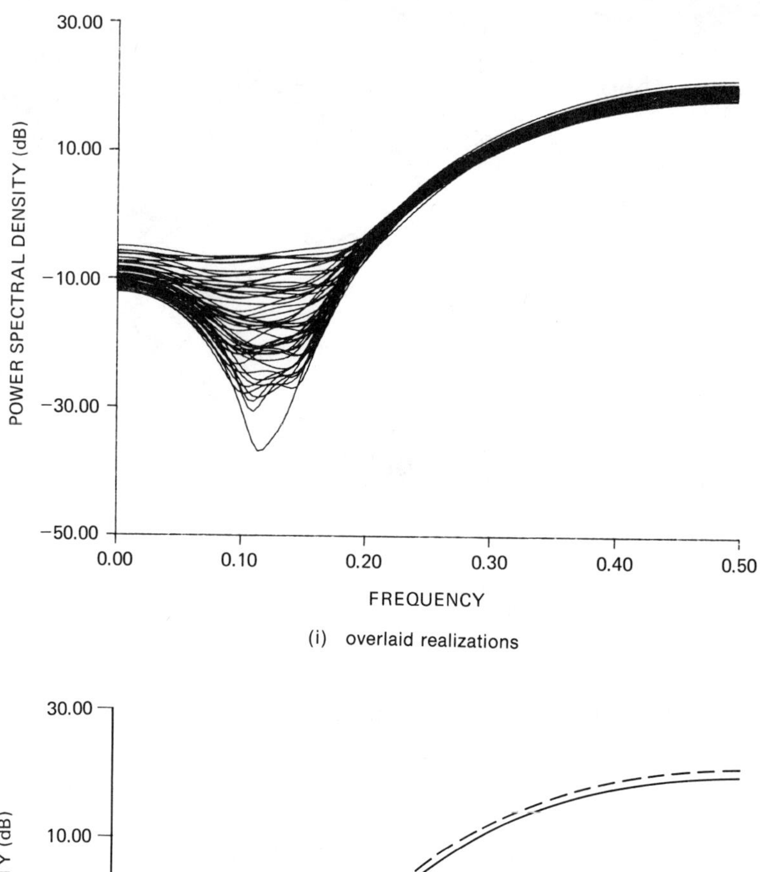

(i) overlaid realizations

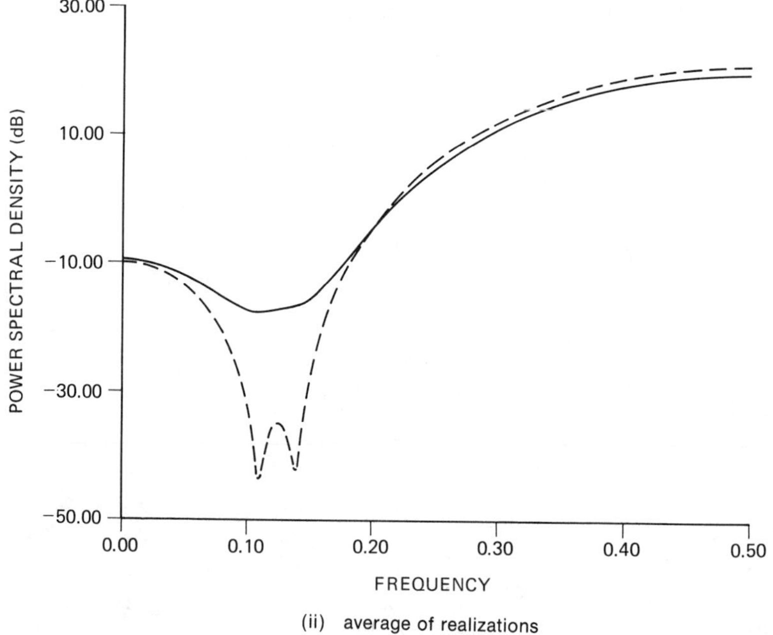

(ii) average of realizations

Figure 8.4 Spectral estimation performance of Durbin's method for process 2.

REFERENCES

Box, G. E. P., and D. A. Pierce, "Distribution of Residual Autocorrelations in Autoregressive–Integrated Moving Average Time Series Models," *J. Am. Stat. Assoc.*, Vol. 64, 1970.

Chow, J. C., "On the Identification of Linear Dynamic System," Ph.D. dissertation, Harvard University, 1971.

Chow, J. C., "On the Estimation of the Order of a Moving-Average Process," *IEEE Trans. Autom. Control,* Vol. AC17, pp. 386–387, June 1972.

Cleveland, W. S., "The Inverse Autocorrelations of a Time Series and Their Applications," *Technometrics,* Vol. 14, pp. 277–293, 1972.

Clevenson, M. L., "Asymptotically Efficient Estimates of the Parameters of a Moving Average Time Series," Tech. Rep. 15, Department of Statistics, Stanford University, Stanford, Calif., July 31, 1970.

Durbin, J., "Efficient Estimation of Parameters in Moving-Average Models," *Biometrika,* Vol. 46, pp. 306–316, 1959.

Kendall, M., and A. Stuart, *The Advanced Theory of Statistics,* Vol. 2, Macmillan, New York, 1979.

PROBLEMS

8.1. For a real MA(1) process, prove that the determinant of $\mathbf{R}_{xx}^{(L)}$, where $\mathbf{R}_{xx}^{(L)}$ is the $L \times L$ autocorrelation matrix, satisfies

$$\det(\mathbf{R}_{xx}^{(L)}) = \sigma^2(1 + b^2[1]) \det(\mathbf{R}_{xx}^{(L-1)}) - b^2[1]\sigma^4 \det(\mathbf{R}_{xx}^{(L-2)}).$$

Use the difference equation to find $\det(\mathbf{R}_{xx}^{(L)})$ explicitly and verify that for large L

$$\det(\mathbf{R}_{xx}^{(L)}) \approx \frac{\sigma^{2L}}{1 - b^2[1]}$$

and hence for $|b[1]|$ not too close to 1,

$$\det(\mathbf{R}_{xx}^{(L)}) \approx \sigma^{2L}.$$

8.2. Use Durbin's method to estimate the parameters of a real MA(1) process. The data are given by

$$x[n] = \begin{cases} \sigma & \text{for } n = 0 \\ \sigma b[1] & \text{for } n = 1 \\ 0 & \text{for } n = 2, 3, \ldots, N-1. \end{cases}$$

Find the approximate solution for large L. Explain your results. *Hint:* Show that the first column of the inverse of

$$\overline{\mathbf{R}}_{xx}^{(L)} = \frac{\mathbf{R}_{xx}^{(L)}}{\sigma^2} = \begin{bmatrix} 1 + b^2[1] & b[1] & 0 & 0 & \cdots & 0 \\ b[1] & 1 + b^2[1] & b[1] & 0 & \cdots & 0 \\ \vdots & \vdots & \vdots & \ddots & \ddots & \vdots \\ 0 & 0 & \cdots & b[1] & 1 + b^2[1] & b[1] \\ 0 & 0 & \cdots & 0 & b[1] & 1 + b^2[1] \end{bmatrix}$$

is given by

$$[\overline{\mathbf{R}}_{xx}^{(L)^{-1}}]_{k1} = \frac{(-b[1])^{k-1} \det (\overline{\mathbf{R}}_{xx}^{(L-k)})}{\det (\overline{\mathbf{R}}_{xx}^{(L)})} \qquad k = 1, 2, \ldots, L$$

and use the results from Problem 8.1.

8.3. Find the asymptotic variance of the MLE of $b[1]$ for a real MA(1) process using (8.11) and compare it to the asymptotic variance of the MLE of $a[1]$ for a real AR(1) process using (6.97). Comment on the difficulty of estimating $b[1]$ when the zero of the process is close to the unit circle. Does this result make intuitive sense?

8.4. For an MA(q) process determine the optimal linear predictor of $x[n]$ based on the infinite past

$$\hat{x}[n] = -\sum_{k=1}^{\infty} \alpha_k x[n-k]$$

by finding the prediction coefficients α_k's. Then show that an equivalent predictor is

$$\hat{x}[n] = \sum_{k=1}^{q} b[k]u[n-k].$$

8.5. Show that the reflection coefficient sequence for a real MA(1) process is given by

$$k_n = \frac{(1 - b^2[1])(-b[1])^n}{1 - (b[1])^{2n+2}}.$$

Comment on the use of the linear prediction error power as a means of model order determination.

APPENDIX 8A

Computer Program for Durbin Method (Provided on Floppy Disk)

```
      SUBROUTINE DURBIN(X,N,IQ,L,B,SIG2)
C
C     This program implements Durbin's method of estimating
C     the MA parameters of a complex time series. See Section
C     8.4 for a description based on real data and the summary
C     at the end of that section. The modifications for complex
C     data are only in the definition of the autocorrelation
C     estimator for complex data.
C
C     Input Parameters:
C
C       X             -Complex array of dimension N×1 of data points
C       N             -Number of complex data points
C       IQ            -Desired MA model order
C       L             -Large AR model order (N/5 usually works well)
C
C     Output Parameters:
```

```
C
C      B             -Complex array of dimension IQ×1 of MA filter
C                     parameter estimates
C      SIG2          -Excitation white noise variance estimate
C
C     External Subroutines:
C
C      CORRELATION            -see Appendix 4D
C      LEVINSON               -see Appendix 6B
C      AUTOCORR               -see Appendix 7B
C
C     Notes:
C
C      The calling program must dimension arrays X,B. The arrays
C      A,AP must be dimensioned greater than or equal to the
C      variable dimensions shown. The array R,AA,RHO in AUTOCORR
C      must be dimensioned greater than or equal to L+1,L×L, and
C      L, respectively. The array AA in Levinson must be dimensioned
C      to agree with dimensions of AA in AUTOCORR.
C
C     Verification Test Case:
C
C      If the complex test data listed in Appendix 1A is input
C      as X, with N=32, IQ=10, and L=15, then the
C      output should be
C        B(1)  = (-4.70956E-02,  4.61267E-02)
C        B(2)  = ( 0.56001     ,  5.76182E-02)
C        B(3)  = ( 0.22638     ,  2.82898E-02)
C        B(4)  = (-1.46033E-02,  8.90383E-03)
C        B(5)  = ( 0.36459     ,  4.33258E-03)
C        B(6)  = (-0.27950     ,  2.51802E-02)
C        B(7)  = ( 0.14134     ,  1.90398E-02)
C        B(8)  = (-0.10563     , -3.48616E-02)
C        B(9)  = (-5.12997E-02, -4.19190E-02)
C        B(10) = ( 1.87891E-02, -1.10760E-02)
C        SIG2  = 1.83942
C      for a DEC VAX 11/780.
C
       COMPLEX X(1),B(1),A(L),AP(L+1)
C     Fit a large order AR model to the data using autocorrelation method
       CALL AUTOCORR(X,N,L,A,SIG2)
C     Use large order AR parameter estimates as new data
       L1=L+1
       AP(1)=(1.,0.)
       DO 20 I=2,L1
20     AP(I)=A(I-1)
C     Use autocorrelation method to generate MA parameter estimates (8.9)
       CALL AUTOCORR(AP,L1,IQ,B,PB)
       RETURN
       END
```

Chapter 9

Autoregressive Moving Average Spectral Estimation: General

9.1 INTRODUCTION

ARMA spectral estimation holds the promise of extending the practical utility of AR spectral estimation to processes which cannot be modeled as purely AR. Since nearly all data are corrupted by some amount of observation noise, the ARMA model is nearly always the appropriate one. Unfortunately, in stark contrast to the multitude of methods available to efficiently estimate parameters of an AR process, no comparable algorithms exist for ARMA estimation. This is because nearly optimal estimators based on maximum likelihood require the solution of a set of highly nonlinear equations. As a result, suboptimal but easily implementable algorithms have been emphasized. In this chapter general background material on ARMA estimation is provided, while practical algorithms are described in Chapter 10.

9.2 SUMMARY

An approximate MLE for the parameters of a real ARMA process is derived in Section 9.3. To obtain the MLE explicitly the function given by (9.4) in the time domain or by (9.6) in the frequency domain must be minimized over the ARMA filter parameters. This function is highly nonlinear in the MA parameters, requiring a computationally burdensome search routine. Next the statistics of the MLE are given for large data records. The MLE of the ARMA parameters is approximately distributed according to a Gaussian PDF with mean equal to the true parameter value and covariance matrix given by (9.8). Similarly, the MLE of the ARMA PSD is approximately Gaussianly distributed with mean equal to the true PSD and covariance matrix given by (9.12). Some methods for estimating the MA parameters of an ARMA process assuming that the AR parameters are known a priori are given in Section 9.5. Model order selection rules are described in Section 9.6, the most popular one being the AIC (9.15). Finally, in Section 9.7 a time series consisting of pure sinusoids in white noise is shown to be approximately modeled by a special ARMA process. In this model the AR and MA parameters are nearly identical.

9.3 MAXIMUM LIKELIHOOD ESTIMATION

The MLE of the ARMA PSD is

$$\hat{P}_{\text{ARMA}}(f) = \frac{\hat{\sigma}^2 \,|\, 1 + \hat{b}[1] \exp(-j2\pi f) + \cdots + \hat{b}[q] \exp(-j2\pi fq) \,|^2}{|\, 1 + \hat{a}[1] \exp(-j2\pi f) + \cdots + \hat{a}[p] \exp(-j2\pi fp) \,|^2} \tag{9.1}$$

where $\{\hat{a}[1], \hat{a}[2], \ldots, \hat{a}[p], \hat{b}[1], \hat{b}[2], \ldots, \hat{b}[q], \hat{\sigma}^2\}$ are the MLEs of the ARMA parameters. This follows from the invariance principle (see Section 3.3.1). To obtain these MLEs we must maximize the likelihood function

$$p(x[0], x[1], \ldots, x[N-1]; a[1], a[2], \ldots, a[p], b[1], b[2], \ldots, b[q], \sigma^2) \tag{9.2}$$

over the unknown parameters. This maximization will involve solving a set of highly nonlinear equations, even with several simplifying assumptions.

To derive the exact likelihood function is somewhat involved and lends little insight into a practical estimation procedure. The reader is referred to the works by Akaike [1974A], Newbold [1974], Ansley [1979], Harvey and Phillips [1979], Gueguen and Scharf [1980], and Perlman [1980] for further details of exact likelihood functions and MLEs. An approximate likelihood function will be derived based on the following assumptions:

1. The data are real and Gaussian.
2. The data record N is large.
3. The poles and zeros are not close to the unit circle.

The basic approach to determining an expression for the approximate likelihood function is to use an AR(∞) model of the ARMA process as described in Section 5.3.2. Then the likelihood function already derived for an AR process can be applied to an ARMA process. Let the ARMA process be modeled as an AR(∞) process with filter coefficients $\{c[1], c[2], \ldots\}$. A finite model order approximation [i.e., an AR(L) model] will be a good approximation to the infinite order model if $c[i] \approx 0$ for $i > L$. This requirement is equivalent to requiring that the impulse response of the filter with system function $1/\mathcal{B}(z)$ be approximately zero for $i > L$. Making this assumption allows us to use the results of Section

6.5.1 for the AR(p) model by letting $p = L$. There it was shown that the approximate likelihood [actually, the conditional likelihood (6.86)] function could be written as

$$p(\mathbf{x} \mid x[0], x[1], \ldots, x[L-1]; \mathbf{a}, \mathbf{b}, \sigma^2)$$
$$= \frac{1}{(2\pi\sigma^2)^{(N-L)/2}} \exp\left[-\frac{1}{2\sigma^2} \sum_{n=L}^{N-1} \left(x[n] + \sum_{j=1}^{L} c[j]x[n-j]\right)^2\right] \quad (9.3)$$

where $\mathbf{x} = [x[0] \;\; x[1] \;\; \cdots \;\; x[N-1]]^T$. $\mathbf{a}$, $\mathbf{b}$ are the vectors of the AR and MA filter coefficients, respectively, which depend on the $c[j]$'s. Note that for (9.3) to apply it was required that N be large and that the poles not be close to the unit circle. Now to maximize the likelihood function over $\mathbf{a}$, $\mathbf{b}$ we must minimize

$$S_2(\mathbf{a}, \mathbf{b}) = \sum_{n=L}^{N-1} \left(x[n] + \sum_{j=1}^{L} c[j]x[n-j]\right)^2. \quad (9.4)$$

S_2 is highly nonlinear in $\mathbf{b}$ but a quadratic function of $\mathbf{a}$. As an example, consider an ARMA(1, 1) process. Then as shown in Section 5.3.2,

$$c[j] = (a[1] - b[1])(-b[1])^{j-1} \qquad j \geq 1$$

so that

$$S_2(\mathbf{a}, \mathbf{b}) = \sum_{n=L}^{N-1} \left[x[n] + \sum_{j=1}^{L} (a[1] - b[1])(-b[1])^{j-1}x[n-j]\right]^2.$$

Because S_2 is quadratic in $\mathbf{a}$, differentiation with respect to $\mathbf{a}$ and substitution of that unique value of $\mathbf{a}$ into S_2 will reduce S_2 to only a function of $\mathbf{b}$ (see Problem 9.1). The resultant S_2 will be nonlinear in $\mathbf{b}$ and hence differentiation will produce a set of equations which if solved may only produce a local minimum. It is also possible to differentiate S_2 with respect to $\mathbf{a}$ and $\mathbf{b}$ and solve the resulting nonlinear equations using a Newton–Raphson approach. This method is discussed in Chapter 10 as the Akaike estimator.

Assuming that S_2 can be minimized to produce $\hat{\mathbf{a}}$, $\hat{\mathbf{b}}$, then in a similar manner to the derivation in Section 6.5.1 the estimate of σ^2 is [see (6.90)]

$$\hat{\sigma}^2 = \frac{1}{N} \sum_{n=L}^{N-1} \left(x[n] + \sum_{j=1}^{L} \hat{c}[j]x[n-j]\right)^2 \quad (9.5)$$

where the $\hat{c}[j]$'s are found as the impulse response of the filter with system function $\hat{\mathcal{A}}(z)/\hat{\mathcal{B}}(z)$. It is seen that unfortunately in the ARMA case no simple set of Yule–Walker type equations result for the MLE. The use of the modified Yule–Walker equations for the estimation of the AR parameters as discussed in Chapter 5 bears no resemblance to the MLE and hence cannot be expected to yield good estimates. This approach is discussed in more detail in Chapter 10.

Alternative expressions for $S_2(\mathbf{a}, \mathbf{b})$ and $\hat{\sigma}^2$ using frequency domain concepts form the basis for many approximate MLE estimation procedures. Assuming that $N \gg L$ (9.4) can be written

$$S_2(\mathbf{a}, \mathbf{b}) \approx \sum_{n=-\infty}^{\infty} \left(\sum_{j=0}^{L} c[j]x[n-j] \right)^2$$

where it is assumed that $x[n] = 0$ for $n < 0$ and $n > N - 1$. Let $e[n] = \sum_{j=0}^{L} c[j]x[n-j]$ and $C(f), E(f)$ be the Fourier transforms of the $e[n], c[n]$ sequences, respectively. Then, by Parseval's theorem,

$$\begin{aligned} S_2(\mathbf{a}, \mathbf{b}) &= \int_{-1/2}^{1/2} | E(f) |^2 \, df = \int_{-1/2}^{1/2} | X(f) |^2 \, | C(f) |^2 \, df \\ &\approx N \int_{-1/2}^{1/2} I(f) \frac{| A(f) |^2}{| B(f) |^2} \, df \end{aligned} \tag{9.6}$$

where

$$I(f) = \frac{1}{N} | X(f) |^2 = \frac{1}{N} \left| \sum_{n=0}^{N-1} x[n] \exp(-j2\pi f n) \right|^2 .$$

$I(f)$ is recognized as the periodogram [see (4.2)]. The different notation is to emphasize that the periodogram is to be considered as a function of the data and not necessarily a spectral estimator. Also, for $N \gg L$, (9.5) becomes

$$\hat{\sigma}^2 = \int_{-1/2}^{1/2} I(f) \frac{| \hat{A}(f) |^2}{| \hat{B}(f) |^2} \, df. \tag{9.7}$$

It is interesting to note that S_2 depends on the data only via the periodogram or equivalently via its inverse Fourier transform $\{\hat{r}_{xx}[0], \hat{r}_{xx}[1], \ldots, \hat{r}_{xx}[N-1]\}$, where $\hat{r}_{xx}[k]$ is the biased ACF estimator. This is a direct consequence of the Gaussian assumption for the data. In (9.6) it is observed that to determine the MLE we must minimize the power at the output of a *pole–zero* or inverse filter $\mathcal{A}(z)/\mathcal{B}(z)$. That the inverse filter now contains poles, whereas in the AR case it contained only zeros, leads to the nonlinearity present in the MLE equations.

9.4 STATISTICS OF THE MAXIMUM LIKELIHOOD ESTIMATOR

As in the AR and MA cases, no statistics for the MLEs of the ARMA parameters and PSD are available for finite data records. For large data records, however, the asymptotic theory of the MLE as described in Section 3.3.1 can provide some approximate statistical descriptions. Based on this theory the ARMA parameter estimator $[\hat{\mathbf{a}}^T \;\; \hat{\mathbf{b}}^T \;\; \hat{\sigma}^2]^T$ for a real ARMA process will be asymptotically (for large data records) distributed according to a multivariate Gaussian PDF. The mean is given by the true values

$$\boldsymbol{\mu} = \mathcal{E} \begin{bmatrix} \hat{\mathbf{a}} \\ \hat{\mathbf{b}} \\ \hat{\sigma}^2 \end{bmatrix} = \begin{bmatrix} \mathbf{a} \\ \mathbf{b} \\ \sigma^2 \end{bmatrix}$$

and the covariance matrix is the inverse of the Fisher information matrix (see Appendix 9A)

$$\mathbf{C}_{a,b,\sigma^2} = \mathcal{E}\left\{\begin{bmatrix} \hat{\mathbf{a}} - \mathcal{E}(\hat{\mathbf{a}}) \\ \hat{\mathbf{b}} - \mathcal{E}(\hat{\mathbf{b}}) \\ \hat{\sigma}^2 - \mathcal{E}(\hat{\sigma}^2) \end{bmatrix}\begin{bmatrix} \hat{\mathbf{a}} - \mathcal{E}(\hat{\mathbf{a}}) \\ \hat{\mathbf{b}} - \mathcal{E}(\hat{\mathbf{b}}) \\ \hat{\sigma}^2 - \mathcal{E}(\hat{\sigma}^2) \end{bmatrix}^T\right\}$$

$$= \begin{bmatrix} \frac{\sigma^2}{N}\mathbf{R}^{-1} & \mathbf{0} \\ \mathbf{0}^T & \frac{2\sigma^4}{N} \end{bmatrix} \tag{9.8}$$

where

$$\mathbf{R} = \begin{bmatrix} \mathbf{R}_{yy} & -\mathbf{R}_{yz} \\ -\mathbf{R}_{yz}^T & \mathbf{R}_{zz} \end{bmatrix} = \begin{bmatrix} p \times p & p \times q \\ q \times p & q \times q \end{bmatrix}$$

$$[\mathbf{R}_{yy}]_{ij} = r_{yy}[i - j] \qquad i = 1, 2, \ldots, p; j = 1, 2, \ldots, p$$

$$[\mathbf{R}_{yz}]_{ij} = r_{yz}[i - j] \qquad i = 1, 2, \ldots, p; j = 1, 2, \ldots, q$$

$$[\mathbf{R}_{zz}]_{ij} = r_{zz}[i - j] \qquad i = 1, 2, \ldots, q; j = 1, 2, \ldots, q$$

and $\mathbf{0}$ is a $(p + q) \times 1$ vector of zeros. $y[n]$ is an AR(p) process with parameters $\{a[1], a[2], \ldots, a[p], \sigma^2\}$ and $z[n]$ is an AR(q) process with parameters $\{b[1], b[2], \ldots, b[q], \sigma^2\}$. These processes are assumed to have been generated as shown in Figure 9.1. As an illustration, for an ARMA(1, 1) process it can be shown that (see Problem 9.2)

$$\mathbf{R} = \begin{bmatrix} r_{yy}[0] & -r_{yz}[0] \\ -r_{yz}[0] & r_{zz}[0] \end{bmatrix}$$

$$= \begin{bmatrix} \dfrac{\sigma^2}{1 - a^2[1]} & \dfrac{-\sigma^2}{1 - a[1]b[1]} \\ \dfrac{-\sigma^2}{1 - a[1]b[1]} & \dfrac{\sigma^2}{1 - b^2[1]} \end{bmatrix} \tag{9.9}$$

so that

$$\mathbf{C}_{a,b,\sigma^2} =$$

$$\begin{bmatrix} \dfrac{(1 - a[1]b[1])^2(1 - a^2[1])}{N(a[1] - b[1])^2} & \dfrac{(1 - a^2[1])(1 - b^2[1])(1 - a[1]b[1])}{N(a[1] - b[1])^2} & 0 \\ \dfrac{(1 - a^2[1])(1 - b^2[1])(1 - a[1]b[1])}{N(a[1] - b[1])^2} & \dfrac{(1 - a[1]b[1])^2(1 - b^2[1])}{N(a[1] - b[1])^2} & 0 \\ 0 & 0 & \dfrac{2\sigma^4}{N} \end{bmatrix}. \tag{9.10}$$

Observe that if $a[1] \approx b[1]$ the variances of $\hat{a}[1]$, $\hat{b}[1]$ are extremely large. This

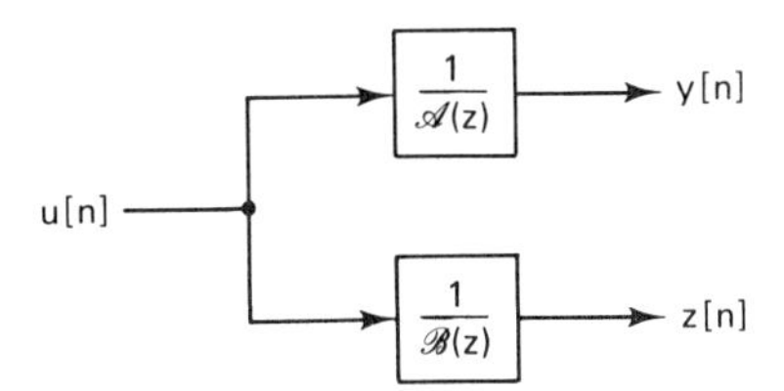

Figure 9.1 Definitions of processes required in Fisher information matrix.

occurs because the ARMA(1, 1) process is nearly white noise, resulting in an overfitting of the model. Caution must be exercised in any estimation procedure in which the model order is not known a priori (see also Problem 9.3). Even if (9.8) is not a good approximation to the covariance matrix of the ARMA parameter estimators as may occur for shorter length data records, it still provides a lower bound in that

$$\begin{aligned} \operatorname{var}(\hat{a}[i]) &\geq \frac{\sigma^2}{N}[\mathbf{R}^{-1}]_{ii} \qquad & i = 1,2,\ldots,p \\ \operatorname{var}(\hat{b}[i]) &\geq \frac{\sigma^2}{N}[\mathbf{R}^{-1}]_{p+i,p+i} \qquad & i = 1,2,\ldots,q \\ \operatorname{var}(\hat{\sigma}^2) &\geq \frac{2\sigma^4}{N} \end{aligned} \tag{9.11}$$

assuming that $\hat{\mathbf{a}}$, $\hat{\mathbf{b}}$, $\hat{\sigma}^2$ are unbiased estimators.

The statistics of the MLE spectral estimator (9.1) can be found by applying the asymptotic properties of the MLE as described in Section 3.3.1. Let

$$\mathbf{p} = [\hat{P}_{\text{ARMA}}(f_1) \quad \hat{P}_{\text{ARMA}}(f_2) \quad \cdots \quad \hat{P}_{\text{ARMA}}(f_M)]^T.$$

Then for large data records $\mathbf{p}$ is distributed according to a multivariate Gaussian PDF with mean equal to the true value and covariance matrix [see (3.23)]

$$\mathbf{C}_p = \mathbf{D}\mathbf{C}_{a,b,\sigma^2}\mathbf{D}^T \tag{9.12}$$

where $\mathbf{D}$ is the $M \times (p + q + 1)$ Jacobian matrix of the transformation from $\mathbf{a}$, $\mathbf{b}$, σ^2 to $\mathbf{p}$:

$$[\mathbf{D}]_{mn} = \begin{cases} \dfrac{\partial P_{\text{ARMA}}(f_m)}{\partial a[n]} & \text{for } m = 1,2,\ldots,M; n = 1,2,\ldots,p \\ \dfrac{\partial P_{\text{ARMA}}(f_m)}{\partial b[n-p]} & \text{for } m = 1,2,\ldots,M; n = p+1, p+2, \ldots, p+q \\ \dfrac{\partial P_{\text{ARMA}}(f_m)}{\partial \sigma^2} & \text{for } m = 1,2,\ldots,M; n = p+q+1. \end{cases}$$

The Jacobian may be evaluated to yield

$$[\mathbf{D}]_{mn} = \begin{cases} -2P_{\text{ARMA}}(f_m)\text{Re}\left[\dfrac{\exp(-j2\pi f_m n)}{A(f_m)}\right] \\ \qquad \text{for } m = 1,2,\ldots,M; n = 1,2,\ldots,p \\ 2P_{\text{ARMA}}(f_m)\text{Re}\left[\dfrac{\exp(-j2\pi f_m(n-p))}{B(f_m)}\right] \\ \qquad \text{for } m = 1,2,\ldots,M; n = p+1, p+2,\ldots,p+q \\ \dfrac{P_{\text{ARMA}}(f_m)}{\sigma^2} \\ \qquad \text{for } m = 1,2,\ldots,M; n = p+q+1 \end{cases}$$

and $\mathbf{C}_{a,b,\sigma^2}$ is given by (9.8). In particular, at any frequency the variance is

$$\text{var}[\hat{P}_{\text{ARMA}}(f)] = \frac{4\sigma^2}{N}P^2_{\text{ARMA}}(f)\mathbf{e}^T\mathbf{R}^{-1}\mathbf{e} + \frac{2}{N}P^2_{\text{ARMA}}(f) \tag{9.13}$$

where

$$\mathbf{e} = \begin{bmatrix} \mathbf{e}_{\text{AR}} \\ \mathbf{e}_{\text{MA}} \end{bmatrix}$$

$$[\mathbf{e}_{\text{AR}}]_n = -\text{Re}\left[\frac{\exp(-j2\pi fn)}{A(f)}\right] \qquad n = 1,2,\ldots,p$$

$$[\mathbf{e}_{\text{MA}}]_n = \text{Re}\left[\frac{\exp(-j2\pi fn)}{B(f)}\right] \qquad n = 1,2,\ldots,q.$$

(See also Problem 9.4). Note that (9.13) may be used to establish a confidence interval for the spectral estimate.

9.5 NUMERATOR DETERMINATION FOR KNOWN AUTOREGRESSIVE PARAMETERS

It was shown in Chapter 5 that the AR parameters of an ARMA model may be determined using the modified Yule–Walker equations [see (5.22)]. Although these equations with the theoretical ACF replaced by an estimated one do not always produce good estimates, as shown in Chapter 10, they do demonstrate the possibility of estimating the AR parameters independent of the MA parameters. It is now assumed that the AR parameters are known so that the denominator of $P_{\text{ARMA}}(f)$ is known. To obtain the MA parameters, we can filter $x[n]$ with $\mathcal{A}(z)$ to obtain a pure MA process and then use any of the MA parameter estimators of Chapter 8. Alternatively, since for spectral estimation only $|B(f)|^2$ is necessary and not explicitly the MA parameters, we can apply any of the Fourier based spectral estimation methods to the time series at the output of $\mathcal{A}(z)$ [Thomson 1977, Kay 1980]. Finally, we may use some theoretical relationships between the numerator and denominator ACFs to obtain the numerator spectral estimator. One example of this approach is now given. Another is described in Problem 9.5, and still others can be found in the works by Kinkel et al. [1979], Kaveh [1979], and Cadzow [1982]. Let $r_{MA}[k] = \mathcal{Z}^{-1}[\sigma^2\mathcal{B}(z)\mathcal{B}^*(1/z^*)]$, where $\mathcal{Z}^{-1}$ denotes the inverse z transform. Then the ARMA PSD is

$$P_{ARMA}(f) = \frac{\sum_{k=-q}^{q} r_{MA}[k]\exp(-j2\pi fk)}{|A(f)|^2}$$

where $r_{MA}[-k] = r_{MA}^*[k]$. To obtain $r_{MA}[k]$, observe that

$$r_{MA}[k] = \mathcal{Z}^{-1}[\mathcal{A}(z)\mathcal{A}^*(1/z^*)\mathcal{P}_{ARMA}(z)].$$

Letting $a_k = \mathcal{Z}^{-1}[\mathcal{A}(z)\mathcal{A}^*(1/z^*)]$ which is assumed known and evaluating the inverse z-transform, $r_{MA}[k]$ may be determined from

$$r_{MA}[k] = \sum_{l=-p}^{p} a_l r_{xx}[k-l] \qquad k = 0, 1, \ldots, q. \tag{9.14}$$

To ensure a nonnegative PSD estimate, $r_{MA}[k]$ must be a positive semidefinite sequence. To estimate $r_{MA}[k]$ we need estimates of $\{r_{xx}[0], r_{xx}[1], \ldots, r_{xx}[p+q]\}$ for use in (9.14). Note that $r_{MA}[k]$ is the ACF of the MA process obtained by filtering $x[n]$ with $\mathcal{A}(z)$ and so is similar in spirit to the filtering approach previously described.

9.6 MODEL ORDER SELECTION

For an ARMA time series the reflection coefficient sequence is infinite in extent so that the prediction error power is always decreasing. This is in contrast to an AR time series, in which the prediction error power first reaches its minimum at the correct model order. Hence model order determination approaches based on the linear prediction error power cannot be used for an ARMA process. Some methods that have been proposed for ARMA model order estimation are now described. The AIC as described in Chapter 7 for AR model order determination and in Chapter 8 for MA model order selection can also be used for the real ARMA case if we define [Akaike 1974B]

$$\text{AIC}(i, j) = N \ln \hat{\sigma}_{ij}^2 + 2(i + j) \tag{9.15}$$

where i is the assumed AR model order, j is the assumed MA model order, and $\hat{\sigma}_{ij}^2$ is the MLE of σ^2 obtained under the assumption that $x[n]$ is an ARMA(i, j) process. As usual, the AIC is computed for all model orders of interest and the orders that minimize it are chosen.

Another approach is to filter $x[n]$ with the estimated inverse filter $\hat{\mathcal{A}}(z)/\hat{\mathcal{B}}(z)$ to generate an estimate $\hat{u}[n]$ of the white noise process. If the correct order has been chosen, $\hat{u}[n]$ will be approximately white noise and hence the estimated ACF should be approximately zero for all lags except the zeroth one. It can be shown that if an ARMA(i, j) model is correct [Box and Pierce 1970], then for a real process

$$Q = N \sum_{k=1}^{M} \left(\frac{\hat{r}_{uu}[k]}{\hat{r}_{uu}[0]} \right)^2 \qquad (9.16)$$

is distributed according to a χ^2_{M-i-j} random variable (see Section 3.2). $\hat{r}_{uu}[k]$ is the biased estimator of the ACF of $\hat{u}[n]$ given by

$$\hat{r}_{uu}[k] = \frac{1}{N} \sum_{n=0}^{N-1-k} \hat{u}[n]\hat{u}[n + \mathrm{k}].$$

M should be the effective impulse response length of the filter with system function $\mathcal{B}(z)/\mathcal{A}(z)$. If the model is incorrect, Q will be large. As an example, for $M = 20$, $i = 2$, $j = 1$, a χ^2_{17} random variable will be less than or equal to 27.59 with 95% probability. A value of Q larger than 27.59 would indicate an incorrect model. We might compute Q over several possible model orders and discard models that had inflated Q's. If all the models but one had inflated values, then by the process of elimination the remaining model could be chosen. Otherwise, further tests would be necessary. Other tests for whiteness of the filtered time series can be found in the book by Jenkins and Watts [1968].

Finally, a model order selection rule based on the modified Yule–Walker equations has been proposed for AR model order determination of an ARMA process by Chow [1972]. If we examine the $i \times i$ matrix $\mathbf{R}'_{xx}$, where

$$\mathbf{R}'_{xx} = \begin{bmatrix} r_{xx}[q+1] & r_{xx}[q] & \cdots & r_{xx}[q-i+2] \\ r_{xx}[q+2] & r_{xx}[q+1] & \cdots & r_{xx}[q-i+3] \\ \vdots & \vdots & \ddots & \vdots \\ r_{xx}[q+i] & r_{xx}[q+i-1] & \cdots & r_{xx}[q+1] \end{bmatrix} \qquad (i \times i) \qquad (9.17)$$

for an assumed model order of i, then for $i > p$, the true AR model order, the matrix will be singular. This follows from the modified Yule–Walker equations (5.16)

$$\sum_{l=1}^{p} a[l] r_{xx}[k-l] = -r_{xx}[k] \qquad k \geq q + 1$$

which imply that the columns of R'_{xx} will be linearly dependent. As an example, if $i = p + 1$, then $\mathbf{R}'_{xx} = [\mathbf{r}'_{q+1} \ \ \mathbf{r}'_q \ \ \cdots \ \ \mathbf{r}'_{q+1-p}]$, where $\mathbf{r}'_k = [r_{xx}[k] \ r_{xx}[k+1] \ \cdots \ r_{xx}[k+p]]^T$. The columns $\mathbf{r}'_i$ are linearly dependent since

$$\sum_{i=0}^{p} a[i] \mathbf{r}'_{q+1-i} = 0$$

which follows from the modified Yule–Walker equations (see Problem 9.6). We can monitor det (R'_{xx}) until it becomes sufficiently small for some i. Note that we need to know q or at least be able to assume that q is not larger than some value q_{MAX}. In the latter case q_{MAX} is used in (9.17) and the actual value of q can be determined by filtering $x[n]$ by $\hat{\mathcal{A}}(z)$ once p has been determined and the AR parameters estimated. To do so we use any of the MA model order selection rules of Chapter 8 applied to the filtered time series.

9.7 A SPECIAL ARMA MODEL

As described in Sections 5.6 and 6.7, a useful model is that of an AR process to which white noise has been added. It was shown there that an ARMA(p, p) model is appropriate if the AR and MA parameters are linked via the relationship

$$\sigma_\eta^2 \mathcal{B}(z)\mathcal{B}^*(1/z^*) = \sigma^2 + \sigma_w^2 \mathcal{A}(z)\mathcal{A}^*(1/z^*). \tag{9.18}$$

Also, it was shown in Section 5.6 that a sum of sinusoids can be modeled in a limiting sense as an AR process with poles on the unit circle. For the special case of *sinusoids in white noise* the PSD of the time series may be approximately modeled by the PSD of an ARMA process. Clearly, if $\sigma^2 \to 0$, which is necessary

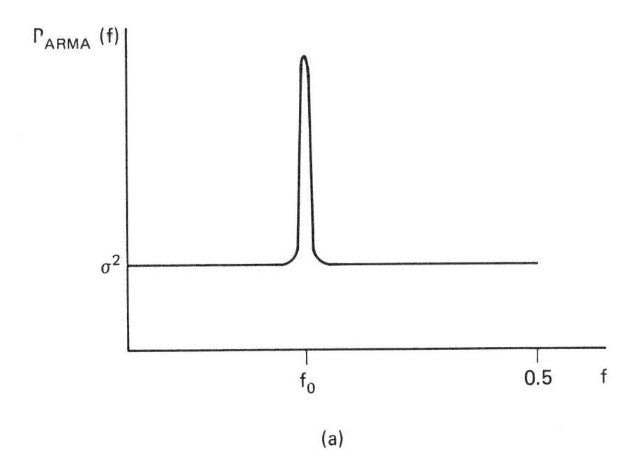

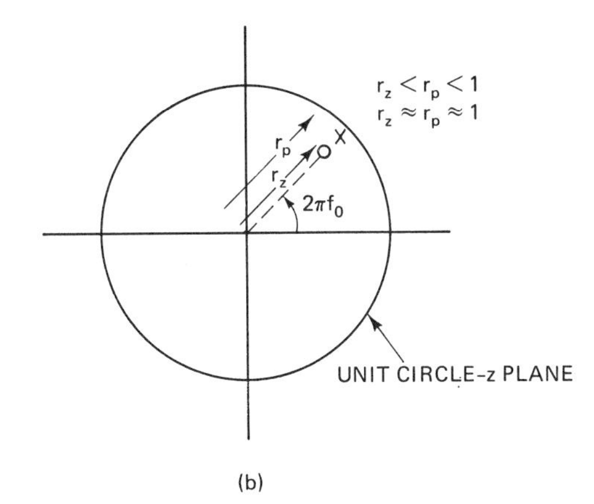

Figure 9.2 ARMA modeling for a complex sinsuoid in white noise. (a) Power spectral density. (b) Pole–zero plot.

in modeling the sinusoids as an AR process (see Section 5.6), then from (9.18),

$$\sigma_\eta^2 \rightarrow \sigma_w^2$$

and

$$\mathcal{B}(z) \rightarrow \mathcal{A}(z)$$

or $b[k] \rightarrow a[k]$. The MA parameters of the ARMA(p, p) model become nearly identical to the AR parameters. It would appear that pole–zero cancellation occurs and that a purely flat PSD is the result. That this is not the case is explained by the fact that the poles are infinitesimally close to the unit circle, producing nearly impulses in the PSD, while the zeros are infinitesimally close to the poles but displaced slightly inward. Figure 9.2 depicts this situation (see Problem 9.7). This special ARMA model was first noted by Ulrych and Clayton [1976]. It demonstrates the possibility of using ARMA spectral estimation techniques for processes consisting of sinusoids in white noise.

REFERENCES

Akaike, H., "Markovian Representation of Stochastic Processes and Its Application to the Analysis of Autoregressive Moving Average Processes," *Ann. Inst. Statist. Math.*, Vol. 26, pp. 363–387, 1974A.

Akaike, H., "A New Look at the Statistical Model Identification," *IEEE Trans. Autom. Control*, Vol. AC19, pp. 716–723, Dec. 1974B.

Ansley, C. F., "An Algorithm for the Exact Likelihood of a Mixed Autoregressive–Moving Average Process," *Biometrika*, Vol. 66, pp. 59–65, 1979.

Box, G. E. P., and D. A. Pierce, "Distribution of Residual Autocorrelations in Autoregressive–Integrated Moving Average Time Series Models," *J. Am. Statist. Assoc.*, Vol. 64, 1970.

Cadzow, J., "Spectral Estimation: An Overdetermined Rational Model Equation Approach," *Proc. IEEE*, Vol. 70, pp. 907–938, Sept. 1982.

Chow, J. C., "On Estimating the Orders of an Autoregressive Moving-Average Process with Uncertain Observations," *IEEE Trans. Autom. Control*, Vol. AC17, pp. 707–709, Oct. 1972.

Guegen, C., and L. Scharf, "Exact Maximum Likelihood Identification of ARMA Models: A Signal Processing Perspective," in *Signal Processing Theory and Applications*, M. Kunt and F. de Coulon, eds, North-Holland, New York, 1980, pp. 759–769.

Harvey, A. C., and G. D. A. Phillips, "Maximum Likelihood Estimation of Regression Models with Autoregressive Moving Average Disturbances," *Biometrika*, Vol. 66, pp. 49–58, 1979.

Jenkins, G. M., and D. G. Watts, *Spectral Analysis and Its Applications*, Holden-Day, San Francisco, 1968.

Kaveh, M., "High Resolution Spectral Estimation for Noisy Signals," *IEEE Trans. Acoust. Speech Signal Process.*, Vol. ASSP27, pp. 286–287, June 1979.

Kay, S., "A New ARMA Spectral Estimator," *IEEE Trans. Acoust. Speech Signal Process.*, Vol. ASSP28, pp. 585–588, Oct. 1980.

Kinkel, J. F., et al., "A Note on Covariance-Invariant Digital Filter Design and Autoregressive Moving Average Spectral Analysis," *IEEE Trans. Acoust. Speech Signal Process.*, Vol. ASSP27, pp. 200–202, Apr. 1979.

Newbold, P., "The Exact Likelihood Function for a Mixed Autoregressive–Moving Average Process," *Biometrika*, Vol. 61, pp. 423–426, 1974.

Perlman, J. G., "An Algorithm for the Exact Likelihood of a High-Order Autoregressive–Moving Average Process," *Biometrika*, Vol. 67, pp. 232–233, 1980.

Thomson, D. J., "Spectrum Estimation Techniques for Characterization and Development of WT4 Waveguide-Part 1," *Bell Syst. Tech. J.*, Vol. 56, pp. 1769–1815, Nov. 1977.

Ulrych, T. J., and R. W. Clayton, "Time Series Modelling and Maximum Entropy," *Phys. Earth Planet. Inter.*, Vol. 12, pp. 188–200, Aug. 1976.

PROBLEMS

9.1. For the MLE of the parameters of a real ARMA(p, q) process we must minimize

$$\frac{1}{N} S_2(\mathbf{a}, \mathbf{b}) = \int_{-1/2}^{1/2} I(f) \frac{|A(f)|^2}{|B(f)|^2} df$$

where $I(f)$ is the periodogram of the data. Show how to minimize this over the AR filter parameters by proceeding as follows:

(a) First show that

$$\frac{1}{N} S_2(\mathbf{a}, \mathbf{b}) \approx \frac{1}{N} \sum_{n=-\infty}^{\infty} \left(\sum_{k=0}^{p} a[k]y[n-k] \right)^2$$

where $y[n]$ is obtained by filtering $x[n]$ (which is nonzero over $[0, N-1]$) with the causal filter $1/\mathcal{B}(z)$. Although $y[n]$ will be nonzero for $0 \le n < \infty$, we choose to ignore $y[n]$ for $n > N-1$ by setting it equal to zero. This approximation will be a good one for large N. Thus, $y(n)$ is nonzero only over $[0, N-1]$.

(b) Then show that the minimum value of $(1/N)S_2(\mathbf{a}, \mathbf{b})$ when minimized over $\mathbf{a}$ is

$$\frac{1}{N} S_2(\hat{\mathbf{a}}, \mathbf{b}) = \hat{r}_{yy}[0] + \sum_{k=1}^{p} \hat{a}[k]\hat{r}_{yy}[k]$$

where

$$\hat{r}_{yy}[k] = \frac{1}{N} \sum_{n=0}^{N-1-k} y[n]y[n+k]$$

and $\hat{a}[k]$ is obtained by using the autocorrelation method of AR estimation. Note that $\hat{\mathbf{a}}$ will be a function of $\mathbf{b}$. For an ARMA(1, 1) process, what is $(1/N)S_2(\hat{a}[1], b[1])$? Can you minimize $(1/N)S_2(\hat{a}[1], b[1])$ over $b[1]$?

9.2. Verify (9.9) used to derive the Cramer–Rao bound of the ARMA parameters of a real ARMA(1, 1) process by making use of the definitions in Figure 9.1.

9.3. Compare the asymptotic variances of the MLE of the AR filter parameter for a real AR(1) process and that of the AR filter parameter for a real ARMA(1, 1) process. Prove that the variance for the AR parameter of the ARMA process is always larger and explain why.

9.4. Evaluate var $[\hat{P}_{\text{ARMA}}(f)]$ for the MLE spectral estimator of a real ARMA(1, 1) process. How does it compare to the variance of the MLE spectral estimator for a real AR(1) process if $b[1] = 0$?

9.5. Assuming $p \geq q$, show that

$$\mathcal{P}_{\text{ARMA}}(z) = \frac{\sigma^2 \mathcal{B}(z)\mathcal{B}^*(1/z^*)}{\mathcal{A}(z)\mathcal{A}^*(1/z^*)}$$

can be written as

$$\mathcal{P}_{\text{ARMA}}(z) = \frac{\mathcal{G}(z)}{\mathcal{A}(z)} + \frac{\mathcal{G}^*(1/z^*)}{\mathcal{A}^*(1/z^*)}$$

where $\mathcal{G}(z) = \sum_{n=0}^{p} g[n]z^{-n}$ by finding $g[n]$ in terms of the AR parameters and the ACF of $x[n]$. Assuming that the AR parameters are known, how could the PSD be estimated based on this representation?

9.6. Consider an ARMA(1, 1) process. Prove that

$$\mathbf{R}'_{xx} = \begin{bmatrix} r_{xx}[2] & r_{xx}[1] \\ r_{xx}[3] & r_{xx}[2] \end{bmatrix}$$

is a singular matrix.

9.7. A complex ARMA(1, 1) process has its pole at $z = r_p \exp(j2\pi f_0)$ and its zero at $z = r_z \exp(j2\pi f_0)$, where $r_z < r_p$. Plot the PSD for $r_p = 0.999$, $r_z = 0.99$, and $\sigma^2 = 1$.

APPENDIX 9A

Derivation of the Cramer-Rao Bounds for ARMA Parameter Estimators

The CR bound that is derived assumes a real ARMA(p, q) process. It is approximate in that the PDF is replaced by the conditional PDF (9.3) and in the exponent of the conditional PDF (9.6) is used. For large data records the approximation will be a good one. The Fisher information matrix $\mathbf{I}_\theta$ of dimension $(p + q + 1) \times (p + q + 1)$ is given by (see Section 3.3.1)

$$[\mathbf{I}_\theta]_{ij} = -\mathcal{E}\left[\frac{\partial^2 \ln p(\mathbf{x}; \mathbf{a}, \mathbf{b}, \sigma^2)}{\partial\theta_i\, \partial\theta_j}\right]$$

where $\boldsymbol{\theta} = [\mathbf{a}^T \;\; \mathbf{b}^T \;\; \sigma^2]^T$. If the exponent in (9.3) is replaced by the approximate expression given by (9.6), then for $N \gg L$ the likelihood function becomes

$$p(\mathbf{x}; \mathbf{a}, \mathbf{b}, \sigma^2) = \frac{1}{(2\pi\sigma^2)^{N/2}} \exp\left[-\frac{N}{2\sigma^2}\int_{-1/2}^{1/2} I(f)\frac{|A(f)|^2}{|B(f)|^2}df\right]$$

where $I(f)$ is the periodogram. The first order partial derivatives with respect to the AR parameters are

$$\frac{\partial \ln p(\mathbf{x};\mathbf{a},\mathbf{b},\sigma^2)}{\partial a[k]} = -\frac{N}{2\sigma^2}\int_{-1/2}^{1/2} I(f)\frac{A(f)\exp(j2\pi fk) + A^*(f)\exp(-j2\pi fk)}{|B(f)|^2}df.$$

Since $A^*(f) = A(-f)$, this reduces to

$$\frac{\partial \ln p(\mathbf{x};\mathbf{a},\mathbf{b},\sigma^2)}{\partial a[k]} = -\frac{N}{2\sigma^2}\int_{-1/2}^{1/2} I(f)\frac{A(f)\exp(j2\pi fk) + A(-f)\exp(-j2\pi fk)}{|B(f)|^2}df$$

and because $I(-f) = I(f)$, $|B(-f)| = |B(f)|$, it follows that

$$\frac{\partial \ln p(\mathbf{x};\mathbf{a},\mathbf{b},\sigma^2)}{\partial a[k]} = -\frac{N}{\sigma^2}\int_{-1/2}^{1/2} I(f)\frac{A(f)\exp(j2\pi fk)}{|B(f)|^2}df.$$

Similarly, it can be shown that

$$\frac{\partial \ln p(\mathbf{x};\mathbf{a},\mathbf{b},\sigma^2)}{\partial b[l]} = \frac{N}{\sigma^2}\int_{-1/2}^{1/2} I(f)\frac{|A(f)|^2 B(f)\exp(j2\pi fl)}{|B(f)|^4}df$$

and

$$\frac{\partial \ln p(\mathbf{x};\mathbf{a},\mathbf{b},\sigma^2)}{\partial \sigma^2} = -\frac{N}{2\sigma^2} + \frac{1}{2\sigma^4}S_2(\mathbf{a},\mathbf{b})$$

where

$$S_2(\mathbf{a},\mathbf{b}) = N\int_{-1/2}^{1/2} I(f)\frac{|A(f)|^2}{|B(f)|^2}df.$$

The second order partial derivatives are

$$\frac{\partial^2 \ln p(\mathbf{x};\mathbf{a},\mathbf{b},\sigma^2)}{\partial a[k]\,\partial a[l]} = -\frac{N}{\sigma^2}\int_{-1/2}^{1/2} I(f)\frac{\exp[j2\pi f(k-l)]}{|B(f)|^2}df$$

$$\frac{\partial^2 \ln p(\mathbf{x};\mathbf{a},\mathbf{b},\sigma^2)}{\partial b[l]\,\partial b[k]} = -\frac{N}{\sigma^2}\int_{-1/2}^{1/2} I(f)\frac{|A(f)|^2\exp(j2\pi fl)}{|B(f)|^4 B^*(f)^2}$$

$$\left\{|B(f)|^2\exp(j2\pi fk) + B^*(f)[B(f)\exp(j2\pi fk) + B^*(f)\exp(-j2\pi fk)]\right\}df$$

$$= -\frac{2N}{\sigma^2}\int_{-1/2}^{1/2} I(f)\frac{|A(f)|^2\exp[j2\pi f(k+l)]}{|B(f)|^2 B^*(f)^2}df$$

$$-\frac{N}{\sigma^2}\int_{-1/2}^{1/2} I(f)\frac{|A(f)|^2\exp[j2\pi f(l-k)]}{|B(f)|^4}df$$

and

$$\frac{\partial^2 \ln p(\mathbf{x}; \mathbf{a}, \mathbf{b}, \sigma^2)}{\partial \sigma^2 \, \partial \sigma^2} = \frac{N}{2\sigma^4} - \frac{1}{\sigma^6} S_2(\mathbf{a}, \mathbf{b})$$

$$\frac{\partial^2 \ln p(\mathbf{x}; \mathbf{a}, \mathbf{b}, \sigma^2)}{\partial a[k] \, \partial b[l]} = \frac{N}{\sigma^2} \int_{-1/2}^{1/2} I(f)$$

$$\left\{ \frac{A(f)[B^*(f) \exp(-j2\pi f l) + B(f) \exp(j2\pi f l)] \exp(j2\pi f k)}{|B(f)|^4} \right\} df$$

$$\frac{\partial^2 \ln p(\mathbf{x}; \mathbf{a}, \mathbf{b}, \sigma^2)}{\partial a[k] \, \partial \sigma^2} = \frac{N}{\sigma^4} \int_{-1/2}^{1/2} I(f) \frac{A(f) \exp(j2\pi f k)}{|B(f)|^2} df$$

$$\frac{\partial^2 \ln p(\mathbf{x}; \mathbf{a}, \mathbf{b}, \sigma^2)}{\partial b[l] \, \partial \sigma^2} = -\frac{N}{\sigma^4} \int_{-1/2}^{1/2} I(f) \frac{|A(f)|^2 B(f) \exp(j2\pi f l)}{|B(f)|^4} df.$$

The expected value of all the second order partials is now taken. Since for large N (see Section 4.3)

$$\mathscr{E}[I(f)] = P_{\text{ARMA}}(f) = \frac{\sigma^2 |B(f)|^2}{|A(f)|^2}$$

by the convergence of the average periodogram to the true PSD, then

$$\mathscr{E}\left[\frac{\partial^2 \ln p(\mathbf{x}; \mathbf{a}, \mathbf{b}, \sigma^2)}{\partial a[k] \, \partial a[l]}\right] = -N \int_{-1/2}^{1/2} \frac{\exp[j2\pi f(k - l)]}{|A(f)|^2} df$$

$$\mathscr{E}\left[\frac{\partial^2 \ln p(\mathbf{x}; \mathbf{a}, \mathbf{b}, \sigma^2)}{\partial b[l] \, \partial b[k]}\right] = -2N \int_{-1/2}^{1/2} \frac{\exp[j2\pi f(k + l)]}{B^*(f)^2} df$$

$$- N \int_{-1/2}^{1/2} \frac{\exp[j2\pi f(l - k)]}{|B(f)|^2} df$$

$$= -2N \mathscr{Z}^{-1}\left[\frac{1}{\mathscr{B}^2(z^{-1})}\right]\Bigg|_{n=k+l} - N \int_{-1/2}^{1/2} \frac{\exp[j2\pi f(l - k)]}{|B(f)|^2} df.$$

But $\mathscr{Z}^{-1}[1/\mathscr{B}^2(z^{-1})]$ is an anticausal sequence, so that for $n = k + l > 0$ the sequence is zero. Since $1 \le k, l \le q$, the first term in the preceding expression is zero and

$$\mathscr{E}\left[\frac{\partial^2 \ln p(\mathbf{x}; \mathbf{a}, \mathbf{b}, \sigma^2)}{\partial b[l] \, \partial b[k]}\right] = -N \int_{-1/2}^{1/2} \frac{\exp[j2\pi f(l - k)]}{|B(f)|^2} df$$

$$\mathscr{E}\left[\frac{\partial^2 \ln p(\mathbf{x}; \mathbf{a}, \mathbf{b}, \sigma^2)}{\partial a[k] \, \partial b[l]}\right] = N \int_{-1/2}^{1/2} \frac{\exp[j2\pi f(k - l)]}{A^*(f)B(f)} df$$

$$+ N \int_{-1/2}^{1/2} \frac{\exp[j2\pi f(k + l)]}{A^*(f)B^*(f)} df$$

$$= N \int_{-1/2}^{1/2} \frac{\exp[j2\pi f(k - l)}{A^*(f)B(f)} df,$$

the second term in the expression being zero due to the anticausality of the sequence. Similarly,

$$\mathcal{E}\left[\frac{\partial^2 \ln p(\mathbf{x}; \mathbf{a}, \mathbf{b}, \sigma^2)}{\partial a[k]\, \partial \sigma^2}\right] = \frac{N}{\sigma^2}\int_{-1/2}^{1/2} \frac{\exp(j2\pi f k)}{A^*(f)}\, df = 0$$

$$\mathcal{E}\left[\frac{\partial^2 \ln p(\mathbf{x}; \mathbf{a}, \mathbf{b}, \sigma^2)}{\partial b[l]\, \partial \sigma^2}\right] = -\frac{N}{\sigma^2}\int_{-1/2}^{1/2} \frac{\exp(j2\pi f l)}{B^*(f)}\, df = 0$$

$$\mathcal{E}\left[\frac{\partial^2 \ln p(\mathbf{x}; \mathbf{a}, \mathbf{b}, \sigma^2)}{\partial \sigma^2\, \partial \sigma^2}\right] = -\frac{N}{2\sigma^4}.$$

To simplify the results two processes as shown in Figure 9.1 are defined. $y[n]$ is an AR(p) process with parameters $\{a[1], a[2], \ldots, a[p], \sigma^2\}$ and $z[n]$ is an AR(q) process with parameters $\{b[1], b[2], \ldots, b[q], \sigma^2\}$. Both processes are driven by the same white noise process $u[n]$ with variance σ^2. Letting $r_{yy}[k]$, $r_{zz}[k]$, and $r_{yz}[k]$ denote the ACF and CCF, respectively,

$$-\mathcal{E}\left[\frac{\partial^2 \ln p(\mathbf{x}; \mathbf{a}, \mathbf{b}, \sigma^2)}{\partial a[k]\, \partial a[l]}\right] = \frac{N}{\sigma^2} r_{yy}[k-l]$$

$$-\mathcal{E}\left[\frac{\partial^2 \ln p(\mathbf{x}; \mathbf{a}, \mathbf{b}, \sigma^2)}{\partial b[k]\, \partial b[l]}\right] = \frac{N}{\sigma^2} r_{zz}[k-l]$$

$$-\mathcal{E}\left[\frac{\partial^2 \ln p(\mathbf{x}; \mathbf{a}, \mathbf{b}, \sigma^2)}{\partial a[k]\, \partial b[l]}\right] = -\frac{N}{\sigma^2} r_{yz}[k-l].$$

Finally, the Fisher information matrix is given by

$$\mathbf{I}_\theta = \frac{N}{\sigma^2}\begin{bmatrix} \mathbf{R}_{yy} & -\mathbf{R}_{yz} & \mathbf{0} \\ -\mathbf{R}_{yz}^T & \mathbf{R}_{zz} & \mathbf{0} \\ \mathbf{0}^T & \mathbf{0}^T & \dfrac{1}{2\sigma^2} \end{bmatrix}$$

where

$$[\mathbf{R}_{yy}]_{ij} = r_{yy}[i-j] \qquad i = 1, 2, \ldots, p; \quad j = 1, 2, \ldots, p$$

$$[\mathbf{R}_{yz}]_{ij} = r_{yz}[i-j] \qquad i = 1, 2, \ldots, p; \quad j = 1, 2, \ldots, q$$

$$[\mathbf{R}_{zz}]_{ij} = r_{zz}[i-j] \qquad i = 1, 2, \ldots, q; \quad j = 1, 2, \ldots, q.$$

Chapter 10

Autoregressive Moving Average Spectral Estimation: Methods

10.1 INTRODUCTION

ARMA parameter estimation via maximum likelihood techniques has been investigated thoroughly by statisticians. As discussed in Chapter 9, the MLE requires that we minimize a highly nonlinear function even with many simplifying assumptions. For this reason the MLE has not proven to be successful in practice, although much research is ongoing to devise computationally efficient algorithms to implement the MLE. As an alternative, many suboptimal but easily implementable algorithms have been proposed. These are based for the most part on equation error modeling approaches, in which a theoretical relationship between the unknown ARMA parameters and the ACF is first established and then by substituting estimates of the ACF into the equation and solving, ARMA parameter estimates may be found. In the statistical literature this method is referred to as the method of moments. It is known that in general the estimators derived in this manner do not attain the CR bound, hence they are suboptimal. In this chapter some specific ARMA estimators are described. Each estimator to be discussed is a typical algorithm within a large class of similar algorithms. For example, the Akaike estimator is typical of the ARMA estimator based on an approximate maximization of the likelihood function. It therefore displays the advantages and disadvantages of algorithms within that class. The particular choice of the Akaike algorithm is not meant to imply that the estimation performance is superior to all other existing estimators within that class. On the contrary, it is expected that the performance is typical for an estimator within that class. The interested reader should consult the references for the proposed variations on the estimators to be described.

10.2 SUMMARY

Section 10.3 contains a description of the Akaike estimator. Based on an approximate maximization of the likelihood function (see Chapter 9) it attempts to minimize the function given by (10.1) over the ARMA filter parameters. The optimization method used is the Newton–Raphson iteration. The estimator can be used only for real data. A computer subroutine entitled AKAIKE is given in Appendix 10C, which implements the algorithm. The use of the modified Yule–Walker equations as described in Chapter 5 is discussed in Section 10.4 for ARMA estimation. The AR parameters may be estimated first by solving the set of linear equations given by (10.7) using the recursive algorithm of (10.8)–(10.13). Once the AR parameter estimates have been obtained, the data are filtered with the estimate of $\mathcal{A}(z)$ to generate an approximate MA process. Then, any of the techniques described in Chapter 8 can be employed to estimate the MA parameters. A computer subroutine entitled MYWE, which is listed in Appendix 10D, implements the estimator with Durbin's method used to estimate the MA parameters. Some of the statistical properties of the estimator are discussed, a salient point being that the model orders should not be chosen to be larger than the true ones. In Section 10.5 an extension of the modified Yule–Walker equation estimator is described. The modified Yule–Walker equations are used with more equations than unknowns, resulting in a least squares solution. This estimator also first finds the AR filter parameter estimates and then uses the filtered sequence to determine the MA parameters. The AR parameters are estimated from (10.18). This modification has been reported to yield higher resolution spectral estimates. A key decision in the use of this approach is the determination of the number of equations. Few guidelines, unfortunately, are available. A computer subroutine entitled LSMYWE which implements this method is contained in Appendix 10E. Again the Durbin MA estimator is used. Next estimators based on input–output identification techniques are examined in Section 10.6. The unobservable white noise excitation is estimated so that standard techniques in identification of linear systems from input and output data can be applied to the ARMA problem. A two-stage least squares estimator is given by (10.20), while a modified approach which has been claimed to improve the estimator is detailed in (10.23)–(10.27). The latter estimator employs three stages of least squares. A computer subroutine implementation of the three-stage least squares estimator entitled MAYNEFIR is given in Appendix 10F. In Section 10.7 we compare the various estimators by computer simulation. The results indicate that the estimator based on the modified Yule–Walker equations performs poorest, the approximate MLE performs best for ARMA processes with poles sufficiently displaced from the unit circle, and the estimator based on solving an overdetermined set of Yule–Walker equations performs best for processes with poles near the unit circle.

The application of the ARMA spectral estimation methods to the test case data described in Section 1.5 produces the spectral estimates shown in Figure 10.1. The computer programs AKAIKE (Appendix 10C), MYWE (Appendix 10D), LSMYWE (Appendix 10E), and MAYNEFIR (Appendix 10F) were used to obtain the results. For all the methods model orders of $p = 7$ and $q = 3$ were chosen. The AKAIKE method, which is valid only for real data, used the real data set, while the remaining methods were applied to the complex test data. The results are discussed in Chapter 12.

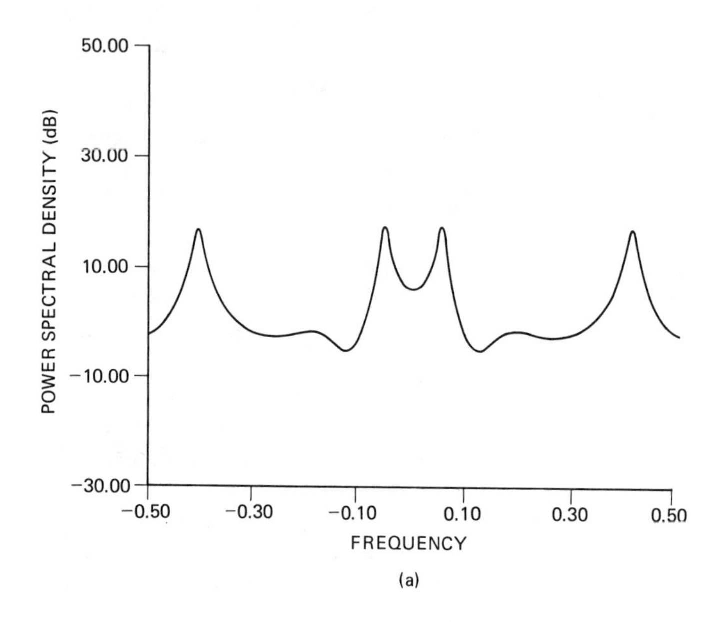

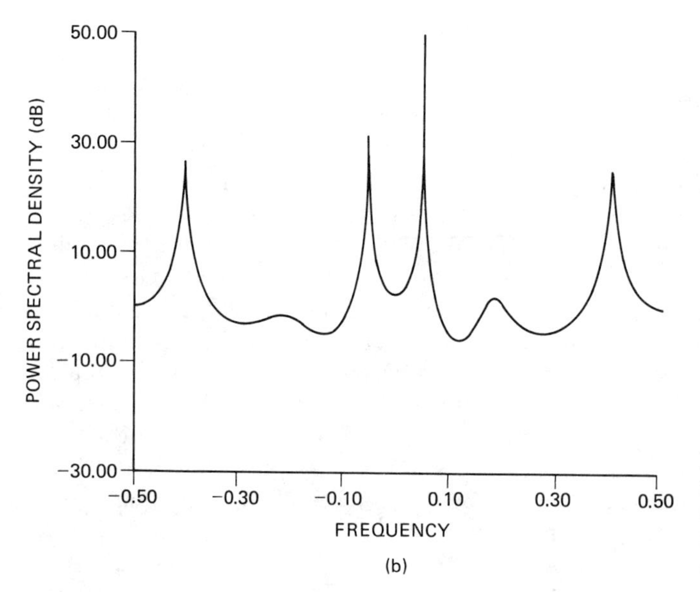

Figure 10.1 ARMA spectral estimates for test case data. (a) Akaike MLE. (b) MYWE ($L = 6$). (c) LSMYWE ($M = 15$, $L = 6$). (d) Mayne–Firoozan ($L = 21$).

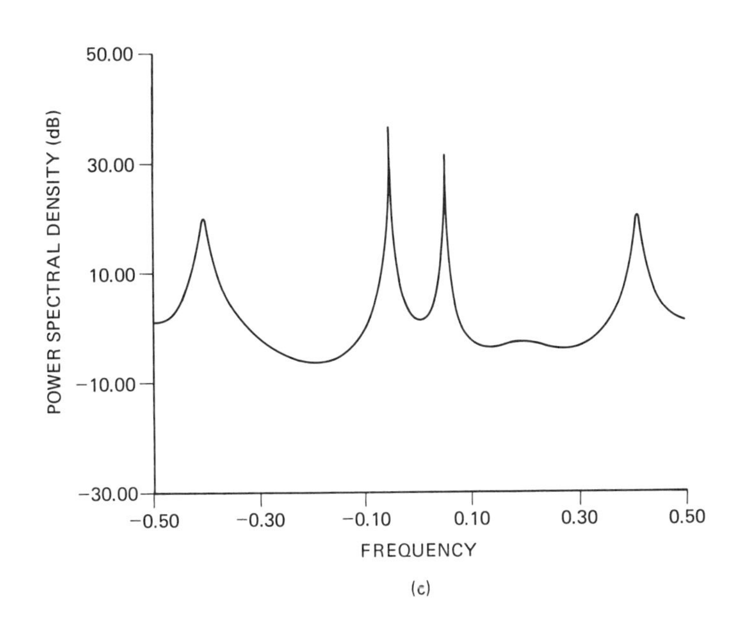

(c)

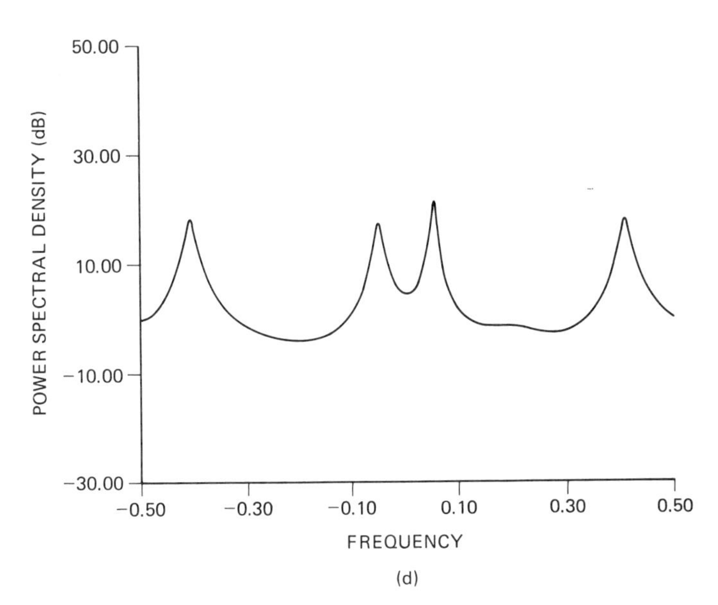

(d)

Figure 10.1 *(Continued)*

10.3 AKAIKE APPROXIMATE MLE

It was shown in Chapter 9 that the approximate MLE of the parameters of a real ARMA process can be determined as the minimum of a highly nonlinear function. In this section a Newton–Raphson iteration is employed to minimize this function. This approach, which was originally proposed by Akaike [1973], is, like all non-linear optimization schemes, iterative in nature and therefore not guaranteed to converge. If convergence does occur, the minimum found may not be the global minimum. It is important to begin the iteration with an estimate that is close to the true parameter value, so that hopefully the global minimum will be found (see Problem 10.1). For large data records local minima are not a problem [Astrom and Soderstrom 1974], in that the log-likelihood function is approximately quadratic in the ARMA parameters and therefore characterized by a single minimum. The simulation of the Akaike algorithm described in Section 10.7 illustrates the good performance obtained.

Recall from Chapter 9 that the approximate MLE of σ^2 is found from (9.7). The approximate MLE of the ARMA filter parameters is obtained as the values that minimize [see (9.6)]

$$Q(\mathbf{a}, \mathbf{b}) = \frac{1}{N} S_2(\mathbf{a}, \mathbf{b}) = \int_{-1/2}^{1/2} I(f) \frac{|A(f)|^2}{|B(f)|^2} df. \tag{10.1}$$

Akaike proposed using a Newton–Raphson iteration to find a zero of $[(\partial Q/\partial \mathbf{a})^T \; (\partial Q/\partial \mathbf{b})^T]^T$, or

$$\begin{bmatrix} \mathbf{a}_{k+1} \\ \mathbf{b}_{k+1} \end{bmatrix} = \begin{bmatrix} \mathbf{a}_k \\ \mathbf{b}_k \end{bmatrix} - \mathbf{H}^{-1}(\mathbf{a}_k, \mathbf{b}_k) \begin{bmatrix} \dfrac{\partial Q}{\partial \mathbf{a}} \\ \\ \dfrac{\partial Q}{\partial \mathbf{b}} \end{bmatrix} \Bigg|_{\substack{\mathbf{a}=\mathbf{a}_k \\ \mathbf{b}=\mathbf{b}_k}} \tag{10.2}$$

$\mathbf{a}_k$, $\mathbf{b}_k$ are the kth iterates of the AR and MA filter parameter vectors, respectively. $\mathbf{H}(\mathbf{a}, \mathbf{b})$ is the Hessian of Q, which is defined as

$$\mathbf{H}(\mathbf{a}, \mathbf{b}) = \begin{bmatrix} \dfrac{\partial^2 Q}{\partial \mathbf{a}\, \partial \mathbf{a}^T} & \dfrac{\partial^2 Q}{\partial \mathbf{a}\, \partial \mathbf{b}^T} \\ \\ \dfrac{\partial^2 Q}{\partial \mathbf{b}\, \partial \mathbf{a}^T} & \dfrac{\partial^2 Q}{\partial \mathbf{b}\, \partial \mathbf{b}^T} \end{bmatrix} = \begin{bmatrix} p \times p & p \times q \\ q \times p & q \times q \end{bmatrix}.$$

In Appendix 10A the required partial derivatives are shown to be approximately

$$\begin{aligned} \frac{\partial Q}{\partial a[k]} &= 2 \sum_{i=0}^{p} a[i] \hat{r}_{yy}[k-i] \qquad && k = 1, 2, \ldots, p \\ \frac{\partial Q}{\partial b[l]} &= -2 \sum_{i=0}^{q} b[i] \hat{r}_{zz}[l-i] \qquad && l = 1, 2, \ldots, q \end{aligned} \tag{10.3}$$

$$\begin{aligned} \frac{\partial^2 Q}{\partial a[k]\, \partial a[l]} &= 2\hat{r}_{yy}[k-l] \qquad && \begin{matrix} k = 1, 2, \ldots, p \\ l = 1, 2, \ldots, p \end{matrix} \\ \frac{\partial^2 Q}{\partial b[k]\, \partial b[l]} &= 2\hat{r}_{zz}[k-l] \qquad && \begin{matrix} k = 1, 2, \ldots, q \\ l = 1, 2, \ldots, q \end{matrix} \\ \frac{\partial^2 Q}{\partial a[k]\, \partial b[l]} &= -2\hat{r}_{yz}[k-l] \qquad && \begin{matrix} k = 1, 2, \ldots, p \\ l = 1, 2, \ldots, q \end{matrix} \end{aligned} \tag{10.4}$$

where

$$\hat{r}_{yy}[k] = \frac{1}{N} \sum_{k=0}^{N-|k|-1} y[n]y[n + | k |]$$

$$\hat{r}_{zz}[k] = \frac{1}{N} \sum_{n=0}^{N-|k|-1} z[n]z[n + | k |]$$

and

$$\hat{r}_{yz}[k] = \begin{cases} \dfrac{1}{N} \displaystyle\sum_{n=0}^{N-k-1} y[n]z[n + k] & \text{for } k \geq 0 \\ \dfrac{1}{N} \displaystyle\sum_{n=-k}^{N-1} y[n]z[n + k] & \text{for } k < 0. \end{cases}$$

The sequences $y[n]$, $z[n]$ are defined as

$$\begin{aligned} y[n] &= \mathcal{Z}^{-1}\left\{\frac{\mathcal{X}(z)}{\mathcal{B}(z)}\right\} \\ z[n] &= \mathcal{Z}^{-1}\left\{\frac{\mathcal{X}(z)\mathcal{A}(z)}{\mathcal{B}^2(z)}\right\} \end{aligned} \tag{10.5}$$

where $\mathcal{X}(z) = \sum_{n=0}^{N-1} x[n]z^{-n}$. (See Appendix 10B for the matrix form of the Hessian.) It is of interest to observe that the $y[n]$, $z[n]$ sequences are estimates of the processes depicted in Figure 9.1 which arose in the derivation of the CR bound. Of course, this is not purely coincidence but can be shown to be a property of MLEs. Since the $y[n]$, $z[n]$ sequences are generated as the outputs of recursive filters, the initial conditions need to be specified. Akaike's approach sets these initial conditions equal to zero on the premise that for large data records any transient introduced will be negligible. Clearly, this will not be the case when the zeros are near the unit circle since then the impulse response will be long.

The Akaike estimator may not yield minimum-phase filter estimates during the course of the iteration. If any iterate of the MA parameters causes $\mathcal{B}(z)$ to have a zero outside the unit circle, then due to the instability of $1/\mathcal{B}(z)$, the $y[n]$, $z[n]$ sequences will grow large. We must therefore monitor the stability of the $1/\mathcal{B}(z)$ filter. An approach to this problem would be to replace any zero outside the unit circle, say z_i, by its conjugate reciprocal or $1/z_i^*$. However, a non-minimum-phase filter would appear to be a deficiency of the algorithm, so that any ad hoc measure might lead to questionable results. Note that without the minimum-phase constraint it is possible to drive $Q(\mathbf{a}, \mathbf{b})$ to zero by making $\mathcal{B}(z)$ arbitrarily large. Assuming that $\hat{\mathcal{B}}(z)$ is minimum-phase, an alternative means of computing $\hat{\sigma}^2$, rather than to use $Q(\hat{\mathbf{a}}, \hat{\mathbf{b}})$, is to use the approximately equivalent expression

$$\hat{\sigma}^2 = \frac{1}{N} \sum_{n=0}^{N-1} \hat{u}^2[n]$$

where

$$\hat{u}[n] = \mathcal{Z}^{-1}\left\{\frac{\mathcal{X}(z)\hat{\mathcal{A}}(z)}{\hat{\mathcal{B}}(z)}\right\}$$

and the initial conditions of the recursive filter are arbitrarily assumed to be zero.

In computing the new iterate of the ARMA parameters as per (10.2), we can avoid the inversion of the Hessian by rewriting the equations as

$$\mathbf{H}(\mathbf{a}_k, \mathbf{b}_k)\begin{bmatrix} \mathbf{a}_{k+1} \\ \mathbf{b}_{k+1} \end{bmatrix} = \mathbf{H}(\mathbf{a}_k, \mathbf{b}_k)\begin{bmatrix} \mathbf{a}_k \\ \mathbf{b}_k \end{bmatrix} - \left.\begin{bmatrix} \dfrac{\partial Q}{\partial \mathbf{a}} \\ \\ \dfrac{\partial Q}{\partial \mathbf{b}} \end{bmatrix}\right|_{\substack{\mathbf{a}=\mathbf{a}_k \\ \mathbf{b}=\mathbf{b}_k}} \tag{10.6}$$

and solving a set of simultaneous linear equations for the new iterate. The Hessian is assured to be invertible since it is positive definite, as shown in Appendix 10B. A Cholesky decomposition [see (2.53)–(2.55)] can be used to solve (10.6). The computer subroutine entitled AKAIKE, which is given in Appendix 10C, implements the entire algorithm just described. As mentioned previously, for good results it is necessary to provide a good set of initial estimates of the ARMA filter parameters. Any of the techniques described in Sections 10.4 to 10.6 can be used for this purpose.

The extension of the Akaike method presented here to complex data has not appeared in the literature, although a similar algorithm that is applicable to complex data has been presented by Ooe [1978]. A closely related ARMA estimator is due to Hannan [1969], which motivated Akaike's work. Also, the interested reader should see the paper by Kohn [1977] for comments on the Akaike and Hannan algorithms. Other iterative ARMA estimators can be found in the works of Durbin [1960A], Walker [1962], Tretter and Stieglitz [1967], Box and Jenkins [1970], Murthy and Kronauer [1973], Anderson [1977], Gutowski et al. [1978], Tuan [1979], Upadhyaya [1979], and the references in Chapter 9.

10.4 MODIFIED YULE–WALKER EQUATIONS

The ARMA estimation methods described in this section and the next are ad hoc in nature. They have arisen from the difficulties associated with the highly nonlinear MLE. Unlike the iterative techniques these methods are direct, relying on the modified Yule–Walker equations, but suboptimal. They do have the advantage that they are computationally simple.

The first approach uses the modified Yule–Walker equations (MYWE) described in Chapter 5. Since these relationships hold when the ACF is known exactly, a reasonable approach is to replace the theoretical ACF samples by estimates and then solve the equations for the AR filter parameters. The MA parameters are subsequently found in a separate step. This leads to the following estimator for the AR filter parameters:

$$\underbrace{\begin{bmatrix} \hat{r}_{xx}[q] & \hat{r}_{xx}[q-1] & \cdots & \hat{r}_{xx}[q-p+1] \\ \hat{r}_{xx}[q+1] & \hat{r}_{xx}[q] & \cdots & \hat{r}_{xx}[q-p+2] \\ \vdots & \vdots & \ddots & \vdots \\ \hat{r}_{xx}[q+p-1] & \hat{r}_{xx}[q+p-2] & \cdots & \hat{r}_{xx}[q] \end{bmatrix}}_{\hat{\mathbf{R}}'_{xx}} \begin{bmatrix} \hat{a}[1] \\ \hat{a}[2] \\ \vdots \\ \hat{a}[p] \end{bmatrix} = -\begin{bmatrix} \hat{r}_{xx}[q+1] \\ \hat{r}_{xx}[q+2] \\ \vdots \\ \hat{r}_{xx}[q+p] \end{bmatrix}. \tag{10.7}$$

The ACF estimator may be either the biased or unbiased estimator. In general, $\hat{\mathbf{a}}$ will not be minimum-phase (see Problem 10.2). Note that the matrix is Toeplitz since the elements along any NW to SE diagonal are the same, although not hermitian. Also, the matrix is not guaranteed to be nonsingular. Once the AR parameters have been estimated and $x[n]$ filtered by $\hat{\mathcal{A}}(z)$ to produce an approximate MA process, any of the methods of Chapter 8 may be used to estimate the MA parameters.

The MYWE can be solved in an efficient manner using an extension of the Levinson recursion. The extension is implicit in the work of Trench [1964], who showed how to invert a nonhermitian Toeplitz matrix recursively. The computational complexity is $O(p^2)$ operations. The recursive algorithm is initialized by

$$\begin{aligned} a_1[1] &= -\frac{r_{xx}[q+1]}{r_{xx}[q]} \\ b_1[1] &= -\frac{r_{xx}[q-1]}{r_{xx}[q]} \\ \rho_1 &= (1 - a_1[1]b_1[1])r_{xx}[q] \end{aligned} \tag{10.8}$$

with the recursion for $k = 2, 3, \ldots, p$ given by

$$a_k[k] = -\frac{r_{xx}[q+k] + \sum_{l=1}^{k-1} a_{k-1}[l]r_{xx}[q+k-l]}{\rho_{k-1}} \tag{10.9}$$

$$a_k[i] = a_{k-1}[i] + a_k[k]b_{k-1}[k-i] \qquad i = 1, 2, \ldots, k-1. \tag{10.10}$$

If $k = p$, exit; if not, continue.

$$b_k[k] = -\frac{r_{xx}[q-k] + \sum_{l=1}^{k-1} b_{k-1}[l]r_{xx}[q-k-l]}{\rho_{k-1}} \tag{10.11}$$

$$b_k[i] = b_{k-1}[i] + b_k[k]a_{k-1}[k-i] \qquad i = 1, 2, \ldots, k-1 \tag{10.12}$$

$$\rho_k = (1 - a_k[k]b_k[k])\rho_{k-1} \tag{10.13}$$

The solution is $a[k] = a_p[k]$, $k = 1, 2, \ldots, p$. It is interesting to note that if $q = 0$ so that the MYWE reduce to the Yule–Walker equations, the algorithm reduces to the Levinson recursion [see (6.46)–(6.48)]. In this case, $b_k[i] = a_k^*[i]$, making the computation of $b_1[1]$ and the recursions (10.11) and (10.12) redundant. Upon examination of (10.9) and (10.11), it is apparent that for the solution to exist it is required that $\rho_i \neq 0$ for $i = 0, 1, \ldots, p-1$, where $\rho_0 = r_{xx}[q]$. This is also obvious if we note that [Trench 1964]

$$\det(\mathbf{R}'_{xx}) = \prod_{i=0}^{p-1} \rho_i. \tag{10.14}$$

A computer subroutine entitled MYWE, which is given in Appendix 10D, implements this algorithm. It employs Durbin's method for MA parameter estimation.

The statistics of the AR filter parameter estimator obtained from the MYWE have been derived for large data records and for real Gaussian data by Gersch [1970]. He has shown that the estimator is asymptotically (as $N \to \infty$) unbiased or

$$\mathcal{E}[\hat{\mathbf{a}}] = \mathbf{a}$$

and that the covariance matrix is

$$\begin{aligned} \mathbf{C}_a &= \mathcal{E}[(\hat{\mathbf{a}} - \mathcal{E}(\hat{\mathbf{a}}))(\hat{\mathbf{a}} - \mathcal{E}(\hat{\mathbf{a}}))^T] \\ &= \frac{\sigma^2}{N-p-q} \mathbf{R}'^{-1}_{xx} \mathcal{R} \mathbf{R}'^{-T}_{xx} \end{aligned} \tag{10.15}$$

where $\mathbf{R}'_{xx}$ is given in (10.7) with the ACF estimates replaced by their true values and

$$\mathcal{R} = \sum_{k=-q}^{q} \left(1 - \frac{|k|}{N-p-q}\right) c_k \mathbf{R}_{xx}[k]$$

$$c_k = \sum_{i=0}^{q-|k|} b[i]b[i+|k|]$$

$$\mathbf{R}_{xx}[k] = \begin{bmatrix} r_{xx}[k] & r_{xx}[k-1] & \cdots & r_{xx}[k-p+1] \\ r_{xx}[k+1] & r_{xx}[k] & \cdots & r_{xx}[k-p] \\ \vdots & \vdots & \ddots & \vdots \\ r_{xx}[k+p-1] & r_{xx}[k+p-2] & \cdots & r_{xx}[k] \end{bmatrix}.$$

Note that in the pure AR case in which $q = 0$, it follows that

$$\mathcal{R} = c_0 \mathbf{R}_{xx}[0] = \mathbf{R}_{xx}$$

$$\mathbf{R}'_{xx} = \mathbf{R}_{xx}$$

so that (10.15) becomes

$$\mathbf{C}_a = \frac{\sigma^2}{N-p} \mathbf{R}_{xx}^{-1} \approx \frac{\sigma^2}{N} \mathbf{R}_{xx}^{-1}$$

which agrees with (6.97). Also, it can be shown that for a pure AR process the use of the MYWE, which involves higher order samples of the ACF, produces poorer estimates than those obtained using the usual Yule–Walker equations [Kay 1980].

The performance of the MYWE approach varies greatly. For some processes the estimates will be quite accurate, while for others they will be very poor. To understand this behavior, consider the example of a real ARMA(1, 1) process. The AR parameter estimator as given by (10.7) is

$$\hat{a}[1] = -\frac{\hat{r}_{xx}[2]}{\hat{r}_{xx}[1]}.$$

The variance is found from (10.15) as

$$\text{var}(\hat{a}[1]) = \frac{\sigma^2}{N-2}\,\frac{(1 + b^2[1])r_{xx}[0] + 2b[1]r_{xx}[1]}{r_{xx}^2[1]}.$$

Using (5.29) for the ACF of an ARMA(1, 1) process this becomes

$$\text{var}(\hat{a}[1]) = \frac{1}{N-2}\,\frac{1 + \left[\dfrac{(a[1] - b[1])^2}{(1 - a^2[1])}\right]^2 + \dfrac{2b^2[1]}{1 - a^2[1]}}{\dfrac{1}{1 - a^2[1]}\left[b[1] - a[1]\left(1 + \dfrac{(a[1] - b[1])^2}{1 - a^2[1]}\right)\right]^2}. \tag{10.16}$$

For a reasonably accurate estimate of $a[1]$ we might require the root-mean-square (RMS) error to be no greater than 0.1 or $\sqrt{\text{var}(\hat{a}[1])} \leq 0.1$. To meet this requirement assuming that $b[1] = 0.5$, the minimum number of samples versus $a[1]$ is plotted in Figure 10.2. It is observed that the statistical fluctuation will be highly

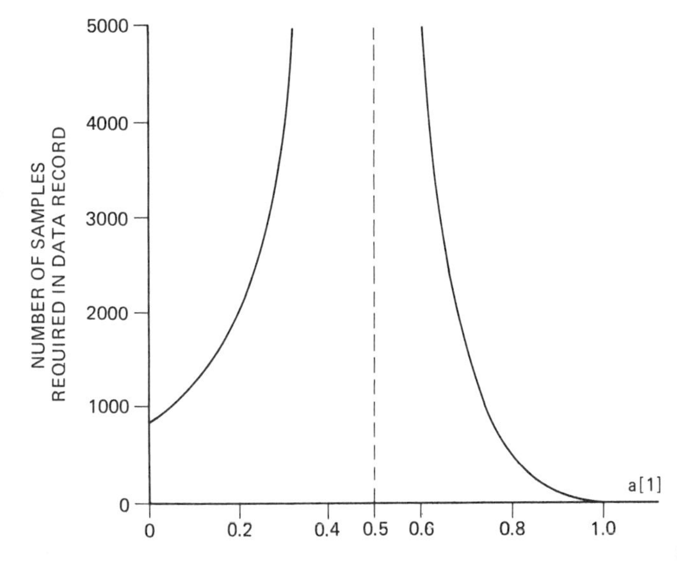

Figure 10.2 Required number of samples for accurate estimation of AR filter parameter of ARMA(1, 1) process when $b[1] = 0.5$ using MYWE. (© 1981 IEEE)

dependent on spectral shape, and in fact, as $a[1] \to 0.5$ the variability will be extremely large since the ARMA(1, 1) model is inappropriate. For $a[1] = 0.5$ the pole and zero cancel, producing a white noise process. Since for a white noise process $\mathcal{E}(\hat{r}_{xx}[1]) = 0$, the estimate given by $-\hat{r}_{xx}[2]/\hat{r}_{xx}[1]$ will be subject to large variations for small perturbations in $\hat{r}_{xx}[1]$. This example illustrates that care must be taken when using the MYWE approach, especially when the model orders are unknown. In general, if the dimensions of $\mathbf{R}'_{xx}$ are greater than the true AR model order and the assumed value for q is larger than the true value, the matrix will be singular (see Problem 9.6). The covariance matrix as given by (10.15) will then be infinite. See Problem 10.3 for a comparison of the MYWE method to the MLE and also Problem 10.4 for estimation of the AR parameters of a noise corrupted AR process.

The statistical properties of the spectral estimator based on the MYWE have been derived by Sakai and Tokumaru [1980]. The results indicate that large variabilities are to be expected for frequencies where the actual PSD is small. Some computer simulation examples of the performance of the MYWE ARMA estimator are provided in Section 10.7.

10.5 LEAST SQUARES MODIFIED YULE–WALKER EQUATIONS

In an attempt to reduce the variance of the MYWE estimator, Mehra [1971] has suggested utilizing more of the available equations. Later, Cadzow [1980] applied this idea to the spectral estimation problem. Since for an ARMA(p, q) process

$$r_{xx}[k] = -\sum_{l=1}^{p} a[l]r_{xx}[k-l] \qquad k \geq q+1 \tag{5.16}$$

the choice of the p equations corresponding to $k = q+1, q+2, \ldots, q+p$ in (10.7) is an arbitrary one. It can be shown that there is information in the ACF at higher order samples [Arato 1961]. To use this information, assume that the highest sample of the ACF that can be accurately estimated is $r_{xx}[M]$, and consider the following theoretical equations:

$$\begin{bmatrix} r_{xx}[q+1] \\ r_{xx}[q+2] \\ \vdots \\ r_{xx}[M] \end{bmatrix} = -\begin{bmatrix} r_{xx}[q] & r_{xx}[q-1] & \cdots & r_{xx}[q-p+1] \\ r_{xx}[q+1] & r_{xx}[q] & \cdots & r_{xx}[q-p+2] \\ \vdots & \vdots & \vdots & \vdots \\ r_{xx}[M-1] & r_{xx}[M-2] & \cdots & r_{xx}[M-p] \end{bmatrix} \begin{bmatrix} a[1] \\ a[2] \\ \vdots \\ a[p] \end{bmatrix}$$

or

$$\mathbf{r} = -\mathbf{R}\mathbf{a}.$$

$\mathbf{R}$ is of dimension $(M-q) \times p$. Assuming that $M - q > p$, there will be more equations than unknowns. When the theoretical ACF is replaced by an estimate, the equations will no longer be satisfied. To account for estimation errors in the ACF the equations should be expressed as

$$\hat{\mathbf{r}} = -\hat{\mathbf{R}}\mathbf{a} + \mathbf{e} \tag{10.17}$$

where $\hat{\mathbf{r}}$, $\hat{\mathbf{R}}$ correspond to the estimators of $\mathbf{r}$, $\mathbf{R}$ and the error term $\mathbf{e}$ is zero if $\hat{\mathbf{r}} = \mathbf{r}$, $\hat{\mathbf{R}} = \mathbf{R}$. It is recommended that the unbiased ACF estimator be used in (10.17) since then the average equation error is zero [Durbin 1960B], or

$$\mathscr{E}(\mathbf{e}) = \mathscr{E}(\hat{\mathbf{r}}) + \mathscr{E}(\hat{\mathbf{R}})\mathbf{a} = \mathbf{r} + \mathbf{R}\mathbf{a} = \mathbf{0}.$$

The form of (10.17) immediately suggests the use of a least squares (LS) estimator. It follows from (3.28) that the LS estimator of the AR parameters is

$$\hat{\mathbf{a}} = -(\hat{\mathbf{R}}^H\hat{\mathbf{R}})^{-1}\hat{\mathbf{R}}^H\hat{\mathbf{r}} \tag{10.18}$$

where

$$\hat{\mathbf{R}} = \begin{bmatrix} \hat{r}_{xx}[q] & \hat{r}_{xx}[q-1] & \cdots & \hat{r}_{xx}[q-p+1] \\ \hat{r}_{xx}[q+1] & \hat{r}_{xx}[q] & \cdots & \hat{r}_{xx}[q-p+2] \\ \vdots & \vdots & \vdots & \vdots \\ \hat{r}_{xx}[M-1] & \hat{r}_{xx}[M-2] & \cdots & \hat{r}_{xx}[M-p] \end{bmatrix}$$

$$\hat{\mathbf{r}} = [\hat{r}_{xx}[q+1] \quad \hat{r}_{xx}[q+2] \quad \cdots \quad \hat{r}_{xx}[M]]^T.$$

This estimator of the AR parameters of an ARMA process is sometimes referred to as an *equation error* modeling approach [Hsia 1977]. Henceforth it will be termed the least squares modified Yule–Walker equation (LSMYWE) estimator. It should be emphasized that *no optimality properties of the LS approach* [Graybill 1976] apply to this problem. This is because $\hat{\mathbf{R}}$ is not a constant matrix nor does $\mathbf{e}$ have the statistical properties necessary to claim optimality (see also Section 3.3.2). Because $\hat{\mathbf{R}}^H\hat{\mathbf{R}}$ is usually positive definite it is invertible by typical routines such as the Cholesky decomposition [see (2.53)–(2.55)] or a myriad of other techniques developed for LS problems [Lawson and Hanson 1974]. In general, $\hat{\mathbf{a}}$ will not be minimum-phase. A computer program subroutine entitled LSMYWE which implements this approach in conjunction with Durbin's method for estimation of the MA parameters is contained in Appendix 10E.

The set of equations given by (10.17)

$$\hat{r}_{xx}[n] = -\sum_{k=1}^{p} a[k]\hat{r}_{xx}[n-k] + e[n] \qquad n \geq q+1$$

are similar in structure to the AR time series model

$$x[n] = -\sum_{k=1}^{p} a[k]x[n-k] + u[n].$$

The LSMYWE estimator can therefore be interpreted as the implementation of the covariance method of linear prediction applied to the "data" sequence $\{\hat{r}_{xx}[q-p+1], \hat{r}_{xx}[q-p+2], \ldots, \hat{r}_{xx}[M]\}$. This suggests application of other AR techniques to ARMA estimators. As an example, see the paper by Friedlander and Porat [1984] for a description of model order selection based on this analogy.

The statistics of $\hat{\mathbf{a}}$ for real data and large data records have been derived by Stoica and Soderstrom [1983]. It has been shown that the estimator is asymptotically unbiased and has a variance which usually decreases as $M - q$, the number of equations, increases [Friedlander and Sharman 1985], as long as $N \gg M - q$. In some instances the CR bound is achieved. It should be cautioned, however, that the same problems encountered with the MYWE for too large an AR model order also apply to the LSMYWE approach. If the model orders chosen are larger than the true ones, $\mathbf{R}$ will not be full rank and hence $\hat{\mathbf{R}}^T\hat{\mathbf{R}}$ will tend to be singular. The asymptotic covariance matrix for the LSMYWE estimator, which is of the form $(\mathbf{R}^T\mathbf{R})^{-1}\mathbf{S}(\mathbf{R}^T\mathbf{R})^{-1}$ (see [Stoica and Soderstrom 1983]), will then be infinite. Nonetheless, the performance of the LSMYWE appears to be superior in general to that of the MYWE as long as the number of equations chosen is not too large. The optimal choice for the number of equations will be highly dependent on the PSD under consideration and so unfortunately is not known a priori. For finite data records no statistical results are available. Based on computer simulations it has been observed that the LSMYWE estimator improves resolution for sharply peaked processes [Kaveh and Bruzzone 1981]. The increased resolution is attributable to the inclusion of the higher order samples of the ACF. For broadband processes, however, the spectral estimate may be poorer in that the variance has been observed to increase. These results have been quantified by Bruzzone and Kaveh [1984], who described the performance of the LSMYWE as being highly dependent on the pole locations. The computer simulation results in Section 10.7 support this conclusion. Other simulation results are available in the works by Friedlander and Porat [1984] and Cadzow [1982]. Some analytical expressions describing the performance of the LSMYWE estimator for sinusoidal signals in white noise can be found in the paper by Israelevitz and Lim [1983]. See Problem 10.5 for a variation of this approach.

10.6 INPUT–OUTPUT IDENTIFICATION APPROACHES

10.6.1 Two-Stage Least Squares

A class of suboptimal ARMA estimation algorithms have been proposed which rely on estimation of the driving white noise $u[n]$. If $u[n]$ were known, we would have knowledge of the input as well as the output. Then the many estimators developed for system identification which require only the solution of linear equations could be used [Hsia 1977, Astrom and Eykhoff 1971]. Historically, it was Durbin [1960A] who first proposed the estimator to be described as the first step in an iterative procedure.

Specifically, if we examine (5.21), it becomes clear that the nonlinear nature of the Yule–Walker equations is due to the unknown cross-correlation between the input and output. Since $u[n]$ is unobservable, $r_{xu}[k]$ cannot be estimated. If, however, we knew $u[n]$, the ARMA parameters could be estimated as the solution of a set of linear equations. In practice, $u[n]$ is estimated from $x[n]$ as the output of a large order prediction error filter. To set up the necessary equations, rewrite the definition of an ARMA(p, q) process with an estimator of the excitation noise replacing $u[n]$ to yield

$$x[n] = -\sum_{k=1}^{p} a[k]x[n-k] + \sum_{k=1}^{q} b[k]\hat{u}[n-k] + \hat{u}[n] \qquad n = 0, 1, \ldots, N-1.$$

In matrix notation this can be written as

$$\mathbf{x} = \mathbf{H}\boldsymbol{\theta} + \hat{\mathbf{u}} \tag{10.19}$$

where

$$\mathbf{x} = [x[0] \quad x[1] \quad \cdots \quad x[N-1]]^T$$

$$\hat{\mathbf{u}} = [\hat{u}[0] \quad \hat{u}[1] \quad \cdots \quad \hat{u}[N-1]]^T$$

$$\boldsymbol{\theta} = [-a[1] \quad -a[2] \quad \cdots \quad -a[p] \quad b[1] \quad b[2] \quad \cdots \quad b[q]]^T$$

$$\mathbf{H} = \begin{bmatrix} x[-1] & x[-2] & \cdots & x[-p] & \hat{u}[-1] & \hat{u}[-2] & \cdots & \hat{u}[-q] \\ x[0] & x[-1] & \cdots & x[-p+1] & \hat{u}[0] & \hat{u}[-1] & \cdots & \hat{u}[-q+1] \\ \vdots & \vdots & \vdots & \vdots & \vdots & \vdots & \vdots & \vdots \\ x[N-2] & x[N-3] & \cdots & x[N-p-1] & \hat{u}[N-2] & \hat{u}[N-3] & \cdots & \hat{u}[N-q-1] \end{bmatrix}.$$

$\mathbf{H}$ has dimension $N \times (p+q)$. Any elements of $\mathbf{H}$ that are not known, such as $x[n]$, $\hat{u}[n]$ for $n < 0$, are set to zero. (10.19) has the structure, although not the properties, of the linear model as discussed in Section 3.3.2. The LS solution is

$$\hat{\boldsymbol{\theta}} = (\mathbf{H}^H\mathbf{H})^{-1}\mathbf{H}^H\mathbf{x}. \tag{10.20}$$

The correlation matrix $(1/N)\mathbf{H}^H\mathbf{H}$ has for its entries estimates of the ACFs $r_{xx}[k]$, $r_{uu}[k]$ and the CCFs $r_{xu}[k]$, $r_{ux}[k]$. As an example, for $p = q = 2$,

$$\mathbf{H} = \begin{bmatrix} x[-1] & x[-2] & \hat{u}[-1] & \hat{u}[-2] \\ x[0] & x[-1] & \hat{u}[0] & \hat{u}[-1] \\ \vdots & \vdots & \vdots & \vdots \\ x[N-2] & x[N-3] & \hat{u}[N-2] & \hat{u}[N-3] \end{bmatrix}$$

which results in

$$\frac{1}{N}\mathbf{H}^H\mathbf{H} = \frac{1}{N}$$

$$\begin{bmatrix} \sum_{n=-1}^{N-2} |x[n]|^2 & \sum_{n=-1}^{N-2} x^*[n]x[n-1] & \sum_{n=-1}^{N-2} x^*[n]\hat{u}[n] & \sum_{n=-1}^{N-2} x^*[n]\hat{u}[n-1] \\ \sum_{n=-2}^{N-3} x^*[n]x[n+1] & \sum_{n=-2}^{N-3} |x[n]|^2 & \underline{\sum_{n=-2}^{N-3} x^*[n]\hat{u}[n+1]} & \sum_{n=-2}^{N-3} x^*[n]\hat{u}[n] \\ \sum_{n=-1}^{N-2} \hat{u}^*[n]x[n] & \underline{\sum_{n=-1}^{N-2} \hat{u}^*[n]x[n-1]} & \sum_{n=-1}^{N-2} |\hat{u}[n]|^2 & \underline{\sum_{n=-1}^{N-2} \hat{u}^*[n]\hat{u}[n-1]} \\ \sum_{n=-2}^{N-3} \hat{u}^*[n]x[n+1] & \sum_{n=-2}^{N-3} \hat{u}^*[n]x[n] & \underline{\sum_{n=-2}^{N-3} \hat{u}^*[n]\hat{u}[n+1]} & \sum_{n=-2}^{N-3} |\hat{u}[n]|^2 \end{bmatrix}. \tag{10.21}$$

This approach does not utilize the additional information that $r_{xu}[k] = r_{ux}^*[-k] = 0$ for $k > 0$ and $r_{uu}[k] = 0$ for $k \neq 0$. If it did, the underlined elements in (10.21) would be zero (see also Problem 10.6). To estimate $u[n]$ we model $x[n]$ by a large order AR model and set $\hat{u}[n]$ equal to the prediction error time series. The estimator for σ^2 can be found once the ARMA filter parameters have been estimated as

$$\hat{\sigma}^2 = \frac{1}{N} \sum_{n=0}^{N-1} |\tilde{u}[n]|^2$$

where $\tilde{u}[n]$ is the estimate of $u[n]$ obtained by filtering $x[n]$ with $\hat{\mathcal{A}}(z)/\hat{\mathcal{B}}(z)$, assuming zero initial conditions for the recursive filter. Note that if $\hat{\mathcal{B}}(z)$ is not minimum-phase, the inverse filter will be unstable. In this case we can either reflect the non-minimum-phase zeros to their conjugate reciprocal locations inside the unit circle [and accounting for the scaling of $\hat{\mathcal{B}}(z)$ which occurs] or compute the sum of squares approximately in the frequency domain as

$$\begin{aligned} \hat{\sigma}^2 &= \frac{1}{N} \int_{-1/2}^{1/2} \frac{|X(f)|^2 \, |\hat{A}(f)|^2}{|\hat{B}(f)|^2} \, df \\ &\approx \frac{1}{NM} \left[\sum_{i=0}^{M-1} \frac{|X(i/M)|^2 \, |\hat{A}(i/M)|^2}{|\hat{B}(i/M)|^2} \right] \end{aligned} \tag{10.22}$$

where $M \gg N$ by using M point FFTs.

The name *two-stage least squares* is due to the use of a large order AR least squares estimate followed by the least squares solution for $\boldsymbol{\theta}$. Although easy to implement, this estimator is known to be inefficient [Durbin 1960A], in that the large data record covariance matrix is not given by the CR bound (see also Problem 10.7 for a similar approach). In an attempt to reduce the variance of the estimator, Mayne and Firoozan [1977] have proposed a modification of the algorithm. This modification, which yields a three-stage LS, is discussed next.

10.6.2 Three-Stage Least Squares

The algorithm due to Mayne and Firoozan [1977] (see also modifications due to Konvalinka and Matausek [1979]) proceeds as follows:

Step 1. Fit a large order AR model to the data $x[n]$ using any of the AR estimators of Chapter 7. Let $\hat{a}_L[k]$ be the estimated AR parameters.

Step 2. Find the estimate of the white noise sequence as the output of the prediction error filter as computed in step 1 or

$$\hat{u}[n] = \sum_{k=0}^{L} \hat{a}_L[k] x[n-k] \qquad n = -q, -q+1, \ldots, N-1 \tag{10.23}$$

where, as usual, the initial conditions $\{x[-q-L], x[-q-L+1], \ldots, x[-1]\}$ are assumed equal to zero.

Step 3. Use (10.20) to generate estimates of the MA filter parameters. Call these estimates $\tilde{b}[k]$.

Step 4. Form the following filtered sequences:

$$y[n] = -\sum_{k=1}^{q} \tilde{b}[k]y[n-k] + x[n] \qquad n = -p, -p+1, \ldots, N-1 \tag{10.24}$$

$$z[n] = -\sum_{k=1}^{q} \tilde{b}[k]z[n-k] + \hat{u}[n] \qquad n = -q, -q+1, \ldots, N-1. \tag{10.25}$$

Set the initial conditions $\{y[-p-q], y[-p-q+1], \ldots, y[-(p+1)]\}$ and $\{z[-2q], z[-2q+1], \ldots, z[-(q+1)]\}$ equal to zero.

Step 5. Determine the final estimates of the ARMA filter parameters by minimizing

$$\sum_{n=0}^{N-1} \left| \sum_{k=0}^{p} a[k]y[n-k] - \sum_{k=0}^{q} b[k]z[n-k] \right|^2 \tag{10.26}$$

over $a[k]$, $b[k]$. This is the same LS problem encountered in (10.19). The solution is

$$\hat{\boldsymbol{\theta}} = (\mathbf{H}^H\mathbf{H})^{-1}\mathbf{H}^H(\mathbf{y} - \mathbf{z}) \tag{10.27}$$

where

$$\hat{\boldsymbol{\theta}} = [-\hat{a}[1] \quad -\hat{a}[2] \quad \cdots \quad -\hat{a}[p] \quad \hat{b}[1] \quad \hat{b}[2] \quad \cdots \quad \hat{b}[q]]^T$$

$$\mathbf{y} = [y[0] \quad y[1] \quad \cdots \quad y[N-1]]^T$$

$$\mathbf{z} = [z[0] \quad z[1] \quad \cdots \quad z[N-1]]^T$$

and

$$\mathbf{H} = \begin{bmatrix} y[-1] & y[-2] & \cdots & y[-p] & z[-1] & z[-2] & \cdots & z[-q] \\ y[0] & y[-1] & \cdots & y[-p+1] & z[0] & z[-1] & \cdots & z[-q+1] \\ \vdots & \vdots & \vdots & \vdots & \vdots & \vdots & \vdots & \vdots \\ y[N-2] & y[N-3] & \cdots & y[N-p-1] & z[N-2] & z[N-3] & \cdots & z[N-q-1] \end{bmatrix}.$$

Steps 1–3 correspond to the two-stage LS estimator, while steps 4 and 5 attempt to generate an efficient estimator. The claim is that the final estimator attains the CR bound. This method works well as long as the zeros of $\mathcal{B}(z)$ do not lie too close to the unit circle [Mayne 1979]. However, the latter restriction appears to apply to almost all ARMA estimation algorithms. Also, it is possible that in step 3 the estimated MA filter may not be minimum-phase. If this occurs, step 4 cannot be implemented since the recursive filter will be unstable. The algorithm should then be terminated, which is a clear deficiency of the method.

As in the two-stage LS method an estimate of σ^2 may be found using (10.22). A computer subroutine entitled MAYNEFIR, which is listed in Appendix 10F, implements this algorithm.

10.7 COMPUTER SIMULATION EXAMPLES

To illustrate the performance of the ARMA estimators described, a computer simulation was conducted. Four different real ARMA(4, 2) processes as summarized in Table 10.1 were realized. The pole locations were chosen to generate wideband as well as narrowband spectral features. Also, the zeros were chosen to yield sharp nulls as well as no nulls. The four ARMA processes have pole–zero plots given in Figure 10.3 and corresponding PSDs given in Figure 10.4. ARMA1 has a broadband spectrum, while ARMA2 is also characterized by a broadband spectrum but with a sharp null. ARMA3 has a narrowband PSD with a sharp null and ARMA4 is also narrowband but without a null. Hence the four processes have PSDs that are representative of a wide class of PSDs. The ARMA algorithms tested were:

1. Akaike MLE (10.2)
2. Modified Yule–Walker equations (MYWE) (10.7) for the AR filter parameters and Durbin's method for the MA parameters (see Chapter 8)
3. Least squares modified Yule–Walker equations (LSMYWE) (10.18) for the AR filter parameters and Durbin's method for the MA parameters (see Chapter 8)
4. Mayne–Firoozan (MAYNEFIR) (10.23)–(10.27)

TABLE 10.1 ARMA(4,2) PROCESSES USED IN COMPUTER SIMULATIONS

	Coefficient						
Process	$a[1]$	$a[2]$	$a[3]$	$a[4]$	$b[1]$	$b[2]$	σ^2
ARMA1	−1.352	1.338	−0.662	0.240	−0.200	0.040	1
ARMA2	−1.352	1.338	−0.662	0.240	−0.900	0.810	1
ARMA3	−2.760	3.809	−2.654	0.924	−0.900	0.810	1
ARMA4	−2.760	3.809	−2.654	0.924	−0.200	0.040	1

	Pole–zero locations	
Process	Poles	Zeros
ARMA1	$0.7 \exp[\pm j2\pi(0.12)]$ $0.7 \exp[\pm j2\pi(0.21)]$	$0.2 \exp[\pm j2\pi(0.17)]$
ARMA2	$0.7 \exp[\pm j2\pi(0.12)]$ $0.7 \exp[\pm j2\pi(0.21)]$	$0.9 \exp[\pm j2\pi(0.17)]$
ARMA3	$0.98 \exp[\pm j2\pi(0.11)]$ $0.98 \exp[\pm j2\pi(0.14)]$	$0.9 \exp[\pm j2\pi(0.17)]$
ARMA4	$0.98 \exp[\pm j2\pi(0.11)]$ $0.98 \exp[\pm j2\pi(0.14)]$	$0.2 \exp[\pm j2\pi(0.17)]$

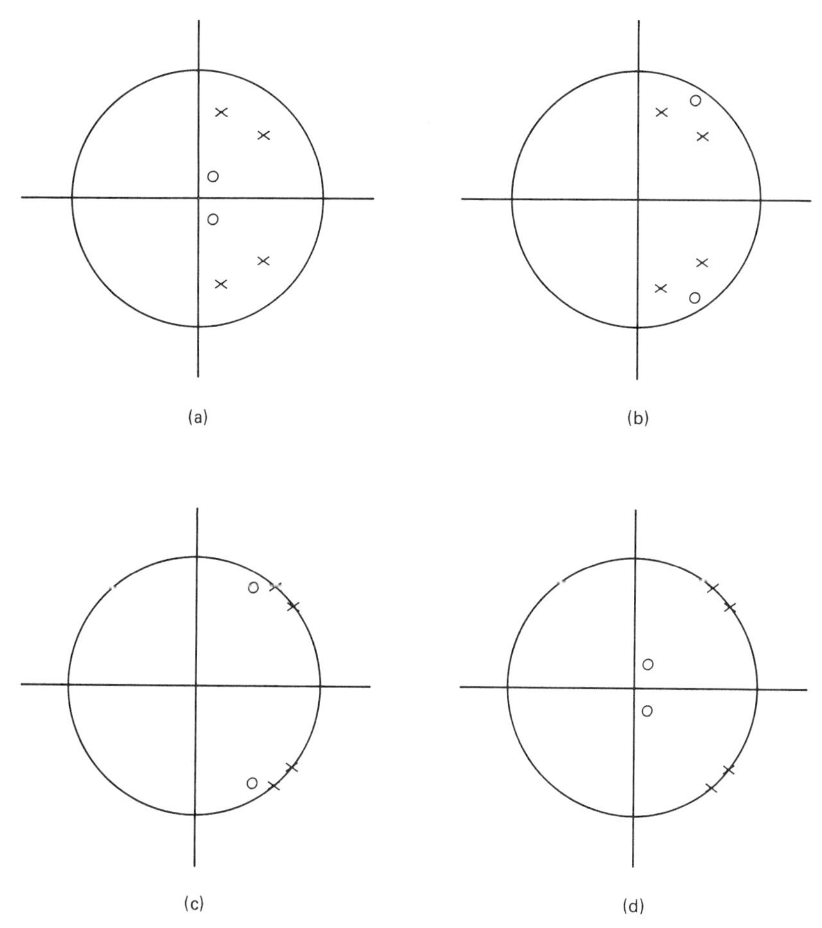

Figure 10.3 Pole–zero diagrams for ARMA test cases. (a) ARMA1. (b) ARMA2. (c) ARMA3. (d) ARMA4.

For all the simulations the data record length is $N = 256$ real data points and the ARMA model orders chosen are the true ones. For Durbin's method and Mayne–Firoozan's method, which require a large AR model order, $L = 50$ was used. The results of each simulation are displayed in two parts. In each figure the average of 50 realizations of the spectral estimator is displayed together with the true PSD (shown as a dashed curve) to indicate the bias. Also, in each figure the same 50 realizations are shown overlaid to indicate the variability of the estimator. The results are given in Figures 10.5 to 10.8.

The Akaike MLE results are given in Figure 10.5. The number of iterations used was 25. At times the estimate of the MA parameters produced a non-minimum-phase filter so that $1/\mathcal{B}(z)$ became unstable. When this occurred, the iteration was halted. For the ARMA4 process almost half of the realizations resulted in non-minimum-phase $\mathcal{B}(z)$ filters, while for the ARMA3 process a non-minimum-phase filter was never observed. These results indicate that the likelihood of obtaining a non-minimum-phase filter during the iteration is greater for processes with zeros near the origin. For the purposes of the simulation the first 50 realizations which resulted in stable $1/\mathcal{B}(z)$ filters were retained. Also, $Q(\mathbf{a}, \mathbf{b})$ increased rather than decreased for some iterations. This effect is presumably due to the approximations of the gradient and the Hessian made in the algorithm. It is seen from Figure 10.5 that the MLE approach yields an accurate spectral estimate for processes with poles sufficiently displaced from the unit circle (assuming that the iteration converges). A further point is that the MLE needed to be initialized or the first iterate needed to be specified. For the simulation the first iterate was obtained by perturbing the ARMA filter parameters slightly from the

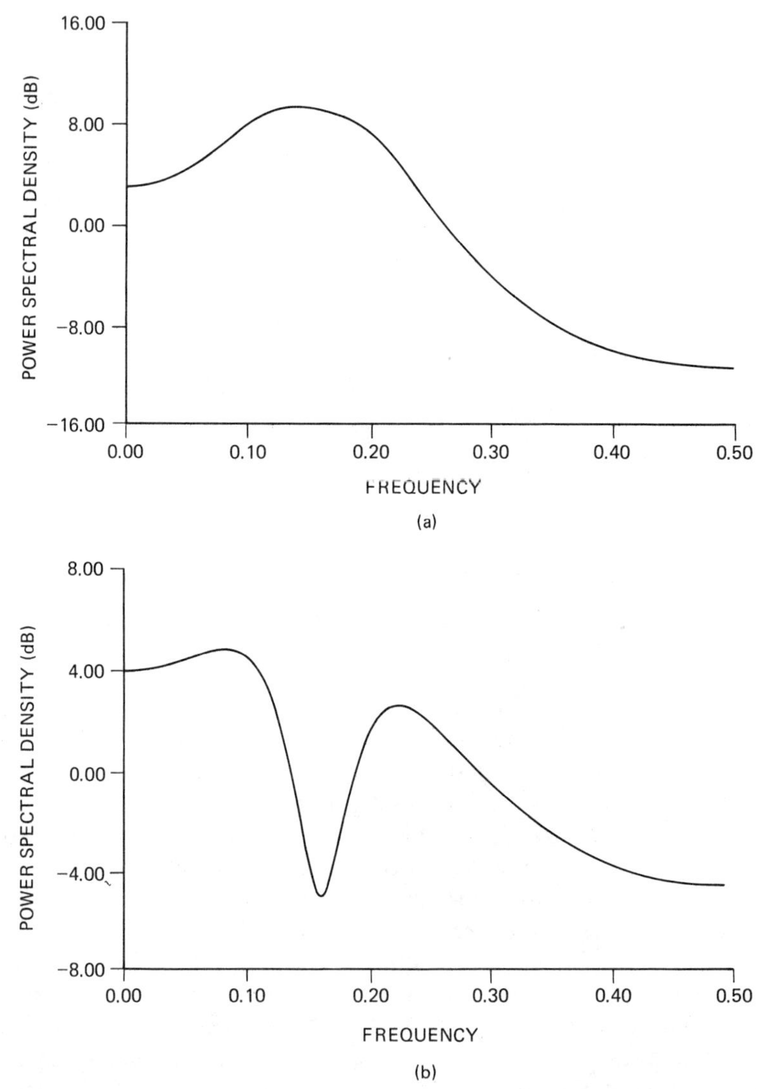

Figure 10.4 Power spectral densities for ARMA test cases. (a) ARMA1. (b) ARMA2. (c) ARMA3. (d) ARMA4.

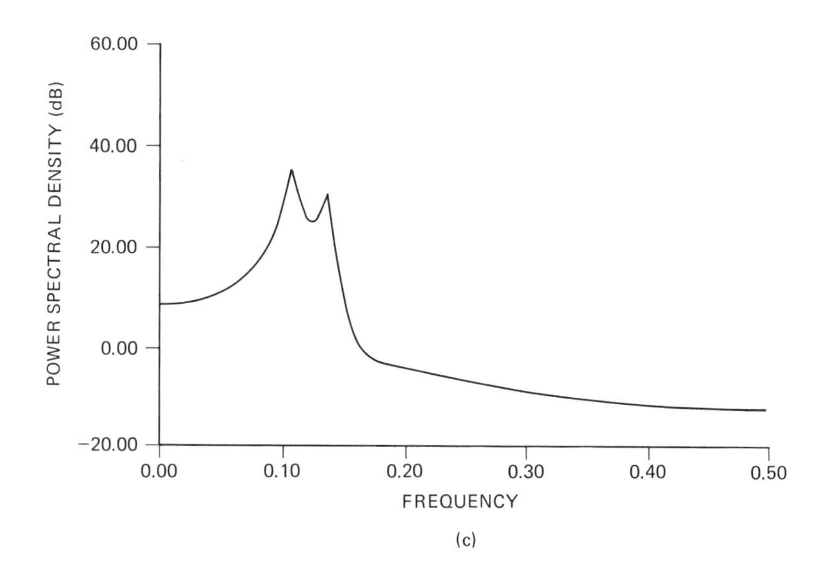

(c)

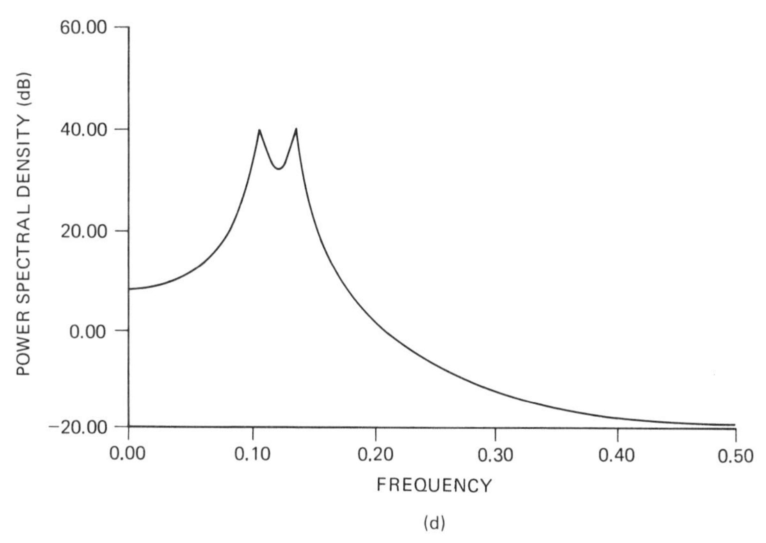

(d)

Figure 10.4 *(Continued)*

true values. Clearly, in practice we would need to use some other ARMA estimator to provide initial estimates of the parameters. Finally, further simulations showed that increasing the data record length improved the convergence of the MLE as well as the quality of the resultant spectral estimates.

The MYWE produced the spectral estimates shown in Figure 10.6. The estimates are in general poor, displaying a large bias. Only the PSD of ARMA1 was estimated with any accuracy.

The LSMYWE spectral estimates are displayed in Figure 10.7. The value of M was 50, so that $M - q = 48$ equations were used. This estimator also exhibits

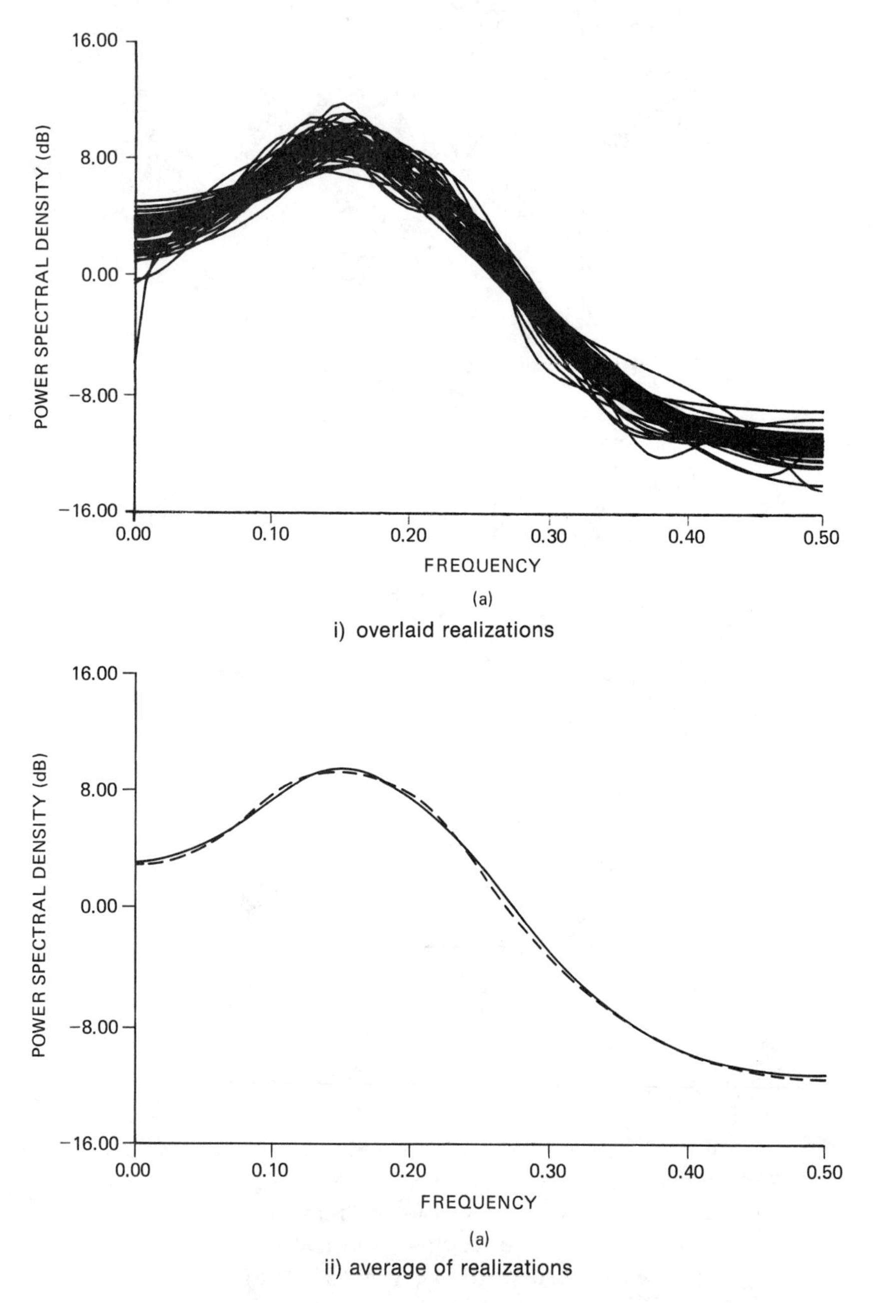

Figure 10.5 Computer simulation results for Akaike MLE. (a) ARMA1. (b) ARMA2. (c) ARMA3. (d) ARMA4.

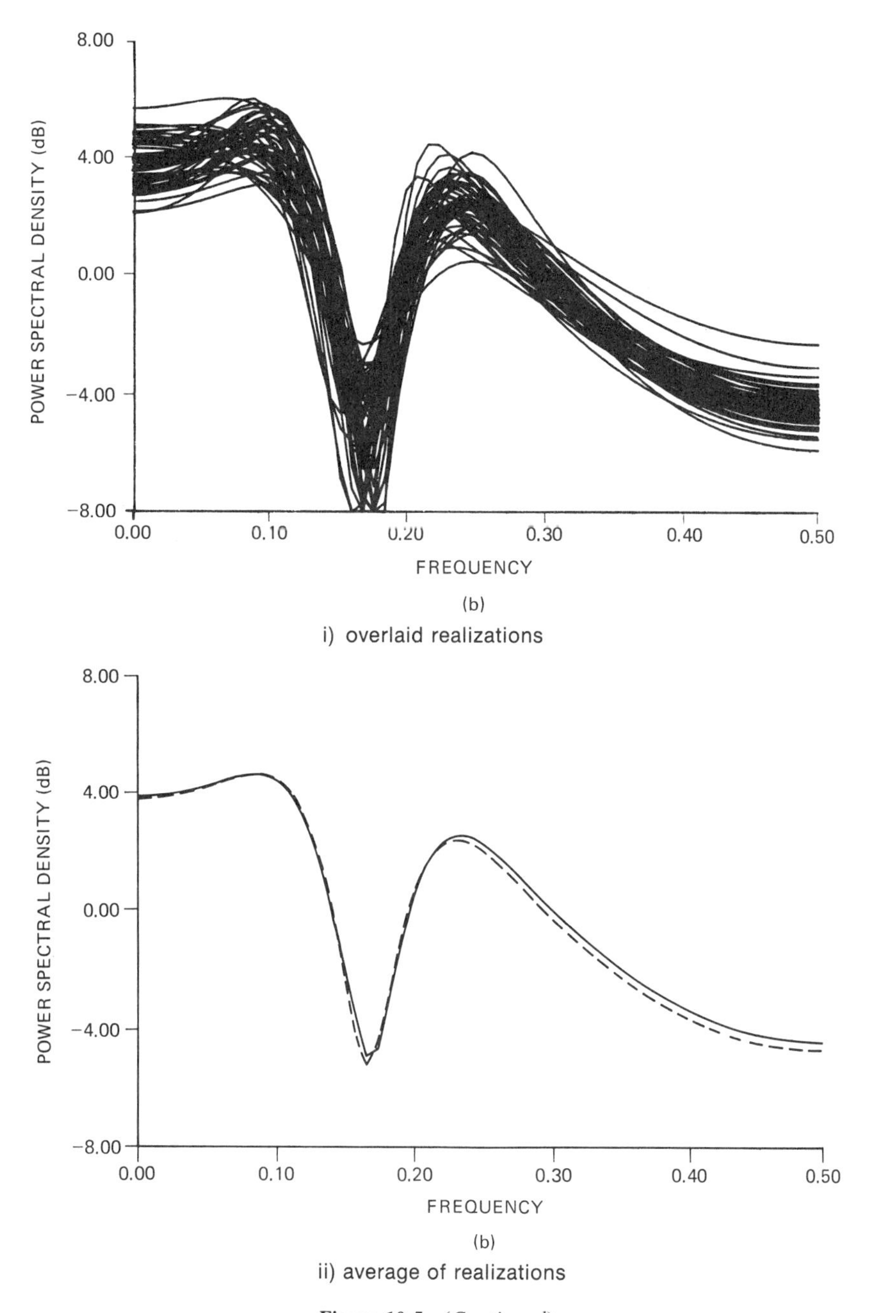

Figure 10.5 *(Continued)*

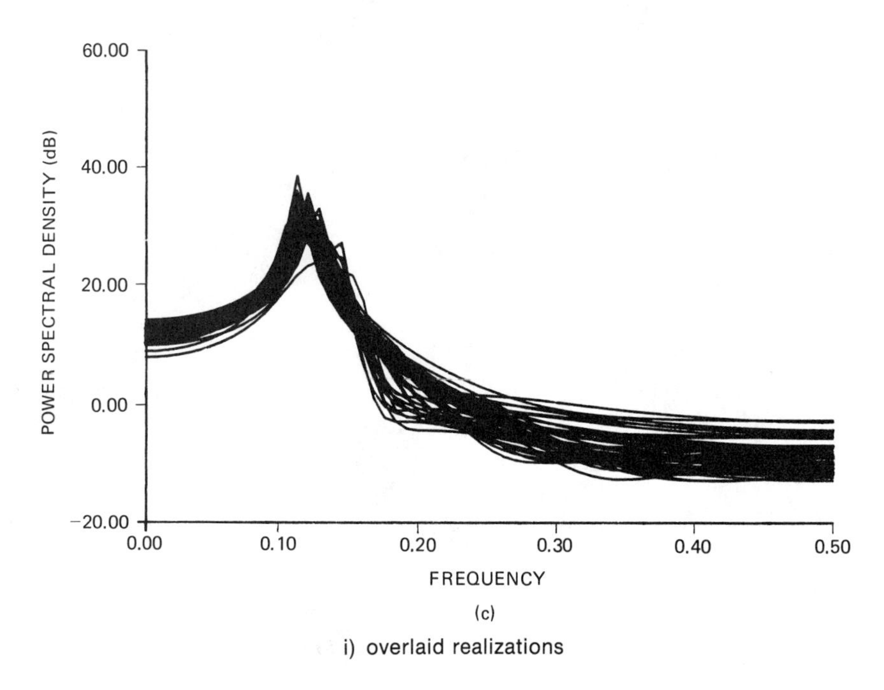

(c)

i) overlaid realizations

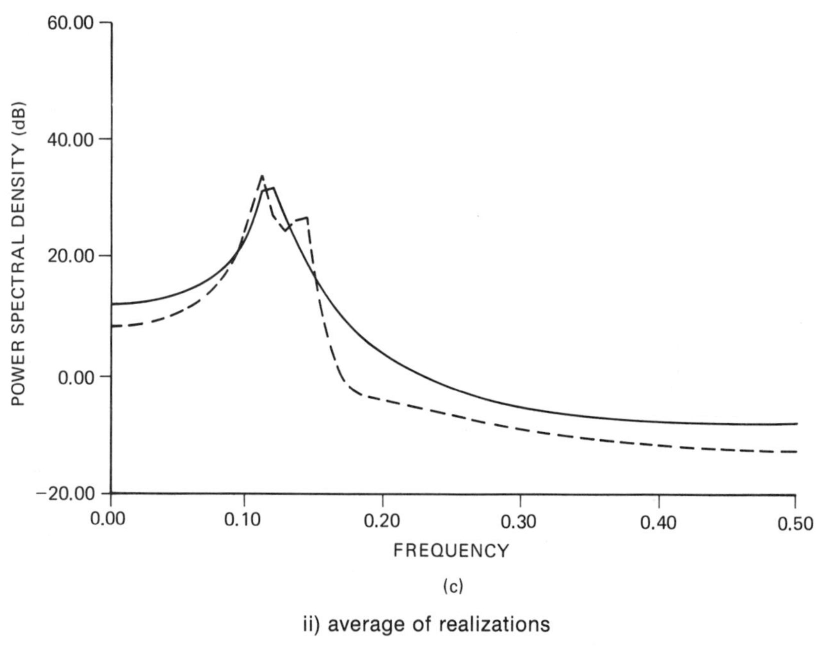

(c)

ii) average of realizations

Figure 10.5 *(Continued)*

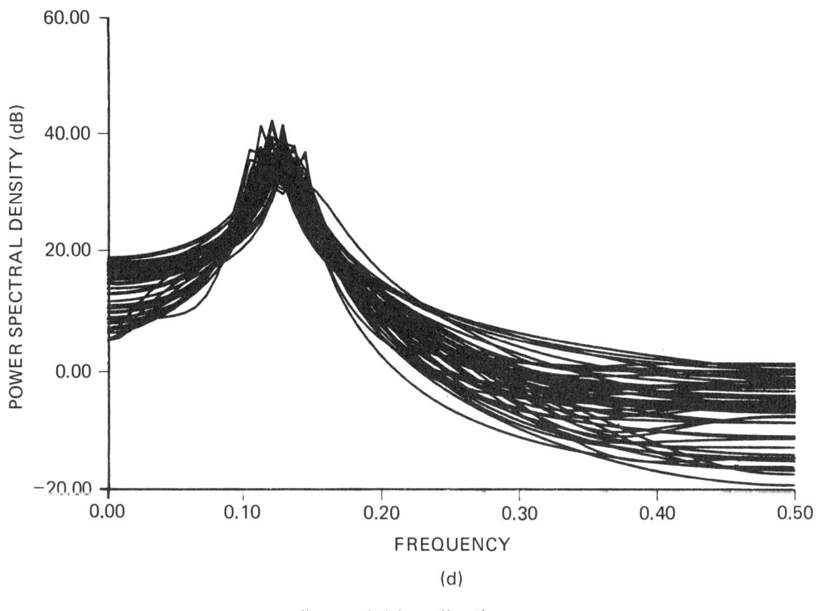

i) overlaid realizations

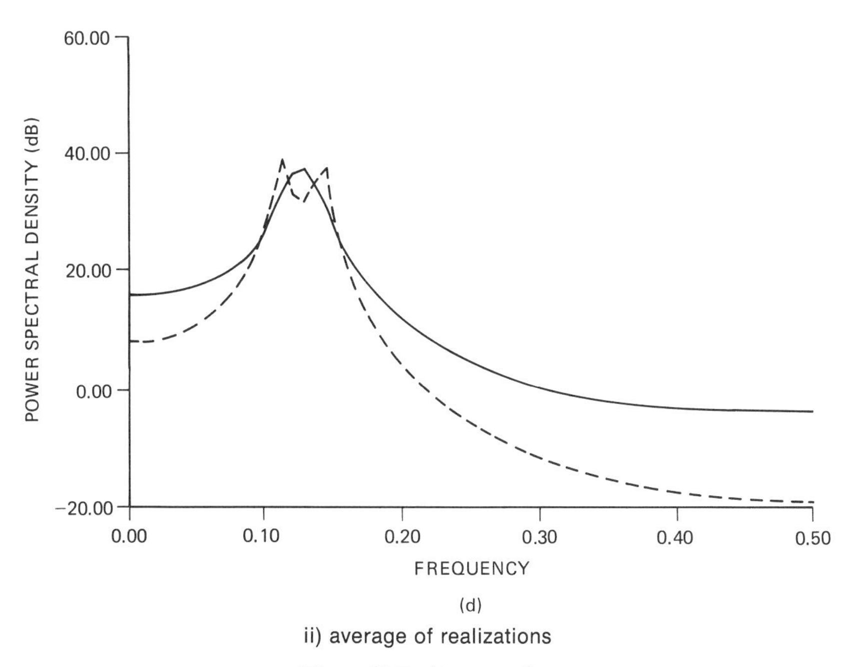

ii) average of realizations

Figure 10.5 *(Continued)*

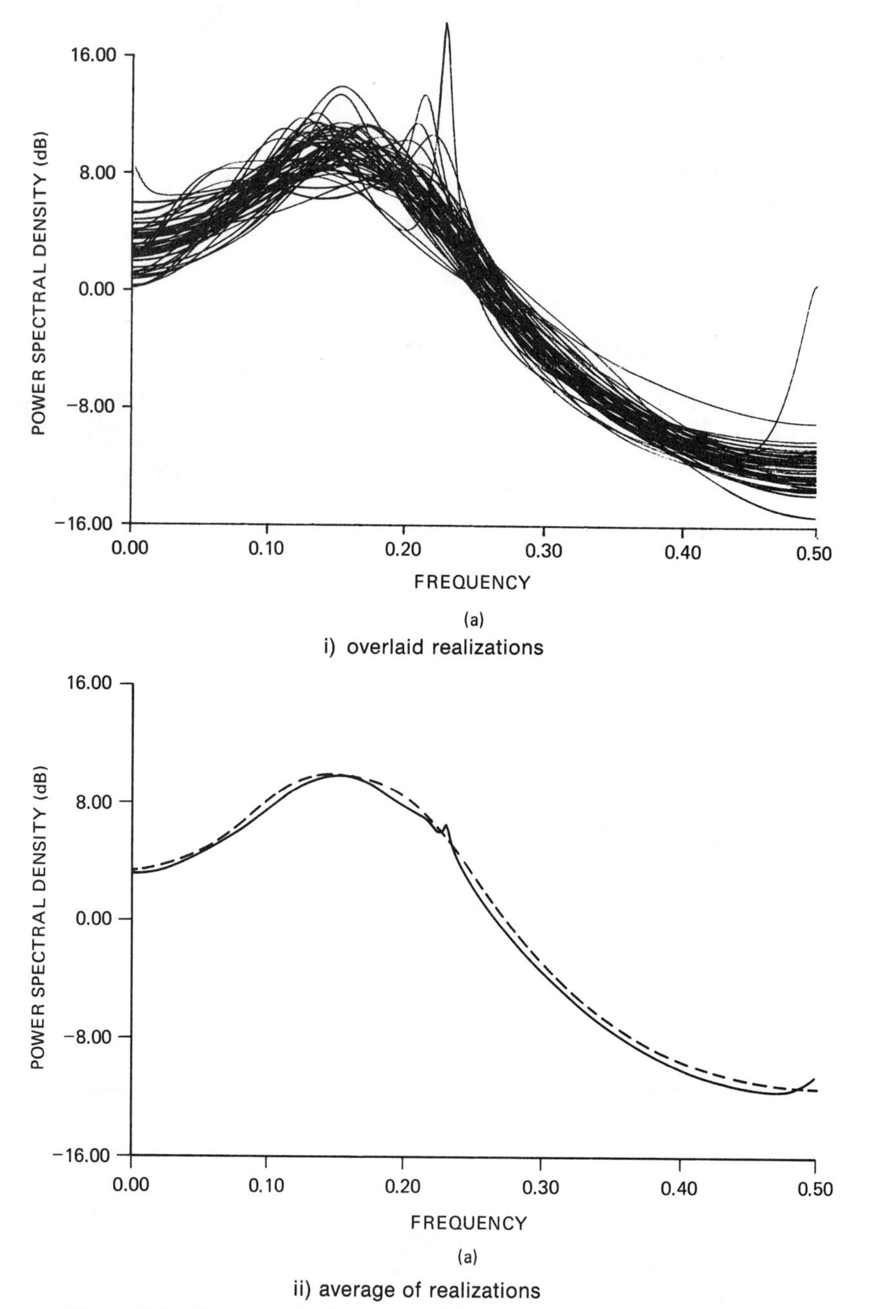

Figure 10.6 Computer simulation results for MYWE. (a) ARMA1. (b) ARMA2. (c) ARMA3. (d) ARMA4.

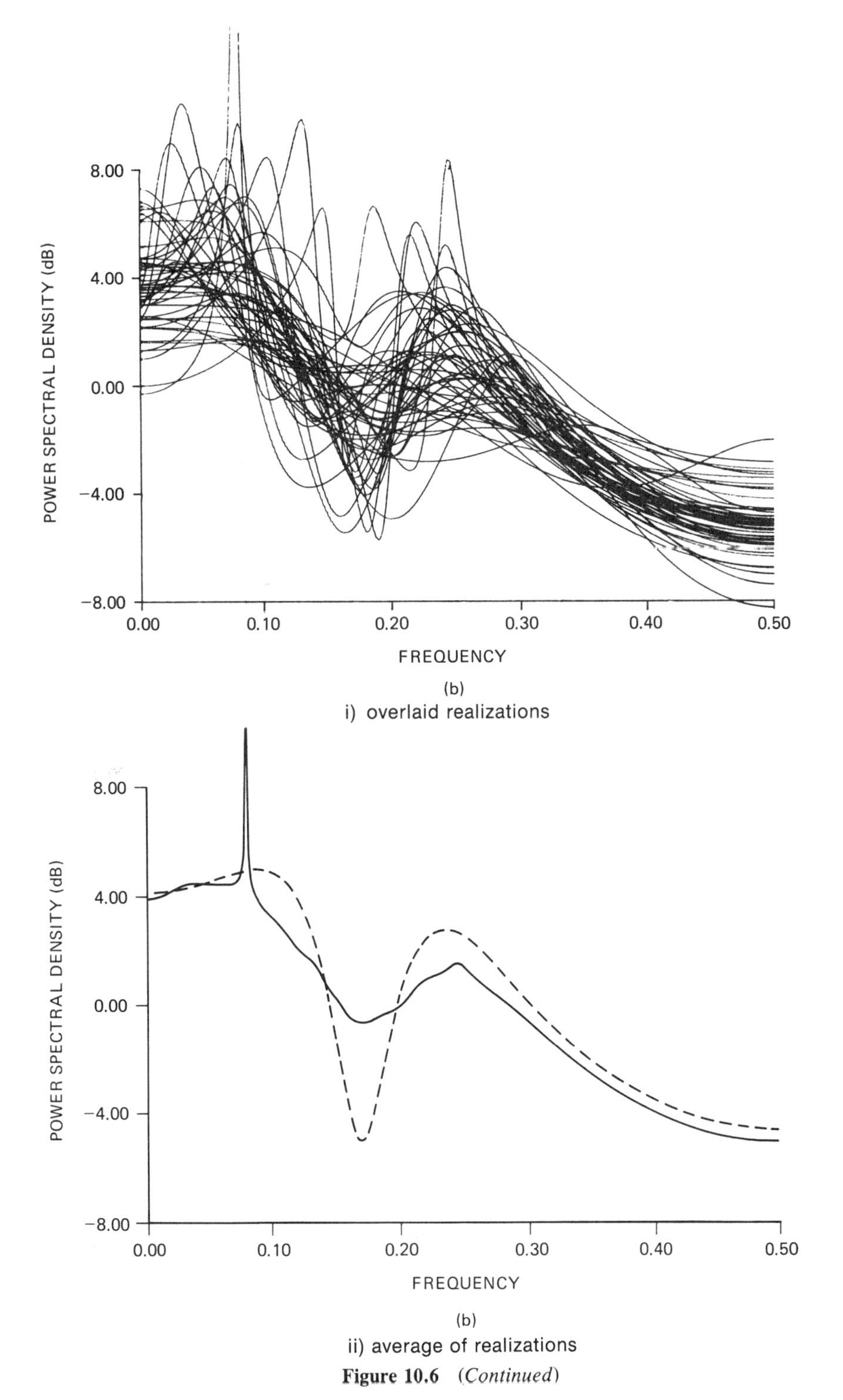

Figure 10.6 *(Continued)*

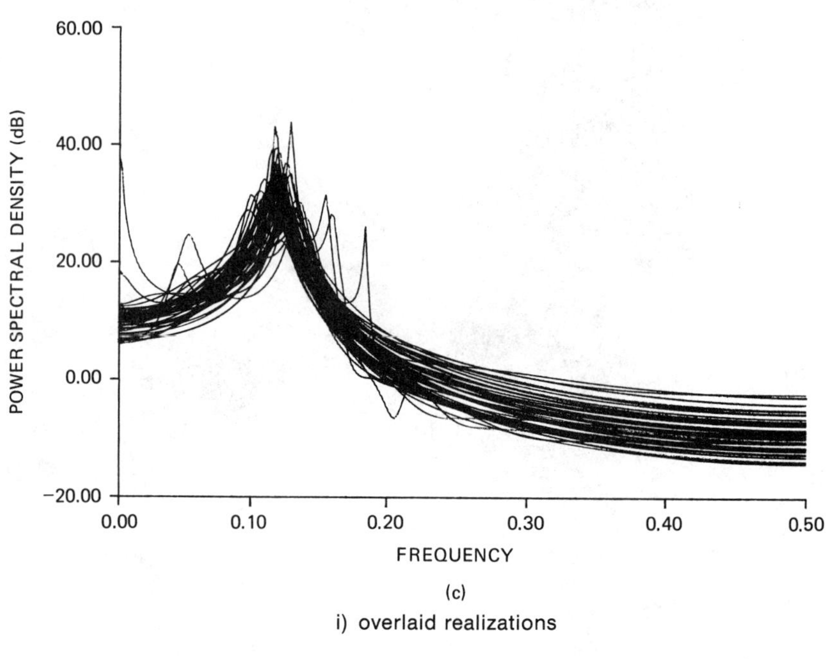

(c)

i) overlaid realizations

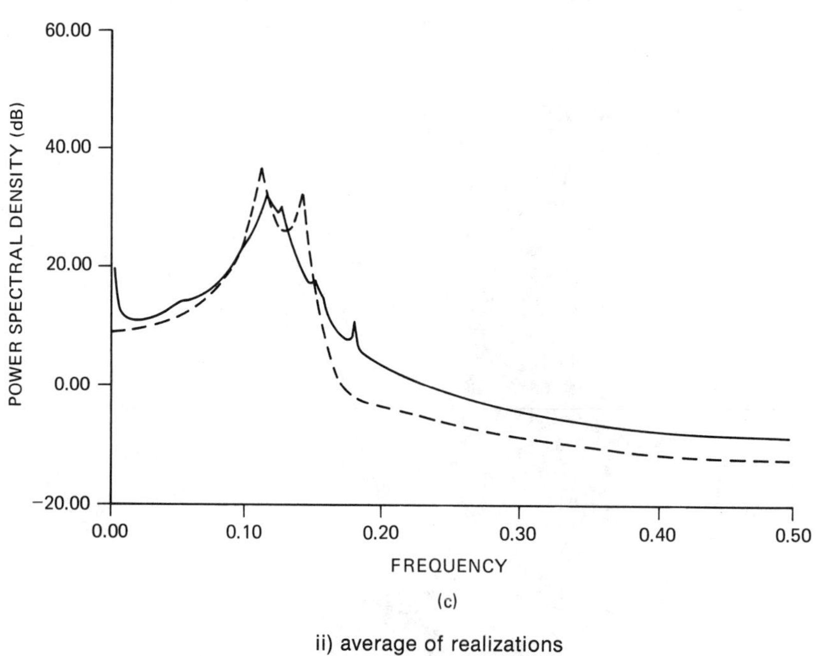

(c)

ii) average of realizations

Figure 10.6 *(Continued)*

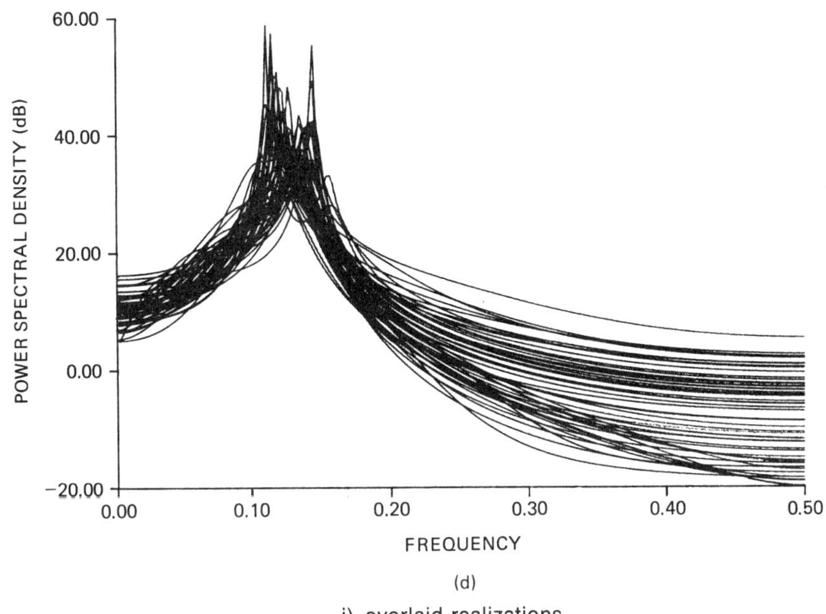

(d)

i) overlaid realizations

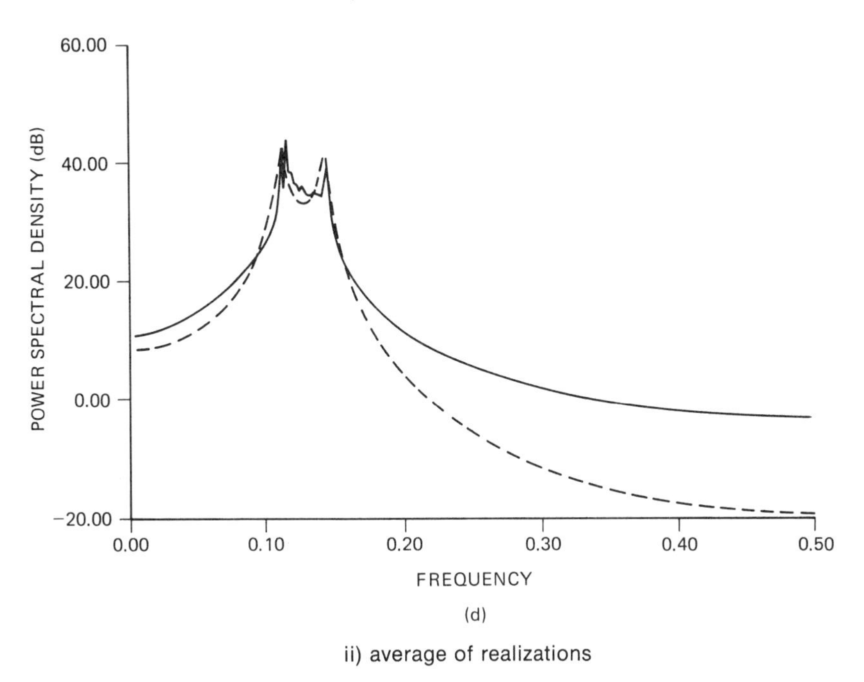

(d)

ii) average of realizations

Figure 10.6 *(Continued)*

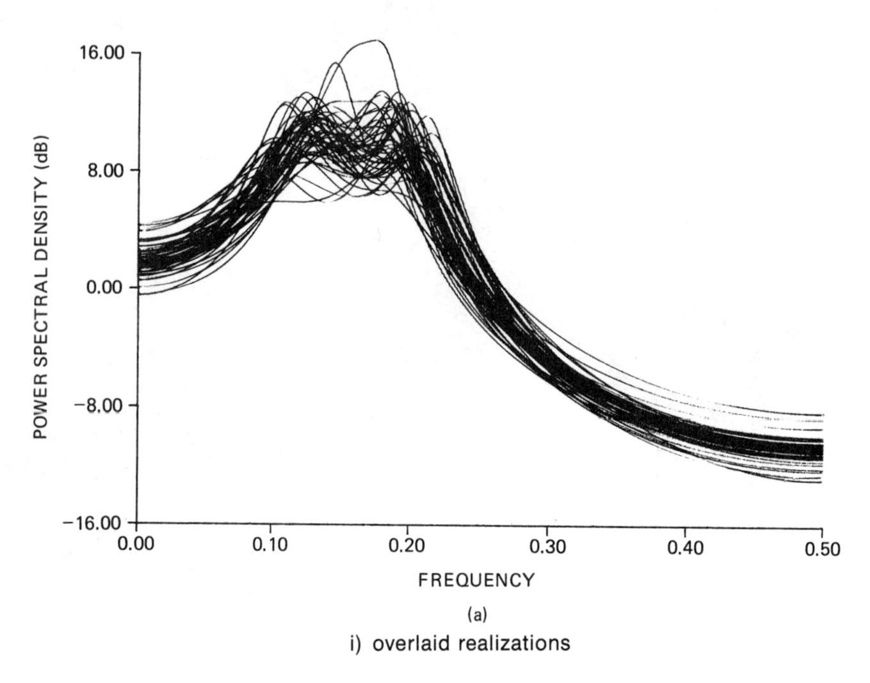

(a)

i) overlaid realizations

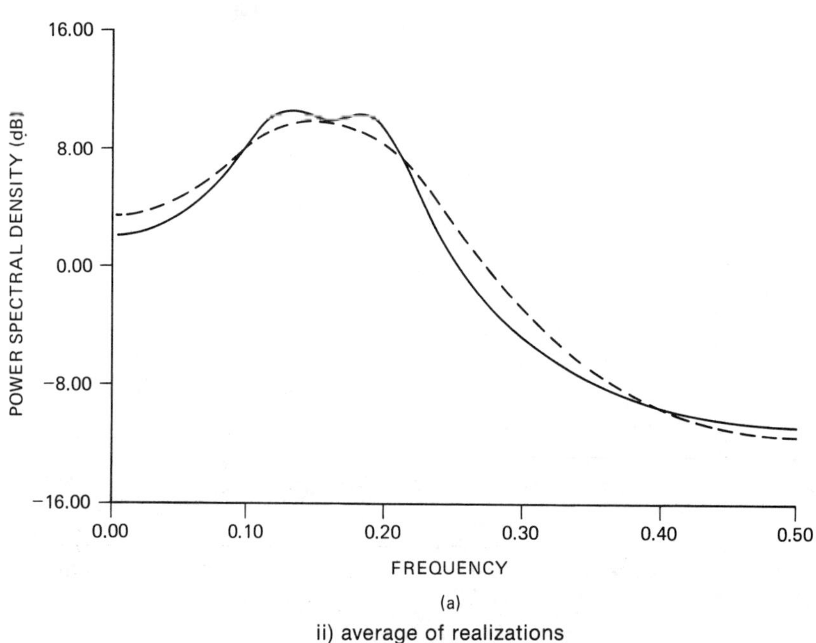

(a)

ii) average of realizations

Figure 10.7 Computer simulation results for LSMYWE. (a) ARMA1. (b) ARMA2. (c) ARMA3. (d) ARMA4.

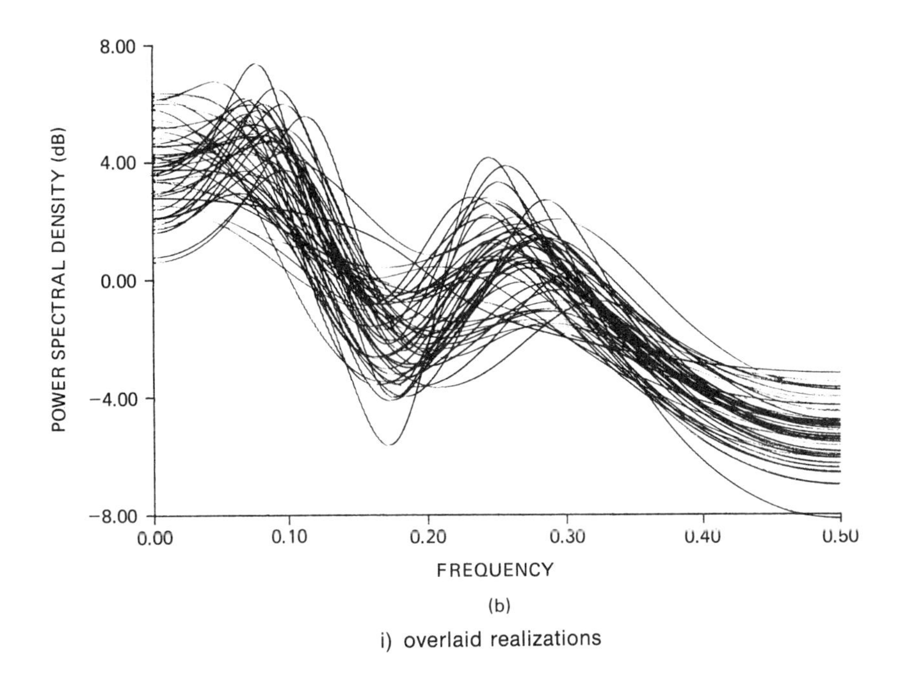

(b)

i) overlaid realizations

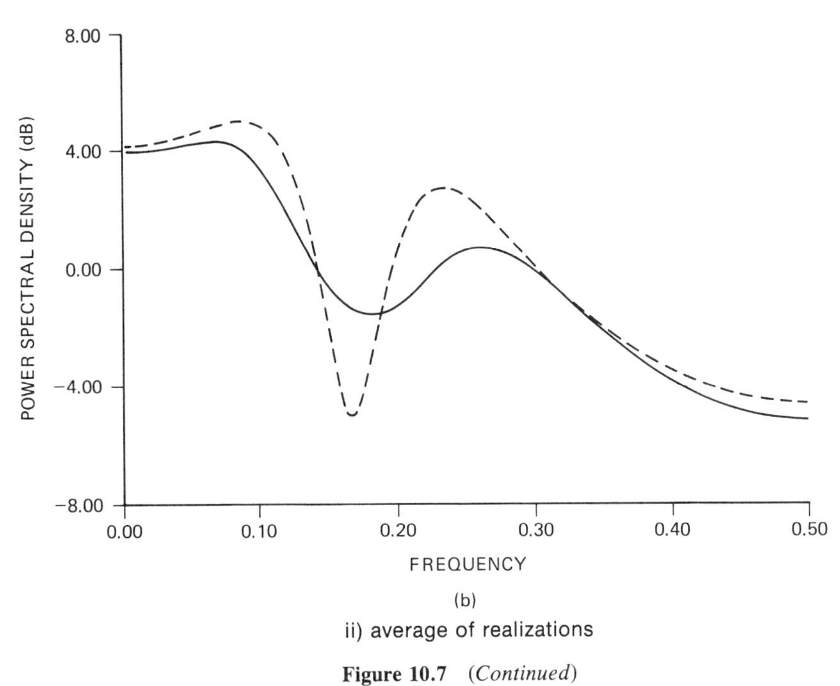

(b)

ii) average of realizations

Figure 10.7 *(Continued)*

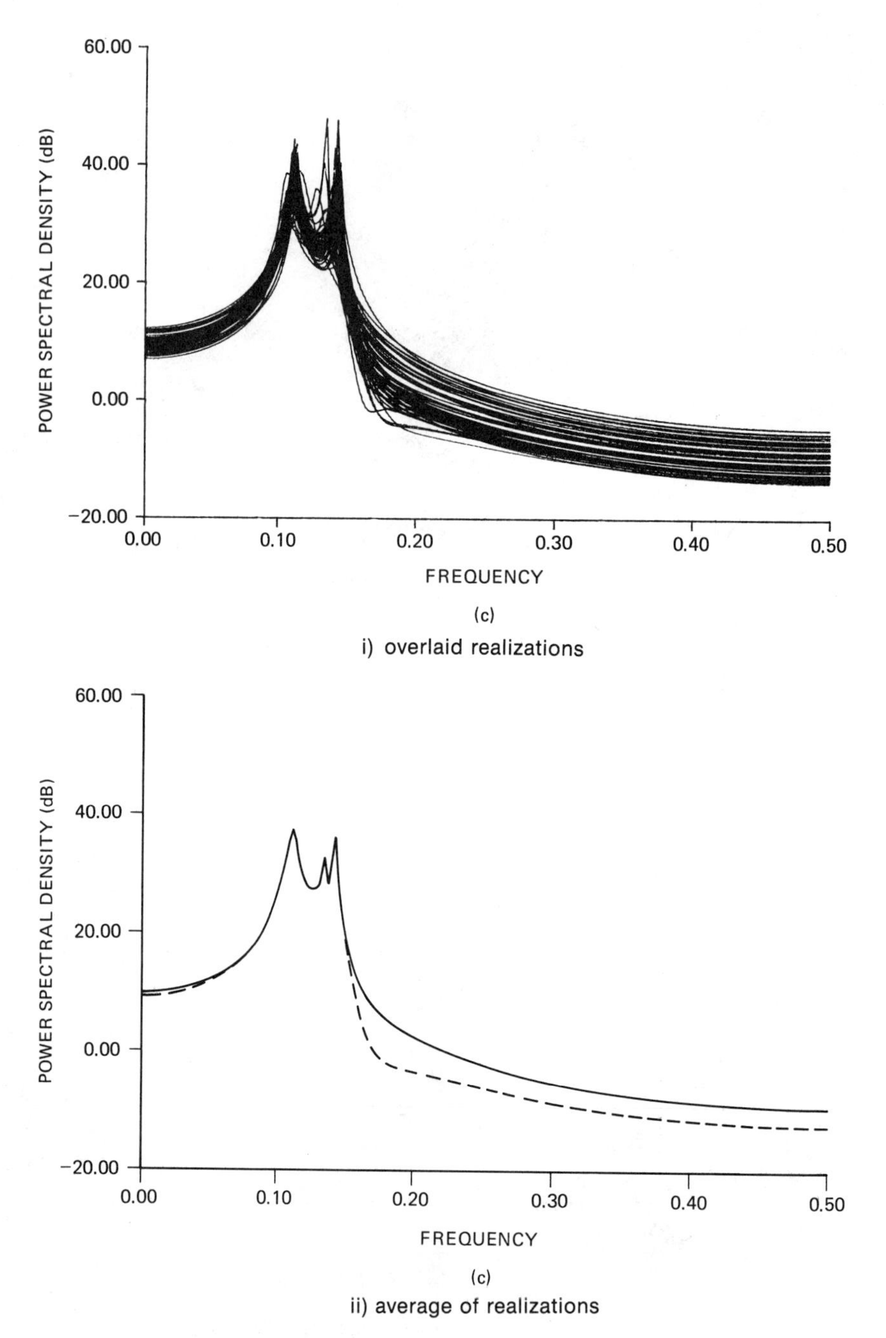

(c)
i) overlaid realizations

(c)
ii) average of realizations

Figure 10.7 (*Continued*)

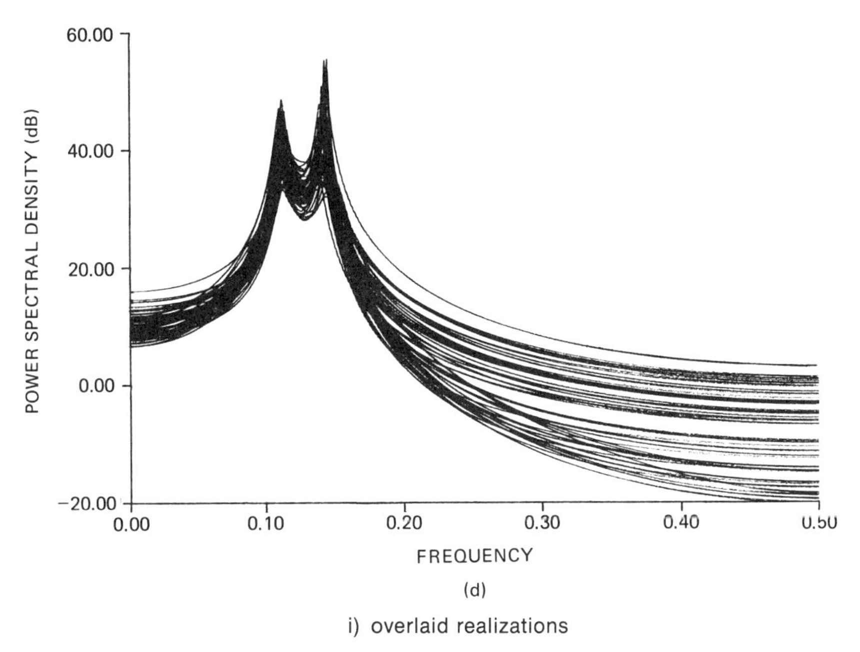

(d)

i) overlaid realizations

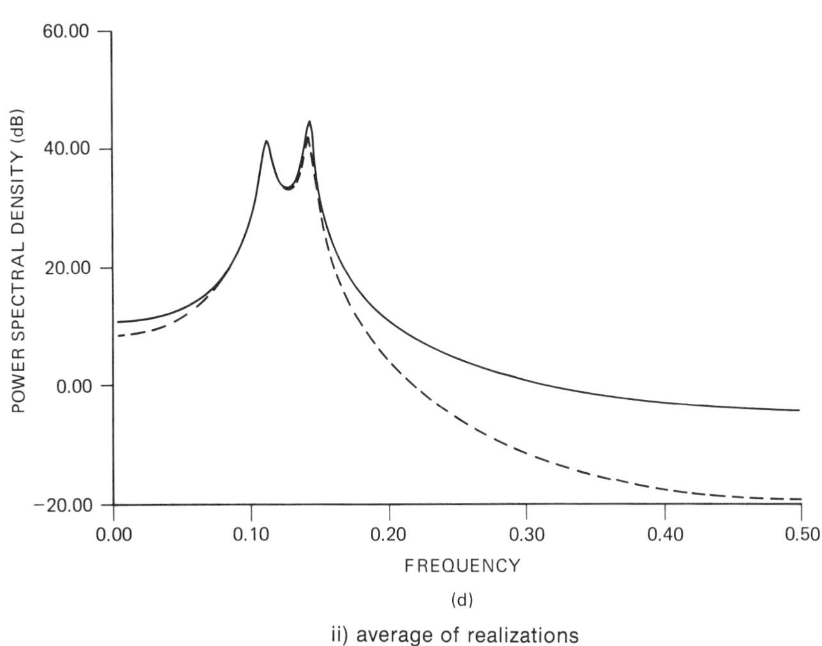

(d)

ii) average of realizations

Figure 10.7 (*Continued*)

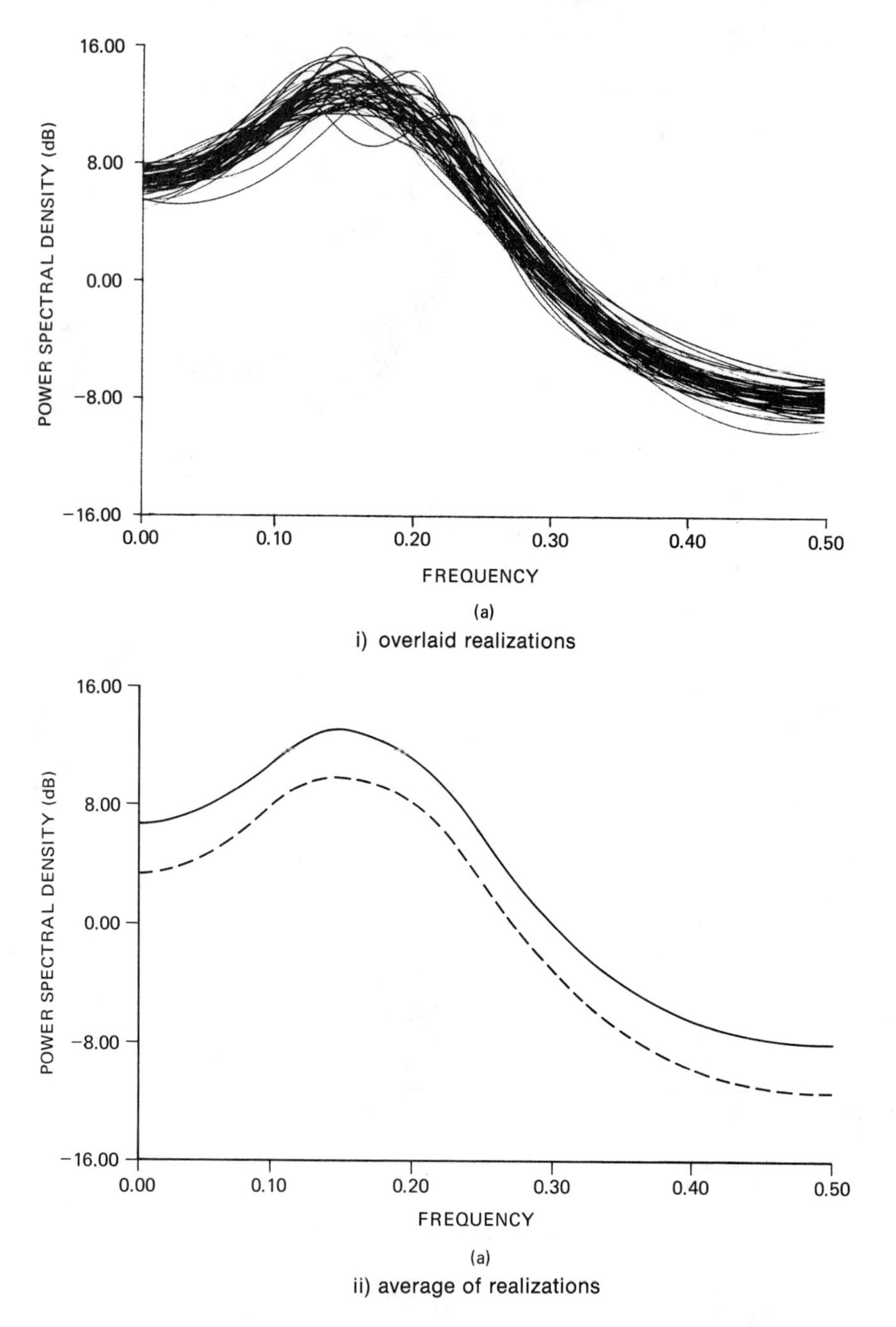

Figure 10.8 Computer simulation results for Mayne–Firoozan. (a) ARMA1. (b) ARMA2. (c) ARMA3. (d) ARMA4.

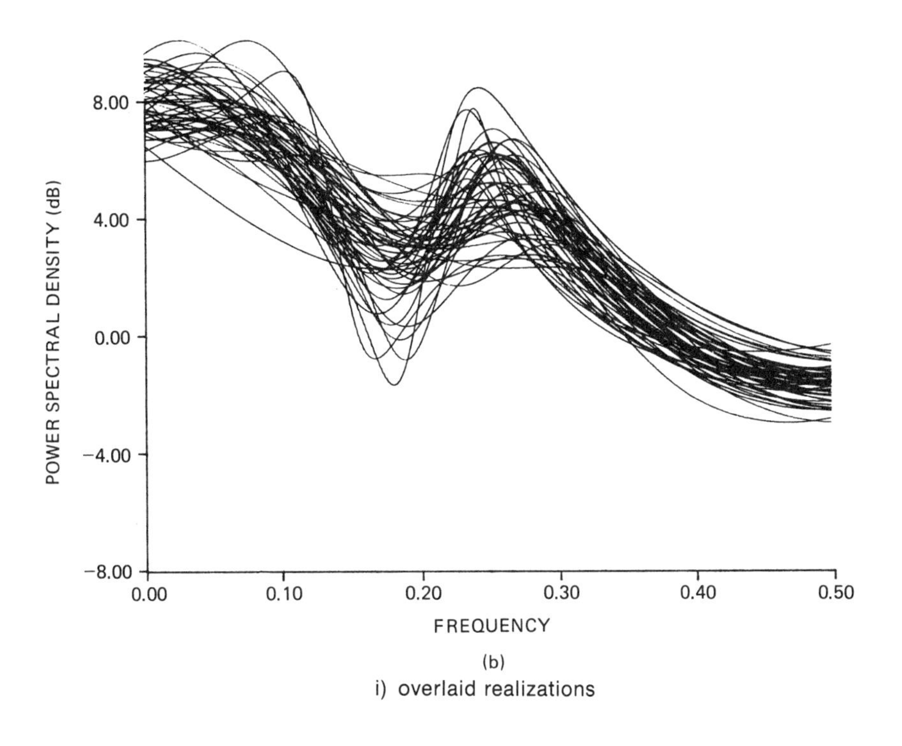

(b)
i) overlaid realizations

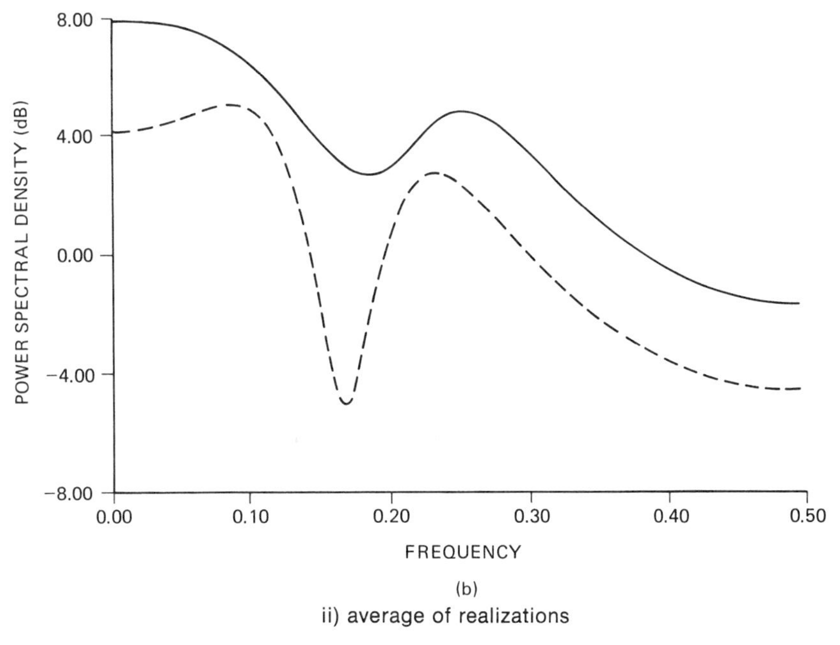

(b)
ii) average of realizations

Figure 10.8 *(Continued)*

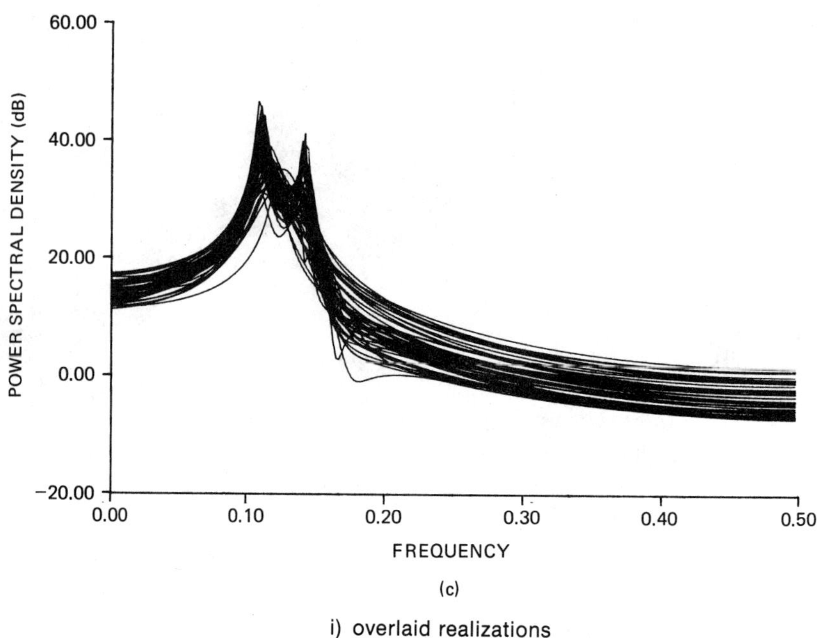

(c)

i) overlaid realizations

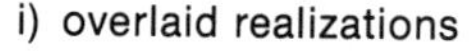

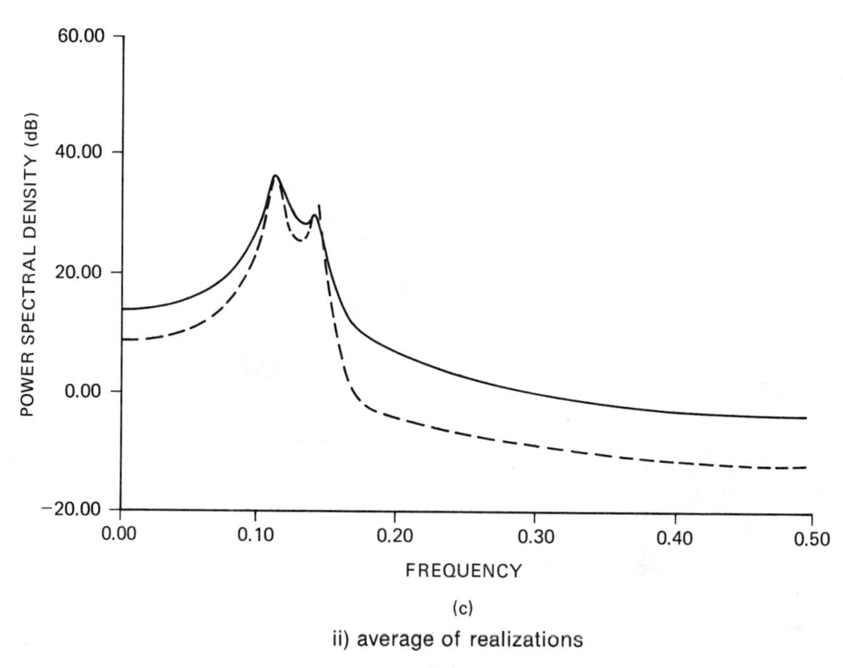

(c)

ii) average of realizations

Figure 10.8 *(Continued)*

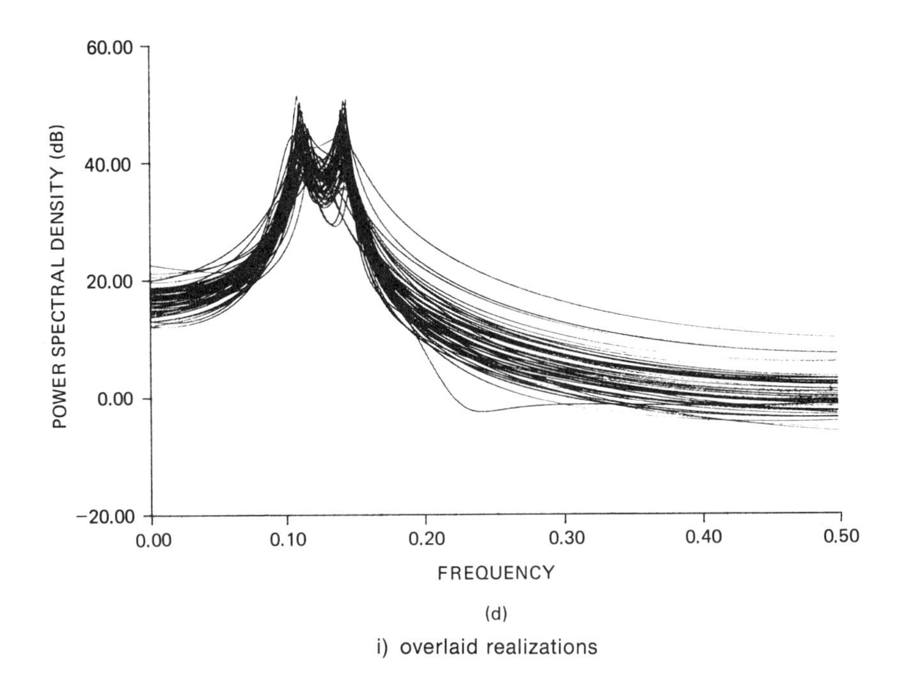

(d)

i) overlaid realizations

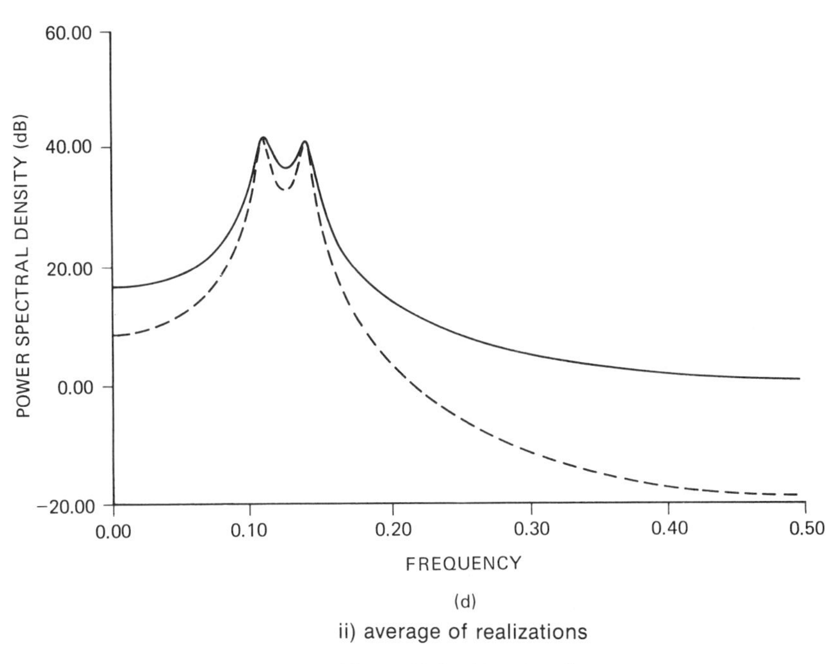

(d)

ii) average of realizations

Figure 10.8 (*Continued*)

TABLE 10.2 CONDITIONS FOR BEST PERFORMANCE OF EACH SPECTRAL ESTIMATOR

	Dominant poles	Weak poles
Dominant zeros	ARMA3 LSMYWE[a]	ARMA2 MLE[a]
Weak zeros	ARMA4 Mayne–Firoozan LSMYWE[a]	ARMA1 Mayne–Firoozan MLE[a]

[a] Appears to provide best performance of methods tested.

some bias, with the bias being most severe for the ARMA2 process. This is attributable to the fact that the large order samples of the ACF are very small unless the poles are near the unit circle. Hence the estimation error will be large relative to the value of the ACF sample being estimated. The AR filter parameter estimates are then poor, causing poor MA parameter estimates since filtering of the data by $\hat{\mathcal{A}}(z)$ will not produce an MA process. Additionally, Durbin's method, which is unable to estimate sharp nulls, probably also contributes to the poor estimate. For the narrowband processes, ARMA3 and ARMA4, the results are quite good with again some bias observed.

The MAYNEFIR spectral estimates are shown in Figure 10.8. The estimates all exhibit a bias which is most pronounced for the ARMA2 process, which has its zeros near the unit circle. This is not unexpected, since the first step of the MAYNEFIR algorithm is to fit a large-order AR model to the data. For zeros near the unit circle the input noise sequence estimate will be severely biased, leading us to conclude that the AR model order needs to be larger.

In comparing all four methods we may conclude that the MLE yields the most accurate estimates for poles sufficiently displaced from the unit circle (i.e., weak poles) if convergence is attained. For poles near the unit circle (i.e., dominant poles) the LSMYWE appears to yield the most accurate estimates. Since the performance of each method is critically dependent on the pole–zero locations, a summary of the conditions under which each method performs well is given in Table 10.2. The MYWE method has been omitted due to its overall poor performance. Also, the entries in the table with an "a" appear to provide the most accurate estimates.

REFERENCES

Akaike, H., "Maximum Likelihood Identification of Gaussian Autoregressive Moving Average Models," *Biometrika,* Vol. 60, pp. 255–265, 1973.

Anderson, T. W., "Estimation for Autoregressive Moving Average Models in the Time and Frequency Domains," *Ann. Statist.,* Vol. 5, pp. 842–865, 1977.

Arato, M., "On the Sufficient Statistics for Stationary Gaussian Random Processes," *Theory Probab. Its Appl.*, Vol. 6, pp. 199–201, 1961.

Astrom, K. J., and P. Eykhoff, "System Identification—A Survey," *Automatica*, Vol. 7, pp. 123–162, 1971.

Astrom, K. J., and T. Soderstrom, "Uniqueness of the Maximum Likelihood Estimates of the Parameters of an ARMA Model," *IEEE Trans. Autom. Control*, Vol. AC19, pp. 769–773, Dec. 1974.

Box, G. E. P., and G. M. Jenkins, *Time Series Analysis: Forecasting and Control*, Holden-Day, San Francisco, 1970.

Bruzzone, S. P., and M. Kaveh, "Information Tradeoffs in Using the Sample Autocorrelation Function in ARMA Parameter Estimation," *IEEE Trans. Acoust. Speech Signal Process.*, Vol. ASSP32, pp. 701–715, Aug. 1984.

Cadzow, J., "High Performance Spectral Estimation—A New ARMA Method," *IEEE Trans. Acoust., Speech, Signal Process.*, Vol. ASSP28, pp. 524–529, Oct. 1980.

Cadzow, J., "Spectral Estimation: An Overdetermined Rational Model Equation Approach," *Proc. IEEE*, Vol. 70, pp. 907–939, Sept. 1982.

Durbin, J., "The Fitting of Time-Series Models," *Revue Inst. Int. Statist.*, Vol. 28, pp. 233–243, 1960A.

Durbin, J., "Estimation of Parameters in Time-Series Regression Models," *J. Roy. Statist. Soc.*, Vol. 22, pp. 139–153, 1960B.

Friedlander, B., and B. Porat, "The Modified Yule-Walker Method of ARMA Spectral Estimation," *IEEE Trans. Aerosp. Electron. Syst.*, Vol. AES20, pp. 158-172, Mar. 1984.

Friedlander, B., and K. C. Sharman, "Performance Evaluation of the Modified Yule–Walker Estimator," *IEEE Trans. Acoust. Speech Signal Process.*, Vol. ASSP33, pp. 719–724, June 1985.

Gersch, W., "Estimation of the Autoregressive Parameters of a Mixed Autoregressive Moving-Average Time Series," *IEEE Trans. Autom. Control*, Vol. AC15, pp. 583–588, Oct. 1970.

Graupe, D., D. J. Krause, and J. B. Moore, "Identification of Autoregressive Moving-Average Parameters of Time Series," *IEEE Trans. Autom. Control*, Vol. AC20, pp. 104–107, Feb. 1975.

Graybill, F. A., *Theory and Application of the Linear Model*, Duxbury Press, Belmont, Calif., 1976.

Gutowski, P. R., E. A. Robinson, and S. Treitel, "Spectral Estimation: Fact or Fiction?," *IEEE Trans. Geosci. and Electron.*, Vol. GE16, pp. 80–84, Apr. 1978.

Hannan, E. J., "The Estimation of Mixed Moving Average Autoregressive Systems," *Biometrika*, Vol. 56, pp. 579–593, 1969.

Hsia, T. C., *System Identification*, D. C. Heath, Lexington, Mass., 1977.

Izraelevitz, D., and J. S. Lim, "Spectral Characteristics of the Overdetermined Normal Equation Method for Spectral Estimation," *Proc. 2nd ASSP Spectrum Estimation Workshop*, pp. 49–54, Nov. 1983.

Kaveh, M., and S. P. Bruzzone, "A Comparative Overview of ARMA Spectral Estimators," *Proc. 1st ASSP Spectrum Estimation Workshop*, pp. 2.4.1–2.4.8, Aug. 1981.

Kay, S., "Noise Compensation for Autoregressive Spectral Estimates," *IEEE Trans. Acoust. Speech Signal Process.*, Vol. ASSP28, pp. 292–303, June 1980.

Kendall, M. and A. Stuart, *The Advanced Theory of Statistics,* Vol. 2, Macmillan, New York 1979.

Kohn, R., "Note Concerning the Akaike and Hannan Estimation Procedures for an Autoregressive-Moving Average Process," *Biometrika,* Vol. 64, pp. 622–625, 1977.

Konvalinka, I. S., and M. R. Matausek, "Simultaneous Estimation of Poles and Zeros in Speech Analysis and ITIF-Iterative Inverse Filtering Algorithm," *IEEE Trans. Acoust. Speech Signal Process.,* Vol. ASSP27, pp. 485–492, Oct. 1979.

Lawson, R. J., and C. L. Hanson, *Solving Least Squares Problems,* Prentice-Hall, Englewood Cliffs, N.J., 1974.

Mayne, D. Q., personal communication of D. Q. Mayne with S. Kay, 1979.

Mayne, D. Q., and F. Firoozan, "An Efficient, Multistage, Linear Identification Method for ARMA Processes," *Proc. IEEE Conf. Decision and Control,* New Orleans, Vol. 1, pp. 435–438, Dec. 1977.

Mehra, R. K., "On-Line Identification of Linear Dynamic Systems with Applications to Kalman Filtering," *IEEE Trans. Autom. Control,* Vol. AC16, pp. 12–21, Feb. 1971.

Murthy, D. N., and R. E. Kronauer, "Methods for Identifying Linear Dynamic Systems," *IEEE Trans. Autom. Control,* Vol. AC18, pp. 549–551, Oct. 1973.

Ooe, M., "An Optimal Complex ARMA Model of the Chandler Wobble," *Geophys. J. Roy. Astron. Soc.,* Vol. 53, pp. 445–457, 1978.

Sakai, H., and H. Tokumaru, "Statistical Analysis of a Spectral Estimator for ARMA Processes," *IEEE Trans. Autom. Control,* Vol. AC25, pp. 122–124, Feb. 1980.

Soderstrom, T., and P. Stoica, "Comparison of Some Instrumental Variable Methods—Consistency and Accuracy Aspects," *Automatica,* Vol. 17, pp. 101–115, 1981.

Stoica, P., and T. Soderstrom, "Optimal Instrumental Variable Estimation and Approximate Implementations," *IEEE Trans. Autom. Control,* Vol. AC28, pp. 757–772, July 1983.

Trench, W. F., "An Algorithm for the Inversion of Finite Toeplitz Matrices," *J. Soc. Ind. Appl. Math.,* Vol. 12, pp. 515–522, Sept. 1964.

Tretter, S. A., and K. Stieglitz, "Power-Spectrum Identification in Terms of Rational Models," *IEEE Trans. Autom. Control,* Vol. AC12, pp. 185–188, Apr. 1967.

Tuan, P., "The Estimation of Parameters for Autoregressive–Moving Average Models from Sample Autocovariances," *Biometrika,* Vol. 66, pp. 555–560, 1979.

Upadhyaya, B. R., "Spectral Domain Estimation of Autoregressive Moving-Average Models Using Shuttle Iterative Methods," *Comput. Electr. Engrg.,* Vol. 6, pp. 163–173, 1979.

Walker, A. M., "Large-Sample Estimation of Parameters of Autoregressive Processes with Moving-Average Residuals," *Biometrika,* Vol. 49, pp. 117–131, 1962.

PROBLEMS

10.1. Plot the function

$$g(x) = \exp\left(-\frac{x^2}{2}\right) + 0.1 \exp\left[-\frac{(x-10)^2}{2}\right]$$

over the domain $-3 \leq x \leq 13$. Find the maximum of the function from the graph. Now use a Newton–Raphson iteration $[x_{k+1} = x_k - g'(x_k)/g''(x_k)]$ to find the maximum. $g'(x)$ and $g''(x)$ denote the first and second derivatives, respectively. Use as the initial estimate $x_0 = 0.5$ and then repeat the experiment for $x_0 = 9.5$. What can you say about the importance of the initial estimate in the Newton–Raphson iteration?

10.2. Let $\hat{a}[1] = -\hat{r}_{xx}[2]/\hat{r}_{xx}[1]$ be the MYWE estimator of the AR filter parameter of a real ARMA(1, 1) process. Assuming an unbiased ACF estimator, find a numerical example to demonstrate that the magnitude of $\hat{a}[1]$ can be greater than 1.

10.3. Compare the variance of the MYWE estimator of the AR filter parameter of a real ARMA(1, 1) process to the CR bound as given by (9.10). Do so by plotting the ratio of the variances for $b[1] = 0.75$ and $-1 < a[1] < 1$.

10.4. Set up the MYWE estimator to estimate the AR filter parameters of a noise corrupted AR process, that is,

$$y[n] = x[n] + w[n]$$

where $x[n]$ is an AR(p) process and $w[n]$ is white observation noise uncorrelated with $x[n]$. Which samples of the ACF need to be estimated to substitute into the equations?

10.5. Consider the real linear model of Section 3.3.2:

$$\mathbf{x} = \mathbf{H}\boldsymbol{\theta} + \mathbf{z}$$

where $\mathbf{z}$ is a zero-mean Gaussian random vector of dimension $N \times 1$, $\boldsymbol{\theta}$ is an unknown constant vector of dimension $k \times 1$, $\mathbf{H}$ is a known constant matrix of dimension $N \times k$ and $\mathbf{x}$ is an observed random vector of dimension $N \times 1$. If the covariance matrix of $\mathbf{z}$ is $\sigma_z^2\mathbf{I}$, where $\mathbf{I}$ is the identity matrix, prove that the MLE of $\boldsymbol{\theta}$ is the usual LS estimator

$$\hat{\boldsymbol{\theta}} = (\mathbf{H}^T\mathbf{H})^{-1}\mathbf{H}^T\mathbf{x}.$$

If now the covariance matrix is $\mathbf{C}$, prove that the MLE is given by the *generalized* LS estimator [Kendall and Stuart 1979]

$$\hat{\boldsymbol{\theta}} = (\mathbf{H}^T\mathbf{C}^{-1}\mathbf{H})^{-1}\mathbf{H}^T\mathbf{C}^{-1}\mathbf{x}.$$

$\hat{\boldsymbol{\theta}}$ may also be viewed as a weighted LS estimator if the necessary assumptions required for the MLE are not justified. *Hint:* Let $\mathbf{C}^{-1} = \mathbf{A}^T\mathbf{A}$, where $\mathbf{A}$ is $N \times N$, so that

$$\begin{aligned}(\mathbf{x} - \mathbf{H}\boldsymbol{\theta})^T\mathbf{C}^{-1}(\mathbf{x} - \mathbf{H}\boldsymbol{\theta}) &= (\mathbf{A}\mathbf{x} - \mathbf{A}\mathbf{H}\boldsymbol{\theta})^T(\mathbf{A}\mathbf{x} - \mathbf{A}\mathbf{H}\boldsymbol{\theta})\\ &= (\mathbf{x}' - \mathbf{H}'\boldsymbol{\theta})^T(\mathbf{x}' - \mathbf{H}'\boldsymbol{\theta}).\end{aligned}$$

Based on these results, how could one improve the LSMYWE estimator? What a priori knowledge is needed? (See the paper by Soderstrom and Stoica [1981].)

10.6. Verify the following Yule–Walker equations for an ARMA(p, q) process:

$$r_{xx}[k] = -\sum_{l=1}^{p} a[l]r_{xx}[k-l] + \sum_{l=1}^{q} b[l]r_{xu}[k-l] \qquad k \geq 1 \tag{10.28}$$

$$r_{ux}[k] = -\sum_{l=1}^{p} a[l]r_{ux}[k-l] + \sum_{l=1}^{q} b[l]r_{uu}[k-l] \qquad k \geq 1. \tag{10.29}$$

Now rewrite the first equation for $k = 1, 2, \ldots, p$ and the second equation for k

= 1, 2, . . . , q and combine all the equations to form the matrix expression

$$\begin{bmatrix} \mathbf{R}_{xx} & \mathbf{R}_{xu} \\ \mathbf{R}_{ux} & \mathbf{R}_{uu} \end{bmatrix} \begin{bmatrix} -\mathbf{a} \\ \mathbf{b} \end{bmatrix} = \begin{bmatrix} \mathbf{r}_{xx} \\ \mathbf{r}_{ux} \end{bmatrix}$$

where

$$[\mathbf{R}_{xx}]_{ij} = r_{xx}[i-j] \qquad \begin{array}{l} i = 1, 2, \ldots, p \\ j = 1, 2, \ldots, p \end{array}$$

$$[\mathbf{R}_{uu}]_{ij} = r_{uu}[i-j] \qquad \begin{array}{l} i = 1, 2, \ldots, q \\ j = 1, 2, \ldots, q \end{array}$$

$$[\mathbf{R}_{xu}]_{ij} = r_{xu}[i-j] \qquad \begin{array}{l} i = 1, 2, \ldots, p \\ j = 1, 2, \ldots, q \end{array}$$

$$[\mathbf{R}_{ux}]_{ij} = r_{ux}[i-j] \qquad \begin{array}{l} i = 1, 2, \ldots, q \\ j = 1, 2, \ldots, p \end{array}$$

and

$$\begin{array}{ll} [\mathbf{r}_{xx}]_i = r_{xx}[i] & i = 1, 2, \ldots, p \\ [\mathbf{r}_{ux}]_i = r_{ux}[i] & i = 1, 2, \ldots, q. \end{array}$$

Show that the two-stage least squares estimator is an equation error modeling approach where (10.28) and (10.29) are solved with estimated correlations replacing the true ones.

10.7. An ARMA(p, q) process is fitted with a large order AR model, resulting in the AR parameters $c[k]$, so that

$$\frac{\mathcal{B}(z)}{\mathcal{A}(z)} \approx \frac{1}{\mathcal{C}(z)}$$

where $\mathcal{C}(z) = 1 + \sum_{k=1}^{L} c[k]z^{-k}$. How can the ARMA filter parameters be recovered from the large-order AR parameters? See Section 5.3.2 for some hints. What type of performance would be expected using this method? (See the paper by Graupe et al. [1975].)

APPENDIX 10A

Evaluation of Partial Derivatives for Akaike MLE

The gradient of $Q(\mathbf{a}, \mathbf{b})$ is first evaluated making use of results in Appendix 9A. Recall that

$$\begin{aligned} p(\mathbf{x}; \mathbf{a}, \mathbf{b}, \sigma^2) &= \frac{1}{(2\pi\sigma^2)^{N/2}} \exp\left[-\frac{N}{2\sigma^2}\int_{-1/2}^{1/2} I(f)\frac{|A(f)|^2}{|B(f)|^2}\,df\right] \\ &= \frac{1}{(2\pi\sigma^2)^{N/2}} \exp\left[-\frac{N}{2\sigma^2} Q(\mathbf{a}, \mathbf{b})\right]. \end{aligned}$$

Hence

$$\frac{\partial \ln p(\mathbf{x}; \mathbf{a}, \mathbf{b}, \sigma^2)}{\partial a[k]} = -\frac{N}{2\sigma^2}\frac{\partial Q(\mathbf{a}, \mathbf{b})}{\partial a[k]}$$

and

$$\frac{\partial \ln p(\mathbf{x}; \mathbf{a}, \mathbf{b}, \sigma^2)}{\partial b[l]} = -\frac{N}{2\sigma^2}\frac{\partial Q(\mathbf{a}, \mathbf{b})}{\partial b[l]} .$$

It follows from the expressions developed in Appendix 9A for $\partial \ln p(\mathbf{x}; \mathbf{a}, \mathbf{b}, \sigma^2)/\partial a[k]$ and $\partial \ln p(\mathbf{x}; \mathbf{a}, \mathbf{b}, \sigma^2)/\partial b[l]$ that

$$\frac{\partial Q}{\partial a[k]} = 2\int_{-1/2}^{1/2} I(f)\frac{A(f)\exp(j2\pi fk)}{|B(f)|^2}\,df$$

$$\frac{\partial Q}{\partial b[l]} = -2\int_{-1/2}^{1/2} I(f)\frac{|A(f)|^2 B(f)\exp(j2\pi fl)}{|B(f)|^4}\,df.$$

The second order partials $\partial^2 \ln p(\mathbf{x}; \mathbf{a}, \mathbf{b}, \sigma^2)/\partial a[k]\,\partial a[l]$, $\partial^2 \ln p(\mathbf{x}; \mathbf{a}, \mathbf{b}, \sigma^2)/\partial b[l]\,\partial b[k]$, $\partial^2 \ln p(\mathbf{x}; \mathbf{a}, \mathbf{b}, \sigma^2)/\partial a[k]\,\partial b[l]$ are also evaluated in Appendix 9A. From those results we obtain

$$\frac{\partial^2 Q}{\partial a[k]\,\partial a[l]} = 2\int_{-1/2}^{1/2} I(f)\frac{\exp[j2\pi f(k-l)]}{|B(f)|^2}\,df$$

$$\frac{\partial^2 Q}{\partial b[l]\,\partial b[k]} = 4\int_{-1/2}^{1/2} I(f)\frac{|A(f)|^2\exp[j2\pi f(k+l)]}{|B(f)|^2 B^*(f)^2}\,df + 2\int_{-1/2}^{1/2} I(f)\frac{|A(f)|^2\exp[j2\pi f(l-k)]}{|B(f)|^4}\,df$$

$$\frac{\partial^2 Q}{\partial a[k]\,\partial b[l]} = -2\int_{-1/2}^{1/2} I(f)\frac{|A(f)|^2\exp[j2\pi f(k+l)]}{|B(f)|^2 A^*(f)B^*(f)}\,df - 2\int_{-1/2}^{1/2} I(f)\frac{|A(f)|^2\exp[j2\pi f(k-l)]}{|B(f)|^2 B(f)A^*(f)}\,df.$$

$\partial^2 Q/\partial b[l]\,\partial b[k]$ and $\partial^2 Q/\partial a[k]\,\partial b[l]$ can be approximated by setting the first term of each expression equal to zero. To justify this step, assume that the partial derivatives are evaluated near the true ARMA parameters and that N is large. Let

$$|U(f)|^2 = I(f)\frac{|A(f)|^2}{|B(f)|^2}$$

which is the periodogram of the sequence obtained by filtering $x[n]$ with $\mathcal{A}(z)/\mathcal{B}(z)$. If $\mathcal{A}(z)$ and $\mathcal{B}(z)$ are the true ARMA filters, then $|U(f)|^2$ is the periodogram of $u[n]$, the white noise excitation. For large N

$$\mathcal{E}(|U(f)|^2) = P_{uu}(f) = \sigma^2$$

by the convergence of the mean of the periodogram to the true PSD [see (4.1)].

The average value of the first term of $\partial^2 Q/\partial b[l]\,\partial b[k]$ is

$$\mathscr{E}\left[4\int_{-1/2}^{1/2} I(f)\frac{|A(f)|^2 \exp[j2\pi f(k+l)]}{|B(f)|^2 B^*(f)^2}\,df\right]$$

$$= 4\int_{-1/2}^{1/2} \frac{\sigma^2 \exp[j2\pi f(k+l)]}{B^*(f)^2}\,df$$

$$= 4\sigma^2 \mathscr{Z}^{-1}\left[\frac{1}{\mathscr{B}^2(z^{-1})}\right]\Bigg|_{n=k+l} = 0.$$

The last equation follows because $1/\mathscr{B}^2(z^{-1})$ is the z-transform of an anticausal sequence and $n = k + l > 0$ since $k = 1, 2, \ldots, q$; $l = 1, 2, \ldots, q$. Also, the average value of the first term of $\partial^2 Q/\partial a[k]\,\partial b[l]$ is

$$-2\mathscr{E}\left[\int_{-1/2}^{1/2} I(f)\frac{|A(f)|^2 \exp[j2\pi f(k+l)]}{|B(f)|^2 A^*(f)B^*(f)}\,df\right]$$

$$= -2\int_{-1/2}^{1/2} \frac{\sigma^2 \exp[j2\pi f(k+1)]}{A^*(f)B^*(f)}\,df$$

$$= -2\sigma^2 \mathscr{Z}^{-1}\left[\frac{1}{\mathscr{A}(z^{-1})\mathscr{B}(z^{-1})}\right]\Bigg|_{n=k+l} = 0$$

by the same reasoning. Neglecting these terms produces the approximate partial derivatives

$$\frac{\partial^2 Q}{\partial b[l]\,\partial b[k]} = 2\int_{-1/2}^{1/2} I(f)\frac{|A(f)|^2 \exp[j2\pi f(l-k)]}{|B(f)|^4}\,df$$

$$\frac{\partial^2 Q}{\partial a[k]\,\partial b[l]} = -2\int_{-1/2}^{1/2} I(f)\frac{|A(f)|^2 \exp[j2\pi f(k-l)]}{|B(f)|^2 B(f)A^*(f)}\,df.$$

For the purposes of computation the partial derivatives can be expressed as correlations of filtered sequences. Define the sequences

$$y[n] = \mathscr{Z}^{-1}\left[\frac{\mathscr{X}(z)}{\mathscr{B}(z)}\right]$$

$$z[n] = \mathscr{Z}^{-1}\left[\frac{\mathscr{X}(z)\mathscr{A}(z)}{\mathscr{B}^2(z)}\right]$$

where $\mathscr{X}(z) = \sum_{n=0}^{N-1} x[n]z^{-n}$. The initial conditions for the causal recursive filters $1/\mathscr{B}(z)$ and $1/\mathscr{B}^2(z)$ are assumed to be zero since $x[n]$ is known only for $0 \le n \le N-1$. Also, to reduce the end effects the filter outputs are computed only for $n = 0$ to $n = N - 1$, so that we do not "run off the data." Values of $y[n]$, $z[n]$ outside this interval are assumed equal to zero. Then

$$\frac{\partial Q}{\partial a[k]} = \frac{2}{N}\int_{-12}^{1/2} \frac{|X(f)|^2 A(f) \exp(j2\pi fk)}{|B(f)|^2}\, df$$

$$= \frac{2}{N}\int_{-1/2}^{1/2} |Y(f)|^2 A(f) \exp(j2\pi fk)\, df$$

$$= \frac{2}{N}\sum_{i=0}^{p} a[i] \int_{-1/2}^{1/2} |Y(f)|^2 \exp[j2\pi f(k-i)]\, df.$$

By Parseval's theorem

$$\int_{-1/2}^{1/2} V^*(f)W(f) \exp(j2\pi fk)\, df = \sum_{n=-\infty}^{\infty} v[n]w[n+k]$$

so that

$$\frac{\partial Q}{\partial a[k]} = \frac{2}{N}\sum_{i=0}^{p} a[i] \sum_{n=-\infty}^{\infty} y[n]y[n+(k-i)].$$

Because $y[n]$ is zero outside the interval $0 \le n \le N-1$,

$$\frac{\partial Q}{\partial a[k]} = 2\sum_{i=0}^{p} a[i] \frac{1}{N} \sum_{n=0}^{N-|k-i|-1} y[n]y[n+|k-i|]$$

$$= 2\sum_{i=0}^{p} a[i]\hat{r}_{yy}[k-i]$$

where $\hat{r}_{yy}[k]$ is the biased ACF estimator of $y[n]$. In a similar fashion we can show that

$$\frac{\partial Q}{\partial b[l]} = -2\sum_{i=0}^{q} b[i]\hat{r}_{zz}[l-i]$$

where $\hat{r}_{zz}[k]$ is the biased ACF estimator of $z[n]$.

The second order partial derivatives are

$$\frac{\partial^2 Q}{\partial a[k]\,\partial a[l]} = \frac{2}{N}\int_{-1/2}^{1/2} |Y(f)|^2 \exp[j2\pi f(k-l)]\, df$$

$$= 2\hat{r}_{yy}[k-l]$$

$$\frac{\partial^2 Q}{\partial b[l]\,\partial b[k]} = \frac{2}{N}\int_{-1/2}^{1/2} |Z(f)|^2 \exp[j2\pi f(l-k)]\, df$$

$$= 2\hat{r}_{zz}[k-l]$$

$$\frac{\partial^2 Q}{\partial a[k]\,\partial b[l]} = -\frac{2}{N}\int_{-1/2}^{1/2} Y^*(f)Z(f) \exp[j2\pi f(k-l)]\, df$$

$$= -\frac{2}{N}\sum_{n=-\infty}^{\infty} y[n]z[n+k-l].$$

Since $y[n]$, $z[n]$ are assumed to be zero for n outside the interval $0 \le n \le N - 1$,

$$\sum_{n=-\infty}^{\infty} y[n]z[n+k-l] = \begin{cases} \displaystyle\sum_{n=0}^{N-(k-l)-1} y[n]z[n+k-l] & \text{for } k \ge l \\ \displaystyle\sum_{n=l-k}^{N-1} y[n]z[n+k-l] & \text{for } k < l \end{cases}$$

which leads to the final result

$$\frac{\partial^2 Q}{\partial a[k]\, \partial b[l]} = -2\hat{r}_{yz}[k-l]$$

where

$$\hat{r}_{yz}[k] = \begin{cases} \displaystyle\frac{1}{N}\sum_{n=0}^{N-k-1} y[n]z[n+k] & \text{for } k \ge 0 \\ \displaystyle\frac{1}{N}\sum_{n=-k}^{N-1} y[n]z[n+k] & \text{for } k < 0. \end{cases}$$

APPENDIX 10B

Positive Definite Property of Approximate Hessian

The approximate Hessian used in the Akaike MLE is given by (10.4). In matrix form it is

$$\mathbf{H}(\mathbf{a}, \mathbf{b}) = 2$$

$$\begin{bmatrix} \hat{r}_{yy}[0] & \hat{r}_{yy}[-1] & \cdots & \hat{r}_{yy}[-(p-1)] & -\hat{r}_{yz}[0] & -\hat{r}_{yz}[-1] & \cdots & -\hat{r}_{yz}[-(q-1)] \\ \hat{r}_{yy}[1] & \hat{r}_{yy}[0] & \cdots & \hat{r}_{yy}[-(p-2)] & -\hat{r}_{yz}[1] & -\hat{r}_{yz}[0] & \cdots & -\hat{r}_{yz}[-(q-2)] \\ \vdots & \vdots & \ddots & \vdots & \vdots & \vdots & \vdots & \vdots \\ \hat{r}_{yy}[p-1] & \hat{r}_{yy}[p-2] & \cdots & \hat{r}_{yy}[0] & -\hat{r}_{yz}[p-1] & -\hat{r}_{yz}[p-2] & \cdots & -\hat{r}_{yz}[p-q] \\ -\hat{r}_{yz}[0] & -\hat{r}_{yz}[1] & \cdots & -\hat{r}_{yz}[p-1] & \hat{r}_{zz}[0] & \hat{r}_{zz}[-1] & \cdots & \hat{r}_{zz}[-(q-1)] \\ -\hat{r}_{yz}[-1] & -\hat{r}_{yz}[0] & \cdots & -\hat{r}_{yz}[p-2] & \hat{r}_{zz}[1] & \hat{r}_{zz}[0] & \cdots & \hat{r}_{zz}[-(q-2)] \\ \vdots & \vdots & \vdots & \vdots & \vdots & \vdots & \ddots & \vdots \\ -\hat{r}_{yz}[-(q-1)] & -\hat{r}_{yz}[-(q-2)] & \cdots & -\hat{r}_{yz}[p-q] & \hat{r}_{zz}[q-1] & \hat{r}_{zz}[q-2] & \cdots & \hat{r}_{zz}[0] \end{bmatrix}$$

$\mathbf{H}(\mathbf{a}, \mathbf{b})$ is symmetric since $\hat{r}_{yy}[-k] = \hat{r}_{yy}[k]$ and $\hat{r}_{zz}[-k] = \hat{r}_{zz}[k]$. To show that it is positive definite, observe that the Hessian can be written as (assuming that $p \geq q$)

$$\mathbf{H}(\mathbf{a}, \mathbf{b}) = \frac{2}{N}\begin{bmatrix} \mathbf{Y}^T\mathbf{Y} & -\mathbf{Y}^T\mathbf{Z} \\ -\mathbf{Z}^T\mathbf{Y} & \mathbf{Z}^T\mathbf{Z} \end{bmatrix}$$

where

$$\mathbf{Y}^T = \begin{bmatrix} y[0] & y[1] & \cdots & y[N-2] & y[N-1] & 0 & 0 & \cdots & 0 \\ 0 & y[0] & \cdots & y[N-3] & y[N-2] & y[N-1] & 0 & \cdots & 0 \\ \vdots & \vdots & \ddots & \ddots & \ddots & \ddots & \ddots & \ddots & \vdots \\ 0 & 0 & \cdots & 0 & y[0] & y[1] & y[2] & \cdots & y[N-1] \end{bmatrix}$$

which is of dimension $p \times (N + p - 1)$ and

$$\mathbf{Z}^T = \begin{bmatrix} z[0] & z[1] & \cdots & z[N-2] & z[N-1] & 0 & 0 & \cdots & 0 \\ 0 & z[0] & \cdots & z[N-3] & z[N-2] & z[N-1] & 0 & \cdots & 0 \\ \vdots & \vdots & \ddots & \ddots & \ddots & \ddots & \ddots & \ddots & \vdots \\ 0 & 0 & \cdots & 0 & z[0] & z[1] & z[2] & \cdots & z[N-1] \end{bmatrix}$$

which has dimension $q \times (N + p - 1)$. Since the Hessian can be expressed as

$$\mathbf{H}(\mathbf{a}, \mathbf{b}) = \frac{2}{N}[\mathbf{Y} \quad -\mathbf{Z}]^T[\mathbf{Y} \quad -\mathbf{Z}]$$

by a standard theorem in linear algebra (see Theorem 5 of Section 2.5), the Hessian is positive definite since the columns of the matrix $[\mathbf{Y} \quad -\mathbf{Z}]$ are linearly independent.

APPENDIX 10C

Computer Program for Akaike MLE (Provided on Floppy Disk)

```
      SUBROUTINE AKAIKE(X,N,A0,B0,IP,IQ,NIT,EPS,PER,A,B,SIG2,IFLAG)
C
C     This program implements the Akaike MLE algorithm for
C     ARMA parameter estimation of real data. A complete
C     description of the method is contained in Section 10.3.
C
C     Input Parameters:
C
```

```
C         X          -Real array of dimension N×1 of real data points
C         N          -Number of real data points
C         A0         -Real array of dimension IP×1 containing initial
C                     estimates of AR parameters for Newton-Raphson
C                     iteration arranged as A0(1) to A0(IP)
C         B0         -Real array of dimension IQ×1 containing initial minimum-phase
C                     estimates of MA parameters for Newton-Raphson
C                     iteration arranged as B0(1) to B0(IQ)
C         IP         -AR model order (must be one or greater)
C         IQ         -MA model order (must be one or greater)
C         NIT        -Maximum number of iterations allowed
C         EPS        -Small positive number to test for ill-conditioning
C                     of Hessian
C         PER        -Stopping criterion, if percentage change from previous
C                     iteration in value of function to be minimized is
C                     less than PER, iteration is halted
C
C       Output Parameters:
C
C         A          -Real array of dimension IP×1 of AR parameter
C                     estimates arranged as A(1) to A(IP)
C         B          -Real array of dimension IQ×1 of MA parameter
C                     estimates arranged as B(1) to B(IQ)
C         SIG2       -Excitation white noise variance estimate
C         IFLAG      -This is a flag indicating conditions of termination
C                     of iteration:
C                        = -1, if one of the di's in CHOLESKY is less than
C                             EPS causing premature termination (Hessian
C                             is ill-conditioned)
C                        =  0, percentage error change in function
C                             less than PER causing termination
C                        =  1, MA filter estimate not minimum-phase
C                             causing premature termination
C                        =  2, all NIT iterations completed causing
C                             termination
C
C       External Subroutines:
C
C         CHOLESKY      -see Appendix 2B
C         STEPDOWN      -see Appendix 6C
C
C       Notes:
C
C         The calling program must dimension arrays X,A0,
C         B0,A,B. The arrays BVC,HC,BC,BBC,Y,Y1,Z,RY,RZ,
C         RYZ,RZY,G,H,BV,P must be dimensioned greater than
C         or equal to the variable dimensions shown. The arrays
C         Y,E in VAR must be dimensioned greater than or equal
C         to N and the array H in HESS, BVECT must be dimensioned
C         to agree with the array H in AKAIKE. The arrays XL,Y,D,
```

```
C          in CHOLESKY must be dimensioned to be greater than or
C          equal to (IP+IQ)×(IP+IQ), IP+IQ, and IP+IQ, respectively.
C          The array AA in STEPDOWN must be dimensioned to agree with
C          BBC in AKAIKE.
C
C        Verification Test Case:
C
C          If the real test data listed in Appendix 1A is input
C          as X, with N=32, IP=7, IQ=3, NIT=100, EPS=1.0E-15, PER=0.1,
C          and with the initial estimates set as
C            A0(1) =   0.22197
C            A0(2) = -1.09315
C            A0(3) = -0.59794
C            A0(4) =   0.62299
C            A0(5) =   0.28311
C            A0(6) =   0.13978
C            A0(7) =   0.13255
C            B0(1) =   3.57056E-02
C            B0(2) = -0.41063
C            B0(3) =   6.39064E-02
C          then the output should be
C            A(1)   = -0.38280
C            A(2)   = -0.85412
C            A(3)   =   0.25719
C            A(4)   =   0.59831
C            A(5)   = -0.43314
C            A(6)   =   9.61728E-02
C            A(7)   =   0.21083
C            B(1)   = -0.44578
C            B(2)   = -0.24197
C            B(3)   =   0.46824
C            SIG2   =   1.67210
C            IFLAG  =   0
C          for a DEC VAX 11/780.
C
           COMPLEX BVC(IP+IQ),HC((IP+IQ)*(IP+IQ+1)/2),BC(IP+IQ),BBC(IQ,IQ)
           DIMENSION A(1),B(1),A0(1),B0(1),X(1),Y(N),Y1(N),
          * Z(N),RY(IP+1),RZ(IQ+1),RYZ(IP),RZY(IQ),G(IP+IQ),
          * H(IP+IQ,IP+IQ),BV(IP+IQ),P(IQ)
C        Compute estimate of white noise variance or Q(a,b) using initial
C        ARMA filter parameter estimates (10.1)
           CALL VAR(X,N,IP,IQ,A0,B0,SIG2P)
           IFLAG=3
           DO 10 I=1,IP
10         A(I)=A0(I)
           DO 20 I=1,IQ
20         B(I)=B0(I)
           DO 100 K=1,NIT
C        Filter data with 1/B(z) to produce y[n] sequence
C        and with A(z)/B^2(z) to produce z[n] sequence (10.5)
```

```
      CALL POLEFILT(X,N,IQ,B,Y)
      CALL POLEFILT(Y,N,IQ,B,Y1)
      CALL ZEROFILT(Y1,N,IP,A,Z)
C     Compute estimates of autocorrelations and cross-
C     correlations, ryy[n],rzz[n],ryz[n],rzy[n]
      CALL CORRCOMP(N,Y,Z,IP,IQ,RY,RZ,RYZ,RZY)
C     Compute gradient vector (10.3)
      CALL GRAD(A,B,IP,IQ,RY,RZ,G)
C     Compute Hessian matrix (10.4)
      CALL HESS(IP,IQ,RY,RZ,RYZ,RZY,H)
C     Compute right-hand-side vector of linear equations
C     to be solved for next iterate (10.6)
      CALL BVECT(A,B,IP,IQ,G,H,BV)
C     Set right-hand-side vector and Hessian equal to
C     complex quantities for use in CHOLESKY
      L=0
      DO 40 J=1,IP+IQ
      BVC(J)=CMPLX(BV(J),0.)
      DO 30 I=1,J
      L=L+1
30    HC(L)=CMPLX(H(I,J),0.)
40    CONTINUE
C     Solve equations to produce next iterate (10.6)
      M=IP+IQ
      CALL CHOLESKY(HC,BVC,M,EPS,IFLAG)
C     Terminate program if equations are ill-conditioned
      IF(IFLAG.EQ.-1)RETURN
      IFLAG=3
C     Check to see if estimate of B(z) is minimum-phase by
C     checking the reflection coefficients
      DO 50 I=1,IQ
50    BC(I)=BVC(IP+I)
      IF(IQ.EQ.1)BBC(1,1)=BC(1)
      PMA=1.
      IF(IQ.GT.1)CALL STEPDOWN(IQ,BC,PMA,BBC,P0,P)
C     If B(z) estimate is not minimum-phase, terminate program
      DO 60 I=1,IQ
      IF(CABS(BBC(I,I)).LT.1.0)GO TO 60
      IFLAG=1
      RETURN
60    CONTINUE
C     Set up new estimates of ARMA parameters
      DO 70 I=1,IP
70    A(I)=REAL(BVC(I))
      DO 80 I=1,IQ
80    B(I)=REAL(BVC(I+IP))
C     Compute new estimate of white noise variance or Q(a,b) and
C     if the change in the variance (error function to be
C     minimized) is less than PER, terminate program (10.1)
```

```
      CALL CHKDN(SIG2P,N,X,A,B,IP,IQ,PER,SIG2,IFLAG)
      IF(IFLAG.EQ.0)RETURN
C     Check to see if all iterations completed and if so
C     terminate program
      IF(K.LT.NIT)GO TO 90
      IFLAG=2
      RETURN
90    SIG2P=SIG2
100   CONTINUE
      RETURN
      END
      SUBROUTINE VAR(X,N,IP,IQ,A,B,SIG2)
C     This program computes the estimate of the white
C     noise variance, which is the error function to
C     be minimized.
      DIMENSION X(1),Y(N),E(N),A(1),B(1)
      CALL POLEFILT(X,N,IQ,B,Y)
      CALL ZEROFILT(Y,N,IP,A,E)
      SUM=0.
      DO 10 I=1,N
10    SUM=SUM+E(I)*E(I)
      SIG2=SUM/N
      RETURN
      END
      SUBROUTINE POLEFILT(X,N,IQ,B,Y)
C     This program implements an all-pole filter, 1/B(z), with
C     the initial conditions of the filter set equal to zero.
      DIMENSION X(1),Y(1),B(1)
      Y(1)=X(1)
      IF(IQ.EQ.1)GO TO 25
      DO 20 I=2,IQ
      Y(I)=X(I)
      DO 10 J=1,I-1
10    Y(I)=Y(I)-B(J)*Y(I-J)
20    CONTINUE
25    DO 40 I=IQ+1,N
      Y(I)=X(I)
      DO 30 J=1,IQ
30    Y(I)=Y(I)-B(J)*Y(I-J)
40    CONTINUE
      RETURN
      END
      SUBROUTINE ZEROFILT(X,N,IP,A,Y)
C     This program implements an FIR filter with coefficients
C     given by A. The values of the input X(I) for I < 1 are
C     set equal to zero.
      DIMENSION X(1),Y(1),A(1)
      Y(1)=X(1)
      IF(IP.EQ.1)GO TO 25
```

```
      DO 20 I=2,IP
      Y(I)=X(I)
      DO 10 J=1,I-1
10    Y(I)=Y(I)+A(J)*X(I-J)
20    CONTINUE
25    DO 40 I=IP+1,N
      Y(I)=X(I)
      DO 30 J=1,IP
30    Y(I)=Y(I)+A(J)*X(I-J)
40    CONTINUE
      RETURN
      END
      SUBROUTINE CORRCOMP(N,Y,Z,IP,IQ,RY,RZ,RYZ,RZY)
C     This program computes the autocorrelations and cross-
C     correlations needed in the Akaike algorithm.
      DIMENSION Y(1),Z(1),RY(1),RZ(1),RYZ(1),RZY(1)
      DO 10 K=0,IP
      CALL CORR(N,K,Y,Y,R)
10    RY(K+1)=R
      DO 20 K=0,IQ
      CALL CORR(N,K,Z,Z,R)
20    RZ(K+1)=R
      DO 30 K=0,IP-1
      CALL CORR(N,K,Y,Z,R)
30    RYZ(K+1)=R
      DO 40 K=1,IQ-1
      CALL CORR(N,K,Z,Y,R)
40    RZY(K+1)=R
      RETURN
      END
      SUBROUTINE CORR(N,LAG,X,Y,R)
C     This program computes a cross (auto) - correlation
C     from two input data sequences.
      DIMENSION X(1),Y(1)
      SUM=0.
      DO 10 J=1,N-LAG
10    SUM=SUM+X(J)*Y(J+LAG)
      R=SUM/N
      RETURN
      END
      SUBROUTINE GRAD(A,B,IP,IQ,RY,RZ,G)
C     This program computes the gradient vector
C     needed in the Akaike algorithm.
      DIMENSION A(1),B(1),RY(1),RZ(1),G(1)
      DO 20 K=1,IP
      G(K)=2.*RY(K+1)
      DO 10 I=1,IP
10    G(K)=G(K)+2.*A(I)*RY(IABS(K-I)+1)
20    CONTINUE
```

```
      DO 40 L=1,IQ
      G(IP+L)= -2.*RZ(L+1)
      DO 30 I=1,IQ
30    G(IP+L)=G(IP+L)-2.*B(I)*RZ(IABS(L-I)+1)
40    CONTINUE
      RETURN
      END
      SUBROUTINE HESS(IP,IQ,RY,RZ,RYZ,RZY,H)
C     This program computes the Hessian needed in the
C     Akaike algorithm.
      DIMENSION RY(1),RZ(1),RYZ(1),RZY(1),H(IP+IQ,IP+IQ)
      DO 20 J=1,IP
      DO 10 K=1,IP
10    H(J,K)=2.*RY(IABS(J-K)+1)
20    CONTINUE
      DO 40 J=IP+1,IP+IQ
      DO 30 K=IP+1,IP+IQ
30    H(J,K)=2.*RZ(IABS(J-K)+1)
40    CONTINUE
      DO 60 J=1,IP
      DO 50 K=IP+1,IP+IQ
      IF(J.GE.(K-IP))H(J,K)= -2.*RYZ(J-(K-IP)+1)
50    IF(J.LT.(K-IP))H(J,K)= -2.*RZY((K-IP)-J+1)
60    CONTINUE
      DO 80 J=IP+1,IP+IQ
      DO 70 K=1,IP
70    H(J,K)=H(K,J)
80    CONTINUE
      RETURN
      END
      SUBROUTINE BVECT(A,B,IP,IQ,G,H,BV)
C     This program computes the right-hand-side vector
C     of the linear equations to be solved for next iterate.
      DIMENSION A(1),B(1),G(1),H(IP+IQ,IP+IQ),BV(1)
      DO 20 I=1,IP+IQ
      BV(I)= -G(I)
      DO 10 J=1,IP+IQ
      IF(J.LE.IP)T=A(J)
      IF(J.GT.IP)T=B(J-IP)
10    BV(I)=BV(I)+H(I,J)*T
20    CONTINUE
      RETURN
      END
      SUBROUTINE CHKDN(SIG2P,N,X,A,B,IP,IQ,PER,SIG2,IFLAG)
C     This program computes the estimate of the white noise
C     variance (error function to be minimized) given the
C     ARMA filter parameter estimates.
      DIMENSION X(1),A(1),B(1)
      CALL VAR(X,N,IP,IQ,A,B,SIG2)
```

```
ERR=100.*(SIG2-SIG2P)/SIG2P
IF(ABS(ERR).LT.PER)IFLAG=0
RETURN
END
```

APPENDIX 10D

Computer Program for Modified Yule-Walker Equations Method (Provided on Floppy Disk)

```
      SUBROUTINE MYWE(X,N,IP,IQ,EPS,A,B,SIG2,IFLAG)
C
C     This program implements the modified Yule-Walker equation
C     estimator for the AR parameters of an ARMA process (10.7).
C     The unbiased autocorrelation estimator is used. The equations
C     are solved using the recursive algorithm of (10.8)-(10.13).
C     The MA parameters and the white noise variance are estimated
C     by filtering the data by an estimate of A(z) and applying
C     Durbin's method (see Section 8.4) to the data at the output
C     of the filter. The large order AR model in Durbin's method
C     is chosen to have an order of N/5.
C
C     Input Parameters:
C
C       X        -Complex vector of dimension N×1 of data points
C       N        -Number of complex data points
C       IP       -Desired AR model order (must be one or greater)
C       IQ       -Desired MA model order (must be one or greater)
C       EPS      -A small positive number to test for ill-conditioning
C                 of matrix
C
C     Output Parameters:
C
C       A        -Complex vector of dimension IP×1 of AR parameter
C                 estimates arranged as A(1) to A(IP)
C       B        -Complex vector of dimension IQ×1 of MA parameter
C                 estimates arranged as B(1) to B(IQ)
C       SIG2     -Excitation white noise variance estimate
C       IFLAG    -Ill-conditioning indicator for solution of
C                 modified Yule-Walker equations
```

```
C                    =   0 for normal termination
C                    =  -1 if magnitude of one of complex rho's is
C                          less than EPS causing premature termination
C
C     External Subroutines:
C
C       CORRELATION     -see Appendix 4D
C       LEVINSON        -see Appendix 6B
C       AUTOCORR        -see Appendix 7B
C       DURBIN          -see Appendix 8A
C
C     Notes:
C
C       The calling program must dimension arrays X,A,B.
C       The arrays R,Y must be dimensioned greater
C       than or equal to the variable dimensions shown.
C       The arrays AA,BB in TOEPEQN must be dimensioned
C       to both be greater than or equal to IP×IP. The arrays
C       A,AP in DURBIN must be dimensioned greater than or equal
C       to ⌊N/5⌋ and ⌊N/5⌋+1, respectively. (⌊M⌋ denotes the
C       largest integer less than or equal to M). The arrays
C       R,AA,RHO in AUTOCORR must be dimensioned greater than or
C       equal to [N/5]+1, [N/5]×[N/5], and [N/5], respectively.
C       The array AA in LEVINSON must be dimensioned to agree with
C       the dimensions of AA in AUTOCORR.
C
C     Verification Test Case:
C
C       If the complex test data listed in Appendix 1A is input
C       as X, with N=32, IP=7, IQ=3, EPS=1.0E-15, then the output
C       should be
C         A(1)   = ( 1.66754     ,-0.13065     )
C         A(2)   = (-2.26033     ,-0.25969     )
C         A(3)   = (-0.71092     ,-6.01127E-02)
C         A(4)   = ( 0.96275     , 0.41967     )
C         A(5)   = ( 0.25055     , 0.22180     )
C         A(6)   = (-0.90021     ,-0.24728     )
C         A(7)   = ( 1.43252     ,-0.21511     )
C         B(1)   = (-0.49970     ,-0.10093     )
C         B(2)   = (-3.66813E-02,-7.14685E-02)
C         B(3)   = ( 0.16940     , 4.09755E-02)
C         SIG2   = 9.21820
C         IFLAG  = 0
C       for a DEC VAX 11/780.
C
        COMPLEX X(1),A(1),B(1),R(IP+IQ+1),Y(N)
C     Compute unbiased autocorrelation estimates
        LAG=IP+IQ+1
        CALL CORRELATION(N,LAG,0,X,X,R)
C     Solve modified Yule-Walker equations using recursive
```

```
C     algorithm of (10.8)–(10.13)
        CALL TOEPEQN(R,IP,IQ,EPS,A,IFLAG)
        IF(IFLAG.EQ.-1)RETURN
C     Filter data by estimated A(z) to yield approximate
C     MA time series (set initial conditions equal to zero)
20      CALL CZEROFLT(X,N,IP,A,Y)
C     Apply Durbin's method with large order AR model order of
C     N/5 to estimate the MA parameters and white noise variance
        L=N/5
        CALL DURBIN(Y,N,IQ,L,B,SIG2)
        RETURN
        END
        SUBROUTINE CZEROFLT(X,N,IP,A,Y)
C     This program implements an FIR filter with coefficients
C     given by A. The values of the input X(I) for I < 1 are
C     set equal to zero.
        COMPLEX X(1),Y(1),A(1)
        Y(1)=X(1)
        IF(IP.EQ.1)GO TO 25
        DO 20 I=2,IP
        Y(I)=X(I)
        DO 10 J=1,I-1
10      Y(I)=Y(I)+A(J)*X(I-J)
20      CONTINUE
25      DO 40 I=IP+1,N
        Y(I)=X(I)
        DO 30 J=1,IP
30      Y(I)=Y(I)+A(J)*X(I-J)
40      CONTINUE
        RETURN
        END
        SUBROUTINE TOEPEQN(R,IP,IQ,EPS,A,IFLAG)
C     This program solves the modified Yule-Walker
C     equations using a generalized Levinson recursion.
        COMPLEX R(1),A(1),AA(IP,IP),BB(IP,IP),C,D,RHO,CORR
        IFLAG=0
C     Initialize recursion (10.8)
        AA(1,1)=-R(IQ+2)/R(IQ+1)
        IF(IP.EQ.1)GO TO 60
        BB(1,1)=-R(IQ)/R(IQ+1)
        RHO=(1.-AA(1,1)*BB(1,1))*R(IQ+1)
C     Check to see if 2×2 matrix is singular
        IF(CABS(RHO).GT.EPS)GO TO 5
        IFLAG=-1
        RETURN
C     Start main recursion
5       DO 50 K=2,IP
C     Compute ak[k] (10.9)
        C=-R(IQ+1+K)
        DO 10 L=1,K-1
```

```
      LAG=IQ+K-L
      IF(LAG.GE.0)CORR=R(LAG+1)
      IF(LAG.LT.0)CORR=CONJG(R(-LAG+1))
10    C=C-AA(L,K-1)*CORR
      AA(K,K)=C/RHO
C     Update in order the ak[i]'s (10.10)
      DO 20 I=1,K-1
20    AA(I,K)=AA(I,K-1)+AA(K,K)*BB(K-I,K-1)
      IF(K.EQ.IP)GO TO 60
C     Compute bk[k] (10.11)
      LAG=IQ-K
      IF(LAG.GE.0)D=-R(LAG+1)
      IF(LAG.LT.0)D=-CONJG(R(-LAG+1))
      DO 30 L=1,K-1
      LAG=IQ-K+L
      IF(LAG.GE.0)CORR=R(LAG+1)
      IF(LAG.LT.0)CORR=CONJG(R(-LAG+1))
30    D=D-BB(L,K-1)*CORR
      BB(K,K)=D/RHO
C     Update in order the bk[i]'s (10.12)
      DO 40 I=1,K-1
40    BB(I,K)=BB(I,K-1)+BB(K,K)*AA(K-I,K-1)
C     Check to see if (k+1)x(k+1) matrix is singular (10.14)
      RHO=(1.-AA(K,K)*BB(K,K))*RHO
      IF(CABS(RHO).GT.EPS)GO TO 50
      IFLAG=-1
      RETURN
50    CONTINUE
C     Find final solution
60    DO 70 I=1,IP
70    A(I)=AA(I,IP)
      RETURN
      END
```

APPENDIX 10E

Computer Program for Least Squares Modified Yule-Walker Equations Method (Provided on Floppy Disk)

```
      SUBROUTINE LSMYWE(X,N,IP,IQ,M,EPS,A,B,SIG2,IFLAG)
C     This program implements the least squares modified
C     Yule-Walker equation estimator (10.18) for the AR
```

```
C     parameters of an ARMA process. The unbiased autocorre-
C     lation estimator is used. The MA parameters and
C     the white noise variance are estimated by filtering the
C     data by an estimate of A(z) and applying Durbin's method
C     to the data at the output of the filter (see Section 8.4).
C     The large order AR model in Durbin's method is chosen
C     to have an order of N/5.
C
C     Input Parameters:
C
C       X          -Complex array of dimension N×1 of data points
C       N          -Number of complex data points
C       IP         -Desired AR model order (must be one or greater)
C       IQ         -Desired MA model order (must be one or greater)
C       M          -(M-IQ) is the number of equations used in LSMYWE
C                   estimator
C       EPS        -Small positive number to test for ill-conditioning
C                   of matrix, required in CHOLESKY used to solve
C                   linear equations
C
C     Output Parameters:
C
C       A          -Complex array of dimension IP×1 of AR parameter
C                   estimates arranged as A(1) to A(IP)
C       B          -Complex array of dimension IQ×1 of MA parameter
C                   estimates arranged as B(1) to B(IQ)
C       SIG2       -Excitation white noise variance estimate
C       IFLAG      -Ill-conditioning indicator for solution of least
C                   squares equations
C                     =   0 for normal termination
C                     = -1 if one of the di's in Cholesky decomposition
C                          is less than EPS causing premature termination
C
C     External Subroutines:
C
C       CHOLESKY        -see Appendix 2B
C       CORRELATION     -see Appendix 4D
C       LEVINSON        -see Appendix 6B
C       AUTOCORR        -see Appendix 7B
C       DURBIN          -see Appendix 8A
C
C     Notes:
C
C       The calling program must dimension arrays X,A,B.
C       The arrays R,RR,AA,Y must be dimensioned greater
C       than or equal to the variable dimensions shown.
C       The arrays XL,Y,D in CHOLESKY must be dimensioned
C       greater than or equal to IP×IP, IP, and IP, respectively.
C       The arrays A,AP in DURBIN must be dimensioned greater
C       than or equal to [N/5] and [N/5]+1, respectively. ([M]
```

```
C        denotes the largest integer less than or equal to M).
C        The arrays R,AA,RHO in AUTOCORR must be dimensioned
C        greater than or equal to [N/5]+1, [N/5]x[N/5], and [N/5],
C        respectively. The array AA in LEVINSON must be dimensioned
C        to agree with the dimensions of AA in AUTOCORR.
C
C      Verification Test Case:
C
C        If the complex test data listed in Appendix 1A is input
C        as X, with N=32, IP=7, IQ=3, M=15, and EPS=1.0E-15, then
C        the output should be
C           A(1)    = ( 0.37106      ,-0.10668      )
C           A(2)    = (-1.09528      ,-0.35745      )
C           A(3)    = (-0.78537      , 9.54137E-02)
C           A(4)    = ( 0.54507      , 0.47929      )
C           A(5)    = ( 0.41105      , 8.75429E-02)
C           A(6)    = ( 0.18216      ,-0.33953      )
C           A(7)    = ( 0.13420      ,-0.10521      )
C           B(1)    = ( 0.15233      ,-0.10394      )
C           B(2)    = (-0.43829      ,-0.26486      )
C           B(3)    = (-4.66616E-02,  1.92027E-02)
C           SIG2    = 1.52461
C           IFLAG   = 0
C        for a DEC VAX 11/780.
C
         COMPLEX X(1),A(1),B(1),R(M+1),RR(M-IQ,IP),AA(IP*(IP+1)/2),Y(N)
C      Compute unbiased autocorrelation estimates
         LAG=M+1
         CALL CORRELATION(N,LAG,0,X,X,R)
C      Fill in the (M-IQ)xIP autocorrelation matrix used
C      in the least squares solution
         DO 20 I=1,M-IQ
         DO 10 J=1,IP
         IF(IQ+I-J.GE.0)RR(I,J)=R(IQ+1+I-J)
         IF(IQ+I-J.LT.0)RR(I,J)=CONJG(R(J-I-IQ+1))
10       CONTINUE
20       CONTINUE
C      Compute R^H R matrix and store in symmetric mode (upper
C      triangular part only) (10.18)
         L=1
         DO 50 J=1,IP
         DO 40 I=1,J
         AA(L)=(0.,0.)
         DO 30 K=1,M-IQ
30       AA(L)=AA(L)+CONJG(RR(K,I))*RR(K,J)
         L=L+1
40       CONTINUE
50       CONTINUE
C      Compute -R^H r or right-hand-side vector (10.18)
         DO 70 I=1,IP
```

```
      A(I)=(0.,0.)
      DO 60 K=1,M-IQ
60    A(I)=A(I)-CONJG(RR(K,I))*R(IQ+1+K)
70    CONTINUE
C   Solve least squares equations (10.18)
      CALL CHOLESKY(AA,A,IP,EPS,IFLAG)
C   Terminate program if equations to be solved are ill-conditioned
      IF(IFLAG.EQ.-1)RETURN
C   Filter data with estimated A(z) to yield
C   approximate MA time series
      CALL CZEROFILT(X,N,IP,A,Y)
C   Apply Durbin's algorithm to estimate MA filter
C   parameters and white noise variance
      LL=N/5
      CALL DURBIN(Y,N,IQ,LL,B,SIG2)
      RETURN
      END
      SUBROUTINE CZEROFILT(X,N,IP,A,Y)
C   This program implements an FIR filter with coefficients
C   given by A. The values of the input X(I) for I < 1 are
C   set equal to zero.
      COMPLEX X(1),Y(1),A(1)
      Y(1)=X(1)
      IF(IP.EQ.1)GO TO 25
      DO 20 I=2,IP
      Y(I)=X(I)
      DO 10 J=1,I-1
10    Y(I)=Y(I)+A(J)*X(I-J)
20    CONTINUE
25    DO 40 I=IP+1,N
      Y(I)=X(I)
      DO 30 J=1,IP
30    Y(I)=Y(I)+A(J)*X(I-J)
40    CONTINUE
      RETURN
      END
```

APPENDIX 10F

Computer Program for Mayne-Firoozan Method (Provided on Floppy Disk)

```
      SUBROUTINE MAYNEFIR(X,N,IP,IQ,L,EPS,NEXP,A,B,SIG2,IFLAG)
C   This program implements the Mayne-Firoozan method for the
C   estimation of the parameters of a complex ARMA process
```

```
C     (10.23) - (10.27).
C
C     Input Parameters:
C
C       X          -Complex vector with first N locations containing
C                   data points (see Notes below)
C       N          -Number of complex data points
C       IP         -Desired AR model order (must be one or greater)
C       IQ         -Desired MA model order (must be one or greater)
C       L          -AR model order for large order AR model
C       EPS        -Small positive number to test for ill-conditioning
C                   of matrix, required in CHOLESKY used to solve
C                   linear equations
C       NEXP       -2**NEXP is the FFT length used to approximate
C                   integral in computation of white noise variance
C
C     Output Parameters:
C
C       A          -Complex vector with first p locations containing
C                   AR parameter estimates arranged as A(1) to A(IP)
                   (see Notes below)
C       B          -Complex vector of dimension IQ×1 of MA parameter
C                   estimates arranged as B(1) to B(IQ)
C       SIG2       -Excitation white noise variance estimate
C       IFLAG      -This is a flag indicating condition of termination.
C                       = -1, if one of di's in CHOLESKY is less than EPS,
C                             causing premature termination
C                       = 0, normal termination
C                       = 1, MA filter estimate not minimum-phase causing
C                             premature termination
C
C     External Subroutines:
C
C       FFT                -see Appendix 2A
C       CHOLESKY           -see Appendix 2B
C       CORRELATION        -see Appendix 4D
C       LEVINSON           -see Appendix 6B
C       STEPDOWN           -see Appendix 6C
C       AUTOCORR           -see Appendix 7B
C
C     Notes:
C
C       The calling program must dimension arrays X,A,B.
C       X and A must be dimensioned greater than or equal to
C       2**NEXP and L, respectively. The arrays U,Y,Z,YZ,BC,P
C       must be dimensioned greater than or equal to the variable
C       dimensions shown. The arrays HH,THETA in LS must be
C       dimensioned greater than or equal to (IP+IQ)×(IP+IQ+1)/2
C       and IP+IQ, respectively. The arrays AA,BB in VARM must
C       each be dimensioned greater than or equal to 2**NEXP.
```

```
C        The arrays XL,Y,D in CHOLESKY must be dimensioned
C        greater than or equal to (IP+IQ)×(IP+IQ), IP+IQ, and
C        IP+IQ, respectively. The arrays R,AA,RHO in AUTOCORR must be
C        dimensioned greater than or equal to L+1,L×L, and L, respectively.
C        The array AA in LEVINSON must be dimensioned to agree with
C        dimensions of AA in AUTOCORR. The array AA in STEPDOWN must
C        be dimensioned to agree with the dimensions of BC in MAYNEFIR.
C        The array Y in FFT must be dimensioned to be greater than or
C        equal to 2**NEXP.
C
C      Verification Test Case:
C
C        If the complex test data listed in Appendix 1A is input
C        as X, with N=32, IP=7, IQ=3, L=21, EPS=1.0E-15, NEXP=9,
C        then the output should be
C          A(1)   = ( 0.31355     ,-0.11364     )
C          A(2)   = (-1.09737     ,-0.17999     )
C          A(3)   = (-0.58869     , 9.76058E-02)
C          A(4)   = ( 0.59138     , 0.22864     )
C          A(5)   = ( 0.24318     ,-1.20419E-02)
C          A(6)   = ( 5.27812E-02,-0.15302     )
C          A(7)   = ( 0.14779     ,-3.40991E-02)
C          B(1)   = ( 0.25299     ,-7.42263E-02)
C          B(2)   = (-0.44655     ,-9.62517E-02)
C          B(3)   = (-7.72834E-03, 1.32956E-02)
C          SIG2   = 1.91462
C          IFLAG  = 0
C        for a DEC VAX 11/780.
C
        COMPLEX X(1),A(1),B(1),U(N)
        COMPLEX Y(N),Z(N),YZ(N),BC(IQ,IQ)
        DIMENSION P(IQ)
        IFLAG=0
C     Apply autocorrelation method to data, x[n] (step 1)
        CALL AUTOCORR(X,N,L,A,SIG2)
C     Filter data x[n] with estimated A(z) to obtain estimate of
C     input white noise sequence, u[n] (step 2 - (10.23))
        CALL CCZEROFLT(X,N,A,L,U)
C     Perform second least squares solution to estimate AR and
C     MA parameters as per step 3
        CALL LS(X,U,X,N,IP,IQ,EPS,A,B,IFLAG)
C     Terminate program if equations to be solved are ill-conditioned
        IF(IFLAG.EQ.-1)RETURN
C     Check to make sure that MA filter, 1/B(z), is minimum-phase
C     by checking magnitudes of reflection coefficients and if
C     not abort
        IF(IQ.EQ.1)BC(1,1)=B(1)
        PMA=1.
        IF(IQ.GT.1)CALL STEPDOWN(IQ,B,PMA,BC,P0,P)
        DO 15 I=1,IQ
```

```
      IF(CABS(BC(I,I)).LT.1.0)GO TO 15
      IFLAG=1
      RETURN
15    CONTINUE
C     Filter data, x[n], with estimated 1/B(z) (10.24)
      CALL CPOLEFILT(X,N,B,IQ,Y)
C     Filter estimated input sequence with 1/B(z) (10.25)
      CALL CPOLEFILT(U,N,B,IQ,Z)
C     Perform last least squares solution for final ARMA
C     parameter estimates (step 5 - (10.27))
      DO 20 I=1,N
20    YZ(I)=Y(I)-Z(I)
      CALL LS(Y,Z,YZ,N,IP,IQ,EPS,A,B,IFLAG)
C     Terminate program if equations are ill-conditioned
      IF(IFLAG.EQ.-1)RETURN
      CALL VARM(X,N,A,IP,B,IQ,NEXP,SIG2)
      RETURN
      END
      SUBROUTINE CPOLEFILT(X,N,B,IQ,Y)
C     This program implements an all-pole filter, 1/B(z), with
C     the initial conditions of the filter set equal to zero.
      COMPLEX X(1),Y(1),B(1)
      Y(1)=X(1)
      IF(IQ.EQ.1)GO TO 25
      DO 20 I=2,IQ
      Y(I)=X(I)
      DO 10 J=1,I-1
10    Y(I)=Y(I)-B(J)*Y(I-J)
20    CONTINUE
25    DO 40 I=IQ+1,N
      Y(I)=X(I)
      DO 30 J=1,IQ
30    Y(I)=Y(I)-B(J)*Y(I-J)
40    CONTINUE
      RETURN
      END
      SUBROUTINE CCZEROFLT(X,N,A,IP,Y)
C     This program implements an FIR filter with coefficients
C     given by A. The values of the input X(I) for I < 1 are
C     set equal to zero.
      COMPLEX X(1),Y(1),A(1)
      Y(1)=X(1)
      IF(IP.EQ.1)GO TO 25
      DO 20 I=2,IP
      Y(I)=X(I)
      DO 10 J=1,I-1
10    Y(I)=Y(I)+A(J)*X(I-J)
20    CONTINUE
25    DO 40 I=IP+1,N
      Y(I)=X(I)
```

```
      DO 30 J=1,IP
30    Y(I)=Y(I)+A(J)*X(I-J)
40    CONTINUE
      RETURN
      END
      SUBROUTINE LS(Y,Z,YZ,N,IP,IQ,EPS,A,B,IFLAG)
C     This program sets up the matrices and vectors needed
C     to compute a least squares solution and solves them
C     using a Cholesky decomposition.
      COMPLEX Y(1),Z(1),A(1),B(1),HKI,HKJ
      COMPLEX HH((IP+IQ)*(IP+IQ+1)/2),THETA(IP+IQ),YZ(1)
      IPQ=IP+IQ
      L=1
      DO 30 J=1,IP+IQ
      DO 20 I=1,J
      HH(L)=(0.,0.)
      DO 10 K=1,N
      IF(I.LE.IP.AND.K-I.GE.1)HKI=Y(K-I)
      IF(I.LE.IP.AND.K-I.LE.0)HKI=(0.,0.)
      IF(I.GT.IP.AND.K-(I-IP).GE.1)HKI=Z(K-(I-IP))
      IF(I.GT.IP.AND.K-(I-IP).LE.0)HKI=(0.,0.)
      IF(J.LE.IP.AND.K-J.GE.1)HKJ=Y(K-J)
      IF(J.LE.IP.AND.K-J.LE.0)HKJ=(0.,0.)
      IF(J.GT.IP.AND.K-(J-IP).GE.1)HKJ=Z(K-(J-IP))
      IF(J.GT.IP.AND.K-(J-IP).LE.0)HKJ=(0.,0.)
10    HH(L)=HH(L)+CONJG(HKI)*HKJ
      L=L+1
20    CONTINUE
30    CONTINUE
      DO 50 I=1,IP+IQ
      THETA(I)=(0.,0.)
      DO 40 K=1,N
      IF(I.LE.IP.AND.K-I.GE.1)HKI=Y(K-I)
      IF(I.LE.IP.AND.K-I.LE.0)HKI=(0.,0.)
      IF(I.GT.IP.AND.K-(I-IP).GE.1)HKI=Z(K-(I-IP))
      IF(I.GT.IP.AND.K-(I-IP).LE.0)HKI=(0.,0.)
40    THETA(I)=THETA(I)+CONJG(HKI)*YZ(K)
50    CONTINUE
      CALL CHOLESKY(HH,THETA,IPQ,EPS,IFLAG)
      DO 60 I=1,IP+IQ
      IF(I.LE.IP)A(I)=-THETA(I)
60    IF(I.GT.IP)B(I-IP)=THETA(I)
      RETURN
      END
      SUBROUTINE VARM(X,N,A,IP,B,IQ,NEXP,SIG2)
C     This program computes the estimate of the white
C     noise variance.
      COMPLEX X(1),A(1),B(1),AA(2**NEXP),BB(2**NEXP)
      M=2**NEXP
      AA(1)=(1.,0.)
```

```
      BB(1)=(1.,0.)
      DO 10 I=2,M
      IF(I.GT.N)X(I)=(0.,0.)
      IF(I.LE.IP+1)AA(I)=A(I-1)
      IF(I.GT.IP+1)AA(I)=(0.,0.)
      IF(I.LE.IQ+1)BB(I)=B(I-1)
10    IF(I.GT.IQ+1)BB(I)=(0.,0.)
      INVRS=1
      CALL FFT(X,NEXP,INVRS)
      CALL FFT(AA,NEXP,INVRS)
      CALL FFT(BB,NEXP,INVRS)
      SIG2=0.
      DO 20 I=1,M
20    SIG2=SIG2+(CABS(X(I))**2)*(CABS(AA(I))**2)/(CABS(BB(I))**2)
      SIG2=SIG2/(N*M)
      RETURN
      END
```

Chapter 11

Minimum Variance Spectral Estimation

11.1 INTRODUCTION

The minimum variance spectral estimator (MVSE), originally developed for seismic array frequency–wavenumber analysis [Capon 1969], estimates the PSD by effectively measuring the power out of a set of narrowband filters [Lacoss 1971]. The popular use of the name maximum likelihood spectral estimator (MLSE) or maximum likelihood method (MLM) is actually a misnomer in that the spectral estimator is not a maximum likelihood spectral estimator *nor does it possess any of the properties of a maximum likelihood estimator*. The motivation for the name is discussed in this chapter. Even the name MVSE that has been chosen to describe this estimator is not meant to imply that the spectral estimator is one that possesses minimum variance but is only used to describe the origins of the estimator. The MVSE is also referred to as the Capon spectral estimator [McDonough 1983]. The difference between the MVSE and the periodogram spectral estimator is that the shape of the narrowband filters in the MVSE is, in general, dependent on the frequency under consideration, whereas for the periodogram the shape of the narrowband filters is the same for all frequencies. In the MVSE the filters adapt to the process for which the PSD is sought with the advantage that the filter sidelobes can be adjusted to reduce the response to PSD components outside the band of interest.

11.2 SUMMARY

The problem of estimating the amplitude of a complex sinusoid in complex Gaussian noise is discussed in Section 11.3. The MLE is shown to be given by (11.5).

Coincidentally, the estimator which is constrained to be a linear function of the data and to be unbiased and which has minimum variance is identical to the MLE. The minimum variance of this estimator is given by (11.12). In Section 11.4 the latter estimator is shown to have a filtering interpretation. The filter is constrained to have its frequency response equal to unity at the frequency under consideration and is designed to minimize the noise power at its output. Thus the filter adaptively "nulls out" strong noise components outside the band of interest. Based on this interpretation the MVSE is defined in Section 11.5 as the power out of this optimal filter for each frequency under consideration. The MVSE is given by (11.17). It exhibits more resolution than Fourier estimators but less than that of an AR spectral estimator. Some analytical expressions are given for one complex sinusoid of an arbitrary frequency in white noise (11.19) and two complex sinusoids with certain frequencies in white noise (11.21). In Section 11.6 the MVSE and the AR spectral estimators are compared. The MVSE exhibits less resolution but also less variance. The lower resolution is explained by the analytical expression (11.26), which relates the MVSE to an average of AR spectral estimators of different orders. Also, the correlation matching property of AR spectral estimators does not hold for the MVSE. Finally, Section 11.7 illustrates the performance of the MVSE based on computer simulations. It is found that for short data records the autocorrelation matrix required to compute the MVSE should be constructed using the modified covariance method. For longer data records a Toeplitz autocorrelation matrix can be used with either the biased or unbiased ACF estimator. Finally, the definition of the MVSE as given by (11.17) should be modified by multiplying it by p, the dimension of the autocorrelation matrix, if the scale of the spectral estimate is of importance.

The application of the MVSE to the complex test case data listed in Appendix 1A results in the spectral estimate of Figure 11.1. The MVSE as defined by (11.17) has been multiplied by p, which was chosen to be $p = 11$. The computer program

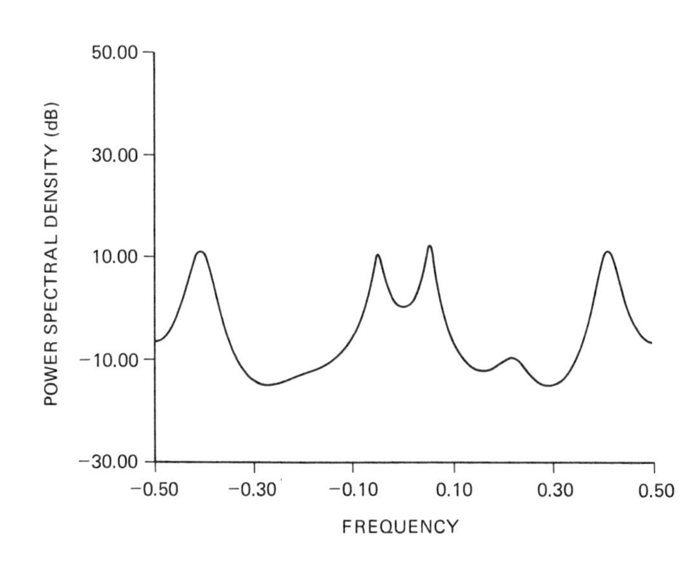

Figure 11.1 MVSE for complex test case data.

MVSE, which is provided in Appendix 11A, was used to obtain the spectral estimate. The results are discussed in Chapter 12.

11.3 MAXIMUM LIKELIHOOD ESTIMATION OF SIGNAL AMPLITUDE

To understand the motivation for the MVSE and the name MLM it is necessary to examine the maximum likelihood estimator (MLE) for the complex amplitude of a complex sinusoid in complex Gaussian noise. Assume that the data are observed as

$$x[n] = A \exp[j(2\pi f_0 n + \phi)] + z[n] \qquad n = 0, 1, \ldots, N-1. \quad (11.1)$$

The signal frequency f_0 is assumed to be known and the amplitude A and phase ϕ are unknown real constants. $z[n]$ is complex Gaussian noise with zero mean and covariance matrix $\mathbf{R}_{zz}$ (see Section 3.5.1) where

$$[\mathbf{R}_{zz}]_{ij} = \mathscr{E}(z^*[n]z[n+i-j]) = r_{zz}[i-j].$$

The probability density function (PDF) of $\mathbf{z} = [z[0] \;\; z[1] \;\; \cdots \;\; z[N-1]]^T$ is [Goodman 1963] (see also Section 3.2.2)

$$p(\mathbf{z}) = \frac{1}{\pi^N \det(\mathbf{R}_{zz})} \exp(-\mathbf{z}^H \mathbf{R}_{zz}^{-1} \mathbf{z}). \quad (11.2)$$

The problem is to estimate A and ϕ given $\{x[0], x[1], \ldots, x[N-1]\}$. Equivalently, if $A_c = A \exp(j\phi)$ denotes the complex amplitude of the sinusoid, the MLE is found by maximizing $p(\mathbf{x} - A_c\mathbf{e})$ over A_c, where $\mathbf{e} = [1 \;\; \exp[j2\pi f_0] \;\; \exp[j4\pi f_0] \;\; \cdots \;\; \exp[j2\pi f_0(N-1)]]^T$. From (11.2),

$$p(\mathbf{x} - A_c\mathbf{e}) = \frac{1}{\pi^N \det(\mathbf{R}_{zz})} \exp[-(\mathbf{x} - A_c\mathbf{e})^H \mathbf{R}_{zz}^{-1} (\mathbf{x} - A_c\mathbf{e})] \quad (11.3)$$

so that the MLE is found by minimizing the hermitian form

$$S = (\mathbf{x} - A_c\mathbf{e})^H \mathbf{R}_{zz}^{-1} (\mathbf{x} - A_c\mathbf{e})$$

over A_c. To effect the minimization, note that S can be written as [see Problem 11.1 and (2.77)]

$$S = (\mathbf{x} - \tilde{A}_c\mathbf{e})^H \mathbf{R}_{zz}^{-1} (\mathbf{x} - \tilde{A}_c\mathbf{e}) + |A_c - \tilde{A}_c|^2 \mathbf{e}^H \mathbf{R}_{zz}^{-1} \mathbf{e} \quad (11.4)$$

where

$$\tilde{A}_c = \frac{\mathbf{e}^H \mathbf{R}_{zz}^{-1} \mathbf{x}}{\mathbf{e}^H \mathbf{R}_{zz}^{-1} \mathbf{e}}$$

The first term does not depend on A_c and the second term is always nonnegative since $\mathbf{R}_{zz}$ is positive definite and hence $\mathbf{R}_{zz}^{-1}$ is positive definite. S is minimized by setting $A_c = \tilde{A}_c$. The MLE of A_c is therefore

$$\hat{A}_c = \frac{\mathbf{e}^H \mathbf{R}_{zz}^{-1} \mathbf{x}}{\mathbf{e}^H \mathbf{R}_{zz}^{-1} \mathbf{e}}. \quad (11.5)$$

The MLE is easily shown to be unbiased, since from (11.1),

$$\mathbf{x} = A_c\mathbf{e} + \mathbf{z}$$

which upon substitution in (11.5) yields

$$\hat{A}_c = \frac{\mathbf{e}^H\mathbf{R}_{zz}^{-1}(A_c\mathbf{e} + \mathbf{z})}{\mathbf{e}^H\mathbf{R}_{zz}^{-1}\mathbf{e}}$$

and hence

$$\mathscr{E}(\hat{A}_c) = A_c + \frac{\mathbf{e}^H\mathbf{R}_{zz}^{-1}\mathscr{E}(\mathbf{z})}{\mathbf{e}^H\mathbf{R}_{zz}^{-1}\mathbf{e}} = A_c.$$

It so happens that the MLE is also the solution to the following problem. Find a linear minimum variance unbiased (LMVU) estimator of A_c. This is to say that we wish to minimize var $(\hat{A}_c)$ for an estimator of the form

$$\hat{A}_c = \mathbf{w}^H\mathbf{x} \tag{11.6}$$

where $\mathbf{w}$ is the vector of weights $[w[0] \quad w[1] \quad \cdots \quad w[N-1]]^T$ and $\hat{A}_c$ is an unbiased estimator. To be unbiased, $\mathbf{w}$ must be constrained to satisfy

$$\mathbf{w}^H\mathbf{e} = 1 \tag{11.7}$$

so that

$$\mathscr{E}(\hat{A}_c) = \mathbf{w}^H\mathscr{E}(\mathbf{x}) = \mathbf{w}^H A_c\mathbf{e} = A_c.$$

The variance can be expressed as the hermitian form

$$\begin{aligned} \text{var}(\hat{A}_c) &= \mathscr{E}[|\hat{A}_c - A_c|^2] = \mathscr{E}[|\mathbf{w}^H\mathbf{x} - \mathbf{w}^H\mathbf{e}A_c|^2] \\ &= \mathscr{E}[\mathbf{w}^H(\mathbf{x} - A_c\mathbf{e})(\mathbf{x} - A_c\mathbf{e})^H\mathbf{w}] = \mathbf{w}^H\mathbf{R}_{zz}\mathbf{w}. \end{aligned} \tag{11.8}$$

In summary, to find the LMVU estimator, we need to minimize var $(\hat{A}_c)$ as given by (11.8) subject to the constraint of (11.7) (see also Problem 11.2). This constrained minimization may be accomplished by making use of the identity

$$\mathbf{w}^H\mathbf{R}_{zz}\mathbf{w} = (\mathbf{w} - \tilde{\mathbf{w}})^H\mathbf{R}_{zz}(\mathbf{w} - \tilde{\mathbf{w}}) + \tilde{\mathbf{w}}^H\mathbf{R}_{zz}\tilde{\mathbf{w}} \tag{11.9}$$

where

$$\tilde{\mathbf{w}} = \frac{\mathbf{R}_{zz}^{-1}\mathbf{e}}{\mathbf{e}^H\mathbf{R}_{zz}^{-1}\mathbf{e}}$$

and which holds if $\mathbf{w}^H\mathbf{e} = 1$. To verify this identity, note that

$$\begin{aligned} &(\mathbf{w} - \tilde{\mathbf{w}})^H\mathbf{R}_{zz}(\mathbf{w} - \tilde{\mathbf{w}}) + \tilde{\mathbf{w}}^H\mathbf{R}_{zz}\tilde{\mathbf{w}} \\ &\quad = \mathbf{w}^H\mathbf{R}_{zz}\mathbf{w} + (2\tilde{\mathbf{w}}^H\mathbf{R}_{zz}\tilde{\mathbf{w}} - \tilde{\mathbf{w}}^H\mathbf{R}_{zz}\mathbf{w} - \mathbf{w}^H\mathbf{R}_{zz}\tilde{\mathbf{w}}) \\ &\quad = \mathbf{w}^H\mathbf{R}_{zz}\mathbf{w} + \frac{1}{\mathbf{e}^H\mathbf{R}_{zz}^{-1}\mathbf{e}}(2 - \mathbf{e}^H\mathbf{w} - \mathbf{w}^H\mathbf{e}). \end{aligned}$$

Since $\mathbf{w}^H\mathbf{e} = 1$, it follows that $\mathbf{e}^H\mathbf{w} = 1$, and hence the identity is proven. To minimize the variance observe from (11.9) that $\tilde{\mathbf{w}}^H\mathbf{R}_{zz}\tilde{\mathbf{w}}$ does not depend on $\mathbf{w}$ and $(\mathbf{w} - \tilde{\mathbf{w}})^H\mathbf{R}_{zz}(\mathbf{w} - \tilde{\mathbf{w}}) \geq 0$ because $\mathbf{R}_{zz}$ is a positive definite matrix. Therefore, the minimum value is obtained by letting $\mathbf{w} = \tilde{\mathbf{w}}$, or

$$\mathbf{w} = \frac{\mathbf{R}_{zz}^{-1}\mathbf{e}}{\mathbf{e}^H\mathbf{R}_{zz}^{-1}\mathbf{e}}. \tag{11.10}$$

Finally, substituting (11.10) into (11.6) the LMVU estimator becomes

$$\hat{A}_c = \frac{\mathbf{e}^H\mathbf{R}_{zz}^{-1}\mathbf{x}}{\mathbf{e}^H\mathbf{R}_{zz}^{-1}\mathbf{e}} \tag{11.11}$$

which is identical to the MLE (11.5). The minimum variance of $\hat{A}_c$ can be found by substituting (11.10) into (11.8) to yield

$$\text{var}\,(\hat{A}_c)_{\text{MIN}} = \frac{1}{\mathbf{e}^H\mathbf{R}_{zz}^{-1}\mathbf{e}}. \tag{11.12}$$

Note that to compute the minimum variance the *noise covariance matrix must be known*. In summary, the MLE of the complex amplitude of a sinusoid in complex Gaussian noise is given by (11.5). Coincidentally, the MLE is also the LMVU estimator. Although it is the LMVU that provides the basis for the MVSE, the connection with the MLE is responsible for the misnomers MLSE and MLM. Henceforth the estimator will be referred to as the LMVU estimator.

11.4 FILTERING INTERPRETATION OF THE LINEAR MINIMUM VARIANCE UNBIASED ESTIMATOR

The LMVU estimator has a filtering interpretation which forms the motivational basis for the MVSE. To simplify the discussion, consider first the case of white noise, so that $\mathbf{R}_{zz} = \sigma^2\mathbf{I}$. Then, using (11.5) yields

$$\hat{A}_c = \frac{\mathbf{e}^H\mathbf{x}}{\mathbf{e}^H\mathbf{e}} \tag{11.13}$$

or

$$\hat{A}_c = \frac{1}{N}\sum_{n=0}^{N-1} x[n] \exp(-j2\pi f_0 n).$$

$\hat{A}_c$ is recognized as the Fourier transform of the data evaluated at $f = f_0$ and scaled by $1/N$. Alternatively, $\hat{A}_c$ is the output of an FIR filter with impulse response

$$h[n] = \begin{cases} \dfrac{1}{N}\exp(j2\pi f_0 n) & \text{for } n = -(N-1), \ldots, -1, 0 \\ 0 & \text{otherwise} \end{cases}$$

at time $n = 0$. The frequency response of the filter is

$$H(f) = \sum_{n=-(N-1)}^{0} h[n] \exp(-j2\pi f n)$$
$$= \frac{\sin[N\pi(f - f_0)]}{N \sin[\pi(f - f_0)]} \exp[j(N - 1)\pi(f - f_0)]. \tag{11.14}$$

The filter response is a sinc function centered at $f = f_0$. Also, the response at $f = f_0$ is

$$H(f_0) = 1$$

so that the sinusoid is passed by the filter undistorted or the signal component at the filter output is just A_c. Note that this is the *identical* filter used to form the periodogram spectral estimate [see (4.3)]. The distortionless nature of the LMVU filter is a direct consequence of the unbiased property of the estimator; that is, the scale factor $\mathbf{e}^H \mathbf{R}_{zz}^{-1} \mathbf{e} = N/\sigma^2$ leads to the unbiased estimator. Consequently, the LMVU estimator when considered as a linear filter is constrained to have unity frequency response at $f = f_0$. In addition, the estimator is chosen to minimize the variance. In filtering terms this is equivalent to minimizing the noise power out of the filter. This is because the variance is the mean square value of $\hat{A}_c - \mathscr{E}(\hat{A}_c) = \hat{A}_c - A_c =$ (output of filter $-$ signal output) $=$ noise output of filter [see also (11.8)]. Thus the minimum variance is identical to the output noise power of the optimal linear filter (see Problem 11.2). For the white noise case it is easily shown from (11.12) to be

$$\text{var}(\hat{A}_c)_{\text{MIN}} = \frac{\sigma^2}{N}.$$

Note that the power out of the filter is the same for all frequencies. Also, the shape of the filter frequency response does not depend on the frequency of the sinusoid. Equation (11.14) shows that the filter response is only shifted with f_0 to ensure a peak response at $f = f_0$. We might have expected this result since the noise background is the same independent of where the filter is centered. However, if the noise is not white, the effect of $\mathbf{R}_{zz}$ in (11.10) is to produce a frequency response with high-level attenuation in the frequency bands where the noise level is large, at the expense of other bands where the noise level is small. As an example, assume that $f_0 = 0$ and $z[n]$ is an AR(1) process with parameter $a[1] = -r \exp(j2\pi f_z)$. Then from (11.10),

$$\mathbf{w} = \frac{\mathbf{R}_{zz}^{-1}\mathbf{e}}{\mathbf{e}^H \mathbf{R}_{zz}^{-1} \mathbf{e}}$$

with $\mathbf{e} = [1 \quad 1 \quad \cdots \quad 1]^T$ since $f_0 = 0$. From (6.64),

$$\mathbf{R}_{zz}^{-1} = \frac{1}{\sigma^2} \begin{bmatrix} 1 & a^*[1] & 0 & 0 & \cdots & 0 \\ a[1] & 1 + |a[1]|^2 & a^*[1] & 0 & \cdots & 0 \\ \vdots & \ddots & \ddots & \ddots & \ddots & \vdots \\ 0 & \cdots & 0 & a[1] & 1 + |a[1]|^2 & a^*[1] \\ 0 & \cdots & 0 & 0 & a[1] & 1 \end{bmatrix}$$

so that

$$\mathbf{R}_{zz}^{-1}\mathbf{e} = \frac{1}{\sigma^2}\begin{bmatrix} 1 + a^*[1] \\ a[1] + 1 + |a[1]|^2 + a^*[1] \\ \vdots \\ a[1] + 1 + |a[1]|^2 + a^*[1] \\ 1 + a[1] \end{bmatrix}$$

and

$$\mathbf{e}^H\mathbf{R}_{zz}^{-1}\mathbf{e} = \frac{1}{\sigma^2}\left[N + (N-2)\,|a[1]|^2 + (N-1)(a[1] + a^*[1])\right].$$

The frequency response of the filter is defined as

$$H(f) = \sum_{n=-(N-1)}^{0} h[n] \exp(-j2\pi fn) \tag{11.15}$$

where $h[n] = w^*[-n]$. Using this definition we have, after some algebra,

$$H(f) = \frac{1}{K}\Big[NB_N(f) - (N-1)r[\exp(j2\pi f_z) + \exp(j2\pi(f - f_z))]B_{N-1}(f) + (N-2)r^2 \exp(j2\pi f)B_{N-2}(f)\Big] \tag{11.16}$$

where

$$B_N(f) = \frac{1}{N}\sum_{n=0}^{N-1} \exp(j2\pi fn)$$

and

$$K = N - 2(N-1)r\cos 2\pi f_z + (N-2)r^2.$$

The magnitude of the frequency response is plotted in Figure 11.2 for $N = 10$, $f_z = 0.25$ for various values of r. For $r = 0$ (i.e., $a[1] = 0$) the noise is white and the filter has the usual sinc type of response (11.14) (shown as a dashed curve). For other values of r the optimal filter attempts to reject the noise by adjusting the response so as to attenuate that region of the frequency band where the noise PSD is the largest. In this case the PSD of the noise is

$$P_{\mathrm{AR}}(f) = \frac{\sigma^2}{|1 + a[1]\exp(-j2\pi f)|^2} = \frac{\sigma^2}{|1 - r\exp[-j2\pi(f - f_z)]|^2}.$$

The PSD has a peak at $f = f_z = 0.25$ and the sharpness of the peak increases with r. This is reflected in the frequency response in which the attenuation in the frequency band centered about $f = f_z = 0.25$ increases with r. In fact, as $r \to 1$ so that the noise approaches a sinusoidal signal, it can be verified from (11.16) that $H(f_z) \to 0$. In effect, a null is placed at the frequency location where the noise power is greatest (see also Problem 11.3). As expected, the gain of the filter at $f = f_0 = 0$ is unity.

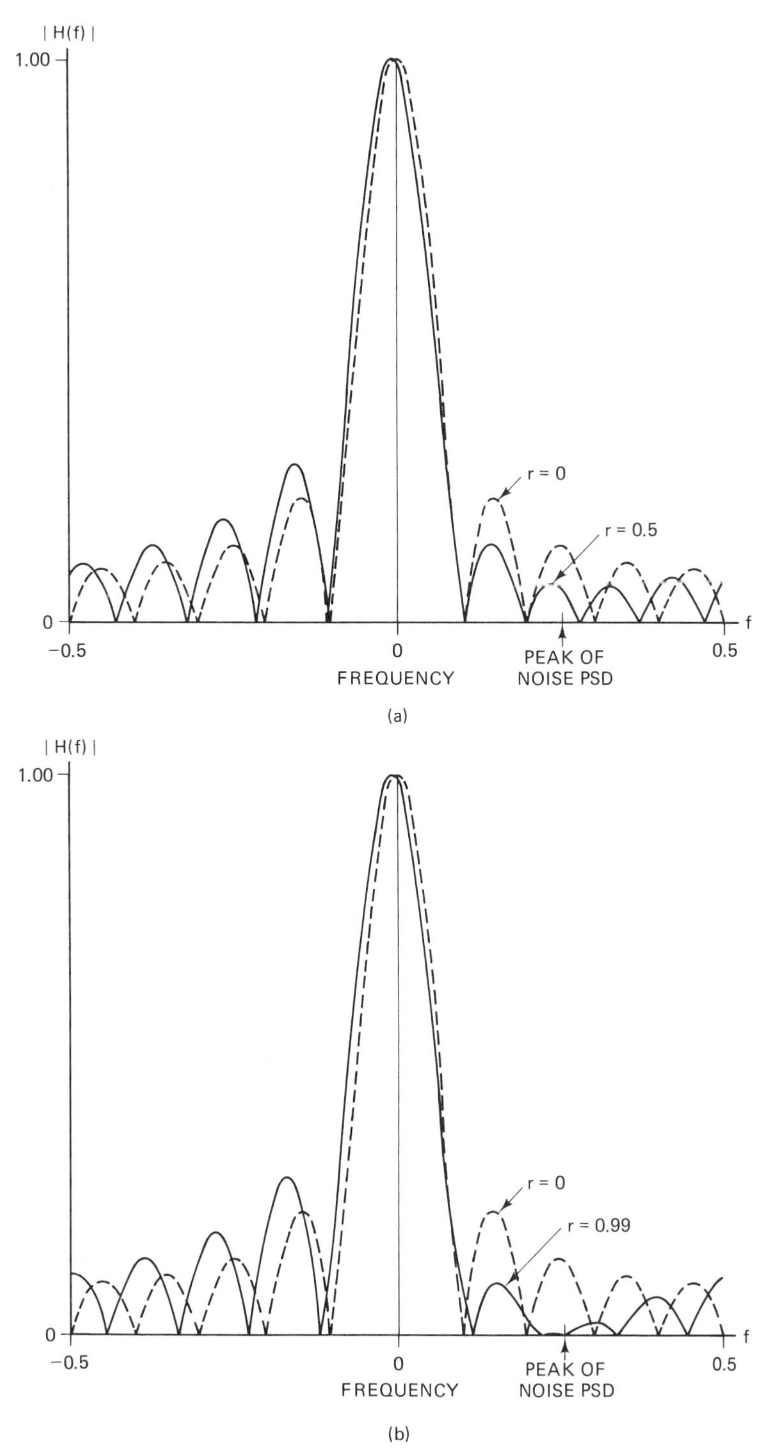

Figure 11.2 Frequency response magnitude of LMVU filter.

It is now apparent that the filter shape will depend on the noise background near the center frequency $f = f_0$. For a given noise covariance $\mathbf{w}$ will form a different filter for each assumed value of f_0. The filter will adjust itself to reject noise components with frequencies not near $f = f_0$ and pass noise components at and near $f = f_0$.

11.5 THE MINIMUM VARIANCE SPECTRAL ESTIMATOR

It was seen in the preceding section that the LMVU estimator forms a filter which only allows through frequency components near f_0 and rejects others in an optimal fashion. The power out of the filter is therefore a good indication of the power in the process in the vicinity of f_0. If we now assume that the data $x[n]$ do not consist of a sinusoid in noise, but more generally a WSS random process, then the power out of the filter for each assumed value of f_0 can be used to indicate spectral content. A spectral estimator can be defined as

$$\hat{P}_{MV}(f) = \frac{1}{\mathbf{e}^H \hat{\mathbf{R}}_{xx}^{-1} \mathbf{e}} \tag{11.17}$$

where $\mathbf{e} = [1 \quad \exp(j2\pi f) \quad \exp(j4\pi f) \quad \cdots \quad \exp(j2\pi f(p-1))]^T$. Note that for a true PSD estimator $\hat{P}_{MV}(f)$ should be divided by $1/p$, the "bandwidth" of the optimal filter, since $\hat{P}_{MV}(f)$ indicates *power* and not power per cycles/samples (see Problem 11.4). As an example, if $x[n]$ is white noise with variance σ^2, then $\hat{P}_{MV}(f) = \sigma^2/p$, which is incorrect by the factor $1/p$. On the other hand, with the definition of (11.17) the MVSE will indicate the true power of a sinusoid in white noise by its peak, as will be shown shortly. If scaling of the PSD is of importance, the definition of $\hat{P}_{MV}(f)$ should be modified by multiplying it by p. (See also the work of Lagunas-Hernandez and Gasull-Llampalas [1984] for an alternative definition of the MVSE which accounts for the filter bandwidth.) To compute the spectral estimate we need an estimate of the autocorrelation matrix. $\hat{P}_{MV}(f)$ has been referred to as the MLM or MLSE or Capon method. For reasons already discussed, (11.17) will be referred to as the minimum variance spectral estimator (MVSE). The connection with maximum likelihood is nonexistent but $\hat{P}_{MV}(f)$ has clearly been motivated by the problem described in Section 11.3. See the paper by Marzetta [1983] for another interpretation of the MVSE.

In practice, the MVSE exhibits more resolution than the periodogram and BT spectral estimators but less than an AR spectral estimator. As an example, consider two real sinusoids in white noise so that the ACF is

$$r_{xx}[k] = 5.33 \cos 0.3\pi k + 10.66 \cos 0.4\pi k + \delta[k].$$

In Figure 11.3 the AR spectral estimator, the MVSE, and the BT spectral estimator with a Bartlett window are compared. The known ACF has been used for all three spectral estimates. For the AR spectral estimator a model order of $p = 10$ has been used, for the MVSE $p = 11$ has been selected, and for the BT spectral estimator the ACF samples up to and including $k = 10$ were used (see (13.41)). All three spectral estimators thus use the same ACF samples. As observed from the figure, the resolution of the MVSE is lower than that of the AR spectral estimator but

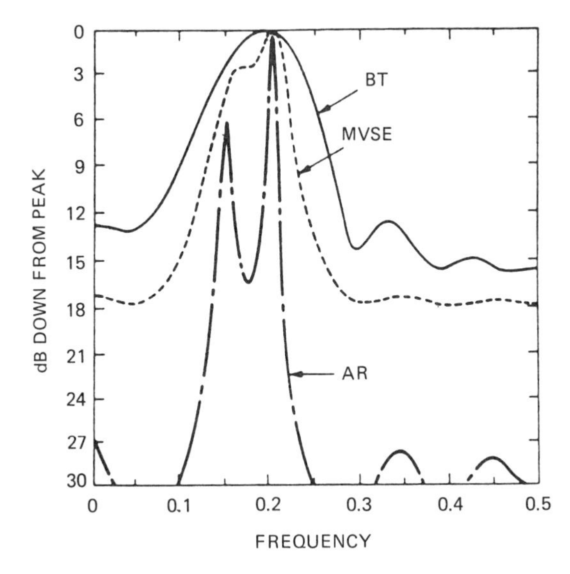

Figure 11.3 Comparison of various spectral estimators. (After Lacoss [1971].)

higher than the BT spectral estimator. The reason for the poorer resolution of the MVSE is given in Section 11.6.

Some further insight into the properties of the MVSE may be gained by examining the analytical expression for the spectral estimate corresponding to one complex sinusoid in complex white noise. Assume that the ACF is given by

$$r_{xx}[k] = P \exp(j2\pi f_0 k) + \sigma_w^2 \delta[k]. \tag{11.18}$$

Then we can show that (see Problem 11.5)

$$P_{\text{MV}}(f) = \frac{\sigma_w^2}{p\left[1 - \dfrac{pP/\sigma_w^2}{1 + pP/\sigma_w^2} \mid B_p(f - f_0) \mid^2\right]} \tag{11.19}$$

where

$$\mid B_p(f) \mid = \left| \frac{\sin p\pi f}{p \sin \pi f} \right| .$$

The peak of the PSD occurs at the sinusoidal frequency and is equal to

$$P_{\text{MV}}(f_0) = \frac{\sigma_w^2}{p}\left(1 + \frac{pP}{\sigma_w^2}\right) . \tag{11.20}$$

For high SNR or $pP/\sigma_w^2 \gg 1$,

$$P_{\text{MV}}(f_0) \approx P$$

which is just the power of the sinusoid. This result should be contrasted with the AR spectral estimator, in which the peak is proportional to the *square* of the power of the sinusoid [see (6.124)]. It can further be shown that the area under the peak of the MVSE is proportional to the square root of the sinusoidal power [Lacoss 1971]. Hence an integrated spectrum would not produce the correct jump size at the sinusoidal frequency location. For two complex sinusoids in white noise the ACF is given by

$$r_{xx}[k] = P_1 \exp(j2\pi f_1 k) + P_2 \exp(j2\pi f_2 k) + \sigma_w^2 \delta[k]$$

which produces a very complicated expression for the inverse of the autocorrelation matrix. In general, it can be shown that the peaks are not at the correct sinusoidal frequencies [Lacoss 1971]. A simple expression can be found for $P_{MV}(f)$ for the special case when $f_1 = k/p$, $f_2 = l/p$ for k,l distinct integers in the range $[-p/2, p/2 - 1]$ for p even and $[-(p-1)/2, (p-1)/2]$ for p odd. Under these conditions $\mathbf{R}_{xx}$ becomes a circulant matrix [see (2.27)] which is easily inverted. The final result is

$$P_{MV}(f) = \frac{\sigma_w^2}{p\left[1 - \dfrac{pP_1/\sigma_w^2}{1 + pP_1/\sigma_w^2} \,|\, B_p(f - f_1) \,|^2 - \dfrac{pP_2/\sigma_w^2}{1 + pP_2/\sigma_w^2} \,|\, B_p(f - f_2) \,|^2\right]}. \tag{11.21}$$

It is of interest to examine the values of the MVSE at the sinusoidal frequency locations. Since $B_p(f_1 - f_2) = B_p(f_2 - f_1) = 0$,

$$P_{MV}(f_1) = \frac{\sigma_w^2}{p}\left(1 + \frac{pP_1}{\sigma_w^2}\right)$$

$$P_{MV}(f_2) = \frac{\sigma_w^2}{p}\left(1 + \frac{pP_2}{\sigma_w^2}\right)$$

and for $pP_1/\sigma_w^2 \gg 1$, $pP_2/\sigma_w^2 \gg 1$,

$$P_{MV}(f_1) \approx P_1$$

$$P_{MV}(f_2) \approx P_2.$$

The peaks of the MVSE can be used to indicate the power of each sinusoid. When the frequencies are not multiples of $1/p$, experiments still indicate that this result is valid. Figure 11.3 illustrates this behavior since the peak amplitudes are separated by 3 dB in accordance with the actual powers. Capon [1969] has shown that for sinusoids in white noise the MVSE exhibits a pseudolinearity property near the peaks. This is to say that the spectral estimate near the peaks is the same as the sum of the spectral estimates for each individual sinusoid.

11.6 COMPARISON OF THE MVSE AND AR SPECTRAL ESTIMATORS

11.6.1 Statistical Properties

The statistical properties of the MVSE for time series data are not known. However, by using statistical results derived for spectral estimation of array data [Capon and Goodman 1970, Capon 1971], it has been argued that the MVSE exhibits less variance than the AR spectral estimator for large data records [Lacoss 1971]. This observation is borne out by simulations comparing the two estimators.

As an example, consider the following model of the estimated ACF:

$$\hat{r}_{xx}[k] = 2\cos 0.4\pi k + \delta[k] + r_k \tag{11.22}$$

where r_k represents statistical estimation error of the ACF estimator. r_k is assumed to be a uniformly distributed random variable with zero mean and variance σ_r^2. Also, r_i is assumed independent of r_j for all i and j. The results of a simulation are shown in Figure 11.4, where the heavy line indicates the spectral estimate with $\sigma_r = 0$ and the other lines are different realizations for $\sigma_r = 0.1$. The AR spectral estimator uses $p = 10$, while the MVSE uses $p = 11$, so that each estimator uses the same samples of the ACF. It is observed that the AR spectral estimator exhibits slightly more variability. However, this increased variability must be traded off against the higher resolution afforded by the AR spectral estimator.

11.6.2 Resolution

A comparison of the resolution between the MVSE and the AR spectral estimator revealed a higher resolution for the latter. Some insight may be gained into this apparent advantage by analytically relating the two spectral estimators [Burg 1972]. To do so, recall that the Cholesky decomposition of $\mathbf{R}_{xx}^{-1}$ is, from Section 6.3.6,

$$\mathbf{R}_{xx}^{-1} = \mathbf{B}\mathbf{P}^{-1}\mathbf{B}^H \tag{11.23}$$

where $\mathbf{R}_{xx}^{-1}$ is of dimension $(p + 1) \times (p + 1)$. Substituting (11.23) into (11.17) and dropping the hat to indicate the theoretical autocorrelation matrix gives us

$$P_{\mathrm{MV}}(f) = \frac{1}{\mathbf{e}^H\mathbf{B}\mathbf{P}^{-1}\mathbf{B}^H\mathbf{e}} \tag{11.24}$$

but

$$\mathbf{B}^H\mathbf{e} = \begin{bmatrix} 1 & 0 & 0 & \cdots & 0 \\ a_1[1] & 1 & 0 & \cdots & 0 \\ a_2[2] & a_2[1] & 1 & \cdots & 0 \\ \vdots & \vdots & \vdots & \ddots & \vdots \\ a_p[p] & a_p[p-1] & a_p[p-2] & \cdots & 1 \end{bmatrix} \begin{bmatrix} 1 \\ \exp(j2\pi f) \\ \exp(j4\pi f) \\ \vdots \\ \exp(j2\pi fp) \end{bmatrix}$$

$$= \begin{bmatrix} 1 \\ a_1[1] + \exp(j2\pi f) \\ a_2[2] + a_2[1]\exp(j2\pi f) + \exp(j4\pi f) \\ \vdots \\ a_p[p] + a_p[p-1]\exp(j2\pi f) + \cdots + \exp(j2\pi fp) \end{bmatrix}$$

$$= \begin{bmatrix} 1 \\ \exp(j2\pi f)A_1(f) \\ \exp(j4\pi f)A_2(f) \\ \vdots \\ \exp(j2\pi fp)A_p(f) \end{bmatrix}$$

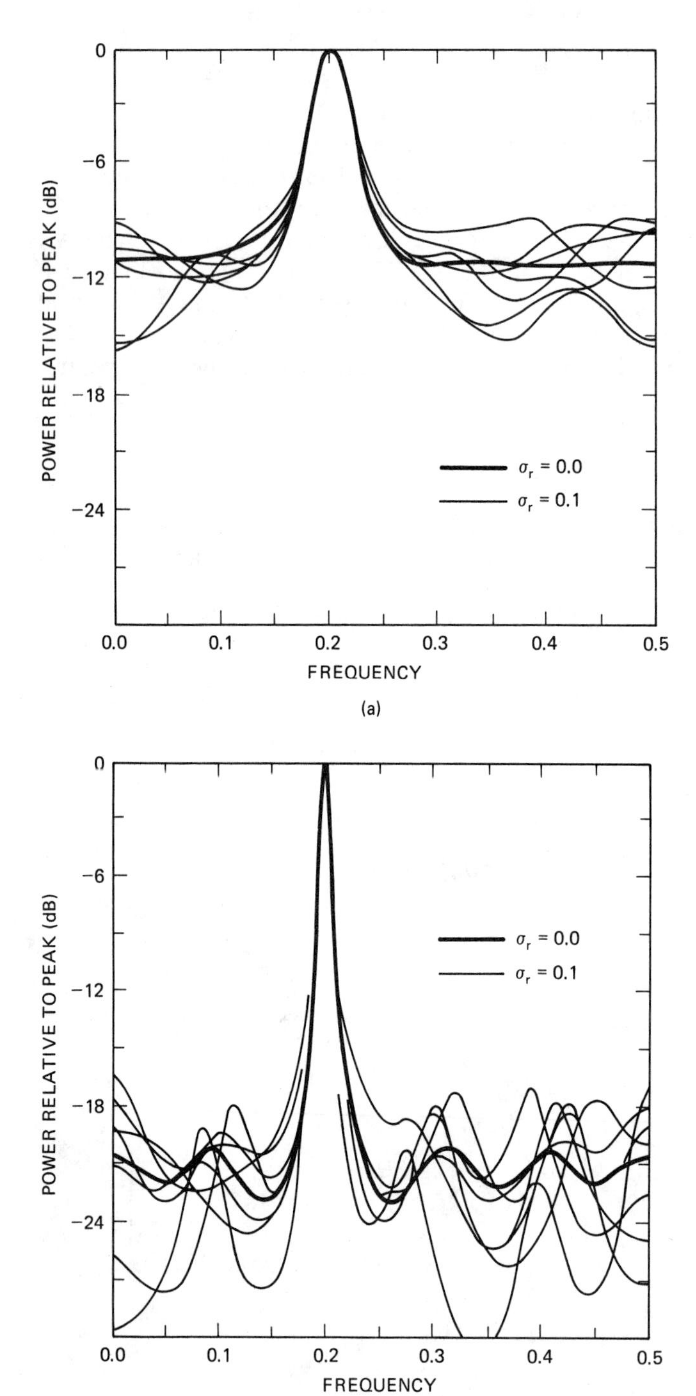

Figure 11.4 Comparison of statistical properties of (a) MVSE and (b) AR spectral estimators. (After Lacoss [1971].)

where

$$A_k(f) = 1 + \sum_{i=1}^{k} a_k[i] \exp(-j2\pi f i).$$

Now (11.24) can be rewritten as

$$P_{\text{MV}}(f) = \frac{1}{\sum_{i=0}^{p} |\mathbf{B}^H\mathbf{e}|_i^2/\rho_i} \tag{11.25}$$

where ρ_i is the prediction error power for the ith order predictor and the elements of $\mathbf{B}^H\mathbf{e}$ are indexed from $i = 0$ to $i = p$. Rewriting the denominator of (11.25) yields

$$\frac{|\mathbf{B}^H\mathbf{e}|_i^2}{\rho_i} = \frac{|A_i(f)|^2}{\rho_i} = \frac{1}{P_{\text{AR}}^{(i)}(f)}$$

where $P_{\text{AR}}^{(i)}(f)$ is the AR spectral estimator for the ith order AR model ($P_{\text{AR}}^{(0)}(f) = \rho_0$). It follows from (11.25) that

$$\frac{1}{P_{\text{MV}}(f)} = \sum_{i=0}^{p} \frac{1}{P_{\text{AR}}^{(i)}(f)}. \tag{11.26}$$

Equation (11.26) explains the lower resolution of the MVSE as due to the "parallel resistor network averaging" effect of combining the low order AR spectra of least resolution with the high order AR spectra of highest resolution. Also, some indication of the lesser variability of the MVSE is given by this averaging effect. Further analytical expressions describing the resolution properties of the MVSE can be found in the works by Cox [1973] and Musicus [1985] (see also Problem 11.6).

11.6.3 Implied Autocorrelation Function

It was shown in Section 6.4.1 that the inverse Fourier transform of the AR spectral estimator yields the identical ACF over the range of known ACF values as the given ACF samples. This property is referred to as the correlation matching property. In contrast, the inverse Fourier transform of $\hat{P}_{\text{MV}}(f)$ does not yield the same ACF values used to compute the PSD. The MVSE may not be viewed as an extrapolation of a sequence of known ACF samples. Of course, the increased resolution of the MVSE over Fourier based spectral estimators is due to the nonzero ACF samples $\{r_{xx}[p], r_{xx}[p+1], \ldots\}$ implicit in the estimator.

11.7 COMPUTER SIMULATION EXAMPLES

The four real ARMA processes described in Section 10.7 were used to test the spectral estimation performance of the MVSE. As noted earlier, for nonsinusoidal processes it is best to use $p\hat{P}_{\text{MV}}(f)$ as the spectral estimator so that this modified

MVSE was used for the simulations. The data record consisted of $N = 256$ real data points and the estimated autocorrelation matrix had dimensions $p \times p = 50 \times 50$. The results of each simulation are displayed in two parts. In each figure the average of 50 realizations of the spectral estimator is displayed together with the true PSD (shown as a dashed curve) to indicate the bias. Also, in each figure the same 50 realizations are shown overlaid to indicate the variability of the estimator. To compute the spectral estimate the autocorrelation matrix was estimated using the implied ACF estimator of the modified covariance method for real data (see Section 7.5):

$$[\hat{\mathbf{R}}_{xx}]_{ij} = \frac{1}{2(N-p)}\left[\sum_{n=p}^{N-1} x[n-i]x[n-j] + \sum_{n=0}^{N-1-p} x[n+i]x[n+j]\right]$$

for $i,j = 1, 2, \ldots, p$. It was found experimentally that the use of a Toeplitz matrix with either the biased or unbiased ACF estimator produced nearly identical results (see Problem 11.7). Since the data record is moderately long, this is expected. For shorter data records the modified covariance method has been found to be more accurate for AR spectral estimation of narrowband processes in white noise. Simulations confirmed this behavior for sinusoids in white noise. The results are shown in Figures 11.5 to 11.8. Observe that the scaling of the vertical axes is not the same for all the figures but has been chosen to be identical to the scaling used to display the ARMA methods (see Figures 10.5 to 10.8). From the figures it appears that the performance of the MVSE is generally good. Some bias is exhibited for the regions of the PSD with sharp peaks, as shown in Figure 11.8. A comparison with the better-performing ARMA spectral estimates indicates a generally larger variance for the MVSE (e.g., compare Figure 11.6 with Figure 10.5b). This is due to the large number of parameters estimated, in this case the $p = 50$ autocorrelation lags, for the MVSE as opposed to the $p + q + 1 = 7$ ARMA parameters. One could reduce the dimension of the autocorrelation matrix but the resolution would decrease as demonstrated in the next example.

Consider two complex sinusoids in complex white noise as the test data. Specifically,

$$x[n] = \sqrt{10}\exp[j2\pi(0.15)n] + \sqrt{20}\exp[j2\pi(0.2)n] + z[n]$$

where $z[n]$ is complex white Gaussian noise (see Section 3.5.1) with variance 1 so that the SNRs of the sinusoids are 10 and 13 dB. The form of the MVSE given by (11.17) was used. The autocorrelation matrix estimator based on the modified covariance method is, for complex data,

$$[\hat{\mathbf{R}}_{xx}]_{ij} = \frac{1}{2(N-p)}\left[\sum_{n=p}^{N-1} x^*[n-i]x[n-j] + \sum_{n=0}^{N-1-p} x[n+i]x^*[n+j]\right] \tag{11.27}$$

for $i,j = 1, 2, \ldots, p$. The results are shown in Figure 11.9 for $N = 256$ complex data points and $p = 50$. Note that the peaks are well resolved due to a large value of p and that the peaks are approximately at the correct locations. The amplitudes correctly indicate the power of the sinusoids in accordance with the analytical

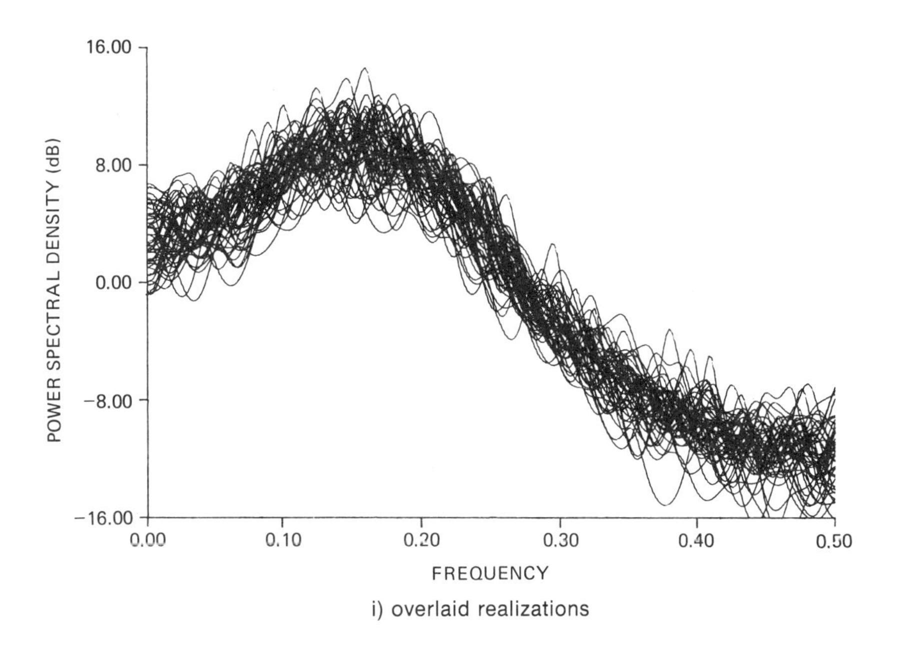

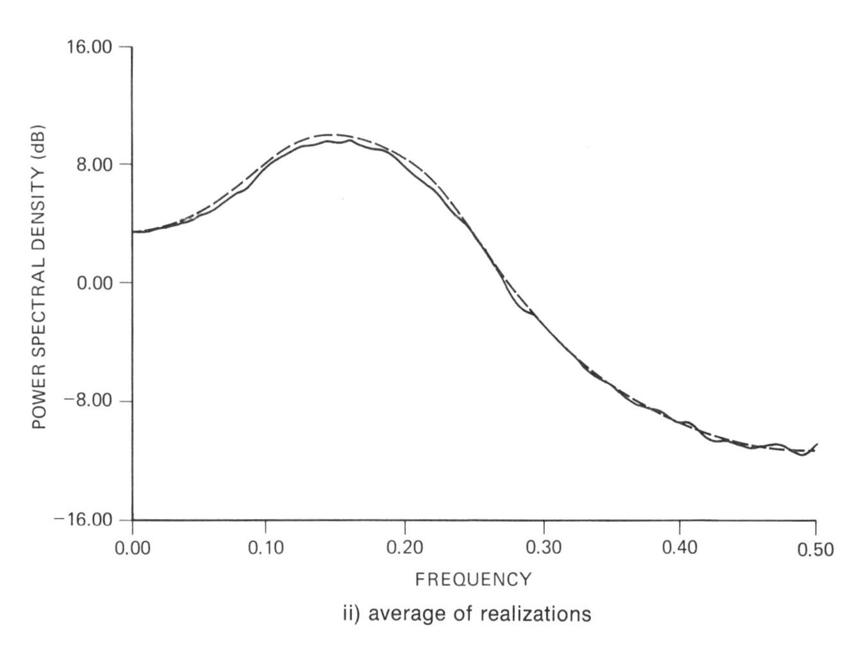

Figure 11.5 Spectral estimates for ARMA1 process.

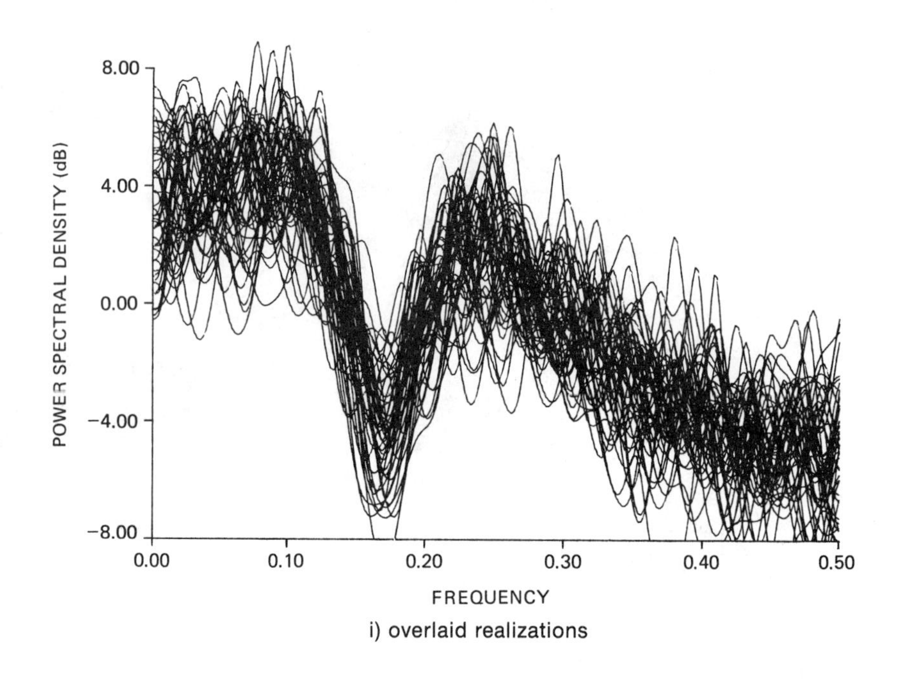

i) overlaid realizations

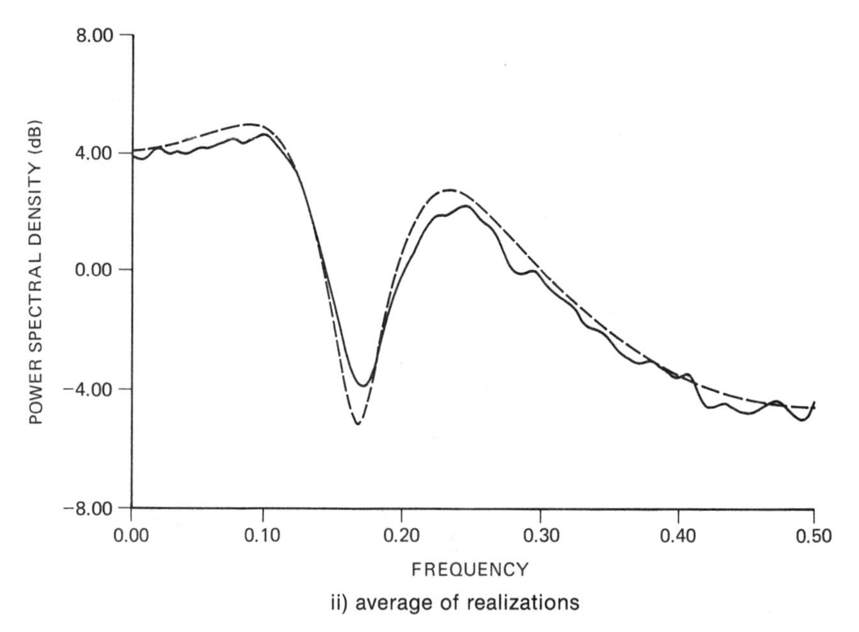

ii) average of realizations

Figure 11.6 Spectral estimates for ARMA2 process.

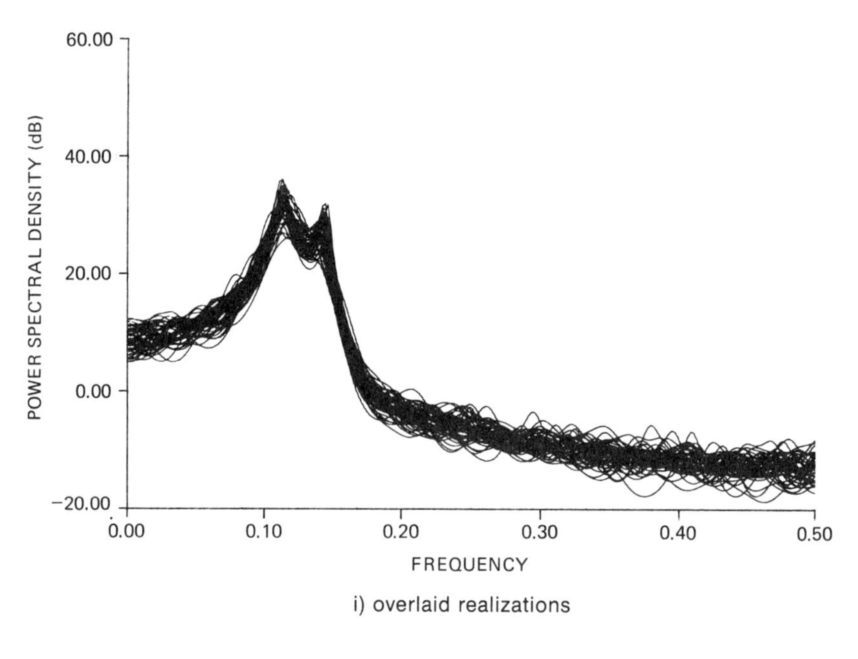

i) overlaid realizations

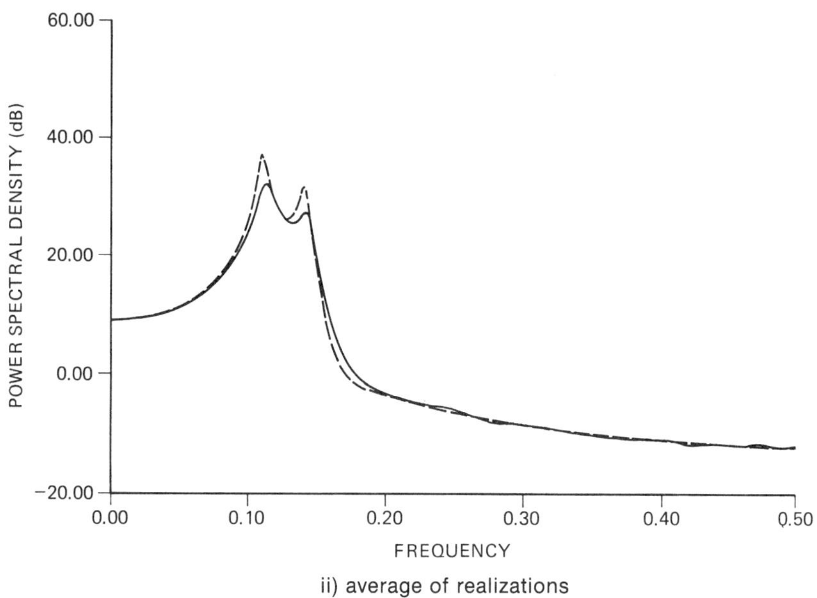

ii) average of realizations

Figure 11.7 Spectral estimates for ARMA3 process.

results. Also, the variance of the spectral estimator is largest for frequency regions having a small PSD. This is because in these regions the outputs of the narrowband filters are due mainly to the noise, which is inherently more random than the sinusoidal signals, which are fixed from realization to realization. If the data record length is decreased to $N = 25$ and the dimension of the covariance matrix is reduced to $p = 10$, the peaks are barely resolved as shown in Figure 11.10. In conclusion, the critical choice in using the MVSE is in the value of p. As in AR

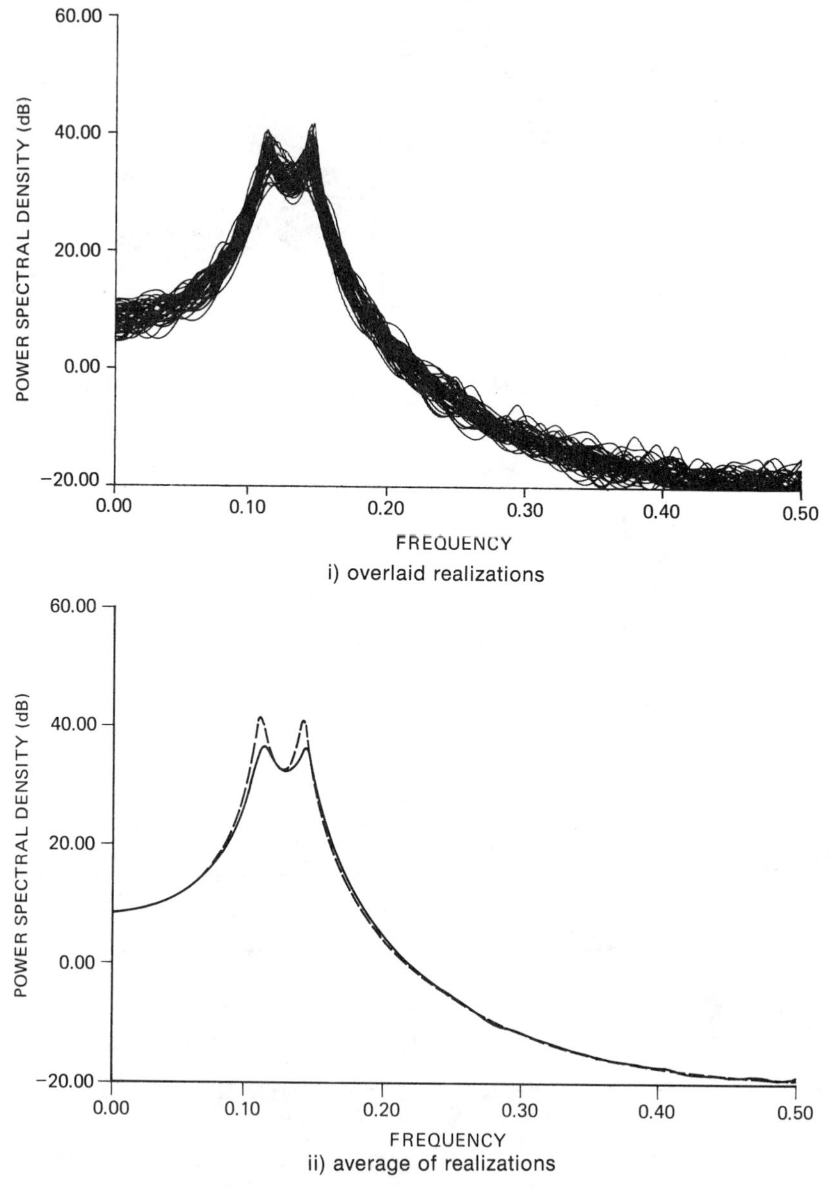

Figure 11.8 Spectral estimates for ARMA4 process.

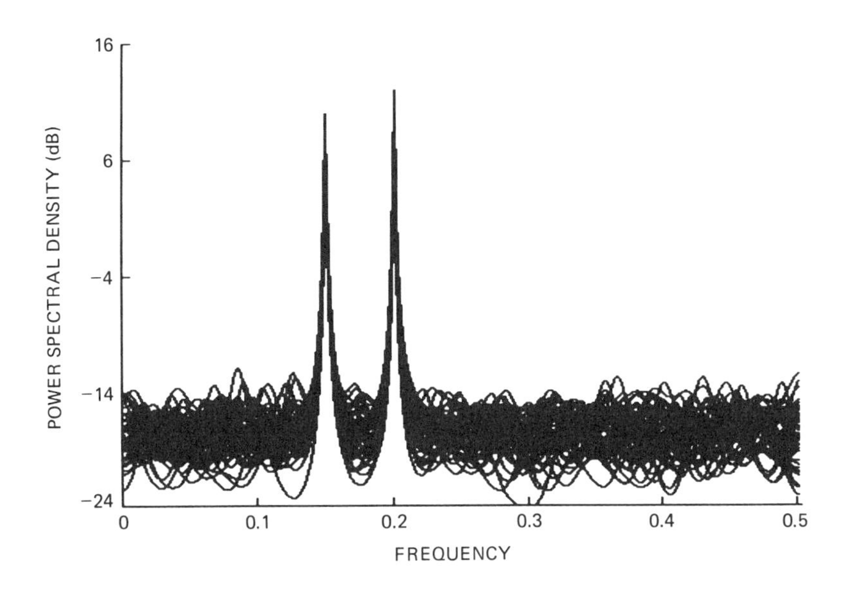

i) overlaid realizations

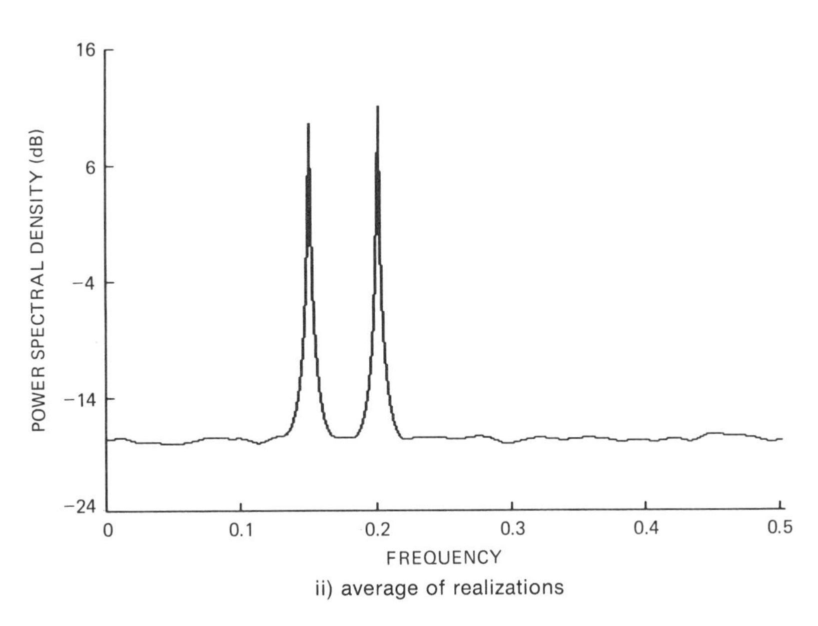

ii) average of realizations

Figure 11.9 Spectral estimates for sinusoidal data ($N = 256$, $p = 50$).

spectral estimation p needs to be chosen large enough for adequate resolution but not so large as to cause excessive spectral variability.

A computer subroutine entitled MVSE is provided in Appendix 11A to implement the MVSE using the modified covariance method for the autocorrelation

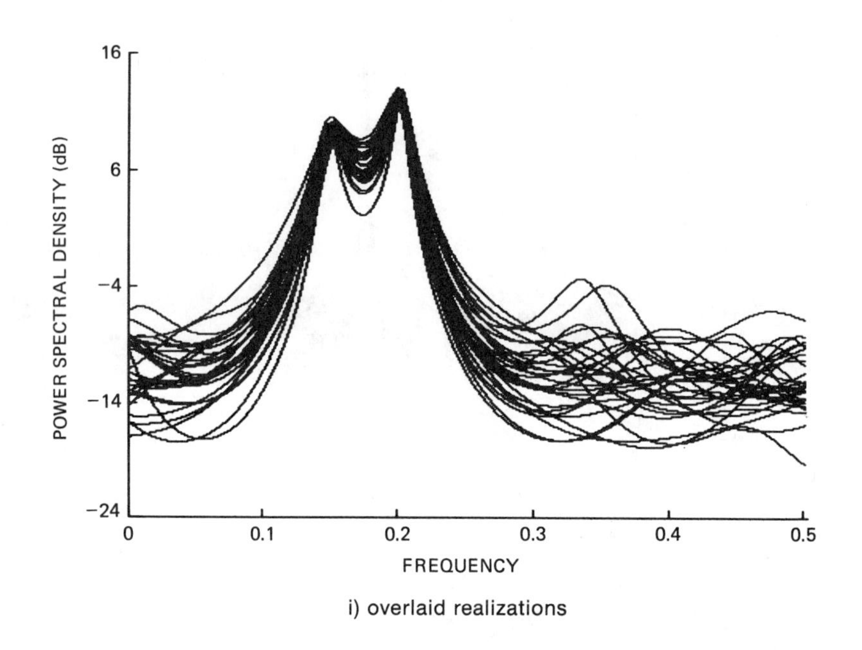

i) overlaid realizations

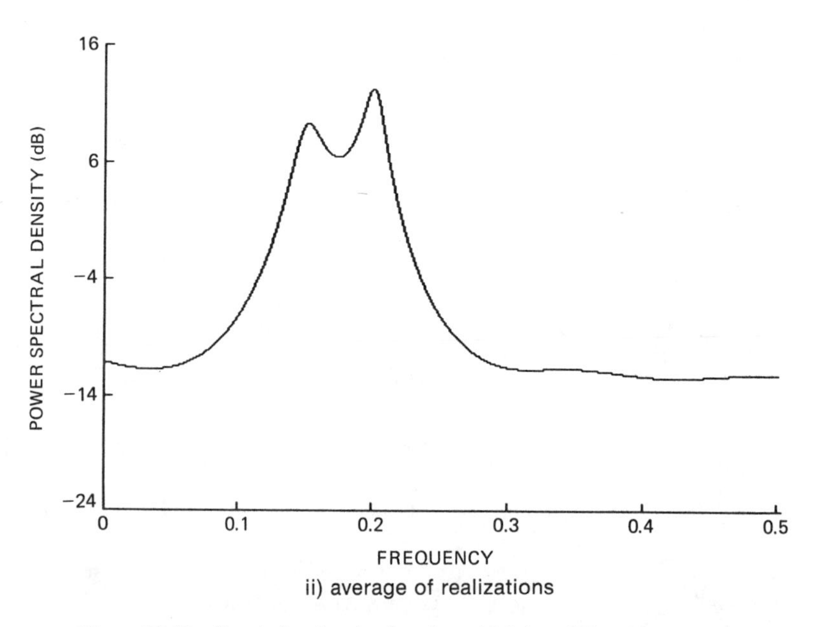

ii) average of realizations

Figure 11.10 Spectral estimates for sinusoidal data ($N = 25$, $p = 10$).

matrix estimator. Note that a fast algorithm has been developed for the case where a Toeplitz autocorrelation matrix with a biased ACF estimator is used [Musicus 1985]. A computer program implementation is given in the book by Marple [1987].

REFERENCES

Burg, J. P., "The Relationship between Maximum Entropy Spectra and Maximum Likelihood Spectra," *Geophysics*, Vol. 37, pp. 375–376, Apr. 1972.

Capon, J., "High-Resolution Frequency–Wavenumber Spectrum Analysis," *Proc. IEEE*, Vol. 57, pp. 1408–1418, Aug. 1969.

Capon, J., "Correction to 'Probability Distributions for Estimators of the Frequency–Wavenumber Spectrum,'" *Proc. IEEE*, Vol. 59, p. 112, Jan. 1971.

Capon, J., and N. R. Goodman, "Probability Distributions for Estimators of the Frequency–Wavenumber Spectrum," *Proc. IEEE*, Vol. 58, pp. 1785–1786, Oct. 1970.

Cox, H., "Resolving Power and Sensitivity to Mismatch of Optimum Array Processors," *J. Acoust. Soc. Am.*, Vol. 54, pp. 771–785, Sept. 1973.

Goodman, N. R., "Statistical Analysis Based on a Certain Multivariate Complex Gaussian Distribution," *Ann. Math. Stat.*, Vol. 34, pp. 152–157, 1963.

Gray, R. M., "Toeplitz and Circulant Matrices: A Review," Tech. Rep. 6502-1, Information Systems Laboratory, Center for Systems Research, Stanford University, June 1971.

Grenander, U., and G. Szego, *Toeplitz Forms and Their Applications*, University of California Press, Berkeley, Calif., 1958.

Kay, S., and C. Demeure, "The High Resolution Spectral Estimator: A Subjective Entity," *Proc. IEEE*, Vol. 72, pp. 1815–1816, Dec. 1984.

Lacoss, R. T., "Data Adaptive Spectral Analysis Methods," *Geophysics*, Vol. 36, pp. 661–675, Aug. 1971.

Lagunas-Hernandez, M. A., and A. Gasull-Llampalas, "An Improved Maximum Likelihood Method for Power Spectral Density Estimation," *IEEE Trans. Acoust. Speech Signal Process.*, Vol. ASSP32, pp. 170–172, Feb. 1984.

Marple, S. L., Jr., *Digital Spectral Analysis with Applications,* Prentice-Hall, Englewood Cliffs, N.J., 1987.

Marzetta, T., "A New Interpretation for Capon's Maximum Likelihood Method of Frequency-Wavenumber Spectrum Estimation," *IEEE Trans. Acoust. Speech Signal Process.*, Vol. ASSP31, pp. 445-449, Apr. 1983.

McDonough, R. N., "Application of the Maximum-Likelihood Method and the Maximum-Entropy Method to Array Processing," in *Nonlinear Methods of Spectral Analysis,* S. Haykin, ed., Springer-Verlag, New York, 1983.

Musicus, B., "Fast MLM Power Spectrum Estimation from Uniformly Spaced Correlations," *IEEE Trans. Acoust. Speech Signal Process.*, Vol. ASSP33, pp. 1333–1334, Oct. 1985.

Pisarenko, V. F., "On the Estimation of Spectra by Means of Non-linear Functions of the Covariance Matrix," *Geophys. J. Roy. Astron. Soc.*, Vol. 28, pp. 511–531, 1972.

PROBLEMS

11.1. Verify (11.4). *Hint:* Let $\mathbf{y} = \mathbf{x} - \tilde{A}_c\mathbf{e}$, $\mathbf{z} = (A_c - \tilde{A}_c)\mathbf{e}$, so that $\mathbf{x} - A_c\mathbf{e} = \mathbf{y} - \mathbf{z}$. Show that $\mathbf{y}^H\mathbf{R}_{zz}^{-1}\mathbf{z} = 0$.

11.2. Show that

$$\text{var}(\hat{A}_c) = \mathbf{w}^H\mathbf{R}_{zz}\mathbf{w}$$
$$= \int_{-1/2}^{1/2} |H(f)|^2 P_{zz}(f)\,df$$

where

$$H(f) = \left[\sum_{n=0}^{p-1} w[n] \exp(-j2\pi fn)\right]^*$$

is the frequency response as defined by (11.15). Next prove that the constraint $\mathbf{w}^H\mathbf{e} = 1$ is equivalent to constraining the frequency response to be unity at $f = f_0$.

11.3. Using the results from Problem 11.2, prove that if

$$P_{zz}(f) = P_1\delta(f - f_1)$$

then for $f_1 \neq f_0$, the LMVU filter will place a notch at $f = f_1$ or equivalently, the $H(f)$ that minimizes the variance subject to the unbiased constraint satisfies $H(f_1) = 0$.

11.4. As $p \to \infty$ it has been proven [Grenander and Szego 1958, Gray 1971] that the eigenvalues $\{\lambda_0, \lambda_1, \ldots, \lambda_{p-1}\}$ and eigenvectors $\{\mathbf{v}_0, \mathbf{v}_1, \ldots, \mathbf{v}_{p-1}\}$ of the $p \times p$ hermitian Toeplitz autocorrelation matrix $\mathbf{R}_{xx}$ are

$$\lambda_i \to P_{xx}\left(\frac{i}{p}\right) \qquad i = 0, 1, \ldots, p-1$$

$$\mathbf{v}_i \to \frac{1}{\sqrt{p}}\left[1 \quad \exp\left[j\left(\frac{2\pi}{p}\right)i\right] \quad \exp\left[j\left(\frac{2\pi}{p}\right)2i\right] \quad \cdots \quad \exp\left[j\left(\frac{2\pi}{p}\right)(p-1)i\right]\right]^T.$$

Use these properties to show that

$$P_{\text{MV}}(f_n) = \frac{p}{\mathbf{e}_n^H\mathbf{R}_{xx}^{-1}\mathbf{e}_n} \to P_{xx}(f_n) \qquad n = 0, 1, \ldots, p-1$$

where $f_n = n/p$ and

$$\mathbf{e}_n = [1 \quad \exp(j2\pi f_n) \quad \exp(j4\pi f_n) \quad \cdots \quad \exp[j2\pi f_n(p-1)]]^T.$$

Note that the PSD has been defined on the frequency interval [0,1] for convenience. What happens if the factor of p is omitted from the definition of the MVSE?

11.5. Verify the analytical expression (11.19) for the MVSE for one complex sinusoid in complex white noise. To do so, use Woodbury's identity (2.32):

$$(\mathbf{A} + \mathbf{u}\mathbf{u}^H)^{-1} = \mathbf{A}^{-1} - \frac{\mathbf{A}^{-1}\mathbf{u}\mathbf{u}^H\mathbf{A}^{-1}}{1 + \mathbf{u}^H\mathbf{A}^{-1}\mathbf{u}}$$

where $\mathbf{A}$ is $p \times p$ and $\mathbf{u}$ is $p \times 1$.

11.6. Consider the modified MVSE to be

$$P_{\text{MMV}}(f) = \frac{p}{\mathbf{e}^H(\mathbf{R}_{xx}^2)^{-1}\mathbf{e}}.$$

As $p \to \infty$, what does this estimator converge to for $f_n = n/p$, $n = 0, 1, \ldots, p - 1$? *Hint:* Use the properties given in Problem 11.4. If the process consists of one complex sinusoid in complex white noise of unit variance, determine the peak amplitude of $P_{\text{MMV}}(f)$ and compare it to the peak amplitude of $pP_{\text{MV}}(f)$. Is the modified MVSE a better frequency estimator? See the works by Pisarenko [1972] and Kay and Demeure [1984].

11.7. Assume that a Toeplitz autocorrelation matrix is used in (11.17) with an unbiased ACF estimator. For $p = 2$ and $f = 0$ find a real data sequence to illustrate the possibility of a negative spectral estimate. Can you also find an example for the case when the biased ACF estimator is used?

APPENDIX 11A

Computer Program for Minimum Variance Spectral Estimator (Provided on Floppy Disk)

```
      SUBROUTINE MVSE(X,N,IP,L,EPS,MODE,PMV,IFLAG)
C
C     This program implements the minimum variance spectral estimator
C     (MVSE) (11.17) using the modified covariance method estimator of
C     the autocorrelation matrix (11.27).
C
C     Input Parameters:
C
C       X          -Complex array of dimension N×1 of data points
C       N          -Number of complex data points
C       IP         -Dimension of autocorrelation matrix used in estimator
C       L          -Number of samples of spectral estimate desired
C       EPS        -Small positive number to test for ill-conditioning
C                   of autocorrelation matrix to be inverted, required
C                   in CHOLESKY to solve linear equations
C       MODE       -If MODE=0, then numerator used in definition of MVSE
C                   is one; if MODE=1, then numerator is given by IP.
C
C     Output Parameters:
C
```

```
C       PMV       -Real array of dimension L×1 of samples of the MVSE, where
C                  PMV(I) corresponds to the spectral estimate at the
C                  frequency F= -1/2+(I-1)/L
C       IFLAG     -Ill-conditioning indicator for solution of linear equations
C                    =   0 for normal termination
C                    = -1 if one of the di's in Cholesky decomposition
C                          is less than EPS causing premature termination
C
C     External Subroutines:
C
C     CHOLESKY     -see Appendix 2B
C
C     Notes:
C
C       The calling program must dimension arrays X,PMV. The
C       arrays B,E,RMAT,A,RINV must be dimensioned greater than or equal
C       to the variable dimensions shown. The arrays XL,Y,D in
C       CHOLESKY must be dimensioned to be greater than or equal
C       to IP×IP, IP, and IP, respectively.
C
C     Verification Test Case:
C
C       If the complex test data listed in Appendix 1A is input
C       as X, with N=32, IP=11, L=512, EPS=1.0E-15, MODE=1,
C       then the output should be
C         PMV(100)=0.04379
C         PMV(200)=0.18816
C         PMV(300)=0.47820
C         PMV(400)=0.03157
C         PMV(500)=0.30242
C         IFLAG=0
C       for a DEC VAX 11/780.
C
        COMPLEX X(1),B(IP),E(IP),RMAT(IP*(IP+1)/2),SUM,A(IP*(IP+1)/2),
      * RINV(IP,IP)
        DIMENSION PMV(1)
        PI2=8.*ATAN(1.)
C     Compute autocorrelation matrix estimate using modified
C     covariance method (11.27)
        CALL CORRMAT(X,N,IP,RMAT)
        IMAX=IP*(IP+1)/2
        DO 5 I=1,IMAX
5       A(I)=RMAT(I)
C     Begin inversion of autocorrelation matrix
C     Invert autocorrelation matrix by solving IP sets of linear
C     equations for X or Rxx X = I, where I is the IP×IP identity matrix
        DO 30 J=1,IP
        DO 10 I=1,IP
        B(I)=(0.,0.)
```

```
10        IF(I.EQ.J)B(I)=(1.,0.)
          CALL CHOLESKY(A,B,IP,EPS,IFLAG)
C       Terminate program if matrix ill-conditioned
          IF(IFLAG.EQ.-1)RETURN
          DO 20 I=1,IP
20        RINV(I,J)=B(I)
30        CONTINUE
C       Computation of hermitian form
          DO 70 K=1,L
          F=-0.5+(K-1.)/L
          DO 40 J=1,IP
40        E(J)=CEXP(CMPLX(0.,PI2*(J-1.)*F))
          SUM=(0.,0.)
          DO 60 I=1,IP
          DO 50 J=1,IP
          SUM=SUM+CONJG(E(I))*RINV(I,J)*E(J)
50        CONTINUE
60        CONTINUE
C       Compute PSD values (11.17)
          IF(MODE.EQ.0)PMV(K)=REAL(1./SUM)
          IF(MODE.EQ.1)PMV(K)=REAL(IP/SUM)
70        CONTINUE
          RETURN
          END
          SUBROUTINE CORRMAT(X,N,IP,RMAT)
C       This program computes an estimate of the IP×IP autocorrelation
C       matrix by using the modified covariance method.
          COMPLEX X(1),RMAT(1)
          L=0
          DO 30 J=1,IP
          DO 20 I=1,J
          L=L+1
          RMAT(L)=(0.,0.)
          DO 10 K=IP+1,N
10        RMAT(L)=RMAT(L)+CONJG(X(K-I))*X(K-J)+X(K-IP+I)*
         * CONJG(X(K-IP+J))
          RMAT(L)=RMAT(L)/(2.*(N-IP))
20        CONTINUE
30        CONTINUE
          RETURN
          END
```

Chapter 12

Summary of Spectral Estimators

12.1 INTRODUCTION

A summary of the Fourier, AR, MA, ARMA, and minimum variance spectral estimators is given in this chapter. All comparisons which are drawn between the various spectral estimators are qualitative in nature and should only be used as an informal guideline in choosing a particular method for an application (see also Section 1.4). The spectral estimation practitioner should utilize the information presented here but should temper his or her choice of an estimator by the constraints of the problem and the expected character of the data. It is usually prudent to apply *several* spectral estimators to the data set to serve as a consistency check of the results.

12.2 TEST CASE DATA COMPARISON

The results of applying the various spectral estimators to the test case data sets have been presented in the summary sections of Chapters 4, 7, 8, 10, and 11. The spectral estimates for each method were obtained by running the computer programs listed in Table 12.1 and contained in the appendices of the various chapters. The specific choices made for the input parameters of each program are given in the verification test case description contained within the documentation of each program. The test case data sets are listed in Appendix 1A and are described in Section 1.5. The spectral estimation methods which are constrained to operate on real data (recursive MLE for AR and Akaike for ARMA) use the real data set

TABLE 12.1 SUMMARY OF SPECTRAL ESTIMATION METHODS USED IN COMPUTER SUBROUTINE IMPLEMENTATIONS

Spectral estimator type	Method	Defining equations	Assumed data type	User-defined parameters	Relative computational complexity	Subroutine listing[a]
Fourier	Periodogram	(4.2)	Complex or real	Data window	Low	PERIODOGRAM (Appendix 4C)
	Blackman–Tukey	(4.23)	Complex or real	Number lags (M), lag window ($w[k]$), ACF estimator	Low	BLACKTUKEY (Appendix 4E)
Autoregressive	Autocorrelation (Yule–Walker)	(5.9), (7.3), (7.5)	Complex or real	Model order (p)	Medium	AUTOCORR[b] (Appendix 7B)
	Covariance	(5.9), (7.8), (7.9)	Complex or real	Model order (p)	Medium	COVMCOV[b] (Appendix 7C)
	Modified covariance (forward-backward)	(5.9), (7.23), (7.24)	Complex or real	Model order (p)	Medium	COVMCOV[b] (Appendix 7C)
	Burg	(5.9), (7.39)–(7.41)	Complex or real	Model order (p)	Medium	BURG[b] (Appendix 7D)
	Recursive MLE	(5.9), (7.46)–(7.52)	Real	Model order (p)	Medium	RMLE[b] (Appendix 7E)
Moving average	Durbin	(5.7), (8.7), (8.9)	Complex or real	Model order (q), large AR model order (L)	Medium	DURBIN[b] (Appendix 8A)
Autoregressive moving average	Akaike MLE	(5.5), (10.2)–(10.5)	Real	Model orders (p, q), initial estimates, iteration stopping criterion	High	AKAIKE[b] (Appendix 10C)
	Modified Yule–Walker equations (high-order Y-W)	(5.5), (8.7), (8.9), (10.7)	Complex or real	Model orders (p, q)	Medium	MYWE[b] (Appendix 10D)
	Least-squares modified Yule–Walker equations	(5.5), (8.7), (8.9), (10.18)	Complex or real	Model orders (p, q), number of equations ($M - q$)	Medium	LSMYWE[b] (Appendix 10E)
	Mayne–Firoozan (three-stage least squares)	(5.5), (10.22)–(10.27)	Complex or real	Model orders (p, q), large AR model order (L)	Medium	MAYNEFIR[b] (Appendix 10F)
Minimum variance	Minimum variance (Capon or maximum likelihood method)	(11.17), (11.27)	Complex or real	Autocorrelation matrix dimension (p)	Medium	MVSE (Appendix 11A)

[a] See also Appendix 1 for brief description as well as supporting subroutines.

[b] All AR, MA, and ARMA methods may use ARMA PSD (Appendix 5C) to compute PSD samples given estimated parameters.

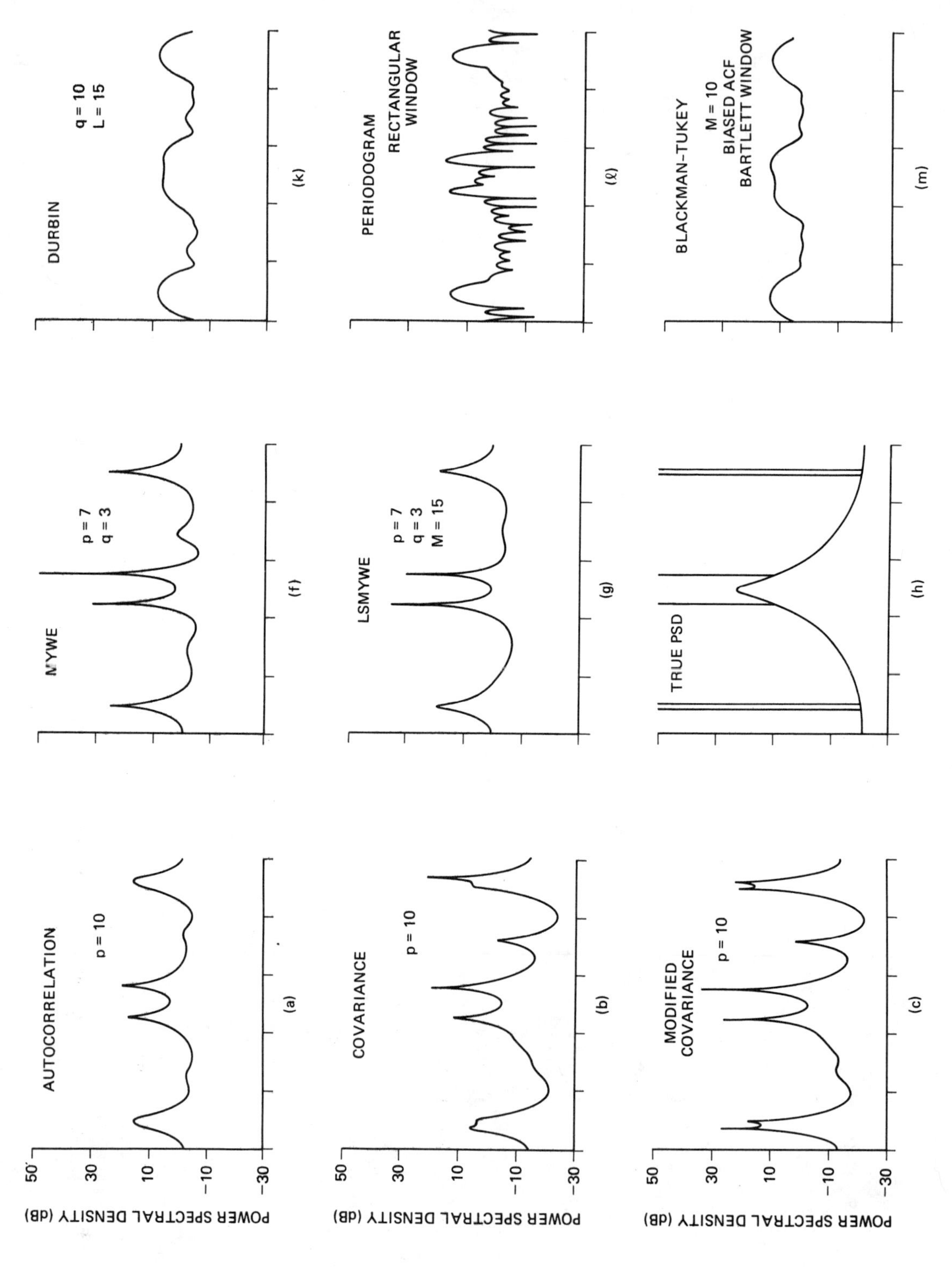
POWER SPECTRAL DENSITY (dB)
50
30
10
−10
−30
AUTOCORRELATION
p = 10
(a)
COVARIANCE
p = 10
(b)
MODIFIED COVARIANCE
p = 10
(c)
MYWE
p = 7
q = 3
(f)
LSMYWE
p = 7
q = 3
M = 15
(g)
TRUE PSD
(h)
DURBIN
q = 10
L = 15
(k)
PERIODOGRAM
RECTANGULAR WINDOW
(ℓ)
BLACKMAN-TUKEY
M = 10
BIASED ACF
BARTLETT WINDOW
(m)

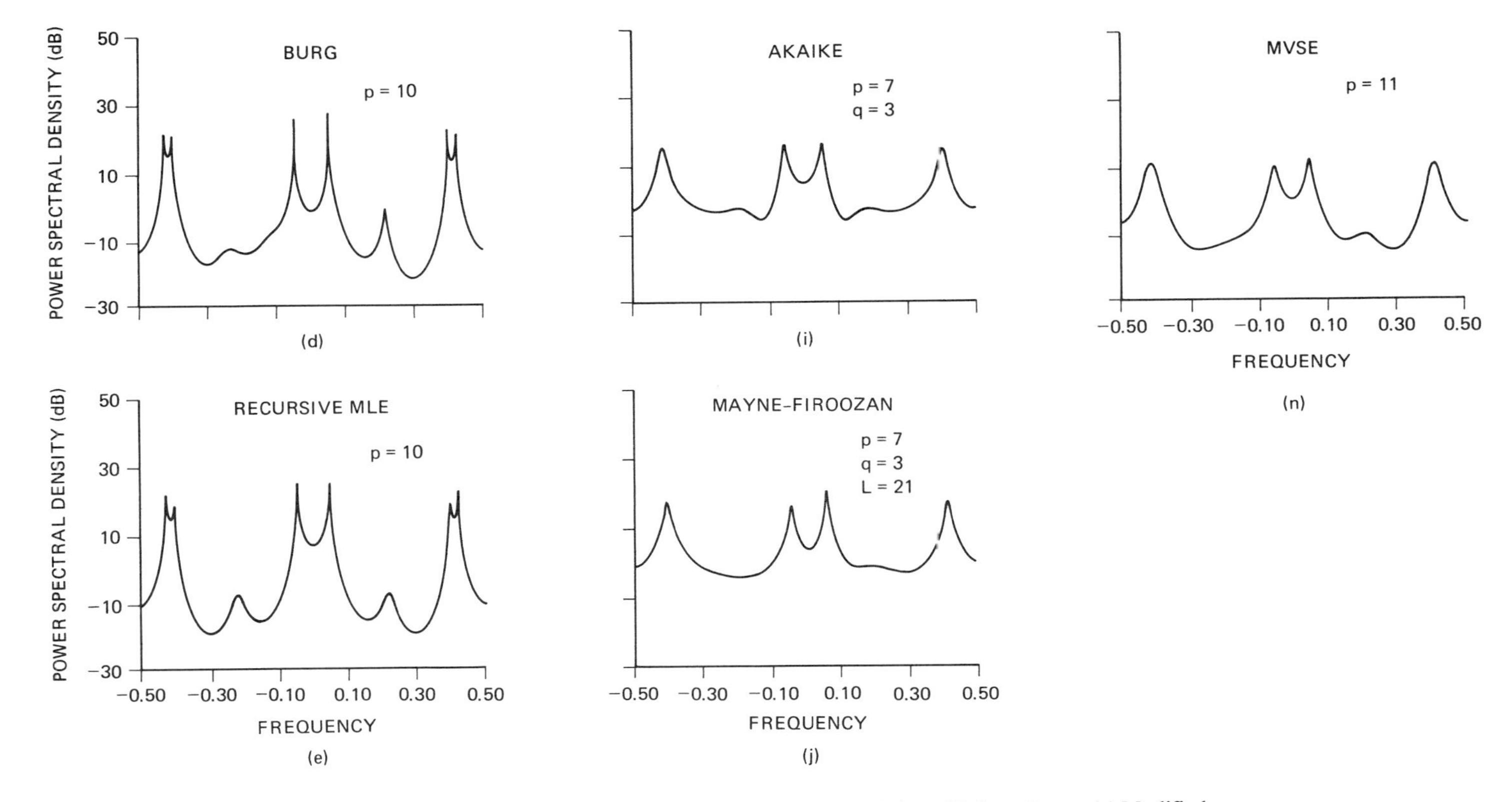

Figure 12.1 Summary of test case results. (a) Autocorrelation. (b) Covariance. (c) Modified covariance. (d) Burg. (e) Recursive MLE. (f) MYWE. (g) LSMYWE. (h) True PSD. (i) Akaike. (j) Mayne–Firoozan. (k) Durbin. (l) Periodogram. (m) Blackman–Tukey. (n) MVSE.

and all others use the complex data set. Each test case has the same PSD, which is shown in Figure 12.1h. There are several narrowband sinusoidal components as well as a broadband AR(1) process component. The closely spaced sinusoids are not resolvable by a Fourier estimator since they are spaced 0.02 cycles/sample apart and the data record length is only 32 points. The sinusoids centered about $f = 0$ are spaced 0.1 cycles/sample apart and so are Fourier resolvable. Because the broadband process is an AR(1) process, the overall data process may be modeled as the limiting form of an ARMA(7, 7) process, where six of the poles lie on the unit circle at the sinusoidal frequency locations and the remaining pole is identical to the AR(1) process pole.

Each spectral estimation method, other than the periodogram, is constrained to estimate the same number of parameters, so that a qualitative comparison would be fair. The exact choices for the model order, lag windows, and so on, were not made to yield the best results but only to illustrate the typical characteristics of each estimator. Other choices will yield similar although not identical results.

Referring to Figure 12.1, the AR spectral estimates based on a model order of $p = 10$ are shown in parts (a)–(e). At most 10 peaks may be present in the spectral estimates. It is observed that the sinusoidal components are resolved by all the estimators except for the autocorrelation method (a), which has the lowest resolution. All the methods produce a spurious peak at about $f = 0.2$, although it is less pronounced for the autocorrelation method (a). The recursive MLE (e) produces a symmetric spectral estimate since it operates on real data and is thus constrained to yield a symmetric estimate. In general, the AR methods are able to resolve closely spaced narrowband components but tend to exhibit peaky spectra even for broadband processes.

The ARMA spectral estimates based on model orders of $p = 7$ and $q = 3$ are shown in parts (f), (g), (i), and (j). None of the methods is able to resolve the spectral components centered about $f = 0.4$. If the AR model order p is increased, it is possible that the peaks may have been resolved. The MYWE (f) and LSMYWE (g) estimates are similar in appearance, as are the approximate MLEs of Akaike (i) and Mayne–Firoozan (j). The LSMYSE (g) used $M = 15$ but did not appear to be overly sensitive to the number of equations $M - q$ used. The Akaike estimate (i) is symmetric since it is contrained to operate on real data. The initial estimates for the Akaike method were obtained using the LSMYWE estimator for real data with $p = 7$, $q = 3$, $M = 15$. It was observed that the iteration did not converge for many other initial estimates, but when convergence did occur, it was rapid. The Mayne–Firoozan estimate (j) used a large AR model order of $L = 21$ but did not appear to be overly sensitive to the value of L.

The MA spectral estimate based on a model order of $q = 10$ and which employs Durbin's algorithm with a large AR model order of $L = 15$ is shown in part (k). It is unable to resolve any of the narrowband spectral components. The choice of the large AR model order L did not significantly affect the spectral estimate.

The Fourier spectral estimates are displayed in parts (l) and (m). The periodogram (l) did not use a data window (rectangular window). As expected, it is unable to resolve the closely spaced sinusoids which are less than $1/N = 0.33$

cycles/sample apart and exhibits the usual sidelobe structure. Note that a peak is visible near $f = 0.2$ which may be responsible for the spurious peak observed in the AR spectral estimates. The Blackman–Tukey estimate (m) was based on a biased autocorrelation estimator for lags $k = 0, 1, \ldots, M = 10$ and employed a Bartlett window. It is observed to be nearly identical to Durbin's estimate (k). This is also expected since the forms of the PSD are identical (see Section 8.3), with only the estimates of the ACF being different.

The minimum variance spectral estimate based on an autocorrelation matrix of dimension $p \times p = 11 \times 11$ is shown in part (n). The factor of p in the numerator of the spectral estimate required for a consistent estimator has been included (see Section 11.5). The closely spaced sinusoids are not discernible and the overall estimate is similar in appearance to the autocorrelation estimate (a). This is probably due to the relationship between the two estimates as given by (11.26). A larger dimension autocorrelation matrix results in a spectral estimate which resolves the closely spaced sinusoids but which also gives rise to many spurious peaks.

Table 12.1 is a descriptive summary of the methods that appear in Figure 12.1. The type of spectral estimator or class to which it belongs and the defining equations are given. The data type expected by each spectral estimator is listed together with the parameters that the user must specify. An indication of the relative computational complexity is given. A quantitative computational comparison requires one to specify an exact implementation and will be critically dependent on any "tricks" used to reduce computation. The subroutine implementations referred to in Table 12.1 *have not been* optimized for minimum computations but rather have been written for pedagogical simplicity. Further discussions of implementational issues may be found in the book by Marple [1987].

12.3 GENERAL COMPARISON

A general comparison of the spectral estimation approaches is given in Table 12.2. The assumed model for each class as well as the form of the spectral estimator is listed. The assumed PSD form imposes certain constraints on the ability of the estimator to estimate the PSD of an arbitrary process with negligible bias. For example, the Fourier estimators are only able to estimate broadband or low dynamic range PSDs. Features with bandwidths less than $1/N$ are not able to be observed. ARMA spectral estimators, on the other hand, have the potential for estimation of both broadband and narrowband or high dynamic range spectra. That potential will only be realized, however, if more accurate estimators of the ARMA parameters are found. A qualitative assessment of the resolution capabilities of the various types of spectral estimators is also given in Table 12.2. For large enough data records the Fourier spectral estimators are preferable since they do not make any additional assumptions about the data other than wide sense stationarity. Also, the statistics for moderately sized data records are known [Jenkins and Watts 1968], which allows us to accurately interpret the spectral estimation results. For the other classes of spectral estimators the statistical per-

TABLE 12.2 COMPARISON OF SPECTRAL ESTIMATION APPROACHES

Spectral estimator type	Assumed model	PSD estimator form	Representable PSDs	Resolution capability
Fourier	None	$\frac{1}{N}\left\lvert \sum_{n=0}^{N-1} x[n]e^{-j2\pi fn}\right\rvert^2$ $\sum_{k=-M}^{M} w[k]\hat{r}_{xx}[k]e^{-j2\pi fk}$	Broadband only	Low
Autoregressive	AR	$\frac{\hat{\sigma}^2}{\left\lvert 1+\sum_{k=1}^{p}\hat{a}[k]e^{-j2\pi fk}\right\rvert^2}$	Broadband or narrow-band but not both	High
Moving average	MA	$\hat{\sigma}^2\left\lvert 1+\sum_{k=1}^{q}\hat{b}[k]e^{-j2\pi fk}\right\rvert^2$	Broadband only	Low
Autoregressive moving average	ARMA	$\frac{\hat{\sigma}^2\left\lvert 1+\sum_{k=1}^{q}\hat{b}[k]e^{-j2\pi fk}\right\rvert^2}{\left\lvert 1+\sum_{k=1}^{p}\hat{a}[k]e^{-j2\pi fk}\right\rvert^2}$	Broadband or narrow-band or both	High
Minimum variance	None	$\frac{1}{\mathbf{e}^H\hat{\mathbf{R}}_{xx}^{-1}\mathbf{e}}$	Broadband or narrow-band but not both	Medium

formance is unknown and the use of the limited results based on asymptotic arguments (large data records) is risky. For short data records the "modern" spectral estimators such as the AR, ARMA, and minimum variance methods offer the promise of higher resolution. Their principal shortcomings are that in the case of AR and ARMA spectral estimation, if the assumed model is inappropriate or if the model orders chosen are incorrect, and in the case of minimum variance spectral estimation, if the dimension of the autocorrelation matrix is inappropriate, then poor spectral estimates will result. Heavy biases and/or large variabilities may be exhibited. In employing a particular spectral estimator we must always be mindful of the potential pitfalls as well as the potential gains. A spectral estimator that combines the best features of all the different approaches would be ideal. Future research in spectral estimation may well result in such an "expert system" [Gaby and Hayes 1984].

REFERENCES

Gaby, J. E., and M. H. Hayes, "Artificial Intelligence Applied to Spectrum Estimation," *Rec. 1984 IEEE Int. Conf. Acoust Speech Signal Process.*, pp. 13.5.1–13.5.4.

Jenkins, G. M., and D. G. Watts, *Spectral Analysis and Its Applications,* Holden-Day, San Francisco, 1968.

Marple, S. L., Jr., *Digital Spectral Analysis with Applications,* Prentice-Hall, Englewood Cliffs, N.J., 1987.

PROBLEMS

12.1. Explain why a comparison of different spectral estimates based on a single data record is not a valid means of comparing the performance of the spectral estimators. How would you make a *quantitative* comparison of several spectral estimators?

12.2. Why might we suspect that there are two sinusoidal components near $f = 0.4$ from examination of the periodogram (Figure 12.1l)?

12.3. The AR spectral estimates (Figure 12.1a–e) do not exhibit much power at $f = 0$, although the true PSD (Figure 12.1h) does. Can you explain the reason for this behavior?

12.4. The Burg spectral estimate (Figure 12.1d) indicates strong peaks at six frequency locations. Discuss how to subtract the sinusoidal components from the data so that the spectral estimate of the broadband PSD may be obtained. Could you use this approach to determine if the peak observed at $f = 0.2$ is a spurious peak?

PART II

ADVANCED CONCEPTS

Chapter 13

Sinusoidal Parameter Estimation

13.1 INTRODUCTION

The estimation of the frequencies of sinusoidal components embedded in white noise is a problem that arises in many fields. For a single sinusoid or multiple sinusoids which are well resolved by Fourier methods, we generally use a periodogram as a frequency estimator. In this instance, the periodogram is optimum. However, for unresolvable sinusoids the periodogram cannot be used, and furthermore, optimal approaches are not practical. This has led to the introduction of many suboptimal methods based on approximate maximum likelihood techniques and eigenvector properties of the autocorrelation matrix. The approximate maximum likelihood techniques regard the sinusoidal process as a deterministic signal with unknown frequencies, while the eigenanalysis approaches employ a WSS random process model so that the frequencies appear as unknown parameters in an autocorrelation matrix. Although the latter approach relies heavily on spectral estimators described in Part I, the methods should be regarded as *frequency estimators,* not spectral estimators. They should not be employed for general spectral estimation since the development of these methods *assumes the data to consist of pure sinusoids in white noise*. The application to narrowband processes in white noise or sinusoids in colored noise may result in poor performance. Although in this chapter we discuss frequency estimation for p complex sinusoids in complex white noise, the same methods generally apply to real sinusoids in real white noise if p is chosen to be *twice* the number of real sinusoids.

13.2 SUMMARY

In Section 13.3 we discuss the problem of frequency estimation based on maximum likelihood methods. It is proven that for a single sinusoid the MLE of frequency is given as the location of the largest peak of the periodogram (13.7). For multiple sinusoids the MLE of the frequencies is found by maximizing (13.13), which unfortunately cannot be done analytically. If, however, the sinusoidal frequencies are well resolved by a periodogram, then the maximization of (13.13) results in frequency estimates which are the locations of the peaks of the periodogram. In Section 13.4 the Cramer–Rao bounds on the variances of the sinusoidal parameters are given. For one sinusoid the bounds are given by (13.16), while for multiple sinusoids they are given by (13.17). Some approximate MLE methods which can be simply implemented are described in Section 13.5. In particular, the iterative filtering algorithm is derived and summarized in Figure 13.2. The use of AR spectral estimation for sinusoidal frequency estimation is shown in Section 13.6 to produce poor results at low SNR. In an effort to improve the performance, some properties of the autocorrelation matrix as described in Section 13.7 can be employed. The important properties are that the signal part of the theoretical autocorrelation matrix has a rank equal to the number of sinusoids, which prompts us to replace the estimated autocorrelation matrix, which in general will be full rank, by one that forces the rank to be equal to the number of sinusoids. Also, it is shown that the signal vectors are orthogonal to the noise subspace eigenvectors as expressed in (13.34). Both these properties lead to improved frequency estimators as described in Section 13.8. The salient frequency estimators are the principal component AR method as given by (13.36), which discards the nonprincipal eigenvectors of the estimated autocorrelation matrix; the Pisarenko harmonic decomposition, which finds the eigenvector of the autocorrelation matrix associated with the minimum eigenvalue and then roots a prediction error filter for the frequencies as given by (13.43); and the MUSIC algorithm, which chooses the frequency estimates as the peaks of a function (13.50) which is theoretically infinite at the sinusoidal frequencies. In Section 13.10 some approaches to the problem of determining the number of sinusoids present are described, while in Section 13.11 various estimators are compared via a computer simulation.

13.3 MAXIMUM LIKELIHOOD ESTIMATION

The problem of estimation of the parameters of complex sinusoids in complex white Gaussian noise is addressed. As will be seen shortly, the MLE for this problem is hopelessly complicated and impractical to implement. To begin the discussion, assume that the data consist of p complex sinusoids in complex white Gaussian noise [Goodman 1963] (see also Section 3.2.2)

$$x[n] = \sum_{i=1}^{p} A_i \exp(j2\pi f_i n + \phi_i) + z[n] \tag{13.1}$$

for $n = 0, 1, \ldots, N-1$. $z[n]$ is complex white Gaussian noise with zero mean and variance σ_z^2. The sinusoidal parameters $\{A_1, A_2, \ldots, A_p, \phi_1, \phi_2, \ldots, \phi_p, f_1, f_2, \ldots, f_p\}$, which consist of amplitudes, phases, and frequencies, are assumed to be constant but unknown and are to be estimated. It is important to remember that the number of sinusoids, p, is assumed to be known. Methods to estimate p are discussed in Section 13.10. Also, the modeling of the sinusoidal process is as a *deterministic* signal with unknown parameters.

13.3.1 Single Sinusoid

The first case to be examined is that of a single sinusoid. The MLE will be shown to be easily obtained by basing the estimate on the Fourier transform of the data. In particular, the MLE of frequency is the frequency where the periodogram attains its maximum. The PDF of $\mathbf{x} = [x[0] \quad x[1] \quad \cdots \quad x[N-1]]^T$ for a signal in complex Gaussian noise is, from (3.13),

$$p(\mathbf{x}-\mathbf{s}) = \frac{1}{\pi^N \det(\mathbf{R}_{zz})} \exp\left[-(\mathbf{x}-\mathbf{s})^H \mathbf{R}_{zz}^{-1}(\mathbf{x}-\mathbf{s})\right] \tag{13.2}$$

where $\mathbf{R}_{zz}$ is the $N \times N$ autocorrelation matrix of the noise and $\mathbf{s} = [s[0] \quad s[1] \quad \cdots \quad s[N-1]]^T$ is the vector of signal samples. Since the noise is assumed to be white,

$$\mathbf{R}_{zz} = \sigma_z^2 \mathbf{I}$$

and the signal is assumed to be a single sinusoid,

$$\mathbf{s} = A_1 \exp(j\phi_1)[1 \quad \exp(j2\pi f_1) \quad \cdots \quad \exp[j2\pi f_1(N-1)]]^T.$$

To find the MLE of A_1, ϕ_1, and f_1, we need to maximize

$$p(\mathbf{x}-\mathbf{s}) = \frac{1}{\pi^N \det(\sigma_z^2 \mathbf{I})} \exp\left[-\frac{1}{\sigma_z^2}(\mathbf{x}-\mathbf{s})^H(\mathbf{x}-\mathbf{s})\right] \tag{13.3}$$

over A_1, ϕ_1, and f_1. Alternatively, we must minimize

$$S = (\mathbf{x}-\mathbf{s})^H(\mathbf{x}-\mathbf{s}). \tag{13.4}$$

Let A_{c_1} denote the complex amplitude or $A_{c_1} = A_1 \exp(j\phi_1)$. Also, let $\mathbf{e}_1 = [1 \quad \exp(j2\pi f_1) \quad \cdots \quad \exp[j2\pi f_1(N-1)]]^T$. Then

$$S(A_{c_1}, f_1) = (\mathbf{x} - A_{c_1}\mathbf{e}_1)^H(\mathbf{x} - A_{c_1}\mathbf{e}_1)$$

where the dependence of S on A_{c_1} and f_1 is made explicit by the notation. If f_1 were known, this problem would be identical to the one encountered in Chapter 11. There the minimizing solution for a known frequency was shown to be

$$\begin{aligned}\hat{A}_{c_1} &= \frac{\mathbf{e}_1^H \mathbf{R}_{zz}^{-1}\mathbf{x}}{\mathbf{e}_1^H \mathbf{R}_{zz}^{-1}\mathbf{e}_1} = \frac{\mathbf{e}_1^H \mathbf{x}}{\mathbf{e}_1^H \mathbf{e}_1} \\ &= \frac{1}{N}\sum_{n=0}^{N-1} x[n] \exp(-j2\pi f_1 n).\end{aligned} \tag{13.5}$$

$\hat{A}_{c_1}$ will be the MLE of A_{c_1} for the problem at hand if f_1 is replaced by its MLE in (13.5). To continue the minimization, $\hat{A}_{c_1}$ as given by (13.5) may be substituted into $S(A_{c_1}, f_1)$ and minimized over f_1.

$$S(\hat{A}_{c_1}, f_1) = (\mathbf{x} - \hat{A}_{c_1}\mathbf{e}_1)^H(\mathbf{x} - \hat{A}_{c_1}\mathbf{e}_1)$$
$$= \mathbf{x}^H(\mathbf{x} - \hat{A}_{c_1}\mathbf{e}_1) - \hat{A}^*_{c_1}\mathbf{e}_1^H(\mathbf{x} - \hat{A}_{c_1}\mathbf{e}_1).$$

The second term is easily shown to be zero by using (13.5), which results in

$$S(\hat{A}_{c_1}, f_1) = \mathbf{x}^H\mathbf{x} - \hat{A}_{c_1}\mathbf{x}^H\mathbf{e}_1 = \mathbf{x}^H\mathbf{x} - \frac{1}{N}|\mathbf{e}_1^H\mathbf{x}|^2. \tag{13.6}$$

Finally, to minimize $S(\hat{A}_{c_1}, f_1)$ we must maximize

$$\frac{1}{N}|\mathbf{e}_1{}^H\mathbf{x}|^2 = \frac{1}{N}\left|\sum_{n=0}^{N-1} x[n] \exp(-j2\pi f_1 n)\right|^2 \tag{13.7}$$

over f_1. Equation (13.7) is recognized as the periodogram [see (4.2)]. Hence *the MLE of the frequency of a single complex sinusoid in complex white Gaussian noise is found by choosing the frequency at which the periodogram attains its maximum* (see also Problem 13.1). A similar result is obtained if the signal is a real sinusoid in real white Gaussian noise (see Problem 13.2). Once $\hat{f}_1$ is found, the MLEs of the amplitude and phase follow from (13.5) as

$$\hat{A}_1 = \left|\frac{1}{N}\sum_{n=0}^{N-1} x[n] \exp(-j2\pi\hat{f}_1 n)\right|$$
$$\hat{\phi}_1 = \arctan\left[\frac{\operatorname{Im}\left(\sum_{n=0}^{N-1} x[n] \exp(-j2\pi\hat{f}_1 n)\right)}{\operatorname{Re}\left(\sum_{n=0}^{N-1} x[n] \exp(-j2\pi\hat{f}_1 n)\right)}\right]. \tag{13.8}$$

Note that the frequency estimate is found as the result of a one-dimensional search. This is usually implemented using an FFT [Palmer 1974, Rife and Boorstyn 1974]. However, to reduce errors due to frequency quantization the data should be zero padded so that a large number of frequency samples are computed. Alternatively, a coarse FFT followed by a Newton–Raphson search can be implemented.

13.3.2 Multiple Sinusoids

To find the MLE for the parameters of multiple complex sinusoids in complex white Gaussian noise, we proceed as before except that now the function to be minimized is

$$S = \left(\mathbf{x} - \sum_{i=1}^{p} A_{c_i}\mathbf{e}_i\right)^H\left(\mathbf{x} - \sum_{i=1}^{p} A_{c_i}\mathbf{e}_i\right) \tag{13.9}$$

where $A_{c_i} = A_i \exp(j\phi_i)$ and $\mathbf{e}_i = [1 \quad \exp(j2\pi f_i) \quad \cdots \quad \exp[j2\pi f_i(N-1)]]^T$. Let $\mathbf{E} = [\mathbf{e}_1 \quad \mathbf{e}_2 \quad \cdots \quad \mathbf{e}_p]$, $\mathbf{A}_c = [A_{c1} \quad A_{c2} \quad \cdots \quad A_{c_p}]^T$, $\mathbf{f} = [f_1 \quad f_2 \quad \cdots \quad f_p]^T$. Then,

$$S(\mathbf{A}_c, \mathbf{f}) = (\mathbf{x} - \mathbf{E}\mathbf{A}_c)^H(\mathbf{x} - \mathbf{E}\mathbf{A}_c). \tag{13.10}$$

The latter equation is in the form of a standard linear least squares problem if $\mathbf{E}$ is assumed to be a known matrix. $S(\mathbf{A}_c, \mathbf{f})$ is minimized over $\mathbf{A}_c$ by [see the discussion of (3.32)]

$$\hat{\mathbf{A}}_c = (\mathbf{E}^H\mathbf{E})^{-1}\mathbf{E}^H\mathbf{x}. \tag{13.11}$$

As before, $\hat{\mathbf{A}}_c$ will be the MLE if $\mathbf{E}$ is replaced by $\hat{\mathbf{E}}$ or the unknown frequencies are replaced by their MLEs. Substituting (13.11) into (13.10) yields

$$\begin{aligned} S(\hat{\mathbf{A}}_c, \mathbf{f}) &= \mathbf{x}^H(\mathbf{x} - \mathbf{E}\hat{\mathbf{A}}_c) \\ &= \mathbf{x}^H\mathbf{x} - \mathbf{x}^H\mathbf{E}(\mathbf{E}^H\mathbf{E})^{-1}\mathbf{E}^H\mathbf{x}. \end{aligned} \tag{13.12}$$

To find the MLE of the frequencies we must maximize

$$L(\mathbf{f}) = \mathbf{x}^H\mathbf{E}(\mathbf{E}^H\mathbf{E})^{-1}\mathbf{E}^H\mathbf{x}. \tag{13.13}$$

$L(\mathbf{f})$ is a *highly nonlinear function of the unknown frequencies* and therein lies the central problem. Note that in general the MLE is not based on a spectral estimator. To maximize this function will require a search over a p-dimensional space, which for several sinusoids is clearly impractical. Alternatively, an iterative optimization can be used, but convergence to the global maximum is not guaranteed. An example to be presented shortly will show that $L(\mathbf{f})$ can indeed have local maxima and minima. If the MLE of the frequencies can be obtained, then the MLE of the amplitudes and phases are easily found from (13.11). See the paper by Marquardt [1963] for a discussion of nonlinear least squares problems.

As an example of the difficulty encountered in maximizing $L(\mathbf{f})$, consider the case of two sinusoids. Then

$$\mathbf{E}^H\mathbf{E} = \begin{bmatrix} \mathbf{e}_1^H\mathbf{e}_1 & \mathbf{e}_1^H\mathbf{e}_2 \\ \mathbf{e}_2^H\mathbf{e}_1 & \mathbf{e}_2^H\mathbf{e}_2 \end{bmatrix}.$$

Let $\beta(f_1, f_2) = \mathbf{e}_1^H\mathbf{e}_2$ and note that $\mathbf{e}_1^H\mathbf{e}_1 = N$, $\mathbf{e}_2^H\mathbf{e}_2 = N$, so that

$$L(\mathbf{f}) = \frac{\mathbf{x}^H[\mathbf{e}_1 \quad \mathbf{e}_2] \begin{bmatrix} N & -\beta(f_1, f_2) \\ -\beta^*(f_1, f_2) & N \end{bmatrix} \begin{bmatrix} \mathbf{e}_1^H \\ \mathbf{e}_2^H \end{bmatrix} \mathbf{x}}{N^2 - |\beta(f_1, f_2)|^2}$$

which after some simplification yields

$$L(\mathbf{f}) = \frac{N[|\mathbf{e}_1^H\mathbf{x}|^2 + |\mathbf{e}_2^H\mathbf{x}|^2] - 2\,\mathrm{Re}\,[\mathbf{x}^H\mathbf{e}_1\mathbf{e}_2^H\mathbf{x}\beta(f_1, f_2)]}{N^2 - |\beta(f_1, f_2)|^2}.$$

Letting $X(f) = \sum_{n=0}^{N-1} x[n] \exp(-j2\pi fn)$, this becomes

$$L(\mathbf{f}) = \frac{N[|X(f_1)|^2 + |X(f_2)|^2] - 2\,\mathrm{Re}\,[X^*(f_1)X(f_2)\beta(f_1, f_2)]}{N^2 - |\beta(f_1, f_2)|^2}. \tag{13.14}$$

This is plotted in Figure 13.1 for the case of no noise for the data

$$x[n] = \exp(j2\pi f_1 n) + \exp(j2\pi f_2 n + \phi_2)$$

where $f_1 = 0.50$, $f_2 = 0.52$, $\phi_2 = \pi/4$ and $N = 25$ data points. Note that the frequencies are assumed to be in the range $0 \leq f \leq 1$ for consistency with many of the reported research results. Even without any noise the function displays

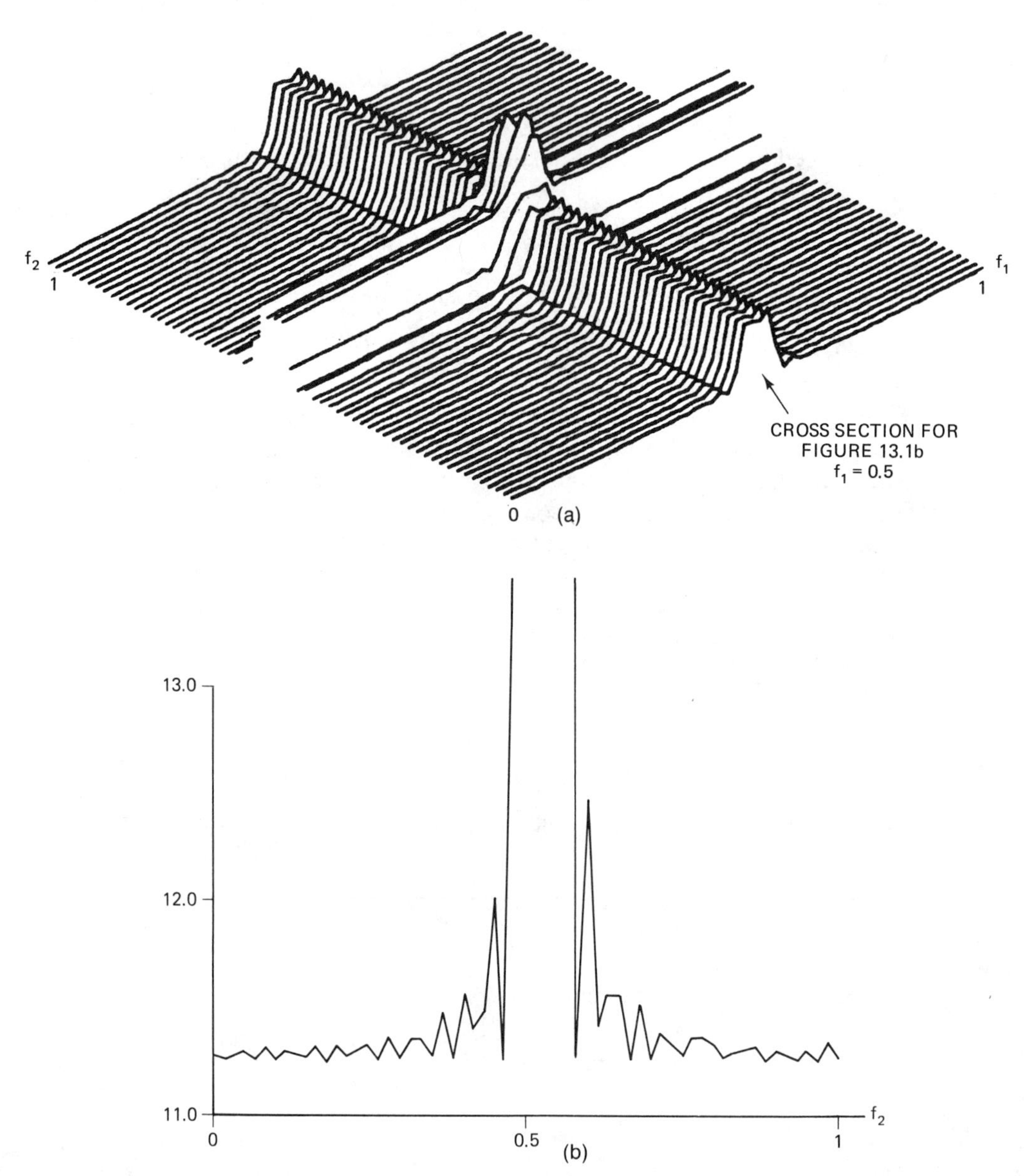

Figure 13.1 (a) MLE function surface for two complex sinusoids (no noise). (b) Cross section of MLE surface at $f_1 = 0.5$.

multiple minima and maxima. Clearly, any iteration approach will require a good initial estimate of the peak of the function to yield the MLE at convergence.

There are, however, some special cases in which the MLE may easily be obtained. As an illustration, consider two sinusoids. The special cases are when:

1. $f_1 = k/N$, $f_2 = l/N$ for k, l distinct integers in the range

$$\left[-\frac{N}{2}, \frac{N}{2} - 1\right] \text{ for } N \text{ even} \quad \text{and} \quad \left[-\left(\frac{N-1}{2}\right), \left(\frac{N-1}{2}\right)\right] \text{ for } N \text{ odd}$$

2. $| f_1 - f_2 | \gg 1/N$

In the first case $\beta(f_1, f_2) = 0$ so that we need only maximize

$$L(\mathbf{f}) = \frac{1}{N}[| X(f_1) |^2 + | X(f_2) |^2] \tag{13.15}$$

by choosing the two frequencies that yield the two largest values of the periodogram when sampled at frequencies that are multiples of $1/N$. Similarly, if we know that the two frequencies are separated by much more than $1/N$, then $\beta(f_1, f_2) \approx 0$. Again, (13.15) must be maximized. To do so we choose the two frequencies as the ones that correspond to the two largest periodogram peaks subject to the constraint that the frequencies are widely separated [Walker 1971]. In both cases the function decouples so as to simplify the maximization. These special conditions on the frequencies are equivalent to requiring the periodogram to yield resolvable peaks. For example, when the frequencies are at multiples of $1/N$, an N point FFT will reveal the presence of the sinusoids because the interference between the sinusoids will be zero. Also, for widely spaced sinusoidal frequencies a periodogram will clearly display two discernible peaks. For closely spaced sinusoids, however, the periodogram is unable to resolve the two sinusoids. We must therefore maximize the exact expression as given by (13.13) to obtain the MLE. Because this task is difficult, if not impossible, researchers have noted the connection between frequency estimation and spectral estimation and proposed the use of various spectral estimation techniques. Their motivation has been to overcome the problems of the periodogram by using "high-resolution" spectral estimators. This approach is discussed in Sections 13.6 and 13.8.

13.4 CRAMER–RAO BOUNDS

Since the methods to be described in the following sections will attempt to estimate sinusoidal parameters, it is useful to be able to compare the various approaches to the Cramer–Rao (CR) bound (see Section 3.3.1). The CR bound has been derived by Rife [1973], a summary of which is contained in the paper by Rife and Boorstyn [1976]. These bounds apply only to unbiased estimators. They are useful in that the best possible performance (smallest variance) for an unbiased estimator may be determined. Estimators whose variance is close to or equals the bound can then be said to be optimal. The cases of a single sinusoid and multiple sinusoids are described separately.

13.4.1 Single Sinusoid

It has been shown that for the data given by (13.1) with $p = 1$ that the CR bounds for the sinusoidal parameters are

$$\begin{aligned} \operatorname{var}(\hat{f}_1) &\geq \frac{6\sigma_z^2}{A_1^2 N(N^2 - 1)(2\pi)^2} \\ \operatorname{var}(\hat{A}_1) &\geq \frac{\sigma_z^2}{2N} \\ \operatorname{var}(\hat{\phi}_1) &\geq \frac{\sigma_z^2(2N - 1)}{A_1^2 N(N + 1)}. \end{aligned} \tag{13.16}$$

As expected, the CR bound for the frequency estimator is inversely proportional to SNR (A_1^2/σ_z^2). Also, the bound on variance decreases as the cube power of data record length, indicating the importance of an adequate number of data points.

13.4.2 Multiple Sinusoids

The CR bounds for the parameters of multiple sinusoids cannot be explicitly expressed but may be numerically evaluated via a computer. Assuming p sinusoids, define

$$\boldsymbol{\theta} = [f_1\ A_1\ \phi_1\ f_2\ A_2\ \phi_2\ \cdots\ f_p\ A_p\ \phi_p]^T.$$

Then the bounds can be shown to be

$$\operatorname{var}(\hat{\theta}_i) \geq \frac{\sigma_z^2[\mathbf{M}^{-1}]_{ii}}{2B_i^2} \tag{13.17}$$

where $B_i = A_i$ if θ_i corresponds to a phase parameter, $B_i = 1$ if θ_i corresponds to an amplitude parameter, and $B_i = 2\pi A_i$ if θ_i corresponds to a frequency parameter. $[\mathbf{M}^{-1}]_{ii}$ is the $[i, i]$ element of the inverse of the $3p \times 3p$ matrix $\mathbf{M}$, which is defined as

$$\mathbf{M} = \begin{bmatrix} \mathbf{M}_{11} & \mathbf{M}_{12} & \cdots & \mathbf{M}_{1p} \\ \mathbf{M}_{21} & \mathbf{M}_{22} & \cdots & \mathbf{M}_{2p} \\ \vdots & \vdots & \ddots & \vdots \\ \mathbf{M}_{p1} & \mathbf{M}_{p2} & \cdots & \mathbf{M}_{pp} \end{bmatrix} \tag{13.18}$$

where $\mathbf{M}_{ij}$ is the 3×3 matrix

$$\mathbf{M}_{ij} = \begin{bmatrix} \sum_{n=0}^{N-1} n^2 \cos \Delta_n[i,j] & -\sum_{n=0}^{N-1} n \sin \Delta_n[i,j] & \sum_{n=0}^{N-1} n \cos \Delta_n[i,j] \\ \sum_{n=0}^{N-1} n \sin \Delta_n[i,j] & \sum_{n=0}^{N-1} \cos \Delta_n[i,j] & \sum_{n=0}^{N-1} \sin \Delta_n[i,j] \\ \sum_{n=0}^{N-1} n \cos \Delta_n[i,j] & -\sum_{n=0}^{N-1} \sin \Delta_n[i,j] & \sum_{n=0}^{N-1} \cos \Delta_n[i,j] \end{bmatrix}$$

and $\Delta_n[i,j] = 2\pi(f_i - f_j)n + (\phi_i - \phi_j)$. The single sinusoid case follows from (13.17) for $p = 1$ and $\Delta_n[1,1] = 0$. Also, note that if $\mathbf{M}_{ij} = 0$ for $i \neq j$, so that $\mathbf{M}$ is block diagonal, then the CR bounds will be the same as for a single sinusoid (see Problem 13.3). This decoupling will hold approximately if all the frequencies are spaced much more than $1/N$ apart from their nearest neighbor. In this case $\mathbf{M}_{ij}/N \approx 0$ for $i \neq j$ since for $k = 0, 1, 2$

$$\begin{aligned} \frac{1}{N} \sum_{n=0}^{N-1} n^k \exp\,[j\Delta_n[i,j]] &= \frac{1}{N} \sum_{n=0}^{N-1} n^k \exp\,(j2\pi\,\Delta f n + \Delta\phi) \\ &= \frac{\exp\,(j\Delta\phi)}{(j2\pi)^k} \left[\frac{d^k}{d\,\Delta f^k} \frac{1}{N} \sum_{n=0}^{N-1} \exp\,(j2\pi\,\Delta f n) \right] \end{aligned} \tag{13.19}$$

where $\Delta f = f_i - f_j$ and $\Delta\phi = \phi_i - \phi_j$. The term in brackets will be approximately zero for $|\Delta f| \gg 1/N$ and N large. For this special case peak picking of the periodogram will result in the MLE, and for large data records this estimator will attain the CR bound.

The explicit CR bounds for the frequencies of two sinusoids will be useful later. It has been shown [Rife 1973] that by using (13.17) the bounds are

$$\begin{aligned} \text{var}\,(\hat{f}_1) &\geq \frac{\sigma_z^2}{2A_1^2(2\pi)^2} [\mathbf{C}]_{11} \\ \text{var}\,(\hat{f}_2) &\geq \frac{\sigma_z^2}{2A_2^2(2\pi)^2} [\mathbf{C}]_{11} \end{aligned} \tag{13.20}$$

where $[\mathbf{C}]_{11}$ is the $[1,1]$ element of the 3×3 matrix $\mathbf{C}$ and

$$\mathbf{C} = (\mathbf{M}_{11} - \mathbf{M}_{12}\mathbf{M}_{22}^{-1}\mathbf{M}_{21})^{-1}.$$

Some interesting observations which result from careful examination of (13.20) are that:

1. Equiamplitude sinusoids have the same CR bounds.
2. The bounds do not depend on the particular frequencies and phases but only on $\Delta_n[1,2] = 2\pi(f_1 - f_2)n + (\phi_1 - \phi_2)$ or the differences.
3. For widely spaced frequencies the bounds reduce to the single sinusoid bound [let $\mathbf{M}_{12} = \mathbf{M}_{21} = \mathbf{0}$ in (13.20)].

4. The bounds for two sinusoids are always greater than or equal to the bound for one sinusoid (see Problems 13.3 and 13.4).

13.5 APPROXIMATE MLE METHODS

It was shown in the Section 13.3 that the MLE frequency estimator is found by minimizing (13.12):

$$S(\hat{\mathbf{A}}_c, \mathbf{f}) = \mathbf{x}^H(\mathbf{I} - \mathbf{E}(\mathbf{E}^H\mathbf{E})^{-1}\mathbf{E}^H)\mathbf{x}.$$

This function is highly nonlinear in the frequencies. In an attempt to replace $S(\hat{\mathbf{A}}_c, \mathbf{f})$ by a more manageable function several researchers have proposed transforming the parameter space from frequencies to AR parameters. Since a sum of sinusoids can be modeled as the impulse response of a causal pole–zero filter with poles on the unit circle, we can equivalently determine the frequencies from the pole locations. Furthermore, the pole locations are functions of the AR parameters. To see how this transformation is effected, note that if the poles of the filter $\mathcal{B}(z)/\mathcal{A}(z)$ are at $z = \exp(j2\pi f_i)$ for $i = 1, 2, \ldots, p$, then

$$\mathcal{A}[\exp(j2\pi f_i)] = \sum_{k=0}^{p} a[k] \exp(-j2\pi f_i k) = 0 \qquad i = 1, 2, \ldots, p. \tag{13.21}$$

Defining the $(N - p) \times N$ matrix

$$\mathbf{A}^H = \begin{bmatrix} a[p] & a[p-1] & \cdots & 1 & 0 & \cdots & 0 \\ 0 & a[p] & \cdots & a[1] & 1 & \cdots & 0 \\ \vdots & \vdots & \ddots & \ddots & \ddots & \ddots & \vdots \\ 0 & 0 & \cdots & a[p] & a[p-1] & \cdots & 1 \end{bmatrix}, \tag{13.22}$$

we have upon using (13.21) that $\mathbf{A}^H\mathbf{E} = \mathbf{0}$ or that the columns of $\mathbf{E}$ are orthogonal to the columns of $\mathbf{A}$. From projection theory (see Problems 13.5 and 13.6) it can be shown that

$$\mathbf{I} = \mathbf{E}(\mathbf{E}^H\mathbf{E})^{-1}\mathbf{E}^H + \mathbf{A}(\mathbf{A}^H\mathbf{A})^{-1}\mathbf{A}^H \tag{13.23}$$

where $\mathbf{I}$ is the $N \times N$ identify matrix. The operator $\mathbf{E}(\mathbf{E}^H\mathbf{E})^{-1}\mathbf{E}^H$ projects a vector onto the subspace spanned by the columns of $\mathbf{E}$ while $\mathbf{A}(\mathbf{A}^H\mathbf{A})^{-1}\mathbf{A}^H$ projects a vector onto the orthogonal complement subspace which is spanned by the columns of $\mathbf{A}$. The sum of these two operators yields the identity operator. Using (13.23) in (13.12) results in

$$S_1(\mathbf{a}) = \mathbf{x}^H\mathbf{A}(\mathbf{A}^H\mathbf{A})^{-1}\mathbf{A}^H\mathbf{x}$$

where $\mathbf{a} = [a[1] \;\; a[2] \;\; \cdots \;\; a[p]]^T$. $S_1(\mathbf{a})$ must be minimized over $\mathbf{a}$ subject to the constraint that the poles lie on the unit circle to yield the MLE of the frequencies. Once the minimizing value of $\mathbf{a}$ is found, the frequencies are obtained by rooting the $\mathcal{A}(z)$ polynomial. Unfortunately, the constrained minimization is difficult. It is easily verified that $\mathbf{A}(\mathbf{A}^H\mathbf{A})^{-1}\mathbf{A}^H$ is a hermitian and idempotent matrix (see Section 2.3 and Problem 2.4), so that

$$S_1(\mathbf{a}) = \| \mathbf{A}(\mathbf{A}^H\mathbf{A})^{-1}\mathbf{A}^H\mathbf{x} \|^2$$

where $\| \mathbf{c} \|^2 = \mathbf{c}^H\mathbf{c} = \sum_{n=0}^{N-1} | c_n |^2$. $S_1(\mathbf{a})$ is still nonlinear in $\mathbf{a}$, but by writing it in this form iterative approaches are suggested. One last manipulation involves replacing $\mathbf{A}^H\mathbf{x}$ by $\mathbf{X}\mathbf{a}'$, where

$$\mathbf{a}' = [1 \ \mathbf{a}^T]^T$$

and $\mathbf{X}$ is the $(N - p) \times (p + 1)$ matrix

$$\mathbf{X} = \begin{bmatrix} x[p] & x[p-1] & \cdots & x[0] \\ x[p+1] & x[p] & \cdots & x[1] \\ \vdots & \vdots & \ddots & \vdots \\ x[N-1] & x[N-2] & \cdots & x[N-1-p] \end{bmatrix}.$$

Then

$$S_1(\mathbf{a}) = \| \mathbf{A}(\mathbf{A}^H\mathbf{A})^{-1}\mathbf{X}\mathbf{a}' \|^2. \tag{13.24}$$

To minimize $S_1(\mathbf{a})$ iteratively, let $\mathbf{A}_k$ = kth iterate of $\mathbf{A}$ and $\mathbf{a}'_{k+1}$ = $(k + 1)$st iterate of $\mathbf{a}'$. To find $\mathbf{a}'_{k+1}$ we minimize

$$\| \mathbf{A}_k(\mathbf{A}_k^H\mathbf{A}_k)^{-1}\mathbf{X}\mathbf{a}' \|^2 \tag{13.25}$$

over $\mathbf{a}'$. This minimization is a simple one since with $\mathbf{A}_k$ fixed the function is hermitian in $\mathbf{a}'$ (see Problem 13.7). Once $\mathbf{a}'_{k+1}$ is found, $\mathbf{A}_k$ is replaced by $\mathbf{A}_{k+1}$ using the new iterate. This approach was first proposed by Evans and Fischl [1973] for least squares estimation of the impulse response in a filter design problem. They point out that even if convergence is attained, the function $S_1(\mathbf{a})$ may not be minimized. A second step in their algorithm is to force the gradient of $S_1(\mathbf{a})$ to be zero using a succeeding iteration. For the sinusoidal frequency estimation problem the AR parameters need to be constrained to reflect the fact that the poles are on the unit circle. These constraints, however, are not easily mapped into constraints on the AR parameters. One attempt to do so constrains the AR parameters to satisfy [Kumaresan and Shaw 1985, Bresler and Macovski 1985]

$$a[p - i] = a^*[i] \qquad i = 0, 1, \ldots, p$$

which when used in (13.25) forces the pole to lie in conjugate reciprocal pairs about the unit circle (see Problem 13.8). Simulation results indicate that this approach works quite well but may be plagued by convergence problems. Also, the computational complexity is high. In an effort to reduce the computations and improve convergence, an algorithm termed the iterative filtering algorithm (IFA) has been proposed [Kay 1984]. It allows us to replace the matrix operations in (13.25) by efficient filtering operations and appears always to converge in a few iterations. The function

$$S_1(\mathbf{a}) = \mathbf{x}^H\mathbf{A}(\mathbf{A}^H\mathbf{A})^{-1}\mathbf{A}^H\mathbf{x}$$

is approximated in the IFA by an asymptotic expression valid for large data records. Define $\hat{\mathbf{A}}$ as the previous iterate so that

$$S_2(\mathbf{a}) = \mathbf{x}^H\mathbf{A}(\hat{\mathbf{A}}^H\hat{\mathbf{A}})^{-1}\mathbf{A}^H\mathbf{x}. \tag{13.26}$$

$S_2(\mathbf{a})$ is to be minimized over $\mathbf{a}$ with $\hat{\mathbf{A}}$ fixed. Using an eigendecomposition of $(\hat{\mathbf{A}}^H\hat{\mathbf{A}})^{-1}$ [see (2.40)],

$$(\hat{\mathbf{A}}^H\hat{\mathbf{A}})^{-1} = \sum_{i=0}^{N-p-1} \frac{1}{\lambda_i}\mathbf{v}_i\mathbf{v}_i^H$$

(13.26) becomes

$$S_2(\mathbf{a}) = \sum_{i=0}^{N-p-1} \frac{|\mathbf{v}_i^H\mathbf{A}^H\mathbf{x}|^2}{\lambda_i}.$$

But $\hat{\mathbf{A}}^H\hat{\mathbf{A}}$ is equal to the autocorrelation matrix of an MA(p) process with parameters $\{\hat{a}[1] \quad \hat{a}[2] \quad \cdots \quad \hat{a}[p]\}$ and $\sigma^2 = 1$. This is because for $i \geq j$,

$$\begin{aligned}
[\hat{\mathbf{A}}^H\hat{\mathbf{A}}]_{ij} &= \sum_{k=1}^{N} \hat{a}[p+i-k]\hat{a}^*[p+j-k] \\
&= \sum_{k=i}^{j+p} \hat{a}[p+i-k]\hat{a}^*[p+j-k] \\
&= \sum_{m=0}^{p-(i-j)} \hat{a}^*[m]\hat{a}[m+i-j]
\end{aligned}$$

since $\hat{a}[k] = 0$ for $k < 0$ or $k > p$. This property allows us to find an approximate expression for the eigenvalues and eigenvectors of the hermitian Toeplitz autocorrelation matrix $\hat{\mathbf{A}}^H\hat{\mathbf{A}}$. It can be shown that as $N \to \infty$ [Grenander and Szego 1958],

$$\lambda_i \to P_{\mathrm{MA}}(f_i) = |\hat{A}(f_i)|^2 \tag{13.27}$$

and

$$\mathbf{v}_i \to \frac{1}{\sqrt{N-p}}[1 \quad \exp(j2\pi f_i) \quad \exp(j\pi 4 f_i) \quad \cdots \quad \exp[j2\pi f_i(N-p-1)]]^T \tag{13.28}$$

for $i = 0, 1, \ldots, N-p-1$. $\hat{A}(f)$ is the frequency response of the MA filter based on the previous iterate or $\hat{A}(f) = 1 + \sum_{k=1}^{p} \hat{a}[k]z^{-k}$ and $f_i = i/(N-p)$. Now let $\mathbf{y} = [y[p] \quad y[p+1] \quad \cdots \quad y[N-1]]^T$, so that

$$\mathbf{A}^H\mathbf{x} = \mathbf{y}$$

where $y[n] = \sum_{k=0}^{p} a[k]x[n-k]$ or $y[n]$ is the output of a PEF. Also, from (13.28),

$$\mathbf{v}_i^H\mathbf{A}^H\mathbf{x} = \mathbf{v}_i^H\mathbf{y} = \frac{\exp(j2\pi f_i p)}{\sqrt{N-p}} \sum_{n=p}^{N-1} y[n]\exp(-j2\pi f_i n)$$

which is the scaled Fourier transform of the $y[n]$ sequence. Ignoring end effects due to the assumption of a large data record, we can use standard properties of Fourier transforms to establish that

$$\mathbf{v}_i^H \mathbf{A}^H \mathbf{x} \approx \frac{\exp(j2\pi f_i p)}{\sqrt{N}} A(f_i) X(f_i)$$

where $X(f)$ is the Fourier transform of the $x[n]$ sequence. Hence

$$\begin{aligned} S_2(\mathbf{a}) &\approx \frac{1}{N} \sum_{i=0}^{N} \frac{|A(f_i)|^2 |X(f_i)|^2}{|\hat{A}(f_i)|^2} \\ &\approx \int_{-1/2}^{1/2} \frac{|A(f)|^2 |X(f)|^2}{|\hat{A}(f)|^2} df \end{aligned} \tag{13.29}$$

which is the desired result. It is clear from (13.29) that to approximately minimize $S_2(\mathbf{a})$ we first filter the data with $1/\hat{\mathcal{A}}(z)$ and then apply any of the AR estimation methods to the data at the output of the filter. The Burg algorithm (see Chapter 7) has been found to work well since it constrains the poles to be within the unit circle, rendering a stable $1/\hat{\mathcal{A}}(z)$ filter. In summary, the IFA proceeds as follows: For $k = 0, 1, \ldots$:

1. Filter $x[n]$ with $1/\hat{\mathcal{A}}_k(z)$ to yield $y_k[n]$, where $\hat{\mathcal{A}}_k(z) = 1 + \sum_{m=1}^{p} \hat{a}_k[m] z^{-m}$ is the kth iterate.
2. Based on $y_k[n]$, estimate the AR parameters using the Burg algorithm to yield $\hat{\mathcal{A}}_{k+1}(z)$.
3. Go to step 1 if convergence has not been attained.

To start the iteration, use $\hat{\mathcal{A}}_0(z) = 1$ and hence $y_0[n] = x[n]$. Also, the initial conditions of the $1/\hat{\mathcal{A}}_k(z)$ filter are always set to zero. Convergence is attained when the zeros of $\hat{\mathcal{A}}_k(z)$ are close to the unit circle. The frequency estimates are the angles of the zeros of $\hat{\mathcal{A}}_k(z)$ at convergence. They may be found using polynomial rooting or by picking the peaks in $1/\hat{A}_k(f)$. The IFA is summarized in Figure 13.2. Simulation results are given in Section 13.11.

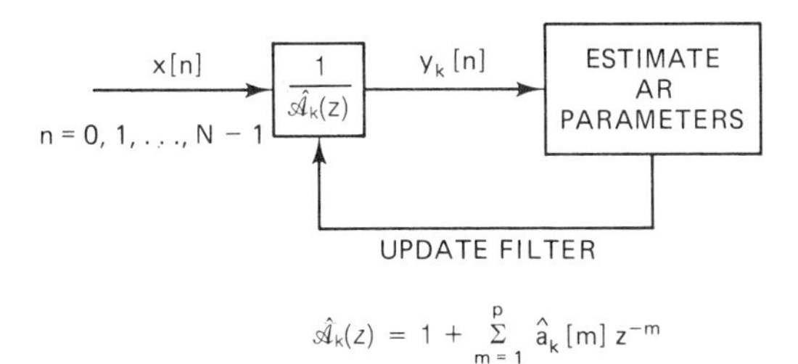

Figure 13.2 Summary of iterative filtering algorithm.

13.6. FREQUENCY ESTIMATION BY SPECTRAL ESTIMATION

Since the periodogram cannot be used to estimate frequencies of sinusoids that are not well resolved, much effort has been directed toward using the "high resolution" techniques of spectral estimation for this purpose. This approach is purely ad hoc in that for sinusoids spaced closely in frequency the MLE is not based on anything that may be interpreted as a spectral estimator. However, in light of the connection of the MLE with spectral estimation for well resolved sinusoids as described in Section 13.3, it is a reasonable alternative to the intractable MLE. In particular, the AR spectral estimator has been extensively studied for use as a frequency estimator. In using it, we must first choose a method for estimating the AR parameters and then a means for estimating the frequencies. Of all the parameter estimation methods described in Chapter 7, the modified covariance method appears to yield the best frequency estimator. This is because it provides estimates which are least sensitive to sinusoidal phases, and spectral peak shifting due to noise effects is less than with other AR methods. In the absence of noise, it can be shown that the modified covariance method yields the true frequencies as the peaks of the spectral estimate. This is not the case with other AR estimators (see Problem 13.9). In assigning frequency estimates based on the AR spectral estimator, one can either choose the estimates as the locations of the p largest spectral peaks or root the prediction error filter polynomial and choose the frequency estimates as the angles of the p zeros closest to the unit circle. The latter approach has been found to be slightly more accurate. It has also been found that for high enough SNR the modified covariance method with polynomial rooting yields frequency estimates which are unbiased and nearly attain the CR bound [Lang and McClellan 1980, Tufts and Kumaresan 1982]. However, at lower SNRs poor estimates result. They display bias and a large variance. This is not unexpected since as discussed in Section 6.7, the zero positions of the prediction error filter based on a known ACF do not lie at the correct frequency locations in the presence of observation noise. To alleviate these problems we may increase the order of the AR model employed. This has the effect of decreasing the modeling error, that is, of reducing the bias of the spectral estimator but also of increasing the variance that may give rise to spurious peaks.

As an example of the effect of increasing the model order on the frequency estimates, consider the case of two closely spaced equiamplitude complex sinusoids in complex white noise with $f_1 = 0.50$, $\phi_1 = 0$, $f_2 = 0.52$, $\phi_2 = \pi/4$, and only $N = 25$ data points. The SNR is 20 dB for each sinusoid. The sinusoids for this case are unresolvable using a periodogram. Assuming a model order of 2, which is only appropriate for the noiseless case, the estimated zeros obtained from 50 realizations using the modified covariance method are shown in Figure 13.3. It is observed that for $p = 2$ one of the zeros models the two sinusoids, while the other zero models the white noise. Clearly, only one peak will be observed in the spectral estimate, so that any reasonable approach would assign the same estimate to each frequency. If the model order is increased to $p = 12$, there appear two zeros at approximately the correct locations. However, for any individual realization it is quite probable that the wrong zero, one that models the

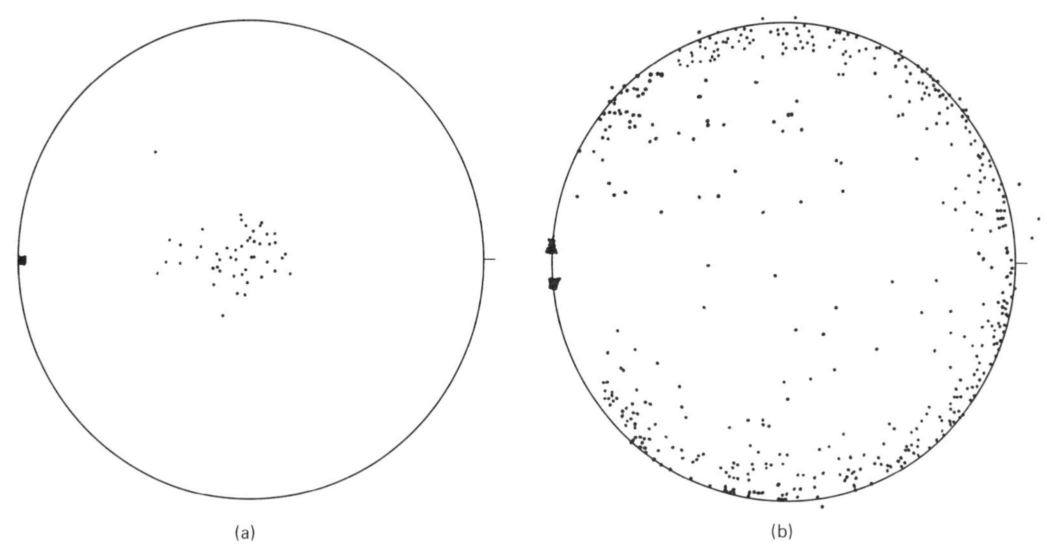

Figure 13.3 Estimated zeros for multiple realizations of modified covariance method for two complex sinusoids in white noise. (a) $p = 2$. (b) $p = 12$. (After Kumaresan [1982].)

noise (i.e., a "noise" zero) would be chosen. This is because the theoretical noise zeros as determined from the known ACF are close to the unit circle, and therefore, slight perturbations due to estimation errors will cause them to be nearer the unit circle than the signal zeros. As a result, there will be a large frequency error. The spectral estimate will be observed to contain a spurious peak. To quantify this behavior further, the frequency estimation performance of the modified covariance method with zero rooting was determined by plotting $10 \log_{10} 1/$(mean square error) versus $\text{SNR} = 10 \log_{10} A^2/\sigma_z^2$. For the frequencies and phases and data record length previously described, the results for model orders of $p = 8$ and $p = 12$ are displayed in Figure 13.4. The model orders correspond to approximately $M/3$ and $N/2$, which have been recommended as the best range for reliable spectral estimates [Ulrych and Bishop 1975]. Also, the CR bound as given by (13.17) is plotted assuming the estimator is unbiased so that the mean square

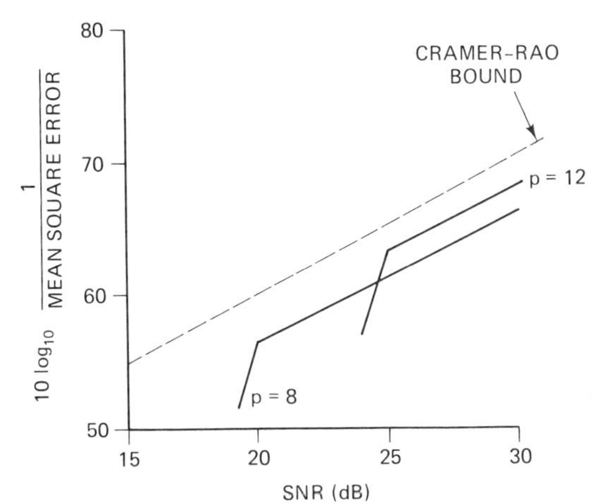

Figure 13.4 Frequency estimation performance of AR spectral estimator using modified covariance method. (After Kumaresan [1982].)

error is equal to the variance. The frequency performance nearly attains the CR bound above some threshold SNR (about 20 dB for $p = 8$ and about 24 dB for $p = 12$). Unfortunately, below the threshold the mean square error increases rapidly, which renders the estimator useless. Methods to extend this threshold are discussed in Section 13.8.

13.7 PROPERTIES OF THE AUTOCORRELATION MATRIX

It has been shown that AR spectral estimation provides reliable frequency estimates for high enough SNR. At lower SNRs the model order needs to be increased to the point where spurious peaks become a problem. Since almost all spectral estimation methods employ an estimated autocorrelation matrix, it is of interest to examine the properties of the theoretical matrix for the specific case of sinusoids in white noise. Hopefully, knowledge of the properties of the autocorrelation matrix will allow us to improve on currently existing techniques as well as to suggest new frequency estimators. Implicit in the use of an autocorrelation matrix is the modeling of the p sinusoids as a WSS random process, which is only possible if the phases are assumed to be random (see Section 3.5.2). As expected, phase estimates are not available from methods based on this type of modeling. Nonetheless, once the frequency estimates have been obtained, the phases and amplitudes may be estimated using (13.11).

Key properties of the $M \times M$ theoretical autocorrelation matrix to be exploited in the estimation of sinusoidal frequencies are as follows:

1. The matrix is composed of a signal autocorrelation matrix and a noise autocorrelation matrix.
2. The signal autocorrelation matrix is not full rank if $M > p$.
3. The p principal eigenvectors of the signal autocorrelation matrix (i.e., those eigenvectors with nonzero eigenvalues) may be used to extract the sinusoidal frequencies.
4. The p principal eigenvectors of the signal autocorrelation matrix are identical to those of the total autocorrelation matrix (i.e., those eigenvectors with the p largest eigenvalues).

To understand the theory behind these properties, consider the data as given by (13.1), except that to ensure a WSS process, assume the phases to be independent random variables uniformly distributed on $[0, 2\pi)$ (see Section 3.5.2). Then the ACF for the process given by (13.1) is easily shown to be

$$r_{xx}[k] = \sum_{i=1}^{p} A_i^2 \exp(j2\pi f_i k) + \sigma_z^2 \delta[k]. \tag{13.30}$$

Let $P_i = A_i^2$ denote the power of the ith sinusoid so that the $M \times M$ autocorrelation matrix for $M > p$ is

$$\mathbf{R}_{xx} = \sum_{i=1}^{p} P_i \mathbf{e}_i \mathbf{e}_i^H + \sigma_z^2 \mathbf{I} = \mathbf{R}_{ss} + \mathbf{R}_{zz}. \tag{13.31}$$

$\mathbf{R}_{xx}$ is seen to be the sum of a signal autocorrelation matrix $\mathbf{R}_{ss}$ and a noise autocorrelation matrix $\mathbf{R}_{zz}$. $\mathbf{R}_{ss}$ is not full rank since it is of dimension $M \times M$ and the rank is only p (see Appendix 13B and Problem 2.9). The total autocorrelation matrix $\mathbf{R}_{xx}$ is, however, full rank due to the inclusion of the $\sigma_z^2\mathbf{I}$ term. The frequency information is contained in the signal vectors, the $\mathbf{e}_i$'s. If one is given $\mathbf{R}_{xx}$ or can estimate it, it would be desirable to be able to effect the decomposition of (13.31). Then the frequency estimates could be obtained from the signal vectors. This does not appear to be possible in practice, although a related decomposition based on the eigenvectors $\mathbf{v}_i$ and eigenvalues λ_i of $\mathbf{R}_{xx}$ is feasible. As an example, consider the case of $p = 1$. Then the autocorrelation matrix may be expressed as

$$\mathbf{R}_{xx} = (MP_1 + \sigma_z^2)\mathbf{v}_1\mathbf{v}_1^H + \sigma_z^2 \sum_{i=2}^{M} \mathbf{v}_i\mathbf{v}_i^H \tag{13.32}$$

where $\mathbf{v}_1 = (1/\sqrt{M})\mathbf{e}_1$ and $\{\mathbf{v}_2, \mathbf{v}_3, \ldots, \mathbf{v}_M\}$ have been chosen to be orthonormal to $\mathbf{v}_1$ and to each other, or $\mathbf{v}_i^H\mathbf{v}_j = \delta[i - j]$. Equation (13.32) relies on the decomposition of the identity matrix as [see (2.39)]

$$\mathbf{I} = \sum_{i=1}^{M} \mathbf{v}_i\mathbf{v}_i^H.$$

It is clear from (13.32) that the eigenvalues of $\mathbf{R}_{xx}$ are $\{MP_1 + \sigma_z^2, \sigma_z^2, \sigma_z^2, \ldots, \sigma_z^2\}$, with corresponding eigenvectors $\{\mathbf{v}_1, \mathbf{v}_2, \ldots, \mathbf{v}_M\}$. The eigenvector corresponding to the one eigenvalue $MP_1 + \sigma_z^2$ that is much larger than the others is termed the *principal eigenvector*. Also, note that since the signal autocorrelation matrix can be written as

$$\mathbf{R}_{ss} = MP_1\mathbf{v}_1\mathbf{v}_1^H$$

there is only one nonzero eigenvalue MP_1. The eigenvector corresponding to this nonzero eigenvalue is the principal eigenvector of $\mathbf{R}_{ss}$ and is identical to the principal eigenvector of $\mathbf{R}_{xx}$. Hence the sinusoidal frequency information that is contained in the principal eigenvector of $\mathbf{R}_{ss}$ is also contained in the principal eigenvector of $\mathbf{R}_{xx}$. Only the latter matrix, however, is readily estimated. One way to extract this information is to make use of the orthogonality of the eigenvectors. Because

$$\mathbf{v}_1^H \sum_{i=2}^{M} \alpha_i\mathbf{v}_i = \frac{1}{\sqrt{M}} \mathbf{e}_1^H \sum_{i=2}^{M} \alpha_i\mathbf{v}_i = 0 \tag{13.33}$$

we can take the Fourier transform of components of the vector formed as an arbitrary linear combination of the *nonprincipal eigenvectors* and choose the frequency that yields zero as per (13.33). [Note that $\mathbf{e}_1^H \mathbf{v}_i = \sum_{n=0}^{M-1} [\mathbf{v}_i]_{n+1} \exp(-j2\pi f_1 n)$.] The nonprincipal eigenvectors are chosen to be those $M - 1$ eigenvectors of $\mathbf{R}_{xx}$ that have the smallest eigenvalues. Since $\mathbf{v}_1$ spans the same subspace as the signal vector $\mathbf{e}_1$ and $\{\mathbf{v}_2, \mathbf{v}_3, \ldots, \mathbf{v}_M\}$ span the orthogonal subspace, it is said that $\mathbf{v}_1$ spans the *signal subspace* and the remaining eigenvectors span the *noise subspace*. This terminology, although in common use, is not entirely accurate since some of $\mathbf{R}_{zz} = \sigma_z^2 \mathbf{I}$ contributes to the signal component in the form of $\sigma_z^2 \mathbf{v}_1 \mathbf{v}_1^H$, as seen from (13.32).

The preceding properties of the autocorrelation matrix are easily generalized for multiple sinusoids. For p complex sinusoids in complex white noise,

$$\mathbf{R}_{xx} = \sum_{i-1}^{p} P_i \mathbf{e}_i \mathbf{e}_i^H + \sigma_z^2 \mathbf{I}.$$

The first term represents the signal autocorrelation matrix, which is of rank p. An eigendecomposition of $\mathbf{R}_{xx}$ produces

$$\mathbf{R}_{xx} = \sum_{i=1}^{p} (\lambda_i + \sigma_z^2) \mathbf{v}_i \mathbf{v}_i^H + \sum_{i=p+1}^{M} \sigma_z^2 \mathbf{v}_i \mathbf{v}_i^H$$

where the signal autocorrelation matrix has been replaced by $\sum_{i=1}^{p} \lambda_i \mathbf{v}_i \mathbf{v}_i^H$ and the decomposition of $\mathbf{I}$ has been used. Furthermore, it is proven in Appendix 13A that the principal eigenvectors $\{\mathbf{v}_1, \mathbf{v}_2, \ldots, \mathbf{v}_p\}$ span the same subspace as the signal vectors $\{\mathbf{e}_1, \mathbf{e}_2, \ldots, \mathbf{e}_p\}$. As a consequence, (13.33) generalizes to

$$\mathbf{e}_i^H \sum_{j=p+1}^{M} \alpha_j \mathbf{v}_j = 0 \qquad i = 1, 2, \ldots, p. \tag{13.34}$$

Hence *the signal vectors are orthogonal to all vectors in the noise subspace.* This property forms the basis for the frequency estimation algorithms described in Section 13.9. A special case occurs when $M = p + 1$, so that the noise subspace has dimension 1. Then

$$\mathbf{e}_i^H \mathbf{v}_{p+1} = 0 \qquad i = 1, 2, \ldots, p$$

or the eigenvector $\mathbf{v}_{p+1}$ associated with the unique minimum eigenvalue σ_z^2 has the property

$$\sum_{n=0}^{p} [\mathbf{v}_{p+1}]_{n+1} \exp(-j2\pi f_i n) = 0 \qquad i = 1, 2, \ldots, p.$$

This says that given $\mathbf{v}_{p+1}$ we can determine the sinusoidal frequencies by finding the zeros of the polynomial

$$\sum_{n=0}^{p} [\mathbf{v}_{p+1}]_{n+1} z^{-n}$$

which will all lie on the unit circle. The angles will be $2\pi f_i$ for $i = 1, 2, \ldots, p$. This result is central to the Pisarenko decomposition described in Section 13.9.1. See Appendix 13B for a formal proof of this property.

To summarize, the important results are the decomposition of the autocorrelation matrix into a signal component and a noise component, the low-rank nature of the signal autocorrelation matrix, and the extraction of the frequency information from the noise subspace eigenvectors. Two general approaches to frequency estimation have been proposed which rely on an eigendecomposition of the estimated autocorrelation matrix. The first class of methods use the standard AR, MVSE, and Blackman–Tukey spectral estimators but impose the constraint on the autocorrelation matrix that it be low rank to match the rank of the theoretical signal autocorrelation matrix. These methods may be generally called principal component techniques and are discussed in Section 13.8. The second class of frequency estimators are based on the orthogonality of the signal vectors to the noise eigenvectors. They include the Pisarenko and MUSIC methods, as well as some variants, and are described in Section 13.9.

13.8 PRINCIPAL COMPONENT FREQUENCY ESTIMATION

Many spectral estimation methods have been applied to data consisting of sinusoids in white noise. The general conclusion is that the peaks of the spectral estimates provide good estimates of the frequencies as long as the SNR is high enough. A specific example that used the modified covariance method for AR spectral estimation was presented in Section 13.6. The sharpness of the peaks and the apparent resolvability tend to increase with the dimension of the autocorrelation matrix estimated. However, if the dimension is too high, spurious peaks or outliers appear which limit the utility of the spectral estimator at low SNR. If we examine the autocorrelation matrix as given by (13.32), it is apparent that the effect of noise is to introduce noise eigenvectors with small eigenvalues. It is known, for example, that these nonprincipal eigenvectors cannot be estimated reliably since they exhibit a large variance [Wilkinson 1965]. Additionally, they bias the autocorrelation matrix estimate by adding components that are not present in the noiseless case. Hence, to gain some noise immunity it is reasonable to retain only the principal eigenvector components in the estimate of the autocorrelation matrix. Principal component analysis, which was originally applied to Fourier spectral estimation by Owsley [1973], has been found to yield improved frequency estimators based on spectral estimators [Tufts and Kumaresan 1982]. It may be viewed as an improvement due to the use of a priori knowledge about the specific theoretical structure of the autocorrelation matrix for the special case of sinusoids in white noise. There have been attempts to gain some noise reduction by subtracting an estimate of $\sigma_z^2\mathbf{I}$ from the estimate of $\mathbf{R}_{xx}$. These approaches have met with only moderate success due to the ever-present estimation errors in σ_z^2 and $\mathbf{R}_{xx}$. The specifics of the principal component methods are now described. The reader is also referred to the many similar approaches in multivariate statistics [Kendall and Stuart 1976] and a generalized spectral estimator proposed by Pisarenko [1972].

13.8.1 AR Frequency Estimation

Many of the results in this section as well as computer simulation results can be found in the works by Kumaresan [1982] and Tufts and Kumaresan [1982]. In AR spectral estimation the AR parameter estimates are [see (7.3) as an example]

$$\hat{\mathbf{a}} = -\hat{\mathbf{R}}_{xx}^{-1}\hat{\mathbf{r}}_{xx}$$

where $\hat{\mathbf{R}}_{xx}$ is some suitable estimate of the $M \times M$ autocorrelation matrix and $\hat{\mathbf{r}}_{xx}$ is an estimate of $[r_{xx}[1] \quad r_{xx}[2] \quad \cdots \quad r_{xx}[M]]^T$. Let the eigenvalues and eigenvectors of $\hat{\mathbf{R}}_{xx}$ be denoted as $\{\hat{\lambda}_1, \hat{\lambda}_2, \ldots, \hat{\lambda}_M\}$ and $\{\hat{\mathbf{v}}_1, \hat{\mathbf{v}}_2, \ldots, \hat{\mathbf{v}}_M\}$, respectively. Recall that for a positive definite hermitian matrix the eigenvalues are real and positive and M eigenvectors can be found which are orthonormal. Then the AR parameter estimate may be written as

$$\hat{\mathbf{a}} = -\sum_{i=1}^{M} \frac{1}{\hat{\lambda}_i} \hat{\mathbf{v}}_i \hat{\mathbf{v}}_i^H \hat{\mathbf{r}}_{xx}.$$

Since the eigenvectors span the entire space, $\hat{\mathbf{r}}_{xx}$ may be expressed as

$$\hat{\mathbf{r}}_{xx} = \sum_{i=1}^{M} \alpha_i \hat{\mathbf{v}}_i$$

for some α_i's. This yields

$$\begin{aligned}\hat{\mathbf{a}} &= -\sum_{i=1}^{M} \frac{1}{\hat{\lambda}_i} \hat{\mathbf{v}}_i \hat{\mathbf{v}}_i^H \sum_{j=1}^{M} \alpha_j \hat{\mathbf{v}}_j \\ &= -\sum_{i=1}^{M} \frac{\alpha_i}{\hat{\lambda}_i} \hat{\mathbf{v}}_i \qquad (13.35) \\ &= -\sum_{i=1}^{p} \frac{\alpha_i}{\hat{\lambda}_i} \hat{\mathbf{v}}_i - \sum_{i=p+1}^{M} \frac{\alpha_i}{\hat{\lambda}_i} \hat{\mathbf{v}}_i.\end{aligned}$$

The solution is a linear combination of all the eigenvectors of $\hat{\mathbf{R}}_{xx}$ [Ulrych and Clayton 1976]. A principal component solution $\hat{\mathbf{a}}_{PC}$ would omit the second term in the last equation of (13.35), so that

$$\hat{\mathbf{a}}_{\mathrm{PC}} = -\sum_{i=1}^{p} \frac{\alpha_i}{\hat{\lambda}_i} \hat{\mathbf{v}}_i. \tag{13.36}$$

This can also be written as

$$\hat{\mathbf{a}}_{\mathrm{PC}} = -\hat{\mathbf{R}}_{xx}^{\#}\hat{\mathbf{r}}_{xx} \tag{13.37}$$

where

$$\hat{\mathbf{R}}_{xx}^{\#} = \sum_{i=1}^{p} \frac{1}{\hat{\lambda}_i} \hat{\mathbf{v}}_i \hat{\mathbf{v}}_i^H$$

is the pseudoinverse of $\hat{\mathbf{R}}_{xx}$. Note from (13.36) that as M is increased, only p

eigenvalues and eigenvectors are used to form the estimate of $\mathbf{a}$. This allows us to increase the AR model order without incurring the spurious peaks which are due to the unstable noise eigenvector components normally present in standard AR spectral estimators. Equation (13.36) has some interesting interpretations. If no noise is present, then for $M > p$ the theoretical autocorrelation matrix $\mathbf{R}_{ss} = \sum_{i=1}^{p} P_i \mathbf{e}_i \mathbf{e}_i^H$ would be singular, being of rank p. The solution for the AR parameter estimates is then not unique. However, all solutions yield PEFs with p zeros on the unit circle at the sinusoidal frequency locations $\exp(j2\pi f_i)$ and with the remaining $M - p$ zeros lying elsewhere (see Problem 13.10). The solution which has the minimum norm or the one for which $\mathbf{a}^H\mathbf{a}$ is minimum is given by

$$\mathbf{a} = -\mathbf{R}_{ss}^{\#}\mathbf{r}_{ss} \tag{13.38}$$

where

$$\mathbf{R}_{ss}^{\#} = \sum_{i=1}^{p} \frac{1}{\lambda_i} \mathbf{v}_i \mathbf{v}_i^H$$

is the pseudoinverse of $\mathbf{R}_{ss}$. It can be shown that the minimum norm solution constrains the $M - p$ noise zeros to lie within the unit circle (see Problem 13.11). The noise zeros will be approximately uniformly distributed about the interior of the unit circle and displaced inward. Equation (13.37) may thus be viewed as an approximate realization of (13.38). Estimation errors and noise will perturb the noise zeros away from their noiseless positions but they will still tend to be uniformly distributed within the unit circle. As a result, there is less chance of mistaking a noise zero for a signal zero or equivalently fewer spurious peaks. It can also be shown that the pseudoinverse solution is obtained in the limit as the noise variance approaches zero [Kay 1980]. A computer program that implements this frequency estimator may be found in the book by Marple [1987].

In practice (13.36) is used with the modified covariance method of AR estimation, as discussed in Chapter 7. The AR spectral estimator based on a principal component decomposition is formed and the sinusoidal frequencies chosen to be the angles of the p zeros closest to the unit circle. This technique with some modifications can also be applied to data consisting of exponentially damped sinusoids in white noise [Kumaresan 1982]. The choice of model order M is a critical one. It has been found empirically that $M = 3/4N$ produces the best performance. The computation of the principal eigenvalues and eigenvectors of an autocorrelation matrix is intensive. To reduce the amount of computation required, we can determine either the p principal components or the $M - p$ nonprincipal components, whichever is less, since

$$\begin{aligned}\hat{a}_{\text{PC}} &= -\sum_{i=1}^{p} \frac{1}{\hat{\lambda}_i} \hat{\mathbf{v}}_i \hat{\mathbf{v}}_i^H \hat{\mathbf{r}}_{xx} \\ &= -\left(\hat{\mathbf{R}}_{xx}^{-1} - \sum_{i=p+1}^{M} \frac{1}{\hat{\lambda}_i} \hat{\mathbf{v}}_i \hat{\mathbf{v}}_i^H\right) \hat{\mathbf{r}}_{xx}.\end{aligned}$$

As a special case if $M = N - p/2$, then the autocorrelation matrix estimate in the modified covariance method may be shown to be of rank p. $\hat{\mathbf{R}}_{xx}$ will have only p nonzero eigenvalues or principal components. The principal component solution can then be found simply by inverting a $p \times p$ matrix as

$$\hat{\mathbf{a}}_{\mathrm{PC}} = -\mathbf{H}^H[\mathbf{H}\mathbf{H}^H]^{-1}\mathbf{h} \tag{13.39}$$

where

$$\mathbf{H} = \begin{bmatrix} x[M-1] & x[M-2] & \cdots & x[0] \\ x[M] & x[M-1] & \cdots & x[1] \\ \vdots & \vdots & \vdots & \vdots \\ x[N-2] & x[N-3] & \cdots & x[N-M-1] \\ x^*[1] & x^*[2] & \cdots & x^*[M] \\ x^*[2] & x^*[3] & \cdots & x^*[M+1] \\ \vdots & \vdots & \vdots & \vdots \\ x^*[N-M] & x^*[N-M+1] & \cdots & x^*[N-1] \end{bmatrix}$$

which has dimension $2(N - M) \times M$ and

$$\mathbf{h} = [x[M] \quad x[M+1] \quad \cdots \quad x[N-1] \quad x^*[0] \quad x^*[1] \quad \cdots \quad x^*[N-M-1]]^T$$

which has dimension $2(N - M) \times 1$. The frequency estimation performance for this value of M is poorer than that obtained for $M = 3/4N$, but the reduced computational burden may make this loss a good trade-off. Simulation results of the principal component AR method are contained in Section 13.11.

13.8.2 Minimum Variance Frequency Estimation

The MVSE was defined in Chapter 11 as

$$\hat{P}_{\mathrm{MV}}(f) = \frac{1}{\mathbf{e}^H\hat{\mathbf{R}}_{xx}^{-1}\mathbf{e}}$$

where $\hat{\mathbf{R}}_{xx}$ is an $M \times M$ positive definite matrix estimator of the autocorrelation matrix. In accordance with the idea of principal components, a frequency estimator may be defined as [Johnson 1982]

$$\begin{aligned} \hat{P}_{\mathrm{MV-PC}}(f) &= \frac{1}{\mathbf{e}^H \displaystyle\sum_{i=1}^{p} \frac{1}{\hat{\lambda}_i} \hat{\mathbf{v}}_i\hat{\mathbf{v}}_i^H \mathbf{e}} \\ &= \frac{1}{\displaystyle\sum_{i=1}^{p} \frac{1}{\hat{\lambda}_i} |\mathbf{e}^H\hat{\mathbf{v}}_i|^2}. \end{aligned} \tag{13.40}$$

The minima of the spectral estimate are used as estimates of the sinusoidal frequencies (See Problem 13.13). Owing to the poorer resolution of the MVSE relative to the AR spectral estimator, few investigations into the frequency estimation performance have been made. The same may be said of the principal component Bartlett spectral estimator described in the next section. See also Owsley [1985] for a similar estimator.

13.8.3 Bartlett Frequency Estimation

The Blackman–Tukey spectral estimator was defined as [see (4.23)]

$$\hat{P}_{\mathrm{BT}}(f) = \sum_{k=-M}^{M} w[k]\hat{r}_{xx}[k] \exp(-j2\pi fk)$$

where $\hat{r}_{xx}[k]$ is the biased ACF estimator. If a Bartlett window is used, then

$$\hat{P}_{\mathrm{BT}}(f) = \hat{P}_{\mathrm{BAR}}(f) = \frac{1}{M} \sum_{k=-(M-1)}^{M-1} (M - |k|)\hat{r}_{xx}[k] \exp(-j2\pi fk).$$

Using the identity

$$\sum_{k=-(M-1)}^{M-1} (M - |k|)g[k] = \sum_{m=0}^{M-1} \sum_{n=0}^{M-1} g[m - n]$$

this becomes

$$\begin{aligned}\hat{P}_{\mathrm{BAR}}(f) &= \frac{1}{M} \sum_{m=0}^{M-1} \sum_{n=0}^{M-1} \hat{r}_{xx}[m - n] \exp[-j2\pi f(m - n)] \\ &= \frac{1}{M} \mathbf{e}^H \hat{\mathbf{R}}_{xx} \mathbf{e}\end{aligned} \tag{13.41}$$

and is termed the *Bartlett spectral estimator*. $\hat{\mathbf{R}}_{xx}$, which has dimension $M \times M$, can be any suitable estimator of the autocorrelation matrix. To ensure a nonnegative spectral estimate the biased ACF estimator should be used with a Toeplitz autocorrelation matrix. A principal component version is

$$\hat{P}_{\mathrm{BAR-PC}}(f) = \frac{1}{M} \sum_{i=1}^{p} \hat{\lambda}_i |\mathbf{e}^H \hat{\mathbf{v}}_i|^2. \tag{13.42}$$

This type of estimator was first proposed by Owsley [1973] for array processing.

13.9 NOISE SUBSPACE FREQUENCY ESTIMATION

The frequency estimation methods to be described in this section all rely on the property that the noise subspace eigenvectors of a Toeplitz autocorrelation matrix are orthogonal to the signal vectors [see (13.34)]. The earliest application of this result is the Pisarenko harmonic decomposition (PHD), which although it does not, itself, provide reliable frequency estimates, has promoted a heavy interest in eigenanalysis approaches to frequency estimation. The PHD, as well as some of the ensuing techniques, are now discussed.

13.9.1 Pisarenko Harmonic Decomposition

In the PHD the eigenvector associated with the minimum eigenvalue of the estimated autocorrelation matrix $\mathbf{R}_{xx}$ is used for frequency estimation [Pisarenko 1973]. The method is based on a trigonometric theorem by Caratheodory [Grenander and Szego 1958]. As pointed out in Section 13.7 and proven in general in Appendix 13B, if $\mathbf{v}_{p+1}$ is the eigenvector of the $(p + 1) \times (p + 1)$ autocorrelation matrix $\mathbf{R}_{xx}$ associated with the minimum eigenvalue, then the zeros of

$$\sum_{n=0}^{p} [\mathbf{v}_{p+1}]_{n+1} z^{-n} \tag{13.43}$$

are at $z = \exp(j2\pi f_i)$ for $i = 1, 2, \ldots, p$. In practice, we replace the theoretical autocorrelation matrix by an estimated one which is Toeplitz and whose elements are the biased ACF estimates. This is necessary to ensure that the estimated autocorrelation matrix retains the properties of the theoretical matrix and in particular that the zeros of (13.43) are on the unit circle. Note that the same approach can be used for real data if the dimension of the autocorrelation matrix is $(2p + 1) \times (2p + 1)$ for p real sinusoids. In this case the zeros will occur in complex-conjugate pairs.

The PHD can also be derived from an AR modeling viewpoint. In Section 5.6 it was shown that p complex sinusoids may be modeled as the limiting form of an AR(p) process. Recall that the limiting process involved letting the poles approach the unit circle as the excitation noise variance approached zero. Based on this AR model the Yule–Walker equations for a purely sinusoidal process are, from (5.19),

$$\underbrace{\begin{bmatrix} r_{ss}[0] & r_{ss}[-1] & \cdots & r_{ss}[-p] \\ r_{ss}[1] & r_{ss}[0] & \cdots & r_{ss}[-p+1] \\ \vdots & \vdots & \ddots & \vdots \\ r_{ss}[p] & r_{ss}[p-1] & \cdots & r_{ss}[0] \end{bmatrix}}_{\mathbf{R}_{ss}} \underbrace{\begin{bmatrix} 1 \\ a[1] \\ \vdots \\ a[p] \end{bmatrix}}_{\mathbf{a}'} = \begin{bmatrix} \sigma^2 \\ 0 \\ \vdots \\ 0 \end{bmatrix} \tag{13.44}$$

where $r_{ss}[k]$ is the ACF of the sinusoids and $\{a[1], a[2], \ldots, a[p], \sigma^2\}$ are the AR parameters. Now implicit in the use of the AR model is the limiting operation $\sigma^2 \to 0$ as the poles of $1/\mathcal{A}(z)$ approach the locations $\exp(j2\pi f_i)$ for $i = 1, 2, \ldots, p$. As a result, (13.44) becomes

$$\mathbf{R}_{ss}\mathbf{a}' = \mathbf{0}. \tag{13.45}$$

But $\mathbf{R}_{xx} = \mathbf{R}_{ss} + \sigma_z^2\mathbf{I}$, so that

$$\mathbf{R}_{xx}\mathbf{a}' = \sigma_z^2\mathbf{a}'. \tag{13.46}$$

Thus $\mathbf{a}'$ is the eigenvector (not normalized) of $\mathbf{R}_{xx}$ associated with the smallest eigenvalue σ_z^2. Clearly, rooting the polynomial formed from this eigenvector is equivalent to rooting the AR filter, which has already been shown to have its zeros on the unit circle at the sinusoidal frequency locations. Another viewpoint is that the PHD is actually a noise compensation technique. If we had knowledge of $\mathbf{R}_{ss}$, then (13.44) with $\sigma^2 = 0$ could be used to solve for the AR parameters. By solving for the eigenvector with the eigenvalue σ_z^2, we are equivalently solving

$$(\mathbf{R}_{xx} - \sigma_z^2\mathbf{I})\mathbf{a}' = 0 \tag{13.47}$$

which is a form of noise compensation. The estimate of the noise variance to be subtracted from the measured autocorrelation matrix is determined by the minimum eigenvalue. As with most noise compensation methods, the PHD performs poorly, as described in Section 13.11.

The classical power method can be applied to $\hat{\mathbf{R}}_{xx}$ to compute the eigenvector of $\hat{\mathbf{R}}_{xx}$ associated with the minimum eigenvalue [Wilkinson 1965]. It should be noted that if the number of sinusoids is unknown, caution must be exercised. Pisarenko's method will produce p zeros on the unit circle even if there are actually fewer than p sinusoids in the data. Statistical properties of the PHD are described in the paper by Sakai [1984]. Some simulation results on the use of the PHD for frequency estimation are given in Section 13.11.

Some extensions to the PHD have been proposed. It may be shown by a similar proof as given in Appendix 13B that if the dimension of $\mathbf{R}_{xx}$ is $M \times M$, where $M > p$, the minimum eigenvalue is still σ_z^2 but now has multiplicity $M - p$. One can therefore average the $M - p$ smallest eigenvalues and use this, hopefully, improved estimate in (13.47), where $\mathbf{R}_{xx}$ is chosen to be $(p + 1) \times (p + 1)$. Another approach relies on the noise subspace orthogonality property. Theoretically, any vector given by [Reddi 1979, Kumaresan 1982]

$$\mathbf{g}' = \sum_{i=p+1}^{M} \alpha_i \mathbf{v}_i \tag{13.48}$$

where $\mathbf{g}' = [g[0] \quad g[1] \quad \cdots \quad g[M-1]]^T$ and the α_i's are arbitrary, lies within the noise subspace. The eigenvectors have been obtained from the $M \times M$ autocorrelation matrix. Therefore, the zeros formed from the polynomial $\mathcal{G}(z) = \sum_{k=0}^{M-1} g[k]z^{-k}$ will have p of its zeros on the unit circle at the sinusoidal frequency locations. It has been proposed to choose the vector in the noise subspace which constrains $g[0] = 1$ and minimizes $\sum_{i=1}^{M-1} | g[i] |^2$. This minimization yields for $\mathbf{g} = [g[1] \quad g[2] \quad \cdots \quad g[M-1]]^T$ the result

$$\mathbf{g} = -\frac{\mathbf{C}^H \mathbf{c}}{1 - \mathbf{c}^H \mathbf{c}} \tag{13.49}$$

where

$$\begin{bmatrix} \mathbf{v}_1^H \\ \mathbf{v}_2^H \\ \vdots \\ \mathbf{v}_p^H \end{bmatrix} = \left[\begin{array}{c|cccc} v_{11}^* & v_{12}^* & \cdots & v_{1M}^* \\ v_{21}^* & v_{22}^* & \cdots & v_{2M}^* \\ \vdots & \vdots & \vdots & \vdots \\ v_{p1}^* & v_{p2}^* & \cdots & v_{pM}^* \end{array}\right] = [\mathbf{c} \mid \mathbf{C}]$$

and $[\mathbf{v}_i]_j = v_{ij}$. Estimates of the eigenvectors are actually used in (13.49). This technique has been shown to outperform the PHD and appears to work well [Kumaresan 1982]. Questions still remain as to the appropriate choice of M.

13.9.2 MUSIC

The multiple signal classification (MUSIC) algorithm is also based on the orthogonality relation of (13.34). It estimates the sinusoidal frequencies as the peaks of the "spectral estimator" [Schmidt 1981]

$$\hat{P}_{\text{MUSIC}}(f) = \frac{1}{\sum_{i=p+1}^{M} |\mathbf{e}^H \hat{\mathbf{v}}_i|^2}. \tag{13.50}$$

Theoretically, when $f = f_i$, the ith sinusoidal frequency, so that $\mathbf{e} = \mathbf{e}_i$, $\hat{P}_{\text{MUSIC}}(f) \to \infty$. Due to estimation errors a peak will be exhibited at or near the sinusoidal frequency. The frequency estimates are found as the frequencies corresponding to the p largest peaks of (13.50). Some simulation results of the MUSIC algorithm are described in Section 13.11. A computer program that implements this frequency estimator may be found in the book by Marple [1987].

The maxima of (13.50) can also be computed by using the signal subspace eigenvectors. To do so, note that $\hat{P}_{\text{MUSIC}}(f)$ is maximized when the denominator is minimized. But the denominator may be expressed as

$$\begin{aligned} \sum_{i=p+1}^{M} |\mathbf{e}^H \hat{\mathbf{v}}_i|^2 &= \mathbf{e}^H \sum_{i=p+1}^{M} \hat{\mathbf{v}}_i \hat{\mathbf{v}}_i^H \mathbf{e} \\ &= \mathbf{e}^H \left(\mathbf{I} - \sum_{i=1}^{p} \hat{\mathbf{v}}_i \hat{\mathbf{v}}_i^H \right) \mathbf{e} \end{aligned} \tag{13.51}$$

where $\mathbf{I} = \sum_{i=1}^{M} \hat{\mathbf{v}}_i \hat{\mathbf{v}}_i^H$ has been used. Maximizing (13.50) is then equivalent to maximizing

$$\sum_{i=1}^{p} |\mathbf{e}^H \hat{\mathbf{v}}_i|^2. \tag{13.52}$$

It is seen that either the noise subspace eigenvectors as originally proposed or the signal subspace vectors as per (13.52) can be used to determine the sinusoidal frequency estimates. It is interesting to note that by a similar analysis (see Problem 13.12) we can show that [Kay and Demeure 1984]

$$\hat{P}'_{\text{MUSIC}}(f) = \frac{1}{1 - \hat{P}(f)} \tag{13.53}$$

where $\hat{P}'_{\text{MUSIC}}(f)$ is identical to (13.50) except that $\mathbf{e}$ is replaced by $\mathbf{e}' = 1/\sqrt{M}\mathbf{e}$ and

$$\hat{P}(f) = \sum_{i=1}^{p} |\mathbf{e}'^{H} \hat{\mathbf{v}}_i|^2 .$$

$\hat{P}(f)$ can be shown to satisfy $0 \le \hat{P}(f) \le 1$. Because (13.53) represents a monotonic transformation from $\hat{P}(f)$ to $\hat{P}'_{\text{MUSIC}}(f)$, the maxima of $\hat{P}'_{\text{MUSIC}}(f)$ or equivalently $\hat{P}_{\text{MUSIC}}(f)$ are identical to the maxima of $\hat{P}(f)$. As an example, for two complex sinusoids a typical spectral plot is shown in Figure 13.5. Although the MUSIC "spectral estimator" would be termed a high-resolution estimator, it is clear from (13.53) that either plot contains the same information and hence either one may be peak picked for use as a frequency estimator. The peaks in Figure 13.5a are present, although not clearly visible. This example serves to illustrate that it is the peak *location*, not the peak sharpness, which is of importance (see

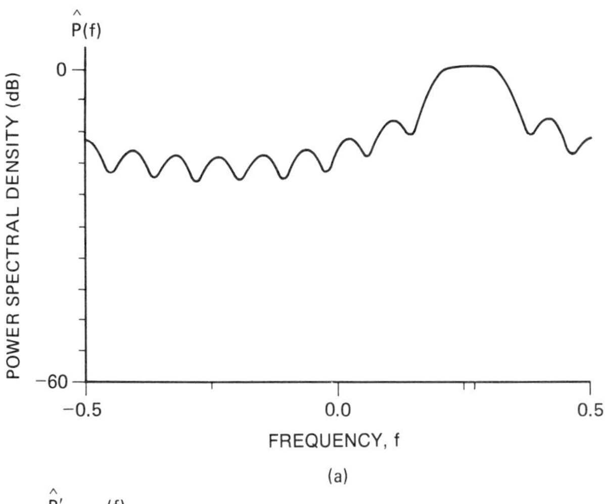

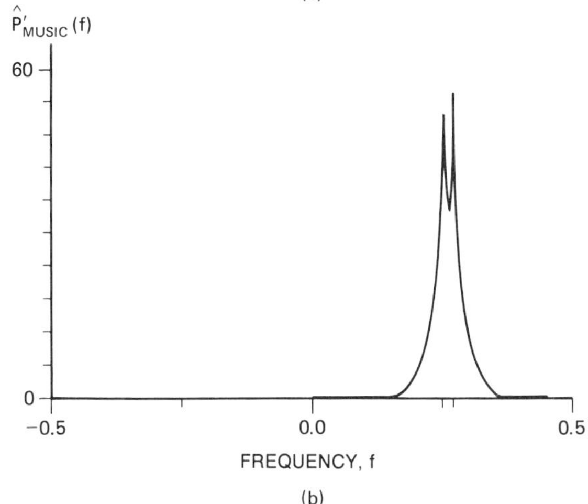

Figure 13.5 Illustration of equivalent spectral plots for frequency estimation. (After Kay and Demeure [1984].) (© 1984 IEEE)

also Section 1.4). Note also that $\hat{P}(f)$ is identical to the principal component Bartlett spectral estimator (13.42) if $\hat{\lambda}_i = 1$.

13.9.3 Generalized Frequency Estimators

The Pisarenko and MUSIC frequency estimators are only two members of a large class of estimators. In general, these estimators estimate the frequencics as the peaks of the function

$$\frac{1}{g(|\mathbf{e}^H\hat{\mathbf{v}}_{p+1}|^2, |\mathbf{e}^H\hat{\mathbf{v}}_{p+2}|^2, \ldots, |\mathbf{e}^H\hat{\mathbf{v}}_M|^2)}.$$

The only requirement on the function which is defined for $\xi_i \geq 0$, $i = p + 1, p + 2, \ldots, M$ is that

$$g(\xi_{p+1}, \xi_{p+2}, \ldots, \xi_M) = 0$$

TABLE 13.1 SUMMARY OF EIGENANALYSIS METHODS

Estimator	Form	Comments
Bartlett-PC	$\frac{1}{M}\sum_{i=1}^{p} \hat{\lambda}_i \mid \mathbf{e}^H\hat{\mathbf{v}}_i \mid^2$	Find peaks
MVSE-PC	$\left(\sum_{i=1}^{p} \frac{1}{\hat{\lambda}_i} \mid \mathbf{e}^H\hat{\mathbf{v}}_i \mid^2\right)^{-1}$	Find minima
AR-PC	$(\mid 1 + \mathbf{e}^H\hat{\mathbf{a}} \mid^2)^{-1}$ $\hat{\mathbf{a}} = -\sum_{i=1}^{p} \frac{1}{\hat{\lambda}_i} \hat{\mathbf{v}}_i\hat{\mathbf{v}}_i^H\hat{\mathbf{r}}_{xx}$	Find peaks or root polynomial
Pisarenko	$(\mid \mathbf{e}^H\hat{\mathbf{v}}_{p+1} \mid^2)^{-1}$	$M = p + 1$; find peaks or root polynomial
MUSIC	$\left(\sum_{i=p+1}^{M} \mid \mathbf{e}^H\hat{\mathbf{v}}_i \mid^2\right)^{-1}$	Find peaks
Generalized	$[g(\mid \mathbf{e}^H\hat{\mathbf{v}}_{p+1} \mid^2, \ldots, \mid \mathbf{e}^H\hat{\mathbf{v}}_M \mid^2)]^{-1}$	Find peaks

Definitions: $\hat{\lambda}_i$, $\hat{\mathbf{v}}_i$ $i = 1, 2, \ldots, M$ are eigenvalues and eigenvectors of $\hat{\mathbf{R}}_{xx}$ $(M \times M)$; $\hat{\lambda}_1 \geq \hat{\lambda}_2 \geq \cdots \geq \hat{\lambda}_M$; $\mathbf{e} = [1 \quad \exp[j2\pi f] \quad \cdots \quad \exp[j2\pi f(M-1)]]^T$

if and only if $\xi_{p+1} = \xi_{p+2} = \cdots = \xi_M = 0$. Clearly, many functions have this property. An example is

$$g(\xi_{p+1}, \xi_{p+2}, \ldots, \xi_M) = \sum_{i=p+1}^{M} \ln(1 + \xi_i)$$

which leads to the frequency estimator

$$\frac{1}{\ln\left[\prod_{i=p+1}^{M} (1 + \mid \mathbf{e}^H\hat{\mathbf{v}}_i \mid^2)\right]}.$$

It should now be apparent that eigenanalysis approaches can yield many different frequency estimators. Table 13.1 summarizes the most popular approaches.

13.10 MODEL ORDER SELECTION

It has been assumed in all the frequency estimation algorithms that the number of sinusoids is known a priori. In practice, this knowledge is seldom available, requiring a means of determining the number of sinusoids from the given data. Some techniques are now briefly discussed.

If the sinusoidal frequencies can be assumed to be spaced farther apart than $1/N$, a normalized periodogram (which would also be the MLE) can be used. The number of sinusoids is determined by the number of peaks exceeding a threshold.

We first compute for each frequency at which a peak occurs, the normalized periodogram

$$l = \frac{\frac{1}{N}\left|\sum_{n=0}^{N-1} x[n] \exp(-j2\pi f_i n)\right|^2}{\sigma_z^2/2}. \tag{13.54}$$

The white noise variance σ_z^2 is assumed to be known and f_i is the frequency of the ith peak. It can be shown (see Problem 13.14) that if noise only is present at and near f_i, then [Fisher 1929]

$$l \sim \chi_2^2 \tag{13.55}$$

so that the probability that l exceeds some threshold γ is $\exp(-\gamma/2)$. If, for example, l were found to be 9.2, the hypothesis that a sinusoid is present at $f = f_i$ would be accepted. This is because if noise only were present, then about 99% of the time l would not exceed this threshold. In this way false noise peaks can be discarded. For sinusoids that are not well resolved, this approach is not valid since the periodogram will display less than p peaks.

For the MLE and approximate MLE methods the AIC as discussed in Appendix 7A can be used to indicate the number of sinusoids. It chooses this number to minimize

$$\text{AIC}(i) = 2N \ln S_i(\hat{\mathbf{A}}_c, \hat{\mathbf{f}}) + 6i \tag{13.56}$$

where $S_i(\hat{\mathbf{A}}_c, \hat{\mathbf{f}})$ is the value of (13.10) after minimization with respect to the complex amplitudes $\mathbf{A}_c$ and frequencies $\mathbf{f}$, assuming that i sinusoids are present. The second term $6i$ in the AIC arises by letting $k = 3i$ in (7A.7) to account for the three parameters (amplitude, phase, and frequency) estimated for each sinusoid. The first term of (13.56) may be verified by maximizing the PDF of the observed data

$$p(\mathbf{x}; \mathbf{A}_c, \mathbf{f}, \sigma_z^2) = \frac{1}{\pi^N \det(\sigma_z^2 \mathbf{I})} \exp\left[-\frac{1}{\sigma_z^2} S_i(\mathbf{A}_c, \mathbf{f})\right]$$

over $\mathbf{A}_c$, $\mathbf{f}$, and σ_z^2 for use in (7A.7) and neglecting the constant terms.

Finally, for the eigenanalysis techniques the eigenvalues of the autocorrelation matrix can be used to indicate the number of sinusoidal components. Since for p sinusoids the eigenvalues are $\{\lambda_1 + \sigma_z^2, \lambda_2 + \sigma_z^2, \ldots, \lambda_p + \sigma_z^2, \sigma_z^2, \sigma_z^2, \ldots, \sigma_z^2\}$, the number of sinusoids can be found as the number of eigenvalues which exceed a given threshold (see Problem 13.15). For the best performance we would like the dimension of the autocorrelation matrix to be large so that the majority of the eigenvalues are equal to σ_z^2. Doing so runs the risk of statistically unstable eigenvalue estimates. It is not known how best to choose the dimension of the autocorrelation matrix, since the statistics of the estimated eigenvalues are unknown. Similar tests based on eigenvalues can be found in multivariate statistics [Kendall and Stuart 1976, Wax and Kailath 1984].

None of the methods thus far described have been extensively tested and therefore their relative advantages and disadvantages are not known. Other approaches to model order selection based on multiple hypothesis testing are given in the works by Anderson [1971] and Stuller [1975].

13.11 COMPUTER SIMULATION EXAMPLES

The data set chosen for the computer simulation is that of two closely spaced complex sinusoids in complex white Gaussian as given by (13.1) for $p = 2$. Specifically,

$$x[n] = \exp(j2\pi f_1 n) + \exp\left[j\left(2\pi f_2 n + \frac{\pi}{4}\right)\right] + z[n] \qquad n = 0, 1, \ldots, 24$$

where $f_1 = 0.50$, $f_2 = 0.52$ and $z[n]$ is complex white noise with variance σ_z^2. The two sinusoids are spaced closer than $1/N = 0.04$, the Fourier resolution. The SNR is defined as

$$\mathrm{SNR} = 10 \log_{10} \frac{1}{\sigma_z^2} \qquad \mathrm{dB}.$$

A comparison is now made between the following frequency estimators:

1. MLE, frequencies maximizing (13.13)
2. IFA (see Figure 13.2)
3. AR spectral estimation using the modified covariance method (see Section 7.5)
4. Principal component AR spectral estimator (13.36)
5. Pisarenko (see Section 13.9.1)
6. MUSIC (13.50).

The details of the various implementations are as follows. The MLE assumes that $p = 2$ and maximizes (13.13) to find the frequency estimates. The IFA uses the Burg algorithm (see Section 7.6) with $p = 2$ and for each SNR 10 iterations are computed. The AR spectral estimator uses a model order of $p = 8$ which from Figure 13.4 produces the best results. The angles of the two zeros closest to the unit circle are chosen as the frequency estimates. The principal component AR spectral estimator uses a model order of $M = 18$, which has been shown to achieve the best performance, and $p = 2$ principal eigenvectors. As before, the frequency estimates are given by the angles of the two zeros closest to the unit circle. The Pisarenko method uses a $(p + 1) \times (p + 1)$ autocorrelation matrix, where $p = 2$, with a biased ACF estimator. Finally, the MUSIC algorithm employs $M = 12$ and $p = 2$ since for smaller values of M the spectral peaks are not resolved. To actually locate the peaks, (13.50) is evaluated on a fine grid of frequencies. For all the methods the mean square error (MSE) of f_1 is computed. Results for f_2 are similar. In calculating the MSE, the lower frequency estimate is assigned to $f_1 = 0.50$. Also, the CR bound as computed from (13.17) is determined. If, however, the frequency estimator is biased, the CR bound is not directly applicable.

It still provides a bound on the MSE or, equivalently, the variance of an unbiased estimator. The frequency estimation performance for the MLE and the principal component version of the modified covariance method as well as the CR bound have been extracted from the work by Kumaresan [1982]. The performance of the IFA has been extracted from the paper by Kay [1984].

Figure 13.6 summarizes the frequency estimation accuracy of all the methods described. As expected, the MLE has the best performance, attaining the CR bound down to a threshold of 3 dB. The IFA has the next-best performance at low SNRs with a threshold of 5 dB. Curiously, the IFA exceeds the CR bound at low SNRs due to an inherent bias. The same bias causes the performance to worsen at higher SNRs where the variance is small but the bias persists. The modified covariance method performs poorly at SNRs less than about 20 dB, while the principal components version attains the CR bound for all SNRs in excess of 7 dB. The use of the principal components has the effect of lowering the threshold by about 13 dB. The Pisarenko method performs extremely poorly for all SNRs. This may be attributed to the use of a Toeplitz autocorrelation matrix with a biased ACF estimator, which even for no noise does not produce the true frequencies. Finally, the MUSIC algorithm gives acceptable performance for

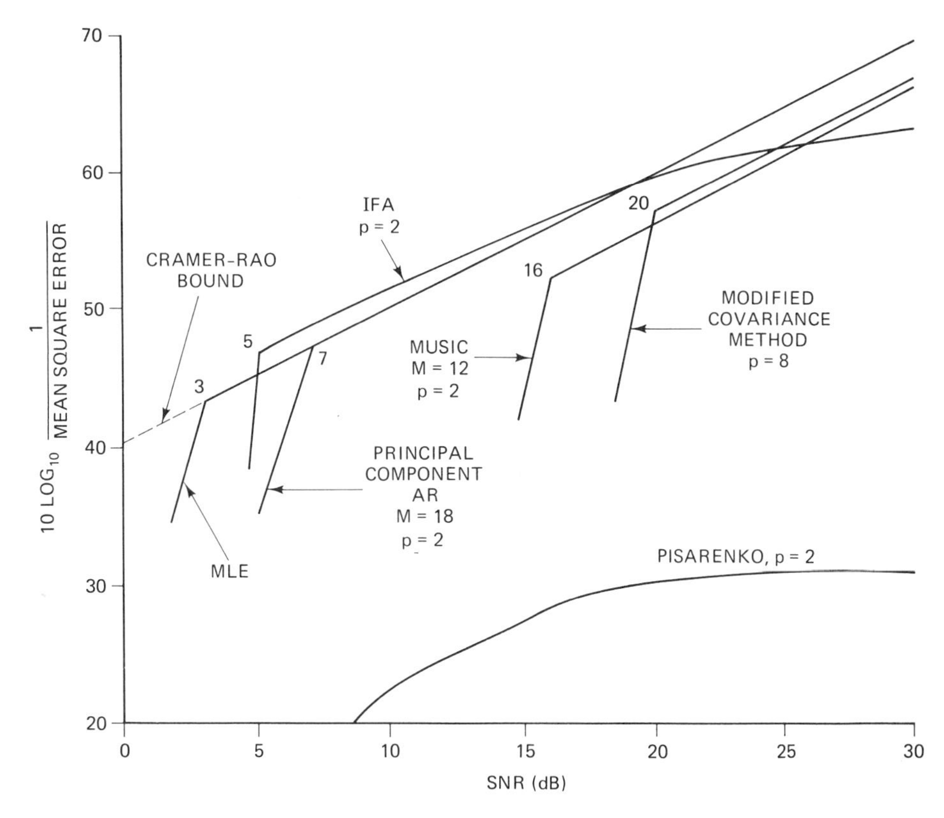

Figure 13.6 Comparison of several frequency estimators.

SNRs above 16 dB. The interested reader should also consult the work by Evans et al. [1982] for a comprehensive comparison of many of these algorithms.

In comparing the various approaches we must also keep in mind the computational complexity. Generally, the eigenanalysis techniques are computationally very intensive, while the modified covariance method and the IFA require a more modest amount of computation as described in Chapter 7. A discussion of some of the computational issues for eigenanalysis frequency estimation can be found in the work by Owsley [1985].

REFERENCES

Anderson, T. W., *The Statistical Analysis of Time Series,* Wiley, New York, 1971.

Bresler, Y., and A. Macovski, "Exact Maximum Likelihood Estimation of Superimposed Exponential Signals in Noise," *Rec 1985 Int. Conf. Acoust. Speech Signal Process.,* Tampa, Fla., pp. 1824–1827.

Evans, A. G., and R. Fischl, "Optimal Least Squares Time-Domain Synthesis of Recursive Digital Filters," *IEEE Trans. Audio Electroacoust.,* Vol. AU21, pp. 61–65, Feb. 1973.

Evans, J. E., J. R. Johnson, and D. F. Sun, "Applications of Advanced Signal Processing Techniques to Angle of Arrival Estimation in ATC Navigation and Surveillance Systems," Tech. Rep. 582, MIT Lincoln Laboratory, June 23, 1982.

Fisher, R. A., "Tests of Significance in Harmonic Analysis," *Proc. Roy. Soc.,* Vol. 125, pp. 54–59, 1929.

Goodman, N. R., "Statistical Analysis Based on a Certain Multivariate Complex Gaussian Distribution," *Ann. Math. Stat.,* Vol. 34, pp. 152–157, 1963.

Grenander, U., and G. Szego, *Toeplitz Forms and Their Applications,* University of California Press, Berkeley, Calif., 1958.

Johnson, D. H., "The Application of Spectral Estimation Methods to Bearing Estimation Problems," *Proc. IEEE,* Vol. 70, pp. 1018–1028, Sept. 1982.

Kay, S., "Autoregressive Spectral Analysis of Narrowband Processes in White Noise with Application to Sonar Signals," Ph.D. dissertation, Georgia Institute of Technology, 1980.

Kay, S., "Accurate Frequency Estimation at Low Signal-to-Noise Ratio," *IEEE Trans. Acoust. Speech Signal Process.,* Vol. ASSP32, pp. 540–547, June 1984.

Kay, S., and C. Demeure, "The High Resolution Spectrum Estimator: A Subjective Entity," *Proc. IEEE,* Vol. 72, pp. 1815–1816, Dec. 1984.

Kendall, Sir M., and A. Stuart, *The Advanced Theory of Statistics,* Vol. 3, Charles Griffin, London, 1976.

Kumaresan, R., "Estimating the Parameters of Exponentially Damped or Undamped Sinusoidal Signals in Noise," Ph.D. dissertation, University of Rhode Island, 1982.

Kumaresan, R., and A. K. Shaw, "High Resolution Bearing Estimation without Eigen Decomposition," *Rec. 1985 IEEE Int. Conf. Acoust. Speech Signal Process.,* Tampa, Fla., pp. 576–579.

Lang, S. W., and J. H. McClellan, "Frequency Estimation with Maximum Entropy Spectral Estimators," *IEEE Trans. Acoust., Speech Signal Process.*, Vol. ASSP28, pp. 716–724, Dec. 1980.

Marple, S. L., Jr., *Digital Spectral Analysis with Applications,* Prentice-Hall, Englewood Cliffs, N.J., 1987.

Marquardt, D. W., "An Algorithm for Least Squares Estimation of Non-linear Parameters," *J. Soc. Ind. Appl. Math.,* Vol. 11, pp. 431–444, June 1963.

Owsley, N. L., "A Recent Trend in Adaptive Spatial Processing for Sensor Arrays: Constrained Adaptation," in *Signal Processing,* J. W. R. Griffiths et al., eds., Academic Press, New York, 1973.

Owsley, N. L., "Sonar Array Processing," in *Array Signal Processing,* S. Haykin et al., eds., Prentice-Hall, Englewood Cliffs, N.J., 1985.

Palmer, L. C., "Coarse Frequency Estimation Using the Discrete Fourier Transform," *IEEE Trans. Inf. Theory,* Vol. IT20, pp. 104–109, Jan. 1974.

Pisarenko, V. F., "On the Estimation of Spectra by Means of Nonlinear Functions of the Covariance Matrix," *Geophys. J. Roy. Astron. Soc.,* Vol. 28, pp. 511–531, 1972.

Pisarenko, V. F., "The Retrieval of Harmonics from a Covariance Function," *Geophys. J. Roy. Astron. Soc.,* Vol. 33, pp. 347–366, 1973.

Reddi, S. S., "Multiple Source Location—A Digital Approach," *IEEE Trans. Aerosp. and Electron. Syst.,* Vol. AES15, pp. 95–105, Jan. 1979.

Rife, D. C., "Digital Tone Parameter Estimation in the Presence of Gaussian Noise," Ph.D. dissertation, Polytechnic Institute of New York, June 1973.

Rife, D. C., and R. R. Boorstyn, "Single Tone Parameter Estimation from Discrete-Time Observations," *IEEE Trans. Inf. Theory,* Vol. IT20, pp. 591–598, Sept. 1974.

Rife, D. C., and R. R. Boorstyn, "Multiple Tone Parameter Estimation from Discrete-Time Observations," *Bell Syst. Tech. J.,* Vol. 55, pp. 1389–1410, Nov. 1976.

Sakai, H., "Statistical Analysis of Pisarenko's Method for Sinusoidal Frequency Estimation," *IEEE Trans. Acoust. Speech Signal Process.,* Vol. ASSP32, pp. 95–101, Feb. 1984.

Schmidt, R. O., "A Signal Subspace Approach to Multiple Emitter Location and Spectral Estimation," Ph.D. dissertation, Stanford University, 1981.

Stuller, J. A., "Generalized Likelihood Signal Resolution," *IEEE Trans. Inf. Theory,* Vol. IT21, pp. 276–282, May 1975.

Tufts, D. W., and R. Kumaresan, "Estimation of Frequencies of Multiple Sinusoids: Making Linear Prediction Perform Like Maximum Likelihood," *Proc. IEEE,* Vol. 70, pp. 975–989, Sept. 1982.

Ulrych, T. J., and T. N. Bishop, "Maximum Entropy Spectral Analysis and Autoregressive Decomposition," *Rev. Geophys. Space Phys.,* Vol. 13, pp. 183–200, Feb. 1975.

Ulrych, T. J., and R. W. Clayton, "Time Series Modelling and Maximum Entropy," *Phys. Earth Planet. Inter.,* Vol. 12, pp. 188–200, Aug. 1976.

Van Trees, H. L., *Detection, Estimation, and Modulation Theory,* Vol. 3, Wiley, New York, 1971.

Walker, A. M., "On the Estimation of a Harmonic Component in a Time Series with Stationary Independent Residuals," *Biometrika,* Vol. 58, pp. 21–36, 1971.

Wax, M., and T. Kailath, "Determining the Number of Signals by Information Theoretic Criteria," *Rec. 1984 IEEE Int. Conf. Acoust. Speech Signal Process., San Diego,* pp. 6.1.3–6.3.4.

Wilkinson, J. H., *The Algebraic Eigenvalue Problem,* Clarendon Press, Oxford, 1965.

PROBLEMS

13.1. For one complex sinusoid in complex white Gaussian noise, assume that the complex amplitude A_{c_1} is a complex Gaussian random variable with mean zero and variance P_1. With this modeling the observed data as given by (13.1) is a complex Gaussian random process with autocorrelation matrix (see Section 3.5.2 and [Van Trees 1971])

$$\mathbf{R}_{xx} = P_1 \mathbf{e}_1 \mathbf{e}_1^H + \sigma_z^2 \mathbf{I}.$$

Show that the MLE of frequency can be found by minimizing $\mathbf{x}^H \mathbf{R}_{xx}^{-1} \mathbf{x}$. Next use Woodbury's identity (2.32) for $\mathbf{R}_{xx}^{-1}$ and perform the minimization. Compare this MLE with the one in Section 13.3.1 in which the signal was assumed to be deterministic.

13.2. Assume that the data consist of a real sinusoid in real white Gaussian noise

$$x[n] = A_1 \cos(2\pi f_1 n + \phi_1) + z[n] \qquad n = 0, 1, \ldots, N-1.$$

Find the function that must be maximized to obtain the MLE of the frequency f_1 by using a derivation similar to the one presented in Section 13.3.1. Show that for large N the MLE of frequency is obtained as the location of the peak of the periodogram. *Hint:* Let $A_1 \cos(2\pi f_1 n + \phi_1) = B_1 \cos(2\pi f_1 n) + B_2 \sin(2\pi f_1 n)$, where $B_1 = A_1 \cos \phi_1$, $B_2 = -A_1 \sin \phi_1$ and note that the MLE of B_1, B_2 is found as the least squares solution (see Section 3.3.2).

13.3. For $N = 25$ plot the Cramer–Rao lower bound for the frequency of a single complex sinusoid as given by (13.16) versus SNR with both axes in decibels. Compare your results with that of Figure 13.4, which plots the frequency bound for either of two equiamplitude complex sinusoids. Explain the difference.

13.4. Prove that the CR bounds for the frequencies of two sinusoids with amplitudes of one are always greater than or equal to the bound for the frequency of a single sinusoid of amplitude one by first showing that (13.16) can be rewritten as

$$\operatorname{var}(\hat{f}_1) \geq \frac{\sigma_z^2}{2(2\pi)^2} [\mathbf{M}_{11}^{-1}]_{11}$$

where $\Delta_n[1, 1] = 0$ and then proving that

$$[\mathbf{M}_{11}^{-1}]_{11} \leq [(\mathbf{M}_{11} - \mathbf{M}_{12}\mathbf{M}_{22}^{-1}\mathbf{M}_{21})^{-1}]_{11}.$$

Hint: Use the matrix inversion lemma

$$(\mathbf{A} + \mathbf{BCD})^{-1} = \mathbf{A}^{-1} - \mathbf{A}^{-1}\mathbf{B}(\mathbf{DA}^{-1}\mathbf{B} + \mathbf{C}^{-1})^{-1}\mathbf{DA}^{-1}$$

where $\mathbf{A} = \mathbf{M}_{11}$, $\mathbf{B} = -\mathbf{M}_{12}$, $\mathbf{C} = \mathbf{M}_{22}^{-1}$, $\mathbf{D} = \mathbf{M}_{21}$ as well as properties of positive definite matrices. Note that the matrix $\mathbf{M}$ can be shown to be positive definite.

13.5. Prove that $\mathbf{A}(\mathbf{A}^H\mathbf{A})^{-1}\mathbf{A}^H\mathbf{x}$ projects $\mathbf{x}$ onto the subspace spanned by the columns of $\mathbf{A}$ by expressing $\mathbf{x}$ as

$$\mathbf{x} = \sum_{i=1}^{N-p} \xi_i \mathbf{a}_i + \sum_{i=N-p+1}^{N} \xi_i \mathbf{e}_i$$

where $\mathbf{a}_i$ is the ith column of $\mathbf{A}$. The $\mathbf{e}_i$'s span the space orthogonal to the space spanned by the columns of $\mathbf{A}$ or $\mathbf{e}_i^H \mathbf{a}_j = 0$. The ξ_i's are arbitrary complex constants (see also Problem 2.4).

13.6. Verify that (13.23) is valid by first letting $\mathbf{F} = [\mathbf{E}\ \mathbf{A}]$ so that $\mathbf{F}$ is full rank and hence invertible and then computing $\mathbf{F}(\mathbf{F}^H\mathbf{F})^{-1}\mathbf{F}^H = \mathbf{I}$.

13.7. Show that (13.25) is minimized for

$$\mathbf{a} = -(\mathbf{X}'^H\mathbf{X}')^{-1}\mathbf{X}'^H\mathbf{x}'$$

where $[\mathbf{x}'\ \mathbf{X}'] = \mathbf{A}(\mathbf{A}^H\mathbf{A})^{-1}\mathbf{X}$ and $\mathbf{x}'$ is $N \times 1$, $\mathbf{X}'$ is $N \times p$.

13.8. Show that the constraint

$$a[p - i] = a^*[i] \qquad i = 0, 1, \ldots, p$$

forces the zeros of $\mathcal{A}(z) = \sum_{k=0}^{p} a[k]z^{-k}$ to occur in conjugate reciprocal pairs or at z_i and $1/z_i^*$.

13.9. Assume that the data consist of p complex sinusoids only. If we use the covariance method with a model order equal to the number of sinusoids, prove that the zeros of the PEF lie on the unit circle at the sinusoidal frequency locations. Does this result also hold for the modified covariance method? Why doesn't this property hold for the autocorrelation method?

13.10. The $M \times M$ signal autocorrelation matrix

$$\mathbf{R}_{ss} = \sum_{i=1}^{p} P_i \mathbf{e}_i \mathbf{e}_i^H$$

has rank p. Prove that for $M \geq p$ any solution of

$$\mathbf{R}_{ss}\mathbf{a} = -\mathbf{r}_{ss}$$

where

$$\mathbf{r}_{ss} = \sum_{i=1}^{p} P_i \exp(j2\pi f_i)\mathbf{e}_i$$

results in a PEF with p zeros at $z = \exp(j2\pi f_i)$, $i = 1, 2, \ldots, p$. *Hint:* Rewrite the equations as $\mathbf{R}_{ss}\mathbf{a}' = 0$, where $\mathbf{R}_{ss}$ is $(M + 1) \times (M + 1)$ and $\mathbf{a}' = [1\ \ \mathbf{a}^T]^T$ by noting that the minimum prediction error power must be zero.

13.11. In Problem 13.10 the solution $\mathbf{a}$ is not unique. One means of providing uniqueness is to choose the solution that has minimum norm (i.e., $\mathbf{a}^H\mathbf{a}$). Prove that this minimum norm solution is given by (13.38) by making use of the property in Appendix 13A. Next, show that the $M - p$ "noise" zeros lie inside the unit circle. Hint: Use the result that

$$\mathbf{a}'^H\mathbf{a}' = \sum_{k=0}^{M} |\, a[k] \,|^2$$

$$= \int_{-1/2}^{1/2} |\, A(f) \,|^2 \, df.$$

Note that there are p zeros constrained to be at $z = \exp(j2\pi f_i)$, $i = 1, 2, \ldots, p$, and apply the minimum-phase theorem of Section 6.3.2.

13.12. Prove the relationship between the MUSIC algorithm and the Bartlett-type frequency estimator as given by (13.53). Can you find a frequency estimator that will exhibit even sharper peaks than the MUSIC estimator?

13.13. Relate the MUSIC algorithm to the modified principal component MVSE. The latter estimator is defined by (13.40) with $\hat{\lambda}_i = 1$. How should the frequency estimates be determined from the modified principal component MVSE?

13.14. Prove that l as given in (13.54) is distributed as a χ_2^2 random variable when $x[n] = z[n]$ and $z[n]$ is complex white Gaussian noise. To do so, first show that

$$y[n] = z[n] \exp(-j2\pi f_0 n) \qquad n = 0, 1, \ldots, N-1$$

are distributed identically to $z[n] \quad n = 0, 1, \ldots, N-1$. This allows us to assume, for convenience, that $f_0 = 0$.

13.15. Find the eigenvalues of the $M \times M$ autocorrelation matrix

$$\mathbf{R}_{xx} = \sum_{i=1}^{p} P_i \mathbf{e}_i \mathbf{e}_i^H + \sigma_z^2 \mathbf{I}$$

if all the frequencies are distinct and integer multiples of $1/M$. Observe that in this case $\mathbf{e}_i^H \mathbf{e}_j = 0$ for $i \neq j$. Comment on the use of the eigenvalues to determine the number of sinusoids present.

APPENDIX 13A

Proof of Spanning Property of Principal Eigenvectors

It is proven that the principal eigenvectors of the signal autocorrelation matrix span the same space as the signal vectors. Mathematically, if

$$\mathbf{R}_{ss} = \sum_{j=1}^{p} P_j \mathbf{e}_j \mathbf{e}_j^H \tag{13A.1}$$

where $\mathbf{R}_{ss}$ is an $M \times M$ matrix, then it is to be shown that

$$\text{span}\,\{\mathbf{e}_1, \mathbf{e}_2, \ldots, \mathbf{e}_p\} = \text{span}\,\{\mathbf{v}_1, \mathbf{v}_2, \ldots, \mathbf{v}_p\}.$$

$\mathbf{v}_i$ for $i = 1, 2, \ldots, p$ are the eigenvectors of $\mathbf{R}_{ss}$ with nonzero eigenvalues. The span of $\{\mathbf{x}_1, \mathbf{x}_2, \ldots, \mathbf{x}_p\}$ is defined as the subspace composed of all the vectors formed as a linear combination of the $\mathbf{x}_i$'s. To establish that $\mathbf{R}_{ss}$ has only p nonzero eigenvalues, we can show that the rank of $\mathbf{R}_{ss}$ is p (in a similar fashion to the proof presented in Appendix 13B) and then use a standard theorem in linear algebra which asserts that the rank of a positive semidefinite matrix is equal to the number of nonzero eigenvalues.

For the signal vectors and the principal eigenvectors to span the same subspace, any vector in the span of $\{\mathbf{v}_1, \mathbf{v}_2, \ldots, \mathbf{v}_p\}$ must also be in the span of $\{\mathbf{e}_1, \mathbf{e}_2, \ldots, \mathbf{e}_p\}$, and conversely. Assume first that $\mathbf{x}$ lies in the span of the principal eigenvectors. It must first be shown that $\mathbf{x}$ can be represented as

$$\mathbf{x} = \sum_{j=1}^{p} \xi_j \mathbf{e}_j. \tag{13A.2}$$

Since the principal eigenvectors form a basis for the subspace,

$$\mathbf{x} = \sum_{i=1}^{p} \alpha_i \mathbf{v}_i. \tag{13A.3}$$

Using the definition of the principal eigenvectors,

$$\mathbf{R}_{ss}\mathbf{v}_i = \lambda_i \mathbf{v}_i \qquad i = 1, 2, \ldots, p \tag{13A.4}$$

for $\lambda_i \neq 0$. Substituting (13A.1) into (13A.4) yields

$$\sum_{j=1}^{p} P_j \mathbf{e}_j \mathbf{e}_j^H \mathbf{v}_i = \lambda_i \mathbf{v}_i$$

or

$$\mathbf{v}_i = \sum_{j=1}^{p} \left(\frac{P_j}{\lambda_i} \mathbf{e}_j^H \mathbf{v}_i \right) \mathbf{e}_j \qquad i = 1, 2, \ldots, p. \tag{13A.5}$$

Substituting (13A.5) into (13A.3) yields

$$\begin{aligned} \mathbf{x} &= \sum_{i=1}^{p} \alpha_i \left[\sum_{j=1}^{p} \left(\frac{P_j}{\lambda_i} \mathbf{e}_j^H \mathbf{v}_i \right) \mathbf{e}_j \right] \\ &= \sum_{j=1}^{p} \left(\sum_{i=1}^{p} \frac{\alpha_i P_j}{\lambda_i} \mathbf{e}_j^H \mathbf{v}_i \right) \mathbf{e}_j \end{aligned}$$

in accordance with (13A.2). For the converse to hold, any vector in the subspace spanned by the signal vectors must also be representable as a linear combination of the principal eigenvectors. But the principal eigenvectors span the entire p-dimensional subspace since they are orthonormal and therefore, linearly independent. Hence the principal eigenvectors form a basis for the signal subspace and any vector lying in the signal subspace may be represented as a linear combination of the principal eigenvectors.

APPENDIX 13B

Proof of Pisarenko Property

It is proven in this appendix that the eigenvector associated with the minimum eigenvalue of a Toeplitz autocorrelation matrix for p sinusoids in white noise has its zeros on the unit circle at the positions $z = \exp(j2\pi f_i)$ for $i = 1, 2, \ldots, p$.

To begin the proof let the $(p + 1) \times (p + 1)$ signal autocorrelation matrix be

$$\mathbf{R}_{ss} = \sum_{i=1}^{p} P_i \mathbf{e}_i \mathbf{e}_i^H.$$

It is first proven that the minimum eigenvalue of $\mathbf{R}_{ss}$ is unique and equal to zero. Observe that $\mathbf{R}_{ss}$ can be written as

$$\mathbf{R}_{ss} = \mathbf{E}\mathbf{P}\mathbf{E}^H$$

where $\mathbf{E} = [\mathbf{e}_1 \ \ \mathbf{e}_2 \ \ \cdots \ \ \mathbf{e}_p]$ and has dimension $(p + 1) \times p$ and $\mathbf{P} = \text{diag}(P_1, P_2, \ldots, P_p)$. $\mathbf{R}_{ss}$ is positive semidefinite since for an arbitrary $\boldsymbol{\xi}$

$$\boldsymbol{\xi}^H \mathbf{R}_{ss} \boldsymbol{\xi} = \sum_{i=1}^{p} P_i \,|\, \mathbf{e}_i^H \boldsymbol{\xi} \,|^2 \geq 0.$$

Hence all the eigenvalues of $\mathbf{R}_{ss}$ are nonnegative. If an eigenvalue is zero, it must be the minimum eigenvalue. A zero eigenvalue will exist if

$$\mathbf{R}_{ss}\boldsymbol{\xi} = \mathbf{0} \tag{13B.1}$$

for some $\boldsymbol{\xi}$ or the nullity of $\mathbf{R}_{ss}$ is greater than zero. Equivalently, a zero eigenvalue will exist if the rank of $\mathbf{R}_{ss}$ is less than $p + 1$. It is now proven that the rank of $\mathbf{R}_{ss}$ is exactly p, so that the nullity of $\mathbf{R}_{ss}$ is 1 and hence (13B.1) is satisfied for some $\boldsymbol{\xi}$. Furthermore, since the nullity is 1, all vectors satisfying (13B.1) lie in a one-dimensional subspace, which implies that there is a single zero eigenvalue. Consider a set of vectors $\mathbf{y}$ which result from the mapping

$$\mathbf{R}_{ss}\mathbf{x} = \mathbf{y}$$

or

$$\mathbf{E}\mathbf{P}\mathbf{E}^H\mathbf{x} = \mathbf{y}$$

for arbitrary $\mathbf{x}$ in R^{p+1}. Let

$$\boldsymbol{\alpha} = \mathbf{P}\mathbf{E}^H\mathbf{x} \tag{13B.2}$$

so that

$$\mathbf{y} = \sum_{i=1}^{p} \alpha_i \mathbf{e}_i. \tag{13B.3}$$

From (13B.3) it is seen that the dimension of the range space or the rank of $\mathbf{R}_{ss}$ is at most p. The rank will be exactly p if the α_i's can be arbitrarily chosen. Since $\boldsymbol{\alpha}$ is $p \times 1$, this requires $\boldsymbol{\alpha}$ to lie anywhere in R^p. By using (13B.2), we need only prove that $\mathbf{P}\mathbf{E}^H$, which is $p \times (p + 1)$, is full rank or the columns of $(\mathbf{P}\mathbf{E}^H)^H = \mathbf{E}\mathbf{P}$ are linearly independent. But

$$\mathbf{E}\mathbf{P}\boldsymbol{\beta} = \sum_{i=1}^{p} \beta_i P_i \mathbf{e}_i$$

which cannot be zero unless all the β_i's are zero (assuming all the P_i's to be nonzero and the frequencies to be distinct). It has thus been proven that the rank of $\mathbf{R}_{ss}$ is p and that there is a unique eigenvalue of $\mathbf{R}_{ss}$ which is zero. Consequently, the *minimum* eigenvalue of $\mathbf{R}_{ss}$ is unique and equal to zero. Because

$$\mathbf{R}_{xx} = \mathbf{R}_{ss} + \sigma_z^2\mathbf{I} \tag{13B.4}$$

has the same eigenvectors as $\mathbf{R}_{ss}$, the eigenvalues are easily seen to be $\{\lambda_1 + \sigma_z^2, \lambda_2 + \sigma_z^2, \ldots, \lambda_p + \sigma_z^2, \sigma_z^2\}$ since the eigenvalues of $\mathbf{R}_{ss}$ are $\{\lambda_1, \lambda_2, \ldots, \lambda_p, 0\}$. Now consider the eigenvector $\mathbf{v}_{p+1}$ of $\mathbf{R}_{xx}$ associated with the smallest eigenvalue σ_z^2

$$\mathbf{R}_{xx}\mathbf{v}_{p+1} = \sigma_z^2\mathbf{v}_{p+1}.$$

From (13B.4),

$$\mathbf{R}_{ss}\mathbf{v}_{p+1} = \mathbf{0}.$$

or

$$\mathbf{EPE}^H\mathbf{v}_{p+1} = \mathbf{0}.$$

Premultiplying by $\mathbf{v}_{p+1}^H$ yields

$$\mathbf{v}_{p+1}^H\mathbf{EPE}^H\mathbf{v}_{p+1} = (\mathbf{E}^H\mathbf{v}_{p+1})^H\mathbf{P}(\mathbf{E}^H\mathbf{v}_{p+1}) = 0.$$

Because $\mathbf{P}$ is positive definite it follows that

$$\mathbf{E}^H\mathbf{v}_{p+1} = \mathbf{0}$$

or

$$\mathbf{e}_i^H\mathbf{v}_{p+1} = 0 \qquad i = 1, 2, \ldots, p. \tag{13B.5}$$

Finally, (13B.5) may be rewritten in the more familiar form

$$\sum_{n=0}^{p} [\mathbf{v}_{p+1}]_{n+1} \exp(-j2\pi f_i n) = 0 \qquad i = 1, 2, \ldots, p$$

which is the desired result. A slight extension of this proof can be used to prove that if the dimension of $\mathbf{R}_{ss}$ is $M \times M$ for $M > p$, the rank of $\mathbf{R}_{ss}$ is p and the number of zero eigenvalues is $M - p$.

Chapter 14

Multichannel Spectral Estimation

14.1 INTRODUCTION

The spectral estimation methods in Part I all assume that we wish to estimate the PSD of a single time series. In many practical situations the data that are available are not limited to the output of a single channel but may well be the result of observations at the output of several channels. It is quite common in the fields of sonar [Knight et al. 1981], radar [Monzingo and Miller 1980], and seismic exploration [Justice 1985] to record data from multiple sensors. With this additional information it is then possible to estimate cross-spectra as well as auto-spectra. The cross-spectra are important in establishing linear filtering relationships between the time series [Jenkins and Watts 1968]. *The multichannel spectral estimation problem is to estimate the auto-spectra of the individual channels and the cross-spectra between all pairs of channels*. Many of the techniques described in Part I are easily extended to multichannel spectral estimation by replacing scalar functions by suitable matrix functions of a vector valued time series. Some extensions, however, are not possible, and these are highlighted.

The descriptions in this chapter are of necessity brief, and many of the important theoretical discussions of vector random processes or multiple time series have been omitted. A comprehensive treatment of multiple time series, including estimation procedures, has been provided by Hannan [1970]. A more applied treatment with many computer programs may be found in the book by Robinson [1967]. Finally, for descriptions of additional algorithms as well as for computer programs, the reader should consult the book by Marple [1987].

14.2 SUMMARY

A review of linear systems and Fourier transform theory for multichannel sequences is the subject of Section 14.3. Multichannel random processes are described in Section 14.4. The ACF is defined by (14.19) and (14.20), while the cross-spectral matrix, which is the quantity to be estimated, is defined by (14.23). The important relationship relating the cross-spectral matrices at the input and output of a multichannel linear system is given by (14.31). Section 14.5 summarizes classical or Fourier spectral estimation. The multichannel periodogram, averaged periodogram, and Blackman–Tukey spectral estimators are defined by (14.33), (14.35), and (14.41), respectively. The biased estimator of the ACF for the multichannel case is given by (14.39). The rational transfer function models are discussed in Section 14.6. The cross-spectral matrices for the multichannel ARMA, AR, and MA processes are given by (14.43), (14.45), and (14.47), respectively. The Yule–Walker equations, which relate the ACF to the model parameters, are given by (14.49) for an ARMA process, by (14.50) for an AR process, and by (14.51) for an MA process. Spectral estimation of multichannel AR processes is discussed in Section 14.7. The most straightforward approach is to replace the theoretical ACF by an estimate in the Yule–Walker equations (14.52) and solve for the AR parameters. The solution to these equations is accomplished by the efficient multichannel version of the Levinson algorithm. This algorithm is summarized in (14.63)–(14.66). A derivation is included in Appendix 14A. The extension of the covariance method for multichannel processes results in the set of equations (14.72) to be solved for the AR parameter estimates with the white noise variance matrix given by (14.73). Estimation of multichannel ARMA processes is described in Section 14.8. The multichannel MYWE and LSMYWE estimators require one to solve (14.76) and (14.78), respectively, for the AR parameters. The MA parameters may be found by filtering the data with the estimate of the AR filter and then applying (14.79) and (14.80). The multichannel MVSE is defined in Section 14.9 by (14.86) and (14.94). Finally, some computer simulation results are given in Section 14.10.

14.3 REVIEW OF LINEAR SYSTEMS AND FOURIER TRANSFORMS

A complex multichannel sequence $\mathbf{x}[n]$ is defined as the complex $L \times 1$ vector

$$\mathbf{x}[n] = \begin{bmatrix} x_1[n] \\ x_2[n] \\ \vdots \\ x_L[n] \end{bmatrix}. \tag{14.1}$$

Typically, $x_i[n]$ represents the data observed at the output of the ith channel. A multichannel linear shift invariant (LSI) filter transforms the input sequence $\mathbf{u}[n]$ into the output sequence $\mathbf{x}[n]$ by

$$\mathbf{x}[n] = \sum_{k=-\infty}^{\infty} \mathbf{H}[k]\mathbf{u}[n - k]. \tag{14.2}$$

$\mathbf{H}[k]$ is a complex $L \times L$ matrix defined as

$$\mathbf{H}[k] = \begin{bmatrix} h_{11}[k] & h_{12}[k] & \cdots & h_{1L}[k] \\ h_{21}[k] & h_{22}[k] & \cdots & h_{2L}[k] \\ \vdots & \vdots & \ddots & \vdots \\ h_{L1}[k] & h_{L2}[k] & \cdots & h_{LL}[k] \end{bmatrix} \tag{14.3}$$

where $h_{ij}[k]$ is the impulse response of the filter between the jth input and the ith output as illustrated in Figure 14.1 for $L = 2$. Note that the output of the ith channel is, from (14.2),

$$x_i[n] = \sum_{j=1}^{L} \sum_{k=-\infty}^{\infty} h_{ij}[k]u_j[n - k] \tag{14.4}$$

or the output of the ith channel is due to all L inputs in general. A multichannel system is causal and stable if and only if each individual scalar system characterized by its impulse response $h_{ij}[k]$ is causal and stable. Multichannel systems for which the number of inputs and outputs are not the same may be obtained as a special case of $\mathbf{H}[k]$ if the appropriate elements in (14.3) are set equal to zero.

The Fourier transform of a multichannel sequence is defined as the vector of individual Fourier transforms or

$$\mathbf{X}(f) = \begin{bmatrix} X_1(f) \\ X_2(f) \\ \vdots \\ X_L(f) \end{bmatrix}$$

where

$$X_i(f) = \sum_{n=-\infty}^{\infty} x_i[n] \exp(-j2\pi fn).$$

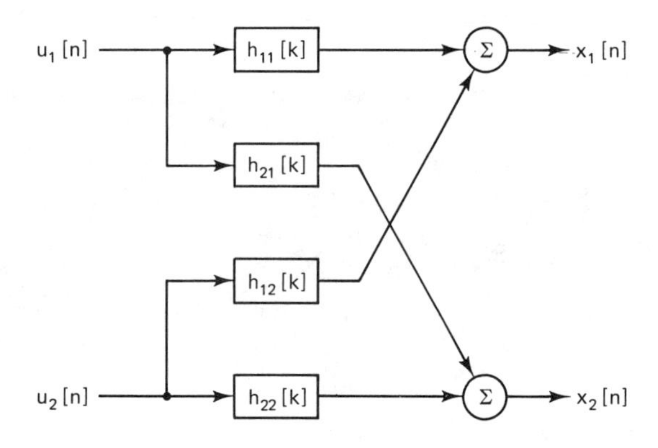

Figure 14.1 Two channel linear shift invariant system.

In vector notation the Fourier transform is the $L \times 1$ vector

$$\mathbf{X}(f) = \sum_{n=-\infty}^{\infty} \mathbf{x}[n] \exp(-j2\pi f n). \tag{14.5}$$

Similarly, the multichannel z transform is defined as the $L \times 1$ vector

$$\mathscr{X}(z) = \sum_{n=-\infty}^{\infty} \mathbf{x}[n] z^{-n}. \tag{14.6}$$

Assuming that the z-transform of $\mathbf{x}[n]$ exists on the unit circle, it is clear from (14.5) and (14.6) that

$$\mathscr{X}[\exp(j2\pi f)] = \mathbf{X}(f).$$

The multichannel z-transform has the usual properties (see Problem 14.1) of linearity

$$\mathscr{Z}(\mathbf{A}\mathbf{x}_1[n] + \mathbf{B}\mathbf{x}_2[n]) = \mathbf{A}\mathscr{X}_1(z) + \mathbf{B}\mathscr{X}_2(z) \tag{14.7}$$

for $\mathbf{A}$, $\mathbf{B}$ being $L \times L$ matrices, and shifting

$$\mathscr{Z}(\mathbf{x}[n - i]) = z^{-i}\mathscr{X}(z). \tag{14.8}$$

Analysis of multichannel linear systems using Fourier and z-transforms is much the same as for scalar systems. If we take the z-transform of $x_i[n]$ as given by (14.4), then (see Problem 14.2)

$$\mathscr{X}_i(z) = \sum_{j=1}^{L} \mathscr{H}_{ij}(z)\mathscr{U}_j(z) \qquad i = 1, 2, \ldots, L$$

which in matrix notation is

$$\mathscr{X}(z) = \mathscr{H}(z)\mathscr{U}(z). \tag{14.9}$$

$\mathscr{H}(z)$ is the multichannel system function and is defined as

$$\mathscr{H}(z) = \begin{bmatrix} \mathscr{H}_{11}(z) & \mathscr{H}_{12}(z) & \cdots & \mathscr{H}_{1L}(z) \\ \mathscr{H}_{21}(z) & \mathscr{H}_{22}(z) & \cdots & \mathscr{H}_{2L}(z) \\ \vdots & \vdots & \ddots & \vdots \\ \mathscr{H}_{L1}(z) & \mathscr{H}_{L2}(z) & \cdots & \mathscr{H}_{LL}(z) \end{bmatrix} \tag{14.10}$$

where

$$\mathscr{H}_{ij}(z) = \mathscr{Z}(h_{ij}[k]).$$

The inverse z-transform of $\mathscr{H}(z)$ is the multichannel impulse response, which is the matrix of impulse responses of the scalar systems connecting $\mathbf{u}[n]$ to $\mathbf{x}[n]$ as given by (14.3).

The multichannel frequency response is obtained as $\mathbf{H}(f) = \mathscr{H}[\exp(j2\pi f)]$. The physical interpretation of the frequency response is much the same as for scalar LSI systems. To be more specific, a multichannel sinusoid is defined as

$$\mathbf{u}[n] = \begin{bmatrix} A_1 \exp[j(2\pi f_0 n + \phi_1)] \\ A_2 \exp[j(2\pi f_0 n + \phi_2)] \\ \vdots \\ A_L \exp[j(2\pi f_0 n + \phi_L)] \end{bmatrix}.$$

Each channel of the multichannel sinusoid is characterized by a complex sinusoid of the same frequency but with a different amplitude and phase in general. Letting the vector of complex amplitudes be denoted by $\mathbf{A}_c$, where

$$\mathbf{A}_c = \begin{bmatrix} A_1 \exp(\phi_1) \\ A_2 \exp(\phi_2) \\ \vdots \\ A_L \exp(\phi_L) \end{bmatrix},$$

the multichannel sinusoid may be written as

$$\mathbf{u}[n] = \mathbf{A}_c \exp(j2\pi f_0 n). \tag{14.11}$$

If this signal is input to an LSI system, the output is found from (14.2) as

$$\begin{aligned} \mathbf{x}[n] &= \sum_{k=-\infty}^{\infty} \mathbf{H}[k]\mathbf{A}_c \exp[j2\pi f_0(n-k)] \\ &= \sum_{k=-\infty}^{\infty} \mathbf{H}[k] \exp(-j2\pi f_0 k)\mathbf{A}_c \exp(j2\pi f_0 n) \\ &= \mathbf{H}(f_0)\mathbf{u}[n]. \end{aligned}$$

As for scalar LSI systems the response of a multichannel LSI system to a complex sinusoid is a complex sinusoid of the same frequency but modified in amplitude and phase by the frequency response.

In contrast to the commutative property of scalar linear systems, the order in which multichannel linear systems are cascaded is important. Consider two linear systems with system functions $\mathcal{H}_1(z)$ and $\mathcal{H}_2(z)$. If the systems are cascaded to produce

$$\mathcal{X}(z) = \mathcal{H}_1(z)\mathcal{U}(z)$$

$$\mathcal{Y}(z) = \mathcal{H}_2(z)\mathcal{X}(z)$$

then the system function relating $\mathcal{U}(z)$ to $\mathcal{Y}(z)$ is $\mathcal{H}_2(z)\mathcal{H}_1(z)$. Since matrix multiplication is not commutative,

$$\mathcal{H}_1(z)\mathcal{H}_2(z) \neq \mathcal{H}_2(z)\mathcal{H}_1(z)$$

in general, so the order in which the systems are cascaded must be preserved.

A multichannel FIR filter is defined as

$$\mathbf{x}[n] = \sum_{k=0}^{N-1} \mathbf{H}[k]\mathbf{u}[n-k] \tag{14.12}$$

or the output of the ith channel is

$$x_i[n] = \sum_{j=1}^{L} \sum_{k=0}^{N-1} h_{ij}[k]u_j[n - k].$$

The system function is

$$\mathcal{H}(z) = \sum_{k=0}^{N-1} \mathbf{H}[k]z^{-k} \tag{14.13}$$

and the system will always be stable and causal. A recursive multichannel filter that is similar in appearance to the scalar all-pole filter is defined by

$$\mathbf{x}[n] = -\sum_{i=1}^{p} \mathbf{A}[i]\mathbf{x}[n - i] + \mathbf{u}[n].$$

The coefficient matrices $\mathbf{A}[i]$ are $L \times L$. To find the system function of the recursive filter, we can make use of (14.7) and (14.8) to yield

$$\mathcal{X}(z) = -\sum_{i=1}^{p} \mathbf{A}[i]z^{-i}\mathcal{X}(z) + \mathcal{U}(z)$$

or

$$\mathcal{H}(z) = \mathcal{A}^{-1}(z) = (\mathbf{I} + \sum_{i=1}^{p} \mathbf{A}[i]z^{-i})^{-1}.$$

In contrast to the scalar case the elements of $\mathcal{H}(z)$ will in general be rational functions of z^{-1} with *zeros* as well as poles. As an example, if $L = 2$, $p = 1$ and

$$\mathbf{A}[1] = \begin{bmatrix} -1 & 1 \\ \frac{1}{2} & -\frac{3}{2} \end{bmatrix}$$

then

$$\mathcal{H}(z) = \begin{bmatrix} 1 - z^{-1} & z^{-1} \\ \frac{1}{2}z^{-1} & 1 - \frac{3}{2}z^{-1} \end{bmatrix}^{-1}.$$

Using (2.30), the matrix may be inverted to yield

$$\begin{aligned} \mathcal{H}(z) &= \frac{\mathbf{C}^T}{\det(\mathcal{A}(z))} \\ &= \frac{\begin{bmatrix} 1 - \frac{3}{2}z^{-1} & -z^{-1} \\ -\frac{1}{2}z^{-1} & 1 - z^{-1} \end{bmatrix}}{1 - \frac{5}{2}z^{-1} + z^{-2}} \\ &= \begin{bmatrix} \dfrac{1 - \frac{3}{2}z^{-1}}{1 - \frac{5}{2}z^{-1} + z^{-2}} & \dfrac{-z^{-1}}{1 - \frac{5}{2}z^{-1} + z^{-2}} \\ \dfrac{-\frac{1}{2}z^{-1}}{1 - \frac{5}{2}z^{-1} + z^{-2}} & \dfrac{1 - z^{-1}}{1 - \frac{5}{2}z^{-1} + z^{-2}} \end{bmatrix} \end{aligned} \tag{14.14}$$

which is observed to contain zeros. Also, the filter is unstable since the poles are at $z = \frac{1}{2}$ and $z = 2$. The denominator polynomial in (14.14), which determines the poles, is equal to det ($\mathcal{A}(z)$). In general, for a stable recursive filter the Lp roots of the polynomial

$$\det\left(\mathbf{I} + \sum_{i=1}^{p} \mathbf{A}[i]z^{-i}\right) \tag{14.15}$$

must lie within the unit circle. As a special case, if $p = 1$ the eigenvalues of $\mathbf{A}[1]$ must be less than 1 in magnitude for stability (see Problem 14.3). This is the multichannel equivalent of requiring $|a[1]| < 1$ for a first order all-pole filter. This is also apparent if we examine the impulse response

$$\mathbf{H}[k] = \begin{cases} (-\mathbf{A}[1])^k & \text{for } k \geq 0 \\ 0 & \text{for } k < 0 \end{cases} \tag{14.16}$$

(see Problem 14.4).

A more general recursive filter is the multichannel equivalent of a pole–zero filter. It is described by the difference equation

$$\mathbf{x}[n] = -\sum_{i=1}^{p} \mathbf{A}[i]\mathbf{x}[n-i] + \mathbf{u}[n] + \sum_{i=1}^{q} \mathbf{B}[i]\mathbf{u}[n-i]. \tag{14.17}$$

The system function may be shown to be (see Problem 14.5)

$$\begin{aligned} \mathcal{H}(z) &= \mathcal{A}^{-1}(z)\mathcal{B}(z) \\ &= \left(\mathbf{I} + \sum_{i=1}^{p} \mathbf{A}[i]z^{-i}\right)^{-1}\left(\mathbf{I} + \sum_{i=1}^{q} \mathbf{B}[i]z^{-i}\right). \end{aligned} \tag{14.18}$$

The system is stable if and only if det $[\mathcal{A}(z)] = 0$ has all its roots within the unit circle.

14.4 REVIEW OF RANDOM PROCESSES

It will be assumed that $\mathbf{x}[n]$ is a multichannel WSS random process. That is, the mean of the process is a constant and the ACF depends only on the lag. Specifically,

$$\mathcal{E}(\mathbf{x}[n]) = \boldsymbol{\mu}[n] = \boldsymbol{\mu}$$

which will henceforth be assumed equal to zero, and the ACF

$$\mathbf{R}_{xx}[k] = \mathcal{E}(\mathbf{x}^*[n]\mathbf{x}^T[n+k]) \tag{14.19}$$

does not depend on n. The kth sample of the ACF is the $L \times L$ matrix

$$\mathbf{R}_{xx}[k] = \begin{bmatrix} r_{11}[k] & r_{12}[k] & \cdots & r_{1L}[k] \\ r_{21}[k] & r_{22}[k] & \cdots & r_{2L}[k] \\ \vdots & \vdots & \ddots & \vdots \\ r_{L1}[k] & r_{L2}[k] & \cdots & r_{LL}[k] \end{bmatrix} \tag{14.20}$$

where the $[i, j]$ element is the CCF between $x_i[n]$ and $x_j[n]$ at lag k or

$$r_{ij}[k] = \mathcal{E}(x_i^*[n]x_j[n + k]). \tag{14.21}$$

Because

$$r_{ji}^*[k] = r_{ij}[-k] \neq r_{ij}[k]$$

the matrix in (14.20) is not hermitian unless $k = 0$. It does, however, have the property that

$$\mathbf{R}_{xx}^H[k] = \mathbf{R}_{xx}[-k] \tag{14.22}$$

which follows from

$$\begin{aligned}
\mathbf{R}_{xx}^H[k] &= \mathcal{E}^H(\mathbf{x}^*[n]\mathbf{x}^T[n + k]) \\
&= \mathcal{E}(\mathbf{x}^*[n + k]\mathbf{x}^T[n]) \\
&= \mathcal{E}(\mathbf{x}^*[n]\mathbf{x}^T[n - k]) \\
&= \mathbf{R}_{xx}[-k].
\end{aligned}$$

As a consequence of (14.22), for $k = 0$, $\mathbf{R}_{xx}^H[0] = \mathbf{R}_{xx}[0]$ or the zeroth lag is hermitian. Furthermore, $\mathbf{R}_{xx}[0]$ is positive semidefinite since for any $\boldsymbol{\xi}$ an $L \times 1$ vector

$$\begin{aligned}
\boldsymbol{\xi}^H\mathbf{R}_{xx}[0]\boldsymbol{\xi} &= \mathcal{E}(\boldsymbol{\xi}^H\mathbf{x}^*[n]\mathbf{x}^T[n]\boldsymbol{\xi}) \\
&= \mathcal{E}(|\boldsymbol{\xi}^H\mathbf{x}^*[n]|^2) \geq 0.
\end{aligned}$$

The power spectral density matrix or *cross-spectral matrix* is defined as

$$\mathbf{P}_{xx}(f) = \begin{bmatrix} P_{11}(f) & P_{12}(f) & \cdots & P_{1L}(f) \\ P_{21}(f) & P_{22}(f) & \cdots & P_{2L}(f) \\ \vdots & \vdots & \ddots & \vdots \\ P_{L1}(f) & P_{L2}(f) & \cdots & P_{LL}(f) \end{bmatrix}. \tag{14.23}$$

The diagonal elements $P_{ii}(f)$ are the PSDs of the individual channels or auto-PSDs, while the off-diagonal elements $P_{ij}(f)$ for $i \neq j$ are the cross-PSDs between $x_i[n]$ and $x_j[n]$, which are defined as

$$P_{ij}(f) = \sum_{k=-\infty}^{\infty} r_{ij}[k] \exp(-j2\pi fk).$$

The magnitude of the cross-PSD describes whether frequency components in $x_i[n]$ are associated with large or small amplitudes at the same frequency in $x_j[n]$, and the phase of the cross-PSD indicates the phase lag or lead of $x_i[n]$ with respect to $x_j[n]$ for a given frequency component. The cross-spectral matrix may also be written as

$$\mathbf{P}_{xx}(f) = \sum_{k=-\infty}^{\infty} \mathbf{R}_{xx}[k] \exp(-j2\pi fk). \tag{14.24}$$

It may be shown that (see Problem 14.6)

$$P_{ji}(f) = P_{ij}^*(f) \tag{14.25}$$

which is a result of the CCF property $r_{ji}^*[k] = r_{ij}[-k]$. Hence the cross-spectral matrix is hermitian. The cross-spectral matrix is also positive semidefinite. To prove this property, define the scalar process $y[n]$ as $y[n] = \boldsymbol{\xi}^T\mathbf{x}[n]$, where $\boldsymbol{\xi}$ is an $L \times 1$ vector. Then

$$\begin{aligned} r_{yy}[k] &= \mathcal{E}(y^*[n]y[n+k]) \\ &= \mathcal{E}(\boldsymbol{\xi}^H\mathbf{x}^*[n]\mathbf{x}^T[n+k]\boldsymbol{\xi}] \\ &= \boldsymbol{\xi}^H\mathbf{R}_{xx}[k]\boldsymbol{\xi}. \end{aligned}$$

Taking the Fourier transform produces

$$P_{yy}(f) = \boldsymbol{\xi}^H\mathbf{P}_{xx}(f)\boldsymbol{\xi} \geq 0$$

so that the cross-spectral matrix is positive semidefinite. As a consequence, the principal minors are nonnegative. This ensures, for instance, that the *magnitude squared coherence* between channels 1 and 2, which is defined as

$$|\gamma_{12}(f)|^2 = \frac{|P_{12}(f)|^2}{P_{11}(f)P_{22}(f)} \tag{14.26}$$

is bounded by 1. Equation (14.26) is easily verified by noting that the determinant of the 2×2 cross-spectral matrix is nonnegative. A physical interpretation of the coherence $\gamma_{12}(f)$ as a correlation coefficient is provided by Jenkins and Watts [1968].

Multichannel white noise is defined as the process whose ACF satisfies

$$\mathbf{R}_{xx}[k] = \boldsymbol{\Sigma}\delta[k] \tag{14.27}$$

so that the cross-spectral matrix is a constant matrix or

$$\mathbf{P}_{xx}(f) = \boldsymbol{\Sigma}. \tag{14.28}$$

For multichannel white noise the individual processes are each white noise processes with variance or PSD $[\boldsymbol{\Sigma}]_{ii}$. The cross-correlation between the processes is zero except at the same time instant where they are correlated with a cross-correlation given by the off-diagonal elements of $\boldsymbol{\Sigma}$.

An important relationship in spectral estimation of a scalar time series is that between the input PSD and output PSD of a LSI filter as given in (3.51). The multichannel version is now derived. From (14.2) the output of a multichannel filter is

$$\mathbf{x}[n] = \sum_{k=-\infty}^{\infty} \mathbf{H}[k]\mathbf{u}[n-k].$$

The ACF becomes

$$\begin{aligned}\mathbf{R}_{xx}[k] &= \mathscr{E}(\mathbf{x}^*[n]\mathbf{x}^T[n+k]) \\ &= \mathscr{E}\left(\sum_{i=-\infty}^{\infty}\sum_{j=-\infty}^{\infty}\mathbf{H}^*[i]\mathbf{u}^*[n-i]\mathbf{u}^T[n+k-j]\mathbf{H}^T[j]\right) \\ &= \sum_{i=-\infty}^{\infty}\sum_{j=-\infty}^{\infty}\mathbf{H}^*[i]\mathbf{R}_{uu}[k+i-j]\mathbf{H}^T[j]. \end{aligned} \tag{14.29}$$

Taking the z-transform yields

$$\mathscr{P}_{xx}(z) = \sum_{i=-\infty}^{\infty}\mathbf{H}^*[i]z^i \sum_{k=-\infty}^{\infty}\mathbf{R}_{uu}[k+i-j]z^{-(k+i-j)}\sum_{j=-\infty}^{\infty}\mathbf{H}^T[j]z^{-j}$$

and finally making use of the definition of the system function and cross-spectral matrices

$$\mathscr{P}_{xx}(z) = \mathscr{H}^*(1/z^*)\,\mathscr{P}_{uu}(z)\mathscr{H}^T(z). \tag{14.30}$$

The cross-spectral matrix is obtained by evaluating (14.30) on the unit circle to yield

$$\mathbf{P}_{xx}(f) = \mathbf{H}^*(f)\mathbf{P}_{uu}(f)\mathbf{H}^T(f). \tag{14.31}$$

14.5 CLASSICAL SPECTRAL ESTIMATION

The single channel methods of classical spectral estimation described in Chapter 4 easily generalize to the multichannel case. An estimator of the cross-PSD $P_{ij}(f)$ is provided by the cross-periodogram [Jenkins and Watts 1968]

$$\hat{P}_{ij}(f) = \frac{1}{N}X_i^*(f)X_j(f) \qquad i,j = 1, 2, \ldots, L \tag{14.32}$$

where

$$X_i(f) = \sum_{n=0}^{N-1} x_i[n]\exp(-j2\pi fn).$$

For the auto-PSD or for $i = j$ this reduces to the usual periodogram. In matrix notation the cross-spectral matrix is estimated as

$$\hat{\mathbf{P}}_{\text{PER}}(f) = \frac{1}{N}\mathbf{X}^*(f)\mathbf{X}^T(f) \tag{14.33}$$

where

$$\mathbf{X}(f) = \sum_{n=0}^{N-1}\mathbf{x}[n]\exp(-j2\pi fn). \tag{14.34}$$

As described in Chapter 4, the statistical properties of the periodogram are poor. The same arguments are valid for the cross-periodograms [Jenkins and Watts 1968]. In particular, $\hat{\mathbf{P}}_{\text{PER}}(f)$ is only a rank 1 matrix and hence is not positive definite (see Problem 2.9). It is easily shown that as a consequence the magnitude squared coherence estimate is always unity (see Problem 14.7). To improve the statistical reliability of the periodogram estimator, we should average several periodograms together. Assuming that K blocks of data are available, the averaged periodogram spectral estimator is defined as

$$\hat{\mathbf{P}}_{\text{AVPER}}(f) = \frac{1}{NK} \sum_{i=0}^{K-1} \mathbf{X}^{(i)^*}(f)\mathbf{X}^{(i)^T}(f) \tag{14.35}$$

where $\mathbf{X}^{(i)}(f)$ is the Fourier transform of the ith block of data. It should be noted that the rank of $\hat{\mathbf{P}}_{\text{AVPER}}(f)$ is K since K dyads have been added together (see Problem 2.9). For a full rank and thus a positive definite matrix, at least $K = L$ periodograms need to be averaged.

As in the single channel case the cross-periodogram can be shown to be the Fourier transform of the estimated CCF. Specifically (see Problem 14.8),

$$\hat{P}_{ij}(f) = \sum_{k=-(N-1)}^{N-1} \hat{r}_{ij}[k] \exp(-j2\pi fk) \tag{14.36}$$

where $\hat{r}_{ij}[k]$ is the biased estimator of the CCF given by

$$\hat{r}_{ij}[k] = \begin{cases} \dfrac{1}{N} \displaystyle\sum_{n=0}^{N-1-k} x_i^*[n]x_j[n+k] \\ \quad \text{for } k = 0, 1, \ldots, N-1 \\ \dfrac{1}{N} \displaystyle\sum_{n=-k}^{N-1} x_i^*[n]x_j[n+k] \\ \quad \text{for } k = -(N-1), -(N-2), \ldots, -1. \end{cases} \tag{14.37}$$

The estimated CCF satisfies the theoretical constraint that $\hat{r}_{ij}[-k] = \hat{r}_{ji}^*[k]$. The computer program CORRELATION provided in Appendix 4D can be used to compute (14.37). A problem which occurs with estimation of the CCF is that of time alignment. From (14.36) it is observed that the samples of the CCF for $|k| \geq N$ are estimated by zeros. We would hope that these samples are indeed small so as not to unduly bias the spectral estimator. In contrast to the ACF, in which the peak value must occur at $k = 0$, no such property exists for the CCF. For example, if $L = 2$ and $x_2[n]$ is a delayed replica of $x_1[n]$ or $x_2[n] = x_1[n - n_0]$, then

$$r_{12}[k] = \mathcal{E}(x_1^*[n]x_2[n+k]) = r_{11}[k - n_0].$$

The CCF will have its maximum value at $k = n_0$. If the delay n_0 is not small relative to $N - 1$, the estimator given by (14.36) will be severely biased. To avoid this problem it is advisable to time align the two sequences prior to spectral estimation [Jenkins and Watts 1968]. The time alignment must be estimated since the peak of the CCF is never known a priori.

The relationship expressed in (14.36) may also be written in matrix notation as

$$\hat{\mathbf{P}}_{\text{PER}}(f) = \sum_{k=-(N-1)}^{N-1} \hat{\mathbf{R}}_{xx}[k] \exp(-j2\pi fk) \tag{14.38}$$

where

$$\hat{\mathbf{R}}_{xx}[k] = \begin{cases} \dfrac{1}{N} \displaystyle\sum_{n=0}^{N-1-k} \mathbf{x}^*[n]\mathbf{x}^T[n+k] \\ \text{for } k = 0, 1, \ldots, N-1 \\ \dfrac{1}{N} \displaystyle\sum_{n=-k}^{N-1} \mathbf{x}^*[n]\mathbf{x}^T[n+k] \\ \text{for } k = -(N-1), -(N-2), \ldots, -1. \end{cases} \tag{14.39}$$

The theoretical relationship given by (14.22) is satisfied by this estimator or $\hat{\mathbf{R}}_{xx}[-k] = \hat{\mathbf{R}}_{xx}^H[k]$. $\hat{\mathbf{R}}_{xx}[k]$ is the multichannel version of the biased ACF estimator given by (4.20).

The multichannel Blackman–Tukey (BT) spectral estimator is defined as [Parzen 1969]

$$\hat{P}_{ij}(f) = \sum_{k=-M}^{M} w_{ij}[k]\hat{r}_{ij}[k] \exp(-j2\pi fk) \qquad i, j = 1, 2, \ldots, L. \tag{14.40}$$

$\hat{r}_{ij}[k]$ is given by (14.37) and the lag windows $w_{ij}[k]$ should satisfy the usual properties of the single channel lag window (see Section 4.5). If all the lag windows are identical, the BT spectral estimator may be written succinctly as

$$\hat{\mathbf{P}}_{\text{BT}}(f) = \sum_{k=-M}^{M} w[k]\hat{\mathbf{R}}_{xx}[k] \exp(-j2\pi fk). \tag{14.41}$$

The same considerations involving the bias–variance trade-off in the choice of lag windows as described for the single channel case apply here as well (see Chapter 4).

14.6 RATIONAL TRANSFER FUNCTION MODELS

The rational transfer function or time series models described in Chapter 5 are easily extended to the multichannel case. The most general difference equation relating a WSS multichannel output process $\mathbf{x}[n]$ to a WSS multichannel input process $\mathbf{u}[n]$ is

$$\mathbf{x}[n] = -\sum_{i=1}^{p} \mathbf{A}[i]\mathbf{x}[n-i] + \mathbf{u}[n] + \sum_{i=1}^{q} \mathbf{B}[i]\mathbf{u}[n-i]. \tag{14.42}$$

where $\mathbf{A}[i]$, $\mathbf{B}[i]$ are $L \times L$ coefficient matrices. $\mathbf{x}[n]$ describes a multichannel ARMA(p, q) process if the input process is white noise or

$$\mathbf{R}_{uu}[k] = \mathbf{\Sigma}\delta[k].$$

The cross-spectral matrix follows from (14.18) and (14.31) as

$$\mathbf{P}_{\mathrm{ARMA}}(f) = \mathbf{A}^{*^{-1}}(f)\mathbf{B}^{*}(f)\mathbf{\Sigma}\mathbf{B}^{T}(f)\mathbf{A}^{T^{-1}}(f) \tag{14.43}$$

where

$$\mathbf{A}(f) = \mathbf{I} + \sum_{i=1}^{p} \mathbf{A}[i] \exp(-j2\pi f i)$$

$$\mathbf{B}(f) = \mathbf{I} + \sum_{i=1}^{q} \mathbf{B}[i] \exp(-j2\pi f i).$$

If the $\mathbf{B}[i]$'s are zero, (14.42) reduces to a multichannel AR(p) process described by

$$\mathbf{x}[n] = -\sum_{i=1}^{p} \mathbf{A}[i]\mathbf{x}[n-i] + \mathbf{u}[n] \tag{14.44}$$

with cross-spectral matrix

$$\mathbf{P}_{\mathrm{AR}}(f) = \mathbf{A}^{*^{-1}}(f)\mathbf{\Sigma}\mathbf{A}^{T^{-1}}(f). \tag{14.45}$$

Similarly, if the $\mathbf{A}[i]$'s are zero, the multichannel MA(q) process results

$$\mathbf{x}[n] = \mathbf{u}[n] + \sum_{i=1}^{q} \mathbf{B}[i]\mathbf{u}[n-i] \tag{14.46}$$

with cross-spectral matrix

$$\mathbf{P}_{\mathrm{MA}}(f) = \mathbf{B}^{*}(f)\mathbf{\Sigma}\mathbf{B}^{T}(f). \tag{14.47}$$

Multichannel Yule–Walker equations that relate the ACF to the parameters of the model may be derived in a similar fashion to the derivation in Chapter 5. To do so, first use (14.42) to yield

$$\begin{aligned}\mathscr{E}(\mathbf{x}^{*}[n]\mathbf{x}^{T}[n+k]) = &-\sum_{i=1}^{p} \mathscr{E}(\mathbf{x}^{*}[n]\mathbf{x}^{T}[n+k-i])\mathbf{A}^{T}[i] + \mathscr{E}(\mathbf{x}^{*}[n]\mathbf{u}^{T}[n+k])\\ &+\sum_{i=1}^{q} \mathscr{E}(\mathbf{x}^{*}[n]\mathbf{u}^{T}[n+k-i])\mathbf{B}^{T}[i]\end{aligned}$$

or

$$\mathbf{R}_{xx}[k] = -\sum_{i=1}^{p} \mathbf{R}_{xx}[k-i]\mathbf{A}^{T}[i] + \mathbf{R}_{xu}[k] + \sum_{i=1}^{q} \mathbf{R}_{xu}[k-i]\mathbf{B}^{T}[i]. \tag{14.48}$$

The multichannel CCF $\mathbf{R}_{xu}[k] = \mathscr{E}(\mathbf{x}^*[n]\mathbf{u}^T[n+k])$ may be evaluated by noting that $\mathbf{x}[n]$ is the output of a causal filter

$$\mathbf{x}[n] = \sum_{i=0}^{\infty} \mathbf{H}[i]\mathbf{u}[n-i].$$

Then

$$\begin{aligned}
\mathbf{R}_{xu}[k] &= \sum_{i=0}^{\infty} \mathbf{H}^*[i]\mathscr{E}(\mathbf{u}^*[n-i]\mathbf{u}^T[n+k]) \\
&= \mathbf{H}^*[-k]\mathbf{\Sigma} \\
&= \begin{cases} \mathbf{\Sigma} & \text{for } k = 0 \\ \mathbf{0} & \text{for } k > 0 \end{cases}
\end{aligned}$$

since $\mathbf{H}[0] = \mathbf{I}$. Finally, the multichannel Yule–Walker equations for an ARMA(p, q) process are

$$\mathbf{R}_{xx}[k] = \begin{cases} -\sum_{i=1}^{p} \mathbf{R}_{xx}[k-i]\mathbf{A}^T[i] + \sum_{i=0}^{q-k} \mathbf{H}^*[i]\mathbf{\Sigma}\mathbf{B}^T[i+k] \\ \text{for } k = 0, 1, \ldots, q \\ \\ -\sum_{i=1}^{p} \mathbf{R}_{xx}[k-i]\mathbf{A}^T[i] \\ \text{for } k \geq q+1. \end{cases} \tag{14.49}$$

$\mathbf{B}[0]$ is defined to be $\mathbf{I}$ in (14.49). The Yule–Walker equations for an AR(p) process are found by setting $\mathbf{B}[i] = \mathbf{0}$ for $i = 1, \ldots, q$ to yield

$$\mathbf{R}_{xx}[k] = \begin{cases} -\sum_{i=1}^{p} \mathbf{R}_{xx}[k-i]\mathbf{A}^T[i] & \text{for } k \geq 1 \\ \\ -\sum_{i=1}^{p} \mathbf{R}_{xx}[-i]\mathbf{A}^T[i] + \mathbf{\Sigma} & \text{for } k = 0. \end{cases} \tag{14.50}$$

An example of the use of these equations to determine the ACF is given in Problems 14.9 and 14.10. Finally, the Yule–Walker equations for an MA(q) process follow from (14.49) by setting $\mathbf{A}[i]$ equal to $\mathbf{0}$ to yield

$$\begin{aligned}
\mathbf{R}_{xx}[k] &= \sum_{i=0}^{q-k} \mathbf{H}^*[i]\mathbf{\Sigma}\mathbf{B}^T[i+k] \qquad k = 0, 1, \ldots, q \\
&= \sum_{i=0}^{q-k} \mathbf{B}^*[i]\mathbf{\Sigma}\mathbf{B}^T[i+k]
\end{aligned}$$

or

$$\mathbf{R}_{xx}[k] = \begin{cases} \sum_{i=0}^{q-k} \mathbf{B}^*[i]\mathbf{\Sigma}\mathbf{B}^T[i+k] & \text{for } k = 0, 1, \ldots, q \\ \mathbf{R}_{xx}^H[-k] & \text{for } k = -q, -q+1, \ldots, -1 \\ \mathbf{0} & \text{for } |k| > q. \end{cases} \tag{14.51}$$

These equations should be compared to the single channel relationships of (5.16), (5.17), and (5.20).

14.7 AUTOREGRESSIVE SPECTRAL ESTIMATION

The estimate of the cross-spectral matrix based on a multichannel AR model is given by

$$\hat{\mathbf{P}}_{\mathrm{AR}}(f) = \hat{\mathbf{A}}^{*^{-1}}(f)\hat{\mathbf{\Sigma}}\hat{\mathbf{A}}^{T^{-1}}(f)$$

where the "hats" denote estimators. The most straightforward means of estimating the unknown parameters is by using the Yule–Walker equations as given by (14.50) with a suitable ACF estimator. In matrix form the Yule–Walker equations become

$$\underbrace{\begin{bmatrix} \mathbf{R}_{xx}[0] & \mathbf{R}_{xx}[-1] & \cdots & \mathbf{R}_{xx}[-(p-1)] \\ \mathbf{R}_{xx}[1] & \mathbf{R}_{xx}[0] & \cdots & \mathbf{R}_{xx}[-(p-2)] \\ \vdots & \vdots & \ddots & \vdots \\ \mathbf{R}_{xx}[p-1] & \mathbf{R}_{xx}[p-2] & \cdots & \mathbf{R}_{xx}[0] \end{bmatrix}}_{\underline{\mathbf{R}}_{xx}} \begin{bmatrix} \mathbf{A}^T[1] \\ \mathbf{A}^T[2] \\ \vdots \\ \mathbf{A}^T[p] \end{bmatrix} = - \begin{bmatrix} \mathbf{R}_{xx}[1] \\ \mathbf{R}_{xx}[2] \\ \vdots \\ \mathbf{R}_{xx}[p] \end{bmatrix}. \tag{14.52}$$

The dimensions of the matrices from left to right are $pL \times pL$, $pL \times L$, and $pL \times L$, respectively. The solution of these equations requires that we solve L sets of simultaneous linear equations, each set containing pL equations. Note that $\underline{\mathbf{R}}_{xx}$ is hermitian since $\mathbf{R}_{xx}[-k] = \mathbf{R}_{xx}^H[k]$ and can be shown to be positive definite for processes that are not perfectly predictable [Nuttall 1976]. It is also block Toeplitz, which allows us to use the multichannel Levinson recursion or LWR algorithm [Wiggins and Robinson 1965] to solve the equations efficiently. A major difference between the Levinson recursion for scalar systems and for multichannel systems is that in the latter case the backward predictor coefficients are not just the time-reversed and complex-conjugated version of the forward predictor coefficients [see (6.37)]. As a consequence, the Levinson recursion produces distinctly different forward and backward prediction error filters and prediction error powers. Also, distinct reflection coefficient matrices arise. To illustrate this difference, consider the problem of determining the optimal pth order forward and backward linear predictors for an AR(p) process. The forward predictor is

$$\hat{\mathbf{x}}^f[n] = -\sum_{i=1}^{p} \mathbf{A}^f[i]\mathbf{x}[n-i]. \tag{14.53}$$

The prediction error power is defined as the sum of prediction error powers for the individual channels $x_1[n], x_2[n] \ldots, x_L[n]$, or

$$\rho^f = \mathscr{E}\{(\mathbf{x}[n] - \hat{\mathbf{x}}^f[n])^T(\mathbf{x}[n] - \hat{\mathbf{x}}^f[n])^*\}. \tag{14.54}$$

Alternatively, ρ^f may be viewed as the sum of the diagonal elements or trace of the covariance matrix

$$\mathbf{\Sigma}^f = \mathscr{E}\{(\mathbf{x}[n] - \hat{\mathbf{x}}^f[n])^*(\mathbf{x}[n] - \hat{\mathbf{x}}^f[n])^T\} \tag{14.55}$$

since for any $L \times 1$ vectors

$$\mathbf{y}^T\mathbf{x}^* = \text{trace}\,(\mathbf{x}^*\mathbf{y}^T).$$

The diagonal elements of $\mathbf{\Sigma}^f$ are the prediction error powers or the variances of the individual prediction errors while the off-diagonal elements are the covariances between the prediction errors. To minimize ρ^f, we can use the orthogonality principle for a vector space composed of *random vectors* [Luenberger 1969]. The inner product between two complex random vectors is defined as

$$\langle \mathbf{x}, \mathbf{y} \rangle = \mathscr{E}(\mathbf{y}^T\mathbf{x}^*) = \text{trace}\,(\mathscr{E}[\mathbf{x}^*\mathbf{y}^T]). \tag{14.56}$$

The orthogonality principle then yields (see Problem 14.11)

$$\mathscr{E}\{\mathbf{x}^*[n-k](\mathbf{x}[n] - \hat{\mathbf{x}}^f[n])^T\} = \mathbf{0} \qquad k = 1, 2, \ldots, p$$

which results in the multichannel Wiener–Hopf equations

$$\mathbf{R}_{xx}[k] = -\sum_{i=1}^{p} \mathbf{R}_{xx}[k-i]\mathbf{A}^{f^T}[i] \qquad k = 1, 2, \ldots, p. \tag{14.57}$$

The prediction error power matrix is

$$\begin{aligned}\mathbf{\Sigma}^f &= \mathscr{E}\{\mathbf{x}^*[n](\mathbf{x}[n] - \hat{\mathbf{x}}^f[n])^T\} \\ &= \mathbf{R}_{xx}[0] + \sum_{i=1}^{p} \mathbf{R}_{xx}[-i]\mathbf{A}^{f^T}[i].\end{aligned} \tag{14.58}$$

As expected, $\mathbf{A}^f[i] = \mathbf{A}[i]$ and $\mathbf{\Sigma}^f = \mathbf{\Sigma}$ or the prediction coefficients are given by the AR filter parameters and the prediction error power matrix is given by the variance matrix of the white noise.

The pth order backward linear predictor is defined as

$$\hat{\mathbf{x}}^b[n] = -\sum_{i=1}^{p} \mathbf{A}^b[i]\mathbf{x}[n+i]. \tag{14.59}$$

The corresponding prediction error power is

$$\rho^b = \mathscr{E}\{(\mathbf{x}[n] - \hat{\mathbf{x}}^b[n])^T(\mathbf{x}[n] - \hat{\mathbf{x}}^b[n])^*\} \tag{14.60}$$

and the prediction error power matrix is

$$\mathbf{\Sigma}^b = \mathscr{E}\{(\mathbf{x}[n] - \hat{\mathbf{x}}^b[n])^*(\mathbf{x}[n] - \hat{\mathbf{x}}^b[n])^T\}. \tag{14.61}$$

Using a similar development the Wiener–Hopf equations become

$$\begin{aligned} \mathbf{R}_{xx}[-k] &= -\sum_{i=1}^{p} \mathbf{R}_{xx}[-k+i]\mathbf{A}^{b^T}[i] \qquad k = 1, 2, \ldots, p \\ \mathbf{\Sigma}^b &= \mathbf{R}_{xx}[0] + \sum_{i=1}^{p} \mathbf{R}_{xx}[i]\mathbf{A}^{b^T}[i]. \end{aligned} \tag{14.62}$$

That the forward and backward prediction coefficients are not related may be observed from an examination of the first-order predictors. From (14.57) and (14.62) we have

$$\begin{aligned} \mathbf{R}_{xx}[0]\mathbf{A}^{f^T}[1] &= -\mathbf{R}_{xx}[1] \\ \mathbf{R}_{xx}[0]\mathbf{A}^{b^T}[1] &= -\mathbf{R}_{xx}[-1] \end{aligned}$$

or

$$\begin{aligned} \mathbf{A}^{f^T}[1] &= -\mathbf{R}_{xx}^{-1}[0]\mathbf{R}_{xx}[1] \\ \mathbf{A}^{b^T}[1] &= -\mathbf{R}_{xx}^{-1}[0]\mathbf{R}_{xx}^{H}[1]. \end{aligned}$$

In the single channel case the coefficients are complex conjugates of each other [see (6.37)]. However, for multichannel prediction no such relationship exists since

$$(\mathbf{A}^{b^T}[1])^H = -\mathbf{R}_{xx}[1]\mathbf{R}_{xx}^{-1}[0] \neq -\mathbf{R}_{xx}^{-1}[0]\mathbf{R}_{xx}[1] = \mathbf{A}^{f^T}[1].$$

The multichannel Levinson algorithm, which is sometimes referred to as the Levinson-Wiggins-Robinson (LWR) algorithm [Wiggins and Robinson 1965], is derived in Appendix 14A. It is summarized below.

1. Initialization:

$$\begin{aligned} \mathbf{A}_1^{f^T}[1] = \mathbf{K}_1^{f^T} &= -\mathbf{R}_{xx}^{-1}[0]\mathbf{R}_{xx}[1] \\ \mathbf{A}_1^{b^T}[1] = \mathbf{K}_1^{b^T} &= -\mathbf{R}_{xx}^{-1}[0]\mathbf{R}_{xx}[-1] \\ \mathbf{\Sigma}_1^f &= \mathbf{R}_{xx}[0][\mathbf{I} - \mathbf{K}_1^{b^T}\mathbf{K}_1^{f^T}] \\ \mathbf{\Sigma}_1^b &= \mathbf{R}_{xx}[0][\mathbf{I} - \mathbf{K}_1^{f^T}\mathbf{K}_1^{b^T}] \end{aligned} \tag{14.63}$$

for $k = 2, 3, \ldots, p$.

2. Reflection coefficients matrices:

$$\begin{aligned} \mathbf{K}_k^{f^T} &= -\mathbf{\Sigma}_{k-1}^{b^{-1}}\mathbf{\Delta}_k^f \\ \mathbf{K}_k^{b^T} &= -\mathbf{\Sigma}_{k-1}^{f^{-1}}\mathbf{\Delta}_k^{f^H} \\ \mathbf{\Delta}_k^f &= \sum_{i=0}^{k-1} \mathbf{R}_{xx}[k-i]\mathbf{A}_{k-1}^{f^T}[i]. \end{aligned} \tag{14.64}$$

3. Predictor coefficient matrices:

$$\mathbf{A}_k^f[i] = \begin{cases} \mathbf{A}_{k-1}^f[i] + \mathbf{K}_k^f \mathbf{A}_{k-1}^b[k-i] & \text{for } i = 1, 2, \ldots, k-1 \\ \mathbf{K}_k^f & \text{for } i = k \end{cases}$$

$$\mathbf{A}_k^b[i] = \begin{cases} \mathbf{A}_{k-1}^b[i] + \mathbf{K}_k^b \mathbf{A}_{k-1}^f[k-i] & \text{for } i = 1, 2, \ldots, k-1 \\ \mathbf{K}_k^b & \text{for } i = k. \end{cases}$$

(14.65)

4. Prediction error power matrices:

$$\begin{aligned} \mathbf{\Sigma}_k^f &= \mathbf{\Sigma}_{k-1}^f[\mathbf{I} - \mathbf{K}_k^{b^T}\mathbf{K}_k^{f^T}] \\ \mathbf{\Sigma}_k^b &= \mathbf{\Sigma}_{k-1}^b[\mathbf{I} - \mathbf{K}_k^{f^T}\mathbf{K}_k^{b^T}]. \end{aligned} \tag{14.66}$$

The solution to the multichannel Yule–Walker equations is

$$\begin{aligned} \mathbf{A}[i] &= \mathbf{A}_p^f[i] \qquad i = 1, 2, \ldots, p \\ \mathbf{\Sigma} &= \mathbf{\Sigma}_p^f. \end{aligned}$$

$\mathbf{A}_k^f[i]$, $\mathbf{A}_k^b[i]$ are the ith prediction coefficients for the kth order multichannel forward and backward linear predictors, respectively. $\mathbf{K}_k^f$, $\mathbf{K}_k^b$ are the multichannel reflection coefficient matrices and $\mathbf{\Sigma}_k^f$, $\mathbf{\Sigma}_k^b$ are the multichannel prediction error power matrices for the kth order predictors. Note that both the forward and backward prediction error filters and associated prediction error powers must be computed at each step of the recursion. A comparison should be made to the scalar Levinson recursion given by (6.46)–(6.48) and summarized in Figure 6.4.

Estimation of the AR parameters may be accomplished by solving the Yule–Walker equations given by (14.52) with an estimated ACF replacing the theoretical one. The biased ACF estimator as given by (14.39) is recommended, as this choice will guarantee a positive definite and hence invertible matrix. The multichannel Levinson recursion may be used to solve the equations. As in the single-channel case, linear prediction approaches may be used (see Chapter 7). The autocorrelation method as well as the covariance method extend readily for multichannel estimation. However, because of the lack of a relationship between the forward and backward prediction coefficient matrices, no such extension is possible for the modified covariance method. Even Burg's method cannot be easily extended, although some proposed extensions may be found in the works by Nuttall [1976], Strand [1977], and Morf et al. [1978].

The multichannel autocorrelation and covariance methods estimate the AR parameters by minimizing the sum of the estimated prediction error powers of the individual channels or

$$\hat{\rho} = \text{trace}\,(\hat{\mathbf{\Sigma}})$$

where

$$\hat{\mathbf{\Sigma}} = \frac{1}{N'} \sum_n (\mathbf{x}[n] - \hat{\mathbf{x}}[n])^*(\mathbf{x}[n] - \hat{\mathbf{x}}[n])^T \tag{14.67}$$

and the predictor is given by

$$\hat{\mathbf{x}}[n] = -\sum_{i=1}^{p} \mathbf{A}[i]\mathbf{x}[n-i].$$

For the autocorrelation method the range of the summation in (14.67) is $-\infty < n < \infty$ and for the covariance method the range is $p \leq n \leq N-1$. Also, the value of N' is N for the autocorrelation method and $N-p$ for the covariance method. Hence, minimizing

$$\hat{\rho} = \operatorname{trace}\left\{\frac{1}{N'}\sum_{n}\left(\sum_{i=0}^{p} \mathbf{A}[i]\mathbf{x}[n-i]\right)^{*}\left(\sum_{i=0}^{p} \mathbf{A}[i]\mathbf{x}[n-i]\right)^{T}\right\} \tag{14.68}$$

results in estimates that are given as the solution of the equations (see Problem 14.12)

$$\frac{1}{N'}\sum_{n} \mathbf{x}^{*}[n-k]\left(\mathbf{x}[n] + \sum_{i=1}^{p} \mathbf{A}[i]\mathbf{x}[n-i]\right)^{T} = \mathbf{0} \qquad k = 1, 2, \ldots, p \tag{14.69}$$

or

$$\sum_{i=1}^{p} \mathbf{C}_{xx}[k, i]\hat{\mathbf{A}}^{T}[i] = -\mathbf{C}_{xx}[k, 0] \qquad k = 1, 2, \ldots, p$$

where

$$\mathbf{C}_{xx}[k, i] = \frac{1}{N'}\sum_{n} \mathbf{x}^{*}[n-k]\mathbf{x}^{T}[n-i].$$

The minimum prediction error power matrix follows from (14.67) and (14.69) as

$$\begin{aligned}\hat{\mathbf{\Sigma}} &= \frac{1}{N'}\sum_{n} \mathbf{x}^{*}[n]\left(\mathbf{x}[n] + \sum_{i=1}^{p} \hat{\mathbf{A}}[i]\mathbf{x}[n-i]\right)^{T} \\ &= \mathbf{C}_{xx}[0, 0] + \sum_{i=1}^{p} \mathbf{C}_{xx}[0, i]\hat{\mathbf{A}}^{T}[i].\end{aligned} \tag{14.70}$$

For the autocorrelation method, if we define $\mathbf{x}[n] = \mathbf{0}$ for $n < 0$ and $n > N-1$, then

$$\begin{aligned}\mathbf{C}_{xx}[k, i] &= \frac{1}{N}\sum_{n=-\infty}^{\infty} \mathbf{x}^{*}[n-k]\mathbf{x}^{T}[n-i] \\ &= \frac{1}{N}\sum_{n=-\infty}^{\infty} \mathbf{x}^{*}[n-k+i]\mathbf{x}^{T}[n] \\ &= \hat{\mathbf{R}}_{xx}[k-i]\end{aligned}$$

where $\hat{\mathbf{R}}_{xx}[k]$ is the biased ACF estimator given by (14.39). As expected, the autocorrelation or Yule–Walker method estimates the AR parameters by substituting the biased multichannel ACF estimator into the Yule–Walker equations and solving. The prediction error power matrix is given by (14.58) with the theoretical ACF replaced by the biased ACF estimate. For the covariance method

$$\mathbf{C}_{xx}[k, i] = \frac{1}{N-p} \sum_{n=p}^{N-1} \mathbf{x}^*[n-k]\mathbf{x}^T[n-i] \tag{14.71}$$

and the equations to be solved are

$$\underbrace{\begin{bmatrix} \mathbf{C}_{xx}[1,1] & \mathbf{C}_{xx}[1,2] & \cdots & \mathbf{C}_{xx}[1,p] \\ \mathbf{C}_{xx}[2,1] & \mathbf{C}_{xx}[2,2] & \cdots & \mathbf{C}_{xx}[2,p] \\ \vdots & \vdots & \ddots & \vdots \\ \mathbf{C}_{xx}[p,1] & \mathbf{C}_{xx}[p,2] & \cdots & \mathbf{C}_{xx}[p,p] \end{bmatrix}}_{\underline{\mathbf{C}}_{xx}} \begin{bmatrix} \hat{\mathbf{A}}^T[1] \\ \hat{\mathbf{A}}^T[2] \\ \vdots \\ \hat{\mathbf{A}}^T[p] \end{bmatrix} = - \begin{bmatrix} \mathbf{C}_{xx}[1,0] \\ \mathbf{C}_{xx}[2,0] \\ \vdots \\ \mathbf{C}_{xx}[p,0] \end{bmatrix} \tag{14.72}$$

and

$$\hat{\mathbf{\Sigma}} = \mathbf{C}_{xx}[0,0] + \sum_{i=1}^{p} \mathbf{C}_{xx}[0,i]\hat{\mathbf{A}}^T[i]. \tag{14.73}$$

Note that $\underline{\mathbf{C}}_{xx}$ is hermitian since $\mathbf{C}_{xx}[i, k] = \mathbf{C}_{xx}^H[k, i]$ but not block Toeplitz. Also, it can be shown to be positive semidefinite (see Problem 14.13), so that the Cholesky decomposition [see (2.53)–(2.55)] may be used to solve the equations (assuming that the matrix is nonsingular). Whereas the autocorrelation method guarantees a stable AR filter, the covariance method does not. Either linear prediction method can be shown to be an approximate MLE.

The determination of the appropriate AR model order is made by use of the AIC (see Appendix 7A). For real multichannel AR processes the AIC becomes [Jones 1974]

$$\mathrm{AIC}(i) = N \ln \det(\hat{\mathbf{\Sigma}}_i) + 2L^2 i \tag{14.74}$$

where $\hat{\mathbf{\Sigma}}_i$ is the estimate of the prediction error covariance matrix assuming an ith order predictor. The latter matrix is most easily found as a by-product of the autocorrelation method when the equations are solved using the Levinson recursion. It has been recommended that the AIC should be computed only for a maximum value of i of $3\sqrt{N}/L$ [Nuttall 1976] in order to produce reliable results.

14.8 AUTOREGRESSIVE MOVING AVERAGE SPECTRAL ESTIMATION

The estimate of the cross-spectral matrix based on a multichannel ARMA model is, from (14.43),

$$\hat{\mathbf{P}}_{\mathrm{ARMA}}(f) = \hat{\mathbf{A}}^{*^{-1}}(f)\hat{\mathbf{B}}^*(f)\hat{\mathbf{\Sigma}}\hat{\mathbf{B}}^T(f)\hat{\mathbf{A}}^{T^{-1}}(f). \tag{14.75}$$

The estimates of the AR and MA coefficient matrices as well as the white noise variance matrix may be obtained using similar procedures as for the scalar case.

Akaike's approximate MLE method for estimation of the parameters of scalar ARMA processes has been extended to the multichannel case [Akaike 1973B]. The problems cited in Chapter 10 for the scalar case become even more severe for the multichannel estimator. Other approximate multichannel MLE ARMA estimators may be found in the works by Hannan [1969A, 1969B], Anderson [1975, 1977], and Nicholls [1976, 1977]. Suboptimal approaches based on the Yule–Walker equations appear to be the only practical means for multichannel ARMA spectral estimation. The multichannel modified Yule–Walker equation (MYWE) estimator is based on (14.49). The AR parameters may be estimated by solving the Yule–Walker equations for $k = q + 1, q + 2, \ldots, q + p$ or

$$\begin{bmatrix} \hat{\mathbf{R}}_{xx}[q] & \hat{\mathbf{R}}_{xx}[q-1] & \cdots & \hat{\mathbf{R}}_{xx}[q-p+1] \\ \hat{\mathbf{R}}_{xx}[q+1] & \hat{\mathbf{R}}_{xx}[q] & \cdots & \hat{\mathbf{R}}_{xx}[q-p+2] \\ \vdots & \vdots & \ddots & \vdots \\ \hat{\mathbf{R}}_{xx}[q+p-1] & \hat{\mathbf{R}}_{xx}[q+p-2] & \cdots & \hat{\mathbf{R}}_{xx}[q] \end{bmatrix} \begin{bmatrix} \hat{\mathbf{A}}^T[1] \\ \hat{\mathbf{A}}^T[2] \\ \vdots \\ \hat{\mathbf{A}}^T[p] \end{bmatrix} = - \begin{bmatrix} \hat{\mathbf{R}}_{xx}[q+1] \\ \hat{\mathbf{R}}_{xx}[q+2] \\ \vdots \\ \hat{\mathbf{R}}_{xx}[q+p] \end{bmatrix}. \quad (14.76)$$

The ACF estimator is given by (14.39) if a biased estimator is desired or by (14.39) with the divisor N replaced by $N - k$ for an unbiased estimator. The multichannel least squares modified Yule–Walker equation (LSMYWE) estimator for the AR parameters is found by solving the set of Yule–Walker equations for $k = q + 1, q + 2, \ldots, M$ in a least squares manner. The theoretical equations are

$$\begin{bmatrix} \mathbf{R}_{xx}[q] & \mathbf{R}_{xx}[q-1] & \cdots & \mathbf{R}_{xx}[q-p+1] \\ \mathbf{R}_{xx}[q+1] & \mathbf{R}_{xx}[q] & \cdots & \mathbf{R}_{xx}[q-p+2] \\ \vdots & \vdots & \vdots & \vdots \\ \mathbf{R}_{xx}[M-1] & \mathbf{R}_{xx}[M-2] & \cdots & \mathbf{R}_{xx}[M-p] \end{bmatrix} \begin{bmatrix} \mathbf{A}^T[1] \\ \mathbf{A}^T[2] \\ \vdots \\ \mathbf{A}^T[p] \end{bmatrix} = - \begin{bmatrix} \mathbf{R}_{xx}[q+1] \\ \mathbf{R}_{xx}[q+2] \\ \vdots \\ \mathbf{R}_{xx}[M] \end{bmatrix}$$

or

$$\underline{\mathbf{R}}\underline{\mathbf{A}}^T = -\underline{\mathbf{r}} \quad (14.77)$$

where $\underline{\mathbf{R}}$ is $(M - q)L \times pL$, $\underline{\mathbf{A}}^T$ is $pL \times L$, and $\underline{\mathbf{r}}$ is $(M - q)L \times L$. Replacing the theoretical ACF by an estimate and solving the equations using least squares to account for the ACF estimation errors results in the AR parameter estimator (see Problem 14.14)

$$\hat{\underline{\mathbf{A}}}^T = -(\hat{\underline{\mathbf{R}}}^H \hat{\underline{\mathbf{R}}})^{-1} \hat{\underline{\mathbf{R}}}^H \hat{\underline{\mathbf{r}}}. \quad (14.78)$$

The equations for the MYWE and LSMYWE estimators may be decomposed into L sets of simultaneous linear equations by partitioning the solution vector and right-hand-side vector into their L columns. Then L sets of simultaneous linear equations can be solved. Furthermore, for the LSMYWE estimator the equations may be solved using a Cholesky decomposition. For either the multichannel MYWE or LSMYWE estimators the methods to find the MA contribution to the cross-spectral matrix are similar to those described for the single-channel case. One possibility is to first filter the data with the estimated AR filter to yield

$$\mathbf{y}[n] = \sum_{k=0}^{p} \hat{\mathbf{A}}[i]\mathbf{x}[n-i] \qquad n = p, p+1, \ldots, N-1$$

which will then approximate a multichannel MA process. In the single channel case Durbin's algorithm was used to estimate the MA parameters. However, this procedure does not appear to have been extended to multichannel MA processes. It is not imperative that the MA parameters themselves be estimated since only the component $\mathbf{B}^*(f)\mathbf{\Sigma}\mathbf{B}^T(f)$ is necessary to determine the cross-spectral matrix. We may thus apply any of the classical spectral estimators described in Section 14.5 to the filtered sequence $\mathbf{y}[n]$ to estimate this factor [Thomson 1977, Kay 1980]. The lower resolution of the Fourier spectral estimators should not degrade the resolution of the overall spectral estimate since it is the AR part that most affects resolution. Alternatively, an expression analogous to (9.14) may be derived for the multichannel case. To do so, first observe from (14.18) and (14.30) that

$$\mathcal{A}^*(1/z^*)\,\mathcal{P}_{xx}(z)\mathcal{A}^T(z) = \mathcal{B}^*(1/z^*)\,\mathbf{\Sigma}\mathcal{B}^T(z).$$

Let

$$\mathbf{R}_{\text{MA}}[k] = \mathcal{Z}^{-1}\left[\mathcal{B}^*(1/z^*)\,\mathbf{\Sigma}\mathcal{B}^T(z)\right]$$

which will be nonzero for $|k| \leq q$. Then

$$\mathbf{R}_{\text{MA}}[k] = \mathcal{Z}^{-1}\left[\mathcal{A}^*(1/z^*)\,\mathcal{P}_{xx}(z)\mathcal{A}^T(z)\right].$$

The inverse z-transform may be shown to be (see Problem 14.15)

$$\mathbf{R}_{\text{MA}}[k] = \sum_{i=0}^{p} \mathbf{A}^*[i] \sum_{j=0}^{p} \mathbf{R}_{xx}[k+i-j]\mathbf{A}^T[j] \qquad k = 0, 1, \ldots, q. \tag{14.79}$$

Hence, given the estimated AR parameters and estimates of $\mathbf{R}_{xx}[k]$ for $0 \leq k \leq p+q$, we can estimate $\mathbf{R}_{\text{MA}}[k]$. Then, using this estimate, we obtain

$$\hat{\mathbf{B}}^*(f)\hat{\mathbf{\Sigma}}\hat{\mathbf{B}}^T(f) = \sum_{k=-q}^{q} \hat{\mathbf{R}}_{\text{MA}}[k]\exp(-j2\pi fk) \tag{14.80}$$

for use in (14.75).

14.9 MINIMUM VARIANCE SPECTRAL ESTIMATION

The single channel minimum variance spectral estimator was described in Chapter 11. The extension to multichannel spectral estimation follows in a straightforward manner. Assume that the data consist of a known frequency sinusoid in noise:

$$x_i[n] = A_{c_i} \exp(j2\pi f_0 n) + z_i[n] \qquad i = 1, 2, \ldots, L. \tag{14.81}$$

The complex amplitudes A_{c_i} differ in general from channel to channel. At each instant of time the data observed for all the channels may be written as

$$\mathbf{x}[n] = \mathbf{A}_c \exp(j2\pi f_0 n) + \mathbf{z}[n] \tag{14.82}$$

where $\mathbf{A}_c = [A_{c_1} \;\; A_{c_2} \;\; \cdots \;\; A_{c_L}]^T$. Note that $\mathbf{A}_c \exp(j2\pi f_0 n)$ is a multichannel sinusoid as discussed in Section 14.3. Assuming that each channel is observed for N samples, the total data set is

$$\mathbf{x} = [\mathbf{x}^T[0] \;\; \mathbf{x}^T[1] \;\; \cdots \;\; \mathbf{x}^T[N-1]]^T \tag{14.83}$$

which is a vector of dimension $NL \times 1$. Using (14.82), this may also be expressed as

$$\mathbf{x} = \begin{bmatrix} \mathbf{A}_c \\ \mathbf{A}_c \exp(j2\pi f_0) \\ \vdots \\ \mathbf{A}_c \exp(j2\pi f_0[N-1]) \end{bmatrix} + \mathbf{z} \tag{14.84}$$

where $\mathbf{z}$ is defined analogously to $\mathbf{x}$. The signal vector is more conveniently written as

$$\begin{bmatrix} \mathbf{I} \\ \mathbf{I} \exp(j2\pi f_0) \\ \vdots \\ \mathbf{I} \exp(j2\pi f_0[N-1]) \end{bmatrix} \mathbf{A}_c \tag{14.85}$$

where each identity matrix has dimension $L \times L$. Defining the $NL \times L$ matrix

$$\mathbf{E} = \begin{bmatrix} \mathbf{I} \\ \mathbf{I} \exp(j2\pi f_0) \\ \vdots \\ \mathbf{I} \exp(j2\pi f_0[N-1]) \end{bmatrix} \tag{14.86}$$

the signal vector becomes $\mathbf{EA}_c$ and the total data set may be written as

$$\mathbf{x} = \mathbf{EA}_c + \mathbf{z}. \tag{14.87}$$

Given this data set, it is desired to find the linear minimum variance unbiased (LMVU) estimator of $\mathbf{A}_c$, or

$$\hat{\mathbf{A}}_c = \mathbf{W}^H \mathbf{x} \tag{14.88}$$

where $\mathbf{W}$ is an $NL \times L$ coefficient matrix. For the estimator to be unbiased

$$\mathscr{E}(\hat{\mathbf{A}}_c) = \mathscr{E}(\mathbf{W}^H(\mathbf{EA}_c + \mathbf{z})) = \mathbf{W}^H\mathbf{EA}_c = \mathbf{A}_c$$

which then yields the constraint that

$$\mathbf{W}^H\mathbf{E} = \mathbf{I}. \tag{14.89}$$

The variance of the estimator is defined as the sum of the variances of the amplitude estimators for each channel or the trace of the covariance matrix

$$\mathbf{C}_{A_c} = \mathscr{E}\{[\hat{\mathbf{A}}_c - \mathscr{E}(\hat{\mathbf{A}}_c)][\hat{\mathbf{A}}_c - \mathscr{E}(\hat{\mathbf{A}}_c)]^H\}. \tag{14.90}$$

Using (14.88), this is easily shown to be

$$\mathbf{C}_{A_c} = \mathbf{W}^H\underline{\mathbf{R}}_{zz}\mathbf{W} \tag{14.91}$$

where $\underline{\mathbf{R}}_{zz} = \mathscr{E}(\mathbf{z}\mathbf{z}^H)$ is the $NL \times NL$ autocorrelation matrix of $\mathbf{z}$. To find the LMVU estimator we must minimize the trace of (14.91) subject to the constraint of (14.89). This minimization produces the optimal weighting matrix [Luenberger 1969]

$$\mathbf{W} = \underline{\mathbf{R}}_{zz}^{-1}\mathbf{E}(\mathbf{E}^H\underline{\mathbf{R}}_{zz}^{-1}\mathbf{E})^{-1} \tag{14.92}$$

which when substituted into (14.91) yields

$$\mathbf{C}_{A_c} = (\mathbf{E}^H\underline{\mathbf{R}}_{zz}^{-1}\mathbf{E})^{-1} \tag{14.93}$$

(see Problem 14.16 for an alternative proof). Note that $\mathbf{C}_{A_c}$ is a matrix of dimension $L \times L$. The same arguments used in the single channel case (see Section 11.5) leads us to propose the multichannel MVSE as

$$\hat{\mathbf{P}}_{\mathrm{MV}}(f) = (\mathbf{E}^H\hat{\underline{\mathbf{R}}}_{xx}^{-1}\mathbf{E})^{-1}. \tag{14.94}$$

$\hat{\underline{\mathbf{R}}}_{xx}$ is an estimator of the $pL \times pL$ hermitian and block Toeplitz matrix given in (14.52). If scaling of the cross-spectral matrix estimate is important, the definition of the MVSE should be modified by multiplying $\hat{\mathbf{P}}_{\mathrm{MV}}(f)$ given in (14.94) by p. To retain the properties of the theoretical autocorrelation matrix the ACF estimator defined in (14.39) should be used to replace the theoretical ACF (principally to ensure that $\hat{\underline{\mathbf{R}}}_{xx}$ is invertible). Then $\hat{\underline{\mathbf{R}}}_{xx}^{-1}$ may be computed using a multichannel version of the Cholesky decomposition of a single channel hermitian Toeplitz matrix as described in Section 6.3.6 [Akaike 1973A]. Note that the order p which determines the number of ACF estimates required should be chosen to be no larger than $N/2$ to ensure statistical stability of the estimated cross-spectral matrix.

14.10 COMPUTER SIMULATION EXAMPLES

A computer simulation example of a multichannel AR spectral estimator is described in this section [Marple 1986]. For examples of Fourier estimators the reader should consult Jenkins and Watts [1968]. The process that is to be considered is a real two channel AR(1) process with parameters

$$\mathbf{A}[1] = \begin{bmatrix} -0.85 & 0.75 \\ -0.65 & -0.55 \end{bmatrix}$$

$$\mathbf{\Sigma} = \mathbf{I}.$$

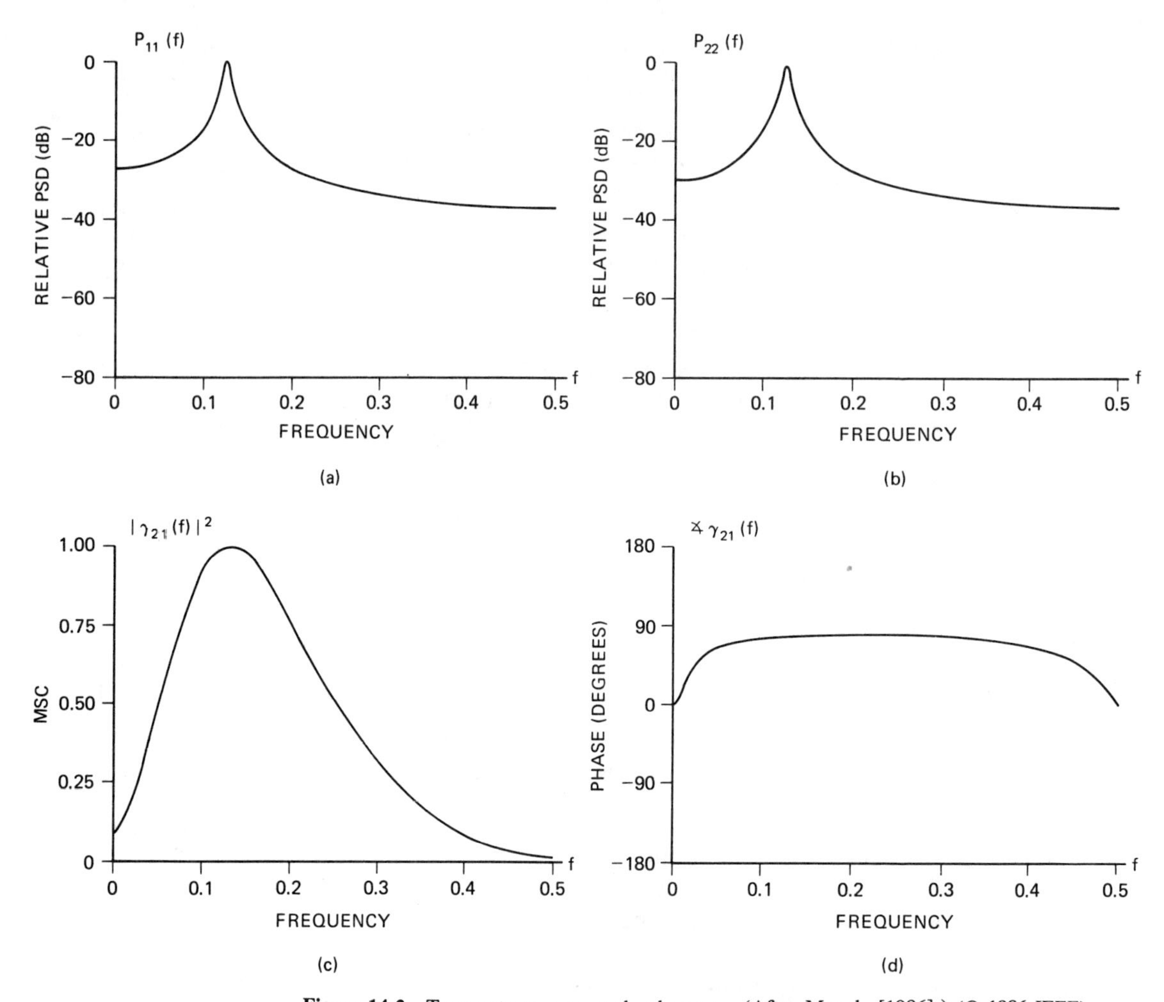

Figure 14.2 True auto-spectra and coherence. (After Marple [1986].) (© 1986 IEEE)

The true autospectra, $P_{11}(f)$, $P_{22}(f)$, as well as the magnitude squared coherence, $|\gamma_{21}(f)|^2$, as given by (14.26), and the phase of the coherence, which is the phase of $P_{21}(f)$, are shown in Figure 14.2. Twenty realizations of this process were generated, with each realization consisting of 100 data points. The various quantities shown in Figure 14.2 were estimated using the covariance method (14.72)–(14.73). The estimates obtained from the 20 realizations are shown overlaid in Figure 14.3. It is observed that the autospectra are estimated well except for the peak which tends to be overly sharp. The coherence estimate appears to be quite accurate except for the phase at $f = 0.5$. Alternative AR estimation algorithms which are generalizations of the Burg method [Nuttall 1976, Strand 1977] and [Morf, et al. 1978] are also compared by Marple [1986]. The estimation performance is similar to that of the covariance method.

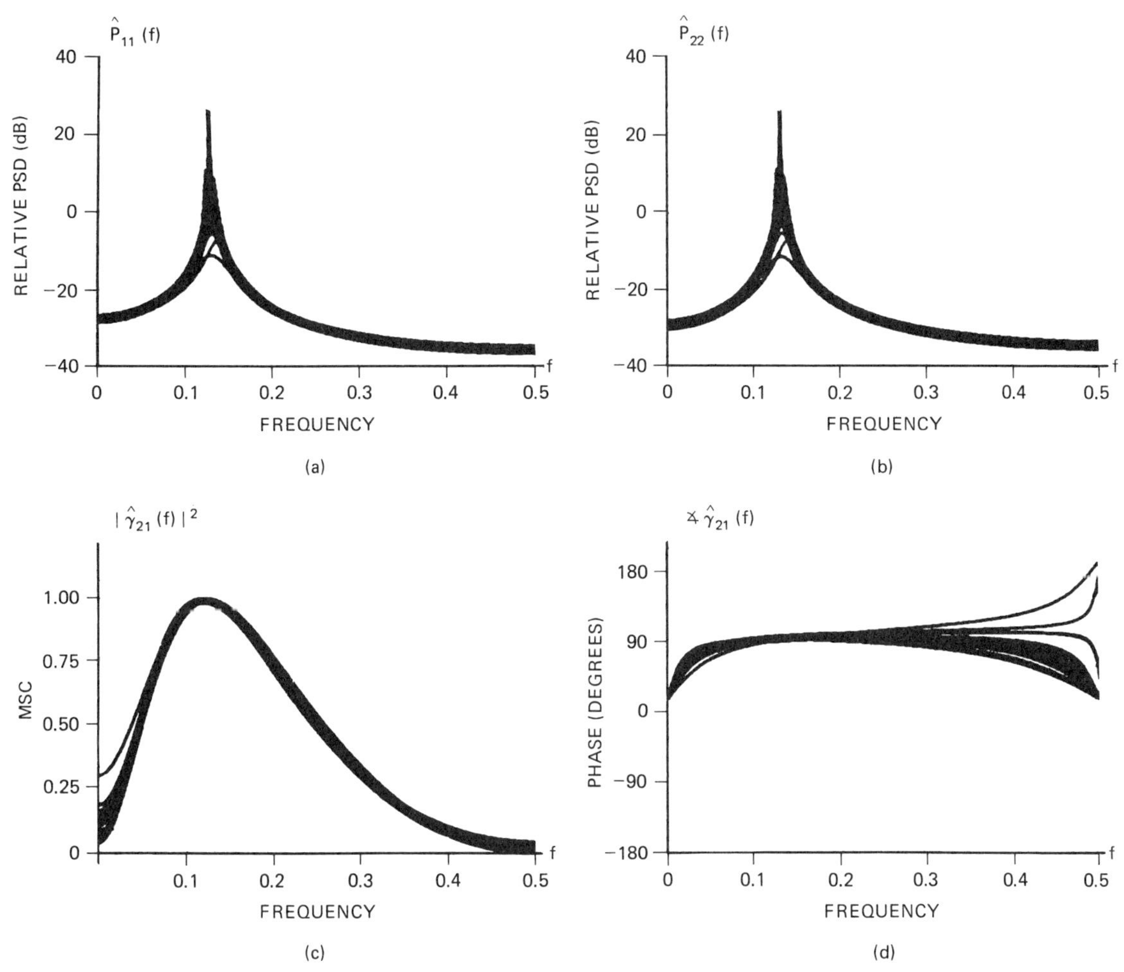

Figure 14.3 Estimated auto-spectra and coherence using covariance method. (After Marple [1986].) (© 1986 IEEE)

REFERENCES

Akaike, H., "Block Toeplitz Matrix Inversion," *SIAM J. Appl. Math.,* Vol. 24, pp. 234–241, Mar. 1973A.

Akaike, H., "Maximum Likelihood Identification of Gaussian Autoregressive Moving Average Models," *Biometrika,* Vol. 60, pp. 255–265, 1973B.

Anderson, T. W., "Maximum Likelihood Estimation of Parameters of Autoregressive Processes with Moving Average Residuals and Other Covariance Matrices with Linear Structure," *Ann. Statist.,* Vol. 3, pp. 1283–1304, 1975.

Anderson, T. W., "Estimation for Autoregressive Moving Average Models in the Time and Frequency Domains," *Ann. Statist.,* Vol. 5, pp. 842–865, 1977.

Burg, J. P., "Maximum Entropy Spectral Analysis," Ph.D. dissertation, Stanford University, May 1975.

Hannan, E. J., "The Identification of Vector Mixed Autoregressive–Moving Average Systems," *Biometrika,* Vol. 56, pp. 223–225, 1969A.

Hannan, E. J., "The Estimation of Mixed Moving Average Autoregressive Models," *Biometrika,* Vol. 56, pp. 579–593, 1969B.

Hannan, E. J., *Multiple Time Series,* Wiley, New York, 1970.

Jenkins, G. M., and D. G. Watts, *Spectral Analysis and Its Applications,* Holden-Day, San Francisco, 1968.

Jones, R. H., "Identification and Autoregressive Spectrum Estimation," *IEEE Trans. Autom. Control,* Vol. AC19, pp. 894–897, Dec. 1974.

Justice, J. H., "Array Processing in Exploration Seismology," in *Array Signal Processing,* S. Haykin et al., eds., Prentice-Hall, Englewood Cliffs, N.J., 1985.

Kay, S., "A New ARMA Spectral Estimator," *IEEE Trans. Acoust. Speech Signal Process.,* Vol. ASSP28, pp. 585–588, Oct. 1980.

Knight, W. C., R. G. Pridham, and S. M. Kay, "Digital Signal Processing for Sonar," *Proc. IEEE,* Vol. 69, pp. 1451–1506, Nov. 1981.

Luenberger, D. G., *Optimization by Vector Space Methods,* Wiley, New York, 1969.

Marple, S. L., Jr., "Performance of Multichannel Autoregressive Spectral Estimators," in Rec 1986 *IEEE Int. Conf. on Acoust., Speech, Signal Process.,* Tokyo, Japan, pp. 197–200.

Marple, S. L., Jr., *Digital Spectral Analysis with Applications,* Prentice-Hall, Englewood Cliffs, N.J., 1987.

Monzingo, R. A., and T. W. Miller, *Introduction to Adaptive Arrays,* Wiley, New York, 1980.

Morf, M., A. Vieira, D. T. Lee, and T. Kailaith, "Recursive Multichannel Maximum Entropy Spectral Estimation," *IEEE Trans. Geosci. Electron.,* Vol. GE16, pp. 85–94, Apr. 1978.

Nicholls, D. F., "The Efficient Estimation of Vector Linear Time Series Models," *Biometrika,* Vol. 63, pp. 381–390, 1976.

Nicholls, D. F., "A Comparison of Estimation Methods for Vector Linear Time Series Models," *Biometrika,* Vol. 64, pp. 85–90, 1977.

Nuttall, A. H., "Multivariate Linear Predictive Spectral Analysis Employing Weighted Forward and Backward Averaging: A Generalization of Burg's Algorithm," *NUSC Tech. Rep. 5501,* Oct. 13, 1976.

Parzen, E., "Multiple Time Series Modeling," in *Multivariate Analysis 2,* P. R. Krishnaiah, ed., Academic Press, New York, 1969.

Robinson, E. A., *Multichannel Time Series Analysis with Digital Computer Programs,* Holden-Day, San Francisco, 1967.

Strand, O. N., "Multichannel Complex Maximum Entropy (Autoregressive) Spectral Analysis," *IEEE Trans. Autom. Control,* Vol. AC22, pp. 634–640, Aug. 1977.

Thomson, D. J., "Spectral Estimation Techniques for Characterization and Development of WT4 Waveguide: Part 1," *Bell Syst. Tech. J.,* Vol. 56, pp. 1769–1815, Nov. 1977.

Wiggins, R. A., and E. A. Robinson, "Recursive Solution to the Multichannel Filtering Problem," *J. Geophys. Res.,* Vol. 70, pp. 1885–1891, Apr. 1965.

PROBLEMS

14.1. Prove that the multichannel z-transform has the properties of linearity (14.7) and shifting (14.8).

14.2. Prove that the z-transform of a sequence at the output of a multichannel LSI system is given by (14.9).

14.3. Consider the multichannel equivalent of a one-pole filter. The difference equation description is

$$\mathbf{x}[n] = -\mathbf{A}[1]\mathbf{x}[n-1] + \mathbf{u}[n].$$

For the filter to be stable the eigenvalues of $\mathbf{A}[1]$ must be less than 1 in magnitude. Prove this by showing that the roots of $\det(\mathbf{I} + \mathbf{A}[1]z^{-1}) = 0$ lie within the unit circle if and only if the eigenvalues are less than 1 in magnitude.

14.4. Show that the impulse response matrix of the multichannel filter described in Problem 14.3 is given by (14.16). Noting that the output can be written as

$$\mathbf{x}[n] = \sum_{k=0}^{\infty} (-\mathbf{A}[1])^k \mathbf{u}[n-k]$$

and assuming that the input may be expressed as

$$\mathbf{u}[n] = \sum_{i=1}^{L} \xi_i[n]\mathbf{v}_i$$

where the $\mathbf{v}_i$'s are the eigenvectors of $\mathbf{A}[1]$, prove that for stability the eigenvalues of $\mathbf{A}[1]$ must be less than 1 in magnitude.

14.5. Verify that the system function for the multichannel equivalent of a pole–zero filter as described by (14.17) is given by (14.18). Find the difference equation representation of the filter with the system function

$$\mathcal{H}(z) = \mathcal{B}(z)\mathcal{A}^{-1}(z).$$

14.6. Prove that $P_{ji}(f) = P_{ij}^*(f)$. What does this imply about the auto-spectra?

14.7. Prove that if the cross-periodogram estimator (14.33) is used for a two channel process, the magnitude squared coherence (14.26) will always be estimated as unity.

14.8. Show that the cross-periodogram may be written as the Fourier transform of the estimated CCF as expressed by (14.36) and (14.37).

14.9. For a multichannel AR(1) process prove that the ACF is

$$\mathbf{R}_{xx}[k] = \mathbf{R}_{xx}[0](-\mathbf{A}^T[1])^k \qquad k \geq 1.$$

Show that $\mathbf{R}_{xx}[0]$ may be found as the solution of

$$\mathbf{R}_{xx}[0] = \mathbf{\Sigma} + \mathbf{A}^*[1]\mathbf{R}_{xx}[0]\mathbf{A}^T[1].$$

Can you solve these equations to find $\mathbf{R}_{xx}[0]$ explicitly?

14.10. Find $\mathbf{R}_{xx}[0]$ for a multichannel AR(1) process by noting that the process may be expressed as

$$\mathbf{x}[n] = \sum_{k=0}^{\infty} \mathbf{H}[k]\mathbf{u}[n-k]$$

where

$$\mathbf{H}[k] = (-\mathbf{A}[1])^k \qquad k \geq 0.$$

14.11. The first-order forward linear predictor is defined as

$$\hat{\mathbf{x}}^f[n] = -\mathbf{A}^f[1]\mathbf{x}[n-1].$$

The orthogonality principle states that the prediction error $\mathbf{x}[n] - \hat{\mathbf{x}}^f[n]$ of the optimal predictor is orthogonal to all vectors in the subspace spanned by $\mathbf{x}[n-1]$, or

$$\langle \mathbf{F}\mathbf{x}[n-1], \mathbf{x}[n] - \hat{\mathbf{x}}^f[n] \rangle = 0.$$

$\mathbf{F}$ is an arbitrary $L \times L$ matrix which when applied to $\mathbf{x}[n-1]$ produces all vectors in the subspace. Using the definition of the inner product (14.56), prove that the optimal predictor satisfies

$$\mathscr{E}\{\mathbf{x}^*[n-1](\mathbf{x}[n] - \hat{\mathbf{x}}^f[n])^T\} = \mathbf{0}.$$

14.12. Show that (14.68) is minimized by the solution of the equations given by (14.69). *Hint:* Extend the results given by (2.79) or, alternatively, define an appropriate inner product for this problem and apply the orthogonality principle.

14.13. Prove that $\underline{\mathbf{C}}_{xx}$ as defined in (14.72) is positive semidefinite for the covariance method. Under what conditions will the matrix be positive definite and hence nonsingular?

14.14. The least squares solution of the modified Yule–Walker equations is given in (14.78). Verify this result by noting that the least squares solution is actually being found for L sets of overdetermined linear equations. Each set of unknowns corresponds to a column of $\mathbf{A}^T$.

14.15. Show that the inverse z-transform of $\mathcal{A}^*(1/z^*)\mathcal{P}_{xx}(z)\mathcal{A}^T(z)$ is given by (14.79).

14.16. Prove that the optimal weighting matrix for the LMVU estimator of the multichannel sinusoidal amplitude is given by (14.92). *Hint:* Verify the identity

$$\mathbf{W}^H\mathbf{R}_{zz}\mathbf{W} = (\mathbf{W} - \tilde{\mathbf{W}})^H\mathbf{R}_{zz}(\mathbf{W} - \tilde{\mathbf{W}}) + (\mathbf{E}^H\mathbf{R}_{zz}^{-1}\mathbf{E})^{-1}$$

where

$$\tilde{\mathbf{W}} = \mathbf{R}_{zz}^{-1}\mathbf{E}(\mathbf{E}^H\mathbf{R}_{zz}^{-1}\mathbf{E})^{-1}$$

and $\mathbf{W}^H\mathbf{E} = \mathbf{I}$. Then note that the trace of a positive semidefinite matrix is greater than or equal to zero.

14.17 Prove that $\mathbf{\Delta}_k^b = \mathbf{\Delta}_k^{f^H}$. To do so, use the result that

$$[\mathbf{0} \;\; \mathbf{A}_{k-1}^{b*}[k-1] \;\; \cdots \;\; \mathbf{A}_{k-1}^{b*}[1] \;\; \mathbf{I}] \begin{bmatrix} \mathbf{R}_{xx}[0] & \mathbf{R}_{xx}[-1] & \cdots & \mathbf{R}_{xx}[-k] \\ \mathbf{R}_{xx}[1] & \mathbf{R}_{xx}[0] & \cdots & \mathbf{R}_{xx}[-(k-1)] \\ \vdots & \vdots & \vdots & \vdots \\ \mathbf{R}_{xx}[k-1] & \mathbf{R}_{xx}[k-2] & \cdots & \mathbf{R}_{xx}[1] \\ \mathbf{R}_{xx}[k] & \mathbf{R}_{xx}[k-1] & \cdots & \mathbf{R}_{xx}[0] \end{bmatrix} \begin{bmatrix} \mathbf{I} \\ \mathbf{A}_{k-1}^{f^T}[1] \\ \vdots \\ \mathbf{A}_{k-1}^{f^T}[k-1] \\ \mathbf{0} \end{bmatrix} = \mathbf{\Delta}_k^f$$

which follows from (14A.3).

APPENDIX 14A

Derivation of Levinson Algorithm for Solution of the Multichannel Yule–Walker Equations

The derivation contained in this appendix follows that of Burg [1975]. Unlike the derivation presented in Chapter 6 for the scalar case, which relied on a vector space approach, the derivation given here is based purely on the algebraic properties of block Toeplitz matrices. The vector space approach is possible with the definition of the inner product given by (14.56), but is more tedious.

It is assumed that the solutions for the $(k-1)$st order forward and backward predictors are already available and that the solutions for the kth order predictors are desired. From (14.57), (14.58), and (14.62), the $(k-1)$st order predictors satisfy the equations

$$\begin{bmatrix} \mathbf{R}_{xx}[0] & \mathbf{R}_{xx}[-1] & \cdots & \mathbf{R}_{xx}[-(k-1)] \\ \mathbf{R}_{xx}[1] & \mathbf{R}_{xx}[0] & \cdots & \mathbf{R}_{xx}[-(k-2)] \\ \vdots & \vdots & \vdots & \vdots \\ \mathbf{R}_{xx}[k-2] & \mathbf{R}_{xx}[k-3] & \cdots & \mathbf{R}_{xx}[1] \\ \mathbf{R}_{xx}[k-1] & \mathbf{R}_{xx}[k-2] & \cdots & \mathbf{R}_{xx}[0] \end{bmatrix} \begin{bmatrix} \mathbf{I} \\ \mathbf{A}_{k-1}^{fT}[1] \\ \vdots \\ \mathbf{A}_{k-1}^{fT}[k-2] \\ \mathbf{A}_{k-1}^{fT}[k-1] \end{bmatrix} = \begin{bmatrix} \mathbf{\Sigma}_{k-1}^{f} \\ \mathbf{0} \\ \vdots \\ \mathbf{0} \\ \mathbf{0} \end{bmatrix} \qquad (14A.1)$$

$$\begin{bmatrix} \mathbf{R}_{xx}[0] & \mathbf{R}_{xx}[-1] & \cdots & \mathbf{R}_{xx}[-(k-1)] \\ \mathbf{R}_{xx}[1] & \mathbf{R}_{xx}[0] & \cdots & \mathbf{R}_{xx}[-(k-2)] \\ \vdots & \vdots & \vdots & \vdots \\ \mathbf{R}_{xx}[k-2] & \mathbf{R}_{xx}[k-3] & \cdots & \mathbf{R}_{xx}[1] \\ \mathbf{R}_{xx}[k-1] & \mathbf{R}_{xx}[k-2] & \cdots & \mathbf{R}_{xx}[0] \end{bmatrix} \begin{bmatrix} \mathbf{A}_{k-1}^{bT}[k-1] \\ \mathbf{A}_{k-1}^{bT}[k-2] \\ \vdots \\ \mathbf{A}_{k-1}^{bT}[1] \\ \mathbf{I} \end{bmatrix} = \begin{bmatrix} \mathbf{0} \\ \mathbf{0} \\ \vdots \\ \mathbf{0} \\ \mathbf{\Sigma}_{k-1}^{b} \end{bmatrix} \qquad (14A.2)$$

where the equation for the prediction error power matrix has been added to form an augmented set of equations [see (5.19)]. The solution for the kth order predictors is found by considering the solution to the equations

$$\begin{bmatrix} \mathbf{R}_{xx}[0] & \mathbf{R}_{xx}[-1] & \cdots & \mathbf{R}_{xx}[-k] \\ \mathbf{R}_{xx}[1] & \mathbf{R}_{xx}[0] & \cdots & \mathbf{R}_{xx}[-(k-1)] \\ \vdots & \vdots & \vdots & \vdots \\ \mathbf{R}_{xx}[k-1] & \mathbf{R}_{xx}[k-2] & \cdots & \mathbf{R}_{xx}[1] \\ \mathbf{R}_{xx}[k] & \mathbf{R}_{xx}[k-1] & \cdots & \mathbf{R}_{xx}[0] \end{bmatrix} \left\{ \begin{bmatrix} \mathbf{I} \\ \mathbf{A}_{k-1}^{f^T}[1] \\ \vdots \\ \mathbf{A}_{k-1}^{f^T}[k-1] \\ \mathbf{0} \end{bmatrix} + \begin{bmatrix} \mathbf{0} \\ \mathbf{A}_{k-1}^{b^T}[k-1] \\ \vdots \\ \mathbf{A}_{k-1}^{b^T}[1] \\ \mathbf{I} \end{bmatrix} \mathbf{K}_k^{f^T} \right\} = \begin{bmatrix} \mathbf{\Sigma}_{k-1}^f \\ \mathbf{0} \\ \vdots \\ \mathbf{0} \\ \mathbf{\Delta}_k^f \end{bmatrix} + \begin{bmatrix} \mathbf{\Delta}_k^b \\ \mathbf{0} \\ \vdots \\ \mathbf{0} \\ \mathbf{\Sigma}_{k-1}^b \end{bmatrix} \mathbf{K}_k^{f^T}. \quad (14A.3)$$

To obtain these equations the forward equations have been increased in dimension by one and added to the backward equations, which have been increased in dimension by one and multiplied by the $L \times L$ matrix $\mathbf{K}_k^{f^T}$, which is yet to be determined. Note that because of the special form of the autocorrelation matrix the middle $k-1$ equations in (14A.3) are satisfied for any choice of $\mathbf{K}_k^{f^T}$. This is because these equations correspond to the equations for the $(k-1)$st order predictors. Only the first and last equations need to be satisfied by an appropriate choice of $\mathbf{K}_k^{f^T}$. Now from (14A.3) $\mathbf{\Delta}_k^f$, $\mathbf{\Delta}_k^b$ are given by

$$\mathbf{\Delta}_k^f = \sum_{i=0}^{k-1} \mathbf{R}_{xx}[k-i]\mathbf{A}_{k-1}^{f^T}[i]$$

$$\mathbf{\Delta}_k^b = \sum_{i=0}^{k-1} \mathbf{R}_{xx}[i-k]\mathbf{A}_{k-1}^{b^T}[i]$$

where $\mathbf{A}_{k-1}^{f^T}[0] = \mathbf{A}_{k-1}^{b^T}[0] = \mathbf{I}$. For (14A.3) to represent the equations for the kth order forward predictor or (14A.1) with $k-1$ replaced by k, it must be true that the last equation is

$$\mathbf{\Delta}_k^f + \mathbf{\Sigma}_{k-1}^b \mathbf{K}_k^{f^T} = \mathbf{0}$$

or

$$\mathbf{K}_k^{f^T} - -\mathbf{\Sigma}_{k-1}^{b^{-1}} \mathbf{\Delta}_k^f.$$

Also, the first equation results in

$$\mathbf{\Sigma}_k^f = \mathbf{\Sigma}_{k-1}^f + \mathbf{\Delta}_k^b \mathbf{K}_k^{f^T}$$

and the solution must satisfy

$$\mathbf{A}_k^{f^T}[i] = \begin{cases} \mathbf{A}_{k-1}^{f^T}[i] + \mathbf{A}_{k-1}^{b^T}[k-i]\mathbf{K}_k^{f^T} & \text{for } i = 1, 2, \ldots, k-1 \\ \mathbf{K}_k^{f^T} & \text{for } i = k \end{cases}$$

or

$$\mathbf{A}_k^f[i] = \begin{cases} \mathbf{A}_{k-1}^f[i] + \mathbf{K}_k^f \mathbf{A}_{k-1}^b[k-i] & \text{for } i = 1, 2, \ldots, k-1 \\ \mathbf{K}_k^f & \text{for } i = k. \end{cases}$$

In a similar fashion we can generate the equations

$$\begin{bmatrix} \mathbf{R}_{xx}[0] & \mathbf{R}_{xx}[-1] & \cdots & \mathbf{R}_{xx}[-k] \\ \mathbf{R}_{xx}[1] & \mathbf{R}_{xx}[0] & \cdots & \mathbf{R}_{xx}[-(k-1)] \\ \vdots & \vdots & \vdots & \vdots \\ \mathbf{R}_{xx}[k-1] & \mathbf{R}_{xx}[k-2] & \cdots & \mathbf{R}_{xx}[1] \\ \mathbf{R}_{xx}[k] & \mathbf{R}_{xx}[k-1] & \cdots & \mathbf{R}_{xx}[0] \end{bmatrix} \left\{ \begin{bmatrix} \mathbf{I} \\ \mathbf{A}_{k-1}^{f^T}[1] \\ \vdots \\ \mathbf{A}_{k-1}^{f^T}[k-1] \\ \mathbf{0} \end{bmatrix} \mathbf{K}_k^{b^T} + \begin{bmatrix} \mathbf{0} \\ \mathbf{A}_{k-1}^{b^T}[k-1] \\ \vdots \\ \mathbf{A}_{k-1}^{b^T}[1] \\ \mathbf{I} \end{bmatrix} \right\} = \begin{bmatrix} \mathbf{\Sigma}_{k-1}^f \\ \mathbf{0} \\ \vdots \\ \mathbf{0} \\ \mathbf{\Delta}_k^f \end{bmatrix} \mathbf{K}_k^{b^T} + \begin{bmatrix} \mathbf{\Delta}_k^b \\ \mathbf{0} \\ \vdots \\ \mathbf{0} \\ \mathbf{\Sigma}_{k-1}^b \end{bmatrix} \tag{14A.4}$$

which should represent the equations for the kth order backward predictor. By the same arguments as before

$$\mathbf{K}_k^{b^T} = -\mathbf{\Sigma}_{k-1}^{f^{-1}} \mathbf{\Delta}_k^b$$

$$\mathbf{\Sigma}_k^b = \mathbf{\Sigma}_{k-1}^b + \mathbf{\Delta}_k^f \mathbf{K}_k^{b^T}$$

$$\mathbf{A}_k^b[i] = \begin{cases} \mathbf{A}_{k-1}^b[i] + \mathbf{K}_k^b \mathbf{A}_{k-1}^f[k-i] & \text{for } i = 1, 2, \ldots, k-1 \\ \mathbf{K}_k^b & \text{for } i = k. \end{cases}$$

Some simplifications to the equations are possible. It may be shown that (see Problem 14.17)

$$\mathbf{\Delta}_k^b = \mathbf{\Delta}_k^{f^H}.$$

Using this result, the multichannel reflection coefficients are then given by

$$\mathbf{K}_k^{f^T} = -\mathbf{\Sigma}_{k-1}^{b^{-1}} \mathbf{\Delta}_k^f$$

$$\mathbf{K}_k^{b^T} = -\mathbf{\Sigma}_{k-1}^{f^{-1}} \mathbf{\Delta}_k^{f^H}.$$

Also, the prediction error power matrices are more conveniently expressed as

$$\begin{aligned} \mathbf{\Sigma}_k^f &= \mathbf{\Sigma}_{k-1}^f + \mathbf{\Delta}_k^b \mathbf{K}_k^{f^T} \\ &= \mathbf{\Sigma}_{k-1}^f - \mathbf{\Sigma}_{k-1}^f \mathbf{K}_k^{b^T} \mathbf{K}_k^{f^T} \\ &= \mathbf{\Sigma}_{k-1}^f [\mathbf{I} - \mathbf{K}_k^{b^T} \mathbf{K}_k^{f^T}] \end{aligned}$$

and similarly,

$$\mathbf{\Sigma}_k^b = \mathbf{\Sigma}_{k-1}^b [\mathbf{I} - \mathbf{K}_k^{f^T} \mathbf{K}_k^{b^T}].$$

The algorithm is initialized by finding the solution for the first order linear predictors. From (14A.1) and (14A.2) the first order predictor coefficients are

$$\begin{aligned} \mathbf{A}_1^{f^T}[1] &= \mathbf{K}_1^{f^T} = -\mathbf{R}_{xx}^{-1}[0]\mathbf{R}_{xx}[1] \\ \mathbf{A}_1^{b^T}[1] &= \mathbf{K}_1^{b^T} = -\mathbf{R}_{xx}^{-1}[0]\mathbf{R}_{xx}[-1]. \end{aligned} \tag{14A.5}$$

The first order prediction error power matrices are from (14A.1) and (14A.2)

$$\begin{aligned}\mathbf{\Sigma}_1^f &= \mathbf{R}_{xx}[0] + \mathbf{R}_{xx}[-1]\mathbf{A}_1^{f^T}[1] \\ &= \mathbf{R}_{xx}[0] - \mathbf{R}_{xx}[0]\mathbf{A}_1^{b^T}[1]\mathbf{A}_1^{f^T}[1] \\ &= \mathbf{R}_{xx}[0][\mathbf{I} - \mathbf{K}_1^{b^T}\mathbf{K}_1^{f^T}]\end{aligned}$$

and similarly,

$$\mathbf{\Sigma}_1^b = \mathbf{R}_{xx}[0][\mathbf{I} - \mathbf{K}_1^{f^T}\mathbf{K}_1^{b^T}].$$

The entire algorithm is summarized in (14.63)–(14.66).

Chapter 15

Two-Dimensional Spectral Estimation

15.1 INTRODUCTION

The "high resolution" performance of some of the one-dimensional (1-D) spectral estimators discussed in Part I has promoted an interest in two-dimensional (2-D) versions of these spectral estimators. In particular, successful extensions will find applications in 2-D digital signal processing, such as image processing [Wernecke and D'Addario 1977, Jain 1981], processing of true array data as in sonar and seismic [Capon 1969], or processing of synthetic array data as in radar and radio astronomy [Wernecke 1977]. Although some of the spectral estimation procedures of Part I have natural extensions to 2-D, many do not. This is due to the differences in 1-D and 2-D discrete systems theory. A key difference is that the mathematics for describing 2-D systems is less complete than for 1-D systems. Polynomials encountered in 1-D rational system models can always be factored, while 2-D polynomials in general cannot, due to a lack of a fundamental theorem of algebra in 2-D. Thus the useful analytic and conceptual tool of isolated poles and zeros is not readily extendible to the 2-D case. Another difference is that 2-D systems have many more degrees of freedom than 1-D systems. For instance, parametric models characterized by recursive difference equations in 1-D can be extended to 2-D models but with many possible difference equation descriptions, not all of which are recursively computable. The high computational load of the 2-D spectral estimation methods has resulted in their application to only small 2-D data sets, which for the most part consist of a few sinusoids in white noise. Only the 2-D periodogram and 2-D MVSE have been widely applied in practice. Suffice it to say that 2-D spectral estimation continues to be an area primarily for research.

A discussion of the issues and mathematical details caused by the significant theoretical differences between 1-D and 2-D spectral estimation would require a lengthy discourse. However, with only one chapter to highlight this topic area, this is not possible. The reader is encouraged to consult the references for many of the details not presented here. A development of multidimensional digital signal processing may be found in the text by Dudgeon and Mersereau [1984]. Additional details on multidimensional spectral estimation are available in the survey paper by McClellan [1982]. The specific application of 2-D spectral estimation for high resolution beamforming or direction finding for a linear spatial array is considered separately in Chapter 16.

15.2 SUMMARY

Sections 15.3 and 15.4 are review sections on 2-D linear systems and Fourier transform theory, and 2-D random process theory, respectively. Classical spectral estimation is discussed in Section 15.5 with an emphasis on the periodogram (15.25) and the Blackman–Tukey (15.28) spectral estimators. Autoregressive spectral estimation is described in Section 15.6. The Yule–Walker equations are derived for the nonsymmetric half plane region of support (15.31) and also the quarter plane region of support (15.36). A key result is that only the infinite NSHP region of support will allow any PSD to be represented without any modeling error. Some methods for AR spectral estimation follow directly from the 1-D case. They are the covariance method and the modified covariance method (15.43). Some analytical results are given for the AR spectral estimator based on a QP region of support for one complex sinusoid in white noise (15.47). An AR spectral estimator which appears to yield good results combines the spectral estimates from individual quadrants (15.54). In Section 15.7 the 2-D extension of the MESE is described. Unlike the 1-D case the MESE for 2-D is unwieldy to compute and so must be implemented by approximate iterative methods. Also, it differs from the AR spectral estimator since the latter does not have the correlation matching property. The MVSE is described in Section 15.8. It is defined by (15.60), where the theoretical autocorrelation matrix is given by (15.62). Some 2-D techniques are briefly described for sinusoidal frequency estimation in Section 15.9. Finally, computer simulation examples described in Section 15.10 illustrate that in 2-D the relative resolutions of the various spectral estimators are about the same as in 1-D.

15.3 REVIEW OF LINEAR SYSTEMS AND FOURIER TRANSFORMS

A complex 2-D sequence or 2-D array $x[m, n]$ is a function defined over the integers $-\infty < m, n < \infty$ in the 2-D plane. Practical 2-D sequences have finite extent; that is, they are nonzero only over a specified finite region. The region over which the sequence is nonzero is termed the *region of support*. A typical finite-extent sequence is the measured data array, which is always assumed to have a region of support $0 \leq m \leq M - 1; 0 \leq n \leq N - 1$. Note that the measured data may have more samples in one dimension than the other. A class of sequences that is sometimes useful are the *separable* sequences. Separable sequences may be expressed as products of 1-D sequences

$$x[m, n] = x_1[m]x_2[n]. \tag{15.1}$$

These special sequences frequently allow the use of 1-D processing techniques. *Periodic* 2-D sequences repeat themselves at regularly spaced intervals in the 2-D plane. They are defined by

$$x[m + M, n + N] = x[m, n] \tag{15.2}$$

for all m and n. M and N are termed the *fundamental periods*. Important specific sequences are the 2-D unit impulse $\delta[m, n]$, defined as

$$\delta[m, n] = \begin{cases} 1 & \text{for } m = n = 0 \\ 0 & \text{otherwise} \end{cases}$$

and the 2-D unit amplitude complex sinusoid, defined as

$$x[m, n] = \exp\,[j2\pi(f_1 m + f_2 n)]. \tag{15.3}$$

The frequency of the sinusoid is (f_1, f_2).

The ouput $x[m, n]$ of a linear 2-D system to the input $u[m, n]$ may be expressed as

$$x[m, n] = \sum_{i=-\infty}^{\infty} \sum_{j=-\infty}^{\infty} h_{i,j}[m, n]u[i, j] \tag{15.4}$$

where $h_{i,j}[m, n]$ is the response of the linear system at $[m, n]$ to a unit impulse applied at $[i, j]$. If the 2-D system response to a unit impulse does not change in form for impulses applied for arbitrary $[i, j]$ but is only shifted, the 2-D system is *linear shift invariant* (LSI). The output as given by (15.4) then results in the convolution sum

$$x[m, n] = \sum_{i=-\infty}^{\infty} \sum_{j=-\infty}^{\infty} h[m - i, n - j]u[i, j]. \tag{15.5}$$

As in the 1-D case, convolution is commutative, so that the order in which LSI systems are cascaded is immaterial.

A 2-D system is *stable* if its output is bounded for all bounded inputs (BIBO stability). A necessary and sufficient condition for a 2-D LSI system to be BIBO stable is that its impulse response be absolutely summable

$$\sum_{m=-\infty}^{\infty} \sum_{n=-\infty}^{\infty} |h[m, n]| < \infty.$$

Two-dimensional stability is far more difficult to test for than 1-D stability.

A 1-D system is *causal* if its output depends only on the present and past inputs. This constraint is useful for processing signals whose independent variable is time, since this property is a reasonable physical constraint, and additionally it yields systems that are implementable in real time. For 2-D applications, at most only one independent variable can represent time, so that causality is not a natural constraint for 2-D systems. However, one property of causal systems in 1-D that is useful is that of recursive computability of a difference equation. Recursive computability allows us to compute the output based on a finite set of past outputs as well as a finite set of the past and present inputs. Recursive computability can be extended to 2-D if we define a *causal LSI system to be one whose impulse response has support on the nonsymmetric half plane (NSHP)*, as shown in Figure 15.1a. A special case occurs when the support is confined to the quarter plane (QP) as shown in Figure 15.1b. A general 2-D recursive difference equation that characterizes an LSI system is

$$\sum_i \sum_j a[i, j]x[m - i, n - j] = \sum_i \sum_j b[i, j]u[m - i, n - j] \tag{15.6}$$

where $a[0, 0] = 1$. The range of the various summations defines the *order* of the difference equation. If the region of support of the $a[i, j]$'s and $b[i, j]$'s is in the NSHP, the output may be recursively computed as a function of the "past" outputs and the "past" and "present" inputs. "Past" outputs are shown in the example of Figure 15.2, in which (15.6) is computed for $b[i, j] = \delta[i, j]$. Other important properties of 1-D linear systems generalize to 2-D by adopting this definition of causality. One of these is spectral factorization, which is discussed in Section 15.6.

The impulse response of the LSI system characterized by the recursive difference equation of (15.6) has an infinite region of support. If the system is stable and causal, then the region of support may again be shown to be the NSHP. As a special case, if $a[i, j] = \delta[i, j]$, then

$$\begin{aligned} x[m, n] &= \sum_i \sum_j b[i, j]u[m - i, n - j] \\ &= \sum_i \sum_j b[m - i, n - j]u[i, j] \end{aligned}$$

which is the convolution sum implementation of the 2-D FIR filter as given by (15.5). If the region of support for the coefficients is the NSHP, the system is causal.

As in the 1-D case, 2-D complex sinusoids are the eigenfunctions of 2-D LSI systems. When (15.3) is used as an input to a 2-D LSI system, the output is

$$x[m, n] = H(f_1, f_2) \exp[j2\pi(f_1 m + f_2 n)]$$

where

$$H(f_1, f_2) = \sum_m \sum_n h[m, n] \exp[-j2\pi(f_1 m + f_2 n)] \tag{15.7}$$

is the system frequency response. The frequency response is periodic in both the f_1 and f_2 frequency variables, each with fundamental periods of 1. The frequencies are assumed to lie in the interval $-\frac{1}{2} \le f_1 \le \frac{1}{2}$, $-\frac{1}{2} \le f_2 \le \frac{1}{2}$. Since (15.7) is a 2-

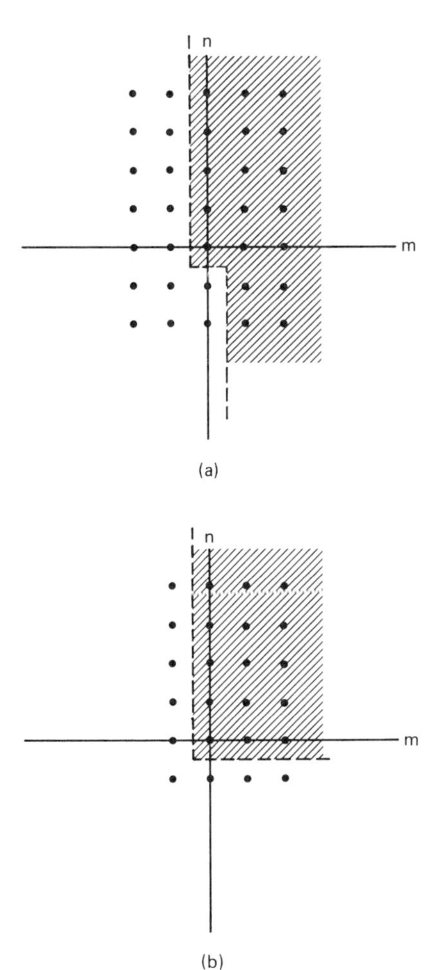

Figure 15.1 Region of support definitions. (a) Nonsymmetric half plane. (b) Quarter plane.

D Fourier transform, the impulse response may be obtained as an inverse 2-D Fourier transform

$$h[m, n] = \int_{-1/2}^{1/2} \int_{-1/2}^{1/2} H(f_1, f_2) \exp [j2\pi(f_1 m + f_2 n)] \, df_1 \, df_2$$

where the integration extends over exactly one period. It is easily verified from (15.7) that if a sequence is separable, the Fourier transform factors into the product of the 1-D Fourier transforms. Other properties of the 2-D Fourier transform are analogous to those of the 1-D Fourier transform and can be found in the book by Dudgeon and Mersereau [1984].

A periodic sequence with fundamental periods M in the horizontal direction and N in the vertical direction was defined in (15.2). It is completely specified by the values in the region $0 \leq m \leq M - 1, 0 \leq n \leq N - 1$. Such a sequence may be represented by a *2-D Fourier series* as

$$x[m, n] = \frac{1}{MN} \sum_{k=0}^{M-1} \sum_{l=0}^{N-1} X[k, l] \exp\left[j2\pi\left(\frac{mk}{M} + \frac{nl}{N}\right)\right] \tag{15.8}$$

where the complex sinusoid is itself periodic with fundamental periods M and N. $X[k, l]$ is termed the *2-D Fourier series coefficient* and is computed as

$$X[k, l] = \sum_{m=0}^{M-1} \sum_{n=0}^{N-1} x[m, n] \exp\left[-j2\pi\left(\frac{mk}{M} + \frac{nl}{N}\right)\right]. \tag{15.9}$$

Note that due to the periodicity of the complex sinusoid the Fourier series coefficients are also *periodic with fundamental periods M and N*. $X[k, l]$ may also be viewed as the Fourier transform of the finite-extent sequence $x[m, n]$ for $0 \leq m$

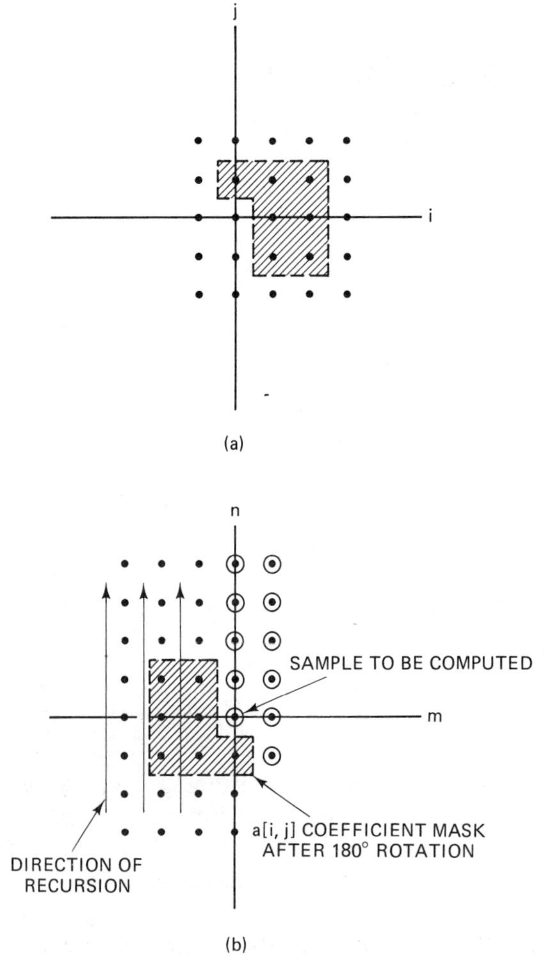

Figure 15.2 Recursive computation of 2-D difference equation. (a) Region of support for a[i,j].(b) Computation of system output.

$\leq M - 1, 0 \leq n \leq N - 1$ when evaluated at $f_1 = k/M$, $f_2 = l/N$ or the Fourier series coefficient is $X(k/M, l/N)$ where

$$X(f_1, f_2) = \sum_{m=0}^{M-1} \sum_{n=0}^{N-1} x[m, n] \exp\left[-j2\pi(f_1 m + f_2 n)\right].$$

(15.9) is then referred to as the 2-D *discrete Fourier transform* (DFT). Direct computation of the DFT requires $4M^2N^2$ real multiplications. It is possible to substantially reduce the amount of computation required by noting that the 2-D DFT may be computed as a series of 1-D DFTs each of which can in turn be computed by a 1-D FFT. The 2-D DFT may be expressed as

$$\begin{aligned} X[k, l] &= \sum_{m=0}^{M-1} G[m, l] \exp\left(\frac{-j2\pi mk}{M}\right) \\ G[m, l] &= \sum_{n=0}^{N-1} x[m, n] \exp\left(\frac{-j2\pi nl}{N}\right). \end{aligned} \tag{15.10}$$

The sequence $G[m, l]$ is obtained by taking the 1-D DFT of each column of $x[m, n]$. Following that operation, which involves M 1-D DFTs, the DFT is taken of each row of the $G[m, l]$ sequence, which involves N 1-D DFTs. Alternatively, we could perform the row DFTs first, followed by the column DFTs. The computational complexity is thus reduced from $4M^2N^2$ real multiplications to only about $2MN \log_2 MN$ real multiplications if 1-D FFTs are employed. The computer program FFT, which is listed in Appendix 2A, may be used to compute the 2-D DFT. As mentioned previously, the frequency samples computed by the DFT are at $f_1 = k/M$, $k = 0, 1, \ldots, M - 1$ and $f_2 = l/N$, $l = 0, 1, \ldots, N - 1$. If the frequency sampling intervals, $1/M$ for f_1 and $1/N$ for f_2, are not fine enough, the sequence may be *zero padded* to yield a finer sampling. This is done by forming a new array

$$x'[m, n] = \begin{cases} x[m, n] & \text{for } 0 \leq m \leq M - 1 \\ & \phantom{\text{for }} 0 \leq n \leq N - 1 \\ 0 & \text{otherwise} \end{cases} \tag{15.11}$$

which has dimensions $M' \times N'$, where $M' > M$, $N' > N$, and by taking the 2-D DFT of $x'[m, n]$. The frequency sample intervals will then be $1/M' < 1/M$ and $1/N' < 1/N$.

The z-transform is a generalization of the Fourier transform which is useful primarily for manipulation of difference equations. The 2-D z-transform of a sequence $x[m, n]$ is defined as

$$\mathscr{X}(z_1, z_2) = \sum_{m=-\infty}^{\infty} \sum_{n=-\infty}^{\infty} x[m, n] z_1^{-m} z_2^{-n}. \tag{15.12}$$

If a sequence composed of the 2-D complex exponential $u[m, n] = z_1^m z_2^n$ is input to an LSI system, then from the convolution sum of (15.5) it is readily seen that the output is

$$x[m, n] = \mathscr{H}(z_1, z_2) z_1^m z_2^n$$

where

$$\mathcal{H}(z_1, z_2) = \sum_{m=-\infty}^{\infty} \sum_{n=-\infty}^{\infty} h[m, n] z_1^{-m} z_2^{-n}.$$

Note that the complex exponential is the eigenfunction of an LSI system, a special case being the complex sinusoid (15.3). $\mathcal{H}(z_1, z_2)$ is the *transfer function* or *system function* of the LSI system and may be defined for a larger class of LSI systems than the frequency response. If $z_1 = \exp(j2\pi f_1)$ and $z_2 = \exp(j2\pi f_2)$, the system function becomes the frequency response

$$\mathcal{H}(\exp[j2\pi f_1], \exp[j2\pi f_2]) = H(f_1, f_2).$$

For LSI systems characterized by recursive difference equations (15.6), the system function can be employed to determine the output as

$$x[m, n] = \mathcal{Z}^{-1}\{\mathcal{H}(z_1, z_2)\mathcal{U}(z_1, z_2)\} \tag{15.13}$$

where $\mathcal{Z}^{-1}$ is the 2-D inverse z-transform (see Dudgeon and Mersereau [1984] for properties) and

$$\mathcal{H}(z_1, z_2) = \frac{\mathcal{B}(z_1, z_2)}{\mathcal{A}(z_1, z_2)}.$$

The 2-D polynomials are defined as

$$\mathcal{A}(z_1, z_2) = \sum_i \sum_j a[i, j] z_1^{-i} z_2^{-j}$$

$$\mathcal{B}(z_1, z_2) = \sum_i \sum_j b[i, j] z_1^{-i} z_2^{-j}$$

which are the z-transforms of the $a[i, j]$ and $b[i, j]$ arrays. It is said that $\mathcal{H}(z_1, z_2)$ possesses a pole at (z_1, z_2) if $\mathcal{A}(z_1, z_2) = 0$ and $\mathcal{B}(z_1, z_2) \neq 0$. Similarly, a zero of the transfer function occurs if $\mathcal{B}(z_1, z_2) = 0$ and $\mathcal{A}(z_1, z_2) \neq 0$. A pole is not a pole in the 1-D sense. Zeros of 1-D polynomials occur at isolated points, while "zeros" of 2-D polynomials are defined by *continuous contours*. As an example, if

$$\mathcal{A}(z_1, z_2) = 1 - z_1^{-1} + 2z_2^{-1} - z_1^{-1}z_2^{-1}$$

then the polynomial is zero whenever

$$z_1 = \frac{1 + z_2^{-1}}{1 + 2z_2^{-1}}$$

which defines a contour.

In 1-D the polynomial $\mathcal{A}(z) = 1 + \sum_{k=1}^{p} a[k]z^{-k}$ can always be factored as

$$\mathcal{A}(z) = \prod_{i=1}^{p} (1 - z_i z^{-1})$$

where the z_i's are the zeros of the polynomial. This factorization is guaranteed by the *fundamental theorem of algebra*. However, in 2-D there is no fundamental theorem of algebra. Thus 2-D polynomials cannot be factored into the product of first order polynomials. As a consequence, many important analytical and implementational tools are not available. An example is stability testing, which becomes much more difficult in 2-D. Some problems illustrating some of the concepts described are given in Problems 15.1–15.5.

15.4 REVIEW OF RANDOM PROCESSES

The properties of a 2-D random process, or more appropriately named a 2-D random field, are now briefly discussed. Let $x[m, n]$ denote the value of the random field at the point $[m, n]$. No particular physical meaning will be attributed to the field, although as discussed in the introduction, there are many physical situations that give rise to such a field. $x[m, n]$ will be assumed to be a WSS or homogeneous field [Yaglom 1962]. Wide sense stationarity in the 2-D case requires that the mean does not depend on the location of the point and the ACF depends only on the distance and orientation between the two points (i.e., the ACF is only a function of the vector joining the two points). More specifically, denote the mean of $x[m, n]$ as $\mu[m, n]$ and the ACF as $r_{xx}[k, l]$. Then, for a WSS random field,

$$\mathcal{E}(x[m, n]) = \mu[m, n] = \mu \tag{15.14}$$

which will henceforth be assumed equal to zero, and

$$r_{xx}[k, l] = \mathcal{E}(x^*[m, n]x[m + k, n + l]) \tag{15.15}$$

does not depend on $[m, n]$ but only on the vector $\mathbf{r} = [k, l]$ from one point to the other. Some special random fields are the isotropic and separable fields. An isotropic field is one whose ACF depends only on the distance $|\mathbf{r}| = \sqrt{k^2 + l^2}$ between the two points [Larimore 1977]. As an example, consider

$$r_{xx}[k, l] = \exp(-\alpha\sqrt{k^2 + l^2}) = \exp(-\alpha|\mathbf{r}|). \tag{15.16}$$

The correlation between any two points that are separated by a distance $|\mathbf{r}|$ is the same, regardless of orientation, hence the name "isotropic." A separable field is one whose ACF can be factored as

$$r_{xx}[k, l] = r_{yy}[k]r_{zz}[l] \tag{15.17}$$

where $r_{yy}[k]$, $r_{zz}[l]$ are valid 1-D ACFs. As an example, if

$$r_{xx}[k, l] = \frac{\sigma^2}{(1 - a^2[1])(1 - b^2[1])}(-a[1])^{|k|}(-b[1])^{|l|} \tag{15.18}$$

for a real field, the field is separable. The significance of this property is that 2-D spectral estimation may be accomplished using 1-D spectral estimators, as will be described shortly.

The 2-D ACF has the hermitian symmetry property

$$r_{xx}[-k, -l] = r_{xx}^*[k, l]$$

which follows from (15.15). Due to this property the ACF need only be estimated on the half plane, one possibility which is shown in Figure 15.1a. Also, as required, $r_{xx}[0, 0]$ is real and from (15.15) $r_{xx}[0, 0]$ is nonnegative. The ACF must be a positive semidefinite sequence in order to guarantee that its Fourier transform will be nonnegative. The Fourier transform of the ACF will be defined shortly to be the PSD. A 2-D positive semidefinite sequence has the property that (see Problem 15.6)

$$\sum_i \sum_j \sum_k \sum_l a[i, j]a^*[k, l]r_{xx}[k - i, l - j] \geq 0 \tag{15.19}$$

for all $a[i, j]$ sequences.

As in the 1-D case, $x[m, n]$ is assumed to be ergodic in the ACF. Making this assumption allows a temporal autocorrelation estimator such as (15.29) to be used as a consistent estimator of the ensemble ACF.

A 2-D PSD can be defined via the 2-D extension of the Wiener–Khinchin theorem as

$$P_{xx}(f_1, f_2) = \sum_{k=-\infty}^{\infty} \sum_{l=-\infty}^{\infty} r_{xx}[k, l] \exp\left[-j2\pi(f_1 k + f_2 l)\right]. \tag{15.20}$$

The PSD is just the 2-D Fourier transform of the ACF. It immediately follows that

$$r_{xx}[k, l] = \int_{-1/2}^{1/2} \int_{-1/2}^{1/2} P_{xx}(f_1, f_2) \exp\left[j2\pi(f_1 k + f_2 l)\right] df_1\, df_2. \tag{15.21}$$

The 2-D PSD is real and nonnegative. If the field is real so that the ACF is real, then $r_{xx}[-k, -l] = r_{xx}[k, l]$ and (15.20) becomes

$$P_{xx}(f_1, f_2) = \sum_{k=-\infty}^{\infty} \sum_{l=-\infty}^{\infty} r_{xx}[k, l] \cos\left[2\pi(f_1 k + f_2 l)\right]. \tag{15.22}$$

It follows that for a real field

$$P_{xx}(-f_1, -f_2) = P_{xx}(f_1, f_2)$$

which is the extension of the even-symmetry property of the PSD for real 1-D time series. Hence, for complex random fields the PSD needs to be estimated over the entire region $-\frac{1}{2} \leq f_1 \leq \frac{1}{2}$, $-\frac{1}{2} \leq f_2 \leq \frac{1}{2}$, while for real random fields a subset of this region which includes any two adjacent quadrants is sufficient (see also Problem 15.7).

If the ACF is separable, the 2-D PSD may be estimated using 1-D spectral estimators. In this case substituting (15.17) into (15.20) yields

$$\begin{aligned} P_{xx}(f_1, f_2) &= \sum_{k=-\infty}^{\infty} \sum_{l=-\infty}^{\infty} r_{yy}[k]r_{zz}[l] \exp\left[-j2\pi(f_1 k + f_2 l)\right] \\ &= P_{yy}(f_1)P_{zz}(f_2) \end{aligned}$$

where

$$P_{yy}(f_1) = \sum_{k=-\infty}^{\infty} r_{yy}[k] \exp(-j2\pi f_1 k)$$

$$P_{zz}(f_2) = \sum_{l=-\infty}^{\infty} r_{zz}[l] \exp(-j2\pi f_2 l).$$

To estimate the 2-D PSD, we need only estimate $P_{yy}(f_1)$ and $P_{zz}(f_2)$ (see Problem 15.8). The 1-D ACFs which are needed are available from the relationships

$$\begin{aligned} r_{yy}[k] &= \frac{r_{xx}[k, 0]}{\sqrt{r_{xx}[0, 0]}} \\ r_{zz}[l] &= \frac{r_{xx}[0, l]}{\sqrt{r_{xx}[0, 0]}} \end{aligned} \tag{15.23}$$

and are unique to within a scale factor. An example of spectral estimation for separable fields is given in Problem 15.19.

Finally, if $x[m, n]$ is assumed to have been generated as the output of a stable LSI filter with system function $\mathcal{H}(z_1, z_2)$ excited at the input by a WSS random field $u[m, n]$ with PSD $P_{uu}(f_1, f_2)$, then the PSD is (see Problem 15.9)

$$P_{xx}(f_1, f_2) = |\mathcal{H}(\exp[j2\pi f_1], \exp[j2\pi f_2])|^2 P_{uu}(f_1, f_2). \tag{15.24}$$

15.5 CLASSICAL SPECTRAL ESTIMATION

Two-dimensional estimation methods based on Fourier methods are straightforward extensions of the 1-D estimators described in Chapter 4. Assuming a data array $x[m, n]$, $m = 0, 1, \ldots, M - 1$; $n = 0, 1, \ldots, N - 1$, the periodogram spectral estimator is

$$\hat{P}_{\text{PER}}(f_1, f_2) = \frac{1}{MN} \left| \sum_{m=0}^{M-1} \sum_{n=0}^{N-1} x[m, n] \exp[-j2\pi(f_1 m + f_2 n)] \right|^2. \tag{15.25}$$

To obtain a statistically stable estimate, an averaged periodogram can be used, in which the data array is sectioned into smaller subarrays, after which the periodogram for each subarray is averaged. Alternatively, the Welch method can be employed. The data are first segmented into *overlapping* subarrays. Then the data in each subarray are windowed. Finally, the periodograms of the windowed subarrays are averaged. Both techniques are described more fully in Chapter 4. To compute the periodogram a 2-D DFT as defined by (15.9) and implemented as an FFT can be used. For adequate frequency sampling the data should be zero padded to form an array of size $M' \times N'$, where $M' > M$, $N' > N$. As explained in Section 15.3, a 1-D FFT is applied to each column of data followed by each row. For an $(M' \times N')$-sized FFT, the computational complexity is about $2M'N' \log_2 M'N'$ real multiplies. This may be reduced further by "pruning" the FFT. This computational complexity is rather modest compared to the "high resolution" methods to be described. The computer program FFT, which is contained in Appendix 2A, can be used to compute the periodogram.

The resolution of the periodogram depends on the data extent in each dimension. Because of the possibility that $M \neq N$, the resolution in one frequency direction may be better than in the other. As an example, for a single complex sinusoid

$$x[m, n] = \exp\{j[2\pi(\nu_1 m + \nu_2 n) + \phi]\}$$

it is easily shown that the periodogram is (see Problem 15.4)

$$\hat{P}_{\mathrm{PER}}(f_1, f_2) = \frac{1}{MN}\left|\frac{\sin[M\pi(f_1 - \nu_1)]}{\sin[\pi(f_1 - \nu_1)]}\right|^2 \left|\frac{\sin[N\pi(f_2 - \nu_2)]}{\sin[\pi(f_2 - \nu_2)]}\right|^2. \tag{15.26}$$

The contours of $\hat{P}_{\mathrm{PER}}(f_1, f_2)$ for a single sinusoid are elliptical, which implies that the resolution will depend on the orientation of the sinusoidal frequencies. An example is given in Figure 15.3. If $M = N$, the contours are approximately circularly symmetric and the resolution does not depend on orientation. The -3 dB contour is approximately at a distance of $0.42/M$ away from the peak (see Problem 15.10). Hence two sinusoids are just resolved if

$$\sqrt{\Delta f_1^2 + \Delta f_2^2} = \frac{0.84}{M} \tag{15.27}$$

where Δf_i is the difference of the sinusoidal frequencies in the f_i direction.

The Blackman–Tukey spectral estimator for 2-D data is defined analogously to that for 1-D data as

$$\hat{P}_{\mathrm{BT}}(f_1, f_2) = \sum_{k=-K}^{K} \sum_{l=-L}^{L} w[k, l]\hat{r}_{xx}[k, l] \exp[-j2\pi(f_1 k + f_2 l)] \tag{15.28}$$

where $w[k, l]$ is a window function used to trade bias for variance exactly as in the 1-D case and $\hat{r}_{xx}[k, l]$ is a positive semidefinite estimator of the ACF. The biased ACF estimator should be used to ensure a nonnegative spectral estimate.

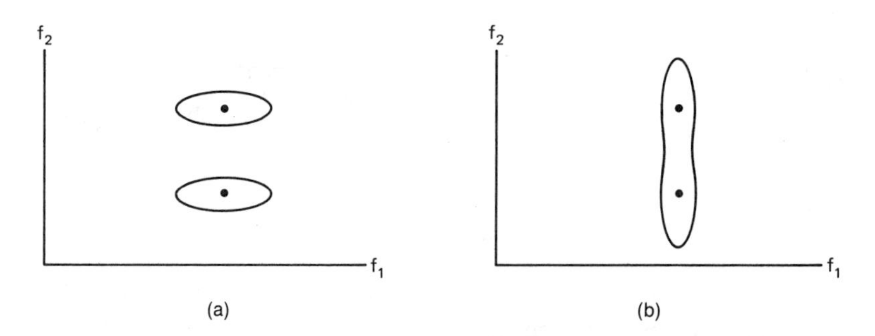

Figure 15.3 -3 dB contour of periodogram for two complex sinusoids. (a) Resolved spectral peaks ($N > M$). (b) Unresolved spectral peaks ($N < M$).

It is defined as

$$\hat{r}_{xx}[k, l] = \begin{cases} \dfrac{1}{MN} \displaystyle\sum_{m=0}^{M-1-k} \sum_{n=0}^{N-1-l} x^*[m, n]x[m + k, n + l] & \text{for } \begin{matrix} k \geq 0 \\ l \geq 0 \end{matrix} \\ \dfrac{1}{MN} \displaystyle\sum_{m=0}^{M-1-k} \sum_{n=-l}^{N-1} x^*[m, n]x[m + k, n + l] & \text{for } \begin{matrix} k > 0 \\ l < 0 \end{matrix}. \end{cases} \tag{15.29}$$

The remaining ACF estimates for $k < 0, l \geq 0$ and $k \leq 0, l < 0$ are found by imposing the hermitian symmetry property

$$\hat{r}_{xx}[-k, -l] = \hat{r}^*_{xx}[k, l].$$

The biased ACF estimator can be shown to be positive semidefinite (see Problem 15.11). The window function $w[k, l]$ should also be positive semidefinite to ensure a nonnegative spectral estimate. One possibility is the 2-D Bartlett window

$$w[k, l] = \begin{cases} \left(1 - \dfrac{|k|}{K}\right)\left(1 - \dfrac{|l|}{L}\right) & \text{for } \begin{matrix} |k| \leq K \\ |l| \leq L \end{matrix} \\ 0 & \text{otherwise.} \end{cases}$$

Since the window is separable, the Fourier transform is the product of the 1-D Fourier transforms, both of which are nonnegative. More generally, 2-D positive semidefinite windows are easily synthesized as products of 1-D positive semidefinite windows. Very little is known about the synthesis of more general 2-D windows and their role in spectral estimation [Huang 1972]. As in the 1-D case the BT spectral estimator is equivalent to the periodogram if $K = M - 1$, $L = N - 1$ and $w[k, l] = 1$.

The statistical performance of the 2-D Fourier spectral estimators is similar to that for the 1-D estimators. Although the extension of analytical results appears to be straightforward, they are not readily available. Some computer simulation examples for the periodogram are given in Section 15.10.

It is also possible to combine Fourier spectral estimation with "high resolution" spectral estimation. A "hybrid" technique for 2-D spectral estimation is valuable if the desired resolution in one dimension is attainable by Fourier methods. A common situation in which this occurs is in frequency–wavenumber estimation for array processing. The $x[m, n]$ data represent the output of the mth sensor at time n. Usually, the temporal resolution is adequate, but the spatial resolution is low due to a limited physical array of sensors [Capon 1969]. We can, in this situation, apply a DFT to each column of time samples, which acts as a narrowband filter, to generate the array $G[m, l]$ given in (15.10). Then a "high resolution" spectral estimator can be applied to each row of $G[m, l]$ to produce a spatial spectral estimate at each temporal frequency. This approach may be interpreted as a *narrowband beamformer*, where the beamforming is effected using other than Fourier techniques [Dudgeon and Mersereau 1984]. An example of this approach is given in the paper by Joyce [1979].

15.6 AUTOREGRESSIVE SPECTRAL ESTIMATION

15.6.1 Two-Dimensional Rational Transfer Function Models

For the purposes of spectral estimation it is convenient to model $x[m, n]$ as the output of a 2-D LSI filter driven by white noise. The white noise excitation, which will be denoted as $u[m, n]$, has an ACF

$$r_{uu}[k, l] = \sigma^2 \delta[k, l].$$

In general, the random field is then represented as

$$x[m, n] = \sum_i \sum_j h[m - i, n - j]u[i, j]$$

where $h[m, n]$ is the impulse response of the LSI filter. The range of summation has purposely been left unspecified. If a "causal" filter is desired, $h[m, n]$ should take on nonzero values only in the NSHP, as discussed in Section 15.3. A causal filter is henceforth assumed. Given this model the PSD is found from (15.24), where the system function is

$$\mathcal{H}(z_1, z_2) = \sum_m \sum_n h[m, n] z_1^{-m} z_2^{-n}$$

and

$$P_{uu}(f_1, f_2) = \sigma^2.$$

The specific rational transfer function models that will be of interest are the 2-D extensions of the time series models introduced in Chapter 5. The system functions for the 2-D AR, MA, and ARMA models are, respectively,

$$\mathcal{H}(z_1, z_2) = \frac{1}{\sum_i \sum_j a[i, j] z_1^{-i} z_2^{-j}}$$

$$\mathcal{H}(z_1, z_2) = \sum_i \sum_j b[i, j] z_1^{-i} z_2^{-j}$$

$$\mathcal{H}(z_1, z_2) = \frac{\sum_i \sum_j b[i, j] z_1^{-i} z_2^{-j}}{\sum_i \sum_j a[i, j] z_1^{-i} z_2^{-j}}$$

where $a[0, 0] = b[0, 0] = 1$. Alternatively, the 2-D AR, MA, and ARMA models are, respectively, given by the difference equation representations

$$
\begin{aligned}
x[m, n] &= -\sum_{\substack{i \quad j \\ [i,j]\neq[0,0]}}\sum a[i, j]x[m-i, n-j] + u[m, n] \\
x[m, n] &= \sum_i \sum_j b[i, j]u[m-i, n-j] \\
x[m, n] &= -\sum_{\substack{i \quad j \\ [i,j]\neq[0,0]}}\sum a[i, j]x[m-i, n-j] \\
&\quad + \sum_i \sum_j b[i, j]u[m-i, n-j].
\end{aligned}
\tag{15.30}
$$

It may be shown that for a stable and causal system the region of support for the AR and MA parameters must be in the NSHP. Then we can compute the sequence $x[m, n]$ in a recursive manner (see Problems 15.12 and 15.13). An example is given in Figure 15.2 for a column-wise raster computation of the AR time series.

From the viewpoint of spectral estimation the constraint of a stable and causal filter is questionable. Jain [1981] as well as Woods [1972] and Pendrel [1979] have proposed several noncausal models for spectral estimation. If the model is motivated by a physical process where there is a natural ordering of the sequence, as in spatial-temporal samples obtained from an array, the constraints are more reasonable. Or if it is necessary to perform filtering of the data in order to estimate the PSD, then a stable causal representation may be preferable. Suffice it to say that the question of stability and causality for spectral estimation has not been resolved. The discussion to follow, which will be restricted to 2-D AR models, will assume causality. (See the paper by Cadzow and Ogino [1981] for spectral estimation methods based on ARMA models.) A strong motivation for doing so is that any PSD may be modeled by an AR process under the following conditions [Whittle 1954]:

1. The region of support for the AR parameters is the NSHP.
2. The number of AR parameters allowed in the model is infinite.

This is the 2-D extension of the property that any 1-D PSD may be modeled by an AR(∞) model as discussed in Section 5.3.2. Further discussions of this property are included in the next section.

15.6.2 Yule–Walker Equations

The Yule–Walker equations for the 2-D AR model are now derived. Analogous equations for the 2-D ARMA model can be found in a similar manner. Let the region of support for the AR parameters (including $a[0, 0] = 1$) be denoted by S_{TNSHP}, where

$$
\begin{aligned}
S_{\text{TNSHP}} &= \{[i, j]: i = 1, 2, \ldots, p_1; j = -p_2, \ldots, 0, \ldots, p_2\} \\
&\quad \cup \{[i, j]: i = 0; j = 0, 1, \ldots, 0, \ldots, p_2\}
\end{aligned}
$$

and TNSHP denotes the truncated NSHP. Figure 15.4 gives some examples of this truncated NSHP region of support. Note that the NSHP, denoted by S_{NSHP}, is obtained by letting $p_1 \to \infty$, $p_2 \to \infty$. Multiplying $x[m, n]$ as given by the first equation in (15.30) by $x^*[m - k, n - l]$ and taking the expected value yields

$$\sum_{[i,j] \in S_{\text{TNSHP}}} \sum a[i, j] r_{xx}[k - i, l - j] = \mathscr{E}(x^*[m - k, n - l] u[m, n])$$

$$= \mathscr{E}(x^*[-k, -l] u[0, 0])$$

since all fields are assumed to be WSS. But from (15.5)

$$x[-k, -l] = \sum_{[i,j] \in S_{\text{NSHP}}} \sum h[i, j] u[-k - i, -l - j].$$

Note that the region of support of the impulse response is the entire NSHP, which follows from the causal and stable filter assumption. Hence

$$\mathscr{E}(x^*[-k, -l] u[0, 0]) = \sum_{[i,j] \in S_{\text{NSHP}}} \sum h^*[i, j] \mathscr{E}(u^*[-k - i, -l - j] u[0, 0])$$

$$= h^*[-k, -l] \sigma^2.$$

Because $h[k, l]$ is the impulse response of a causal filter

$$\mathscr{E}(x^*[-k, -l] u[0, 0]) = \begin{cases} 0 & \text{for } [k, l] \in S'_{\text{NSHP}} \\ h[0, 0] \sigma^2 & \text{for } [k, l] = [0, 0] \end{cases}$$

where S'_{NSHP} denotes the NSHP with the $[0, 0]$ point deleted or $S_{\text{NSHP}} = S'_{\text{NSHP}} \cup [0, 0]$. $h[0, 0]$ may be shown to be unity (see Problem 15.14) so that the 2-D Yule–Walker equations become

$$\sum_{[i,j] \in S_{\text{TNSHP}}} \sum a[i, j] r_{xx}[k - i, l - j] = \begin{cases} 0 & \text{for } [k, l] \in S'_{\text{NSHP}} \\ \sigma^2 & \text{for } [k, l] = [0, 0]. \end{cases} \tag{15.31}$$

These equations should be compared to those for the 1-D case given by (5.17). From the ACF $r_{xx}[k, l]$ for $[k, l] \in S_{\text{TNSHP}}$, we can determine the AR parameters by solving a set of linear equations. As an example, consider the case of $p_1 = p_2$

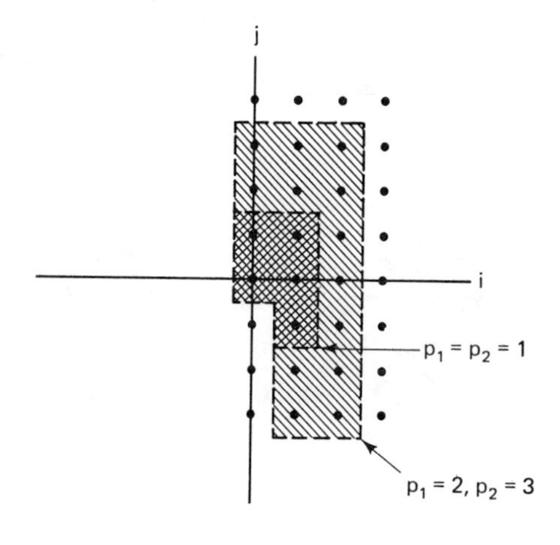

Figure 15.4 Examples of truncated nonsymmetric half plane region of support for AR parameters.

$= 1$ so that the unknown AR parameters are $\{a[0, 1], a[1, -1], a[1, 0], a[1, 1]\}$, as illustrated in Figure 15.4. Then from (15.31) with $[k, l] \in S'_{\text{TNSHP}} = \{[0, 1], [1, -1], [1, 0], [1, 1]\}$,

$$\begin{aligned} r_{xx}[k, l] = &- a[0, 1]r_{xx}[k, l - 1] - a[1, -1]r_{xx}[k - 1, l + 1] \\ &- a[1, 0]r_{xx}[k - 1, l] - a[1, 1]r_{xx}[k - 1, l - 1] \qquad [k, l] \in S'_{\text{TNSHP}} \\ \sigma^2 = &\ r_{xx}[0, 0] + a[0, 1]r_{xx}[0, -1] + a[1, -1]r_{xx}[-1, 1] \\ &+ a[1, 0]r_{xx}[-1, 0] + a[1, 1]r_{xx}[-1, -1] \end{aligned}$$

or in matrix form

$$\begin{bmatrix} r_{xx}[0, 0] & r_{xx}[0, -1] & r_{xx}[-1, 1] & r_{xx}[-1, 0] & r_{xx}[-1, -1] \\ r_{xx}[0, 1] & r_{xx}[0, 0] & r_{xx}[-1, 2] & r_{xx}[-1, 1] & r_{xx}[-1, 0] \\ r_{xx}[1, -1] & r_{xx}[1, -2] & r_{xx}[0, 0] & r_{xx}[0, -1] & r_{xx}[0, -2] \\ r_{xx}[1, 0] & r_{xx}[1, -1] & r_{xx}[0, 1] & r_{xx}[0, 0] & r_{xx}[0, -1] \\ r_{xx}[1, 1] & r_{xx}[1, 0] & r_{xx}[0, 2] & r_{xx}[0, 1] & r_{xx}[0, 0] \end{bmatrix} \begin{bmatrix} 1 \\ a[0, 1] \\ a[1, -1] \\ a[1, 0] \\ a[1, 1] \end{bmatrix} = \begin{bmatrix} \sigma^2 \\ 0 \\ 0 \\ 0 \\ 0 \end{bmatrix}. \tag{15.32}$$

The matrix is hermitian due to the hermitian symmetry property of the ACF and can be shown to be positive semidefinite.

A special case of the NSHP AR model is the quarter plane (QP) model. For this model the region of support of the AR parameters is the quarter plane, examples of which are given in Figure 15.5. For this example, the Yule–Walker equations can be found from (15.32) be deleting the third row and third column to yield

$$\left[\begin{array}{cc|cc} r_{xx}[0, 0] & r_{xx}[0, -1] & r_{xx}[-1, 0] & r_{xx}[-1, -1] \\ r_{xx}[0, 1] & r_{xx}[0, 0] & r_{xx}[-1, 1] & r_{xx}[-1, 0] \\ \hline r_{xx}[1, 0] & r_{xx}[1, -1] & r_{xx}[0, 0] & r_{xx}[0, -1] \\ r_{xx}[1, 1] & r_{xx}[1, 0] & r_{xx}[0, 1] & r_{xx}[0, 0] \end{array}\right] \begin{bmatrix} 1 \\ a[0, 1] \\ a[1, 0] \\ a[1, 1] \end{bmatrix} = \begin{bmatrix} \sigma^2 \\ 0 \\ 0 \\ 0 \end{bmatrix}. \tag{15.33}$$

Note that the matrix is hermitian and block Toeplitz, with the blocks given by the indicated partitions. Also, to solve for the four unknown AR parameters it is necessary to estimate five samples of the ACF. This is in contrast to the 1-D case, in which the number of ACF samples needed is equal to the number of AR parameters to be estimated (see Problem 15.15). In general, the region of support for the QP model is $S_{\text{TQP}} = \{[i, j]: i = 0, 1, \ldots, p_1; j = 0, 1, \ldots, p_2\}$. The Yule–Walker equations of (15.31) then reduce to

$$\sum_{i=0}^{p_1} \sum_{j=0}^{p_2} a[i, j]r_{xx}[k - i, l - j] = \begin{cases} 0 & \text{for } [k, l] \in S'_{\text{QP}} \\ \sigma^2 & \text{for } [k, l] = [0, 0] \end{cases} \tag{15.34}$$

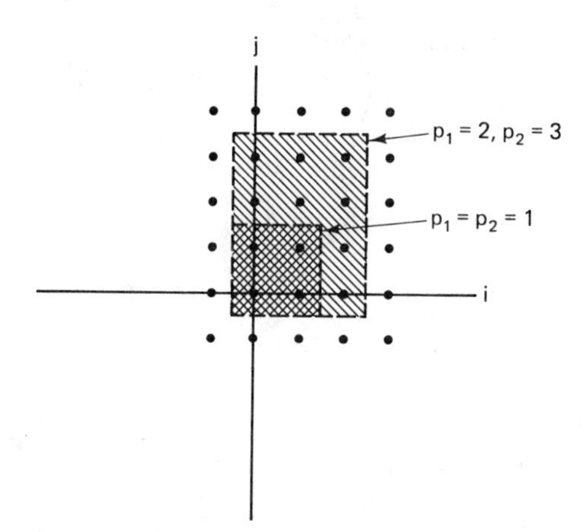

Figure 15.5 Examples of truncated quarter plane region of support for AR model parameters.

where $S_{QP} = \{[i, j]: i \geq 0, j \geq 0\}$ and S'_{QP} denotes the set S_{QP} with the point [0, 0] deleted or $S_{QP} = S'_{QP} \cup [0, 0]$. The Yule–Walker equations for the QP model can be used to determine the AR parameters from the ACF samples. Generalizing (15.33) leads to the equations

$$\underline{\mathbf{R}}_{xx}\underline{\mathbf{a}} = \underline{\mathbf{h}} \tag{15.35}$$

where $\underline{\mathbf{R}}_{xx}$ has dimension $(p_1 + 1)(p_2 + 1) \times (p_1 + 1)(p_2 + 1)$ and $\underline{\mathbf{a}}$ and $\underline{\mathbf{h}}$ both have dimension $(p_1 + 1)(p_2 + 1) \times 1$. More explicitly, (15.35) can be written as (see Problem 15.16)

$$\begin{bmatrix} \mathbf{R}_{xx}[0] & \mathbf{R}_{xx}[-1] & \cdots & \mathbf{R}_{xx}[-p_1] \\ \mathbf{R}_{xx}[1] & \mathbf{R}_{xx}[0] & \cdots & \mathbf{R}_{xx}[-(p_1 - 1)] \\ \vdots & \vdots & \ddots & \vdots \\ \mathbf{R}_{xx}[p_1] & \mathbf{R}_{xx}[p_1 - 1] & \cdots & \mathbf{R}_{xx}[0] \end{bmatrix} \begin{bmatrix} \mathbf{a}[0] \\ \mathbf{a}[1] \\ \vdots \\ \mathbf{a}[p_1] \end{bmatrix} = \begin{bmatrix} \sigma^2\mathbf{h}_1 \\ \mathbf{0} \\ \vdots \\ \mathbf{0} \end{bmatrix} \tag{15.36}$$

where

$$\mathbf{a}[i] = [a[i, 0] \quad a[i, 1] \quad \cdots \quad a[i, p_2]]^T \qquad (p_2 + 1) \times 1$$

$$\mathbf{h}_1 = [1\ 0 \cdots 0]^T \qquad (p_2 + 1) \times 1$$

and

$$\mathbf{R}_{xx}[i] = \begin{bmatrix} r_{xx}[i, 0] & r_{xx}[i, -1] & \cdots & r_{xx}[i, -p_2] \\ r_{xx}[i, 1] & r_{xx}[i, 0] & \cdots & r_{xx}[i, -(p_2 - 1)] \\ \vdots & \vdots & \ddots & \vdots \\ r_{xx}[i, p_2] & r_{xx}[i, p_2 - 1] & \cdots & r_{xx}[i, 0] \end{bmatrix}$$

$$(p_2 + 1) \times (p_2 + 1).$$

In this form the equations are analogous to the augmented Yule–Walker equations for 1-D AR processes [see (5.19)]. Note that $\mathbf{R}_{xx}[-i] = \mathbf{R}^H_{xx}[i]$, so that $\underline{\mathbf{R}}^H_{xx} = \underline{\mathbf{R}}_{xx}$ or $\underline{\mathbf{R}}_{xx}$ is hermitian. It may also be shown to be positive semidefinite. Furthermore, $\underline{\mathbf{R}}_{xx}$ is doubly block Toeplitz; that is, the matrix itself has a block Toeplitz structure and each individual block is Toeplitz since $[\mathbf{R}_{xx}[i]]_{mn} = r_{xx}[i, m - n]$, although not hermitian. The block Toeplitz property enables us to solve (15.36) by a slight modification of the efficient multichannel Levinson algorithm described in Section 14.7 [Therrien 1981]. A computer program to solve (15.36) has been provided by Marple [1987].

In contrast to the 1-D case there is no autocorrelation matching property of AR spectral estimation (either for the QP or NSHP model) in 2-D. This follows from the observation that it requires more ACF samples than AR parameters to be estimated to solve the Yule–Walker equations [see (15.32) and (15.33)]. On the other hand, the 2-D MESE to be discussed in Section 15.7 does have this property. It is immediately obvious that *AR spectral estimation and MESE are not equivalent in 2-D*. Whether a good spectral estimator should be constrained to have the matching property has yet to be determined.

15.6.3 Autoregressive Modeling and Prediction Theory

As in the 1-D case there is a strong connection between AR modeling and linear prediction theory. To predict $x[m, n]$ as a linear combination of "past" data samples, we could use

$$\hat{x}[m, n] = - \sum_{[i,j] \in S'} \sum a[i, j] x[m - i, n - j]. \tag{15.37}$$

S' will denote either the truncated NSHP or the truncated QP with the [0, 0] point omitted. To determine the optimal set of prediction coefficients the prediction error power

$$\begin{aligned} \rho &= \mathscr{E}\left(\,|\, x[m, n] - \hat{x}[m, n] \,|^2\right) \\ &= \mathscr{E}\left(\left| \sum_{[i,j] \in S} \sum a[i, j] x[m - i, n - j] \right|^2 \right) \end{aligned} \tag{15.38}$$

is minimized. Using a similar development to that of Section 6.3.1, it can be shown that

$$\sum_{[i,j] \in S} \sum a[i, j] r_{xx}[k - i, l - j] = \begin{cases} 0 & \text{for } [k, l] \in S' \\ \rho_{\text{MIN}} & \text{for } [k, l] = [0, 0]. \end{cases} \tag{15.39}$$

These equations are identical to the Yule–Walker equations, which implies that if a random field is a 2-D AR process with support S, the optimal linear prediction coefficients of support S are just the AR filter parameters. Also, the minimum prediction error power ρ_{MIN} is equal to the white noise variance. An important consequence of this result is the following. Assume that we choose a NSHP or QP AR model for the data, compute the AR parameter estimates from (15.39) assuming a known ACF $r_{xx}[k, l]$ (not necessarily the ACF of an AR process), and determine the spectral estimate as

$$P_{\text{AR}}(f_1, f_2) = \frac{\rho_{\text{MIN}}}{\left| \sum_{[k,l] \in S} \sum a[k, l] \exp\left[-j2\pi(f_1 k + f_2 l)\right] \right|^2}. \tag{15.40}$$

Then even if the ACF is known exactly and the region of support is infinite, only the NSHP will in general yield the correct PSD [Whittle 1954]. This is attributable to the property that an arbitrary 2-D PSD can only be spectrally factored as

$$\frac{\sigma^2}{\mathcal{A}(z_1, z_2)\mathcal{A}^*(1/z_1^*, 1/z_2^*)}$$

if the region of support for the AR filter coefficients is the infinite NSHP [Ekstrom and Woods 1976]. One way to verify this result is by examining the whiteness of the prediction error for the infinite QP and NSHP. It is shown in Problem 15.18 that if the region of support for the prediction coefficients is the QP, the ACF of the prediction error will be zero on that QP and the one obtained by rotating it 180°. This is illustrated in Figure 15.6. For the NSHP the ACF of the prediction

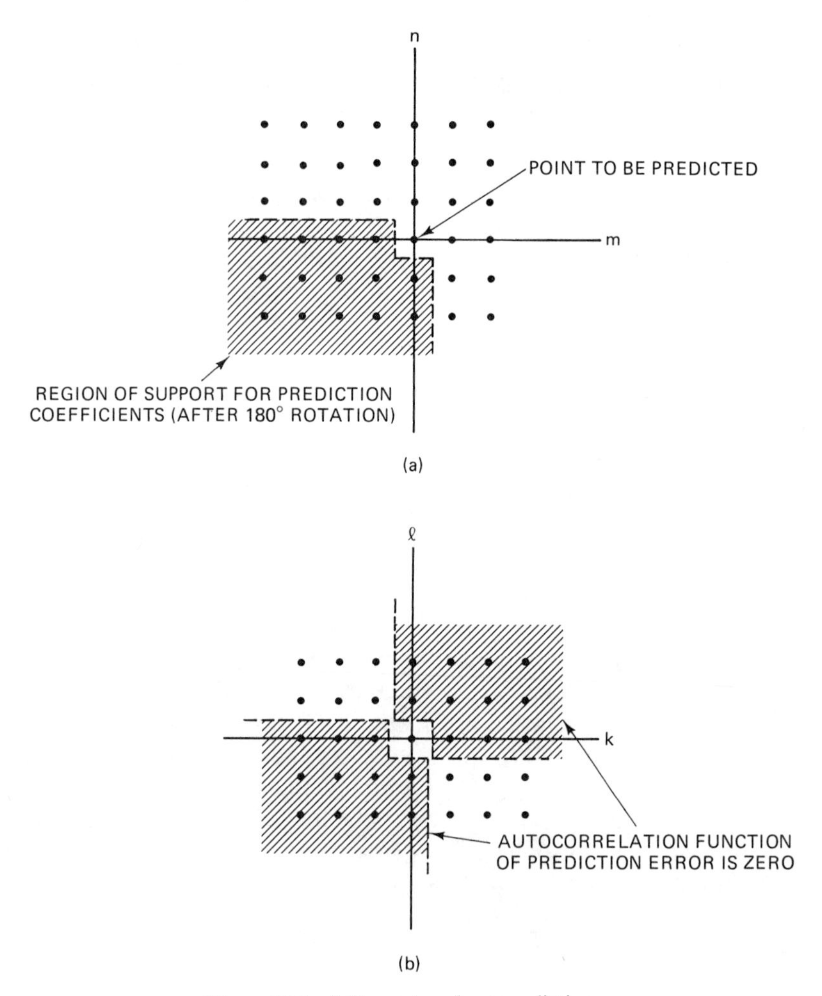

Figure 15.6 2-D quarter plane prediction.

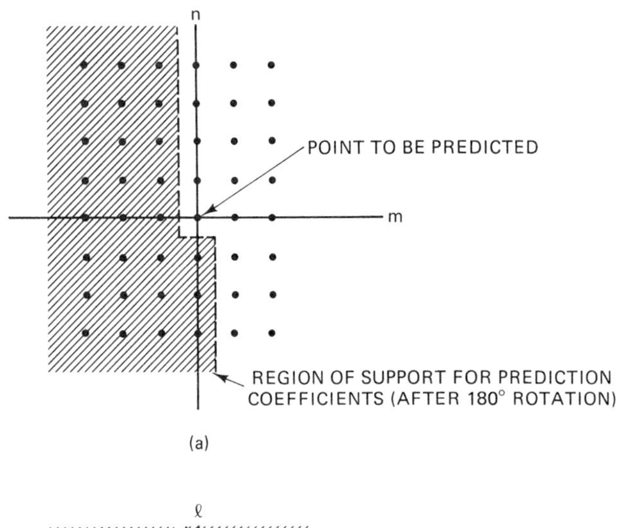

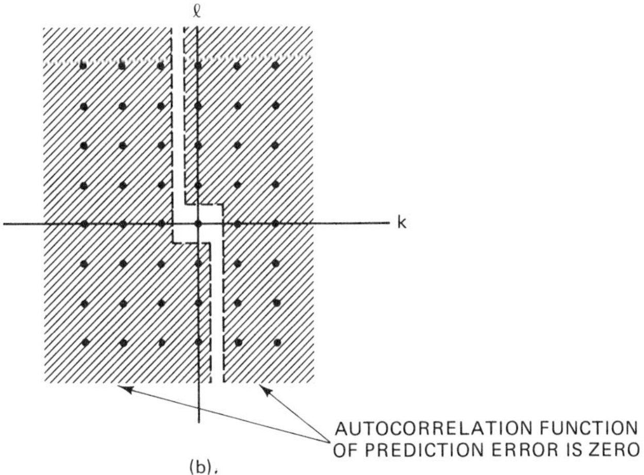

Figure 15.7 2-D nonsymmetric half plane prediction.

error will be zero for all $[k, l] \neq [0, 0]$ as shown in Figure 15.7, and hence the prediction error is white. Only the NSHP will yield the true PSD in general for an infinite region of support [Whittle 1954].

Many other concepts of 1-D linear prediction can be extended to 2-D. Some examples are the Levinson recursion [Justice 1977], the reflection coefficients [Marzetta 1980], and the lattice filter [Parker and Kayran 1984].

15.6.4 Methods for Autoregressive Spectral Estimation

As described in the previous sections the causal AR models are based on the NSHP and the QP. It has been observed from simulations that for sinusoidal signals in noise spectral estimators based on the NSHP perform poorly [Pendrel

1979]. The reason is not known, although it is possible that the NSHP requires too high an order. The discussion to follow will concentrate on the QP AR model whose Yule–Walker equations are given by (15.36). The spectral estimator will be given by

$$\hat{P}_{AR}(f_1, f_2) = \frac{\hat{\sigma}^2}{\left| \sum_{m=0}^{p_1} \sum_{n=0}^{p_2} \hat{a}[m, n] \exp\left[-j2\pi(f_1 m + f_2 n)\right] \right|^2} \quad (15.41)$$

where $\hat{\sigma}^2$, $\hat{a}[m, n]$ are the estimates of the AR parameters. For separable processes 1-D estimation methods may be used, an example of which is given in Problem 15.19. In general, we could substitute an estimate of the ACF into (15.36) and solve for the AR parameters. To preserve the positive semidefinite property of the matrix the ACF estimator given by (15.29) should be used (see Problem 15.20). This results in the 2-D autocorrelation method (see Section 7.3). Another possibility is to estimate the AR parameters using the 2-D version of the covariance method and modified covariance method. The modified covariance method in analogy with the 1-D version discussed in Section 7.5 estimates the AR parameters by minimizing [Ulrych and Walker 1981, Nikias and Raghuveer 1985]

$$\frac{1}{2(M-p_1)(N-p_2)} \sum_{m=p_1}^{M-1} \sum_{n=p_2}^{N-1} \left| \sum_{i=0}^{p_1} \sum_{j=0}^{p_2} a[i,j]x[m-i, n-j] \right|^2 + \frac{1}{2(M-p_1)(N-p_2)} \sum_{m=0}^{M-1-p_1} \sum_{n=0}^{N-1-p_2} \left| \sum_{i=0}^{p_1} \sum_{j=0}^{p_2} a^*[i,j]x[m+i, n+j] \right|^2 \quad (15.42)$$

where $a[0, 0] = 1$. The covariance method finds the AR parameters by minimizing only the first term of (15.42). The minimization is easily accomplished by using (2.79) to yield the result

$$\sum_{i=0}^{p_1} \sum_{j=0}^{p_2} \hat{a}[i, j] \left\{ \begin{array}{l} \frac{1}{2(M-p_1)(N-p_2)} \sum_{m=p_1}^{M-1} \sum_{n=p_2}^{N-1} x[m-i, n-j]x^*[m-k, n-l] \\ + \frac{1}{2(M-p_1)(N-p_2)} \sum_{m=0}^{M-1-p_1} \sum_{n=0}^{N-1-p_2} x^*[m+i, n+j]x[m+k, n+l] \end{array} \right\} = 0 \quad (15.43)$$

for

$$k = 0, 1, \ldots, p_1$$

$$l = 0, 1, \ldots, p_2$$

$$[k, l] \neq [0, 0]$$

and where $\hat{a}[0, 0] = 1$. The estimator of the white noise variance is

$$\hat{\sigma}^2 = \sum_{i=0}^{p_1} \sum_{j=0}^{p_2} \hat{a}[i, j] \left\{ \frac{1}{2(M - p_1)(N - p_2)} \sum_{m=p_1}^{M-1} \sum_{n=p_2}^{N-1} x^*[m, n]x[m - i, n - j] \right.$$

$$\left. + \frac{1}{2(M - p_1)(N - p_2)} \sum_{m=0}^{M-1-p_1} \sum_{n=0}^{N-1-p_2} x^*[m + i, n + j]x[m, n] \right\} . \tag{15.44}$$

The covariance method finds the AR filter parameters by omitting the second term in braces in (15.43) and finds the white noise variance by (15.44) with the second term in braces omitted and the remaining term multiplied by 2. No extensions of the Burg algorithm or the recursive MLE are known (see Chapter 7).

To gain some insight into the performance of the QP AR spectral estimator, analytical results may be obtained for one complex sinusoid in complex white noise, assuming a known ACF and AR parameters given by the solution of (15.36) [Chien 1981]. The ACF is

$$r_{xx}[k, l] = P \exp [j2\pi(\nu_1 k + \nu_2 l)] + \sigma_w^2 \delta[k, l]. \tag{15.45}$$

Using this in (15.36) results in the AR parameters (see Problem 15.21)

$$\begin{aligned} a[m, n] &= - \frac{P}{[(p_1 + 1)(p_2 + 1) - 1]P + \sigma_w^2} \exp [j2\pi(\nu_1 m + \nu_2 n)] \\ \sigma^2 &= \sigma_w^2 \left(1 + \frac{P}{\sigma_w^2 + [(p_1 + 1)(p_2 + 1) - 1]P}\right) . \end{aligned} \tag{15.46}$$

The AR spectral estimator becomes

$$P_{\text{AR}}(f_1, f_2) = \frac{\sigma_w^2 \left(1 + \dfrac{P}{\sigma_w^2 + [(p_1 + 1)(p_2 + 1) - 1]P}\right)}{\left| 1 - \dfrac{P}{[(p_1 + 1)(p_2 + 1) - 1]P + \sigma_w^2} \displaystyle\sum_{\substack{m=0 \\ [m,n]\neq[0,0]}}^{p_1} \sum_{n=0}^{p_2} \exp[-j2\pi(f_1 - \nu_1)m] \exp [- j2\pi(f_2 - \nu_2)n] \right|^2} . \tag{15.47}$$

Note that the peak of the PSD is at $f_1 = \nu_1$, $f_2 = \nu_2$ and has a value

$$\begin{aligned} P_{\text{AR}}(f_1, f_2) \big|_{\text{MAX}} &= P_{\text{AR}}(\nu_1, \nu_2) \\ &= \sigma_w^2[1 + \eta((p_1 + 1) (p_2 + 1) - 1)][1 + \eta(p_1 + 1) (p_2 + 1)] \end{aligned} \tag{15.48}$$

where $\eta = P/\sigma_w^2 = \text{SNR}$. For high SNR or $\eta \gg 1$ and $(p_1 + 1)(p_2 + 1) \gg 1$

$$P_{\text{AR}}(f_1, f_2) \big|_{\text{MAX}} \approx \sigma_w^2[\eta(p_1 + 1)(p_2 + 1)]^2$$

so that the peak is proportional to the square of the power in the sinusoid. These results are all analogous to those obtained for the 1-D AR spectral estimator in Section 6.7. The principal difference is the replacement of p, where p is the order of the 1-D AR model, by $(p_1 + 1)(p_2 + 1) - 1$, the number of AR filter parameters in the 2-D case. Hence, as an example, the peak value of the 2-D AR PSD with $p_1 = p_2 = 10$ is about 100 times as large as that of the 1-D PSD with $p = 10$.

For multiple complex sinusoids in complex white noise an analytical expression similar to (15.47) can be derived. Due to its complexity it is of limited utility. In general, it is found that the spectral peaks will be displaced from the correct frequency locations, as might be expected from the 1-D case. Also, the peaks will be approximately proportional to the square of the sinusoidal power so that, as explained in Section 6.7, lower level sinusoids may be masked by higher level ones. It should be realized that if the QP model fails to yield a good spectral estimate, then increasing the model orders may not improve the quality. This is because the use of a QP model with even an infinite region of support may not be an accurate model for the PSD under consideration.

In deriving the QP AR model it was assumed that the region of support for the AR parameters was

$$S_1 = \{[i, j]: i = 0, 1, \ldots, p_1; j = 0, 1, \ldots, p_2\}$$

examples of which are given in Figure 15.5. The subscript on S now denotes the quadrant of support. There is no reason why we could not have chosen

$$S_4 = \{[i, j]: i = 0, 1, \ldots, p_1; j = 0, -1, \ldots, -p_2\}$$

an example of which is shown in Figure 15.8. Also, estimators based on the second and third quadrants can be used. Due to the hermitian symmetry property of the ACF, the use of the second quadrant will result in the identical spectral estimate as obtained by using the fourth quadrant (see Problem 15.22). Thus, restricting consideration to the first and fourth quadrants, we have the following sets of Yule–Walker equations:

$$\sum_{i=0}^{p_1} \sum_{j=0}^{p_2} a_1[i, j] r_{xx}[k - i, l - j] = \begin{cases} 0 & \text{for } [k, l] \in S_1' \\ \sigma^2 & \text{for } [k, l] = [0, 0] \end{cases} \tag{15.49}$$

which is obtained from (15.34). The subscript on the AR parameters denotes the quadrant of support. Similarly,

$$\sum_{i=0}^{p_1} \sum_{j=-p_2}^{0} a_4[i, j] r_{xx}[k - i, l - j] = \begin{cases} 0 & \text{for } [k, l] \in S_4' \\ \sigma^2 & \text{for } [k, l] = [0, 0]. \end{cases} \tag{15.50}$$

In general, the spectral estimates based on the first and fourth quadrants will be different. As an example, consider a single complex sinusoid in complex white noise so that the ACF is given by (15.45). Then the first quadrant AR model will yield (15.46) while the fourth quadrant model yields

$$\begin{aligned} a_4[m, n] &= -\frac{P}{[(p_1 + 1)(p_2 + 1) - 1]P + \sigma_w^2} \exp\,[j2\pi(\nu_1 m - \nu_2 n)] \\ \sigma^2 &= \sigma_w^2 \left(1 + \frac{P}{\sigma_w^2 + [(p_1 + 1)(p_2 + 1) - 1]P}\right). \end{aligned} \tag{15.51}$$

As a further illustration, consider the spectral estimates given by the first and

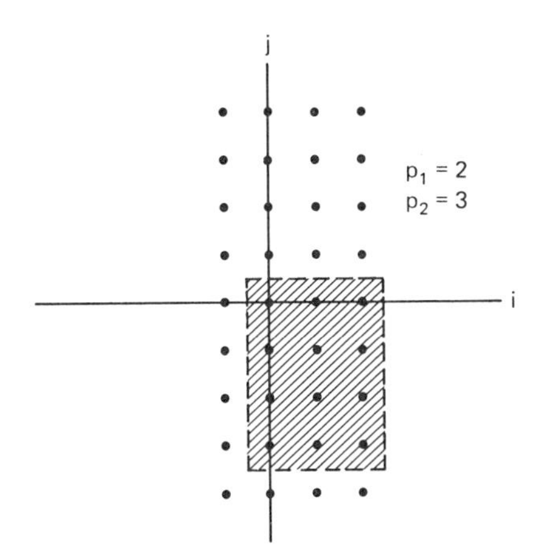

Figure 15.8 Example of fourth-quadrant region of support for quarter plane AR parameters.

fourth quadrant models, that is,

$$P_1(f_1, f_2) = \frac{\sigma^2}{\left| \sum_{m=0}^{p_1} \sum_{n=0}^{p_2} a_1[m, n] \exp\left[-j2\pi(f_1 m + f_2 n)\right] \right|^2} \tag{15.52}$$

and

$$P_4(f_1, f_2) = \frac{\sigma^2}{\left| \sum_{m=0}^{p_1} \sum_{n=-p_2}^{0} a_4[m, n] \exp\left[-j2\pi(f_1 m + f_2 n)\right] \right|^2}. \tag{15.53}$$

For $p_1 = p_2 = 2$, $P = \sigma_w^2 = 1$ and a sinusoid with a frequency in the first quadrant the spectral estimates obtained from (15.52) and (15.53) are shown in Figure 15.9a and b. We see that even though the peak of the spectral estimate is located at the sinusoidal frequency, the contours of constant PSD level are elliptical. This means that the frequency estimate in one direction will tend to be more accurate than the estimate of frequency for the other direction. Furthermore, this preferred direction will depend on whether we employ the first or fourth quadrant.

To alleviate this problem it has been suggested that the following AR spectral estimator be used [Jackson and Chien 1979]:

$$P_{1,4}(f_1, f_2) = \frac{\sigma^2}{\frac{1}{2}\left[\, |A_1(f_1, f_2)|^2 + |A_4(f_1, f_2)|^2 \right]} \tag{15.54}$$

where $|A_1(f_1, f_2)|^2$, $|A_4(f_1, f_2)|^2$ are given by the denominators of (15.52) and (15.53), respectively. For the same example the spectral estimate using (15.54) is shown in Figure 15.9c. The contours are now circular. Another motivation for the use of the combined quadrant estimator is that for sinusoids in white noise more accurate frequency estimates are obtained and spurious peaks are less often observed [Chien 1981]. The latter observation is because a 2-D polynomial is zero

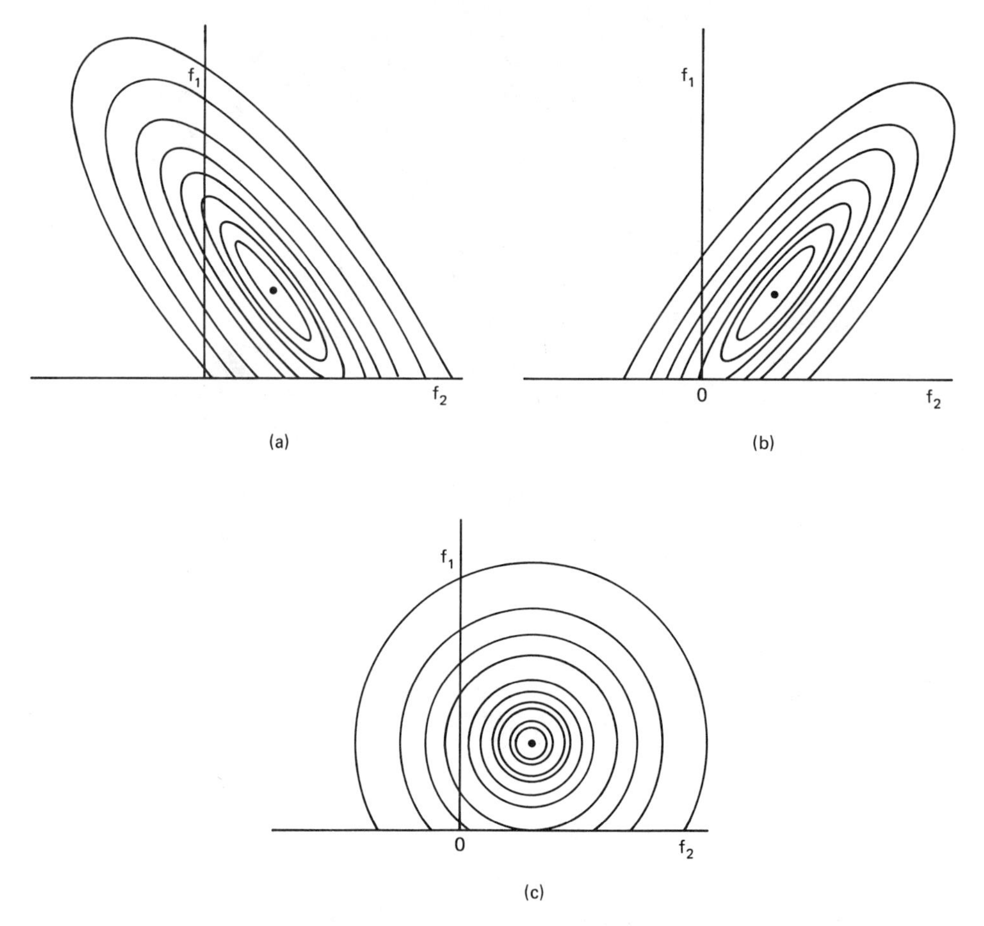

Figure 15.9 Contours of spectral estimate for QP and combined QP AR modeling. (a) First quadrant model. (b) Fourth quadrant model. (c) Combined quadrant model. (After Jackson and Chien [1979].) (© 1979 IEEE)

on a contour instead of at isolated points (i.e., the zeros) as in the 1-D case. Due to estimation errors, (15.52) or (15.53) will produce a spurious peak if the zero contour crosses the surface $z_1 = \exp(j2\pi f_1)$, $z_2 = \exp(j2\pi f_2)$. The use of (15.54) will produce a spurious peak only if A_1 and A_4 are both near zero, which is much less likely. In fact, it has been observed that the use of a single quadrant sometimes produces a ridge of peaks, but not for the combined two-quadrant estimator [Chien 1981]. For processes other than sinusoids in white noise the properties of this estimator have not been studied. In computing the combined quadrant spectral estimate, we can take advantage of a property of the multichannel Levinson recursion in which the first and fourth quadrant AR parameter estimates are simultaneously computed [Therrien 1981] to reduce the computation.

15.7 MAXIMUM ENTROPY SPECTRAL ESTIMATION

The MESE for 2-D data does not easily follow as an extension of the 1-D MESE (see Section 6.4.2). Due to the lack of a fundamental theorem of algebra for 2-D polynomials, even the existence of the MESE is not guaranteed, and when it exists, finding the MESE requires the solution of a set of nonlinear equations. The equivalence between the AR spectral estimator and the MESE in 1-D does not carry over to 2-D. Criticisms of the MESE for either 1-D or 2-D are that the criterion of maximum entropy has not been justified for spectral estimation and that the assumption of known ACF samples is seldom satisfied in practice.

Specifically, the MESE assumes that the ACF $r_{xx}[k, l]$ is known over the region $|k| \le p_1$, $|l| \le p_2$. It is desired to maximize the entropy rate, which is proportional to

$$\int_{-1/2}^{1/2} \int_{-1/2}^{1/2} \ln P_{xx}(f_1, f_2)\, df_1\, df_2 \tag{15.55}$$

subject to the constraint that the inverse Fourier transform of the PSD which maximizes (15.55) is equal to the known ACF over the given region. Thus the MESE effectively extrapolates the ACF by choosing the PSD that maximizes the entropy. It is not obvious that such an extrapolation exists. In the 1-D case Burg [1975] showed that there are an infinite number of possible extrapolations, all of which produce nonnegative PSDs. However, in the 2-D case, even if the known ACF samples are positive definite, it does not follow that even a single valid extrapolation exists [Dickinson 1980]. Forcing the extrapolated PSD to have an ACF that matches the known ACF samples may produce a negative spectral estimate. Hence it must be assumed that the known ACF samples are part of a valid ACF. Under this condition it can be shown that the MESE exists and is unique [Woods 1976]. How one actually determines whether the given ACF samples are part of a valid ACF can be found in the paper by Lang and McClellan [1981]. The test is difficult to implement.

Assuming that the MESE exists, it is found by maximizing (15.55) with respect to the unknown ACF samples. It can be shown that this maximization leads to a PSD of the form [Barnard and Burg 1969]

$$P_{xx}(f_1, f_2) = \frac{1}{\sum_{k=-p_1}^{p_1} \sum_{l=-p_2}^{p_2} \lambda_{k,l} \exp\left[-j2\pi(f_1 k + f_2 l)\right]} \tag{15.56}$$

where the $\lambda_{k,l}$'s are found so that the ACF corresponding to $P_{xx}(f_1, f_2)$ matches the given ACF samples. In the 1-D case the MESE has the form

$$P_{xx}(f) = \frac{1}{\sum_{k=-p}^{p} \lambda_k \exp(-j2\pi f k)} \tag{15.57}$$

and the denominator polynomial in the z plane (i.e., $\sum_{k=-p}^{p} \lambda_k z^{-k}$) is factorable into $\mathcal{A}(z)\ \mathcal{A}^*(1/z^*)/\sigma^2$. This factorization is not possible in general for the denominator of (15.56).

No closed-form solution is available for the $\lambda_{k,l}$ coefficients due to the nonlinear nature of the problem. One iterative algorithm that attempts to find the MESE but does not explicitly find the $\lambda_{k,l}$'s is due to Lim and Malik [1981]. Its performance is studied in the paper by Malik and Lim [1982]. Some other iterative approaches can be found in the papers by Woods [1976] and Jain and Ranganath [1978]. An approach that extrapolates the known autocorrelation matrix has been proposed by Roucos and Childers [1980].

15.8 MINIMUM VARIANCE SPECTRAL ESTIMATION

The 1-D MVSE described in Chapter 11 can easily be extended to two or more dimensions. Although not discussed here, the MVSE is also readily found for nonuniformly spaced data, which makes it quite useful in practice for array processing [Capon 1969]. The filtering interpretation, then, is to construct a 2-D FIR filter that will pass a 2-D complex sinusoid at a given frequency undistorted while simultaneously minimizing the power at the output of the filter. Let the data be given as

$$x[m, n] = A \exp\,[j2\pi(f_1 m + f_2 n) + \phi] + z[m, n] \qquad \begin{matrix} m = 0, 1, \ldots, M-1 \\ n = 0, 1, \ldots, N-1 \end{matrix} \tag{15.58}$$

where $z[m, n]$ is a complex zero mean random noise field. If the data are ordered by columns, then

$$\underline{\mathbf{x}} = A_c \underline{\mathbf{e}} + \underline{\mathbf{z}} \tag{15.59}$$

where $A_c = A \exp(j\phi)$,

$$\underline{\mathbf{x}} = [x[0,0] \quad x[0,1] \quad \cdots \quad x[0, N-1] \quad x[1,0] \quad x[1,1] \quad \cdots \quad x[1, N-1]$$
$$\cdots \quad x[M-1, 0] \quad x[M-1, 1] \quad \cdots \quad x[M-1, N-1]]^T$$

$$\underline{\mathbf{z}} = [z[0,0] \quad z[0,1] \quad \cdots \quad z[0, N-1] \quad z[1,0] \quad z[1,1] \quad \cdots \quad z[1, N-1]$$
$$\cdots \quad z[M-1, 0] \quad z[M-1, 1] \quad \cdots \quad z[M-1, N-1]]^T$$

$$\underline{\mathbf{e}} = [1 \quad z_2 \quad \cdots \quad z_2^{N-1} \quad z_1 \quad z_1 z_2 \quad \cdots \quad z_1 z_2^{N-1}$$
$$\cdots \quad z_1^{M-1} \quad z_1^{M-1} z_2 \quad \cdots \quad z_1^{M-1} z_2^{N-1}]^T$$

and $z_1 = \exp(j2\pi f_1)$, $z_2 = \exp(j2\pi f_2)$. Now filter $\underline{\mathbf{x}}$ with the FIR filter with coefficients $\underline{\mathbf{w}}$, which are column ordered. The output of the filter is $\underline{\mathbf{w}}^H \underline{\mathbf{x}}$. The filter coefficients are constrained to satisfy $\underline{\mathbf{w}}^H \underline{\mathbf{e}} = 1$ so that the sinusoid will be undistorted at the filter output. To minimize the power out of the filter, we need to minimize (see Section 11.3) $\underline{\mathbf{w}}^H \underline{\mathbf{R}}_{zz} \underline{\mathbf{w}}$, where $\underline{\mathbf{R}}_{zz} = \mathcal{E}(\underline{\mathbf{z}}\underline{\mathbf{z}}^H)$. The result, which is obtained in exactly the same manner as in the 1-D case, is that the optimal FIR filter has weights

$$\underline{\mathbf{w}} = \frac{\underline{\mathbf{R}}_{zz}^{-1}\underline{\mathbf{e}}}{\underline{\mathbf{e}}^H\underline{\mathbf{R}}_{zz}^{-1}\underline{\mathbf{e}}}.$$

and the minimum power out of the filter is

$$\frac{1}{\underline{\mathbf{e}}^H\underline{\mathbf{R}}_{zz}^{-1}\underline{\mathbf{e}}}.$$

Finally, by the same arguments as in Section 11.5, the MVSE is defined as

$$\hat{P}_{\mathrm{MV}}(f_1, f_2) = \frac{1}{\underline{\mathbf{e}}^H\hat{\underline{\mathbf{R}}}_{xx}^{-1}\underline{\mathbf{e}}}. \tag{15.60}$$

$\hat{\underline{\mathbf{R}}}_{xx}$ is an estimate of the autocorrelation matrix and is of dimension $p_1p_2 \times p_1p_2$. The dimensions should be chosen to ensure that $p_1 \ll M$, $p_2 \ll N$ so that the autocorrelation estimator will have a sufficient number of lagged products for statistical stability. As discussed in Chapter 11, a more appropriate definition of the MVSE is one which multiplies (15.60) by p_1p_2. The reader is referred to that chapter for further details. The theoretically known autocorrelation matrix is block Toeplitz, which allows us to invert it using the method of Akaike [1973]. To see this, first observe that

$$\underline{\mathbf{R}}_{xx} = \mathscr{E}(\underline{\mathbf{x}\mathbf{x}^H})$$

where $\underline{\mathbf{x}}$ consists of the data $x[m, n]$, $m = 0, 1, \ldots, p_1 - 1$; $n = 0, 1, \ldots, p_2 - 1$ ordered in a columnwise fashion. If $\underline{\mathbf{x}}$ is rewritten as

$$\underline{\mathbf{x}} = [\mathbf{x}[0]^T \quad \mathbf{x}[1]^T \quad \cdots \quad \mathbf{x}[p_1 - 1]^T]^T \tag{15.61}$$

where $\mathbf{x}[i]$ is the ith column of data, then

$$\begin{aligned}
\underline{\mathbf{R}}_{xx} &= \mathscr{E}(\underline{\mathbf{x}\mathbf{x}^H}) \\
&= \mathscr{E}\left\{\begin{bmatrix} \mathbf{x}[0] \\ \mathbf{x}[1] \\ \vdots \\ \mathbf{x}[p_1 - 1] \end{bmatrix} [\mathbf{x}[0]^H \quad \mathbf{x}[1]^H \quad \cdots \quad \mathbf{x}[p_1 - 1]^H]\right\} \\
&= \begin{bmatrix}
\mathscr{E}(\mathbf{x}[0]\mathbf{x}[0]^H) & \mathscr{E}(\mathbf{x}[0]\mathbf{x}[1]^H) & \cdots & \mathscr{E}(\mathbf{x}[0]\mathbf{x}[p_1 - 1]^H) \\
\mathscr{E}(\mathbf{x}[1]\mathbf{x}[0]^H) & \mathscr{E}(\mathbf{x}[1]\mathbf{x}[1]^H) & \cdots & \mathscr{E}(\mathbf{x}[1]\mathbf{x}[p_1 - 1]^H) \\
\vdots & \vdots & \ddots & \vdots \\
\mathscr{E}(\mathbf{x}[p_1 - 1]\mathbf{x}[0]^H) & \mathscr{E}(\mathbf{x}[p_1 - 1]\mathbf{x}[1]^H) & \cdots & \mathscr{E}(\mathbf{x}[p_1 - 1]\mathbf{x}[p_1 - 1]^H)
\end{bmatrix}.
\end{aligned}$$

But

$$\mathcal{E}(\mathbf{x}[i]\mathbf{x}[j]^H) = \mathcal{E}\left\{\begin{bmatrix} x[i,0] \\ x[i,1] \\ \vdots \\ x[i,p_2-1] \end{bmatrix} [x^*[j,0] \quad x^*[j,1] \quad \cdots \quad x^*[j,p_2-1]]\right\}$$

$$= \begin{bmatrix} r_{xx}[i-j,0] & r_{xx}[i-j,-1] & \cdots & r_{xx}[i-j,-(p_2-1)] \\ r_{xx}[i-j,1] & r_{xx}[i-j,0] & \cdots & r_{xx}[i-j,-(p_2-2)] \\ \vdots & \vdots & \ddots & \vdots \\ r_{xx}[i-j,p_2-1] & r_{xx}[i-j,p_2-2] & \cdots & r_{xx}[i-j,0] \end{bmatrix}$$

$$= \mathcal{E}(\mathbf{x}[i-j]\mathbf{x}[0]^H).$$

Letting $\mathbf{R}_{xx}[i] = \mathcal{E}(\mathbf{x}[i]\mathbf{x}[0]^H)$, it follows that

$$\underline{\mathbf{R}}_{xx} = \begin{bmatrix} \mathbf{R}_{xx}[0] & \mathbf{R}_{xx}[-1] & \cdots & \mathbf{R}_{xx}[-(p_1-1)] \\ \mathbf{R}_{xx}[1] & \mathbf{R}_{xx}[0] & \cdots & \mathbf{R}_{xx}[-(p_1-2)] \\ \vdots & \vdots & \ddots & \vdots \\ \mathbf{R}_{xx}[p_1-1] & \mathbf{R}_{xx}[p_1-2] & \cdots & \mathbf{R}_{xx}[0] \end{bmatrix}. \quad (15.62)$$

Thus $\underline{\mathbf{R}}_{xx}$ is doubly block Toeplitz, with each block being of dimension $p_2 \times p_2$. Also, $\underline{\mathbf{R}}_{xx}$ is hermitian since $\mathbf{R}_{xx}[-k] = \mathbf{R}_{xx}^H[k]$ and can easily be shown to be positive semidefinite. Note that $\underline{\mathbf{R}}_{xx}$ is identical to the autocorrelation matrix of (15.36) which arose in the QP AR model (albeit with different dimensions). This is not coincidental since the column-ordering approach used here can also be applied to the QP AR case, as described in Problem 15.16. When $\underline{\mathbf{R}}_{xx}$ is replaced by an estimate, we must be careful to ensure that the matrix is positive definite and hence invertible. Otherwise, large variances in the spectral estimator may be present. An autocorrelation estimator that guarantees this property is given by (15.29) (see Problem 15.20). Also, the implied autocorrelation estimators that resulted from the covariance and modified covariance methods for AR estimation [see (15.43)] can also be used. As in the 1-D case, the MVSE may be related to the AR spectral estimator. See the paper by Dowla and Lim [1984] for details.

15.9 SINUSOIDAL PARAMETER ESTIMATION

As in 1-D, the MLE of the parameters for complex sinusoids in complex white noise is difficult to compute. The amplitudes and phases can be found in an analogous manner to the method described in Section 13.3 once the frequencies have been estimated. If, however, the sinusoidal frequencies are well resolved, the peak locations of the 2-D periodogram provide the MLE of the frequencies. Cramer–Rao bounds on the sinusoidal parameters have been derived by Kumaresan [1980] and are given in the work by Chien [1981] together with some examples.

The principal component AR frequency estimator discussed in Chapter 13 has been extended to 2-D by Kumaresan and Tufts [1981]. To avoid polynomial rooting, which in general is not possible, the peaks of the frequency function (15.54) are used as frequency estimates. The Pisarenko method described in Chapter 13 relies on polynomial rooting and hence cannot be extended to 2-D, but a generalized Pisarenko method is given by Lang and McClellan [1982]. The MUSIC method [Schmidt 1981], which was discussed in Chapter 13 as a 1-D spectral estimation technique, was originally proposed for multiple dimensions. The eigenvectors of the 2-D autocorrelation matrix are computed and the noise subspace eigenvectors chosen to form the 2-D version of the frequency function given by (13.50). As in the 1-D case, the peaks of the frequency function are used as the frequency estimates. Note, however, that the function must be evaluated over a 2-D frequency variable. Another eigenanalysis approach for 2-D frequency estimation can be found in the paper by Durrani and Chapman [1983].

15.10 COMPUTER SIMULATION EXAMPLES

Some simulation examples which illustrate the various spectral estimators are now presented. These examples have been extracted from the work by Ulrych and Walker [1981]. The data consist of two complex sinusoids in complex zero

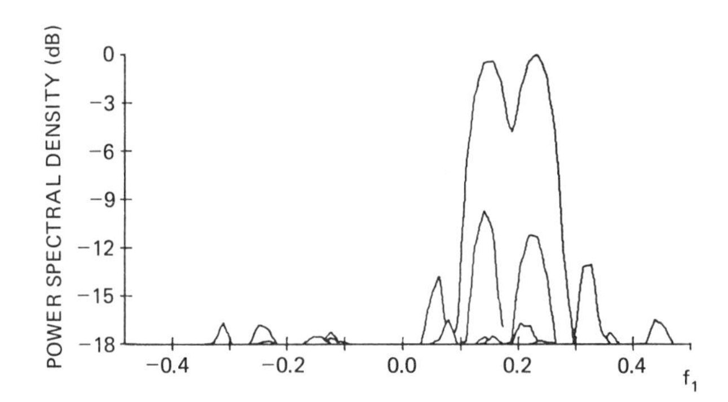

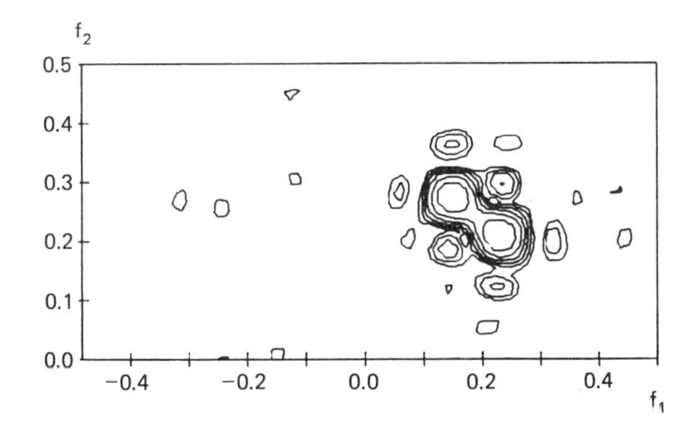

Figure 15.10 Periodogram estimate for two sinusoids in white noise (after Ulrych and Walker [1981].)

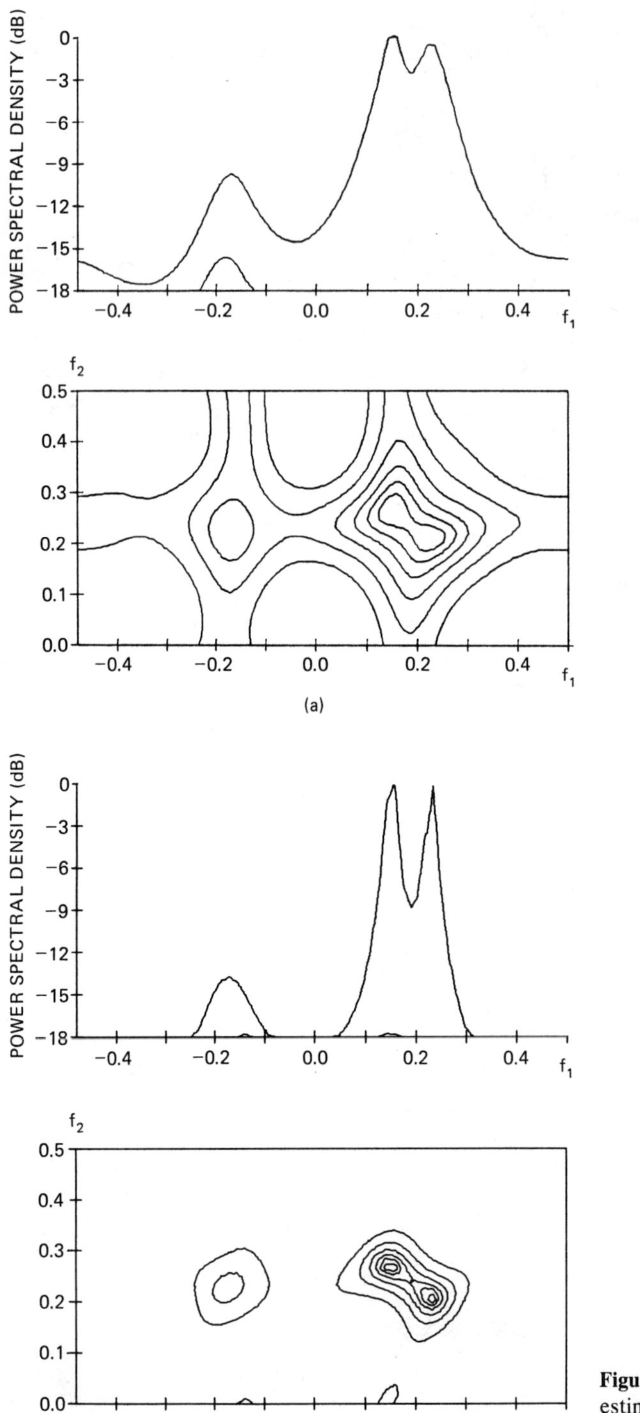

Figure 15.11 Combined QP spectral estimate for two sinusoids in white noise. (a) AR-AUTO. (b) AR-MCM (after Ulrych and Walker [1981].)

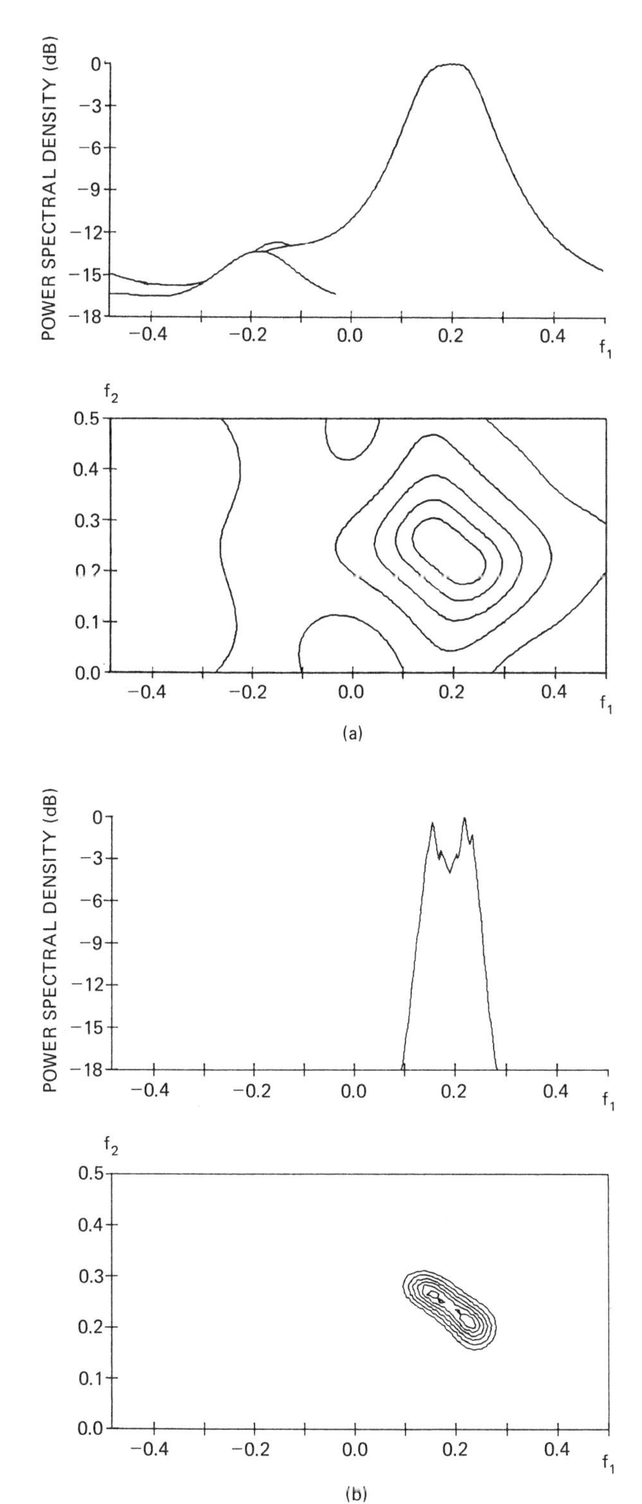

Figure 15.12 MVSE for two sinusoids in white noise. (a) MVSE-AUTO. (b) MVSE-MCM (after Ulrych and Walker [1981].)

mean white noise:

$$x[m, n] = A \exp[j2\pi(0.23m + 0.21n)] + A \exp[j2\pi(0.15m + 0.27n)] + z[m, n] \quad \begin{matrix} m = 0, 1, \ldots, 15 \\ n = 0, 1, \ldots, 15 \end{matrix} \tag{15.63}$$

so that $M = N = 16$. The variance of $z[m, n]$ is σ_z^2 and $A^2/\sigma_z^2 = 10$, so that the SNR is 10 dB. The sinusoids are resolved since the -3 dB circularly symmetric contour of the periodogram has width $0.84/M = 0.05$ and the sinusoids are spaced $\sqrt{(0.23 - 0.15)^2 + (0.27 - 0.21)^2} = 0.1$ apart in frequency as per (15.27). The spectral estimators to be compared are:

1. *Periodogram:* (15.25) with $M = N = 16$.
2. *Combined quadrant AR:* (15.54) with $p_1 = p_2 = 4$. The AR parameters are estimated by using (15.49) and (15.50) with the ACF estimator of (15.29), termed the AR-AUTO to denote the 2-D autocorrelation method, or by using the modified covariance method given by (15.43) for quadrant 1 and a similar expression for quadrant 4, termed the AR-MCM.
3. *MVSE:* (15.60) with $p_1 = p_2 = 5$. The autocorrelation matrix is estimated using either (15.62) with the ACF estimator given by (15.29), termed the MVSE-AUTO, or by using the autocorrelation matrix estimator implicit in the modified covariance method given by (15.43), termed the MVSE-MCM.

The results are shown in Figures 15.10 to 15.12. The periodogram shown in Figure 15.10 is able to resolve the sinusoids and is characterized by the usual sidelobe structure. Windowing of the data will reduce the sidelobes but also the resolution. The AR spectral estimator resolves the sinusoids but also exhibits an extra peak, as seen in Figure 15.11. The AR-MCM appears to produce a more accurate spectral estimate. The MVSE-AUTO spectral estimator does not resolve the sinusoids, while the MVSE-MCM barely does so, as seen in Figure 15.12. The comparison is only illustrative of the various techniques. No attempt has been made to compare the variance of the spectral estimators nor optimize their performance through model order selection. A definitive comparison for arbitrary data sets, then, is still required to evaluate the strengths and weaknesses of each estimator. See also the work by Frey [1982] for a computational comparison of the various techniques as well as some more simulation examples for sinusoids in white noise.

REFERENCES

Akaike, H., "Block Toeplitz Matrix Inversion," *SIAM J. Appl. Math.*, Vol. 24, pp. 234–241, Mar. 1973.

Barnard, T., and J. P. Burg, "Analytical Studies of Techniques for the Computation of High-Resolution Wavenumber Spectra," *Advanced Array Research Rep. 9*, Texas Instruments, May 1969.

Burg, J. P., "Maximum Entropy Spectral Analysis," Ph.D. dissertation, Stanford University, 1975.

Cadzow, J. A., and K. Ogino, "Two-Dimensional Spectral Estimation," *IEEE Trans. Acoust. Speech Signal Process.*, Vol. ASSP29, pp. 396–401, June 1981.

Capon, J., "High Resolution Frequency–Wavenumber Spectrum Analysis," *Proc. IEEE*, Vol. 57, pp. 1408–1418, Aug. 1969.

Chien, H., "Two-Dimensional Spectral Estimation from Autoregressive Models with Various Areas of Support," Ph.D. dissertation, University of Rhode Island, 1981.

Dickinson, B., "Two-Dimensional Markov Spectrum Estimates Need Not Exist," *IEEE Trans. Inf. Theory*, Vol. IT26, pp. 120–121, Jan. 1980.

Dowla, F. U., and J. S. Lim, "Relationship between Maximum-Likelihood-Method and Autoregressive Modeling in Multidimensional Spectrum Estimation," *Rec. 1984 IEEE Int. Conf. Acoust. Speech Signal Processing*, pp. 5.3.1–5.3.4.

Dudgeon, D. E., and R. M. Mersereau, *Multidimensional Digital Signal Processing*, Prentice-Hall, Englewood Cliffs, N.J., 1984.

Durrani, T. S., and R. Chapman, "Eigenfilter Methods for 2D Spectral Estimation," *Rec. 1983 IEEE Int. Conf. Acoust. Speech Signal Process.*, pp. 863–866.

Ekstrom, M. P., and J. W. Woods, "Two-Dimensional Spectral Factorization with Applications in Recursive Digital Filtering," *IEEE Trans. Acoust. Speech Signal Process.*, Vol. ASSP24, pp. 115–128, 1976.

Frey, E. J., "Two-Dimensional Spectral Estimation: A Comparison of Current Techniques," MS thesis, University of Colorado Boulder, 1982.

Huang, T. S., "Two-Dimensional Windows," *IEEE Trans. Audio Electoacoust.,* Vol. AU20, pp. 88–90, Mar. 1972.

Jackson, L. B., and H. C. Chien, "Frequency and Bearing Estimation by Two-Dimensional Linear Prediction," *Rec 1979 IEEE Int. Conf. Acoust. Speech Signal Process.*, pp. 665–668.

Jain, A. K., "Advances in Mathematical Models for Image Processing," *Proc. IEEE*, Vol. 69, pp. 502–528, May 1981.

Jain, A. K., and S. Ranganath, "Two-Dimensional Spectral Estimation," *Proc. RADC Spectrum Estimation Workshop,* Rome, N.Y., pp. 151–157, May 1978.

Joyce, L. S., "A Separable 2-D Autoregressive Spectral Estimation Algorithm," *Rec. 1979 IEEE Int. Conf. Acoust. Speech Signal Process.*, pp. 677–680.

Justice, J. H., "A Levinson-Type Algorithm for Two-Dimensional Wiener Filtering Using Bivariate Szego Polynomials," *Proc. IEEE*, Vol. 65, pp. 882–886, June 1977.

Kumaresan, R., unpublished notes, 1980.

Kumaresan, R., and D. W. Tufts, "A Two-Dimensional Technique for Frequency–Wavenumber Estimation," *Proc. IEEE*, Vol. 69, pp. 1515–1517, Nov. 1981.

Lang, S. W., and J. H. McClellan, "Spectral Estimation for Sensor Arrays," *Proc. 1st ASSP Spectrum Estimation Workshop*, pp. 3.2.1–3.2.7, Aug. 1981.

Lang, S. W., and J. H. McClellan, "The Extension of Pisarenko's Method to Multiple Dimensions," *Rec. 1982 IEEE Int. Conf. Acoust. Speech Signal Process.*, pp. 125–128.

Larimore, W. E., "Statistical Inference on Stationary Random Fields," *Proc. IEEE*, Vol. 65, pp. 961–970, June 1977.

Lim, J. S., and N. A. Malik, "A New Algorithm for Two-Dimensional Maximum Entropy Power Spectrum Estimation," *IEEE Trans. Acoust. Speech Signal Process.*, Vol. ASSP29, pp. 401–413, June 1981.

Malik, N. A., and J. S. Lim, "Properties of Two-Dimensional Maximum Entropy Power

Spectrum Estimates," *IEEE Trans. Acoust. Speech Signal Process.*, Vol. ASSP30, pp. 788–798, Oct. 1982.

Marple, S. L., Jr., *Digital Spectral Analysis with Applications,* Prentice-Hall, Englewood Cliffs, N.J., 1987.

Marzetta, T. L., "Two Dimensional Linear Prediction: Autocorrelation Arrays, Minimum-Phase Prediction Error Filters, and Reflection Coefficient Arrays," *IEEE Trans. Acoust. Speech Signal Process.*, Vol. ASSP28, pp. 725–733, Dec. 1980.

McClellan, J. H., "Multidimensional Spectral Estimation," *Proc. IEEE,* Vol. 70, pp. 1029–1039, Sept. 1982.

Nikias, C. L., and M. R. Raghuveer, "Multi-dimensional Parametric Spectral Estimation," *Signal Process.*, Vol. 9, pp. 191–205, Oct. 1985.

Parker, S. R., and A. H. Kayran, "Lattice Parameter Autoregressive Modeling of Two-Dimensional Fields: Part 1," *IEEE Trans. Acoust. Speech Signal Process.*, Vol. ASSP32, pp. 872–885, Aug. 1984.

Pendrel, J. V., "The Maximum Entropy Principle in Two-Dimensional Spectral Analysis," Ph.D. dissertation, York University, 1979.

Roucos, S. E., and D. G. Childers, "A Two-Dimensional Maximum Entropy Spectral Estimator," *IEEE Trans. Inf. Theory*, Vol. IT26, pp. 554–560, Sept. 1980.

Schmidt, R. O., "A Signal Subspace Approach to Multiple Emitter Location and Spectral Estimation," Ph.D. dissertation, Stanford University, 1981.

Therrien, C. W., "Relations between 2-D and Multichannel Linear Prediction," *IEEE Trans. Acoust. Speech Signal Process.*, Vol. ASSP29, pp. 454–456, June 1981.

Ulrych, T. J., and C. J. Walker, "High Resolution 2-Dimensional Power Spectrum Estimation," in *Applied Time Series II*, D. Findley, ed., Academic Press, New York, 1981, pp. 71–99.

Wernecke, S. J., "Two-Dimensional Maximum Entropy Reconstruction of Radio Brightness," *Radio Sci.*, Vol. 12, pp. 831–844, Sept.–Oct. 1977.

Wernecke, S. J., and L. R. D'Addario, "Maximum Entropy Image Reconstruction," *IEEE Trans. Comput.*, Vol. C26, pp. 351–364, Apr. 1977.

Whittle, P., "On Stationary Processes in the Plane," *Biometrika*, Vol. 41, pp. 434–449, Dec. 1954.

Woods, J. W., "Two-Dimensional Discrete Markovian Fields," *IEEE Trans. Inf. Theory*, Vol. IT18, pp. 232–240, Mar. 1972.

Woods, J. W., "Two-Dimensional Markov Spectral Estimation," *IEEE Trans. Inf. Theory*, Vol. IT22, pp. 552–559, Sept. 1976.

Yaglom, A. M., *Introduction to the Theory of Stationary Random Functions*, Prentice-Hall, Englewood Cliffs, N.J., 1962.

PROBLEMS

15.1. Compute the output of the LSI system with impulse response

$$h[m, n] = \begin{cases} (-a[1])^m(-b[1])^n & \text{for } m \geq 0,\ n \geq 0 \\ 0 & \text{otherwise} \end{cases}$$

to the "unit step" input

$$u[m, n] = \begin{cases} 1 & \text{for } m \geq 0, n \geq 0 \\ 0 & \text{otherwise.} \end{cases}$$

Under what conditions is this system stable? Is this system causal?

15.2. For the first order recursive system

$$x[m, n] = -a[0, 1]x[m, n-1] - a[1, 0]x[m-1, n] + u[m, n]$$

show that the impulse response is given by

$$h[m, n] = \begin{cases} \dfrac{(m+n)!}{m!n!}(-a[1, 0])^m(-a[0, 1])^n & \text{for } m \geq 0, n \geq 0 \\ 0 & \text{otherwise} \end{cases}$$

by explicitly computing the output for an impulse input using the difference equation. Assume that any "initial conditions" are equal to zero.

15.3. Find the system function as well as the frequency response for the system of Problem 15.2. What type of filter can the system be classified as?

15.4. Find the Fourier transform of the sequence

$$x[m, n] = \begin{cases} \exp\,[j2\pi(\nu_1 m + \nu_2 n)] & \text{for } 0 \leq m \leq M-1, 0 \leq n \leq N-1 \\ 0 & \text{otherwise.} \end{cases}$$

Plot the magnitude of the transform.

15.5. Compute the DFT of the sequence

$$x[m, n] = \begin{cases} \exp\left[j2\pi\left(\dfrac{mk_0}{M} + \dfrac{nl_0}{N}\right)\right] & \text{for } 0 \leq m \leq M-1, 0 \leq n \leq N-1 \\ 0 & \text{otherwise.} \end{cases}$$

How is the DFT related to the Fourier transform found in Problem 15.4?

15.6. Prove that a valid ACF must be positive semidefinite by noting that

$$\text{var}\left(\sum_k \sum_l a^*[k, l]x[k, l]\right) \geq 0$$

for an arbitrary sequence $a[k, l]$. Assume $x[k, l]$ is zero mean.

15.7. For an ACF with the property

$$r_{xx}[-k, l] = r^*_{xx}[k, l]$$

in what part of the plane must the ACF be known in order to find the PSD? What symmetry properties does the PSD have?

15.8. Find and plot the PSD for the ACF given by (15.18) with $a[1] = b[1] = \frac{1}{2}$ and also with $a[1] = b[1] = -\frac{1}{2}$.

15.9. Prove the relationship between the input PSD and output PSD of a LSI system as given by (15.24).

15.10. Show that the contours of (15.26) with $M = N$ are approximately circular near $f_1 = \nu_1$, $f_2 = \nu_2$ by assuming that $(\sin M\theta)/(\sin\theta) \approx M - M^3\theta^2/6$ for θ small. Now find the radius of the circle to verify (15.27).

15.11. Prove that the biased ACF as given by (15.29) is positive semidefinite. *Hint:* Define

$$x'[m, n] = \begin{cases} x[m, n] & \text{for } 0 \le m \le M - 1, 0 \le n \le N - 1 \\ 0 & \text{otherwise} \end{cases}$$

so that the ACF estimator is

$$\hat{r}_{xx}[k, l] = \frac{1}{MN} \sum_{m=-\infty}^{\infty} \sum_{n=-\infty}^{\infty} x'^*[m, n]x'[m + k, n + l].$$

15.12. Show that the AR model with a truncated NSHP region of support is recursively computable by referring to Figure 15.2.

15.13. Show that the ACF of the real 2-D AR model

$$x[m, n] = -a[1]x[m - 1, n] - b[1]x[m, n - 1] - a[1]b[1]x[m - 1, n - 1] + u[m, n]$$

is given by (15.18).

15.14. Prove that the impulse response for the truncated NSHP AR model is equal to unity at $[m, n] = [0, 0]$. *Hint:* Extend the so-called initial value theorem to 2-D.

15.15. Explain why it requires more ACF samples than unknown AR parameters to determine the AR parameters for the QP model by using a convolution interpretation of the Yule–Walker equations. Show that for the QP model which consists of $(p_1 + 1)(p_2 + 1)$ parameters, we require $2p_1p_2 + p_1 + p_2 + 1$ ACF lags. Compare this result to the 1-D case.

15.16. In this problem the block Toeplitz property of the matrix in the Yule–Walker equations for the QP AR model is derived. Rewrite the definition of the AR process for the point [0, 0] as

$$\sum_{i=0}^{p_1} \sum_{j=0}^{p_2} a[i, j]x[-i, -j] = u[0, 0]$$

or

$$\sum_{i=0}^{p_1} \mathbf{x}[i]^H \mathbf{a}[i] = u[0, 0]$$

where

$$\begin{aligned} \mathbf{a}[i] &= [a[i, 0]\ a[i, 1] \cdots a[i, p_2]]^T \\ \mathbf{x}[i] &= [x^*[-i, 0]\ x^*[-i, -1] \cdots x^*[-i, -p_2]]^T \end{aligned} \qquad i = 0, 1, \ldots, p_1$$

which is a concatenation by columns. Define $\underline{\mathbf{a}}$ and $\underline{\mathbf{x}}$ as

$$\underline{\mathbf{a}} = \begin{bmatrix} \mathbf{a}[0] \\ \mathbf{a}[1] \\ \vdots \\ \mathbf{a}[p_1] \end{bmatrix}$$

$$\underline{\mathbf{x}} = \begin{bmatrix} \mathbf{x}[0] \\ \mathbf{x}[1] \\ \vdots \\ \mathbf{x}[p_1] \end{bmatrix}$$

so that $\underline{\mathbf{x}}^H\underline{\mathbf{a}} = u[0, 0]$. Now explicitly evaluate $\mathscr{E}(\mathbf{x}\mathbf{x}^H\mathbf{a}) = \mathscr{E}(\mathbf{x}u[0, 0])$ to obtain (15.36). Note that $\mathbf{R}_{xx}[i] = \mathscr{E}[\mathbf{x}[n + i]\mathbf{x}^H[n]]$. *Hint:* Invoke the causality of the model to prove that

$$\mathscr{E}(x^*[-i, -j]u[0, 0]) = 0 \qquad i = 0, 1, \ldots, p_1; j = 0, 1, \ldots, p_2; [i, j] \neq [0, 0]$$

by noting that the impulse response has a QP region of support.

15.17. Consider the infinite length predictor for the 1-D case,

$$\hat{x}[n] = \sum_{l=1}^{\infty} \alpha_l x[n - l].$$

The optimal prediction coefficients are found by employing the orthogonality principle

$$\mathscr{E}(x^*[n - l]e[n]) = 0 \qquad l \geq 1$$

where $e[n] = x[n] - \hat{x}[n]$ is the prediction error. Prove that $r_{ee}[k] = 0$ for $k \geq 1$ for the optimal predictor. Hence $r_{ee}[k] = r^*_{ee}[-k] = 0$ for $k \leq -1$ and $e[n]$ is a white noise process.

15.18. In this problem we extend the results of Problem 15.17 to 2-D by considering the predictor

$$\hat{x}[m, n] = - \sum\sum_{[i,j] \in S'_{\text{NSHP}}} \alpha_{ij} x[m - i, n - j]$$

where S'_{NSHP} denotes the infinite NSHP with the point [0, 0] omitted. The orthogonality principle yields

$$\mathscr{E}(x^*[m - i, n - j]e[m, n]) = 0 \qquad [i, j] \in S'_{\text{NSHP}}$$

where $e[m, n]$ is the prediction error. Prove that $e[m, n]$ is a white noise process by showing that $r_{ee}[k, l] = 0$ for $[k, l] \neq [0, 0]$. What happens if only an infinite QP is used for prediction?

15.19. For the data model of Problem 15.13, we could estimate $a[1]$ by taking any row of $x[m, n]$ and forming

$$\hat{a}^{(n)}[1] = - \frac{\sum_{m=0}^{M-2} x[m, n]x[m + 1, n]}{\sum_{m=0}^{M-1} x^2[m, n]}.$$

Is this a reasonable estimator? How could the estimator be improved? What is the corresponding estimator for $b[1]$?

15.20. Prove that $\underline{\mathbf{R}}_{xx}$ as given in (15.36) is positive definite if the biased ACF estimator as given by (15.29) is substituted.

15.21. Verify that the AR parameter estimates of a QP AR model for a complex sinusoid in complex white noise are given by (15.46). Assume a known ACF.

15.22. Prove that the AR spectral estimates for the QP model are the same if we use the first or third quadrant as the region of support. Assume a known ACF.

Chapter 16

Other Applications of Spectral Estimation Methods

16.1 INTRODUCTION

In the preceding chapters we have discussed the theory and application of modern spectral estimation. Much of the underlying theory has been applied to areas other than spectral estimation. Since these further applications are of sufficient interest, in this chapter we summarize some of these applications. The topics to be discussed are not meant to be an all-inclusive listing of these additional applications, but only a representative sampling of the more common areas.

16.2 TIME SERIES EXTRAPOLATION AND INTERPOLATION

The theoretical foundations of modern spectral estimation have led to applications in the extrapolation and interpolation of time series data. When the PSD is known, the solution to these problems is well known [Papoulis 1965]. However, for processes with unknown PSDs, the optimal solutions cannot be implemented. Consider first the problem of extrapolation (or prediction). If the time series is an AR(p) process, then the optimal linear prediction coefficients are just the AR parameters as discussed in Chapter 6. The parameters can be estimated from the data by any of the methods of Chapter 7. If the process is not an AR process, the number of prediction coefficients of the optimal predictor is, in general, infinite. Theoretically, as the number of prediction coefficients increases, the extrapolation error will decrease. When we are limited to a finite data set from which to estimate the prediction coefficients, the extrapolation error will be minimized

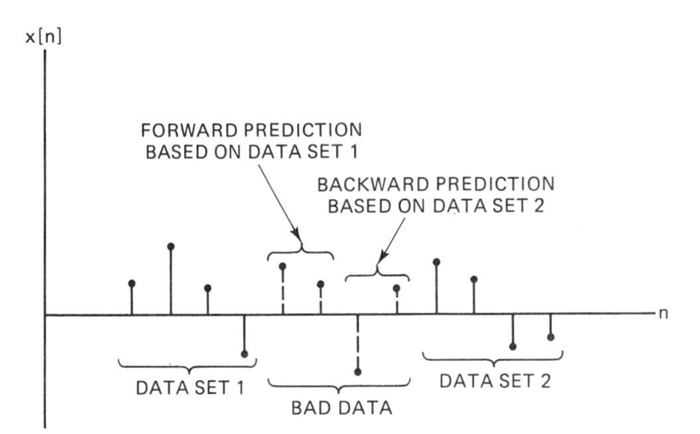

Figure 16.1 Interpolation of bad data points.

by choosing a predictor with as large an order as possible, subject to the constraint that the prediction coefficients can be accurately estimated [Cleveland 1971].

Although the techniques described in Chapter 7 can yield only a one-step predictor, one can use the predicted sample as if it were part of the original data set and continue the extrapolation to the next sample [Bowling and Lai 1979]. It has even been proposed to use the enlarged set of original and extrapolated data with a conventional periodogram or a BT spectral estimator to improve the resolution [Frost and Sullivan 1979]. However, extreme care must be taken in extrapolating too far from the original data set since these extrapolations will tend to be unreliable.

In addition to extrapolation, interpolation may be performed by using a forward and backward predictor as shown in Figure 16.1. This is valuable for replacing bad data points [Nuttall 1976]. For AR(p) processes, interpolation of a bad data point need only be based on the p samples on either side of the bad point [Kay 1983A].

16.3 SIGNAL DETECTION

A prewhitening filter is a natural use of the estimated filter of a rational transfer function based spectral estimate. For example, in AR spectral estimation the prediction error filter $\hat{\mathcal{A}}(z)$ is a whitening filter. The output of the filter (the prediction error) is white noise if the observed process is an AR(p) process. If the time series is not an AR process, the output time series will still be characterized by a "flatter" PSD than at the input (see Section 6.4.3). Prewhitening filters naturally arise in the design of optimal detectors, where it must be assumed that *the power spectral density is known*. When the PSD is unknown, other approaches to detection must be tried. As an example, a prewhitener is an integral part of the detector for a signal in colored noise. A practical application is the detection of radar target returns in a background of clutter. The optimal detector is a prewhitener followed by a matched filter, which is matched to the signal at the prewhitener output [Van Trees 1968]. Since the clutter spectrum is usually time varying, the whitening filter parameters and matched filter must be updated in time. The success of the prewhitening scheme will depend on the time variation of the clutter and the ability to estimate the parameters of the spectrum before they change. AR whiteners naturally suggest themselves and have been found to work well in practice [Bowyer et al. 1979]. Alternatively, optimal detectors that incorporate estimated prewhiteners may be derived. Using an AR model of the clutter, performance close to that of a detector designed with a priori knowledge of the spectrum may be obtained [Gobien 1973, Kay 1983B].

The use of the AR spectral estimator as a basis for the detection of signals in white noise has been shown to have some advantages. Specifically, for a sinusoidal signal whose frequency changes over the duration of the data record, as is frequently encountered in a doppler radar or sonar, a detector that thresholds the AR spectral estimate has been proposed. Whereas the detection performance of a periodogram degrades rapidly for a frequency modulated signal, the performance of an AR spectral estimate detector is relatively unaffected by moderate amounts of frequency modulation [Kay 1982]. This result is due to the semicoherent nature of the ACF used to estimate the spectrum [Lank et al. 1973]. Also, the use of a detector based on an AR(1) spectral estimate may be implemented more efficiently than an FFT.

Finally, the use of AR modeling for detection of broadband random signals in white noise has been proposed. It has been shown that when the PSD of the signal is unknown, detectors based on AR models yield superior performance to conventional energy detectors [Kay 1985].

16.4 BANDWIDTH COMPRESSION

An important problem in speech research is that of bandwidth compression. If the redundancy in speech can be reduced, then more speech signals can be transmitted through a fixed bandwidth channel or stored in some mass storage. A common technique is differential pulse-code modulation (DPCM) [Jayant 1974]. The basis of DPCM is to transmit only information that cannot be predicted, often termed the innovations of the process [Kailath 1974]. In fact, if the speech waveform were perfectly predictable from a set of previous samples, the receiver, once it had knowledge of those samples, could perfectly reconstruct the entire waveform (assuming no transmission errors due to noise). There would be no need for transmission. In practical DPCM systems, speech samples are analyzed at the transmitter to determine the prediction coefficients. Then, only the prediction error time series and the prediction coefficients are transmitted. The speech signal is reconstructed at the receiver. The bandwidth reduction is achieved because the variance of the prediction error time series is less than that of the speech waveform, or

$$\rho = r_{xx}[0] \prod_{i=1}^{p} (1 - k_i^2) \le r_{xx}[0]. \tag{16.1}$$

Hence fewer quantizer levels are necessary to encode the prediction error time series. Note that for maximum bandwidth compression the predictor parameters must be continually updated as the statistical character of speech changes (i.e., from phoneme to phoneme).

The most dramatic technique for bandwidth reduction is linear predictive coding (LPC), in which AR modeling is used to represent the speech spectrum [Markel and Gray 1976]. Assuming that speech can be accurately modeled as the output of an all-pole filter driven by white noise for unvoiced speech, or driven by a train of impulses for voiced speech, the speech waveform may be reduced to a small set of parameters. Thus only the model parameters and the period of the impulse train need be transmitted or stored. Speech synthesis is then accomplished by employing the appropriate model for each speech sound. Also, speech recognition is possible using the same LPC model. We can determine which speech sound was spoken by comparing the measured LPC coefficients (estimated AR parameters) to a set of stored coefficients or templates for a class of speech sounds. The sound whose template is closest to the measured LPC coefficients is chosen [Rabiner and Schafer 1978].

16.5 SPECTRAL SMOOTHING AND MODELING

Conventional periodogram and BT spectral estimators lead to estimates that are characterized by many "hills and valleys." This effect is due to the random nature of the spectral estimate. Autocorrelation windowing or spectral smoothing will substantially reduce the fluctuations but not eliminate them. An AR spectral estimator can be used to smooth these fluctuations since a pth-order estimator is constrained to have at most p peaks (or valleys). For p small a smoothed spectral estimate will result, as discussed in Section 6.8. It is now argued that the AR spectral estimator accurately represents the peaks of a periodogram but not the valleys [Makhoul 1975, Markel and Gray 1976]. Consider the estimate of the AR parameters found using the autocorrelation method (see Chapter 7). We must minimize

$$\hat{\rho} = \frac{1}{N} \sum_{n=-\infty}^{\infty} |\, e[n] \,|^2 \tag{16.2}$$

where $e[n] = x[n] + \sum_{k=1}^{p} a[k]x[n-k]$ and it is assumed that $x[n]$ is nonzero only over the $[0, N-1]$ interval. Then, by Parseval's theorem,

$$\hat{\rho} = \frac{1}{N} \int_{-1/2}^{1/2} |\, E(f) \,|^2 \, df$$

where $E(f)$ is the Fourier transform of $e[n]$. Denoting the Fourier transform of $x[n]$ by $X(f)$, $\hat{\rho}$ becomes

$$\begin{aligned} \hat{\rho} &= \frac{1}{N} \int_{-1/2}^{1/2} |\, X(f) \,|^2 \, |\, A(f) \,|^2 \, df \\ &= \sigma^2 \int_{-1/2}^{1/2} \frac{(1/N) \,|\, X(f) \,|^2}{P_{xx}(f)} \, df \end{aligned} \tag{16.3}$$

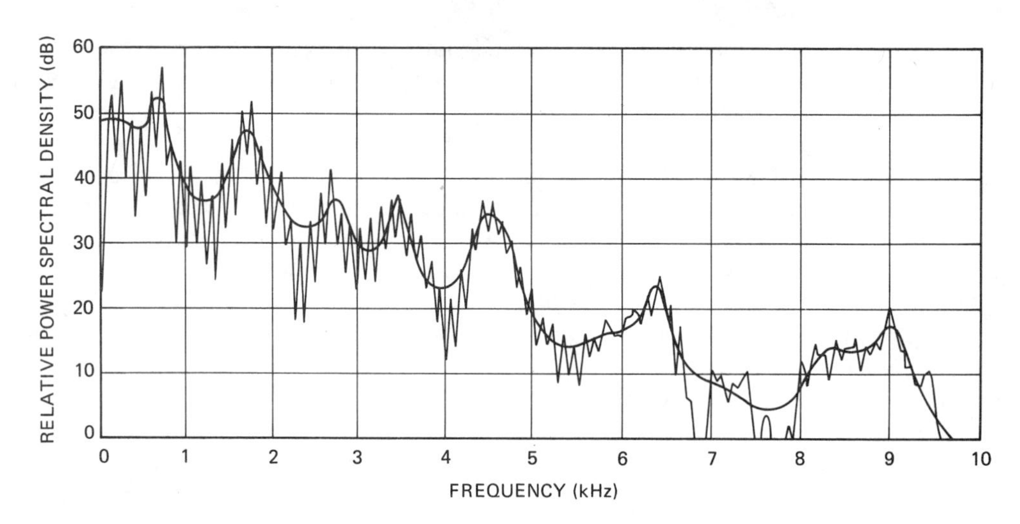

Figure 16.2 Periodogram and smoothed AR(28) spectral estimates of speech data. (After Makhoul [1975].) (© 1975 IEEE)

where $|X(f)|^2/N$ is the periodogram and

$$P_{xx}(f) = \frac{\sigma^2}{|A(f)|^2}$$

is the AR PSD model. To minimize $\hat{\rho}$, the AR PSD $P_{xx}(f)$ should match the periodogram over the frequency regions where the periodogram is large to reduce the large contribution to the error as given by (16.3). Over frequency regions where the periodogram is small, there is only a small contribution to the error so that matching is not necessary. The result is that *the AR spectral estimate matches the peaks but not the valleys of the periodogram*. If we wish to represent or model the peaks of a spectrum, then we need only take the inverse Fourier transform of the spectrum to yield the ACF and then use this information to compute the AR PSD. Viewed in this manner the procedure models a spectrum that is assumed to be the true PSD. An example of AR modeling for a speech signal is shown in Figure 16.2.

16.6 BEAMFORMING/DIRECTION FINDING

In beamforming or direction finding, we are interested in obtaining an estimate of the spatial structure of a physical random field. If we sample in space a random field using a line (linear) array of equally spaced sensors, a vector time series $\{x(t, 0), x(t, \delta_x), \ldots, x(t, [M-1]\delta_x)\}$, is obtained, where $x(t, i\delta_x)$ is the continuous waveform at the ith sensor. Such an array is illustrated in Figure 16.3. The field can be expressed as [Capon 1969]

$$x(t, i\delta_x) = \iint \Psi(f, k_x) \exp[j2\pi(ft - k_x i\delta_x)]\, df\, dk_x \tag{16.4}$$

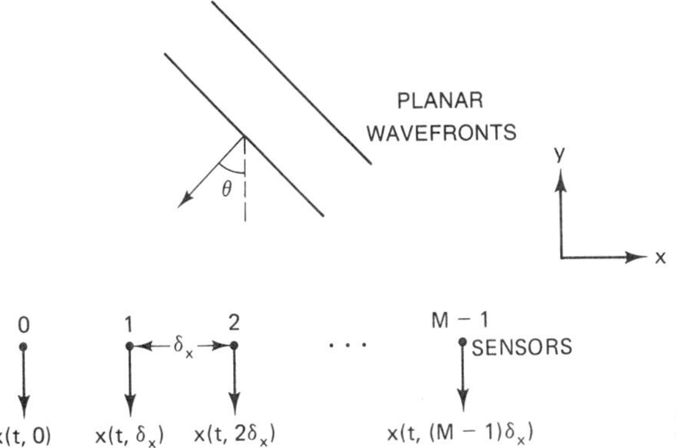

Figure 16.3 Line array geometry.

which represents the field as the sum of an infinite number of monochromatic plane waves with random complex amplitudes $\Psi(f, k_x)$. The temporal frequency is denoted by f, while the spatial frequency along the x direction is denoted by k_x. The spatial frequency or wavenumber component k_x is the reciprocal of the wavelength of a monochromatic plane wave along the x direction. Because $k_x = (f/c) \sin \theta$, where θ is the angle indicated in Figure 16.3, $\mathcal{E}[|\Psi(f, k_x)|^2]$ is the power at frequency f arriving from the θ direction. From (16.4) the inverse Fourier transform is

$$\begin{aligned}\Psi(f, k_x) &= \sum_{i=0}^{M-1} \left[\int x(t, i\delta_x) \exp(-j2\pi f t)\, dt \right] \exp(j2\pi k_x i\delta_x) \\ &= \sum_{i=0}^{M-1} X_f(i\delta_x) \exp(j2\pi k_x i\delta_x) \qquad (16.5)\end{aligned}$$

where $X_f(i\delta_x)$ is the Fourier transform of the continuous process observed at the output of the ith sensor. Note that a summation is present in (16.5) since the spatial field is sampled at the sensor locations. This expression is the Fourier transform of the spatial "time series" $X_f(i\delta_x)$ considered for a fixed temporal frequency f and indexed by the sensor number i. If the spatial field is homogeneous,

$$\mathcal{E}[x^*(t, i\delta_x)x(t, j\delta_x)] = g(t, [j - i]\delta_x) \qquad (16.6)$$

for some function $g(\ )$ or the spatial correlation function depends only on the separation between two points, the spatial "time series" is WSS and estimation of $\mathcal{E}(|\Psi(f, k_x)|^2)$ for a fixed frequency f is analogous to one-dimensional spectral estimation. Any of the one-dimensional techniques described are then applicable if the data record is considered to be $\{X_f(0), X_f(\delta_x), \ldots, X_f([M - 1]\delta_x)\}$ for some f. Generally, this data set is replaced by $\hat{X}_f(i\delta_x)$ for $i = 0, 1, \ldots, M - 1$, where $\hat{X}_f(i\delta_x)$ is the Fourier transform of a set of temporal samples from the ith sensor

$$\hat{X}_f(i\delta_x) = \sum_{n=0}^{N-1} x(n\delta_t, i\delta_x) \exp(-j2\pi f n\delta_t). \qquad (16.7)$$

Then this narrowband Fourier analysis is followed by "high resolution" spectral estimation in the spatial direction (see also Section 15.5 for a further discussion of these "hybrid" spectral estimation approaches). If the temporal data set is too small to allow adequate temporal frequency resolution using Fourier techniques, the 2-D methods of Chapter 15 should be applied to the spatial-temporal data.

It should be noted that some extra averaging is available in the spatial case which is not available in the temporal case. For instance, the spatial "autocorrelation function" at lag k or $\mathcal{E}[x^*(n\delta_t, i\delta_x)x(n\delta_t, (i + k)\delta_x)]$ can be estimated as

$$\frac{1}{N(M - k)} \sum_{n=0}^{N-1} \sum_{i=0}^{M-k-1} x^*(n\delta_t, i\delta_x)x(n\delta_t, (i + k)\delta_x) \qquad k \geq 0$$

where it is assumed that $x(n\delta_t, i\delta_x)$ is temporally WSS over the interval $0 \leq n\delta_t \leq (N - 1)\delta_t$. This estimate includes an extra time-averaging operation. Further details about spatial spectral estimation or, alternatively, beamforming or direction finding can be found in the works by McDonough [1974], Gabriel [1980], Barnard [1982], Evans et al. [1982], and Johnson [1982].

16.7 LATTICE FILTERS

The minimum-phase lattice filter described in Chapter 6 has the property that its coefficients are bounded by 1 in magnitude. This is very desirable when we must quantize the coefficients for transmission or storage [Viswanathan and Makhoul 1975]. The lattice structure may also be used to synthesize stable all-pole filters or stable pole–zero filters.

16.8 OTHER APPLICATIONS

Other applications of the techniques described in this book are:

1. Equalization for digital communications [Makhoul and Viswanathan 1978, Proakis 1983]
2. Transient analysis [Van Blaricum and Mittra 1978, Kumaresan 1982]
3. Digital filter design [Oppenheim and Schafer 1975]
4. Predictive deconvolution [Robinson 1967]
5. Cepstral analysis [Landers 1978]
6. Image processing [Jain 1981].

The interested reader may consult the references for further details.

REFERENCES

Barnard, T. E., "Two Maximum Entropy Beamforming Algorithms for Equally Spaced Line Arrays," IEEE Trans. Acoust. Speech Signal Process., Vol. ASSP30, pp. 175–189, Apr. 1982.

Bowling, S. B., and S. Lai, "The Use of Linear Prediction for the Interpolation and Extrapolation of Missing Data Prior to Spectral Analysis," *Proc. RADC Workshop Spectrum Estimation,* pp. 39–50, Oct. 3–5, 1979.

Bowyer, D. E., P. K. Rajasekaran, and W. W. Gebhart, "Adaptive Clutter Filtering Using Autoregressive Spectral Estimation," *IEEE Trans. Aerosp. Electron. Syst.*, Vol. AES15, pp. 538–546, July 1979.

Capon, J., "High-Resolution Frequency–Wavenumber Spectrum Analysis," *Proc. IEEE*, Vol. 57, pp. 1408–1418, Aug. 1969.

Cleveland, W. S., "Fitting Time Series Models for Prediction," *Technometrics*, pp. 713–723, Nov. 1971.

Evans, J. E., J. R. Johnson, and D. F. Sun, "Applications of Advanced Signal Processing Techniques to Angle of Arrival Estimation in ATC Navigation and Surveillance Systems," Tech. Rep. 582, MIT Lincoln Laboratory, June 23, 1982.

Frost, O. L., and T. M. Sullivan, "High Resolution Two-Dimensional Spectral Analysis," *Rec. 1979 IEEE Int. Conf. Acoust. Speech Signal Process.*, pp. 673–676.

Gabriel, W. F., "Spectral Analysis and Adaptive Array Superresolution Techniques," *Proc. IEEE*, Vol. 68, pp. 654–666, June 1980.

Gobien, J. O., "Simultaneous Detection and Estimation: The Use of Sufficient Statistics and Reproducing Probability Densities," Ph.D. dissertation, University of Michigan, 1973.

Jain, A. K., "Advances in Mathematical Modeling for Image Processing," *Proc. IEEE*, Vol. 69, pp. 502–528, May 1981.

Jayant, N. S., "Digital Coding of Speech Waveforms: PCM, DPCM, and DM Quantizers," *Proc. IEEE*, Vol. 62, pp. 611–632, May 1974.

Johnson, D. H., "The Application of Spectral Estimation Methods to Bearing Estimation," *Proc. IEEE,* Vol. 70, pp. 1018–1028, Sept. 1982.

Kailath, T., "A View of Three Decades of Linear Filtering Theory," *IEEE Trans. Inf. Theory*, Vol. IT20, pp. 146–181, Mar. 1974.

Kay, S., "Robust Detection via Autoregressive Spectrum Analysis," *IEEE Trans. Acoust. Speech Signal Process.*, Vol. ASSP30, pp. 256–268, Apr. 1982.

Kay, S., "Some Results in Linear Interpolation Theory," *IEEE Trans. Acoust. Speech Signal Process.,* Vol. ASSP31, pp. 746–749, June 1983A.

Kay, S. M., "Asymptotically Optimal Detection in Unknown Colored Noise via Autoregressive Modeling," *IEEE Trans. Acoust. Speech Signal Process.*, Vol. ASSP31, pp. 927–940, Aug. 1983B.

Kay, S., "Broadband Detection of Signals with Unknown Spectra," *Rec. 1985 IEEE Int. Conf. Acoust. Speech Signal Process.*, pp. 33.1.1–33.1.3.

Kumaresan, R., "Estimating the Parameters of Exponentially Damped or Undamped Sinusoidal Signals in Noise," Ph.D. dissertation, University of Rhode Island, 1982.

Landers, T., "Maximum Entropy Cepstral Analysis," *Proc. RADC Spectrum Estimation Workshop,* Rome, N.Y., pp. 245–258, May 1978.

Lank, G. W., I. S. Reed, and G. E. Pollon, "A Semicoherent Detection and Doppler Estimation Statistic," *IEEE Trans. Aerosp. Electron.*, Vol. AES9, pp. 151–165, Mar. 1973.

Makhoul, J., "Linear Prediction: A Tutorial Review," *Proc. IEEE*, Vol. 63, pp. 561–580, Apr. 1975.

Makhoul, J., and R. Viswanathan, "Adaptive Lattice Methods for Linear Prediction," *Rec. 1978 IEEE Int. Conf. on Acoustics, Speech and Signal Processing*, pp. 83–86.

Markel, J. D., and A. H. Gray, Jr., *Linear Prediction of Speech*, Springer-Verlag, New York, 1976.

McDonough, R. N., "Maximum Entropy Spatial Processing of Array Data," *Geophysics*, Vol. 39, pp. 843–851, Dec. 1974.

Nuttall, A., "Spectral Analysis of a Univariate Process with Bad Points, via Maximum Entropy, and Linear Predictive Techniques," Tech. Rep. 5303, Naval Underwater Systems Center, New London, Conn., Mar. 26, 1976.

Oppenheim, A. V., and R. W. Schafer, *Digital Signal Processing*, Prentice-Hall, Englewood Cliffs, N.J., 1975.

Papoulis, A., *Probability, Random Variables, and Stochastic Processes*, McGraw-Hill, New York, 1965.

Proakis, J., *Digital Communications*, McGraw-Hill, New York, 1983, Chap. 6.

Rabiner, L. R., and R. W. Schafer, *Digital Processing of Speech Signals*, Prentice-Hall, Englewood Cliffs, N.J., 1978.

Robinson, E. A., "Predictive Decomposition of Time Series with Application to Seismic Exploration," *Geophysics*, Vol. 32, pp. 418–484, June 1967.

Van Blaricum, M. L., and R. Mittra, "Problems and Solutions Associated with Prony's Method for Processing Transient Data," *IEEE Trans. Antennas Propag.*, Vol. AP26, pp. 174–182, Jan. 1978.

Van Trees, H. L., *Detection, Estimation, and Modulation Theory,* Vol. 1, J. Wiley, New York, 1968.

Viswanathan, R., and J. Makhoul, "Quantization Properties of Transmission Parameters in Linear Predictive Systems," *IEEE Trans. Acoust. Speech Signal Process.*, Vol. ASSP23, pp. 309–321, June 1975.

APPENDIX 1
Summary of Computer Programs

The programs described in this appendix which require data as an input may all be used with complex data sets except for RMLE and AKAIKE, which accept only real data. The programs which require complex data may be used for real data sets if the real data are first converted to its complex form by setting the imaginary part equal to zero.

1.1 FFT

1.1.1 Description
This program computes the fast Fourier transform using a decimation-in-time algorithm.

1.1.2 Data Type
Complex

1.1.3 Called Subroutines
None

1.1.4 Where Found
Appendix 2A

1.2 CHOLESKY

1.2.1 Description
This program solves a hermitian positive definite set of linear simultaneous equations using the Cholesky decomposition.

1.2.2 Data Type
Complex

1.2.3 **Called Subroutines**
None

1.2.4 **Where Found**
Appendix 2B

1.3 PERIODOGRAM

1.3.1 **Description**
This program computes a data windowed periodogram.

1.3.2 **Data Type**
Complex

1.3.3 **Called Subroutines**
FFT

1.3.4 **Where Found**
Appendix 4C

1.4 CORRELATION

1.4.1 **Description**
This program computes the autocorrelation or cross-correlation estimates. Either the biased or unbiased estimators may be computed.

1.4.2 **Data Type**
Complex

1.4.3 **Called Subroutines**
None

1.4.4 **Where Found**
Appendix 4D

1.5 BLACKTUKEY

1.5.1 **Description**
This program computes the Blackman–Tukey spectral estimator. Either the biased or unbiased autocorrelation estimator may be used as well as a lag window.

1.5.2 **Data Type**
Complex

1.5.3 **Called Subroutines**
FFT, CORRELATION

1.5.4 **Where Found**
Appendix 4E

1.6 WGN

1.6.1 **Description**
This program generates samples of real white Gaussian noise with zero mean and a specified variance.

1.6.2 **Data Type**
Not applicable

1.6.3 **Called Subroutines**
RAN (Intrinsic function random number generator on DEC VAX 11/780)

1.6.4 **Where Found**
Appendix 5A

1.7 GENDATA

1.7.1 **Description**
This program generates a realization of a real or complex AR, MA, or ARMA time series given the filter parameters, excitation white noise variance, and an array of zero mean, unit variance, uncorrelated random variables.

1.7.2 **Data Type**
Not applicable

1.7.3 **Called Subroutines**
STEPDOWN

1.7.4 **Where Found**
Appendix 5B

1.8 ARMAPSD

1.8.1 **Description**
This program computes a set of power spectral density values given the parameters of a real or complex AR, MA, or ARMA model.

1.8.2 **Data Type**
Not applicable

1.8.3 **Called Subroutines**
FFT

1.8.4 **Where Found**
Appendix 5C

1.9 LEVINSON

1.9.1 **Description**
This program implements the Levinson recursion to solve the Yule–Walker equations given the autocorrelation function.

1.9.2 **Data Type**
Complex

1.9.3 **Called Subroutines**
None

1.9.4 **Where Found**
Appendix 6B

1.10 STEPDOWN

1.10.1 **Description**
This program implements the step-down procedure to find the coefficients and prediction error powers of all the lower order predictors given the coefficients and prediction error power for a pth order linear predictor.

1.10.2 **Data Type**
Complex

1.10.3 Called Subroutines

None

1.10.4 Where Found

Appendix 6C

1.11 AUTOCORR

1.11.1 Description

This program implements the autocorrelation method of linear prediction (Yule–Walker method) for estimation of the parameters of an AR process.

1.11.2 Data Type

Complex

1.11.3 Called Subroutines

CORRELATION, LEVINSON

1.11.4 Where Found

Appendix 7B

1.12 COVMCOV

1.12.1 Description

This program implements either the covariance method of linear prediction or the modified covariance (forward-backward) method to estimate the parameters of an AR process.

1.12.2 Data Type

Complex

1.12.3 Called Subroutines

CHOLESKY

1.12.4 Where Found

Appendix 7C

1.13 BURG

1.13.1 Description

This program implements the Burg method for estimation of the parameters of an AR process.

1.13.2 Data Type

Complex

1.13.3 Called Subroutines

None

1.13.4 Where Found

Appendix 7D

1.14 RMLE

1.14.1 Description

This program implements the recursive MLE method for estimation of the parameters of an AR process.

1.14.2 Data Type

Real

1.14.3 Called Subroutines

None

1.14.4 Where Found

Appendix 7E

1.15 DURBIN

1.15.1 Description

This program implements Durbin's method for estimation of the parameters of an MA process.

1.15.2 Data Type

Complex

1.15.3 Called Subroutines

CORRELATION, LEVINSON, AUTOCORR

1.15.4 Where Found

Appendix 8A

1.16 AKAIKE

1.16.1 Description

This program implements the Akaike MLE algorithm for estimation of the parameters of an ARMA process.

1.16.2 Data Type

Real

1.16.3 Called Subroutines

CHOLESKY, STEPDOWN

1.16.4 Where Found

Appendix 10C

1.17 MYWE

1.17.1 Description

This program implements the modified Yule–Walker equation estimator for the AR filter parameters and Durbin's method for estimation of the MA filter parameters and white noise variance of an ARMA process.

1.17.2 Data Type

Complex

1.17.3 Called Subroutines

CORRELATION, LEVINSON, AUTOCORR, DURBIN

1.17.4 Where Found

Appendix 10D

1.18 LSMYWE

1.18.1 Description

This program implements the least squares modified Yule–Walker equation estimator for the AR filter parameters and Durbin's method for estimation of the MA filter parameters and white noise variance of an ARMA process.

1.18.2 Data Type
Complex

1.18.3 Called Subroutines
CHOLESKY, CORRELATION, LEVINSON, AUTOCORR, DURBIN

1.18.4 Where Found
Appendix 10E

1.19 MAYNEFIR

1.19.1 Description
This program implements the Mayne–Firoozan method (three-stage least squares) estimator for the parameters of an ARMA process.

1.19.2 Data Type
Complex

1.19.3 Called Subroutines
FFT, CHOLESKY, CORRELATION, LEVINSON, STEPDOWN, AUTOCORR

1.19.4 Where Found
Appendix 10F

1.20 MVSE

1.20.1 Description
This program implements the minimum variance spectral estimator [Maximum Likelihood Method (MLM) or Capon's method].

1.20.2 Data Type
Complex

1.20.3 Called Subroutines
CHOLESKY

1.20.4 Where Found
Appendix 11A

APPENDIX 2

Glossary of Symbols, Abbreviations, and Notational Conventions

SYMBOLS†

$*$ — complex conjugation

$a[i]$ $(\mathbf{A}[i], a[i, j])$ — ith AR filter parameter

$a_k[i]$ $(\mathbf{A}_k[i])$ — ith predictor coefficient for kth order predictor

$\mathcal{A}(z)$ $(\boldsymbol{\mathcal{A}}(z), \mathcal{A}(z_1, z_2))$ — AR filter system function

$A(f)$ $(\mathbf{A}(f), A(f_1, f_2))$ — AR filter frequency response

$b[i]$ $(\mathbf{B}[i], b[i, j])$ — ith MA filter parameter

$\mathcal{B}(z)$ $(\boldsymbol{\mathcal{B}}(z), \mathcal{B}(z_1, z_2))$ — MA filter system function

$B(f)$ $(\mathbf{B}(f), B(f_1, f_2))$ — MA filter frequency response

$\mathbf{C}_\theta$ — covariance matrix for estimator of θ

χ^2_N — chi-squared random variable with N degrees of freedom

$\text{cov}(x, y)$ — covariance between random variables x and y

$\delta[n]$ $(\delta[m, n])$ — discrete delta function

$\delta(f)$ — Dirac delta function

$\det(\mathbf{A})$ — determinant of the matrix $\mathbf{A}$

$\text{diag}(d_1, d_2, \ldots, d_n)$ — diagonal matrix with diagonal elements d_i

$\sim$ — "is distributed according to"

$\mathbf{e}$ — vector of complex exponentials

$\mathcal{E}(\ )$ — expectation operator

† Symbols in parentheses are multichannel and/or 2-D versions.

$e_k^f[n],\ e_k^b[n]$	forward and backward prediction errors of order k
f	frequency
$\mathcal{F},\ \mathcal{F}^{-1}$	Fourier transform and inverse Fourier transform operators
$\dfrac{\partial}{\partial \mathbf{x}}$	gradient with respect to vector $\mathbf{x}$
H	complex conjugate and transpose of a matrix
$\hat{}$	denotes an estimator
$h[n]\quad (\mathbf{H}[n],\ h[m, n])$	impulse response
$\mathcal{H}(z)\quad (\mathcal{H}(z),\ \mathcal{H}(z_1, z_2))$	system function
$H(f)\quad (\mathbf{H}(f),\ H(f_1, f_2))$	system frequency response
$\dfrac{\partial^2}{\partial \mathbf{x}^2}$	Hessian with respect to vector $\mathbf{x}$
$\mathbf{I}$	identity matrix
$\mathbf{I}_\theta$	Fisher information matrix for parameter $\boldsymbol{\theta}$
$I(f)$	periodogram of data
$\mathrm{Im}\,(\ )$	imaginary part
$\langle x, y \rangle$	inner product between the vectors x and y
$\mathbf{J}$	exchange matrix
$k_i\quad (\mathbf{K}[i])$	ith reflection coefficient
λ_i	ith eigenvalue of a matrix
μ	mean of a random variable
$N(\mu, \sigma^2)$	Gaussian or normal distribution with mean μ and variance σ^2
$N(\boldsymbol{\mu}, \mathbf{C})$	multivariate Gaussian or normal distribution with mean vector $\boldsymbol{\mu}$ and covariance matrix $\mathbf{C}$
n	time index of sequence
N	number of data points
$\lVert \mathbf{x} \rVert$	norm of a vector $\mathbf{x}$
$O(\ )$	"on the order of"
p	AR model order
$p(\ )$	probability density function
$P_{xx}(f)\quad (\mathbf{P}_{xx}(f),\ P_{xx}(f_1, f_2))$	power spectral density of $x[n]$
$P_{xy}(f)$	cross-power spectral density between processes $x[n]$ and $y[n]$
$\mathcal{P}_{xx}(z)\quad (\mathcal{P}_{xx}(z))$	z-transform of the autocorrelation function
$\mathcal{P}_{xy}(z)$	z transform of the cross-correlation function
$\Pr\{\ \}$	probability of a random variable
$\Phi(\ )$	cumulative distribution function
#	pseudoinverse operator for a matrix
q	MA model order
$\mathrm{Re}\,(\ )$	real part
$r_{xx}[k]\quad (\mathbf{R}_{xx}[k],\ r_{xx}[k, l])$	autocorrelation function of the process $x[n]$

$r_{xy}[k]$ $(\mathbf{R}_{xy}[k], r_{xy}[k, l])$	cross-correlation function between the processes $x[n]$ and $y[n]$
$\mathbf{R}_{xx}$ $(\underline{\mathbf{R}}_{xx})$	autocorrelation matrix
ρ_i $(\mathbf{\Sigma}_i)$	prediction error power for ith order predictor
σ^2 $(\mathbf{\Sigma})$	excitation white noise variance of time series model
$\star$	convolution operator
T	transpose of a matrix
$\boldsymbol{\theta}$	unknown vector parameter
$u[n]$ $(\mathbf{u}[n], u[m, n])$	input white noise of time series model
var ( )	variance of a random variable
$\mathbf{v}_i$	ith eigenvector of a matrix
$x[n]$ $(\mathbf{x}[n], x[m, n])$	observed data process
$\mathcal{X}(z)$ $(\mathcal{X}(z), \mathcal{X}(z_1, z_2))$	z-transform of the sequence $x[n]$
$X(f)$ $(\mathbf{X}(f), X(f_1, f_2))$	Fourier transform of the sequence $x[n]$
$\mathcal{Z}, \mathcal{Z}^{-1}$	z-transform and inverse z-transform operators

ABBREVIATIONS

1-D	one-dimensional
2-D	two-dimensional
ACF	autocorrelation function
AIC	Akaike information criterion
AR	autoregressive
AR(p)	autoregressive process of order p
ARMA	autoregressive moving average
ARMA(p, q)	autoregressive moving average process with AR order p and MA order q
AVPER	averaged periodogram
BAR	Bartlett
BT	Blackman–Tukey
CAT	criterion autoregressive transfer function
CCF	crosscorrelation function
CDF	cumulative distribution function
CR	Cramer–Rao
DAN	Daniell
DEN	denominator
DFT	discrete Fourier transform
FFT	fast Fourier transform
FIR	finite impulse response
FPE	final prediction error
IFA	iterative filtering algorithm
LMVU	linear minimum variance unbiased
LS	least squares

LSI	linear shift invariant
LSMYWE	least squares modified Yule–Walker equations
MA	moving average
MA(q)	moving average process of order q
MAX	maximum
MESE	maximum entropy spectral estimator
MIN	minimum
MLE	maximum likelihood estimator
MLM	maximum likelihood method
MSE	mean square error
MUSIC	multiple signal classification
MV	minimum variance
MVSE	minimum variance spectral estimator
MYWE	modified Yule–Walker equations
NSHP	nonsymmetric half plane
PC	principal components
PDF	probability density function
PEF	prediction error filter
PER	periodogram
PHD	Pisarenko harmonic decomposition
PSD	power spectral density
QP	quarter plane
RMS	root mean square
SD	step-down
SNR	signal-to-noise ratio
TNSHP	truncated nonsymmetric half plane
TQP	truncated quarter plane
WSS	wide sense stationary

NOTATIONAL CONVENTIONS

All vectors are *column* vectors. Matrices and vectors are indicated by boldface type. Elements of matrices and vectors are denoted as follows. $[\mathbf{A}]_{ij}$ denotes the $[i, j]$ element of $\mathbf{A}$, and $[\mathbf{a}]_i$ denotes the ith element of $\mathbf{a}$.

Appendix 3
Description of Floppy Disk Contents

A double sided/double density (DS/DD) 360 Kb floppy disk containing several files has been provided with the book *Modern Spectral Estimation: Theory and Application* by Steven M. Kay, published by Prentice-Hall, Englewood Cliffs, N.J. The files may be read using an IBM PC, PC/XT, or PC/AT with MS-DOS 2.1 or higher. The files are:

1. twenty Fortran subroutines
2. two data files
3. a demonstration program
4. a single text file

The Fortran subroutines which are listed within the book and also briefly described in Appendix 1 are:

1. FFT
2. CHOLESKY
3. PERIODOGRAM
4. CORRELATION
5. BLACKTUKEY
6. WGN
7. GENDATA
8. ARMAPSD
9. LEVINSON
10. STEPDOWN
11. AUTOCORR
12. COVMCOV
13. BURG
14. RMLE
15. DURBIN
16. AKAIKE
17. MYWE
18. LSMYWE
19. MAYNEFIR
20. MVSE

Note that the file names will be truncated to 8 characters (if longer than 8 characters) and given the extension ".FOR".

The two data files CD.DAT and RD.DAT contain the 32 point complex and real test data sequences, respectively, listed in Appendix 1A. The files may be read into a VAX 11/780 using the following Fortran statements:

```
C     For the real data set use "DIMENSION" instead of
C     "COMPLEX"
        COMPLEX X(32)
        CHARACTER*6 FNNNNN
        PRINT*,'ENTER DATA FILENAME (6 CHAR.)'
        READ(5,'(A6)')FNNNNN
        OPEN(UNIT=2,FILE=FNNNNN,STATUS='OLD')
        DO 10 I=1,32
        READ(2,99)X(I)
C     For the real data set use "FORMAT(2X,F9.6)"
99      FORMAT(2X,F10.7,F10.7)
10      CONTINUE
        CLOSE(2)
```

The demonstration program is an abbreviated version of a menu-driven spectral estimation software package which is described more fully in Appendix 4. It may be run by typing "demo" and requires 256 Kb memory, a math coprocessor, and a color graphics adaptor. The text file is a copy of this appendix and is contained in README.TXT. Any questions or comments should be directed to:

Professor Steven M. Kay
Electrical Engineering Dept.
University of Rhode Island
Kingston, R.I. 02881
401-792-2505

Appendix 4

Description of Menu-Driven Spectral Estimation Software

An interactive menu-driven software package entitled Modern Interactive Spectral Analysis (MISA) has been developed to aid the user in understanding the tradeoffs inherent in applying the techniques of modern spectral estimation. It has been designed to complement the book *Modern Spectral Estimation: Theory and Application* by Steven Kay, published by Prentice-Hall and is based on the exact Fortran subroutines given in the book. A demonstration program is included on the floppy disk provided with the book and may be run by typing "demo". MISA will run on an IBM PC, PC/XT, PC/AT or a compatible machine equipped with 512Kb memory, a math coprocessor chip and a color (or B/W) graphics display adapter and monitor. As part of the software package a User Manual and Workbook are included. The workbook consists of numerous exercises to guide the user in developing a critical understanding by experimenting with the spectral estimation methods. Key features of MISA are:

- generation of real or complex data consisting of combinations of sinusoids, AR/MA/ARMA processes, and white noise
- pole/zero input for data generation by on-screen pole-zero plot
- graphing of real or imaginary parts of data
- implementation and graphing of Fourier, AR (maximum entropy, linear prediction), MA, ARMA, and minimum variance (Capon, maximum likelihood method) spectral estimators
- overlaid spectral plots of different estimators
- overlaid spectral plots of different realizations of same estimator for variability determination

- graphing of true power spectral density of generated data
- data file manipulations for storage of time series or spectral data, for inputting external time series data for analysis or spectral data for graphing
- on-line descriptions of all implemented methods and suggestions for user parameter inputs

The specific spectral estimation methods implemented are described in Table 12.1 of the book with the exact subroutine listing given in the appendices throughout the book. They are:

	Method Type	*Method*
1.	Fourier	periodogram
2.	Fourier	Blackman-Tukey
3.	AR*	autocorrelation
4.	AR*	covariance
5.	AR*	modified covariance
6.	AR*	Burg
7.	AR*	recursive MLE
8.	MA	Durbin
9.	ARMA	Akaike
10.	ARMA	modified Yule-Walker equations (MYWE)
11.	ARMA	least squares MYWE
12.	ARMA	Mayne-Firoozan
13.	minimum variance	Capon (''maximum likelihood method'')

* Also referred to as ''maximum entropy'' or ''linear prediction'' spectral estimation

An order form for MISA is included on the last page of the book.

It is anticipated that a version of MISA will be available as a Fortran library for mainframe implementations. The user will be required to provide graphics. Further information is available from:

Spectrasoft
190 Eighth St.
Providence, R.I. 02906
401-274-3793

Index

Copies of MISA and the User Manual/Workbook may be obtained by using the order form below. Site licenses are available for multiple users.

Modern Interactive Spectral Analysis

ORDER FORM

Please send me the MISA software and User Manual/Workbook @ $295 for each copy:

	Quantity		Price
	________		________
Sales tax (R.I. residents add 6%)			________
		Total	$________

Please check one of following payments:

☐ Purchase order enclosed		#________
☐ Check/money order enclosed	Amount	________
☐ Charge VISA	Amount	________

Account #________________________________

Expiration date ____________________________

Signature ________________________________

Ship to: (Please print or type)

Name ______________________________________

Department __________________________________

Organization _________________________________

Address ____________________________________

City, State, Zip ______________________________

Country ____________________________________

Area code/Phone ______________________________

Spectrasoft 190 Eighth St. Providence, R.I. 02906 401-274-3793